DECODING GRAVITY, TIME AND CAUSALITY

DYNAMIC GENERAL RELATIVITY,
A THEORY OF QUANTUM GRAVITY AND
A POTENTIAL THEORY OF EVERYTHING

Dr. Rohit Kale

INDIA · SINGAPORE · MALAYSIA

Notion Press Media Pvt Ltd

No. 50, Chettiyar Agaram Main Road
Vanagaram, Chennai, Tamil Nadu – 600 095

First Published by Notion Press 2022
Copyright © Dr. Rohit Kale 2022
All Rights Reserved.

ISBN

Hardcase 979-8-88555-444-2

Paperback 978-1-68563-948-8

This book has been published with all efforts taken to make the material error-free after the consent of the author. However, the author and the publisher do not assume and hereby disclaim any liability to any party for any loss, damage, or disruption caused by errors or omissions, whether such errors or omissions result from negligence, accident, or any other cause.

While every effort has been made to avoid any mistake or omission, this publication is being sold on the condition and understanding that neither the author nor the publishers or printers would be liable in any manner to any person by reason of any mistake or omission in this publication or for any action taken or omitted to be taken or advice rendered or accepted on the basis of this work. For any defect in printing or binding the publishers will be liable only to replace the defective copy by another copy of this work then available.

COPYRIGHT NOTICE

Caution: The contents of this book are protected by © Copyright owned by Dr. Rohit Kale. The content in this book should not be unlawfully reproduced, downloaded, disseminated, transferred, plagiarised or published in any form (electronic or otherwise) and may not be built upon, renamed, or profited from, without the prior written permission of the author.

Scientific publication in the subject using the ideas presented within the book without collaborating with the author should be avoided.

CONTENTS

ACKNOWLEDGEMENTS 9
WHAT IS THIS BOOK AND HOW TO READ IT? 11
OTHER BOOKS WRITTEN BY ME 13

PART I

CHAPTER 1:	CRISIS IN PHYSICS	17
CHAPTER 2:	CIRCULAR REASONING IN PHYSICS	22
CHAPTER 3:	PSYCHOLOGY OF A BELIEF	25
CHAPTER 4:	AIMS OF A GOOD THEORY IN SCIENCE	32
CHAPTER 5:	HUMAN BELIEFS AND BELIEF BIASES	35
CHAPTER 6:	HOW MY SECOND BOOK LED TO MY THIRD BOOK?	39
CHAPTER 7:	CONVENTIONAL VS AUTHORS METHOD	47
CHAPTER 8:	PROBLEMS OF UNIFYING GR WITH QM	53
CHAPTER 9:	IMPORTANCE OF AN OBSERVER AND THE MEASUREMENT PROBLEM IN QUANTUM MECHANICS	56
CHAPTER 10:	WHY THIS THEORY OVER OTHER POTENTIAL THEORIES?	58
CHAPTER 11:	INSIGHTS THAT EMERGE FROM THE THEORY	62
CHAPTER 12:	SERIOUS QUESTIONS STILL UNANSWERED	66

PART II

CURRENT STATUS OF THE THEORY OF DGR AND VERY RECENT INCITES AND DIFFICULT PROBLEMS IN PHYSICS 67

CURRENT STATUS OF THE THEORY OF DYNAMIC GENERAL RELATIVITY 69

CHAPTER 1:	BACKGROUND THINKING	71
CHAPTER 2:	BASIC ASSUMPTIONS OF THE THEORY OF DGR (DYNAMIC GENERAL RELATIVITY)	73
CHAPTER 3:	WHAT DOES GENERAL RELATIVITY MEAN FUNDAMENTALLY? (DERIVING A MODEL OF SPACE-TIME FROM GR)	79
CHAPTER 4:	EXPLAINING UNIVERSAL AND LOCAL TIME	91
CHAPTER 5:	THE T=0 AGGREGATE CONTINUUM AND THE SIGNIFICANCE OF "E" OR "э"	97
CHAPTER 6:	TIME RESETTING WAVES (TRW) AND THEIR INTERFERENCE	99
CHAPTER 7:	THE PHENOMENON OF RECRUITMENT OF PC'S	110
CHAPTER 8:	THE ENIGMA OF POTENTIAL ENERGY	111

CHAPTER 9:	KINETIC ENERGY	116
CHAPTER 10:	MATHEMATICS OF DGR	118
CHAPTER 11:	CONSTRUCTING A MODEL	120
CHAPTER 12:	DEFINING DISTANCES	121
CHAPTER 13:	DEFINING TIME DILATION	123
CHAPTER 14:	NOMENCLATURE OF LEVELS	125
CHAPTER 15:	DEFINING VOLUME LOSS	129
CHAPTER 16:	NUMBER OF EVENTS PER PLANK TIME	134
CHAPTER 17:	THE INSTANTANEOUS VELOCITY OF INWARD PROPAGATION OF EPCA	135
CHAPTER 18:	ANALYSING DATA FROM COMPUTER SIMULATIONS	139
CHAPTER 19:	GRAVITATIONAL LIMIT	141
CHAPTER 20:	A POTENTIAL EXPLANATION OF MONDIAN FORCES IN DGR	143
CHAPTER 21:	SIGNIFICANCE OF A_o	150

DIFFICULT PROBLEMS IN PHYSICS ... 153

CHAPTER 22:	DGR AND DARK MATTER	155
CHAPTER 23:	DARK ENERGY AND CRISIS IN COSMOLOGY	158
CHAPTER 24:	INFORMATION LOSS WITHIN A BLACKHOLE OR SINGULARITY IN DGR	168
CHAPTER 25:	THE TROUBLE WITH HAWKING RADIATION	170
CHAPTER 26:	DGR AND THE BIG BANG THEORY	172
CHAPTER 27:	IS STRING THEORY COMPLETELY WRONG?	174
CHAPTER 28:	TAKING A CLOSER LOOK AT G, NEWTONIAN UNIVERSAL CONSTANT AND C – DGR AND VARIABLE SPEED OF LIGHT	178
CHAPTER 29:	DGR AND DIRAC'S LARGE NUMBER HYPOTHESIS	185
CHAPTER 30:	THE RATIO OF ELECTROMAGNETIC FORCE TO GRAVITY	188
CHAPTER 31:	MANIFEST AND UN-MANIFEST FORM	190
CHAPTER 32:	AN ELEGANT WAY TO TEST THE THEORY OF DGR	194
CHAPTER 33:	DISCUSSION	199

PART III
EVOLUTION OF THE THEORY AFTER THE CONCEPTUAL IDEA OF EPCAS CAME 207

CHAPTER 1:	UNDERSTANDING LOCAL TIME	209
CHAPTER 2:	IMPLICATIONS OF OUR THEORY ON EINSTEIN'S SPECIAL RELATIVITY	214
CHAPTER 3:	MATTER ANTIMATTER ASYMMETRY	227
CHAPTER 4:	WAVE-PARTICLE DUALITY OF MATTER AND LIGHT	240
CHAPTER 5:	A PARTICLE	254
CHAPTER 6:	MATTER-ANTIMATTER (OR MATTER-EXOTIC NEGATIVE ENERGY ANTIMATTER) MODEL.. 270

CHAPTER 7:	THE CHARGED PARTICLE	279
CHAPTER 8:	A BOND	287
CHAPTER 9:	MAGNETISM AND ELECTRICITY	292
CHAPTER 10:	ENTANGLEMENT**	305
CHAPTER 11:	DESTRUCTION AND COMPENSATION TO AVOID A VOID	320
CHAPTER 12:	CREATION AS A COMPENSATION TO AVOID A VOID	325
CHAPTER 13:	HOW DO "EPCA FORMING TRWS" (TIME RESETTING WAVES) EMANATING FROM DIFFERENT PARTICLES INTERACT?	327
CHAPTER 14:	ATTRACTION AND REPULSION	334
CHAPTER 15:	THE MATTER-WAVE AS PREDICTED BY DGR	343
CHAPTER 16:	STATIC GENERAL RELATIVITY VS DYNAMIC GENERAL RELATIVITY	344
CHAPTER 17:	EXPLAINING PUSH-PULL BANDS	348
CHAPTER 18:	ENEA PARTICLES, THEIR PROPERTIES, INTERACTIONS AND IMPLICATIONS	349
CHAPTER 19:	WHAT'S THERE IN THE EMPTY SPACE?	355
CHAPTER 20:	WHAT'S THE SHAPE OF THE UNIVERSE?	361
CHAPTER 21:	PHOTON AND TYPES OF EM RADIATION	365
CHAPTER 22:	THERMODYNAMICS - NEWER INSIGHTS	369
CHAPTER 23:	THE ELECTROMAGNETIC BOND	371
CHAPTER 24:	THE CHEMICAL BONDING AND THE ENIGMA OF CHEMICAL POTENTIAL ENERGY	377
CHAPTER 25:	METALLIC BONDS PREDICTED BY DGR	387
CHAPTER 26:	THE STRONG NUCLEAR FORCE	390
CHAPTER 27:	WEAK NUCLEAR FORCE	396
CHAPTER 28:	SUPERPOSITION OF SPIN OF A CHARGED PARTICLE	402
CHAPTER 29:	WHY THERE IS SOMETHING AND NOT NOTHING? - THE PROCESS OF GENESIS	407
CHAPTER 30:	STRING THEORY AND OUR THEORY	417
CHAPTER 31:	CONTEMPORARY PHYSICS AND DGR	419
CHAPTER 32:	FINAL ASSUMPTIONS OF THE THEORY	422

PART IV

VERY EARLY IMMATURE STAGES OF THE THEORY 431

CHAPTER 1:	EVOLUTION OF DGR IN MY MIND (1**-6**)	433
CHAPTER 2:	GRAVITY	439
CHAPTER 3:	EINSTEIN'S GENERAL THEORY OF RELATIVITY, IS IT RIGHT?	441
CHAPTER 4:	BACKGROUND THINKING	446
CHAPTER 5:	BASIC ASSUMPTIONS	449
CHAPTER 6:	HOW TIME GRADIENT LEADS TO DILATION OF SPACE?	455
CHAPTER 7:	THE POSITIVE FEED-BACK LOOPS OR ROLLER COASTER	463

CHAPTER 8:	DYNAMICITY OF THE APPARENTLY STATIC SPACE TIME CHANGES IN GENERAL RELATIVITY AND HOW IT IS DIFFERENT THAN CURRENT THEORY	464
CHAPTER 9:	WHY PLANCK LENGTHS ARE NOT MINIMUM POSSIBLE LENGTHS AND PLANCK TIMES ARE NOT MINIMUM POSSIBLE TIMES?	466
CHAPTER 10:	TIME	467
CHAPTER 11:	COMMUNICATION BETWEEN ADJACENT PLANCK COMPARTMENTS	477
CHAPTER 12:	AN ELECTROMAGNETIC WAVE	485
CHAPTER 13:	GRAVITATIONAL LENSING EXPLAINED1[#]	498
CHAPTER 14:	ENERGY	500
CHAPTER 15:	TEMPERATURE	506
CHAPTER 16:	INERTIA	508
CHAPTER 17:	MOND	522
CHAPTER 18:	BLACK HOLES	533
CHAPTER 19:	CHARGED PARTICLE	536
CHAPTER 20:	ELECTRON	549
CHAPTER 21:	MESON THEORY OF NUCLEUS AND QUANTUM CHROMODYNAMICS	559
CHAPTER 22:	MATTER ANTIMATTER PROBLEM	564

BIBLIOGRAPHY AND REFERENCES *567*

ABBREVIATIONS AND EXPLANATIONS *579*

ACKNOWLEDGEMENTS

I thank the Almighty for guiding me through the journey of writing this book.

Special thanks to my dear wife and my two sons for their love, support and encouragement while I wrote this book. Writing this book would not be possible without the constant support and blessings of my parents and my family.

Writing this book would not have been possible if the relevant knowledge was not freely available on the web. A special thanks to all the Teachers on the web who take significant efforts to teach the concepts of physics to anybody who wants to learn without expecting anything in return. These teachers of mine gave me the pieces of this puzzle and the only thing I did was to put the pieces of the puzzle together. Thus this book stands on the shoulders of these stalwarts of Physics and a special credit for writing this book goes to all of them. These include Sabine Hossenfelder, Don Lincoln, Matt O'Dowd (PBS digital studios), Arvin Ash, Joe Scott, Dibyajyoti Das (For the love of Physics), Parth G., Alexander Unzicker and many others. A special thanks to all these great teachers to make the difficult concepts available within the reach of everyone.

A special thanks to my big brother, Chetan Kathalay, who seeded the love of physics in me and gave a patient ear to all my ugly, sometimes unrealistic and seemingly unscientific chatter and also actively contributed to or gave direction to my thoughts, while I indulged in thinking that was needed in understanding the concepts within the book.

I want to thank the Notion press team including everyone who worked on this book.

I pray to the Almighty that the knowledge assimilated within this book comes to fruition and it helps us, humans, to come closer to the ultimate truth.

WHAT IS THIS BOOK AND HOW TO READ IT?

What is this book?

This book is an attempt to describe a theory of Quantum Gravity and a potential direction towards a Theory of Everything that I call Dynamic General Relativity (DGR). I stumbled upon the idea of DGR while thinking about Gravity. The book employs methods and direction of thinking that is almost unknown to current Physics. Although a lot of physics information is present in the book and by reading the book, a good insight can be obtained as to "How to construct a theory", this book is not a textbook of physics and the theories and arguments discussed are not yet accepted by mainstream Physics. The book (most chapters) can be understood by anyone who has some basic undergraduate knowledge of Physics and knowledge of advanced techniques in Mathematics are not necessary for understanding the concepts in the book. Although some Mathematics is discussed in a few Chapters, it is undergraduate level and not necessary to understand the theory itself.

What's in the name!

The original name of this book was **"Dynamic General Relativity – A Theory of Quantum Gravity and a Potential Theory of Everything"** which, I feel, is still the most appropriate one. This name, however, seems too technical, psychologically burdening and seems appropriate only for the Physics community. For the cognitive ease of the Laymen who might be interested in knowing the truth, the more acceptable name of **"Decoding Gravity, Time and Causality"** was chosen, just to ensure that common man with little knowledge of complex mathematics and still interested in knowing the truth don't shy away from reading the book.

How to read it?

"Part I" is a general description of the problems and fallacies of thinking of current physics, the major problems and crises in current physics and cosmology and essentially why this book is needed? It also tries to answer what, I believe, were the main hurdles due to which the physics community could not decode these arguments or reach the insights that I could reach through DGR.

The rest of the book is the description of how the theory evolved and what is the current state of the theory. As this book was written and the theory evolved over three years, it was quite a challenge to decide how to arrange its contents.

The best way to understand the theory is probably to read Part IV first (after reading Part I) and only then Part III and finally Part II.

Part I – Part IV – Part III – Part II

This way, one can get a complete idea as to how the theory evolved in my mind from the presently accepted physics to the current state of the theory. However, Part IV is the very early part of the theory development. It contains many faulty lines of thinking. To clarify certain insights that emerged later or to point out the error in the direction of thinking, I added text boxes during the editing stage of the book and retained these faulty lines of thinking instead of deleting or significantly editing them but instead underlined them to highlight their inaccuracy.

Thus one should be constantly aware that, throughout the book, wherever text boxes are added, they are added later to give additional information that emerged at a later date and point out mistakes. Also, note that any text which is underlined anywhere in the book is possibly wrong or needs further clarification that is provided in the nearby Text boxes.

Those who are in a hurry can just read the first chapter (Evolution of the Theory) in Part IV and skip the rest of part IV keeping in mind the risk of losing the additional insight which you might get by reading this section.

Part I of the book is general topics of what is wrong with current physics and what is needed to reach the current theory. Part IV is nothing but my notes while thinking about the very early evolution of my theory. Part III can further be divided before and after the idea of EPCA formation heralded. Part III is when the actual theory took shape and started looking promising. Part II is the current status of the theory and how DGR solves a lot of difficult problems in current Physics. Worth noting that the extrapolations of DGR to explain the Strong Nuclear Force and Weak nuclear force (QCD) are merely speculative at present, given the little lack of knowledge and understanding that I had about these. More work is needed especially in this regard.

Caution: When I was writing "Part iv and early part of Part iii", the theory was just taking shape and was not complete (even today the theory remains incomplete and multiple questions remain unanswered). These were the times when I had to fly high in imagination and I often drifted in the wrong direction and the arguments made may feel unsatisfactory. Discussion in chapters in these sections is often loud thinking done by me which at times may be wrong. Wherever possible, I have underlined and added a textbox and left them minimally edited. The end result, which is the current form of the theory, however, appears satisfactory.

OTHER BOOKS WRITTEN BY ME

How the Homo Sapiens Blundered

The book is on Global Warming, Climate change and other crises like biodiversity crisis, extinction crisis facing Humanity and what are the possible blunders made by them that led to this ugly situation. It also discusses many solutions that I believe are inevitably needed.

One of the important messages from this book is "The most important blunder of Humans is the continual destruction of Ecosystems/Green members and continual inhibition of natural cycles all of which reduce the Carbon sink (and not just burning fossil fuels or emitting CO_2 as is the belief propagated everywhere) and that solutions that solely focus of GHG emissions reduction while continuing the Ecocide are not only going to be useless but would be counterproductive." The ugly possibility that even the Climate science community is completely riddled with bias and keeps focussing on GHG emissions reduction which by itself (although needed) will do nothing to Ecology and will add to the Ecocide done by Humans. The answer isn't Eco-socialism or eco-fascism or destroying Capitalism which has become the politically riddled Climate science movement but to take the help of Capitalism (Eco-centric regulated Capitalism) and make big Corporates, Customers and even the Governments accountable. The real answer lies in Protecting Ecology, Preserving Ecosystems and more importantly lending a helping hand in healing the Ecosystems. The focus should shift from "GHG Emissions" to "Ecology and Ecosystems".

Decoding Human Psyche

The realization that Human Psyche plays an important role in what Humans choose and these have lasting consequences, Climate change being one of them led me to my second book. In this book, I intended to decode all the psychological barriers in understanding the importance of "Ecology" and understanding the arguments made in my first book.

I tried to decode how beliefs take shape in humans, how their subconscious mind programming affects their perception and decisions and how these affect Ecology and ecosystems.

The ugly realization from this book was that Humans are "Delusional creatures" and are extremely prone to confirm the preconceived beliefs that make them biased about making some choices rather than others. Even the entire Climate science is riddled with Confirmation bias-Congruence bias and is in a way leading the world into darkness by teaching the world hatred for Capitalism, hatred for CO_2 emissions and making only wealthy Capitalists pay while giving a complete clean chit to Consumers and to those activities which spread ecocides or inhibit ecosystems formation or habitat destruction but do not emit GHG emissions. CO_2 emissions calculation and reduction has become a big red herring in all environmental discussions completely overshadowing other potentially more important issues like the Biodiversity crisis, Extinction crisis, Consumption crisis and Insect Apocalypse. The root cause of all these crises which I collectively call the "CBEC crisis or Consumption Biodiversity Extinction Consumption crisis" is a Psychological crisis due to extreme and widespread cognitive biases like Anthropocentrism bias, Dunning Kruger effect, hundreds of Belief biases, Attribution bias, Confirmation/Congruence bias and most importantly dis-information and complete absence of critical thinking.

The realization from this book about how the "Beliefs" within the human mind sway our thinking away from the right direction led me to my third book "Dynamic General Relativity" where I realized how certain beliefs from General Relativity and Special Relativity were hindering the physicists from thinking in the right direction and were the major hurdles towards understanding the ultimate reality.

(Both the Books are available on Kindle, Amazon and also Notion press website: https://notionpress.com/store.)

PART I

CHAPTER 1

CRISIS IN PHYSICS

It is not an exaggeration to say that physics has come into crisis today. The Standard Model of particle physics and the quantum field theory which is the most accepted theory has been described in the early part of the last century. After that, very little progress has occurred. The last major shift in the understanding of physics occurred because of the theory of relativity (both special and general) and Quantum Mechanics. Although the general theory of relativity has been accepted and has been confirmed by multiple scientific observations, unifying it with Quantum Mechanics has been a big enigma.

In Quantum Mechanics itself, multiple questions remain unanswered. Richard Feynman, a famous physicist has said "nobody understands Quantum Mechanics". The term "Shut up and calculate", which means it is best to stick to the equations given by Quantum Mechanics (as they give extremely accurate results and conform with the experimental results to an extreme degree) without asking many questions, has become popular.

There are many controversial and confusing things/phenomena in Quantum Mechanics which seem unavoidable. Quantum Superposition is one such phenomenon. Non-locality or quantum entanglement or the apparent need for faster than light transfer of information is another one. More recently, the "Delayed Choice Quantum Eraser" experiment has questioned even the arrow of time and suggest that in the quantum realm, future events can affect the past.

As discussed separately in detail, the major hurdle in unifying QM with GR is the problem of background independence and the problem of time. QM assumes a fixed background with fixed coordinate locations and a single constantly moving forward moving entity of time, while in GR, space-time is not a static background but is constantly changing or evolving and time for different points in space may run at a different pace. They are thus incompatible with each other and several attempts to unify them both have failed.

The problem of Flat rotation curves and Velocities of Galaxy Clusters

In Astrophysics, the problem of higher than expected velocities of the peripheral stars in the Galaxies was described by Fritz Zwicky in 1933 *(Zwicky, Fritz – 107/108)* which necessitates either a modification in our understanding of gravity (i.e. Modified Newtonian dynamics – MOND) or more commonly accepted alternative of "the existence of Dark Matter". The enigma of finding evidence of Dark Matter particles or creating one in the particle accelerators continues to date with little success. The knowledge about what these dark matter particles constitute continues to elude the physicists.

Crisis in Cosmology

The observations that suggest that the expansion of the universe is accelerating rather than decelerating has created another conundrum.

The rate of expansion of the Universe is given by a term called the Hubble constant.

There are two methods of calculating Hubble's constant. The first method is by using the Lambda CDM model using the CMB (Cosmic Microwave Background). This method gives a value of 67.66 + 0.42, published by Planck mission *(The Planck Collaboration – 78)* This, however, gives the rate of expansion of the Universe at the time CMB was released.

The second method is to calculate the rate at which the Galaxies recede from us as we observe today using Cepheid variables and Type Ia supernovae as standard candles to measure intergalactic distances. This second method gives a slightly different value of Hubble's constant.

It was initially thought that the difference in these values is due to systematic error in measurement of each of these values and over some time, with improvement in techniques and as more precise measurement techniques for measuring the CMB or observation of standard candles become available, these values would come closer. But despite all the improvements in techniques and advancements, the difference persists.

These findings suggest that there is yet unidentified energy called Dark Energy which is responsible for this accelerated expansion. The trouble is that the findings suggest that the value of the density of this dark energy remains constant instead of falling with the increased volume of the Universe, which goes against intuition.

We don't know the majority of the Universe

What the above findings suggest is that the Universe is made up of 70% Dark energy, 25% Dark matter and what we know or can see I.e. the Baryonic matter constitutes just 5% of the total Universe.

Current theories which we use to describe our Universe, which is General Relativity (GR) and Quantum Field theory (QFT) or just Standard Model (SM) of particle physics do not have a way to explain or accommodate this Dark Matter or Dark Energy.

Enigma of Black holes

Black holes (BH) are regions of the Universe where gravitational pull inwards is so strong that even light cannot escape out. GR supports the existence of such objects. Evidence is accumulating that there are numerous such objects in our Universe. Every Galaxy has a BH at its centre, which may have masses of several million to a billion solar masses. These are called Supermassive Black holes (SMBH) or Ultra-Massive Black holes (UMBH). Current theories cannot predict with certainty what happens within the event horizon.

Data cannot be obtained for events that happen within the event horizon and thus even the presence of a BH has to be made out by its effects on the surrounding stellar objects.

The problem of information loss within a BH has not been solved with any confidence yet and our current theories completely break down at the centre where it is hypothesized that a singularity exists.

The moment of creation

The currently accepted LCDM or Lambda Cold Dark Matter Model of the early Universe can explain what events happened at the start of the Universe with good confidence up to about 10^{-35} seconds. Before this, the Quantum effects of Gravity start becoming important and thus current theories break down.

Crisis in Foundations of Physics

The foundations of physics have GR and QFT I.e. Standard Model of Particle physics. GR is 100 years old and SM was almost fully described by the 1960s. Little has been added in these theories after these.

Problems like Quantum properties of Space-time in GR and the measurement problem in Quantum Mechanics persist with little explanation offered by the current theories for the same.

Crisis in gathering Experimental evidence and failed predictions

There were several successes in the experimental detection of Gravitational waves by the LIGO (Light Interferometer Gravitational-Wave Observatory) and detection of the Higgs Boson by the LHC (Large Hadron Collider) at CERN. However, most theoretical predictions today are at energy ranges that need extremely large machines with high costs to build. Multiple predictions made by theories were not found to have experimental confirmation. The LHC was expected to find evidence of supersymmetric particles, miniature Black holes and extra dimensions, none of which were found. Many large "Dark Matter particle detectors" are functional and have failed to find the particles they were built to detect, despite multiple upgrades. Large experimental setups to detect proton decay to prove "Grand Unified Theories" have failed to detect Proton decay. Many high energy particle detectors built with extremely high costs are in function. A large amount of data collected from them is discarded and only about 1% of data from it is accepted by the scientists for analysis. Thus getting a positive experimental result with reasonable certainty of not being obtained by chance has increasingly reduced.

The sorry state of Quantum Gravity theories

Many contender theories can be called potential theories of Quantum Gravity and still others can be called contenders of the elusive "Theory of Everything".

Out of these, the majority of manpower in Physics is engaged in two theories namely Loop Quantum Gravity (LQG) and String Theory (ST). These two theories are almost completely incompatible with each other. LQG is an extension of the theory of GR and assumes quantization of space-time. String theory on the contrary starts with QFT and is based on two-dimensional Strings vibrating in 11-dimensional hyperspace. Most physicists working on these believe that the theory they work on is right and the other one is wrong. The theoretical physics community is thus divided into "Loopy" physicists and "Stringy" physicists. Even the academic conferences of those working on these stay separate and there are very few who are working to join both or working to make LQG reconcile with ST.

Both LQG and ST are presently trapped in the theoretical realm with their Mathematical complexity increasing by the day with very little hope for a testable prediction from either. LQG has a slightly higher chance of coming up with a testable prediction in the future. With 10^{500} possible shapes of the 11-dimensional manifolds on which the Strings vibrate, there is little hope for String theory to come in a testable range or come anywhere close to falsifiability.

However, one of the predictions of LQG of a slight difference in velocity of light with different energies has not been detected as yet.

Other problems of Quantum Gravity include "the flat rotation curves of Galaxies and Rotation velocities of outer stars and clusters". To explain these problems, the physicists are divided into two groups. Some agree with Modified Newtonian Dynamics and are engaged in disproving the idea of Dark Matter. Others believe in Dark matter and claim that Einstein's GR cannot be wrong even at large distances and small accelerations. Thus the physics community is divided into MONDian physicists and DM or LCDM physicists.

Other less known theories of Quantum Gravity (and possibly TOE) like Emergent gravity, Causal Dynamical Triangulation (CDT), Geometric unity (GU) have very few people working on them and given the small manpower engaged in them, the progress in these has been slow. However, these carry some hope simply because these scientists keep an open mind and are ready to consider out of the box ideas and change their methods instead of repeating the same old mistakes.

The Challenge

The real challenge with the future physicists is thus

- ❖ To get a single theory that unifies QM with GR,
- ❖ A theory that successfully explain the Galactic rotation curves either by explaining MOND or with Dark Matter in which case it has to explain what it is and how to find it.
- ❖ The theory could potentially be a unification of LQG and ST, taking the strengths of the two and removing the weaknesses of each.
- ❖ The theory should provide valid logical explanations to all these problems and more, with a minimum number of assumptions.

- ❖ If possible the theory should (if possible) unify all the forces that are described in SM with each other (Grand Unified Theory) and with Gravity, thus providing us with a single "Theory of Everything" (TOE) that can explain all the phenomena happening.
- ❖ The theory should be such that it gives testable predictions, and other accepted theories should effortlessly emerge from the theory without much fuss.
- ❖ Ideally, the theory should have a minimum number of assumptions and should successfully derive all the constants of nature i.e. it should derive, from the assumptions of the theory itself, the masses of all particles like the mass of Electron, the mass of the Quarks etc., explain the values of constants of nature like the fine structure constant, the ratio of the strength of EM force to Gravity, the ratio of the radius of the Universe to the radius of the Proton etc.

There is presently no theory that comes even close to achieving all this.

Chapter 2

CIRCULAR REASONING IN PHYSICS

Physics is full of this fallacy of logic.

Circular reasoning essentially involves saying A is because of B and B is because of A.

That means the proof of A is B and the proof of B is A thus both A and B have proof. In reality, however, neither A nor B have proof.

Many examples can be quoted.

Light is an electromagnetic wave.

An Electromagnetic wave is nothing but a disturbance in electric and magnetic fields that are perpendicular to each other and are travelling forwards at the speed of light.

One might ask, what exactly is getting disturbed here. The instant answer is "a hypothetical field called the electric field and another hypothetical field called the magnetic field". Both of these are present all over the Universe at every point. Both these are vector fields.

One might then ask, what a vector field is?

The instant answer to this would be, a vector field is something that has a value and a well-defined direction at every point.

However, "what exactly is a field made of?" is unclear. "What is the difference between the electric and the magnetic fields?" is unclear. What exactly is an electric or a magnetic field made of is not known. This means what exactly is an electromagnetic wave made of is unclear. This means what exactly is light made of is also unclear.

In phenomena like refraction, the wave-front of the wave of light changes direction. In reflection, the wave-front reverses the direction. The wave-front also interferes with other similar wave-fronts to an intricate extent both constructively or destructively. Our theories based on light being a wave-front of a wave can accurately describe these phenomena in stark detail. However, "what this wave-front is made up of?" Is the question which is often dodged by modern physics with its circular reasoning. What makes these wave-fronts behave the way they behave is not known although their behaviours are known to remarkable precision.

The current theory on which entire physics is defined today is called the Quantum field theory or QFT. This theory assumes the presence of a separate field for every particle that exists according to the Standard Model (SM) of particle physics. These fields are imaginary undefined entities and nobody can explain their fundamental structure.

The need of the hour is to have a theory that can give proper answers to each of the following questions

- What is the wave of light made of, or what exactly is waving here?
- what is the wave-front made of?
- what is a magnetic and electric field?
- what is the explanation for the weird behaviour of light with wave-particle duality of light?
- why light behaves like a particle as well as a wave?

In QFT

- what are these fields made up of?
- what happens to these fields when one particle gets destroyed and another is created?
- what happens to other fields when a particle of a particular field is formed?

Many such questions remain unanswered or are forbidden from being asked in the modern QFT.

Energy and work

"What exactly is the entity called energy?" is not known. Although energy is arguably one of the most abundant things one can see, "what it is?" remains unclear in modern Physics.

Energy is defined based on work that its presence can accomplish.

"How exactly does the energy produce work?" is not known.

Energy is the "ability to do work".

In this, what is work? It is defined in terms of force applied and movement achieved in a particular body with a particular mass. Applying force itself needs energy. In this manner, energy is defined in terms of work and work is defined in terms of energy.

Abstract entities in Physics

Physics has many abstract entities, the true nature of which is unknown. It probably started with Michael Faraday's description of the Magnetic field and Electric field. After this event, everything that we see is described based on a field while "what these fields are?" cannot be explained. They remain abstract entities. Another recent addition to this abstract entity of fields is the Higgs field which is another all-encompassing field that gives mass to most objects/particles.

Energy, wave and fields as described above are not the only abstract entities. The list is pretty long. The various specialised types of energies are equally abstract with potential energy and kinetic energy in classical mechanics, nuclear energy in nuclear physics, electrical or magnetic potential energy in electronics, chemical potential energy in chemistry just to name a few. Entities like Inertia and Momentum are abstract entities that are necessarily to be included in the assumptions of the theory to be explained. What these are fundamentally made of is unclear and why they behave the way they do are converted into "Laws of Nature" to evade the possible demand for "need for explanation".

Bonds that are formed by various particles are also made possible by the exchange of virtual particles which the Standard model describes as Bosons. The Photon, the Gluon, the W and Z bosons are all force-carrying particles that enable the transfer of forces among individual particles. What these particles are made up of fundamentally is unclear. These have multiple properties which have to be incorporated in assumptions of the theory.

Many phenomena like Superposition, wave-particle duality of light and wave-particle duality of matter and matter-antimatter asymmetry are yet to be explained. The probabilistic nature of Quantum Mechanics and the Uncertainty Principle are accepted despite them seeming highly illogical.

The probabilistic nature of QM is explained by some with Copenhagen interpretation which prohibits asking questions and by some with Multiverse hypothesis. The Uncertainty Principle is explained based on wave-particle duality of matter which itself is a relatively little known or ill-explained entity. Although both these have enough evidence and cannot be proven wrong, the underlying reality which they try to point out is based currently on abstract entities.

Each of these abstract entities could be called "Beliefs". The fact that all these beliefs are generally accepted as science or physics and that the process of cognition or thinking in itself is nothing but "generation of newer beliefs" means that at least the basic understanding about how beliefs form and how our mind works is imperative if one has to know how these beliefs formed and where errors in "belief-formation process" could have occurred and which of these errors could be potentially corrected to arrive at a "Theory of Everything".

It is imperative to understand certain Psychological aspects of the thinking process if one has to overcome the hurdles that lie in between our clearer understanding of the Theory of everything.

Chapter 3

PSYCHOLOGY OF A BELIEF

A belief is a fundamental unit of the thinking process. What constitutes a Belief is difficult to define.

But "to believe" is something that is done fundamentally by our cognition. A belief is something that a conscious person can have in the subconscious mind. It remains stored within, can pop into the conscious mind from time to time and can in fact influence perception of the conscious mind even without being visible.

Typical examples of beliefs are "a tiger is a threat to my life i.e. I should run away from one" or "I like pizza" or "it feels safe at my home".

Each one of these is a subconscious belief stored in a person's mind.

It figures that there are millions or billions of such Beliefs and they keep forming and getting stored in our memory in the subconscious mind as we are exposed to more stimuli, more events and as we choose more actions.

In the process of the creation of beliefs, what stimuli we get are crucial. In the same way, what pre-existing beliefs are present in the subconscious mind alter our perception of a new stimulus.

For example, a person with a belief that there is a tiger in these woods would feel scared of every little gush of wind. A person who has no idea of a tiger would have a different reaction to the same gush of wind.

The psychological glitch

Beliefs are like walls in our minds.

They can get reinforced with life events. A person might have had a belief wall of 1cm before and after multiple reinforcements, the same wall becomes 1 inch thick. The thicker is the wall of beliefs, the thicker it is to break.

Unfortunately, the wall of beliefs makes the person blind to viewing beyond his beliefs. A person's belief forms a sort of a box and the persons thought process gets trapped within that box.

The term lateral thinking or thinking out of the box points out the same psychological mechanism.

When a box of beliefs forms for a person, the "train of thoughts" keeps entering the same box, so that the person gets trapped into thinking in a particular manner. This means that the person keeps giving importance to some aspects and fails to consider other more important aspects as the belief that these are unimportant has probably arisen and has taken a foothold subconsciously.

A psychological glitch is the same phenomenon in which the person gets trapped in a way of thinking and his thinking process is unable to think beyond the walls of his beliefs.

Many examples of the same can be given.

An architect will view the beauty and utility of a building, a businessman would think only in terms of finance and revenue generated by the building while an environmentalist or a doctor might be more capable of thinking about the ecological aspects or healthcare spread or disease spread or ventilation related aspects of the same building.

This is the reason why multiple brains working on the same project helps as different people view the project from their unique perspective and give more importance to variable things.

A very important example related to physics can be quoted here.

Physicists often get trained to think in one particular manner. Physics has many equations, lots of mathematics and a lot of conservation laws. It also has many fields. Thus most theories proposed as candidates for a theory of everything maintain the same line of thinking with multiple fields. The emphasis is on getting the Math right and getting everything down as an equation. A firm belief in minds of physicists is that "a theory must have a good equation with mathematical beauty, even if its implications are showing weird logical results."

Examples of scientific beliefs

Multiple examples can be quoted here.

Democritus believed that matter is made up of indestructible little balls which he called "atoms". He envisioned (i.e. believed) that they probably have little spikes on their surface. According to the arrangement of these spikes, the physical properties of liquids solids and gases was explained.

Socrates and Plato, contrary to this, had different beliefs. They believed that everything is made up of matter and form. The matter is non-conscious but material. While the form is conscious but not material. This concept of "Dualism" was soon given up due to the lack of evidence of this "form" and "materialism" was wholeheartedly accepted by the scientific community.

Newton believed that light is made up of corpuscles i.e. particles. Contrary to that Huygens believed that "light is made up of waves." Phenomena like diffraction, reflection and interference proved that light was a wave. But experiments on the photoelectric effect,

Einsteinian explanation of the photon in his landmark paper and also Compton's scattering experiment showed conclusively that light behaves like particles as well.

Initially, it was believed that there is a universal medium called Ether that pervades the entire Universe. Many experiments were conducted to prove or disprove its presence. The famous Michelson Morley experiment which intended to detect ether wind proved that there is no detectable "ether wind" and established the fundamental nature of the speed of light.

Thus the scientific process is a process of "Formation of Beliefs" and then looking to prove them or refute them.

Eventually, the beliefs are slightly or sometimes profoundly modified to suit the experimental data.

For example, Einstein's landmark papers on Special theory and General theory of relativity were milestones that broke many such pre-existing beliefs or preconceptions about physics. The theories changed the complete picture of Space and time which the scientific community had. He wedded space and time into a single intricately woven fabric of space-time. Given in it was the troubling notion of "no universal simultaneity" and "relativity of time so that every point in space-time can have its version of time and thus time runs differently for different individuals".

Here the long-held preconception that "a second is same measured anywhere in the Universe" was shattered.

The scientific process starts with "a scientific question" which the scientists want to answer or provide an explanation for.

A scientist or a team of scientists attempts to answer this scientific question by the formation of some beliefs. This collection of beliefs is what is called the "hypothesis" or a "theory".

Once the hypothesis is formed, theoretical models are formed and extrapolations are obtained based on previously known knowledge. These extrapolations might involve the prediction of certain experimental results. If these predictions are proven, it gives credibility to the hypothesis.

Often the hypothesis needs to be modified or additional assumptions need to be added to make the predictions of the theory fit observations.

The best examples of such a hypothesis that we can give today are String theory and Loop Quantum gravity. Both these are theories that have certain assumptions. Based on these assumptions, mathematical and physical theoretical models are created and the predictions obtained. The energy needed to test the predictions of these theories is too much for us presently and we may have to wait for their confirmation.

This means that these two theories are still in the "realm of beliefs".

Examples of how some scientific beliefs acted as a psychological glitch and prevented thinking in the right direction in the past.

Almost all scientific advances in foundations of any subject are either the theoretical prediction of "how some beliefs could be wrong" or experimental confirmation of the same.

"Time is absolute and is same everywhere" was a long-held belief in the scientific community. This was broken by Special Relativity. "Gravity is a force which instantaneously acts on a distant object and pulls it towards the Earth" was a long-held belief held post-Newton's description of Gravitation. This was broken down by the description of General Relativity (GR) by Einstein and its experimental confirmation by demonstration of Gravitational lensing and Gravitational time dilation by numerous experiments.

The belief of "the existence of Antimatter particles" which was first described by Paul Dirac was confirmed by the actual detection of Positron by Carl David Anderson in 1932.

The belief of a steady-state of the Universe which was long held by the scientific community was thrashed by the detection of Galactic redshifts by Edwin Hubble and heralded the belief of an expanding Universe. This belief of a "steadily expanding Universe" was thrashed by the calculation of Hubble's constant with CMB which suggested that the Universe is expanding faster today than it was before.

In this manner, many scientists, by their scientific contributions, have broken the previously held beliefs. Multiple examples can be quoted like Darwin (theory of evolution and natural selection), Galileo (Sun centric model), Max Planck (quantum nature of heat loss by a black body which Einstein in his paper on the Photoelectric effect, called "the Photon"), Bohr (Bohr's atomic orbits), Pauli (existence of a third particle in Beta decay later discovered to be the Neutrino). The list is endless.

Beliefs/Facts/Pseudo-facts

Beliefs are things that a person agrees or believes without the need for proof.

For example, "there is a tiger out there" is a belief that doesn't necessarily require a conclusive proof.

A fact however is something that has conclusive proof that can be accurately measured and confirmed.

Facts are rare. Often things that were believed to be facts turn out to be beliefs that persisted for a long time due to lack of evidence.

Thus, "Nitrogen is a gas" is no longer a fact for those who know that in a compressed cold cylinder, nitrogen can remain in a liquid state as Liquid nitrogen.

"Solar system is Earth-centric" was believed to be a fact in the pre-Galilean era. After the Sun-centric solar system was proposed, it turned out to be just a belief that persisted for long as a fact. In short, it turned out to be a pseudo-fact.

A belief can thus be a right belief or a wrong belief.

A fact is always right. But such things are rare. After the edge of our knowledge expands, a previously known fact has a likelihood of turning into another pseudo-fact.

Identifying such Pseudo-facts in our thinking is the key. Because these Pseudo-facts are beliefs that sway our thinking and cause a psychological glitch in our mind and prevent us from thinking in the right direction.

Beliefs in the scientific process

"A belief" is an integral part of the scientific process.

The scientific process aims at making sense of the world we are living in.

This process involves making observations of what constitutes this world and then explaining the same.

The process of observation can be direct observation of physical phenomena or it could be performing simple or complex experiments and obtaining their results.

Once these observations or results of experiments i.e. experimental data are known, what is needed is to offer explanations based on assumptions.

These assumptions or Hypotheses are basically "beliefs".

Thus the scientific process is a process of formation of beliefs in our mind and then looking out for evidence for the same so that one can convert the belief into a fact or a Pseudo-fact.

The process starts with identifying a "Scientific Question or Problem". An assessment is made of the current experimental evidence and current beliefs. The present beliefs are modified or completely new beliefs are thought of. With these newer beliefs, theoretical models are obtained. The theory that these beliefs constitute is described in as much detail as possible. The necessity here is that the model which is obtained should not directly contradict the current experimental data and also should not give contradictory predictions. Thus once a theory (which is a set of beliefs) takes birth, the first mandatory thing to be looked at is that it does not contradict the current experimental evidence. Thus a theory of Gravity that predicts that an apple will rise up instead of falling cannot be entertained any further as it contradicts with most observed phenomena. Most new theories fail in this first step itself and this step indeed filters out most of the possible theories. Once it is clear that the theory is compatible with most of the experimental evidence, predictions are obtained by extrapolating the findings of the theory at conditions not routinely observed. Experiments are then performed to try and find if the predictions of the new theory match with the new experimental findings.

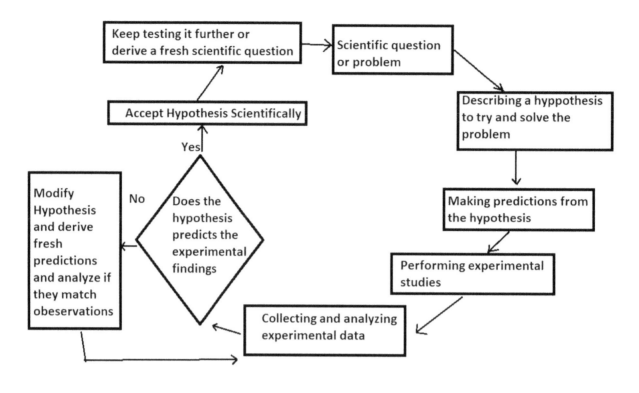

The Scientific Process

Figure i/3.1: Shows how the Scientific Process works

How to spot a good theory from a bad theory?

A good theory can make testable predictions. Thus a completely non-falsifiable theory cannot be a good theory. A good example of such a theory is any theory that suggests a "Multiverse". The probability of testable predictions that confirm such a theory would be extremely low and such a theory should ideally not be a part of the Scientific Process. However, many theories today require an energy range that is beyond the current testable range including String theory.

Another feature of a good theory is the number of assumptions and types of assumptions that a theory requires. A theory that can explain all experimental observations with a minimum number of assumptions is better. A theory that makes assumptions that are within the boundaries of logic is better than a theory that needs assumptions beyond logic. For example, if a theory needs only three dimensions and can explain all findings, it is preferable to a theory that needs higher dimensions that have never been conclusively proven. But adding more assumptions is not necessarily bad if it can explain more experimental data.

A theory, which can explain, majority of the experimental findings with just a handful of assumptions would thus be extremely good.

Current theories like Quantum Field Theory have to assume the presence of at least 37 different fields present throughout the Universe. It also has to add the values of constants in their assumptions as none of these can be derived from the theory itself. String theory requires an 11-dimensional hyperspace that has never been detected.

Thus, if a theory has exotic-looking assumptions, then it automatically becomes a poorer theory than a theory with lesser and straightforward assumptions.

A theory that needs complex manipulations and the addition of multiple assumptions to explain newer experimental findings and which despite decades of research by hordes of physicists could not make ample progress would be poorer. Compared to this, if for a theory, very few additional assumptions are needed to offer explanations of a wide variety of phenomena and most accepted theories can be derived at appropriate locations from it, it should be considered as a better theory. In short, a good theory is one which can not only encompass, but also explain in detail most other accepted theories.

CHAPTER 4

AIMS OF A GOOD THEORY IN SCIENCE

The ultimate aim of science is to come closer to reality. The ultimate theory should thus explain all the phenomena happening with a minimum number of assumptions. This means that the theory should be capable of explaining multiple phenomena with a lesser number of assumptions.

Not only it should confirm all the experimental findings, but capable of incorporating all the newer and old findings with minimal addition of assumptions. It should be capable of incorporating the majority of other theories which are essentially incomplete and based on some aspects of science and form an overarching theory that explains the phenomena happening in these theories.

For example, the theory that materials are made of atoms was useful as it was capable of explaining physical features like state, temperature etc. The theory that they are made of negatively and positively charged sub-particles would explain a lot of things including chemistry and electromagnetic bonds. The plum-pudding model emerged but was unable to explain the findings of Rutherford's experiment and thus came the planetary model of the atom which could successfully explain the findings. Soon it was realized that even it was wrong as the Electron revolving around the positively charged nucleus would keep falling inwards due to electromagnetic radiation within a few milliseconds and is thus incompatible with the then known laws of Electromagnetism. The atomic orbits that were first described by Neil's Bohr also explained the emission spectra and absorption spectra. Even Bohr's orbits failed when it was applied on atoms larger than the hydrogen atom. With the availability of Schrodinger's equation, the s, p, d and f orbitals were described by solving it and the current model of the atom which can explain a lot of phenomena like electromagnetic bonds and atomic spectra became a widely accepted view of the atom. To note in this entire episode in the history of deriving the acceptable model of an atom is that a model which contradicts with the known knowledge of the era was discarded and the better models which were able to recreate the experimental observations were accepted despite having additional assumptions. Nowhere in this was a theory contradicting the current knowledge entertained.

A good theory must make everything simpler rather than making things complicated despite explaining all the experimental findings. A good theory should explain the existence of or derive most of the constants of nature.

A good theory should be specific enough to make predictions that can be tested. It should, thus, be falsifiable. A good theory must reduce the fundamental constants rather than adding more constants.

A bad theory, on the contrary, is just the opposite. A theory that is so vague that it is incapable of making any prediction and it has so many free parameters in its assumptions that any discrepancy in predictions and observations can be explained by some manipulation is a bad scientific theory.

The best example of such a theory is the hypothesis of Dark matter. There is no specified way that the theory describes how this dark matter is distributed. There is also no specified amount for any Galaxy. The physicists can pick up any amount of it and place it wherever needed and in whatever density that is needed in the computer simulations to explain the findings. The theory cannot derive any predictions whatsoever and all its successes are based on more addition of assumptions. However, there is no other theory at least till present which can explain the findings seen in the Cosmic microwave background better than the LCDM model and thus despite its drawbacks, the Dark Matter hypothesis is still accepted by many. This will remain until a better alternative becomes available in the future if it does.

An ideal method of deriving a theory is

- deriving a phenomenological description,
- then deriving mathematical details,
- then creating a model and deriving predictions and
- the final step being testing if the experimental observations match the predictions.

If a theory can match any number of deviations in experimental observations by internal tinkering, it is essentially non-falsifiable and is a bad theory.

A theory where every assumption and every prediction it makes is beyond measurement is also essentially non-falsifiable and is thus just theoretical crap even if one can derive beautiful mathematical equations or scientific computer simulations for it.

A theory that makes things extremely complicated mathematically so that the assumptions needed are not only not reduced, but where the assumptions tend towards infinity, is not a good scientific theory. One such theory is the String theory. To note that, mathematical techniques derived from research in String theory can still be useful and thus the effort of all those working on it isn't going to go waste.

Ideally, the process of theory development and obtaining experimental data should be independent. A good theory is a theory that effortlessly derives experimental observations without the need for additional tinkering.

Thus a theory that is derived with the sole purpose of deriving a single observational finding up to twelve decimal places by adding multiple assumptions to it isn't good even if it agrees with the experimental result. This is because the theory would be different if a different result was obtained.

An ideal candidate for such a theory is Quantum Electrodynamics (QED). Here multiple assumptions were added to the theory to include multiple levels of interaction like emission of a photon, reabsorption of a photon, creation of pairs and destruction of pairs of virtual particles etc. as diagrammatically represented by Feynman diagrams. These interactions were just made up to modify the theoretically predicted result up to twelve decimal places so that it matches with the observations. This mathematical "hocus pocus" (as quoted by Feynman himself) is bad science. The falsifiability of all these additional interactions is zero. And QED can be modified further by adding more such interactions just to modify the theoretically predicted value just in case the experimental observations vary with an increase in precision of measurement in the future. (Note that QED is one of the foundational theories on which DGR itself is based since the idea of Quantum foam and quantum energy fluctuations at the Planck scale is taken from QED. Thus not every part of a theory may be wrong. One has to just learn to distinguish the useful from the useless.)

Another common fallacy in the scientific method today, which is described elsewhere in detail is the Congruence bias.

In this, the physicists keep trying to prove their beliefs right and thus focus only on experimental data and observations that prove them right.

A subtler way this can be done is by looking only at the data that proves them right.

Today, most experiments in physics are extremely complicated and are carried out with huge funding with a big team of experts engaged in achieving a common goal. The mega particle accelerators today produce an astronomical amount of data. The physicists have to literally handpick the data that potentially matches with the predictions of the theory. All the other data which might amount to as high as 95-99% of the data is discarded as background noise. It is not impossible then that the physicists might discard important data which prove them wrong in the zeal to prove themselves right. Acknowledging that the theory on the predictions of which this multimillion-dollar project was built is indeed wrong is a difficult thing for a physicist whose pay-check depends on not acknowledging it.

CHAPTER 5

HUMAN BELIEFS AND BELIEF BIASES

When beliefs form in the Human mind, they constantly affect the way the human mind works.

As discussed before, beliefs are like walls, so that the human mind just cannot think beyond those walls. These walls have a unique influence on the direction of the thinking process. The mind learns to think in a particular direction and it soon becomes the norm so that any other direction of thinking becomes banned. Another way of expressing it is that the mind learns to think in a particular manner or develops cognitive ease in thinking in a particular manner and any deviation in direction of thinking provokes anxiety.

This is like a constant Glitch in the electronics of a Computer created by a Computer virus that makes the computer repeat a mistake again and again.

These glitches created in the human mind are called Cognitive Biases.

Literature on Cognitive biases is growing tremendously fast and cognitive biases are being described in behavioural financing, consumer behaviour, healthcare decisions by healthcare professionals, researchers etc. just to name a few.

In short, the "Thinking Process" of humans is not completely flawless and multiple flaws can be introduced by these biases.

Belief biases

This glitch created by any belief is termed as the "Belief bias". Although details of these biases are not needed here, what is important to understand is that when humans have a belief within their mind and they come across another idea, the mind gets excessively influenced by the belief and relies excessively on the believability of the idea rather than evidence. In research, this bias is critical. When a researcher believes in a hypothesis (say dark matter), when he comes across another article published on the subject of MOND, the underlying belief that MOND is wrong excessively influences the mind and thus the researcher reads the article only to find faults and as soon as he finds one, latches on to it completely forgetting the merit in the article. Thus relying on the believability of the result instead of the quality of research.

Anchoring bias

In this kind of bias, the human mind gives excessive importance to the first information that it comes across. Thus an amateur Physicist during the early days comes across a Mentor good in String Theory and develops a belief that String theory is right and thus LQG is wrong. Note that these initial beliefs keep getting reinforced with advances in his career as he reads more stuff confirming his beliefs (see Confirmation bias)

Ostrich bias

In this bias, the mind firmly believes that "I am unbiased" or I can never be biased. In this, the mind keeps rationalizing the tendency to confirm one's belief and subconsciously keeps ignoring the negatives of the project or theoretical framework he is working on and keeps focussing on positives.

Confirmation bias

In this, the human mind has cognitive ease in interacting with people who confirm its preconceived beliefs. The researcher derives more pleasure reading articles that confirm his beliefs and gets annoyed while reading articles or books that criticise his beliefs. A Loopy physicist attending a Loopy conference and a Stringy physicist attending a Stringy conference is in a way confirmation bias unless they feel equally comfortable attending all types of conferences and presenting their work.

Self-Serving bias

Ideally, scientists should work to find out the truth and not be annoyed or unhappy if proven wrong. This, however, is far from reality and the underlying psychological glitch in it is called the "Self-serving bias". This is an underlying often firm belief that "I have to maintain my self-esteem". Thus any article or paper or book criticizing your ideas and thus trying to disprove "these believes" held by the mind provokes anxiety. People who achieve success in life like hundreds of papers published in reputed journals or positions of authority in reputed institutes are often having an inflated self-serving bias. Science, however, demands that the scientists who engage in it should stay grounded and recognize the gap between what is known and what is unknown.

This cognitive glitch would be the single most important hurdle that makes people in authority from accepting a possible good but less popular direction towards a TOE offered by DGR.

Congruence Bias

The true meaning of this is that while undertaking research, a researcher tries to prove his preconceived ideas right and in the process fails to ask questions or collect data that can prove his hypothesis wrong.

The more hidden meaning of this is that the human mind thinks in a direction that tries to prove the preconceived ideas right instead of proving them wrong. In this, the mind keeps thinking in lines that are comfortable (despite repeated failure) instead of thinking in other more unfamiliar lines for the anxiety of failure or cognitive strain such approaches provoke.

Congruence bias is probably the single most important reason why, I believe, that a hard-core PHD holding physicist would never have reached the theory described in this book. This is because even at the graduate level of physics, excessive stress on mathematics is given. Abstract looking entities and approaches like "Fields" and "Energy" are normalized and it is repeatedly taught not to question what these are and what they mean fundamentally.

Congruence bias is the reason why most physicists would have a hard time giving up the idea of fields from a point where they believe in a theory that requires 37 fields. Congruence bias would make most physicists abhorrent to the idea that there could be something wrong in GR or SR.

Congruence bias of the entire "Physics community" is what I consider as the single most reason why progress in General Relativity and Quantum Mechanics Stalled for such a long time.

It won't be an exaggeration to state that it was the realization and understanding of the Congruence bias among the entire Physics community that heralded within me and made it possible for me to reach this theory.

"Is the entire Physics that we know of Wrong then?"

When Einstein described his theory of SR and GR, he proved a lot of physics known at that time as wrong or partially correct. However, although some aspects of physics changed, many things remain valid.

Thus, the concept of fields that are a very useful mathematical tool will remain valid. The real meaning of what a field means would be known to us if "Dynamic General Relativity" (DGR) is accepted or proved correct in the future.

Thus DGR is a new direction of thinking of physical phenomena and doesn't necessarily prove old approaches wrong. It just gives a better phenomenological description of the physical events.

The BCC Cycle

When a belief forms in the mind of a physicist, the mind follows a unique cycle which I like to call the BCC cycle or the Belief bias - congruence bias - confirmation bias cycle. In this, a belief "B" forms in the mind of a scientist. The scientist (here a physicist) writes a paper with congruence bias in it with the sole intention of upholding the belief or proving it right. This paper is read by another scientist who also had the same belief. The process of reading confirms the belief B of the second scientist. This second scientist under the reinforced belief writes another paper with congruence bias. This second paper, read by a third scientist has the same effect of confirming the belief B of the third scientist. In this manner, scientists keep reinforcing the beliefs of each other.

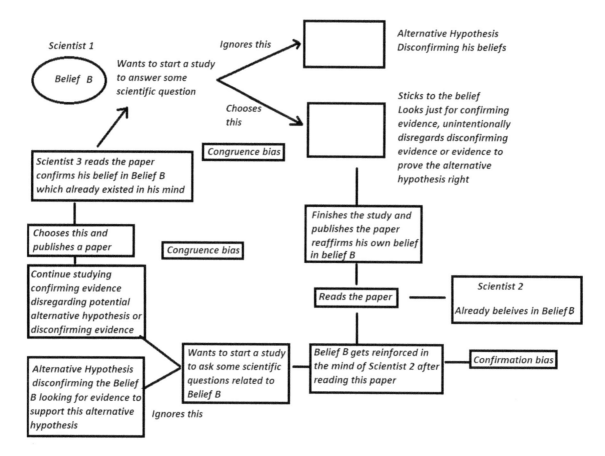

Figure i/5.1: Shows the BBC cycle or Belief bias- Congruence bias- Confirmation bias cycle.

There can be alternatively stated as the ABC cycle or the Anchoring bias- belief bias- confirmation bias cycle.

Note that scientists take great care to avoid the BCC cycle by utilizing multiple methods like double-blinding, incorporating Control cohorts, using sigma to calculate the probability of getting the same finding by chance etc. However, these methods cannot be helpful in every aspect of a theory.

CHAPTER 6

HOW MY SECOND BOOK LED TO MY THIRD BOOK?

One of the first questions which one might ask after reading this book is:

"How can a person without any experience in research in physics or post-graduate education/PhD in physics or higher mathematics write a book on the theory of everything?"

This would indeed cast serious doubts on the credibility of whatever is written. This would indeed be the most important hurdle in acceptance of whatever is given in the book on mainstream physics.

In my second book, I have described how pre-existing beliefs or prejudices skew the way the human mind thinks.

Only a scientist in physics or mathematics indeed has enough knowledge about all the subspecialties of physics, knowledge about all the newer experiments, prevailing theories and scientific advances in physics and other related fields, the conservation laws of physics and how mathematics works for physics and thus can take the present theories forward without conflicting with the present theories and experimental data.

The present physics however has come into a stage where there have been no significant advances in the field since the last century. Multiple mysteries have remained and newer ones have been added. There is no answer to them and there are no prospects of answering them in the near future with current technologies. Expensive experiments involving thousands of physicists working for analysing thousands of gigabytes of data and involving an extreme amount of energy like the Particle accelerators like CERN have failed to make significant progress in the understanding of our theories. Opinions are divided amongst physicists regarding the feasibility of building bigger particle accelerators with higher expenditure and involving higher computing capabilities. Currently, a lot of physicists are engaged in theoretical frameworks most of which provide no prospects of providing experimental evidence. As explained elsewhere, the physicists working on fundamental physics can be divided into those supporting String Theory and those supporting Loop Quantum Gravity.

The number of proposed theories is very high and keeps multiplying. Thus a physicist who knows all the postulated theories in full detail is difficult to find, most of them focus on a small part of physics.

The requirement of complex mathematics in each of these means that modern physics has gone beyond the understanding of a layman. Most of these theories are based on complex mathematical techniques and are full of operators and symbols, none of which make sense to the layman.

It is the opinion of at least some of the physicists that current physics is literally "lost in Mathematics".

The current way forward for theories of physics is to create a mathematical equation, even if making sense of it is difficult. Often the interpretation of equations obtained to explain data is difficult or impossible, creating further hurdles.

Even experimenting in today's world is a problem. Most of the answers to the very basic physics have been answered by myriad experiments. Today, experimenting in physics to answer the questions remaining to be answered cannot be carried by a single person and has to be a team effort of thousands of scientists, physicists and other experts working together.

Physicists of today thus have to specialise in respective sub-specialities. Reading more about one aspect of an extremely small portion of a subject is needed which means that their ability to think holistically involving multiple specialities at the same time is limited.

Despite all these problems, it is still clear that a good knowledge of presently accepted science and discussed theories is needed before one can think of taking such an advanced step as describing a theory of everything.

Beliefs as walls

When I was writing the second book, I came across the unflinching fact that the human mind creates walls of beliefs.

The mind becomes blind to looking beyond the walls of beliefs formed within. This fact was very well exemplified in Einstein's description of space-time.

The belief that "time is absolute and that a second will be a second wherever you go in the universe" was so extremely ingrained in all the physics in the pre-Einstein era that nobody questioned it. It was like a wall beyond which no physicist before Einstein could see. Most theories that came before Einstein had this as an invisible postulate that needed no mention or proof.

Einstein managed to break this prejudice and view beyond it. He shattered the wall of this belief.

Subsequently, many such walls were shattered when Quantum Mechanics was described. One example is the belief that light is made of waves, which belief was shattered and replaced by the "wave-particle duality" of light.

Advances in fundamental disciplines of science like physics are nothing other than shattering pre-existing walls of beliefs of the scientific community.

The funny part of research in physics or for that matter any subject is that not only does it shatter walls, it also forms newer walls of belief.

A researcher can have such a Belief introduced into his brain at a tender stage of graduate or postgraduate training. As more and more reading about these beliefs progresses, the size of these walls keeps getting stronger.

While this "Theory of everything" took shape in my mind, I came across many such beliefs which are, in my opinion, responsible for skewing the minds of physicists away from reality. These are the beliefs that prevented the physicists from thinking in the right direction to reach the theory of everything.

These beliefs are described below along with reasons why these prevented the physicists from thinking straight. I believe that giving up these beliefs was mandatory before one can reach the Unification of all forces and reach the theory of everything.

(Note that what has been reached at the end of this book isn't the complete theory of everything, but just the right way of thinking to potentially achieve the Unification of all fundamental forces of nature and a potential theory of everything.)

It is clear that the modern physics curriculum extensively magnifies each of these beliefs and thus physics curriculum adds to the thickness of walls in the minds of physicists.

This is, I believe, the reason due to which it was unlikely for a hard-core physicist to reach the present theory, as it was mandatory to look beyond these walls to reach it.

1. Fields exist i.e. forces at a distance are caused by fields

   The concept of fields entered physics after Michael Faraday described a field (later called as the magnetic field) formed due to current passing through a circuit. After Maxwell successfully described his equations and was able to show how light is an electromagnetic wave i.e. a wave in the electric and magnetic field, the concept of fields was further strengthened. Subsequently, many fields other than these have been described, one of them being the Higgs field responsible for mass. The currently accepted theory of the minute particles is called the Quantum Field Theory which assumes the presence of more than 37 fields and every fundamental particle is just a vibration in one of its respective fields. The conceptual idea of a field has been deeply rooted in minds of physicists to such an extent that a physics theory where no fields are present is almost bound to be discarded before being understood.

   The idea of fields is compatible with mathematical calculations and has received good experimental support due to its predictive ability. However, a field is nothing but a value given to a physical entity at every point in space. In this manner, imagining fields is helpful. However, imagining how 37 fields can co-exist at every point in space all over the universe is difficult for a layman and indeed if there is a simpler explanation that can successfully account for all the experimental evidence, it is better than imagining 37 different fields.

2. Time and space are not absolute

This belief was started after Einstein described the Special Theory of Relativity. After this came the General Theory of Relativity which proved superior to the Newtonian theory of Gravity. The success of both these theories has established the relativity of space and time in the physicist's mind to such an extent that current physicists firmly believe that there is no absolute coordinate system and space can contract/expand and bend. Even time can dilate and need not be absolute.

Although this was a significant advance and emergence of this belief from Einstein theories can be considered a shattering of previous prejudices, it also constitutes a belief that needs to be shattered.

Instead of this, what seems closer to reality and what is the assumption in our theory is that

-there is an absolute time called Universal Time and there is a relative time called Local Time

-there is an absolute 3-dimensional co-ordinate system in the universe with a single unit being one Planck length and the space has the ability to actively expand or contract, due to the process of active Creation or Destruction of PCs, which makes it relatively flexible and capable of bending despite the absolute coordinate system.

Until we leave the absoluteness of the above "Belief of relativity of space-time" and accept the above two beliefs, we cannot reach our theory.

Although Einstein successfully managed to wed space and time into space-time and was able to see the relativity of space and time, he did not attempt to find out why they behave in this manner.

The ability of space to bend was considered a fundamental property of space-time instead of some underlying mechanism responsible for it.

Popular physics teaches that "Gravity is just an illusion and that there is no force directed downwards acting on the falling apple."

Einstein's equivalence principle implies that gravity is just an illusion and it isn't a force, just an effect caused due to curved space-time.

Einstein was willing to give up the centuries-old observation that things fall due to a downward acting force of gravity and willing to believe the absurd-looking non-intuitive claim that gravity is just an illusion due to the bending of space-time.

How exactly should space bend and enable the apple to fall towards the Earth was extremely difficult to imagine for humans. Einstein was willing to believe that this inability to imagine wasn't due to a flaw in his theory but the inherent flaw in the human mind to imagine it. It is this firm wall in his mind that gravity is solely due to bending of space-time instead of some underlying mechanism of space-time which causes the space-time to bend secondarily that clouded his mind from thinking towards active spatial contraction due to time dilation.

The bombardment of Special and General Relativity to physics graduates with complex mathematics which rarely makes complete sense keeps adding to the wall of belief. It is taught to them never to question these beliefs or be sceptical of them.

We live in a three-dimensional universe. An actively expanding or contracting universe isn't something unacceptable to Physics and expansion of the universe is a known phenomenon.

A differentially contracting or expanding space can lead to a bent space-time leading to bent Geodesics described by General Relativity even with an absolute coordinate system.

3. There is no connection between energy and time.

    There are multiple beliefs related to the connection between energy and space-time in modern physics that need a relook. These include the following -

    - ❖ ***Energy is treated as an independent entity and has no conceivable connection with time.***
    - ❖ ***Energy doesn't need any space, i.e. a huge amount of energy (equal to almost the entire energy in the universe) can be concentrated into an infinitesimally small region of space.***
    - ❖ ***The presence of energy has no Consequence on space.***
    - ❖ ***Energy is regarded as an abstract entity that can do work. Energy is said to exist in itself and is unrelated to time.***
    - ❖ ***The above definition of energy is enough for physicists and the need to further delve deeper to examine its relationship with space and time is unimportant.***

    The relationship between time and energy is probably the biggest prejudice that physicists will have to break before understanding the present theory.

4. Quantum particles, according to currently accepted physics, have a property called Superposition, in which they can be said to exist in more than one state at the same time I.e. for example an electron can and does take all the possible paths from point A to point B.

    Superposition is a property that emerged as an explanation of the weird findings of the Double slit experiment.

    The details of how the wave-particle duality of light and wave-particle duality of particles came into physics and how Superposition emerged as a potential explanation is discussed elsewhere.

The idea of Superposition along with Heisenberg's Uncertainty Principle implies that a Quantum particle cannot possess accurate location and momentum. This isn't just due to our ignorance or limitation of measuring-ability of our instruments but a fundamental property of nature, i.e. it is not true that the electron takes only one path or location at an instant and the nature knows it but it is unknowable to us. Instead, the electron, according to teaching in Quantum Mechanics, does take all the paths which are possible or is present in all the locations around the atom at once and is thus called the Electron cloud.

This weird looking property is due to the wave-particle duality of particles. (A particle that has no connection with a wave need not obey Heisenberg's Uncertainty Principle, the one which is closely associated with a wave has to obey it).

This wave-particle duality however is still abstract and it is unclear what it means even to those who teach it.

It might mean one of the below mentioned things in the case of the electron by using plain simple logic.

- the electron is a particle and a wave both at the same time every moment it exists when it is not detected yet i.e. both at the same time (it collapses into a particle i.e. collapse of the wave function when a measurement is made) (this is the accepted view)
- the electron is always a single well-defined particle with a well-defined location but is associated with a wave that has independent existence to the electron (independent particle interacting with an independent wave)
- the electron is always a well-defined particle but its interaction with its neighbouring space-time creates a wave whose dimensions depend upon the properties of the particle. (Election has independent existence but the wave is generated secondary to the presence of electron) (as suggested in our theory)
- the electron is only a wave when not measured and a particle only when measured (an alternative viewpoint to what is accepted)

Another equally weird explanation of the findings of the double-slit experiment which is often given as an alternative viewpoint to Superposition is called as "Many World's Interpretation", in which every moment, the world splits into two worlds, in one the electron goes through one of the slits, in the other, it goes through the other. This approach avoids Superposition but is equally detrimental to logic.

Although these beliefs are weird and a little away from logic, they have been accepted by physicists and are strongly ingrained in physicists' minds.

5. There is no creation

The "materialism bias" of science is hard to shatter. The occurrence of the word "creation" in a theory will invariably mean its doom in the present state of physics.

In such an abhorrent environment of science, how can an imprisoned mind think of a theory whose foundations are based on the weird assumption that

"At each point in space, Creation and Destruction are going on spontaneously and that everything that exists is due to the variation in the ratio between Creation and Destruction?"

The acceptance of the weird facts of "universe is expanding" "the big bang theory" and "spontaneous formation of a virtual particle and anti-particle pair happening in quantum foam" all of which imply the process of Creation of something from nothing, is acceptable to modern physics. This acceptance, however, is conditional. The condition is not to use the word "Creation" anywhere. The use of "ratio of Creation to Destruction" would indeed be extremely difficult for science to accept. Wherever we need to take the help of something close to Creation, to prevent the entry of "a Creator" any amount of weirdness is acceptable in the theory (including infinite universes).

With so much hatred for Creation, a theory where Creation and Destruction are central to its assumptions would need to break significant thickness of belief walls.

6. Universe and everything in it is purposeless and everything is made of material elements alone, there is nothing more than material elements in the universe.

The first step in breaking this firm bias of materialism was taken when physicists agreed to believe the weirdness of Quantum Mechanics. Thus this belief isn't present in presently accepted theories. But the psyche of individuals still has this belief to a significant extent. Presence of something more than material elements like "information about local time" that can determine space and which can be transferred from one place to another constitutes something beyond the physical realm.

How exactly this information is transferred from place to place is extremely unclear yet. But the implications of the transfer of such information is mind-blowing.

If such information does exist and does get transferred from one point in the universe to another, every point in the universe would need to be capable of understanding the information coming from all directions, interpreting it and transferring forward the result. This implies that every point in the Universe is conscious and capable of interpreting such abstract information. How exactly this happens is not known, but the very thought of a conscious Universe which allows simultaneous Creation and Destruction at every point would be deterrent to any further acceptance by contemporary physics.

It is clear then that these psychological walls of beliefs are too much for a physicist to think beyond.

Thinking out of the box

With the walls of these beliefs firmly in place, physicists of today won't be able to think in the direction which leads to the present theory.

They have to learn to think beyond these beliefs.

The realisation of the reasons why so many experts in physics could not solve the riddle of the theory of everything encouraged me to think out of the box of supposedly firm but not infallible beliefs that haunt physics today.

It is this out of the box thinking which might take the theory away from realms of reality into imagination, into a series of unprovable theories and increasing complexities. But physics is already in this realm and most theories presently presented are complex and beyond experimental evidence. Thus, it was fair to take this risk. The reward was too great.

The limited knowledge of mathematics, instead of being a hurdle, was a help as there was a possibility of getting lost in the complexity of the math without getting any output.

The theory at its present form is promising, however, it has to stand the test of mathematics and science to make it compatible with all the experimental data before one can jump to conclusion that it is useful.

Chapter 7

CONVENTIONAL VS AUTHORS METHOD

The scientific method involves making observations, doing experiments, generating data and ultimately making theoretical models with assumptions that can explain the given observations. Once we have certain theories or models that can successfully account for the observations of the experiments, there comes extrapolation and predictions of these theories which would then be put to test by additional experiments and additional data production.

Once the additional data from the new experiments is obtained, the theories are put to test if they can match the observations.

This relentless cycle keeps going.

It involves the process - "observe/experiment - form theories- test the theories so that they conform with the experimental data- derive theoretical prediction and again do further experiments to see if the new predictions fit experimental data".

Physics is full of conservation laws. While deriving newer theories, one has to ensure that the newer theory respects these conservation laws. In addition, the experimental data set limits to certain variables. Thus the newer theory should not conflict with the previous experimental data. If a new theory conflicts with experiments done a century ago, it isn't a good theory.

One of the ways by which one can test the correctness of a theory is its ability to make the correct theoretical prediction. If a theory can make good predictions, it is at least partially correct.

Another characteristic of a good theory would be its ability to explain all the experimental observations with minimal assumptions. The theory is not better if another theory already exists with lesser assumptions than it which can explain all the findings.

However, if a theory can explain some experimental findings which the previous theory was unable to explain, despite its assumptions being more, it is indeed going to be a better theory.

Another needed characteristic of a theory is its ability to be described mathematically and transformed into an equation. Mathematics is an integral part of a theory.

There are two potential methods of the development of a theory.

1) Describe what is happening first followed by deriving its mathematical expression.

2) Derive a mathematical equation first and based on this derive a theory as to what it could mean.

The Newtonian mechanics, Newtonian gravitation wherein Newton observed scientifically and critically what is happening first and derived mathematical expressions for what is happening using newer terminologies like mass, acceleration, force etc., is an example of first mechanisms of theory formation.

The theory of Electromagnetism described first by Faraday and its mathematics derived later by Maxwell, also represents the first mechanism of theory formation.

Even Einstein's theories of Special and General Relativity are derived with the first mechanism wherein Einstein made observations and formed theories in his mind and later utilised Riemannian tensor calculus to derive mathematics that can describe it.

This mechanism, however, is no longer helpful now as deriving observations for Quantum objects is increasingly difficult.

Today, the second method, wherein mathematics is described first and the theoretical framework to describe it is obtained later is more commonly employed.

The first time this method was utilised was probably by Paul Dirac wherein he solved the Schrödinger's equation to include Special Relativity in it i.e. derived relativistic version of the Schrödinger's equation and secondarily predicted the existence of Antimatter.

The complex mathematical derivation of the String theory and m theory are other examples of the second method.

The two methods are intrinsically related and sometimes both the process of theorising and deriving mathematical expressions of the theory can go simultaneously in the scientist's brains.

After the arrival of Quantum Mechanics, the philosophical method or the first method has somehow taken a back-foot. What actually happens is increasingly difficult and hypothesizing only that without taking help of mathematics increases the chances of getting lost. The possibilities of what happens sometimes approach infinity and thus physicists like Feynman had to take the help of mathematical tricks like renormalization to get rid of the infinities.

The events that take place at the quantum level, presently are bizarre or unclear.

There are many hurdles to the philosophical method of theorising. In Quantum Mechanics, there are multiple concepts that make things uncertain. The Uncertainty Principle says that we cannot have information about a Quantum particle's momentum and location at the same time. The Quantum Superposition makes a Quantum particle capable of being present at multiple locations at the same time. The Wave-particle duality of light and wave-particle duality of matter particles further complicate the picture. It would indeed make it difficult for theorizing what exactly happens to a Quantum particle when we don't know if it is a particle or a wave at that moment and the thing which is trying to move it is a particle of light or a wave of light.

The difficulty in visualising General Relativity and Quantum Mechanics together creates further hurdles in the first method. If time is not absolute and it is unclear what is the shape of the background space, if particles can suddenly appear from nowhere and disappear into nothingness in the quantum foam, the first method becomes almost impossible.

It is imperative given all these facts and given the fact that the modern physics curriculum exacerbates these deeply rooted assumptions further, that one has to think beyond these assumptions and if maintained firmly, these assumptions and the biases they create, make it impossible for progress to occur in the first method of theorising.

To arrive at the present theory, the first thing needed is to think beyond these assumptions.

How the author arrived at the present theory?

The first most important observation that I assumed to be true without a doubt and which has been confirmed countless times is

"The space between the falling apple and the Earth or for that matter any object gravitating down towards the Earth's surface progressively and spontaneously gets smaller."

There is a sort of corollary to this observation that forms the first assumption of the theory.

This is that

"space is capable of spontaneous contraction - this essentially means that spontaneous reduction in the number of units of spatial volume - the Planck Compartments between two bodies is possible."

Note here that "the Planck Compartment" is derived from the assumptions of Loop Quantum Gravity.

The second assumption that has no need to be doubtful about and which comes from the Hubble's observations of distant galaxies which proved "the expansion of the universe"

"Space is capable of spontaneous expansion i.e. spontaneous increase in the number of units of spatial volume -the Planck Compartments can also occur."

The third assumption comes from the Special and General Theory of Relativity which suggests that

"there exists time dilation nearer the Earth and the time dilation reduces as we go away from the Earth - gravitational time dilation"

The fourth assumption comes from the Quantum electrodynamics of Quantum Mechanics.

"The Quantum foam exists and as such spontaneous appearance or disappearance of virtual particles is possible"

As Quantum electrodynamics is one of the most successful theories, this "quantum foam" assumption has little doubt.

There is a small corollary that was added as an additional assumption. As the word virtual particle seemed nonspecific or ill-defined, the additional rather controversial assumption that

"these virtual particles could very well be Planck Compartments or aggregates of Planck Compartments."

With this, we reach the assumption that

"Planck Compartments or their aggregates can spontaneously appear or disappear. The active spontaneous expansion or contraction of space could be due to this the spontaneous appearance or disappearance of these Planck Compartments that form a part of the quantum foam."

With these four relatively non-controversial assumptions in the mind, the journey began.

If space can spontaneously increase in size or decrease in size (i.e. volume) and Planck Compartments can spontaneously appear or disappear, could it solve the riddle of Gravity? Can there be a link between the two?

That link could prove to be "time".

If the space in between the apple and the Earth keeps reducing, and there exists a time gradient, can time gradient be responsible for the active loss of Planck Compartments in between the two?

What was essentially needed was to assume a link between time and Quantum foam.

It occurred to me that if a theory is formed, whose basic assumptions established a link between Quantum foam and time, can it explain at least some aspects of reality.

I realised that if this link exists, it has to be a universal fundamental link that happens at every corner of the Universe.

To make it work, some other stubborn biases needed to be given up or modified like "relativity of time" and "relativity of space" I.e. absence of an absolute coordinate system and absence of absolute time.

Initially, I felt that giving up these implications of Relativity could prove suicidal to a theory. As the theory took shape, it became clear that it is possible to have relativity of time and space despite having an absolute coordinate system and absolute time.

Initially, I would never have imagined that the theory would go as far as it went.

But as I started analysing this imaginary Universe formed with the assumptions made by this new theory, it started occurring to me that, the theory can explain many aspects of the reality which were initially assumed to be just fundamental properties. These include inertia, momentum and kinetic energy. The theory could successfully incorporate and explain in detail, Newtonian gravity and General Relativity and also explain Modified Gravity at extreme distances. When I realized that it can, with just a handful of assumptions of its own, nicely explain wave-particle duality of both particles and light, explain the weird findings of the double-slit experiment without the weirdness of Quantum Superposition and possibly also give us clues about Matter Anti-matter asymmetry and arrow of time, I knew that I had unearthed a treasure. I knew that this could possibly provide a unified force that explains all the four fundamental forces of nature and thus could potentially lead us towards "a Theory of Everything".

Authors methods

What I have utilised here is pure imagination.

I kept on applying the above-mentioned assumptions along with a handful of other assumptions to establish the universal link between "time" and Quantum foam and applied all the emergent extrapolations to all the possible scenarios of physical reality that I could think of.

All the pieces of this jigsaw puzzle were already in readiness to fit properly into their places.

The theory worked remarkably and with such few assumptions, I could explain a host of physical findings. All this was possible without the need for assuming the presence of any field. The presently accepted Quantum Field Theory presumes the existence of at least 37 fields and still cannot explain many findings and presumes many others to be fundamental properties. Thus the present theory certainly has a much higher explanatory capacity than modern-day QFT.

Drawbacks of the approach

The theorising which is done here could be considered the first or the philosophical method of theorising. In essence, presently, the theory does not have mathematics to describe it. It tries to explain the gross observations observed in experiments (like falling apple or repulsion between magnets or electrons) based on time and quantum foam i.e. appearance or disappearance of Planck compartments. However, as it incorporates most theories accepted presently in itself, the mathematics of other theories remains valid.

Even then significant work is needed and if needed significant modifications may have to be poured into the present theory for making it consistent with already proven conservation laws.

Although, the introduction of mathematics is needed in the core theory, more importantly, many fundamental questions remain like how do like charges repel and opposite charges attract, the exact reason why the Electromagnetic force is so strong compared to Gravity etc. A significant amount of computer simulations and mathematics may have to come in to enable us to understand these.

Mathematics by itself would not prove to be a big hurdle and I do not believe we are incapable of introducing proper math for the theory. The mathematics that is needed for the proper description of the theory could be equivalent or higher than General Relativity and would need much higher expertise in mathematics than I possess.

Theorising without any mathematics has a chance of completely going off the track. This means that there may be places where the theory, on detailed analysis comes into serious mathematical or philosophical conflict with other proven theories and thus may need significant modifications.

This is especially true with the Matter-Antimatter model of charged particles described later in the book.

The approach needs to utilise small but non-controversial observations from a lot of smaller disciplines to see if the theory can explain them. A superficial agreement with observations may prove to be inadequate or in reality fallacious on detailed observations. For example, my knowledge and the descriptions of the Weak and Strong Nuclear Forces and events happening inside the nucleus which is called "the Quantum Chromodynamics" may be inadequate to extrapolate my theory onto these situations. This needs a lot of experts including experts in geometry and experts in computer simulations to work in unison to complete this part of the theory.

All in all, the present description of the theory may be considered extremely inadequate and further work including the work of multiple experts in other fields may be needed in collaboration before the theory achieves a true form of a complete well written "theory of everything"

However, I am confident that such a theory that can explain so many aspects with so few assumptions, could not be wrong in its entirety and thus physicists need to give it a fair analysis before jumping to a conclusion and discarding it.

CHAPTER 8

PROBLEMS OF UNIFYING GR WITH QM

Quantum Mechanics is the theory on which the interplay between quantum particles is explained. Quantum particles and Quantum distances are extremely small. These could be to the tune of 10^{-10} or smaller. At larger distances, the laws of Quantum Mechanics somehow do not apply.

In Quantum Mechanics, the particles follow Heisenberg's Uncertainty Principle, so that they cannot have a definite position and momentum at the same time. Their position is just given by the probability of finding a particle at a co-ordinate location. At a given point in time, if a measurement is made, the wave function determining the probability collapses and the particle achieves a definite location.

Both these require the existence of a fixed three-dimensional-ordinate system that is unchanging with time. It also needs a properly defined and fixed time.

The theory is thus said to have a fixed background co-ordinate system or is background independent.

In it, a small unit of time everywhere in the quantum system assumes the same quantity. This means that a nanosecond at one point is the same as a nanosecond at some other point. This means that the theory assumes rigidity of time.

This is in contrast to General Relativity in which both space and time are wedded together and both are relative. That is space can expand or contract with respect to time i.e. there is no rigid background co-ordinate system. This means that the coordinate system itself is variable and can change. The background coordinate system isn't rigid. This means that General Relativity is background dependent. Any theory compatible with General Relativity cannot be background-independent or cannot be with a fixed background co-ordinate system.

The time in General Relativity is also relative and the quantity (or length) of a nanosecond at one point may be different than the quantity of a nanosecond at some other point.

There are thus two problems why Quantum Mechanics and General Relativity are incompatible.

1. ***Background independence problem***

    And

2. ***The time problem.***

This indicates that only one of the two can be right. Only one of the two theories can represent reality. However, both are so successful theories and closely predict the observations in their own realms that it has become difficult to give up one for the other.

There have been several attempts to unify them. None of them was successful.

There are two potential ways of achieving it,

1. ***Convert General Relativity into a Background independent entity***

    Or

2. ***Convert Quantum Mechanics into a relativistic theory***

The first one is somehow unpopular.

Several attempts of achieving the second have been partially successful.

Paul Dirac, while attempting to find out a relativistic version of Schrödinger's equation stumbled upon a prediction of the existence of Antimatter which was later confirmed to be true.

What does logic say?

The logic goes more with the background-independent and rigid constantly forward progressing version of time rather than the difficult to understand bending of space-time.

How our theory achieves this?

In our theory i.e. the Dynamic General Relativity, the "static" General Relativity which is a background dependent theory is being converted into a background independent entity.

Einstein's General Relativity predicts multiple things. All of these predictions can be explained even by Dynamic General Relativity, despite our theory being a background-independent theory. The background dependence and flexibility or variability of time emerge out effortlessly in it despite its rigidity of time and background independence.

Grand Unification

There are 4 fundamental forces of nature- gravity, electromagnetic force, strong and weak nuclear force.

The electromagnetic force evolved into physics from the unification of the electric and the magnetic force by Sir James Clark Maxwell through his Maxwell's equations.

The electromagnetic and weak nuclear forces are said to be arising from the same entity -the electroweak force. This was proposed by Steven Weinberg and others (Weinberg, S – 103/104).

Further unification and thus simplification was hoped by multiple scientists. However, attempts to achieve that have thus far failed.

The term Grand unified theories would be theories that unify the electro-weak force with the strong nuclear force.

A Theory of Everything is a theory that explains all the interactions with a single force. That is, it unifies all the fundamental forces into one single force and thus, in a way, unifies gravity with all the other forces of nature.

Our theory of Dynamic General Relativity explains how the distribution of local time changes around a particle and the resultant changes in space that occur, which can explain all the fundamental forces of nature. It thus successfully unifies all the forces of nature. All the forces of nature, according to it are caused by changes in local time and thus it is in a way a "Theory of everything".

Why it falls short of being a "Theory of everything"

Multiple mysteries and questions persist and although a lot can be explained by our theory, it cannot explain everything as yet if at all.

It cannot explain- Biology, self-organizing systems and consciousness.

Multiple questions about the internal architecture of the theory persist and thus it cannot be called a complete theory in itself and should be considered as a direction of thinking or a way to reaching a theory of everything.

Chapter 9

Importance of an Observer and the Measurement Problem in Quantum Mechanics

In the Copenhagen interpretation of Quantum Mechanics, the quantum particle is supposed to be in a Superposition of all possible quantum states at the same time and the particle collapses into one of the states when a measurement is made. This means that when a measurement is made, the properties like handedness of an Electron, instantaneous location of the electron, and path of an electron collapse from a probabilistic to an actual state.

However, what constitutes "a measurement" isn't defined in Quantum Mechanics. This is the so-called measurement problem of Quantum Mechanics. "Whether a conscious observer is needed to define a measurement or just a measurement made by an artificially intelligent computer that is never read by a conscious observer or an observation made by an unintelligent conscious observer like a cat also constitutes an observation?" Is not defined.

This creates multiple interpretations and confusion. This creates a divide between physicists into three or more camps. The "Copenhagen interpretation camp" are the ones who believe that the state in which the electron was before the observation was made isn't important and should not be asked. The common phrase "shut up and calculate" is used to imply this attitude that one is supposed to defer asking such questions. This is literally like saying that the state of electrons before a conscious observer makes an observation is unimportant. Einstein hated this idea and commented, "Do you really think the Moon isn't there when you are not looking?".

The second camp being the "realists" who believe that the Electron exists irrespective of whether we are observing it or not and that it possesses different measurable properties irrespective of an observer. These properties are independent of an observer.

There are many other alternative interpretations of QM including the more counterintuitive "Many world's interpretations" described by Everett.

An Observer is comparatively unimportant in General Relativity so that the space-time curvature is independent of whether an observer is seeing it or not.

The weird findings of the double-slit experiment and its extensions like the which-way experiment are indeed troubling to explain.

The observer dependence and status of the measurement problem in our theory

DGR assumes a definite coordinate location of every particle at a miniature moment in time which is known to nature up to a coordinate location of a Planck length. This location depends mainly on the status of Local Time and direction and frequency of EPCAs forming near a particle and thus is independent of an observer. Thus DGR is a realist theory.

The Superposition of particles that we observe is an emergent phenomenon due to our inability to decipher events as small as a Planck time or smaller. Multiple other hidden variables contribute to the probabilistic nature of QM.

The wave associated with the particle can easily explain the weird findings of the double-slit experiment without the need for Superposition.

CHAPTER 10

WHY THIS THEORY OVER OTHER POTENTIAL THEORIES?

This is one of the first and the most relevant questions a thinking mind would ask. There are already enough "theories of everything" in the market. Do we really need to waste more time on another theory? What is so good about this theory that it deserves the physicist's time, given that it demands a paradigm shift in how physics works?

The human mind has certain limitations. Due to cognitive incapability or maybe some other more fundamental reason, the human mind can only focus on a part of the reality or know a part of the reality.

I like to call this "Tunnel vision".

The eye of the mind of a conscious individual can visualize only that part of knowledge that is available to it or is known to it.

Different thinkers focus on a different part of reality and in the process fail to focus on some other aspect. This is akin to the blind men and the elephant analogy where the blind men touch several parts of the elephant and assume that the elephant is like a pillar or a fan or a rope or a snake.

This applies to theories of Physics or for that matter every other subject. Every theory looks at or focuses on some aspect of potential reality and often ignores other important aspects of it.

For example, the Earth-centric model of the Universe failed to take into account evidence of motions of the planet that went against it. The Sun-centric model of the Solar system considered that and came closer to reality. Newton took other aspects into consideration like force and acceleration and thus derived Newtonian gravity but failed to consider other aspects like the orbit of Mercury as this information wasn't available to his mind at his time. Einstein's theory took these and other things into consideration including SR. And thus came up with a theory that is closer to reality.

Thus every theory is partially correct and partially wrong. It gets some aspects right and some aspects of it may have to be modified or prove to be incorrect subsequently. Newton got the equations of motion right at human scales but his theory got "instantaneous gravitational influences thousands of light-years away" wrong. Although he did admit this flaw and he was personally uncomfortable with the idea of instantaneous action at a distance which his theory demanded.

This applies to the many theories of everything and gravity that are already there as well. Most of these theories try to derive a theory better than the pre-existing ones. That is a theory that is closer to reality.

One has to realize that the goal here is to reach a theory that is right in all aspects, which possibly is impossible to achieve. Thus any theory which a physicist is working on is partially correct and partially incorrect and thus the only thing needed is to identify where it errs and where it is perfect and which portions of it may be useful.

However, if the physicists have a stubborn belief that every aspect of the theory they believe in is correct and cannot question any aspect of their theory and on top of it, believe that every aspect of the theory that others believe is wrong and is unworthy of a thought, then they will keep falling in the rabbit-hole of failure.

In short, every effort made to reach the ultimate reality is useful. Every theory that is thought of can be useful. Every theory can and will have some right aspects and some errors.

The present theory aims to find a middle ground between GR and QM. In doing so, it manages to find some potential flaws in GR and QM. If these flaws are real and the physicists keep denying them, they cannot reach the ultimate reality.

Successes of DGR

DGR has two foundational and three phenomena-based assumptions.

The two foundational assumptions being

1. *there is an absolute space and*
2. *there is absolute time.*

These are more logical rather than assumptions. But since these were almost given up post-GR, these need to be included as assumptions.

There are some important extensions of this assumption of absolute space and time.

Space is made of an array or stack or bundle of PCs that are 1Planck volume thick. These are not stretchable and two PCs cannot take the same spot at once. They are constantly being created and destroyed every Planck second. This is an assumption based on the idea of Quantum foam. These PCs can be considered the smallest possible indivisible unit of space.

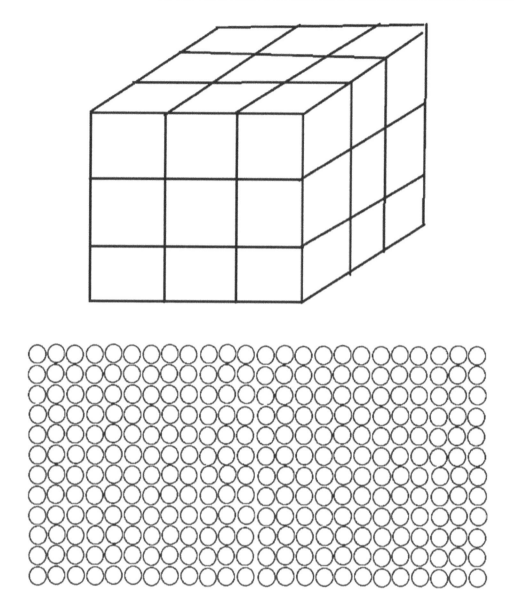

Figure i/10.1: Shows Planck compartments stacked on top of each other with a minimum gap as the proposed most fundamental unit of space-time

The three assumptions of phenomena happening are

1. Creation,
2. Destruction, and
3. Entanglement (or achieving compensation or maintaining balance.)

Entanglement is a series of processes happening to compensate for the Creation or Destruction. Some underlying assumptions need to be taken for these three to coexist. Compensation or balancing happens by entanglement because a PC vacuum created is not allowed and two PCs cannot be forced at a single place.

With these assumptions and just three processes, an extremely complex theory can be constructed. To start with, it can successfully explain Newtonian gravity, Einsteinian GR, and MOND. It can successfully unite QM and GR and provide an understanding of why these appeared different to us. It can explain the weird-looking concepts like variable time, bending space, Quantum Superposition, wave-particle duality. It can also potentially explain other interactions like electromagnetism, Electromagnetic bonds, chemistry, etc. It can provide clues to a lot of unknown abstract entities like energy, momentum, inertia, light waves, time, information, entropy etc. Although presently explaining the Weak Nuclear Force and the Strong Nuclear Force from this theory is not completely clear and more work is needed, prospects that it can subsequently explain these as well is good. It can potentially solve the conundrum of Dark matter and Dark energy and the cosmological crisis.

It has a very high likelihood of providing a testable prediction and it is not completely non-falsifiable. A theory in which so many theories pop out automatically without much effort must be pointing to some underlying aspect of reality.

The theory can prove to be the theory that unites Loop Quantum Gravity and String theory and redefines the word "physical dimensions" or make us understand higher dimensions better.

There is, in my opinion, a high likelihood of it being closer to the truth. Even if a part of it is wrong, there is a high probability that a part of it is right. It can give important clues to the direction of thinking of the physicists and should therefore be given adequate thought.

It is thus a no brainer that a theory like this deserves more attention from the physicists.

CHAPTER 11

INSIGHTS THAT EMERGE FROM THE THEORY

> **This section was written at the end after the current status of the Theory was understood. It is prudent that this section will be understood and thus should be read after the entire theory is known and understood. Although it can be read before, its real significance will be understood only after reading it after understanding the theory.**

The new paradigm shifts in thinking or understanding (like "energy and mass are equivalent" and "Space and time are wedded to form space-time" in SR or "massive objects bend space-time" in GR) that the theory demands or the newer insights that emerge from the theory are as follows:

1. **Quantum foam is linked to time**: There are two different types of times. The time that we measure is the Local Time. This variable entity of time is but an illusion caused by active spatial contraction happening near the Earth. There is a single Universally progressive time called Universal Time.

2. **At first - creation linked to Local Time destruction to UT**: Gravity emerges out from this difference. That deviation in time can cause spatial expansion or contraction. The subsequent realization that this time resetting is not fundamental. What is fundamental is the property of entanglement due to which PC vacuum, formed anywhere, is compensated by appropriate spatial expansion or contraction. The variability in local time that occurs is secondary. (Later this was replaced by the three processes of Creation, Destruction & Entanglement and the concept of the C/D ratio.)

3. **A link between Positive Energy and Volume loss,** the existence of Negative energy and Volume gain link, in short Energy - time deviation link, time dilation - spatial contraction link, time contraction - spatial dilation link, T=0 compartment -as a positive energy quantum or the packet of energy described in conventional physics.

4. **Destruction takes place at the core of every particle**

5. **Phenomenology of EPCA formation to compensate for the volume loss** and the insight that it is not the deviation in local time that causes spatial changes but it is the other way around with spatial changes leading to a stubborn illusion of deviation in local time

6. **The insight that more than one event (e or ϶) of destruction can happen at the core of a particle per PT**. The insight that mass or energy in a particle is equivalent to "e"

7. The insight that a time resetting EPCA forming wave starts at the place of destruction and moves outwards. The insight that this **TRW can travel faster than the speed of light. That information and Gravitational interactions can travel faster than the speed of light**. That higher is the value of e, faster is the pace with which this TRW moves out. That EPCA forming TRWs can explain the wave-particle duality of matter and light

8. **The insight that negative energy particles can't form aggregates and remain separate.** Positive energy particles can form aggregates

9. **The insight that a charged particle may be made of a positive energy core with negative energy particles forming a lattice**. That zones of excessive spatial contraction or expansion develop around them due to **interference of the TRW from them leading to the creation of push-pull bands which can be analogous to electric or magnetic flux lines.**

10. **The insight that the charged particle has asymmetry at poles so there must be an asymmetry in negative energy to positive energy**. The insight that the TRWs emerging out from these particles can travel faster than the speed of light.

11. **The insight that the negative energy lattice cannot stay static and would rotate explaining the spin of charged particles**. The deficiency or discrepancy between positive and negative energy in charged particles may mean an excess of an ENEA or a deficiency of an ENEA. This would mean that one pole has a deficit of ENEA and one has an excess of it explaining the magnetic dipole of charged particles and handedness of spin. **The poles can shift with ENEA particles shifting positions like a pendulum** so that a pole with negative energy excess can turn into a pole with positive energy excess. This explains the Superposition of spin-handedness of charged particles and that a right-handed one can be turned into a left-handed one instantly.

12. **The insight that the edge of the Universe probably has a positive energy excess with an infinite PC vacuum that keeps pulling the edge PCs apart explaining the Expansion of the Universe** (if the Universe is Flat as the current observations indicate). The intergalactic zones may have compensatory negative energy excess where a new constant genesis of particles can take place.

13. **The insight that Dark matter and Dark energy are both made up of ENEA particles.** The insight that "within the Nucleon is positive energy which binds the Quarks together". The negative energy counterpart of this unbalanced positive energy in the nucleons must lie entangled with it at the outskirts of the Galaxy as Dark matter.

14. **Empty space contains active spatial contraction or dilation.** Nearer the Galaxies, it is the spatial contraction in the form of EPCAs which may be circular or linear. Even within a solid substance or liquid, in between molecules and atoms will be the presence of dilated time with EPCAs. The slower velocity of light while travelling through the glass prism explains refraction. In the intergalactic regions in between non-gravitationally bound Galaxies, where Expansion of the Universe happens, spatial expansion type EPCAs would exist and thus local time here would run faster.

15. The insight that the push-pull bands travelling faster than light bend under the influence of high density of positive energy spatial contraction EPCAs within the high positive energy zone of Nucleons well within the zone where Quarks exist. These bent push-pull bands can interact with those coming from other Quarks within the same Nucleon or other Nucleon, explaining Gluons.

16. The insight that the ENEA particles forming the Buckyball lattice around a charged particle, have a slight asymmetry between negative and positive energy particles. This makes the ENEA particles move in a pendulum fashion and thus the charged particle keeps switching poles and thus handedness. At any given point, the particle is in a Superposition of both right and left-handedness, which essentially means it is changing from one to the other handedness in a short interval. This spin becomes fixed if it passes through a magnetic field. The pendulum motion would create a crest and a trough even for the push-pull bands emanating out.

17. These crests and troughs of time dilation or time contraction would form a wave. The push bands emanating from a negatively charged particle, like an electron, can curve around the proton and meet at the other end thus explaining the repulsion between the proton and electron after they come to lie in atomic orbitals.

18. Higher dimensions as described in String theory may have a different counterintuitive interpretation. The evolution of the local state of a cubical space with time may be dependent on multiple factors each of which contributing to the change in the ultimate location of its vertices. Thus the ultimate shape of geodesic of the ever-changing space in DGR would be determined by many of these factors including spin, momentum, TRWs from PEP, ENEA, etc. Each of these influences can be considered as independent dimensions and thus at present, about 9 dimensions (possibly more) can be defined that influence the geodesics around a particle.

19. After a critical distance, the Gravity-related TRW which forms EPCAs would form linear EPCAs and would result in a uniform inwards movement of the EPCAs. This

critical acceleration can be termed as a_0 and gravitational laws behave differently beyond this critical limit so that the MONDian type of influence which is directly related to the mass of the Galaxy is what determines the inwards pull and resultant rotation velocities.

20. **The insight that DGR is a variable speed of light theory** as described by Einstein in 1911 *(Einstein, A – 39)*. The insight that G in Newtonian gravitational force equation is the outward pull due to MOND like forces due to mass of the rest of the Universe. The Insight that DGR is perfectly compatible with Mach's principle as noted by Einstein. The insight that DGR, like the Large number hypothesis of Paul Dirac, predicts a variable G and is consistent with the spontaneous creation of elements from nothing so that mass of the Universe (i.e. the sum of masses of all the Baryons in the Universe) is not a constant but keeps increasing with Age of the Universe and G, is related to the mass of the Universe should also keep increasing. Thus the ratio of Strength of Electromagnetism to Strength of Gravity should keep reducing with the Age of the Universe as per DGR. *(Unzicker, Alexander – 95, 96, 97, 98, 99, 100)*

21. **The insight that Radio waves are different from other EM waves.** All the other waves have positive energy particles called photons which are T=0 compartment aggregates surrounded by compensatory EPCA forming TRW forming the wavefront. So they are only positive energy waves. Radio waves, however, are the Push-pull bands from a charged particle drifting outwards with the speed of light and thus have alternating zones of excessive time contraction i.e. negative energy and excessive time dilation i.e. positive energy.

22. **The insight that Gravitational waves detected by LIGO are different from the EPCA forming TRWs (that move much faster than the speed of light and are responsible for Gravitational interactions).** They are created whenever EPCA formation can no longer compensate for the destruction taking place at the centre. They can be potentially created when any particle or object or heavenly body with mass moves.

Chapter 12

SERIOUS QUESTIONS STILL UNANSWERED

> **To be read after reading the entire book.**

1. What's the significance of c? Why does light travel with speed c, which is one PC per PL? Do higher energy photons have a mass?
2. Why the EM force is 10^{37} times the force of Gravity and the Strong nuclear force is 10^{40} times stronger?
3. What is h? What is the connection between h and volume loss or h and energy?
4. What is the fine structure constant? How to derive it theoretically? How to explain the various orbitals within a Bohr orbit? How to explain the formation of orbits, orbitals, EM bonds, and overall chemistry by time distribution patterns, phenomena, and geodesics?
5. What is the exact equation connecting volume loss and mass, volume loss and energy, volume loss and time dilation, or volume gain and time contraction?
6. The exact flow of PCs and thus defining the exact geometry of the geodesics when the various phenomena described are happening. This includes phenomena like EPCA formation leading to the attraction between two positive energy particles, formation of Push-pull bands, time changes happening around a particle due to motion which we call kinetic energy, and time changes when the motion is restricted in a field that we call potential energy.
7. The exact geometry, time distribution, and geodesics while electrons bond with Protons/Nucleons within an atom.
8. The exact geometry, time distribution, geodesics, and velocity of flow of PCs around the events in the history of the Universe like the recombination epoch, the early Big Bang, and maybe even at the time of the Big Bang.
9. What is the significance of other constants of nature? To derive the details of the extent of volume loss per PT happening at all the fundamental particles in the Standard Model with mass? How to theoretically derive masses of various particles in the Standard Model and explain them? How to explain the Ratio of Proton to Electron?
10. To find out the maximum velocity of entanglement both theoretically as predicted by DGR and practically with experimental observations.

Thus a lot of work remains and it can be said that DGR is probably in its infancy yet.

PART II

CURRENT STATUS OF THE THEORY OF DGR AND VERY RECENT INCITES AND DIFFICULT PROBLEMS IN PHYSICS

CURRENT STATUS OF THE THEORY OF DYNAMIC GENERAL RELATIVITY

Currently, there are two theories to explain gravity, Newtonian mechanics and General Relativity. It is well known that Newtonian gravity has many pitfalls and although at reasonable distances it gives an accurate description of Gravity and allows precise calculations, it is beyond doubt now that General Relativity is closer to reality as it can explain many phenomena like Gravitational lensing, Gravitational time dilation and precession of the orbit of Mercury. (*Einstein Albert- 33,34,35,36,37*)

General Relativity has its own problems, however, and at very large scales like at distances of thousands of light-years, the theory fails to explain the speed of rotation of the peripheral stars in the Galaxies and fails to explain the flat rotation curves described first by Fritz Zwicky (*Zwicky, F-107,108*) and confirmed later by Vera Rubin *(Rubin, Vera-86,87)* and colleagues. Its mathematics also breaks down at extreme situations like within a Blackhole or the start of the Universe i.e. at the time of the Big bang.

There is still no way of combining General Relativity (GR), which deals with great distances and Quantum mechanics or Quantum field theory (QFT) which is an extremely successful theory describing extremely short distances. Several attempts have been made to describe a Quantum theory of Gravity and detect the elusive Graviton with Loop Quantum Gravity (LQG) and String Theory. These theories, however, remain in the realm of hypothesis till now.

The problems of uniting Quantum Field Theory with Relativity include background independence and the problem of time. The space and time in GR are wedded to form a single entity of space-time which means that both space and time are flexible and can bend/curve contract or expand and are essentially active participants in the processes while in QM, time is a background entity which progresses relentlessly and there exists an independent background of fixed coordinates in space.

Most approaches to find common ground between these is to convert present Quantum theory into a relativistic theory with variable time wedded to space like what was attempted by Paul Dirac and others.

Presented below is a theory, which we can call Dynamic General Relativity (DGR) which can, with just a handful of assumptions, successfully pave the way of describing Quantum Gravity and can potentially unite General Relativity with Quantum mechanics.

CHAPTER 1

BACKGROUND THINKING

We consider a Meteor/particle at a point P with mass m at a distance h from the surface of the Earth, actively under influence of Earth's gravitation, accelerating downwards.

General Relativity says that due to Earth's gravitation, at the surface of the Earth just below the meteor at point P, time is dilated or a clock ticks slower. As we move from M towards P, gravitational time dilation (GTD) reduces.

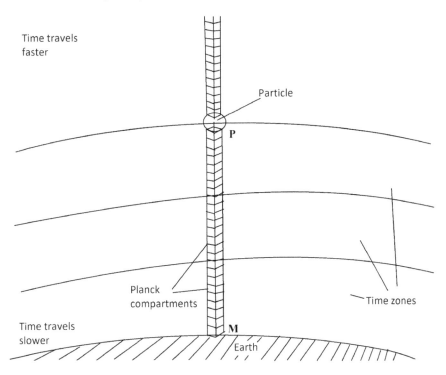

Figure ii/1.1: Shows a particle P suspended in Gravitational time dilation zones of the Earth. A skyscraper of compartments 1mm³ or 1cm³ or 1m³ extending from point P to point M on the Earth's surface is shown. Each of these compartments will lie in different time zones and as per DGR will lose volume actively and contract, thus pulling the particle P downwards.

We can imagine small cubical compartments stacked on top of each other forming a stack starting from M on the Earth's surface and reaching P. Due to GTD, each of these Compartments would have time dilation which progressively reduces as we move from Earth upwards.

We can, if we wish to, have these compartments of any size with each side 1 cm or 1 mm or even a side of 1 Planck length. Logic says that even if each Compartment has a height of 1 Planck length, there will be a finite number of PCs from M to P. As the Meteor accelerates towards the surface, the number of these PCs between the instantaneous position of P and M would keep reducing.

We know, according to Quantum electrodynamics that at Planck scales, there exists Quantum foam wherein spontaneous creation and destruction of virtual particles is constantly taking place at an extremely rapid pace at Planck times.

In QED, which first described Quantum foam, the virtual particles have to be created or destroyed in pairs. Furthermore, these particles are not necessarily real, aptly called virtual, although their presence can have measurable effects. We presume that there is an underlying hidden presumption in QED that these virtual particles have no effect whatsoever on the volume of space or on time.

Our theory assumes that these virtual particles are appearing and disappearing Planck Compartments (PC) and that their appearance adds volume and thus expands space and their destruction contracts space.

Disappearing PCs being positive energy particles and appearing PCs being negative energy particles in our theory, analogous to matter and antimatter particles in QED. These represent the fluctuations in energy that are seen at Planck scales.

Central to our theory thus is a ratio called as C/D ratio. A place where the C/D ratio is 1 has time running at the pace of Universal time(UT), which is an entity of time that constantly flows forwards since the start throughout the Universe.

This Universal time is equivalent to the universally constant unchanging background time of Quantum mechanics.

There is however another entity in DGR called the Local time (LT).

There is a universally present link between local time and this process of creation or destruction so that

- ❖ ***Wherever there is local time dilation, there is active volume loss due to active ongoing destruction of Planck Compartments.***
- ❖ ***Wherever there is a contracted LT, the appearance of excess PCs causes expansion of space.***

This LT is the time equivalent of the time we see in GR and the only time which we can measure.

The assumption here is that the Quantum foam in these 1cm or 1mm-sized compartments will have a predominance of destruction proportional to time dilation within them and this destruction, as exclaimed by Newtonian gravity, pulls the meteor down. The time dilation is because of the EPCAs as explained later.

CHAPTER 2

BASIC ASSUMPTIONS OF THE THEORY OF DGR (DYNAMIC GENERAL RELATIVITY)

1. Space is made up of fundamental units with a volume of 1 Planck length^3 i.e. Cubical or spherical compartments with a side of a Planck length. Each of them has a coordinate location accurate to Planck length. With this assumption, it is established that there is a fixed universal coordinate system.
2. These Planck Compartments lie in close contact with their neighbours. They are incompressible and cannot vary in size.
3. Planck Compartments can be "Created" from nowhere. This Creation explains the addition of new space units in the expansion of the Universe.
4. Planck Compartments can be "Destroyed".

    Both these processes of Creation and Destruction of Planck Compartments are rapidly taking place at Quantum level in every corner of the Universe wherein at some places the process of Destruction dominates while at other places the process of Creation dominates. At most Places, the process of Destruction counter-balances Creation so that there is no net loss or gain of volume. The ugly implication of this assumption is that

    "Every nook and corner of this Universe is being Created and Destroyed and then Re-Created continuously. This cycle of Creation and Destruction keeps going relentlessly"
5. Energy, which is defined in contemporary physics as the capacity to do work, is of two types. Positive energy (PE) is when the process of Destruction predominates and Negative energy (NE) is when the process of Creation of Planck Compartments (PCs) predominates.
6. There are two times- Universal time and Local time

    There exists a Universal time (UT) that is constant throughout the Universe. This started at the time of the Big Bang and is constantly ticking at a universally constant pace throughout the Universe. This time can be considered the time in Quantum mechanics.

    There also exists a Local time (LT) that varies from point to point in the Universe depending on which of the two processes (Creation or Destruction) predominates.

    Wherever Destruction predominates, space is in an active state of contraction, is losing volume and here Local time is dilated relative to the Universal time (UT)

Wherever Creation is in predominance, new PCs are being formed and thus volume is being gained or space is in process of active expansion. Here Local time is in a contracted state relative to the UT.

Local time is the time we mention in Relativity. How the two are related and how the state of space determines the time is explained below.

7. Matter is made up of a high concentration of positive energy. This means that at the core of all matter particles, a high amount of positive energy resides. Positive energy indicates the predominance of Destruction. So at the core of matter particles, Destruction predominates.

8. In the intergalactic regions, Negative Energy predominates probably in the form of negative energy particles.

9. The scales which we will use in the description of this theory would be Planck scales so that most processes would be described with the length of Planck length, the volume of Planck volume and time of Planck time.

10. The PCs are in a state of entanglement with their neighbours.

    Entanglement is another major assumption.

    This means that they are in close proximity to their neighbours and empty space between them or true PC vacuum cannot exist. Any force which pushes PCs together to compensate for volume loss leads to additional Destruction and any force that pulls them apart to compensate for volume gain leads to additional Creation.

11. EPCA formation

    When a positive energy particle exists at a point - i.e. at a PC, there is a predominance of destruction. This could mean millions or billions of PCs being destroyed every Planck time. More is the mass of the particle, more is the energy within it in accordance with the Einsteinian mass-energy equation. More is the energy within, more is the extent of destruction within.

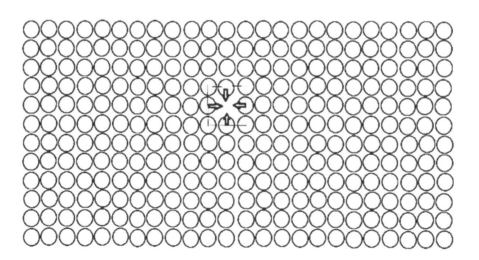

Figure ii/2.1: Shows how Eddy Planck Compartment Aggregates (EPCAs) form around a point where Destruction predominates.

This destruction would create a void or empty space or PC vacuum around the particle. This is compensated by the process of entanglement.

Every PC destroyed creates a PC vacuum which pulls the neighbouring PCs towards it. These PCs get entangled together and local time changes within this aggregate of PCs in such a manner that it loses enough volume to take the place of the destroyed PC. This, however, creates a PC vacuum immediately outside the entangled PCs, which entangles their neighbours. A spherical wave resetting local time starts and moves outwards quickly.

Because they are spherical and they are PC aggregates, we can call them EPCAs (Eddy Planck Compartment Aggregates).

Note that for nature, there is no smallest time.

If 1000 PCs are being destroyed in a Planck time, the Planck time will have to be divided into 1000 equal parts and one event of PC destruction happens at each of these fragments of time. At every destruction, a new EPCA forms. This means that the PCs around this particle are pulled inwards and a cycle of EPCA formation starts and repeats every 1/1000 th of a Planck time. As the first level of immediately adjacent PCs moves inwards PC vacuum forms around it and the inward pull is now shifted to the second level PC, which now aggregates together to form an EPCA and move inwards. This cycle of shifting outwards of PC vacuum and shifting inwards of PC aggregates continues incessantly and a new cycle begins with every event of PC destruction that happens at the centre. The wave that forms and moves outwards inevitably leads to horizontal volume contraction (leading to reduced circumference/surface area of the EPCA) and vertical volume contraction (leading to inwards movement of the EPCA) to compensate for the PC Vacuum created inside. This horizontal volume loss or horizontal spatial contraction leads to secondary resetting of time. The amount of horizontal volume contraction determines how much time dilation will happen. Thus, this outwardly progressing disturbance created by the PC destruction at the centre leads to sequential resetting of time in each outer concentric volume of space.

This local time resetting compensatory wave travels 1000 PCs away from the original point in one Planck time. Which is extremely rapid and is much faster than the speed of light (its velocity is determined by the extent of destruction happening at the centre – more the destruction, more is the velocity of outwards progression of the Time resetting wave. It is unclear whether there is a limit to how fast this can occur and if there is a limit, this will be the limit with which information transfer due to "entanglement" can occur) However, this process doesn't violate GR as what is moving faster than the speed of light is local-time resetting tendency and not a particle with mass. Even the movement of EPCAs which represent spatial expansion or contraction doesn't happen faster than the speed of light beyond a point.

The EPCAs explain the gradually decreasing time dilation as we go from M towards P and it also explains the active contraction of space in between M and P.

In fact, every particle that makes up the Earth's mass contributes to the destruction of PCs. A sphere of PCs immediately around the surface of the Earth would thus be sucked similarly to compensate for the volume loss happening at the core of each of its constituent particles. The immediate neighbours entangle and form another EPCA beyond it. This cycle keeps repeating.

All this process essentially means that entangled aggregates of PCs form around the Earth (or for that matter any heavenly object) and these EPCAs gradually move inwards due to spatial contraction and volume loss within them, like any sphere which has a reducing surface area. An apt analogy is a spherical balloon made of rubber which has air-filled within at high pressure, has a pressure differential inside and outside and due to this, is losing air fast so that due to elasticity of its surface it keeps losing surface area and keeps moving inwards.

The process of destruction keeps going at the core of each particle and so does the process of formation of EPCAs around it.

This process explains the change in the shape of space-time as predicted by GR and thus explains a lot of findings of GR like Gravitational lensing and Gravitational red-shift.

The change in space-time, however, is slightly different and is dynamic instead of what was predicted by GR. The bending of space-time is not static but actively changing, actively losing volume and actively changing the location of points on the space with time.

Arcuate vs linear EPCAs

Closer to any object with mass, the destruction happening at its centre due to positive energy leads to the creation of spherical or curved EPCAs. This disturbance can potentially keep moving indefinitely.

EPCAs formed around the Sun or other stars in the Milky way are also curved at lesser distances like the diameter of the solar system. The time resetting waves travel outwards at Speeds much faster than the speed of light and at distances as big as the Galactic radius, the EPCAs would be linear. Here there is no horizontal volume loss and no curvature. But there are just entangled PC aggregates shifting their position (vertical volume loss will still happen). The EPCAs become linear at distances beyond which the gravitational time dilation caused by a star or a planet goes below the Planck time in sub-Planck time realms.

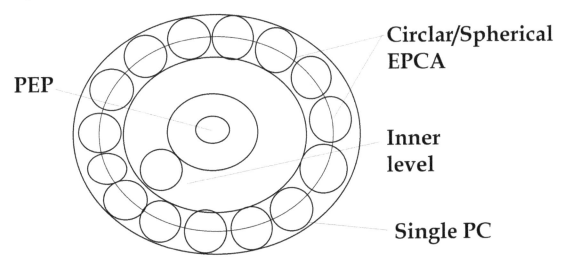

The EPCAs, although not curved are indeed present and would interact with other EPCAs from other stars. This provides the potential explanation for modified Gravity at great distances which is called Modified Newtonian Dynamics (MOND).

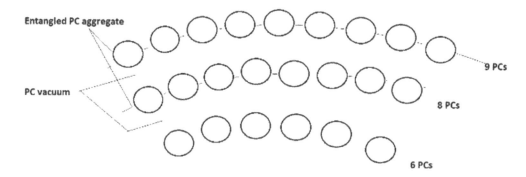

Figure ii/2.2: Shows a spherical/circular EPCA formed around a central point of destruction. Also shown are the various levels where the EPCA would lie in the future as it loses volume and its surface area reduces.

Figure ii/2.3: Shows the formation of a linear EPCA. Here, there is no horizontal volume contraction (i.e. it is negligible). However, there is a shifting of position or level of the EPCA at a relatively constant acceleration.

CHAPTER 3

WHAT DOES GENERAL RELATIVITY MEAN FUNDAMENTALLY? (DERIVING A MODEL OF SPACE-TIME FROM GR)

Einstein's General Relativity describes how an object curves space-time and how this curvature results in many observable phenomena like gravity or gravitational lensing. (*Einstein A. - 33, 34, 35, 36, 37, W. de Sitter – 22,23*)

A typical example that is given to demonstrate it is how the surface of a trampoline bends or curves when a heavy metallic ball is placed over it.

How exactly this is achieved at a fundamental or microscopic level is not explained except that this curvature moves outwards with the speed of light.

One of the well-known properties of a Trampoline that is inevitable and directly responsible for this curvature is elasticity. Elasticity essentially means that the surface or material of the trampoline resists deformation and when the deforming force is removed, has a tendency to bounce back.

The question which would arise is, is space-time equally elastic at a fundamental level?

The elasticity of materials has one prerequisite. The individual components must be connected or bonded and these bonds need to be such that when stretched by pulling the constituent particles or fundamental units apart, the strength of bondage should increase so that the more you pull them apart, the more they pull each other.

Whatever the particle with mass does, one thing is clear in General Relativity. That the mass of this particle gives it an ability with which it can modify time immediately adjacent to it and apply some kind of force on the fundamental units which make up space-time. (Note that how the particle with mass achieves this is not a part of GR)

The trampoline analogy should not be taken too far as a trampoline is two dimensional and changes occurring in space-time due to Gravity are happening in a three-dimensional space and also vary with time.

However, one thing is clear. The presence of a particle with mass at point A in space-time leads to changes in the curvature of space-time around it and these changes travel outwards as a sphere of increasing diameter.

This means that curvature at one point in space-time can affect the curvature of space-time immediately adjacent to it.

That is, if there exists a fundamental unit of space-time, two things are implied in General Relativity. The first is that these fundamental units are connected. The second likely possibility is that there is elasticity.

Demonstrating elasticity

Let's imagine a linear rubber band made of 10 particles which can be assumed to be the fundamental building blocks or fundamental units which form the rubber band. Let us name them A, B, C, D... etc. We know that there exist linear elastic bonds between A and B, B and C, C and D etc. So that when A and J which are the end particles are pulled apart, the pull is evenly distributed so that each of these distances increase and the rubber band gets stretched and the bonds between neighbouring particles become stronger leading to increased pull between them. As soon as this pull is relieved, the particles come back to lie in their original configuration and the distances between them revert back to normal. This property is essentially the elasticity of the material.

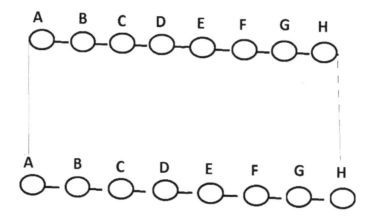

Figure ii/3.1: Demonstrating Elasticity

Elasticity of space-time

We will try to describe a toy model for understanding elasticity of the space-time.

Let's imagine a hundred tennis balls each one of them representing individual building blocks of space-time. We know that space-time is apparently continuous and there is currently no evidence that such fundamental spherical units of space-time exist or empty space in between them exists.

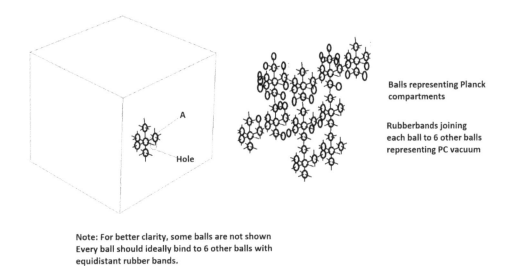

Figure ii/3.2: Shows a toy model off space-time with multiple tennis balls attached with 6 rubber bands within a box.

However, we will assume that this is true for now.

We assume that Einstein's space-time (or more appropriately just space) is made of these individual units. Each of these units or tennis balls is bonded to six other balls with a thick rubber band. All these 100 balls are arranged in a lattice arrangement forming a crystal-like cubical structure similar to a crystal of salt or metals.

The 100 balls are within a box with rigid walls so that the peripheral balls are bonded to the wall with a similar thick rubber band.

Imagine that one of the faces of the cubical box has a small hole matching the diameter of the ball, overlying one of the balls. Let's name this Ball A.

Through this small hole, we can access the rubber band that binds A to the wall and thus we can pull the ball A outwards so that it almost comes out of the box.

Note that there are going to be five balls immediately adjacent to A and the sixth rubber band attaches it to the wall of the box which we have removed and through which we are pulling A out.

The five immediately adjacent balls will be pulled inwards equally towards the missing or exiting ball. This force will travel outwards so that the balls immediately adjacent to the adjacent balls also feel the inward pull towards A.

The proper symmetric equidistant arrangement of the balls would indeed get disturbed and this disturbance keeps travelling outwards till all the balls get disturbed from their original position.

Now imagine if we thrust the ball A back inwards. What will essentially happen is that all the balls will immediately move back to their original position and some might even overshoot.

Now further imagine what would happen if we keep pulling and thrusting inwards the ball A at regular intervals.

Indeed, a wave of disturbance and then a wave of recovery of original position would start from A and keep moving outwards.

Now further imagine that this box is huge, of the size of a 100 Olympic sized swimming pools and there are not only 100 balls but millions or even billions of balls bonded like this.

Indeed, the disturbance and recovery waves can travel significant distances.

Computer simulations of these can give a much better idea.

Planck Compartments as individual units or fundamental building blocks of space-time

If we consider Planck Compartments as individual building blocks of space-time (or more appropriately just space), they need to have certain properties.

At the macro level, this granularity is not visible. Thus if they are real, they should be such that they merge with each other.

The critical question is if they are there, do they have walls?

The answer to them is pretty straightforward. Yes, and no.

One can imagine a drop of water or about 10-15 ml water in deep space. In zero-gravity, the constituent elements of the material that makes up the water keep pulling each other and it is expected to form a floating spherical or irregularly shaped blob of water. Inside it is water. Just outside it is air within the spaceship or the space station. The walls of the blob constitute the surface of the blob.

If such a blob of water is thrown into the vacuum of deep space, what happens to it depends on the outside temperature and also on its mass. If the mass is small and thus its Gravity is negligible, the outward pull of vacuum on individual water molecules just outside the blob would pull the individual constituents apart and the blob cannot stay stable. However, if the blob is as big as a planet and has significant Gravity, such a planet made completely of water can survive without breaking off.

Such a Blob will have water inside it, it will have a wall but it doesn't have a wall. Just outside it, is nothing or a vacuum or empty space.

If two such blobs come together they can merge into one so that they can form a continuous medium

Planck Compartments made of the "elusive ether" with PC vacuum or nothingness or empty space in between

Space in the vacuum of deep space is also not made of "nothing" but is made of this granular space itself through which light and other electromagnetic radiation or gravitational waves can travel.

As LIGO has detected Gravitational waves, the elusive ether or the material that makes up space-time, which MM (Michelson-Morley) were so desperately trying to find out and which Einstein also believed in, becomes likely again.

The vacuum isn't empty but has Planck Compartments at the fundamental level which are themselves made of this "ether or space".

In between these Planck Compartments would be a potential space where the real nothing or the real vacuum or we can call "Planck Compartment vacuum" can be thought to exist.

Forces acting on a Planck Compartment

Imagine two cubical boxes A and B which are joined together by a small pipe. Imagine that there is a small ball within the pipe such that the ball has the same diameter as the pipe. The two ends of the pipe have a net so that the ball cannot escape out into the two boxes. The two boxes are filled with air with equal temperature so that pressure on both sides of the ball would be expected to be equal.

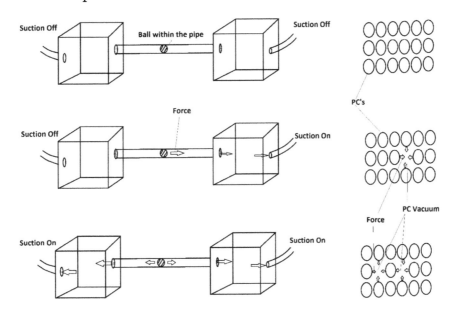

Figure ii/3.3 Demonstrating forces acting on a Planck Compartment.

Now imagine that we close the inlet of the pipe in box B and we remove all the air within it. So that now the box B has a vacuum within it. As soon as we open the opening of the pipe, the air within the pipe towards box B is sucked out and the pressure on the ball towards box B falls. The ball would be pulled towards box B with a force and it would move towards the B end of the pipe quickly.

Now imagine a spherical Planck Compartment named X which is surrounded by 26 Planck Compartments all around it forming a cube. Imagine that the Planck Compartment immediately to the right of X is Y.

What will happen if Y vanishes or is destroyed. There is a formation of Planck Compartment vacuum (PCV) on the right side of X. X will be pulled towards the right. Note that there will be 26 Planck Compartments around Y and all of them will be pulled towards Y. The pull will remain only until they come close and the PC vacuum vanishes.

X will be pulled to right and would move only half Planck length.

Another more likely possibility is that as soon as Y vanishes, all the 26 Planck Compartments around Y get entangled and time instantly changes in this entangled Planck Compartment aggregate or EPCA so that excess Destruction happens within and all the Planck Compartments other than X get destroyed and only X moves to the right a full 1 Planck length. Even this seems unlikely and another possibility that each of the neighbouring Planck Compartments loses some volume due to sub-Planck length destruction and the result is that a total of 26 Planck volumes space is lost and only one Planck Compartment volume space remains. This is more likely as we know that when Y gets destroyed, all the 26 surrounding PCs move inward equally causing a symmetrical spherical wave moving outwards instead of an asymmetrical wave caused by 1 Planck length movement of a single Planck Compartment X.

Now further imagine that in our experiment above, we start removing air from both the boxes A and B instead of only B. Now the pressure distribution on both sides of the ball equalises and the ball remains stable in its position instead of being pulled to one side.

Similarly, if our Planck Compartment X is such that there is Planck Compartment vacuum on both sides, the forces neutralise and it won't move to left or right.

Note that this is oversimplifying an extremely complex matter.

In reality, at the centre of a particle with mass, destruction of Planck Compartments would be going on to the tune of billions or trillions or quadrillions. With such high Destruction, Planck Compartment vacuum is likely to develop at extremely high rates like every 1/billionth or 1/trillionth or 1/quadrillionth of a Planck second.

If the PC vacuum is forming at a faster rate at the right side of X than on the left, it would indeed move on the right. Also, space-time is bound to be extremely complicated with the presence of dilated or contracted local times, especially around a charged particle.

Explaining entanglement and EPCA formation

Imagine a Planck Compartment X where a billion Planck Compartments are being destroyed per Planck time.

At the end of the first Planck time, a PC vacuum of the radius of one billion Planck Lengths would form. (note that this is for understanding purposes as, at sub-Planck times, the corrective EPCAs would be expected to start forming.)

This sphere of Planck Compartment Vacuum with a diameter of 1 billion Planck lengths, is lined by an aggregate of Planck Compartments. Each of these Planck Compartments is pulled inwards towards X. Thus they can be said to be entangled. Due to this pull, they would lose some PCs and reduce in volume and thus reduce in surface area, thus moving inwards in an attempt to compensate for the Destruction.

Because the central destruction continues, the inward progression of this already entangled PC aggregate would continue, this time, however, the extent of destruction would increase and thus the local time dilation would increase in the Eddy PC aggregate or EPCA.

As the PC aggregate moves inwards, a spherical volume of PC vacuum is forming around this entangled PC aggregate sphere. This has another PC aggregate just outside it. All these PC's immediately outside this new PC vacuum again get entangled and the cycle repeats. This newly formed EPCA again loses volume and surface area so that the EPCA moves inwards.

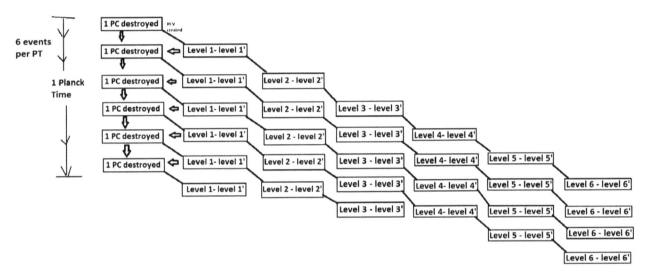

Figure ii/3.4: Shows how fresh cycles of EPCA formation keep repeating after every event of PC Destruction at the centre. Note that the numbering of levels would be arbitrary and not accurate.

Note that with a Destruction of a billion Planck Compartments happening every Planck time, every event described above happens at a 1/billionth of a Planck second. This means that the formation of the EPCA and inward movement of the EPCA due to excessive Destruction and the Creation of a new PC vacuum around it, happen extremely rapidly at sub-Planck time scales. The inwards movement of each EPCA also would happen at sub-Planck length scales for every cycle. This cycle keeps getting repeated outwards.

Note that instead of a single particle with the Destruction of 1 billion PCs per Planck time, there can be a sphere of a thousand bonded particles each of which has a million PCs being destroyed at its core. Even this arrangement would lead to a formation of a PC vacuum around it of 1 billion Planck length diameter at the end of one Planck time.

EPCAs as rubber spheres

The EPCAs are going to be single Planck Compartment thick and having an identical local time within. Because of the constantly developing PC vacuum within this sphere, there is a constant force directed inwards. This means that the PCs entangled together in an EPCA is essentially like a rubber sphere or a balloon that is forcibly inflated with excess air and now the air is passing out from it.

The difference here is that the sphere will eternally keep losing volume until it reaches the centre where extreme destruction is happening, unlike a rubber sphere or a balloon which will stop collapsing once its minimum size is attained. Also, in this example, there won't be just one but a billion PC aggregates being formed per Planck time, each of them moving inwards.

PCs react as if Dominos react to the falling of the first domino

Another useful analogy is that of "The Domino Effect" wherein when hundreds of dominos are arranged standing on one of their sides and the first one falls, a chain reaction starts with one domino leading to the fall of the second domino which causes the fall of the third domino and so on. The individual PCs or the EPCA's behave like dominos with one inwards shifted EPCA leading to the development of the second and the second inwards shifting leading to the creation of the third EPCA. The cycle continues incessantly.

Explaining Pull bands

Imagine a cylindrical region of space-time.

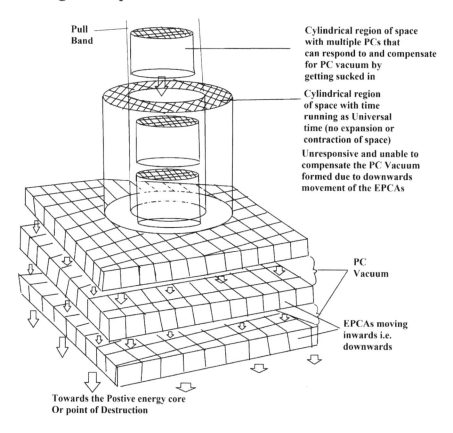

Figure ii/3.5: Shows probable mechanisms underlying the formation of a Pull band. Note that the obliquely shaded cylindrical zone and the uppermost layer of PCs (or 10^6 PC aggregates) have Universal time due to interference between waves from Positive and Negative particles. Thus they cannot expand or contract, responding to the PC Vacuum developing below the uppermost layer. The Destruction happening at the core leads to the constant development of EPCAs which

move inwards causing the constant development of PC vacuum below the uppermost layer. The only way of compensating it is by pulling in PCs from the narrow responsive cylindrical zone which forms the Pull band.

Further, imagine that the PCs lining this cylinder are being pulled outwards due to PC vacuum formation. Let this outward pull be just enough to prevent them from getting entangled and thus if a cylindrical PC vacuum developed just inside this cylinder due to extreme destruction happening due to the presence of a high mass particle at the lower end of the cylinder, these PCs are incapable of entangling and thus incapable of taking part in compensation of the destruction happening at the lower end of the cylinder. (Alternatively one can imagine a cylindrical region of space where the peripheral zone has T=UT i.e. due to the interference of various EPCAs, there forms a non-contracting non-expanding zone of space that is unable to respond to PC vacuum developed below.)

The typical method of EPCA formation would probably be unable to completely compensate for the destruction happening here. The rate of formation of uncompensated PC vacuum would be higher leading to extreme pull on the small cylindrical region which can respond. PC's from other regions probably from the top might have to be pulled within the cylinder, to compensate for the loss. The smaller is this responsive region, the greater is the suction or pull effect. This is akin to the relationship between pressure and surface area, wherein with the same pressure, more force can be applied by reducing the cross-section of the pipe. This could be an important mechanism in explaining why the Electromagnetic force is stronger than Gravity.

This cylindrical region can be of any length. The band thus formed could be curved or it can spiral around.

This constitutes a "Pull band".

For the push bands, we have to understand in detail how excessive creation works as explained below.

Magic-jelly and Pac man

A positive energy particle or a particle with positive mass is a Planck Compartment where the C/D ratio is shifted towards destruction and thus it eats up the volume from all sides.

This eating up is perfectly symmetrical so that PC vacuum is created equally all around the Planck Compartment.

This is true for a static particle.

However, these being quantum particles, are difficult to remain in one position. All particles including quantum particles and particles of light have what we call momentum. This can be linear or angular momentum.

To explain this, another analogy is prudent.

Imagine that there is a large tank or swimming pool filled with magical jelly. The individual particles of the jelly do not tolerate any empty space so that they start attracting each other if there is an empty space created at any place.

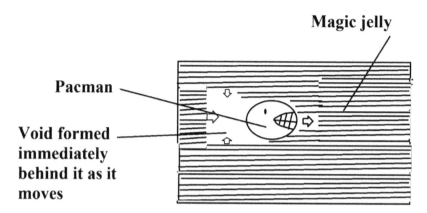

Figure ii/3.6: Shows Magic jelly and Pacman analogy

We know that exactly this happens when a fish or a submarine or a person is swimming deep within the water. As a fish swims forward, there is a void created immediately behind it. The pointed shape of the fish is to create a space in front by pushing the water molecules aside while it moves forward. This void created immediately behind it is instantly filled in by water rushing behind it from sides or the back. Similarly, when swimming forward, the fish has to push the water molecules lying in front aside. This is because the constituent molecules of the fish cannot take the same place as the water molecules. So when the fish swims, the molecules of water lying in front are pushed aside.

The Positive energy particle when it is moving is like a Pacman within this magic jelly.

We know that the Pacman possesses a direction. If the Pacman is directed from left to right, the Pacman keeps eating the jelly and at the same time, it keeps moving forward. As it moves forward, a small gap or zone is created where there is no jelly. Due to the properties of this magic jelly, the surrounding jelly is quickly pulled and the gap is filled in by the expanding jelly.

This Pacman, however, can move with a limited speed ahead.

What if the eating velocity is much higher than the moving velocity? What will happen is much more jelly will be eaten up on the progressing side rather than on the regression side.

As soon as the resting Pacman eats the jelly immediately around it, the particles of jelly immediately around it will be pulled inwards and they will lose contact with the particles beyond this sphere. Essentially, all this repetitive cycle will lead to a wave of movement of the jelly particles. This wave of disturbance starts from the Pacman and moves outwards in a spherical manner.

When the Pacman is moving but with a much slower velocity than the eating up process, the Pacman eats much more in the quadrant at the direction of the movement while the gap forms in the diametrically opposite quadrant which can be called the regression or receding border.

Even in this, the waves would be formed. But the wave would be asymmetric.

Explaining the Push bands

Creation from nothing is a more difficult concept to grasp than destruction.

Imagine two magnets held at a distance facing North poles to each other. They would repel each other. This repulsion can be explained by two means, either a force in between them pushing them away from each other or two independent forces on their far side, pulling them away from each other.

When we have a negative energy particle, a new Planck Compartment is formed every Planck time in it. This new Planck Compartment pushes the existing Planck Compartments away forming negative energy EPCAs.

Here we have to recognise a very important property of the Planck Compartment.

Two Planck Compartments, although they are supposed to contain nothing intuitively, cannot attain the same coordinate location.

This is pretty evident from the box and ball analogy discussed above. The individual balls are kept apart at a constant distance. Any attempt to increase the distance between the balls increases the force between them pulling them together. In the same way, any attempt to push the two balls together would increase repulsion between them so that the two balls remain separate and don't come too close.

When a new cubical Planck Compartment comes into existence at a point P which is shared by 8 pre-existing cubical PC's, the PC's can no longer stay at the same place as before and are pushed apart. As they move apart, new PCs have to appear to compensate for the volume needed to encircle the newly formed PC. From 8PCs, it has to become 26 PCs. When these new PCs form, each of them pushes the outer PCs apart and modify the C/D ratio in them to enable further Creation so that they can move apart and accommodate the outward movement.

Unlike a light wave or a wave in water, the PCs cannot move through each other and would possibly obey a principle similar to Pauli's exclusion principle.

A newly formed PC has the capability to push the PC's apart or create a pushing force. Two negative energy particles coming together would push each other apart due to the excessive Creation happening within and this pushing away effect. Implications are, that, unlike positive energy particles which attract and thus coalesce, negative energy particles remain separate.

At the end of a single second, our compartment produces 10^{43} new Planck Compartments. This is volume bigger than the volume of a linear string of PCs 1meter long. This is because 10^{35} Planck lengths are equal to one meter.

This means that even the baseline negative energy particle can produce significant space volume.

This is like forcing 100 people forcibly into a small box which is a meter in breadth.

It is unclear what happens to positive energy Planck Compartments. In them, multiple T=0 compartments can coalesce to form a compartment with hyper dilated time. Here, the Compartments are getting destroyed but the ability to destroy one compartment per Planck time is being retained. It is not exactly two Planck Compartments coinciding on top of each other.

It is impossible to know if the Compartments coalesce or stay apart and still keep doing their job. The effect on the outward Planck Compartments would be almost the same.

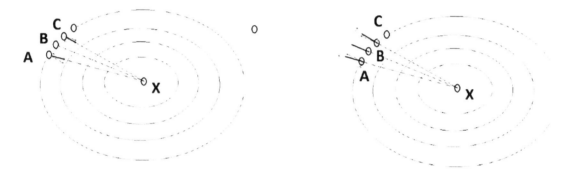

Figure ii/3.7: Shows three particles that form a part of a peripheral EPCA. In the left figure, X is a PEP and ABC are being pulled inwards for compensating the PC vacuum inside (I.e. towards X). As the EPCA moves towards the inner level, A and C are pushed towards each other. Alternatively, after Destruction, the re-creation of all the three PCs is not possible due to the smaller surface area of the inner circle. Thus A pushing towards C causes destruction of B. In the second figure, X is an ENEA particle and the EPCA is negative energy type of EPCA which is being pushed outwards by excessive Creation at X. As A, B and C move to the outer level, more surface area becomes available so that the Creation of additional PC in between A and B and B and C, now becomes possible.

CHAPTER 4

EXPLAINING UNIVERSAL AND LOCAL TIME

Space and time are wedded in GR so that one cannot be separated from the other. In DGR, space is separate and so is time.

The two types of time seem confusing at first. And due to the effect of GR for a century, the concept of relativity of time seems difficult to be given up. This will undoubtedly bring maximum opposition to the theory.

In DGR, there is only one time and that is UT. The LT is but a stubborn illusion created due to actively contracting or expanding space. It is inevitable for us to detect the deviated time i.e. LT by any experiment, the time is still going forward along with UT.

To understand this, it's easier to do a thought experiment.

Imagine a photon clock which is essentially two mirrors fixed together rigidly at a known distance and a single photon bouncing back and forth. All this assembly being in the vacuum of space. Here, we assume that the photon is a point and has no length or breadth. It moves with the speed c which is 1 Planck length per Planck time. We assume that the length between the mirrors is 100 Planck lengths so that it takes 100 Planck times for the photon to reach a mirror and get reflected.

This means that the photon clock ticks at every 100 Planck times or 200 Planck times depending on how you want to measure it.

Now further imagine a pendulum made of iron which is used to measure time.

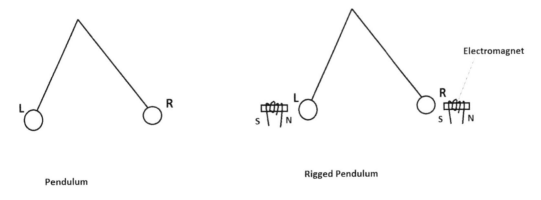

Figure ii/4.1: Shows the rigged Pendulum

Let it be such that it goes from one extreme to the other at time t. Now, what if we corrupt this apparatus by inserting two electro-magnets at the two walls near the extreme points of the Pendulum. Let these two magnets be L and R. Every time the Pendulum goes away from R, R gets activated and pulls the ball of the pendulum in a direction opposite to its swing. Whenever The pendulum reaches near L and starts moving towards R, L starts pulling it towards itself.

The result of these would be to slow down the pendulum. If a researcher who is unaware of this flaw in the pendulum uses it to measure time, his experiments would have an error of excessively dilated time.

Now we come back to the photon clock.

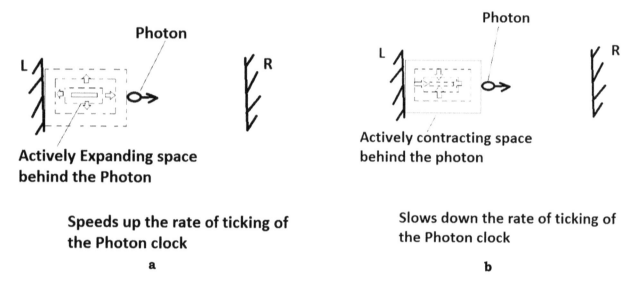

Figure ii/4.2: Showing how actively contracting space just before the photon of the photon clock, delays its journey and slows down its ticking (a) while actively expanding space just before the photon makes it reach the ends faster and hastens up the ticking.

Let's say that the two extremes of the photon be R and L.

What will be the effect of the presence of active spatial contraction in the space immediately behind the photon during its journey is a question worth pondering. The active spatial contraction and loss of volume immediately behind the photon due to the constant presence of EPCAs would inevitably slow down the journey of the photon towards the extreme points. There would be no practical way of knowing how much this slowing down is until we have any way to measure what is the pace of UT. Whatever time we have measured till now has always been LT and we have no means of measuring UT unless we go infinitely away from all the Gravitational bodies.

The same slowing down would be true for any clock including the atomic clock in which the Vibrations of the Caesium atom are slowed down due to the presence of actively contracting space.

If the settings are changed for our rigged pendulum clock so that the electromagnets repel the pendulum to actively push it towards the opposite side (for this we would need a permanent magnet in the ball of the pendulum), this would increase the rapidity with which each swing happens and the clock will run faster or show a contracted time. Similarly, a constant presence of actively expanding space just behind the photon pushes the photon forwards and makes it reach its destination faster and thus gives an erroneous measurement of faster ticks.

Comparing the periodicity of the same photon clocks kept at three different places, one at the surface, one at a satellite and the third at infinite distance from any gravitationally active body would be ideal.

The assumption in the above thought experiment worth noting is that the active spatial expansion or contraction happening due to the presence of the Gravitational body doesn't reduce the distance between the two mirrors.

Also note that the spatial contraction or expansion is present both in front and behind the photon but only the spatial contraction/expansion behind the photon is of relevance, for some yet unknown reason.

Local time

Time within A Planck Compartment

At the start of the evolution of this theory, it was presumed that every PC can have time. So that every PC that forms a part of the PC aggregate has the same time. It means that every individual PC can have a property of time assigned to it.

Now it has become clearer that this isn't the case. A single PC cannot have variable time.

A Planck Compartment can be only in three possible states.

1. A state with T=0 compartment,
2. A T=Tmax compartment and
3. A T=UT compartment.

When Destruction predominates, it is T=0, when Creation predominates it is T=Tmax and when Creation balances Destruction, it is T=UT.

The Local Time would then be decided in a bigger volume of space.

When all the constituent PCs are T=UT, there is no spatial expansion or contraction and time is running in this entire region at UT.

When there are some T=0 compartments while others are P=UT, the Local Time is given by

volume loss ΔV ratio of the total number of T=0 PCs to Total no of PCs.

Here, the numerator is where only Destruction happens, while the denominator is where Creation is happening. Here, there is an imbalance towards destruction which deviates the C/D ratio and causes active spatial contraction.

The true C/D ratio = Total PCs where Creation is happening/total PCs where Destruction is happening

= (Total Number of PCs- Total number of T=0 PCs)/Total number of UT Pcs + Total number of T=0 PCs)

= Total number of UT PCs/Total number of PCs

When there are some T = Tmax compartments and some T=UT compartments, then local time is given by the ratio

ΔV ratio = Total number of T=Tmax compartments/Total no of PCs.

Here, the numerator is where negative volume loss or negative Destruction or (positive) Creation is happening while the denominator is where Destruction is happening. Here Creation predominates and active spatial expansion happens. Here the ΔV ratio would be similar but negative.

The true C/D ratio here would be

= Total number of PCs where creation is happening/Total number of PCs where destruction is happening

= (Total number of UT PCs + Total number of T=T max PCs where only creation happens)/ Total number of UT PCs

= Total number of PCs/Total number of UT PCs

What is the true meaning of Creation, Destruction and Entanglement?

Consider a circular aggregate of PCs that has 100 PCs that are in the state of UT i.e. Creation in them balances Destruction.

What this essentially means is that with every Planck time, 100 PCs are destroyed here and all the 100 PCs are re-created. Thus the number of PCs remains 100 and destruction counterbalances creation. This circle will have a steady circumference and the PCs will not move inward or outwards.

Now imagine that there is a development of PC Vacuum immediately inside these PCs causing them to be pulled inwards uniformly. In the next PT, again all the 100 PCs get destroyed. But while getting created, their inward movement reduces the circumference of the circle. This means that the number of PCs that can be accommodated in this new circumference is smaller. Thus at the time of re-creation, some PCs have no room. For the recreation of a PC, enough volume needs to be available to accommodate it. If enough volume is not available, the creation fails. This means Destruction occurred but Creation failed for these PCs.

Number of PCs destroyed = Circumference of the outer circle -circumference of the inner circle.

Note that in reality, in a 3-dimensional world, an EPCA is like a spherical aggregate of PCs rather than a circular one.

The inverse is also true.

When an extra PC is created in the space between four PCs at their intersection, it cannot be accommodated unless it pushes all the four neighbouring PCs outwards. In the next cycle, all these 4 PCs would be destroyed and while getting recreated, there is more circumference available for them.

This causes a net creation of some PCs.

The cycle of Destruction and then Re-Creation

The fact that many events can happen at a point per Planck time, the cycle of Destruction followed by Re-Creation followed by Re-Destruction followed by Re-Creation at a point can happen any number of times as needed per Planck time.

Possibilities

Local time at a place depends on the C/D ratio within i.e. depends on how much Creation and Destruction is happening within.

Several possibilities arise as discussed below individually.

The T=0 compartment

In this compartment, the local time is zero which means the local time has stopped ticking.

Here, the C/D ratio is 1/2. This means that one PC is created and two are destroyed per PT.

Effectively speaking, from a volume of 1 PC, the volume increases to 2 PCs and then 2 PCs are destroyed. Thus effectively, the PC loses its entire volume or vanishes or is destroyed. To note that there can be temporary T=0 compartments that vanish immediately and just appear due to the process of entanglement and are a part of EPCAs. Also, there are permanent T=0 compartments that are massless particles and move with the speed of light and can coalesce with other T=0 compartments and represent a quantum of energy. These permanent T=0 PCs retain their ability to destroy 1PC volume per PT indefinitely and lead to the formation of EPCA's around them. These constitute the quantum or packet of **Positive energy**.

The PC at UT

This is a PC that is at Universal time (UT). It means that local time is running the same as UT, i.e. is not actively contracting or dilating. In this PC, the C/D ratio is 1 i.e. one Planck volume of space is created and destroyed and there is no change in the net volume.

The Dilated local time

In this, there is a net deviation towards destruction. Again, this depends on the C/D ratio. Another ratio that describes the local time of such a space is the ratio of T=0 PCs to PCs at UT.

In a space where the local time is dilated, there is active spatial contraction, however not in all the PCs and how much spatial contraction will depend on this T=0 by UT ratio.

The T= Tmax compartment

In this compartment, the C/D ratio is 2/1. So that 2 PCs are created but one PC volume of space is destroyed so that one PC volume gets converted to 2 PC volumes or there is an increase in the net volume of 1 PC. This creates a positive energy EPCA around it.

The Contracted Local time

In this, there is a net deviation towards creation. There is the deviation of the C/D ratio towards Creation. It is also given by the number of T=Tmax PCs to the number of PCs at UT.

The more the number of T=Tmax PCs in a given unit of volume, the more is the expansion of space.

Even these can be temporary or permanent T=Tmax compartments.

The Hyper-dilated local time

T=0 compartments attract each other and can merge with other T=0 compartments. There is no theoretical upper limit to how many T=0 compartments can merge. Thus 2 or 3 or 1000 or 10^{30} or 10^{100000} can merge.

At points where multiple T=0 compartments are merging, can be said to have a hyper-dilated time as time is running below zero.

Photons are aggregates of a varied number of T=0 compartments. Even particles with mass have a high amount of T=0 compartments aggregated together depending on the mass. The number of T=0 compartments within a point or a particle is indicative of the amount of energy contained within it.

T=0 compartment being a quantum of positive energy and T= Tmax compartment being a quantum of negative energy.

The Hyper-contracted time

To note that like hyper dilated time, hyper contracted time isn't common as T-Tmax compartments repel each other and do not form aggregates.

However, hyper contracted time was present at the time immediately after the big bang where an insanely high density of T=Tmax compartments was present at a small volume.

CHAPTER 5

THE T=0 AGGREGATE CONTINUUM AND THE SIGNIFICANCE OF "E" OR "ɜ"

More the T=0 compartments aggregated together, more is the density of positive energy and more is the value of "e" (since e is commonly used for a charge of an electron one can use an alternate symbol "ɜ") which is the number of events happening at a point per PT.

More is the number of T=0 compartments aggregated together, more is the number of events happening at that point and thus more is the extent of destruction per PT happening at it.

There is a predicted relationship between the amount of destruction happening at the centre and the amount of energy.

Thus, Photons of electromagnetic waves, which have a quantum of energy within them, would have increasing amounts of T=0 compartments aggregated within and an increasing amount of destruction happening proportional to the energy that is contained within.

Similarly, more is the mass of a particle, more is the value of "e" and more is the destruction per PT that happens at its core. This applies to supermassive objects like stars and also supermassive or ultra-massive Black holes with the value of "e" and the amount of destruction happening per PT at the core being directly proportional to the mass.

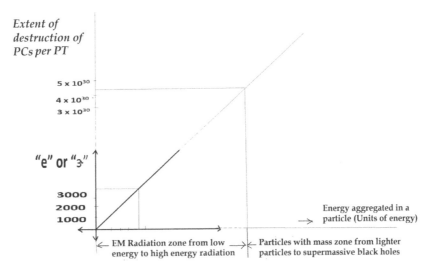

Figure ii/5.1: Shows a diagrammatic graph showing the probable relationship between the extent of Destruction at a point and density of energy concentrated at a point, starting from lesser

energy EM radiations to higher energy radiation to particles with mass to massive and Supra-Massive Black holes.

We are thus interested in finding out the following equations.

F(∋) = Extent of Energy packed together (energy density) in any particle (massive or massless)

F(∋) = Frequency of the resultant EM radiation

Where ∋ is the extent of Destruction happening at the core of the particle, also called the number of events happening at the core of the particle per Planck Time.

A Positive energy particle or PEP or PEM particle (Positive Energy Matter particle)

An aggregate of T=0 compartments forms a PEP.

It represents a quantum of energy and depending upon the amount of destruction involved can be a photon or a massive particle or a supermassive Blackhole.

A PEP can thus be defined by "e".

At the end of one Planck time, the PEP destroys e number of PCs and thus leaves behind a PC vacuum equal to a radius of e Planck Volumes.

In other words, an event happens in a PEP, every 1/e th fraction of a Planck time and sucks in the neighbouring PCs for compensation, thus starting a cycle of inward movement of PC aggregates. Thus at the core of a PEP, e number of cycles start per PT.

Due to this PC vacuum formation, the surrounding PCs belonging to the next level move inwards partially as described in more detail subsequently while defining PC levels around a PEP. These inward moving levels have to lose volume to compensate for the volume loss happening at the PEP and thus have time within them reset.

Thus a time resetting wave (TRW) starts propagating outwards. This wave moves extremely fast, much faster than the speed of light. The speed of outward propagation of the TRW depends on e. it travels "e" PL outwards per PT. This is a "positive energy EPCA forming" time resetting wave, each of these EPCAs having a relative dilation of time within them.

Note that these TRWs are different from the Gravitational waves discussed in the Potential energy section.

A Negative energy particle (can also be called the ENEA i.e. the Exotic Negative Energy Antiparticle)

An ENEA having local time, $T=T_{max}$ remains separate and tends to repel each other. Each of them can be said to have a negative mass.

They also form spatial expansion type "negative energy EPCA forming" time resetting waves around themselves.

CHAPTER 6

TIME RESETTING WAVES (TRW) AND THEIR INTERFERENCE

Constructive Interference of TRW

Like any other wave, TRWs can also positively or negatively interfere. Thus time dilation type TRWs from two PEPs can constructively interfere in the zone in between as shown in the figure.

In the figure ii/6.1, two PEPs A and B are shown with their TRWs which are named 1-6 tentatively. As one moves from A towards B on line AB, the time dilation caused by A reduces while that caused by B would increase. Thus a zone of maximum resultant time dilation and spatial contraction is formed at the line AB. As one considers peripheral points C and D, time dilation in them would be smaller due to their distance from A and B. Resultant spatial contraction here would be much lesser. E and F would have an even smaller time dilation and resultant spatial contraction due to the presence of A and B. The result would be a formation of a maximally time contracted zone from A to B where spatial contraction pulls A and B together and the volume loss in which is compensated by EPCAs as shown in the next figure.

The figure ii/6.2 shows two PEPs P and Q in the vicinity of each other with their respective zones of time dilation shown with tentative numbers 6 to 1 (which can be considered random units of time dilation used here only for understanding the purpose for simplicity and should not be considered real units of representing time). Note how the extreme time dilation at the centre leads to a TRW around each particle.

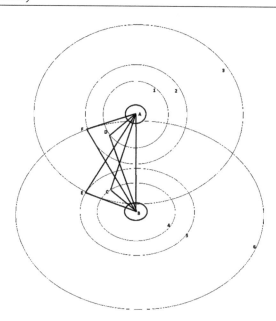

Figure ii/6.1: Shows two PEPs A and B with TRWs 1,2,3 emanating from A and 4,5,6 emanating from B. There will be maximum time dilation along the line AB and time dilation at C, D, E and F depend on the TRWs interfering at these points.

When the two TRWs emanating from each particle meet, their influences can add up or cancel each other depending upon whether their influence is time dilation or time contraction. Thus a time dilation of 3 caused by P and time dilation of 4 caused by Q leads to a net resultant time dilation of 7 and so on. Note that although this is a convenient way of analysing these TRWs, it is unclear whether the resetting of time is the primary event and spatial contraction happens secondary to it or the other way round. The spatial contraction with EPCA formation is likely the primary event and the apparent local time resetting is secondary.

If we analyse in detail, the zone between the particle can be divided into various zones with varying time dilations with maximal time dilation at the line PQ. A maximal spatial contraction would also happen here and this spatial contraction along PQ acts like a rubber band, constantly losing volume and pulling the PEPs towards each other. This volume loss is compensated by spatial contraction happening at the neighbouring zones with time dilation of 6,5,4 etc. The resultant EPCAs would look like the figure.

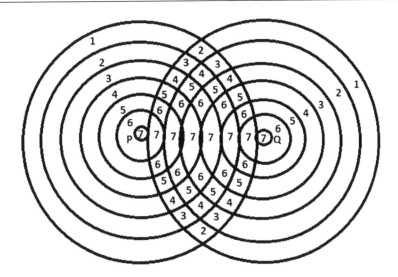

Figure ii/6.2 Shows two PEPs P and Q close to each other with TRW around them with an imaginary unit of time dilation (just used for an understanding purpose – the real time-dilation would not reduce linearly but will follow Einstein's equation for time dilation.)

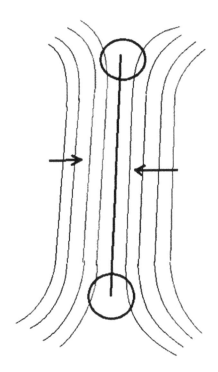

Figure ii/6.3 Shows the probable shape of the resultant EPCAs and how they will move inwards. The line joining the two particles will have the maximum time dilation spatial contraction and thus would progressively attract each other. The layers shown above and below the line joining the two particles are the EPCAs which would move towards the centre.

Subsequently, the figure ii/6.4 and ii/6.5 show three PEPs in the vicinity of each other with a time dilation of 10 units at the centre and shows how the resultant time dilation around them can be calculated. Each of these would be pulled towards each other due to constructive interference until they coalesce to form an aggregate with mass equal to the addition of their masses.

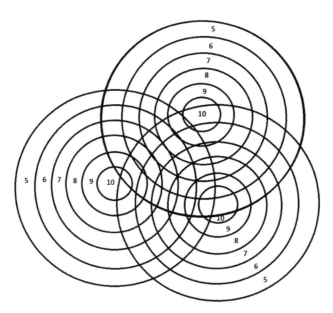

Figure ii/6.4 Three PEPs with hypothetical time dilation of 10 with TRWs starting from each of them interfering in the space in between is shown. The TRWs will interfere constructively and the resultant time dilations can be calculated by the time dilation caused by each wave that arrives at a point.

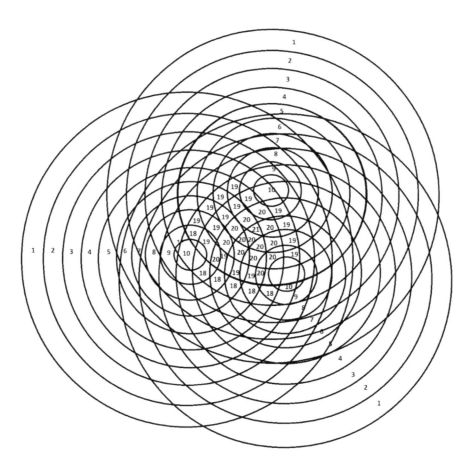

Figure ii/6.5 The figure shows the resultant hypothetical time dilations caused by interference of TRWs emanating from the three PEPs shown in Figure 6.3. The additive time dilation in all the regions means that there will be active spatial contraction between all these particles.

Destructive Interference of TRW

Figure 6.6 shows two particles with their respective TRWs. Here, particle P on the left is a PEP with a time dilation of 7 units at the centre and particle Q has a time dilation of −7 i.e. a time contraction of 7. Both of these are hypothetical particles with hypothetical units. Here, a time dilation of 6 would cancel out time dilation of −6 and a place where time dilation is 2 and −4 would have a net resultant time dilation of −2.

Detailed analysis of these time dilation zones gives the following inferences.

- The zone in-between the particle does not have a consistent amount of time dilation.
- Wherever the positive and negative time dilations cancel each other out would have a 0 local time i.e. local time running as per UT. Here, there will be no spatial contraction or expansion.
- Wherever there is a net positive time dilation, there will be a spatial contraction.
- Wherever there is a net negative time dilation i.e. wherever there is a net local time contraction will have spatial dilation.
- If the two particles come close to each other, the net negative time dilation near the right-sided ENEA particle causes repulsion. The PEP would have a much more spatial contraction in the direction away from the direction of the ENEA. Both these effects make the PEP move away from the ENEA. However, it is bonded to the ENEA due to the spatial contraction between the two which pulls the ENEA towards it as well.
- This behaviour is consistent with the prediction of Bondi et al who said that a pair of positive and negative mass particles would exhibit a perpetual runaway motion (*Bondi, H – 12*).

However, to note that an ENEA cannot form aggregates with other ENEAs unless there is extreme pressure from outside (like at the time of the Big Bang). The maximum possible time contraction is T=Tmax. Thus 7 such ENEAs forming a particle with mass −7 is not possible.

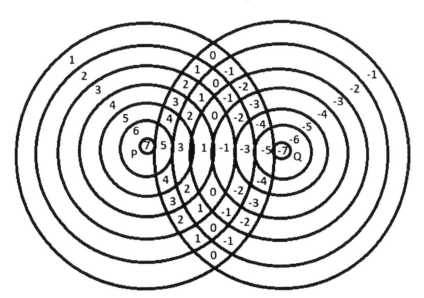

Figure ii/6.6 TRWs from a positive and a negative energy particle named P and Q respectively. The time dilations marked are hypothetical and only for understanding purposes. One can see

how at places the time dilation cancels out time contraction forming a zone of T=UT where there is no spatial contraction or expansion. To the right of the zone of UT, towards the negative energy particle, the space will actively expand pushing the particle Q away from P. On the left side of the UT zone, where resultant time dilation remains positive, active spatial contraction happens which pulls Q towards P. Thus the two particles have a complex love-hate type of interaction which Bondi et al called perpetual motion (*Bondi, H – 12*). **Q will keep running away from P and P will keep pulling Q so that it will keep following Q wherever it goes. Note that the two can never come close to each other contrary to the usual belief that matter and antimatter particles can annihilate each other.**

Taking a closer look at the figure ii/6.6, we can note that in between the two particles of opposite energies would exist a narrow zone where the TRWs meet where Local time becomes equal to UT so that there is no Spatial contraction or expansion.

This is true for particles with equal and opposite negative energies.

But what if we have a higher energy PEP which is at a higher distance thus creating an equivalent time dilation?

Higher is the positive energy within the PEP, higher will be the distance at which the smaller ENEA can remain stably due to the creation of a zero local time. At this level, the ENEA will neither move inwards towards the PEP due to spatial contraction time dilation nor move away due to time contraction spatial expansion.

Imagine a PEP with our hypothetical time dilation of +100 at the centre. As the TRW around it starts, the immediate outer sphere would have time dilation of 99,97,96 and so on. At the hundredth sphere around it, the resultant time dilation would be +1. At this zone, what if a hundred negative energy particles i.e. ENEAs can remain in such a manner that the TRW emanating from them with time dilation of –1 interferes with the one emanating from the central PEP and get cancelled forming a zone with local time zero?

Each of these ENEAs would have an active creation of a PC per PT. This addition of volume, however, can no longer be compensated in all directions. Especially not in the direction where the PEP lies.

The PEP also has a lot of destruction happening at its core which produces TRW and leads to the formation of EPCAs that travel inwards. The sphere just inside the location of the ENEAs will also form EPCAs which move inwards. The PC Vacuum produced here at the 99th sphere is no longer compensated due to the presence of the ENEAs. The negative energy spatial expansion time contraction type of TRW starting from the neighbouring ENEAs can interact and would interfere constructively. At this location as well, i.e. at the equator of the ENEAs along the surface of the sphere that forms the Bucky-ball lattice, there is a likelihood of formation of an LT= Universal time zone so that there is no expansion or contraction of space even here.

It figures that the only way that the extra formation of spatial volume happening at the ENEA can be compensated is by moving aside or pushing the PCs in the direction away from the PEP. Contrary to that, the only way that the PC Vacuum formed immediately inside the sphere which has these ENEAs is compensated is by pulling in, extra PCs from the spaces in between the ENEA particles.

The linear velocity with which a push or a pull band forms would be much faster than 1PL/1PT i.e. the speed of light but once they form, they move away from the charges with the speed of light forming the electromagnetic radiation.

The extra formation of PCs forming the push bands keeps pushing the Bucky-ball Lattice around. The bands will thus not go outwards in a straight line but spiral outwards as shown in Figure ii/6.6.

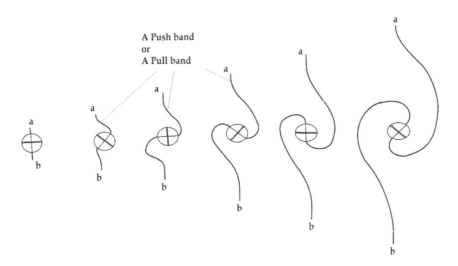

Figure ii/6.7 Shows spiralling out of the Pull or Push band due to the rotation/spin of the charged particle in the centre.

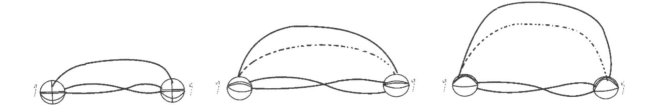

Figure ii/6.8 Pull band once formed, can keep moving away from its origin with the speed of light forming electromagnetic radiation. This is especially when the charges are moving towards or away from each other as further explained in the Potential energy section.

Figure ii/6.8 shows two charges with Push/pull bands in between. The velocity with which the bands move from one charge to the other would be faster than the speed of light, but as the Band moves outwards, this will happen with the speed of light.

Detailed computer simulations of this time distribution and this movement of PCs to compensate for the destruction or creation can be constructed and would be helpful to visualise these interference zones better.

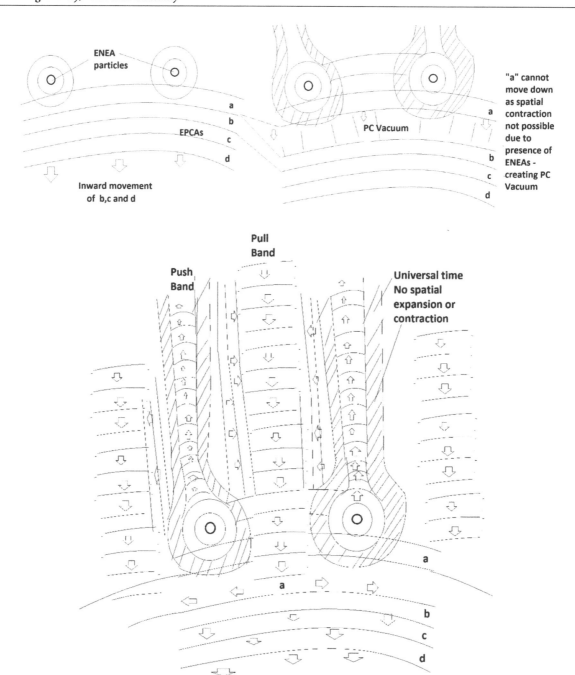

Figure ii/6.9 Showing the time dilation pattern possibly produced when two ENEA particles are lying in the path of a TRW from a higher energy PEP, like in the ENEA lattice around a charged particle. The obliquely shaded region is the T=UT zone and cannot contribute to the compensatory expansion or contraction. Thus the compensation of Volume loss or Volume gain can only occur by movement of PCs in the arrows shown thus forming push and pull bands. Note that the figure is just diagrammatic as the push and pull bands cannot lie in the same plane.

Figure ii/6.9 shows two ENEAs within the spherical EPCAs of a PEP. The Time resetting wave from the central PEP will be impeded from propagating forward due to the existence of these ENEAs. The EPCAs marked as a, b, c, d should ideally have both horizontal volume loss leading to each of them moving inwards. Out of these b, c and d can move inwards, but a cannot develop uniform volume contraction due to the presence of ENEAs. Thus the PC Vacuum formed due to inward movement of b can only be compensated by inward movement of PCs from space in between the ENEA particles. Because much more volume needs compensation, the PCs in the zone between ENEAs are pulled inwards much faster, possibly faster than 1 PC per 1 Planck time.

Note that this figure is just diagrammatic and the push bands and pull bands can never come in a single plane as is shown as the pull bands will form, not in-between two ENEA particles as shown, but in the space exactly between three or four neighbouring ENEAs. The thickness of obliquely shaded zone with Universal time i.e. LT = UT is also shown arbitrarily. The real distribution of time will have to be elucidated after one has the equation equating the number of events of destruction at the core to mass. With this information, the real-time dilations within each of the EPCAs can be calculated. The calculated diameter of the Electron can give the approximate distance between the central PEP core and the peripheral ENEA lattice.

These are just probable ways by which the Push-pull bands can be explained.

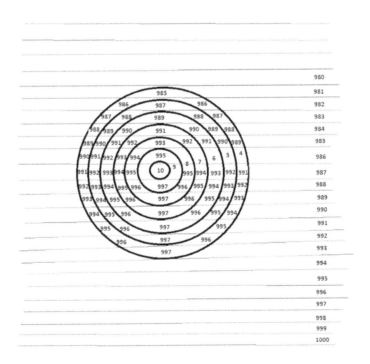

Figure ii/6.10 Shows how TRWs of a small PEP interacts with a TRW of a large PEP like a gravitational field. A band of maximum time dilation forms, as shown, and maximum spatial contraction happening here pulls the particle downwards –note the time dilation of 997 formed downwards. (the time dilations shown here are hypothetical and not real)

TRWs and Gravity

Gravity, as per DGR, is the result of the distribution of time dilation around the particle due to its presence in a time dilation field of a large PEP like the Earth. As shown in the figure ii/6.10, the TRWs from the larger object and the TRWs from the smaller object interfere and the resultant time dilation distribution pattern results in a band of time dilation which pulls the particle deeper into the field.

TRWs in a moving PEP: An Explanation for Inertia

This perfectly spherical shaped TRW emanating outwards from a PEP is just theoretical. Most particles, along with mass possess velocity, momentum and thus kinetic energy.

Momentum, Newton's laws of motion and Inertia are considered fundamental laws of nature and no physical theory has been able to provide a satisfactory explanation for the same.

In DGR, a potential explanation emerges without much effort.

A PEP, with an extreme amount of destruction happening at its core, is like a vacuum cleaner sucking inside and destroying PCs from all directions symmetrically. Thus, it results in a perfectly symmetrical spherical TRW with spherical EPCAS.

However, if the PEP in question is not static but possesses a velocity, (let's presume in the direction of left to right), what would be the result on the TRWs and the resultant EPCAs?

Figure ii/6.11 shows one such PEP. Every PEP, as described elsewhere in the description of levels, has 1 PL thick levels around it which move inwards and form EPCAs to compensate volume loss happening at the particle core. The perfectly spherical shape of the EPCA is because the level 1 PCs in a static particle have to move to equal lengths before they come in contact and compensate for the volume loss.

However, if the PEP is moving, it will change its position so that the point where maximum volume loss happens keeps changing for every subsequent cycle. Massless particles i.e. those who have a comparatively small value of "e" can travel with the velocity of light and thus travel 1 PL per 1 PT.

As shown in figure ii/6.11, as the particle moves from left to right, a PC vacuum develops in the receding border. This PC vacuum would mean that the destruction happening at the particle can no longer be compensated equally from the front and back. The PCs in the preceding border of the PEP have to move much faster and contribute much more to the compensation of destruction happening at the core. The inward movement of the PCS at the receding border would then be predominantly to compensate for the PC vacuum created here and only partially to compensate for the volume loss. This indicates that the EPCAs that form here move inwards much slower than the EPCAs that move inwards in the preceding border. This is true for every cycle so that the same events are repeated "e" times every PT.

One can relate this to a vacuum cleaner tube directed in one direction or a suction effect of rotating fan in an aircraft jet engine. An unbalanced inward pulling of the air by the vacuum cleaner tube causes the tube to move forwards as well unless it is held firmly. This persistent asymmetrical volume loss compensation keeps pulling the PEP forwards thus contributing to Inertia at motion.

This asymmetry is directly proportional to the velocity v and also directly proportional to mass m or the particle. This property of the particle is what we call momentum p.

If there are multiple particles bonded together, each of them having velocity v, the resultant momentum would be a sum of all the individual momenta.

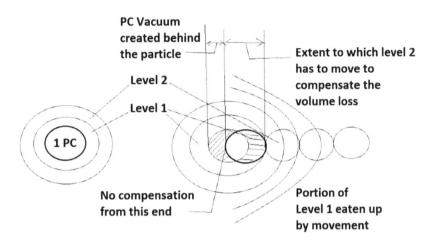

Figure ii/6.11: Shows a PEP moving from left to right. Note the levels 1 and 2 shown and the asymmetry created due to movement.

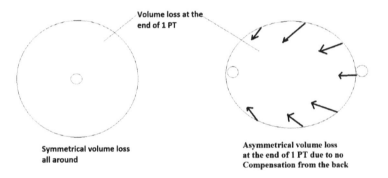

Figure ii/6.12: Shows the symmetrical volume loss vs asymmetrical volume loss at the end of 1 PT happening when the PEP moves from left to right, due to no compensation from the receding i.e. left border.

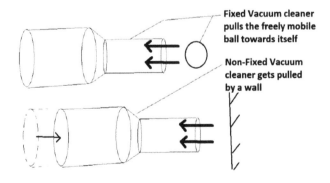

Figure ii/6.13: Shows analogy of a vacuum cleaner

CHAPTER 7

THE PHENOMENON OF RECRUITMENT OF PC'S

As the PEP moves forwards, it eats up much more PCs on the preceding border. This can be termed as "recruitment" of PCs. This phenomenon of "recruitment of PCs" has important implications.

If a perfectly spherical object is rotating, each of the constituent particles revolves around the centre and would keep recruiting different PCs in their preceding borders. The vector sum of the momentum of each of these would give the angular momentum of the spherical object.

The phenomenon of recruitment of PCs offers a different solution to the twin paradox clearly delineating, at least in theory, the difference between the static twin from the moving twin. The constituent particles that make up the spaceship and the twin within it, would recruit PCs when they move away. On the other hand, the Twin that remained on the Earth did not move with respect to the universal coordinate system of the surrounding PCs and no recruitment occurred in the constituent particles that make up the static twin. Only the Local time of the twin that went on the Space ship and returned would age slower. But the Universal time for both the twins would move at the same rate suggesting that the slower movement of time which the moving twin experiences, is just a stubborn illusion and that when the two return, with respect to Universal time, both would be exactly the same age.

CHAPTER 8

THE ENIGMA OF POTENTIAL ENERGY

The word potential energy comes at many places in various subjects like physics, chemistry, atomic and nuclear physics, etc. The term energy itself is a bit abstract with a nonspecific definition. Thus adding the word "potential" makes it even more abstract. Literal meaning or the word Potential means non-existent in reality.

Is the potential energy non-existent in reality?

A probable reason why it is called "potential" is that it is invisible and where and in which form it lies is still unclear.

Whatever is the truth, in most places where potential energy is used, there are well-defined formulae to calculate it and thus at most places, its value can be mathematically obtained.

We know that potential energy is readily converted into other more visible forms of energy like kinetic energy or heat. Thus there is little doubt that it exists.

It exists in various forms like gravitational potential energy, chemical potential energy, electric potential energy, magnetic potential energy, nuclear potential energy just to name a few.

The potential energy in DGR

In DGR, energy is nothing but deviation in local time which leads to a change in C/D ratio leading to active spatial expansion or contraction.

Active contraction or time dilation beyond Universal time is positive energy. Active spatial expansion or time contraction beyond Universal time is negative energy.

It then figures that even all forms of potential energy would be some sort of time changes leading to spatial changes.

How Potential Energy apparently breaks the law of conservation of energy?

Imagine a robotic arm powered by a battery-powered by some fossil fuel like diesel. Further, imagine a very heavy piece of machinery, say 100 kg.

Imagine that the robotic arm utilizes the energy released from the burning diesel to lift this machinery by a height of h. This process done against gravity needs some energy which is provided by the burning of the fuel. Imagine that there is no locking mechanism available in the robotic arm so that after reaching h if one wishes to keep it there, one has to keep supplying that energy, or one has to keep supplying the fuel.

One can see that the above experimental setup can keep consuming diesel. The robotic arm is not moving the machinery but still, energy demand would continue. Where is all this energy going? Note that the robotic arm can keep consuming more fuel and keep the machine for 1 min or 1 hour or 1 day. At the end of 1min, the energy consumed is much less than the energy consumed at the end of 1 hour which would be much less than the energy consumed by the end of one day. However, the potential energy of the machine would continue to be the same at the end of one minute or one hour, or one day.

Where did all this consumed energy go?

Some say that it was probably radiated out as heat. There will be some heat loss while burning the fuel. But there is no reason to believe that the temperature of this setup rises.

Note that this robotic arm and heavy machinery are just to increase the apparent impact. We may as well use a 5kg weight and our deltoid muscle with ATP as a source of energy. Most people would know how much energy needs to be spent to keep a 5kg weight elevated at a height for the entire day

Explanation by DGR

DGR would say that EPCAs are travelling downwards towards the Earth due to the destruction happening at the core of each constituent particle of the Earth.

EPCAs just beneath the heavy machinery at the height h will keep going down forming a PC vacuum just below the machine. If unsupported, the machinery would move down and thus fall freely with the EPCA until it reaches the ground.

When the EPCAs form beneath a machine supported from above, the machinery cannot move down. This causes the persistence of PC Vacuum.

The mass of every particle constituting the machine absorbing and destroying PCs adds to this un-compensated vacuum. This vacuum would then be inevitably compensated by the inward flow of PCs from the sides.

The smooth flow of EPCA, like the flow of a fluid in laminar flow, is disturbed and the vertically downwards flow is replaced by rapid horizontally moving PCs. This creates a Gravitational-wave that keeps on absorbing and taking away the energy lost to keep the machinery up.

The mass of the machinery creates its EPCAs all around it and all these zones add to the un-compensated PC vacuum.

Potential energy is given by the term

PE = mgh

In this term, the "gh" indicates the gravitational potential at the point where the machine lies. And m indicates the contribution of the machine's mass towards the potential energy.

The downward progression of the EPCAs due to Earth's gravitation keeps pulling the machine down. This downward pull has to be counterbalanced by the upward pull of the robotic arm which requires energy. This energy is dissipated away as the gravitational wave.

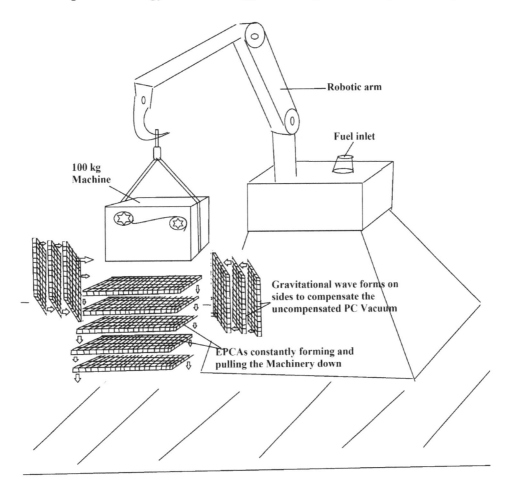

Figure ii/8.1: Shows a Robotic arm lifting a 100 kg machine with help of energy derived from burning fuel. EPCAs constantly form beneath the machine and keep getting pulled downwards, create a PC vacuum underneath the machine and in turn keep pulling the machine down. That constitutes the Potential energy within the machine. The Robotic arm has to constantly counter the pull. The uncompensated PC Vacuum is compensated by the formation of a Gravitational-wave formed from all the sides which carry off or dissipates the energy spent by the robotic arm.

What happens beneath a children's see-saw or a weighing balance?

In a see-saw or weighing balance, the two sides are pulled down by the EPCAs from the Earth's gravitation. This causes an un-compensated PC Vacuum beneath both sides. These can get converted into tension in the substance of the levers on both sides and ultimately compression of the fulcrum. Assuming that both sides are at the same height and the term "gh" i.e. gravitational potential at that place or PC vacuum at that place due to the downward directed EPCAs is the same, the difference in masses on both sides causes an asymmetrical PC vacuum, more at the side with more mass. Thus more time dilation and more downward pull will be beneath the heavier side.

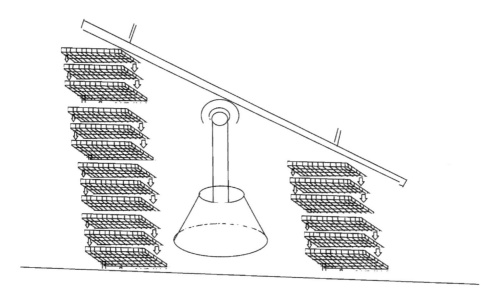

Figure ii/8.2: Shows how the constant formation of EPCAs beneath the two ends of a kids see-saw leading to the constant creation of the PC vacuum beneath keeps pulling the two ends downwards. The overall pull will thus depend on the mass of each side.

Electric Potential and Magnetic potential

Imagine a charge which lies in an electric field of another oppositely charged particle, say an Electron, bonded to a Proton in the outermost filled orbital of an atom. Here the Proton's electrical field pulls the negatively charged Electron inwards. At this orbital, the potential energy of the Electron is the lowest possible one. Now if we start pulling the Electron outwards, the electric attraction starts obstructing our intention to free the Electron. One has to supply enough energy to the Electron to jump to a higher orbital level. At this outer allowable orbit, there exists a constant inwards pull and thus this Electron can be said to possess electrical potential energy.

Similar to this description, we can imagine a small bar magnet that is in contact with a bigger magnet so that the North pole of the smaller and South pole of the large magnet is facing each other and is stuck to it. If we try to pull out the smaller magnet, magnetic flux (pull bands) emanating from them interact and constantly keep pulling the two together and opposing our intervention. The unrestricted pull towards the bigger magnet creates potential energy. If the smaller magnet is kept pulled apart, it will possess potential energy which will be called magnetic potential energy.

Explanation by DGR

The electric and magnetic fields and interactions are explained based on push-pull bands along with spin.

In the case of an Electron, as soon as the push-pull bands of an Electron come in contact with those of a Proton, it rotates and aligns itself in such a way that the north pole of the

Proton aligns with the South pole of the Electron or the other way round. In this configuration, the density of pull bands of the two charges match, and thus the two pull each other. This happens until the push bands of the electron go around the curved geodesic around the Proton and meet at the diametrically opposite side and form a standing wave. Beyond this point, the electron cannot move any closer. In this situation, the inward pull of the pull bands at the NS poles of the two charged particles counter-balances the outward push of the push bands of the electron around the Proton.

Any attempt to move Electron outwards leads to an uncompensated inwards pull by the perfectly aligned pull bands of the two charged particles. These un-compensated pull bands have extremely rapidly flowing EPCAs within them moving towards each of the particles. The un-compensated pull is compensated differently. The PC vacuum created by these un-compensated pull bands is compensated by a sudden inwards rush of PCs from all directions that propagate outwards or drift outwards with the speed of light forming/EM waves/Radio waves. The too and fro motion of an electron from an inner orbit back to an outer orbit and back inwards and the compensation process of the resultant increase and decrease in electrical potential energy leads to the creation of an EM wave.

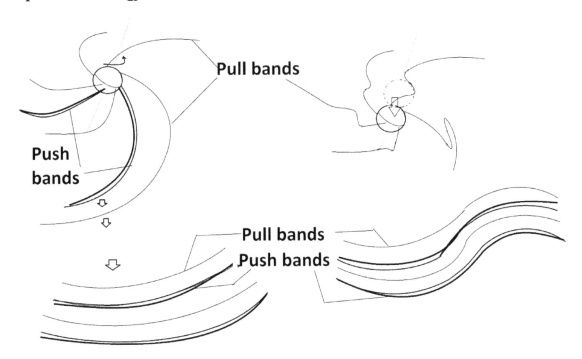

Figure ii/8.3: Showing how Push-pull bands drifting outwards form the Radio waves

The Push-pull bands move outwards with the speed of light and form an electromagnetic wave (radio waves).

CHAPTER 9

KINETIC ENERGY

This is another abstract entity. Nobody knows where the kinetic energy is stored. What we do know is that potential energy can get transformed into Kinetic energy and back.

The well-known expression for KE is

KE $= 1/2\ mv^2$

One can see that KE is directly proportional to mass as well as to the square of velocity. Why the term 1/2 comes in the equation is not known yet.

The Kinetic Energy in DGR

Any particle which moves with a speed v will have destruction within it and an EPCA formation around it, although as explained earlier, at least at some cycles, it can no longer count on the receding border for compensation of the destruction taking place in its core.

In these cycles, where there lies a PC vacuum just behind the particle at the receding border, the particle relies more on the preceding border for compensation of the PC loss. Thus, in the preceding edge, the EPCAs move inwards faster and there is a faster change in time gradient. Because of the excess compensation from the preceding border, the particle is pulled further towards the preceding border, thus pushing the particle further forwards, thus explaining inertia of motion.

At the receding border, the EPCAs that form partly compensate the PC Vacuum formed here and partly contribute to the compensation of the central destruction. Due to the movement, the preceding border in the direction of motion of the particle will have a higher than usual time dilation which will also mean that the inwards motion of EPCAs here is faster. On the contrary, the receding border, exactly opposite to the direction of velocity, will have a much smaller time dilation and a smaller reduction in time dilation

At the receding border or edge, the fast-moving EPCAs are like a spherical rubber surface moving rapidly to compensate for the PC Vacuum forming behind the particle as it moves. The more is the mass of the particle and the more is the velocity of the particle, the more will be this PC vacuum behind it. Thus this PC vacuum at the back is directly proportional to velocity and mass which is the momentum p of the particle. Note that the movement of the particle is also creating or leading to time changes immediately in front of the particle at the preceding border.

These time changes are directly proportional to mass. These time changes happen at the front half and the back half. Also, each of these changes in front and back is directly proportional to the velocity v. If the velocity v is increased, both the time changes in front and back would increase. Thus an entity that describes a combination of all these changes would be proportional to the square of the velocity. This explains the equation of Kinetic energy well.

CHAPTER 10

MATHEMATICS OF DGR

Mathematics of Newtonian mechanics, Special Relativity (SR) and GR

Newtonian mechanics, SR and GR pop out of DGR at appropriate distances without much effort. The mathematics of each of these would, indeed, remain valid.

In addition to the equations suggested in Newtonian and Einsteinian Gravity, other equations are likely in DGR.

New Equivalences suggested

Like the mass-energy equation suggested by SR, DGR points to the existence of other important, yet undescribed, equivalences like loss of volume-mass equivalence or loss of volume - energy equivalence.

In addition, it suggests ***a "loss of spatial volume- time dilation equivalence" or "gain in spatial volume time contraction equivalence"***.

In addition, ***"a positive energy- loss of spatial volume- mass equivalence"*** and ***"a negative energy- gain in spatial volume- negative mass-anti gravity equivalence"*** is suggested by the theory.

$+f(\Delta V) \rightarrow$ Positive Energy

Or

$F(\Delta V) \rightarrow$ Mass

Or

$-f(\Delta V) \rightarrow$ Negative Energy

Where (ΔV) is volume loss per unit of time (I.e. per Planck time)

Do we need Calculus?

In DGR, we start from the extremely small and examine the effects of these small events on the large. The space is already in an extremely divided state and thus further division of space is not necessary. Thus dividing it further and getting an integral would not help. Even time intervals in the assumptions of the theory are extremely small and although time will need to be divided further based on the number of events per Planck time, the "dt" part or the denominator is already per Plank time and thus derivatives like velocity d(distance)/dt can just be expressed in terms of distance per Planck time. Thus velocity in DGR (say the inwards velocity of an EPCA) would be a simple summation of distances travelled during each cycle per Planck time. The real challenge is to carry out these calculations with extremely small individual distances for an extremely large number of times. We have to get used to the exponential function i.e. extremely small distances or times and the extremely large number of times the same simple mathematical operation is repeated if we have to crack the Mathematics of DGR.

One of the fundamental assumptions in DGR would be that

"The Total number of events that can happen at an infinitesimal fraction of time can be infinite."

What this means is that in each small infinitesimal fraction of time, let's arbitrarily say $1/100^{th}$ of a Planck time, an infinite (countable infinity/practically infinite – i.e. not infinite in the true sense but un-countably large) number of events can potentially happen all at the same time. That is a PC can get destroyed at the centre starting a new cycle, all the 26 surrounding PCs can be pulled inwards by an inwards directed force, also causing a resetting of the time. At the same infinitesimal time, the previous cycle causing PC vacuum just outside level 1' and the cycle previous to that PC vacuum around level2' and 3' and 4' and so on. Thus there is no upper limit as to how many simultaneous events can happen at infinitesimal fractions of time.

This assumption is inevitable and is based on intuition.

CHAPTER 11

CONSTRUCTING A MODEL

The first step towards deriving the Mathematics of DGR would indeed be to construct a model of how the Planck Compartments react and form the EPCAs. Given the extremely tiny distances we are talking about and the extremely high number of PCs that would react, this is going to be challenging.

The real challenge is not to make a model. The real challenge is to carry out the computations of this high number of PCs at extremely high speeds, accounting for extremely small fragments of time, making sense of extremely small i.e. mostly sub-Planck length distances and then adding all of them together.

The real challenge is to carry out each of these calculations for every cycle for large distances and then carry out these calculations with increasing rates of cycles happening per Planck time.

Chapter 12

DEFINING DISTANCES

Sub-Planck length distances

We can compare the velocities of a ball and a photon. Let the ball travel from left to right with a meagre velocity (say 10 m/sec) while light travels at c. Both are travelling on parallel tracks of PCs so that when they move, they occupy the next subsequent PC.

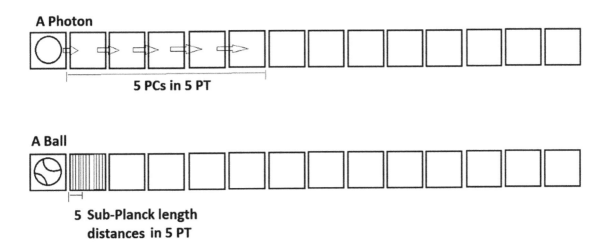

Figure ii/12.1: Shows a comparison of a Photon and a ball. A Photon travels one Planck length per Planck time i.e. it crosses one of the compartments shown in one Planck time. At the same time, a ball or for that matter anything travelling slower than the speed of light would cover an extremely small Sub-Planck length distance. Thus sub-Planck length distances are real.

We know that the photon would travel one PC per Planck time. The ball cannot travel faster than this or equal to this so it must necessarily travel slower than this. This means at the end of the first Planck time, it would have travelled a miniature sub-Planck length distance. The ratio v/c would determine how many Planck times are needed for the stone to cross a Planck Compartment.

Note that most events in our life happen with velocities slower than the speed of light. Thus most of these events happen at sub-Planck length distances per PT. In short, the assertion that Planck length is the smallest possible distance is not true.

Every cycle of destruction of one PC at the centre of a Positive energy particle would create an extremely small sub-Planck length shifting of the surrounding PCs. Measuring any of these experimentally would be impossible. However, even modelling these in theoretical models will be difficult. With multiple cycles happening per Planck time at the centre, we have to keep track of these miniature sub-Planck length shifts caused at each level by each cycle and obtain a sum of the total shift that is obtained per Planck time. At places where the EPCAs are moving faster than the speed of light, at least these sums of shifts of PCs after taking into account all the cycles that happened in a Planck time would go beyond one Planck length. However, as explained above, for EPCAs moving smaller than the speed of light, which will be the case most often, the shifts happening per PT after summing up the effects of all cycles would still be below Planck lengths. Thus only particles with extremely high destruction happening at their core would give any meaningfully calculable distances. As one moves outwards, more is the distance from the central destruction, still smaller in the sub-Planck length realm will be the distances.

Thus we have to get used to having calculations done at such extremely small distances.

CHAPTER 13

DEFINING TIME DILATION

Time dilation

The T=0 compartment

Consider the photon clock in the figure.

Let us consider an instance when the Photon is at point P and moving to the right.

Let's further imagine that the PC immediately before P has LT (local time) of T=0. The Photon will travel towards the right covering one PL per PT. But the T=0 compartment will destroy one PC per PT on the left thus effectively pulling the Photon back to its original position. Thus, the photon in the photon clock stops ticking. A photon entering the T=0 compartment effectively stops moving and is thus lost. (here the local time within the PC to the left of P is 0 and not local time within the whole Planck clock)

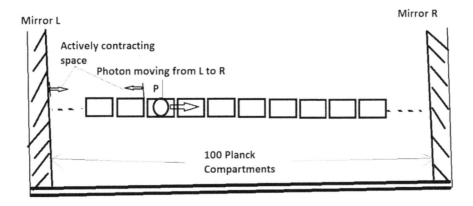

Figure ii/13.1: Shows a Photon clock with a Photon P moving from right to left which is being pulled towards the left due to actively contracting space due to the presence of an EPCA.

Time dilation due to movement

Let's imagine that the whole set-up of the photon clock with its photon is travelling towards the right. If the set-up is travelling with a velocity of light, the rightward movement of the left mirror one PC per PT effectively cancels out the rightward movement of the photon and the photon clock again stops ticking.

What if the velocity of rightward propagation is much less than c? Let's presume it to be half of c. Every two rightward propagations of the photon happening at 2 PT would be partially cancelled by the 1PL rightward movement of the left mirror. The velocity of light would effectively be turned to half and the photon clock will measure a dilated time.

Note that rightward propagation of the photon clock apparatus is effectively equivalent to spatial contraction happening to the left of P.

Gravitational time dilation

If the photon clock is kept in such a manner that there is a constant presence of an EPCA propagating at any direction but effectively causing spatial contraction to the left of P, this will also hinder the rightward movement of the photon and slow down the ticking of the photon clock explaining GTD.

Note that extent of Time dilation is directly proportional to the rate of active Spatial contraction or the other way round. This means that nearest to the Gravitating object, the active volume loss is much more. Essentially, if multiple Photon clocks are kept at different levels starting from the Earth's surface, going upwards, the cancelling effect of the active spatial contraction will be highest at the surface and it would keep reducing as we go towards the higher Planck clocks.

This is the troubling but insightful fact, towards which DGR is pointing at. This is Einstein's VSL or Variable speed of light theory, as noted subsequently. The forward progression of the Photon of light will be hindered much more at the surface and much less as we go higher, thus the speed of light is not a constant (although in true sense it is constant since light travels 1 PL per PT) but is a highly variable entity, depending on active spatial changes caused by the status of Local Time.

CHAPTER 14

NOMENCLATURE OF LEVELS

1. For any point P, there can be defined, a single spherical Planck compartment with radius 0.5 PL i.e. diameter 1 PL around it. This can be considered as level 0.

2. Immediately around it, can be defined a cube of 3*3*3 PCs i.e. a total of 26 PCs. This can be called level 1. Immediately outside this 3*3*3 cube, another cube with 5*5*5 can be easily defined. This would be level 2. Outside this, cubes of 7*7*7, 9*9*9 and 11*11*11 PCs representing level 3, level 4 and level 5 respectively can be easily drawn. In this manner, levels can be defined until infinity. Spheres can then be drawn within each such level which represents the PCs.

3. When the PC, whose centre is P gets destroyed, the PC vacuum can be compensated by all the surrounding 26 PCs to lose some volume (here the first level loses 18 PCs) and form 8 PCs. The point P now lies outside each of these PCS. This 8 PC level could be called the level 1'. Thus level 1 EPCA has been converted to level 1' EPCA here.

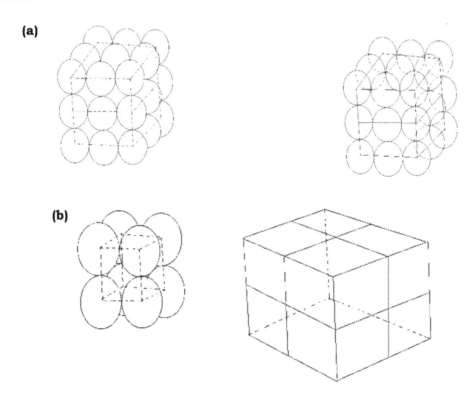

Figure ii/14.1 (a) and (b): (a) Shows a cube made of 3x3x3 Planck Compartments, (b) shows the same 3x3x3 cube after getting converted to a 2x2x2 cube.

4. The volume of 1 PC lost at the centre and the 18 PCs lost at level 1 (i.e. total 18+1=19 PCs) will remain outside level 1' as PC vacuum. This volume of 19 PCs, however, will have to be distributed over a surface area occupied by 26 PCS before. Thus the distance between level 2 and the new level of level 1' will be smaller than one PL, that is level 2 has to move smaller than 1 PL to compensate it. This distance can be given by 19/26=0.730769PL.

5. Another PC lost at the P subsequently will convert the 8 PCs in level 1' to one PC with a loss of 7 PCs at level 1'. This shift from level 1' to level 0 starts another cycle of formation of PC vacuum and EPCA formation. Here, a PC vacuum of 7 PCs lost is distributed in a volume of 8 PCs. The level 2' is sucked in to recreate the 26 PC cube and the cycle continues. Here, Level 1 formed level 1', level 2 formed level2', level 1' formed level 0 and level 2' again formed level 1.

(note that the nomenclatures level 1 –level 1' are variable and not consistent and the figure is just to show the events happening in each PT)

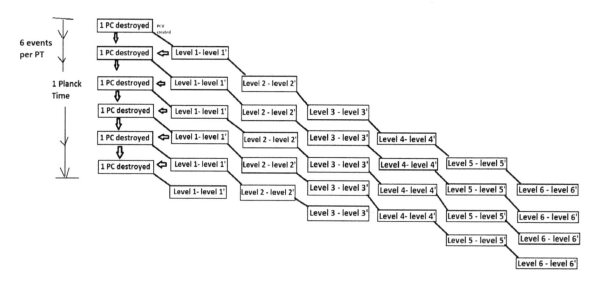

Figure ii/14.2: Shows how an event of PC destruction initiates a cycle of EPCA formation and inward shifting of these EPCAs. Here, in 1 Planck time, 6 events are happening, thus each event happening 1/6th of a Planck time (note that the levels are numbered arbitrarily and will keep changing with every cycle.)

6. Imagine two adjacent PCs forming a part of EPCA at any level. Even if spherical, the neighbouring PCs would have a contact point. Each PC will have four such points. A circle can be defined passing through each of these points having the centre as the centre of the PC (called the central circle of the PC in subsequent discussion). The surface area of this circle is of importance and so is the distance between its centre and P. It is worth noting that very close to the centre P, the curvature of the PC aggregate will be much more and the four contact points would be much lower. The central circle would then not be a planar structure but would have a convexity outwards.

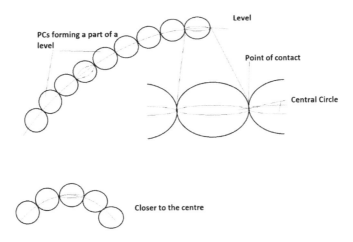

Figure ii/14.3: Showing the central circle of each PC belonging to a PC aggregate. Note the point of contact between adjacent PCs which depends on the curvature of the EPCA. With a more curved EPCA, the point of contact goes down.

One can note that the centre of level 0 is 0 PL from P, the centre of each PC in level 1 is about 1.5 PL from P and so on. In this manner, levels 1.5, 2.5, 3.5, 4.5.... Etc. can be defined till infinity with the help of a computer. Note that these levels and their distances will hold only before the time-resetting EPCA forming wave reaches them. After the EPCAs form, they will move inwards and will form newer yet unknown levels and subsequent cycles will affect them in these newer levels and not the previous levels. As we move outwards, (discussed below) will become extremely small, in the sub-Planck length realm and this error will become smaller unless the number of events happening at P is huge in which case the error would persist and would be significant.

7. Each level will form a sphere. The surface area (SA) of this sphere would be the sum of surface areas of the central circles of each constituent PC.

    i.e. (total number of PCs) * (surface area of each central circle i.e. 0.785 PL square)

    If the PCs are considered cubical, then this SA can be considered equal to the total number of PCs in the level.

8. The distances of these central circles (which represent the levels) from the P are definable and will remain constant. The location of P is irrelevant here. Such levels can be defined around any point irrespective of whether it is near the Sun or Alpha Centauri or in the Andromeda Galaxy.

9. For each level, 3 things can be defined precisely. The total number of PCs in it (which is equivalent to the volume contained within the EPCA and loosely the SA of the sphere), the distance from P (r) and the surface area of the sphere with this centre as its radius. Each of these will be constant for each level.

10. For every level r, the next incomplete level r' can be defined (i.e. computed by computers) where r can go to r'.

11. SA at r and SA at r' would be definable for any value of r. The value of p*0.785 at each level can be defined, thus one can find out the number of PCs that need to be lost to move from level r to r'.

12. With this, the value of δd, which is the distance travelled by the EPCA while crossing from level r to r' can be calculated.

    For this δd = (Total volume lost at the centre + sum of total volume lost in each of the EPCAs up to level r)/SA at r

13. At each movement r to r', the outer EPCA is trying to compensate for the volume lost till now by moving inwards. This means that the PC vacuum formed within an EPCA is equal to the volume lost within the centre and at all the subsequent EPCAs in this cycle. This is compensated by the formation of the inward moving EPCA whose new radius is r.

    Thus the sum of volume lost in all the previous EPCAs and central volume loss = PC vacuum created

14. If the number of events that are happening at P (e) per PT is known and δd at each of these events (δd1, δd2, δd3...... δde) can be calculated, the velocity of propagation of the EPCA at r would be given by

    Velocity of propagation of EPCA at r = sum of δd at each in one Planck time

    Velocity of propagation of EPCA at r $\sum_{n=1}^{e} \delta dn$

15. It is worth noting that r can take any value including sub-Planck length values. But the Surface area of the Sphere can take only integer values.

CHAPTER 15

DEFINING VOLUME LOSS

Imagine a sphere made of 10^6 Planck volumes. Let's imagine that this sphere is actively losing a volume of 100 PV per PT and thus is in a process of active contraction.

Let this volume loss be Δv per Planck time.

This process can be achieved by two different processes.

The first one (which is not in accordance with our theory) is that the volume loss happens uniformly throughout at an infinitesimally small level and so every unit of the sphere loses some volume. In this, division of the sphere into small PCs is not needed. Even if we do divide the space into smaller volumes or compartments, each Compartment must have a variable volume so that each Compartment becomes smaller as time proceeds.

The other process is closer to our theory. In this, the entire sphere is in turn made of PCs that have a fixed volume and cannot change. Thus space contraction is carried at specific locations within it and is not generalized. Out of the 10^6 PCs forming this sphere, 100 or possibly lesser) will lose a PC every PT. (Note that the assumption here is that the rest of the PCs stay on or in them Destruction balances Creation or every Planck second one PC gets created and also destroyed so that their volume remains unchanged.)

These localised areas of volume loss would then create disturbances around them and create additional volume loss for compensation. This volume loss needed for compensation is responsible for the actual spatial contraction.

Theoretically, volume loss can be either uniform or non-uniform.

Uniform volume loss happens within an EPCA while in a bigger volume of space, the volume loss would be non-uniform.

In uniform volume loss, the ratio of lost PCs/Total PCs would remain throughout. So that if we divide it into left and right halves, each of them will have a similar ratio of lost PCs to Total PCs.

There is no necessity that volume loss will be uniform in a bigger space and the volume loss is necessarily non-uniform within a sphere which has a positive energy particle within it.

```
Volume lost per PT Δv= original volume — resultant volume = sum of volume of
total no. PCs — sum of volume of PCs not lost = sum of volume of total no.
PCs — (sum of volumes of total number of PCs - sum of volume of PCs lost)

Δv-ratio=lost PCs/total PCs
```

Consider the denominator of this equation. What will happen if the denominator is trending towards infinity? A finite numerator would then have no significance. This happens in linear EPCAs in which the total number of PCs is trending towards infinity and thus there is hardly any volume loss or curvature. (note that the infinity here is countable infinity and is not infinity in the real sense)

Now, what if the denominator is tending towards 0 or the numerator is tending towards infinity? The equation breaks down simply because of the appearance of infinity. Although this happens, the theory allows aggregation of multiple T=0 PCs. In supermassive Black holes near the centre of Galaxies, a countably infinitely large number of T=0 events can exist within a volume of 0 PCs (i.e. numerator is infinity and denominator is 0) (Note that the term "Countable infinity" is used loosely and rather inappropriately here as in the real mathematical sense, counting these PCs is difficult but isn't impossible and they cannot be called infinite PCs in real sense.)

Explaining this would need an underlying theory that offers an explanation of DGR and will have to wait.

If there is a point where excessive volume loss is occurring like a particle with positive energy or mass and is leading to the formation of EPCAs around, volume loss would be given by

```
Δv occurring within a large volume of space per PT=Sum of Volume loss occurring
in the centre per PT+ volume loss occurring in each EPCA per PT.
```

Another ratio that is possibly equivalent and might have some significance is

Total number of T=0 compartments/Ratio of the total number of PCs having UT (having balanced creation and destruction or total PCs remaining at the end of a PT)

Uniform vs Non-Uniform volume loss

Let's focus on an EPCA at r distance from a central point P where there is active destruction going on. The assumption here is that this is the only particle and thus there is no other EPCA entering the region of space.

Each EPCA will have the following properties

1. The total number of PCs that form it should be a perfect integer so that EPCAs with one or more half PCs or quarter PCs or any other imperfect number are not permissible. This means that it can have 100 PCs but not 100.5 PCs simply because a half PC cannot exist.

2. They have a uniform Δv-ratio throughout the EPCA so that volume loss happening in the EPCA has to be uniformly distributed. This means that the T=0 compartments would be uniformly distributed. The Δv-ratio at an EPCA usually determines how much volume loss per PT is happening within it which determines the extent of LT dilation. (conversely, GR predicted time dilation can give us a clue as to calculate the Δv at any level.)

Let's presume that this EPCA has 10^{10} total PCs and there are 10 T=0 PCs for every 1000 PCs or Δv-ratio = 10/1000 =0.01.

This Δv-ratio remains valid whether we consider the whole of the EPCA or a section or arc of it. So that if an arc of the EPCA has 10,000 PCs, then it will also necessarily contain 100 T=0 PCs. Due to the uniform distribution of the T=0 compartment, the volume loss is going to be uniform and there won't be any distortion of the perfect spherical shape of the EPCA as it moves inwards unless another external influence like the entry of another EPCA from another particle is acting on it.

3. With every Planck time, the EPCA would move inwards after the loss of a fixed number of PCs and becomes the next level EPCA after moving inwards by a distance δd. If multiple events are happening per PT at the centre, the EPCA will move inwards that many times. δd would vary at every level change. If the EPCA is moving slower than light, which is most often the case, δd would be in the sub-Planck length range.

4. If the entire sphere of disturbance created by the presence of positive energy particle at P is considered, there is a non-uniform distribution of Δv. If a large cubical volume made of 10^6 mm^3 is kept at a distance from P, as the EPCAs move inwards, the cubical space also moves inwards.

The outermost layers of the cubical space will lose much less space while the inner layers will lose much more volume, thus distorting the Cube and shrinking the cube significantly by the time it reaches the particle at P.

Note that *the total number of layers of the cube will remain the same and thus height of the cube will not change despite a variable change in the breadth of each side of each layer. The vertical volume loss will be in the form of level change.*

For example, the cube shown in figure ii/15.1, with multiple layers A-L is under the influence of gravity. Then, the individual thickness of A-L would not change and the number of layers would not change. The volume loss happening in the vertical direction would result in each of these levels shifting downwards.

The breadth of each level will keep reducing as the cube moves downwards with the lower layer L showing maximum volume loss as it reaches the surface.

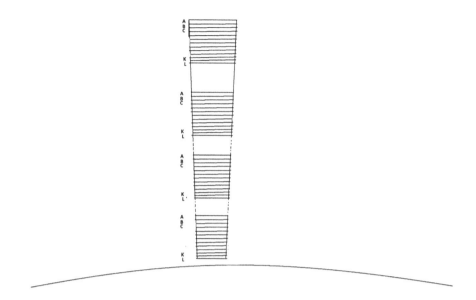

Figure ii/15.1: This shows how a cubical region of space gets distorted as it moves inwards under the influence of Gravity.

5. If the extent of energy at P is more, i.e. mass at P is more, this means the number of events happening per PT at P would be high. This increases the inward velocity and extent of distortion. At a single level at a distance r, more number of EPCA cross over per PT if mass at P is more. This means that the number of T=0 events happening per PT at every level would be more and the effective spatial contraction and time dilation would be more. Thus both volume loss/contraction in a vertical direction, as well as volume loss/contraction in a horizontal direction, increases with an increase in mass at P. (explained in detail later)

Temporary vs Permanent T=0 PCs

The T=0 compartment temporarily appearing in the EPCA causing the volume loss occurring in an EPCA is different from a permanent T=0 compartment which constitutes positive energy. The former T=0 compartment (or volume loss) happens due to entanglement and disappears quickly, while the latter persists relentlessly causing volume destruction and causing EPCAs around it. The temporary T=0 compartment in the EPCA will invariably have a Planck vacuum in two directions, towards the centre in which direction the EPCA is progressing and the opposite direction where PC vacuum is created due to inward movement of this EPCA. Thus this temporary T=0 compartment will not lead to any EPCA formation all around it. Instead, the volume loss happening due to the temporary T=0 compartment will only cause a change in the level of the EPCA from higher to lower level.

A permanent T=0 compartment, on the contrary would lead to PC vacuum all around and would lead to EPCA formation around it. Being a massless particle, it would also keep propagating in any one direction with the speed of light.

ΔV within an EPCA

An EPCA can be defined as an aggregate of Planck compartments forming a planar surface or a sphere that has a number of neighbouring PCs entangled together, usually to compensate for volume loss happening at a point and having a PC vacuum at one or two directions. Within an EPCA, volume loss can happen at horizontal plane alone.

The thickness of EPCA remains same.

The vertical plane volume loss happens in an EPCA as a change in the level of the EPCA.

Following the volume loss, the Surface area and thus the radius of the sphere-forming the EPCA reduces and it moves inwards. When a T=0 event happens in an EPCA, the resultant void would cause a localised EPCA (EPCA within an EPCA) around it in the same plane with its upper and lower portions chopped off due to the PC vacuum.

The localised EPCA would have to lose some PCs to compensate for the volume loss and contract and in the process change the level or moves the EPCA inwards. This cycle continues till all volume loss is compensated and EPCA reaches the next level. The cycle either stops here if the volume loss at the centre has stopped or more commonly continues as the inner EPCA has moved inwards by that time reforming the PC inward vacuum, so another cycle starts.

Alternatively, one can imagine, all the aggregated PCs being pulled inward by the creation of the PC vacuum leading to all the temporary T=0 events happening simultaneously causing loss of volume and loss of surface area.

Volume lost at each EPCA movement from r to r' = (SA at r) − (SA at r')

$$= 4\pi r^2 - 4\pi r'^2 = 4\pi(r^2 - r'^2)$$

The term r' can be defined as $(r-\delta d)$

Thus

Volume lost at each EPCA movement from r to r' $= 4\pi \,(r^2 - (r-\delta d)^2)$

$$= 4\pi \,(r^2 - (r^2 - 2r\delta d + \delta d^2))$$

$$= 4\pi \,(2r\delta d - \delta d^2)$$

Chapter 16

NUMBER OF EVENTS PER PLANK TIME

This depends on the extent of positive energy or mass present at P (according to DGR).

If 10 PC's are being destroyed at a PT, a PT would be divided into 10 equal parts 1/10 Planck times each and each destruction of PC would happen at each of these intervals. The Planck time can be divided as much as is needed like this. The number of events happening at P could be 100 or 1000 or 10^8 or 10^{100} as well. Every time a PC destruction event happens, the 26 PCs in level 1 are sucked inside and move ½ Planck length inwards and start a new cycle. Worth noting is that the PCs here would travel 0.5 PL every sub- Planck time interval. If 100 events are happening at the centre per PT, then level 1 PCs are moving 0.5 PL in 1/100th of a PT, which is 250 times faster than the speed of light.

More the number of events per PT at the centre, more number of cycles initiated per PT.

More is the number of cycles initiated per PT, more the number of times each EPCA moves per PT.

Thus the velocity of movement of EPCAs at each level is directly proportional to the number of events happening at P which is directly proportional to the amount of energy or mass.

Chapter 17

THE INSTANTANEOUS VELOCITY OF INWARD PROPAGATION OF EPCA

Velocity of propagation of an EPCA at a fixed distance r would be directly proportional to m (as suggested by Newtonian equation).

To note that the Volume that needs to be lost depends on the distance δd that the EPCA is supposed to move.

For every cycle, r-r' or δd would be very small in sub-Planck length realms.

When multiple cycles are happening at P, every EPCA would be moved multiple times every PT. If there are 100 events, every EPCA would move 100 times every PT. Thus, for every PT, there can be multiple δd (which can be designated as $δd_1$, $δd_2$, $δd_3$...... etc.).

The sum of all of them would determine the distance every EPCA moved in this PT, which would determine the velocity of inward movement of the EPCA.

Distance moved by an EPCA in a PT i.e. velocity of propagation of EPCA $= \sum_{e=1}^{n} δd$

PC vacuum that needs compensation at level r in one PT or Total volume lost = Volume lost at P in one PT in this cycle + sum of Volume lost at all the inner EPCAs in this cycle

= Volume lost at P in one PT in this cycle $\sum_{r=0}^{r} \Delta V$ (i.e. horizontal Volume lost at each subsequent EPCA)

δd at any level in one event = Total volume lost within it in this cycle till now or PC vacuum created by inward movement of the inner EPCA/Total SA at this level (I.e. number of PCs the volume gets distributed to)

i.e. **PC vacuum created/Volume of PC's it gets distributed to**

δd at any level = {1(e) + $\sum_{r=0}^{r}$ **[(SA at r) − (SA at r')]}/(SA at r)**

Where 1(e) is the PC volume lost in the centre in one PT,

∑ (SA at r) - (SA at r') is the summation of PC volume lost at each EPCA,

Where *(SA at r)* is equivalent to the Surface area in which the volume is distributed.

This equation tells us that δd at any level is proportional to the total volume lost up to r which is directly proportional to the number of events happening at P.

δd is also inversely proportional to the total Volume of PCs it gets distributed to which is essentially the number of PC's that make up the sphere of the EPCA with radius r and is equivalent to SA at r.

$\delta d \; \alpha \; m$

$\delta d \; \alpha \; 1/SA$ i.e. $1/r^2$

The theory predicts that as volume loss happening at each level is distributed at the higher level with a higher surface area, the distance that the outer EPCA has to travel inwards to compensate it i.e. δd keeps reducing as we move outwards for every cycle.

i.e. for a fixed value of m, δd is inversely proportional to the square of r.

Each δd is necessarily sub-Planck length.

However, the sum of all the δds happening per PT can be lower or significantly higher than a PL.

Thus inner EPCAs might move inwards significantly faster than c. As we move out, subsequent EPCAs have a smaller δd and so move slower.

There must be a critical value of r (r_{cric}) beyond which the velocity of inward propagation of EPCAs goes below c so that further increase in r, despite e number of events happening at P will cause such a small δd *per cycle*, that $\sum_{n=1}^{e} \delta dn$ per PT is smaller than one PL.

This value of r is equivalent to the Schwarzschild radius of a Blackhole of mass equal to the mass of the positive energy particle at P.

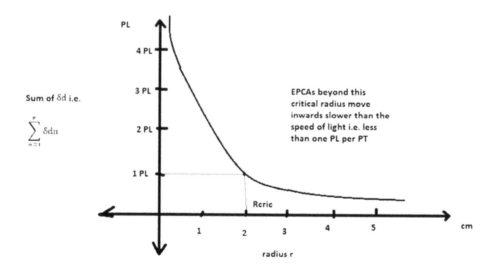

Figure ii/17.1: A diagrammatic representation of r_{cric} i.e. critical radius around the particle P with a positive energy particle at its centre wherein beyond this radius, the value of $\sum_{n=1}^{e} \delta dn$ per PT falls below 1 PL per PT so that the inward propagation of EPCAs beyond this is slower than the speed of light

i.e. $r_{cric} = \dfrac{2Gm}{C^2}$

Solving this equation might give us the elusive relation between mass and the number of events at P per PT.

m→f(e) or m →f(dV)

at $\sum_{n=1}^{e} \delta dn = 1\text{PL}$

(Where m is mass at P, dV is volume lost per PT at P and e is the number of events happening per PT at P)

i.e. we have to find out the relation between mass at P and volume loss at P per PT or the number of events at P per PT.

(the fact is that the symbol "e" used in this book as the number of events happening at P is already used conventionally to denote the charge of an electron. Thus another exotic rarely used symbol like ɜ may also be used to denote this variable)

Inwards velocity of the EPCA is equivalent to velocity with which a random object at that EPCA falls inwards.

Because δd is inversely proportional to Surface area which is proportional to the square of r, the velocity of inward propagation of EPCA would be inversely proportional to r^2

More the number of events at P per PT, more the number of δds, which means more is the inwards movement at radius r.

Thus, the inwards velocity of the EPCA would be directly proportional to the number of events e which is equivalent to mass.

Further work will be needed to construct computer simulations according to theory to accurately test if these relations indeed hold.

Further work is needed to test if the vertical and horizontal volume contractions predicted by DGR matches with the changes in the shape of Space-time predicted by equations of GR.

Obtaining horizontal loss of volume ΔV- *LT time dilation relation*

At every level r, horizontal volume loss ΔV can be calculated by SA at r – SA at r'

With the same value of r applied to the time dilation formula of Einstein's equation given above, the extent of time dilation predicted by GR can be calculated at each level.

This ΔV can be plotted against time dilation. The theory predicts that ΔV should be directly proportional to the time dilation so that more is the horizontal volume loss, more is the time dilation. As one moves outwards, i.e. as the value of r increases, the value of ΔV reduces and so does the value of time dilation caused by the events happening at P.

However, for this, the mass to "e" relation is required so that the effective value of mass for the given value of e can be inserted in time dilation equation of GR.

Further work is needed to test whether the effective horizontal loss of volume ΔV at every level is enough to cause a photon clock to slow down enough to measure/cause time dilation as predicted by GR.

Uniformity of Volume contraction

The horizontal volume contraction or volume loss at r is represented by

ΔV = SA at r − SA at r'

ΔV = number of PCs in SA of sphere r − number of PCs in SA of sphere r' = PCs lost while in transition from r to r'

The vertical volume contraction or volume loss at r is equal to

δd = (SA at r')/(SA at r)

δd = (Number of PCs in SA at r')/(Number of PCs in SA at r)

In theory or by logic, the vertical and horizontal volume contraction at a point in space should be equal. However, it remains to be seen if this is true or can be assumed.

What is clear is that both vertical and horizontal volume loss keep reducing as the value of r increases until a point after which the vertical volume loss becomes a constant and the horizontal volume loss becomes extremely small.

CHAPTER 18

ANALYSING DATA FROM COMPUTER SIMULATIONS

After a computer simulation of the space-time made of PCs is ready, with all the levels as described above, complex tedious calculations need to be undertaken to determine the inward velocity of propagation of EPCAs with e=1.

The value of $δd$ would need to be calculated at each level as the value of r increases. $δd$ will have to be plotted against r.

Relationship of $δd$ as a function of r needs to be found after plotting this graph.

F(r)→$δd$

(for e = 1)

An arbitrary low level of $δd$ can be set up before, say 10^{-6} PL, after which calculating of $δd$ can be halted simply to limit the number of calculations, as such fractions of 1 PL distance are insignificant at bigger scales.

Furthermore, the same calculations can be repeated after increasing the number of events per PT.

Note that the above calculations need to be done for r up to several km if not more or at least 10^{35} PLs i.e. one m. This means the computer algorithm has to run at least 10^{35} times.

These calculations need to be done for very high number of events per PT i.e. e=100, 200,500,1000, 10^6, 10^{10}, 10^{100} events per Planck time. They have to be repeated for every cycle and every outward propagation of every cycle.

Obtaining these data and finding computers that can perform such calculations is going to be the real challenge.

With this data, velocities of EPCAs can be obtained with equation $\sum_{e}^{1} δd$ per PT.[1*]

> 1* - To note here is that the velocity which we will get here is velocity of EPCA when there is no external influence.
>
> This force would be equivalent to $F \approx m_1 m_2 / r^2$
>
> Subsequently, it was realized that the effect of G could be due to MONDian forces due to mass of the rest of the Universe acting on the two gravitating bodies which reduce the overall attraction by a factor of 10^{10} and effectively pulling them apart. How to get the force equivalent to the complete Newtonian equation of $F = Gm_1 m_2 / r^2$ or incorporating the effect of this MONDian outward pull on the velocity of EPCA at r is not yet clear. Although its effect is clear. It will reduce the effective velocity of the inwards movement of the EPCAs.

These EPCAs have to be matched with those predicted by Newtonian gravity for a known mass, say mass of the Earth at a known distance (R + h) with the help of g.

Alternatively, the equation for the Schwarzschild radius of the black hole can be utilised with v = c, to get the relationship between mass and number of events per PT.

CHAPTER 19

GRAVITATIONAL LIMIT

For a given number of events per PT (i.e. for a particle or heavenly object with a known mass say one solar mass), if we put an insanely high value of r like a few light-years, the value of SA at r would be very high i.e. a sphere with a radius of a few light-years will have an extremely high number of PCs (countable infinity). The value of δd here would be in the sub-Planck length range.

ΔV [(horizontal volume loss at a given level of EPCA) which is "(SA at r) – (SA at r')"] would be very small (i.e. – no. of PC's lost/total no of PC's. is extremely small or would tend to zero due to countable infinity at the denominator and a small number at the numerator.

This means that the horizontal volume loss will be negligible and there would be just shifting of aggregated PCs without much volume lost with subsequent shifts of the EPCA.

The instantaneous velocities of EPCAs given by $\sum_{e}^{1} \delta d$ per PT which represents gravitational acceleration would be extremely small.

Because there is no horizontal volume loss, LT dilation would be negligible. No volume loss would mean that our Photon clock would not be slowed down, if kept here.

This is predicted even in GR *(Einstein A.- 33,34)* where an extremely high value of r in the gravitational time dilation equation given below, gives a sub-Planck time value of time dilation.

$$t = t' \left(\frac{1}{1 - \sqrt{2GMr/c^2}} \right)$$

where t is the dilated time and t' is the stationary time, M is the mass of the source of gravity and r is the distance

The value of r, when the above equation gives a difference in dilated and stationary time in sub-Planck time realms I.e. Gravitational time dilation ceases to be significant or is in sub-Planck time realms can be called the Gravitational limit of that object.

Here, the horizontal volume loss is not happening and vertical volume loss is happening at extremely low accelerations.

This is the zone where the spherical EPCAs become linear EPCAs.

This is probably the distance where Newtonian dynamics ceases to be significant and MOND becomes applicable (as per the prediction of DGR).

$$\Delta V = \infty - \infty = 0,$$

$$\delta d = \frac{countable\,\infty}{countable\,\infty}$$

However, there will be a non-zero value of $\sum_{e}^{1} \delta d$ especially if the value of e is high i.e. mass of the gravitating object is high.

In the central region where curved EPCAs are present and significant time dilation spatial contraction exists, there exists a relationship between *δd* and r which can be written as

δd = f (r)

As r takes higher values, *δd* reduces.

δd however is inversely proportional to r^2

For a given value of "e" i.e. for a star of known mass, "$\sum_{e}^{1} \delta d$" which represents the instantaneous velocity of the EPCA at r in one PT or acceleration achieved by the EPCA at r, would also be inversely proportional to r^2.

In MOND, the inward-directed force that keeps the peripheral stars in orbit is stronger than Newtonian gravity and its relationship with r is different. The force should reduce proportionally to r instead of proportional to r^2.

For DGR to be consistent with MOND, after gravitational limit, the new MONDian force should also follow this relation and should reduce proportionally to r. However, it remains to be seen if this is true.

Chapter 20

A POTENTIAL EXPLANATION OF MONDIAN FORCES IN DGR

Let's imagine the region in the periphery of a Galaxy where linear EPCAs from four stars (A, B, C and D) are meeting as shown in figure ii/20.1, A being the peripheral star being held together.

Worth noting is that the distance r for each of these stars would be variable. Even masses of the three stars would be different and so the value of e i.e. number of events happening per PT at the core of each star would be different. This means that the number of linear EPCAs passing in each of the three directions would be different.

It is difficult to imagine how the confluence of the time resetting waves from all the three stars would look like.

A Planck time would be divided into variable fractions and number of cycles per PT and thus distance travelled per PT by each EPCA would be different for each of them. All of them would confluence together creating a complex pattern of LT distribution.

At the core of each of these stars, according to DGR, an extreme amount of destruction is happening with e number of events per PT happening. To compensate for this, a time resetting wave starts and travels outward much faster than the speed of light and causes the formation of EPCAs. These time resetting waves are time dilation spatial contraction waves. At the end of the Gravitational limit, these waves would not stop resetting time, however, reset time in a sub-Planck realm. At such time dilation, there is no volume contraction and thus EPCAs forming here would become linear EPCAs. As these waves cover more distance, the sub-Planck time dilation of time increases with it being proportional to r as per the Einsteinian equation.

When two or more time-dilation type time-resetting waves interfere, they would do it constructively. On the other hand, a time dilation type wave interferes with a time contraction type of wave, there would be destructive interference.

The time dilation waves from all these stars, which are presently causing sub-Planck time dilation of time, meet and interfere in the region shown. The wave from B and D add up such that a uniform zone of very low time dilation in between the two stars exists.) The horizontally placed linear resetting time waves from the stars A and C also interfere constructively with this arrangement of local time. As one moves from the central point towards both B and D, the resultant time dilation receives an equal contribution from the horizontally placed linear time resetting waves from A and B so that the normal variation of time dilation would be disturbed and in such a situation, the resultant time dilation doesn't fall in the same proportion that it would without the existence of A and D.

Note that the Andromeda Galaxy has a trillion stars and the Milky Way Galaxy has an estimated 100 thousand million i.e. 100 billion stars. Each of these stars would produce these time resetting waves moving outwards. Thus if we focus on a peripheral region of the Milky Way, it would receive 100 thousand million such waves, each in different stages of sub-Planck time dilations of time. The resultant time dilation at any point would then be the sum of all the time dilations caused by each of these waves. If the sum of these time dilations, still falls in the sub-Planck time realm, there won't be any spatial contraction. But in zones where this sum results in much larger time dilation, active spatial contraction can result. Note that this kind of spatial contraction would depend only on the mass of the Galaxy and not on any individual stars. More the number of sub-Planck time waves meeting and adding up, more is the contraction. Thus a larger Galaxy can produce a bigger force and thus can tolerate a larger peripheral rotation velocity. This also explains the "plateauing" of the velocity of peripheral stars so that their velocity depends, not on their masses but the overall mass of the Galaxy.

Thus DGR predicts that the asymptotic speed of the peripheral stars

V α Mass of the Galaxy

The MONDian velocity- mass relationship given by Mordehai Milgrom (*Milgrom, M 69-73*) is

$V_\infty^4 = GMa_0$

Where V_∞ is the velocity of outer stars, M is the mass of the Galaxy and a_0 is the resultant constant and extremely low accelerations produced by MOND.

This equation given by MOND explains the Tully Fisher relationship well *(McGaugh- 67,68)*. That is, as the mass or luminosity of Galaxy increases, (the fourth power of) the rotational velocity at which the rotation of peripheral star's plateaus increases in proportion.

More work is needed before it can be said conclusively if DGR can also similarly predict the asymptotic speeds of peripheral stars and get a similar relationship and explain the Tully Fischer relationship.

Another observation is that of $\delta d = \dfrac{countable \infty}{countable \infty}$

Here, we can consider an EPCA at an extremely large distance from a star, say the Sun.

Let's focus on an EPCA at a distance r = 4 light-years.

Theoretically, it is possible to draw a sphere of radius 4 light-years forming this EPCA around our star. But the number of PCs within the surface area of this sphere would be astronomically large. If this EPCA has to move inwards by a sub-Planck length distance, the amount of reduction in SA needed would be equally insignificant. At human scales, this EPCA would have practically no curvature and would behave like a linear EPCA.

Here, the SA at r as well as SA at r' both would be (countable)∞.

In our theory, when the EPCA contracts from a radius r to r' to compensate the PC vacuum inside it, the Volume equivalent to all the PCs destroyed or lost till now for compensating the volume loss happening at the centre and at each of the EPCAs (which is a fraction of the volume of the number of PC's at r' and thus here is close to countable ∞, including the number of PCs destroyed while shifting from r to r'), gets distributed just outside the EPCA at r', thus forming a new PC Vacuum for the formation of another outward EPCA. For linear EPCAs this volume of PCs needed to be destroyed to shift from r to r' is negligible.

The absurdity comes when we have to compensate a volume of ∞ PCs and distribute it to ∞ PCs. This is not absurd in the real sense as here ∞ is not real infinity.

Essentially the inner PC vacuum within a linear EPCA, which is about ∞ PCs, gets shifted outside the shifting EPCA and forms the new PC vacuum pulling the next EPCA inwards.

But the changes in δd that you get with every change in r would be extremely small. This essentially means that beyond such a value of r, the acceleration due to the gravity of a star would not only be extremely small but would become constant instead of reducing with every increment in distance r.

This means that any object at such a distance from the star would move inwards with a constant acceleration. Any value of r beyond this critical limit would have the same acceleration inwards due to the linear EPCAs moving inwards with a constant acceleration.

MOND Bonds predicted by DGR

Let's return to our example of four stars in figure ii/20.1.

Like A, every peripheral star would have some stars which are immediate neighbours and some which are extremely distant.

In our example, the time dilation caused between A and B, which normally is responsible for the attractive force between A and B but at such astronomical distances is extremely small and ineffective, is supplemented by time dilation from all the other stars put together. Thus MOND would be acting mainly to keep these peripheral stars from flying off the edge of the galaxy by bonding them with the neighbouring stars.

One can imagine that B, C and D, all being comparatively peripheral stars are being bound to the inner stars by MOND bonds and the most peripheral star A being bound to these and all the other immediate neighbours as well as to the more massive far away neighbours including the Central Supermassive Blackhole of the Galaxy by such bonds wherein the sub-Planck time waves resetting time supplement their bonds. Note that such bonds would exist even between B and D, B and C and so on.

This is akin to multiple carbon atoms bonded together in a crystal of a diamond with each atom rigidly bonded to their neighbours. This explains why all these stars move with the same velocity and thus peripheral part of the galaxy moves like a disc rather than just being bonded to and orbiting around the Central Supermassive Black hole.

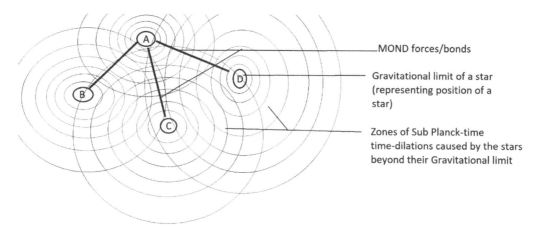

Figure ii/20.1: Shows three stars B, C and D keeping the fourth star A bonded together with MOND bonds.

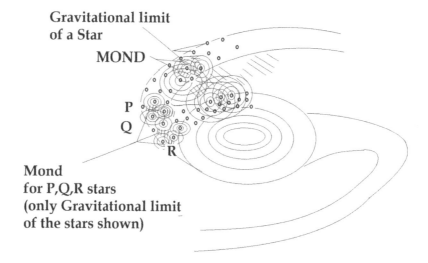

Figure ii/20.2: Shows multiple stars bonded together with MOND bonds in a Galaxy just like Carbon atoms bonded together in a diamond.

Let's focus on a point Q as shown in Figure ii/20.3. An EPCA moving downwards would create a PC vacuum at Q. This needs compensation by the formation of an EPCA just above it. At the same time, another EPCA forms moving towards the left, towards the right and upwards. The PC at Q would be pulled in all directions at the same time, with an acceleration of a_o.

As shown in the figure, there will be the creation of an additional PC vacuum than a_o. This additional PC vacuum leads to additional spatial contraction and is represented by the sum of time dilations caused by each of the stars belonging to the Galaxy. In reality, the EPCAs will form at different frequencies depending on the value of e within every star. The resultant wave that results due to interference of each of the (now linear) time resetting waves starting from each of the constituent stars of the Galaxy and now reaching Q, would ideally give the resultant spatial contraction at Q. Thus the resultant PC vacuum at Q would be roughly na_o where n is the total number of stars in the Galaxy.

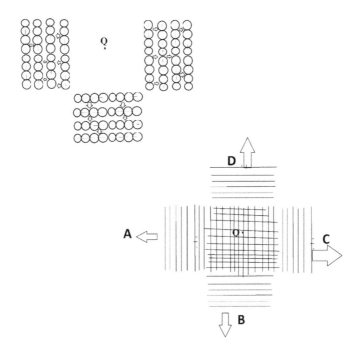

Figure ii/20.3: Shows linear EPCAs at a point in between four stars A, B, C and D. The central shaded portion is almost akin to a Blackhole, given that the void created within this region is difficult to compensate. This gives rise to additional PC vacuum and additional pull, much more than the normal Gravitational pull between the individual stars. (This is almost akin to the creation of a temporary T=0 compartment or akin to a miniature Black hole formation at the square shaded region. When the same phenomenon happens at the centre of the Galaxy, it may explain the central supermassive Blackhole that we see.)

Let's look at the following figure ii/20.4. These show 4 stars A, B, C and D at various places in the Galaxy. We can analyse in what directions will the linear EPCAs arising from each of these would be arranged at two points within the galaxy (X and Y) and four points outside the Galaxy (P, Q, R and S). The lines A', B', C' and D' represent the linear EPCAs arranged at various points. (note that these figures are just diagrammatic representations as real Galactic distances cannot be reproduced in such a small figure)

At X, which is a relatively central point in the Galaxy, the linear EPCAs from A and B move opposite each other while EPCAs from D and C are at an angle. At central locations, the stars would be pulled all around in the plane of the disc of the Galaxy and thus MOND forces would cancel each other out. At the periphery, like in Y, the EPCAs at C, B and D are moving to the right while A' is moving towards the left. For a peripherally located star, most linear EPCAs reaching it from other stars of the Galaxy would be directed inwards pulling the star inwards and a net inward force would exist, which can cancel out the centripetal force of the rotational velocity.

For a point in the plane of the Galaxy, but located out of it like P, the relative position of EPCAs would be additive and all of them, i.e. A', B', C' and D' would point inwards. Here, the inwards pull can be significantly more. For points like Q lying outside the plane of the Galaxy but closer to it, the direction of linear EPCAs is divergent and thus the resultant force/acceleration would be much less or would be the vector sum of all the accelerations. At a much greater distance from the Galactic centre, at R, the EPCAs start aligning and finally at S, the inward movement of all the EPCAs aligns well so that the resultant inward pull would again be additive like at P.

The direction of inward pull by the linear EPCAs at Q, R and S along with the fact that there is a paucity of centripetal forces here due to Galactic rotation could also potentially explain the disc shape of the Galaxy instead of a sphere.

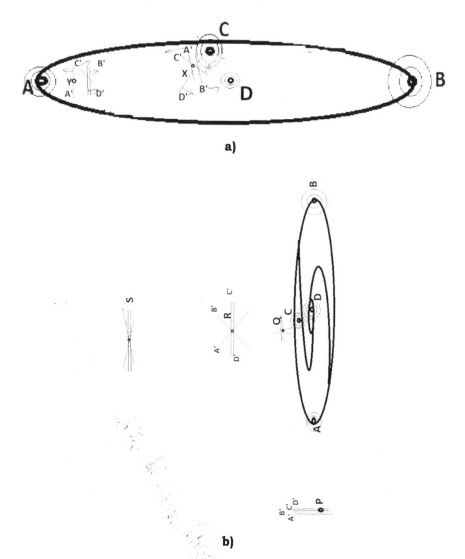

Figure ii/20.4 a) and b): a) Shows the direction of MONDian forces acting on various stars depending upon their location within the Galaxy. The central stars will have MONDian forces

on all sides, thus cancelling each other, while peripheral stars have a net inward force directed towards the direction of the Blackhole/mass of the Galaxy.

b) Shows how the direction of net MONDian forces would change at various locations just outside the Galaxy. Note that in the plane of the Galaxy, just outside the limits of the Galaxy, the MONDian forces are directed in the same direction and just add up. Just above the plane of the Galaxy, the forces are directed in different directions and the resultant force would not be the direct sum but the vector sum of all the forces. This may not add up completely, offering an explanation of why high-velocity stars do not exist above the plane of the Galaxy and the Galaxies are spiral discs or ellipses and not Spherical.

Centre of the Galaxy

In contemporary Physics, we note that the centre of every Galaxy has a Supermassive Blackhole. The discovery of Black holes themselves began when Astrophysicists noted extremely curved trajectories of stars around the centre of the Milky way which is also called the Sagittarius A. In DGR, we know that MONDian linear EPCAs reach this point from every star that forms a constituent part of the Galaxy. These EPCAs would have a similar effect so that the constant outwards pull on the PCs lying here cannot be compensated by inward flow from PCs from any neighbouring region.

This gives us a potential clue. The question to which no answer is known yet but which is extremely interesting and mindboggling is

"Is there really a Blackhole at the Sagittarius A or for that matter at the centre of all the Galaxies? Or is it just the manifestation of the MONDian gravity of the mass of all the Galaxy put together which remains uncompensated and thus leads to constant destruction?"

If most of the Black holes in the centre of the Galaxies aren't Black holes but are just manifestations of MONDian gravity, the next troubling question is "Is there a Black hole in reality or most of the Black hole-like objects or points in the Universe which have properties of a Blackhole are such points where the force of Gravity is focussed and gets exaggerated?

These questions are unanswered yet.

CHAPTER 21

SIGNIFICANCE OF A_o

There are several important consequences of linear EPCA formation and a_o.

1. Every particle with mass counts irrespective of whether it is a part of a star or a planet or is a part of the stellar gas cloud.
2. Galaxies and galaxy clusters can have Gravitational consequences at extremely high distances.

Every particle with mass counts:

Let's imagine a large sphere that contains 1 solar mass stellar gas i.e. Hydrogen. The difference indeed would be that the molecules would have a negligible density compared to that within the Sun. Each of these particles would have destruction going on within them and would give rise to EPCA formation around them. The velocity of inwards propagation would depend on the number of events happening at the centre and the distance from the centre.

After a critical distance from each particle, these EPCAs would become linear EPCAs and would move inward with an acceleration of a_o.

There would exist a critical density for every molecule of a given mass beyond which only time resetting waves with linear EPCAs would interact. They can get positively reinforced and cause a net attraction between the two molecules irrespective of the distance between them. If the temperature of the gas is above a critical limit, the gas molecules would have enough momentum to overcome this pull and stay separate as gas. If enough momentum is not available, the gaseous molecules collapse towards each other and the cloud collapses until their density and their temperature increases beyond such a point that the inward gravitational pull is counterbalanced by the outward pressure and a star forms.

Regardless of whether the mass is distributed as a small sphere of a star or a large low-density sphere of stellar gas, it keeps losing volume and needs a constant supply of in-falling PCs immediately outside it.

The extent of volume loss happening per PT is determined by the number of particles and extent of destruction per PT within each of them. The time resetting waves from each particle forming linear EPCAs coalesce and thus at the edge of our sphere, the resultant PC vacuum created is proportional to the amount of destruction happening.

In short, every particle matters.

This spherical gas cloud, beyond its boundaries, acts very similar to a one solar mass star at its centre regardless of density. This is known before, i.e. Gravity is independent of density. A 4 solar masses Blackhole is gravitationally equivalent to a 4 solar mass star.

The importance of this is that every molecule of gas that contributes to volume destruction within the Galaxy counts.

Galaxies and Galaxy clusters can have Gravitational consequences at extremely high distances

Beyond the edge of any Galaxy, linear EPCA forming time resetting waves coalesce and get positively reinforced.

Each of these time resetting waves leads to a gravitational acceleration of a_o.

The resultant inward pull of EPCAs would be directly proportional to the mass of the Galaxy.

This goes well with the velocity-mass relation given by MOND

$v^4 = GMa_o$

This acceleration due to Gravity will not reduce with distance so that it will remain valid even at large interstellar distances from the Galaxy.

This is valid even with a group of satellite Galaxy clusters with its central Galaxy. Thus Andromeda with all its satellite Galaxies can exert gravitational influence on the Milky way and all its satellite clusters.

All these likely give rise to linear EPCA producing time resetting waves that coalesce when considering distances beyond the local cluster and thus exert Gravitational influence at distant Galaxies. In MOND, this is called the "external field effect". There seems to be no end to this and it is indeed likely that a Galaxy at one edge of the observable Universe has a Gravitational influence at another Galaxy located diametrically opposite to it and thus the mass of the entire Universe can potentially have an influence on every constituent. This goes well in line with the Mach's principle.

There is however one way of removing all the Gravitational influence of a large Galaxy cluster from its far-away neighbour. This can happen if there exists a zone in the intergalactic region where the expansion of the Universe is much faster than the resultant acceleration of linear EPCAs.

Here, what is essentially happening is that all the volume contraction within the mass of the Galaxies that produce the EPCAs is being completely compensated by new PCs being formed. This will need a critical density of ENEA particles in the intergalactic region.

However, the TRWs (time resetting waves) of Gravity do travel faster than the velocity of light depending on the value of e and thus it is more likely that the Gravitational influences might just penetrate through any thickness of ENEAs.

If the former is the case, then the Universe is likely to be made up of islands of positive energy excess completely alienated from gravitational influences of each other by zones of negative energy excess.

This however seems unlikely.

DIFFICULT PROBLEMS IN PHYSICS

CHAPTER 22

DGR AND DARK MATTER

DGR predicts that Gravity behaves more like MOND. So can we say that Dark matter is not needed?

It is a known fact that although MOND can provide an explanation for Gravity within the Galaxy, beyond the Galaxy, it falls short to explain the velocities of Galaxy clusters. Thus although MOND is likely, we cannot give away the idea of Dark matter.

DGR can provide a potential explanation of what Dark matter is and how it behaves. But for this, we need to look into another mystery of Physics, the matter-antimatter asymmetry.

Matter Antimatter asymmetry - how DGR provides a potential explanation?

Paul Dirac's equation suggested the existence of something called anti-matter. *(Dirac, P.-28)* Since the discovery of the Positron *(Anderson, Carl – 6,7)* (which is the antiparticle of the Electron, having the same mass but opposite charge) and following the discovery of antiparticles of many other particles like Charm Quark, enough evidence has accumulated to support the existence of such antiparticles for almost all the particles described in the Standard Model of particle physics. Antimatter is being created frequently within the particle accelerators like in CERN and FERMI labs. These antimatter particles have been combined to create antimatter atoms as well, like an antimatter hydrogen atom.

These discoveries have posed another enigma to physicists, the question which has no answer yet- why is there so much matter and so little antimatter?

At the start, immediately after the Big bang, it is expected that equal quantities of both matter and antimatter particles were created. Many of those particles are annihilated with each other reforming the energy. However, there was a slight asymmetry between matter and antimatter particles that somehow led to a world that is predominated by matter particles. The question, "what happened to all the antimatter particles or why there is matter excess?" has puzzled physicists.

The theory of DGR has an inbuilt explanation for this dilemma.

However, for that, we need to redefine a few terms or visualise them a bit differently.

Like there are matter and antimatter particles, DGR has two types of particles within its basic assumptions, positive energy and negative energy particle. The negative energy particles can thus be considered as the antimatter equivalent of DGR.

DGR predicts that during the Big Bang, an equal amount of positive and negative energy particles which are DGR equivalent of matter and antimatter particles were created. There might be a slight excess of negative energy antimatter particles which is inevitable to explain the expansion of the Universe. The positive energy particles of DGR can be called PEPs and the negative energy equivalents of them could be called Exotic Negative Energy Antiparticles or ENEAs.

In QED, at Planck length scales, minute energy fluctuations exist so that there is a spontaneous formation and destruction of Virtual particle-antiparticle pairs *(Feynman, Richard – 45, E. Fermi. – 44, Paul Dirac – 27)*. In DGR, these are nothing but PCs with a unit of positive energy and a unit of negative energy. When these pairs are created as well as destroyed, no change in volume results, the PC remains with UT.

Once formed, DGR predicts that the ENEAs have active Creation going on within and PEPs have active Destruction happening within. Multiple PEPs can attract each other and form a single large aggregate. However, the ENEAs, being negative energy particles, will repel each other and repel all the other particles, be it positive or negative energy particles, and thus remain separate.

If at a point in space, millions of PEPs form and combine in the centre to form a central positive energy core, the simultaneously formed ENEAs would not be able to annihilate them and would form a lattice around this core and start spinning around it. The positive energy core would keep pulling the lattice inwards and each ENEA within the lattice would keep pushing each other and the positive energy core away. This positive energy core with negative energy lattice in DGR is equivalent to the matter particles in contemporary physics.

If there is absolute symmetry with an equal number of positive and negative energy particles, an uncharged particle forms. But if there is a minute asymmetry with one (or few) excess PEP or one (or few) excess ENEA particle, a charged particle forms. These would represent the Electron and the Positron and for that matter the Quarks and the Antiquarks.

This would imply that within a matter particle, there exists enough Antimatter to refute extreme asymmetry.

Even then, within the Protons and Neutrons, there is a high amount of positive energy trapped within, which constitutes 99% of the mass of these particles. There is no antimatter/negative energy equivalent of this positive energy around.

It is a presumption of DGR, that each of these PEPs when formed, had an antimatter equivalent in the form of an ENEA which formed alongside, but was ejected away.

Thus, each positive energy matter-particle which forms the mass of Protons or Neutrons which forms the entire baryonic matter in the Galaxy must have an antimatter equivalent, still bonded and entangled to it. These ENEA particles, having the active creation of PCs happening within them and thus being repulsive, would move away from the matter and would form layers along the edge of the Galaxy and even form a sphere around the disc. This can be equivalent to Dark matter. These particles are mutually repulsive and interact with each other and also with the matter particles in the Galaxy by repulsion. They are, however, entangled with their original partner and cannot fly away from the Galaxy and stay connected to it gravitationally. They, however, represent negative Gravity instead of true Gravity i.e. they repel instead of attracting. The extra force exerted by these could explain a lot of phenomena like extended MOND i.e. movement of the Galaxy clusters.

Note that the antiparticles of a fermion in contemporary physics which we call antimatter are, as per DGRs predictions, not antimatter but just formed by addition or subtraction of a single (or a few) ENEA particle from a balanced uncharged particle.

The ideal candidate particle for the ENEA particle is the neutrino. This is because Neutrinos and antineutrinos are produced in reactions where charged particles are transformed. Neutrinos travel extremely fast near the speed of light and they repel other particles and are shy particles with little interaction with others.

Further research on them would shed more light. However, Neutrinos, unlike ENEAs, have a positive mass and also are possibly their antiparticle. Thus there is a likelihood that they might be made of a few ENEA particles around a miniature core of positive energy.

The theory thus predicts that a Positron, its neutral counterpart and the Electron differ by the ratio of negative to positive energy within them so that adding a neutrino (which is nothing but a single or a few ENEA particles) to a Positron might give a neutral electron-counterpart and adding an additional neutrino/ENEA particle would lead to the creation of an Electron.

All this, at this stage, is just blind extrapolation and further work is needed in this regard.

At the time of the Big Bang, there was a still-unexplained abundance of the negative energy particles. But the positive and negative energy particles produced and those who took part in the creation of the Galaxy and its elements are still existing in and around the Galaxy.

This explains a lot of things like the poor stability and short half-lives of most Antiparticles. These findings subsequently would pave the way for a potential theory of everything from DGR.

CHAPTER 23

DARK ENERGY AND CRISIS IN COSMOLOGY

The currently accepted theory for cosmology is the Lambda CDM model (*Scott, D – 92*). This model is partly based on the interpretation of the Cosmic Microwave Background (CMB) which was first detected by Penzias and Wilson in 1965. (*Penzias, A.; Wilson, R.-77*)

Currently, an extremely precise map of the CMB is available thanks to the COBE satellite (*Smoot et al, - 94*), WMAP satellite (*Abbott, B – 4*) and more recently Planck telescope *(The Planck Collaboration – 78,79)*. In this map of the CMB, the CMB appears uniform in all directions. However, there are minor Temperature fluctuations of about 2.7 Kelvin. Also, there are Baryonic acoustic oscillations within the map. These are reverberations of Baryonic matter at that age of the Universe producing sound waves. The Baryonic matter then was in the form of a Photon-Proton-Electron plasma as temperatures were too hot for them to bind together and form molecules. The Photons were dominant and were constantly interacting and getting scattered around. This made the space-time opaque. The Universe was expanding faster than the speed of light at that time.

(*Peebles, P – 76, Cirigliano, D – 19*)

At a moment in the evolution of the Universe, the Universe cooled enough for the Protons and Electrons to bond with each other and thus form neutral Hydrogen atoms. The Thomson Scattering of Photons, which the charged particles did, stopped and the Universe immediately turned transparent from an opaque one, and these photons instead of taking curvilinear complex paths started moving straight. This is literally like a photograph of the early Universe which started its journey long back and now due to the extreme redshift is visible in the form of the radio waves/microwaves of the CMB.

The moment in time when the charged particles lost enough energy to start bonding is called the recombination epoch and the release of the photons is called decoupling.

The Baryon acoustic oscillations which were present at the time of this transformation froze and were imprinted at the CMB *(Hu, W-59)*. A detailed analysis of the distribution and sizes of these BAOs and their temperature curves can be carried out and these BAOs can be used as ancient rulers to get an idea about the distribution of Dark energy, Dark Matter, and Baryonic matter at that time in the Universe and it is also possible to get an estimate of how fast the Universe was expanding. This is given by the value of the Hubble constant.

The most detailed analysis of the CMB given by the Planck mission of the European Space Agency gives a value of Hubble constant of 67.74 + –0.46 km/sec/Mpc. (*Planck Collaboration – 78,79*)

The second method of measuring the Hubble constant

There are other methods of getting an estimation of the Hubble constant.

One of them is using the cosmic ladder created by the use of Cepheid variables and Type Ia Supernovae as standard candles to measure intergalactic distances. This method, however, gives a slightly different value of the Hubble constant of 71+ –1.3 km/sec/Mpc.

Crisis in cosmology

Recent advances in these methodologies have improved the accuracy of measurements of both these methods. Initially, it was thought that the disagreement in the Hubble constant values predicted by both these methods is due to some systematic error and further refinements in techniques would eventually make these values converge. The refinements, instead of reducing the gap, confirmed with greater confidence that there exists a significant difference in the value of the Hubble constant of today than that in the early Universe. (*Peebles, P – 76*)

This implies that the Universe is expanding much faster now than it was at the time of the start of the CMB radiation.

This is counterintuitive.

If we assume that Dark energy is responsible for the expansion of the Universe, as the volume of the Universe increases, logic says that its energy density should keep on reducing as the same amount of energy will get distributed in a larger volume. Here there is an underlying assumption that newer Dark energy cannot be created from "nothing" as this violates the principle of conservation of energy.

Instead of going according to scientific logic, the Universe, as always, is choosing to behave anomalously. Expansion of the Universe with a static amount of Dark energy would mean that the energy density should keep falling till a point after which Gravity would take over and the Universe would start re-collapsing. Not only is the Universe expanding, but the above scientific findings suggest that the expansion of the Universe is accelerating. This means that Dark energy density, instead of falling, is remaining constant or increasing.

This constitutes the real Crisis in Cosmology.

Lambda CDM model and DGR. How DGR can provide a potential explanation of the Crisis in Cosmology?

Several questions arise while inferring the predictions of DGR in the Lambda CDM model. The size of the Universe at the time of the release of the CMB was just a 1100th fraction of what it is today. The Universe has grown a thousand times bigger than it was at that time.

As noted in the discussion of the Big Bang theory, no mention of the Ultra Massive Black Holes (UMB) holes and SMBHs comes during the discussion of the BAOs. The increased density zones could represent the zones where these enormous Black holes exist.

If they existed, they had extremely high values of e and would have extremely high velocities of EPCAs around them. The inward pull of EPCAs leading to inward Gravity is being cancelled out by the outwardly directed radiation pressure. This can explain the Baryonic acoustic oscillations well.

If we presume that the SMBH and UMBH existed at the centre of the density fluctuations, another question that might arise is what was their mass at that time and has their mass remained constant, or has it changed now as compared to then.

In DGR, expansion of the Universe happens because there is an apparent imbalance between positive and negative energy. This means that there are zones where there is the Creation of PCs happening but there is no compensatory destruction happening elsewhere.

This observation goes against our observations of the Universe that it is Flat at larger scales.

A Universe with a predominance of Destruction would have negative curvature. A Universe with a predominance of negative energy i.e. predominance of Creation rather than Destruction should have positive curvature. And a Universe with absolutely balanced Destruction and Creation should be a flat one.

In DGR terms, this means that there are more negative energy particles or ENEAs than the PEPs. If the ENEAs and PEPs both are formed in pairs, and we can find more ENEA particles, where did their partner PEPs go?

There is a rather counterintuitive and controversial explanation of this in DGR. This is controversial and is just tentative because there is little reason to believe it is right or wrong, it is completely non-falsifiable and there is a high likelihood that it may be wrong and an alternative explanation might exist.

Why does the Universe Expands and what happens at the edge of the Universe?

"What lies beyond the Universe?' is an enigma.

There are only two possibilities. Either the Universe is finite and it has a well-defined edge or the Universe goes on infinitely and has no edge. The second one comes in conflict with the Big Bang theory and would indicate that there was Universe even before the Big Bang or that the Universe has no starting point. It would also indicate that the expansion of the Universe is happening basically due to excessive creation indicating a positive curvature of the Universe at large. This will come in conflict with multiple observations.

So we assume that the first one is right and there is indeed an edge to the Universe and that it is finite.

If the Universe has an edge and is finite, "what lies at the edge of the Universe?" is a tricky question that has no answer yet.

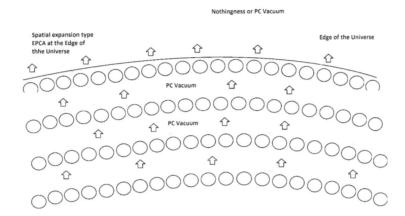

Figure ii/23.1: Edge of the Universe as predicted by DGR. There exists an infinite PC vacuum beyond the edge of the Universe and the PCs that line up the edge have an unbalanced outward-directed pull that constantly creates Negative energy type spatial expansion type of EPCAs.

It figures that DGR can provide a prediction.

According to DGR, what lies beyond the edge of the Universe is "nothingness".

Note that this could be true nothingness or it could be multiple other Universes, forming a part of a Multiverse, that are beyond contact and presently moving away from our Universe thus creating the ever-present PC vacuum between them.

If Universe had a starting point of creation at the Big Bang, then it may have a finite size and it would have a starting point in time. "What would lie before the starting point?" is again nothingness.

When two PCs move away from each other, what remains between them is also this "nothingness" that we call the PC vacuum.

If a huge (possibly infinite) zone of PC vacuum exists beyond our Universe, then the PCs which are present at the edge would have a constantly present unbalanced force directed outwards forcing them to move outwards. All these PCs lining the edge of the Universe would be entangled and would move outwards. This can be accomplished only if the Sphere which they form increases in its Surface area. This means that each pair of neighbouring PCs would be pulled apart and would lead to the creation of new PCs in between. This indicates that they would form a Negative energy EPCA or an EPCA with a contracted time like the one seen around an ENEA particle.

The PCs immediately inside them now become the PCs lying at the outer edge of the Universe. These would get pulled outwards and would form a negative energy EPCA and would move outwards to compensate for the PC Vacuum immediately outside the sphere that is formed by them. As this second EPCA moves outwards, it creates a PC vacuum immediately inside the sphere. This cycle keeps repeating. This tendency for the outwards movement of the EPCA would move inwards and would spread inwards.

Note the similarity between this constantly present outwards pull at the edge of the Universe and G which is a constantly present outward pull due to MONDian Gravity from the mass of the whole Universe. It is possible that this constant outwards pull is due to Gravitational interactions between our Universe with other neighbouring Universes. This might be pointing towards an infinite Universe with us living in a part of the Universe which was created in between multiple Universes during the Big bang or it might be hinting towards a Multiverse with the completely alienated neighbouring Universes having little means of exchanging information.

The zones of positive energy excess which were discussed in the honeycomb model of the Universe would be pulled apart due to the constantly present outward pull.

In essence, this negative energy excess is only an apparent one and the expansion of the Universe as per this model is caused by the ever-present PC Vacuum outside the Universe which represents the missing positive energy.

This, however, only explains the expansion of the Universe. It doesn't explain the accelerated expansion.

However, multiple questions arise like "how much volume or distance of PC vacuum exists outside the Universe and is there a limit to it and if yes what lies beyond that?"

We know that if, at a point, there exists positive energy with a high value of "e", the surrounding spatial contraction EPCAs move much faster than the speed of light to compensate for the volume loss. It is quite likely that the "edge PCs" are also moving faster than the speed of light outwards.

Note that this may be equivalent to saying that there exists an extremely large Blackhole at the outer edge of the Universe and the outward pull of this Blackhole is the cause of the expansion of the Universe. There is nothing to be afraid of, though, as the PCs trying to compensate for it aren't destroyed as they never reach it.

We can call this the "Positive energy dominance at the edge" (PEDE) model in that there is an excessive imbalance of positive energy/negative energy ratio towards positive energy at the edge.

Honey-comb Model of the Universe

We know that although the Universe looks uniform at larger scales, it is patchy and baryonic matter tends to clump at places. Baryonic matter, according to DGR is made of predominantly positive energy and is gravitationally attractive to each other. Every Galaxy has a high amount of positive energy at the centre in the form of SMBH or UMBH. These are the regions of the Universe where positive energy predominates. There is a predominance of destruction happening here which is compensated by the inwards movement of EPCAs. Thus the entire clump of gravitationally bound Baryonic matter including multiple gravitationally bound Galaxies or Galaxy clusters or Nebulae with satellite Galaxies form a zone where positive energy and positive energy EPCAs predominate. Here, space is in a constant state of contraction that is determined by the proximity to a Gravitationally active object.

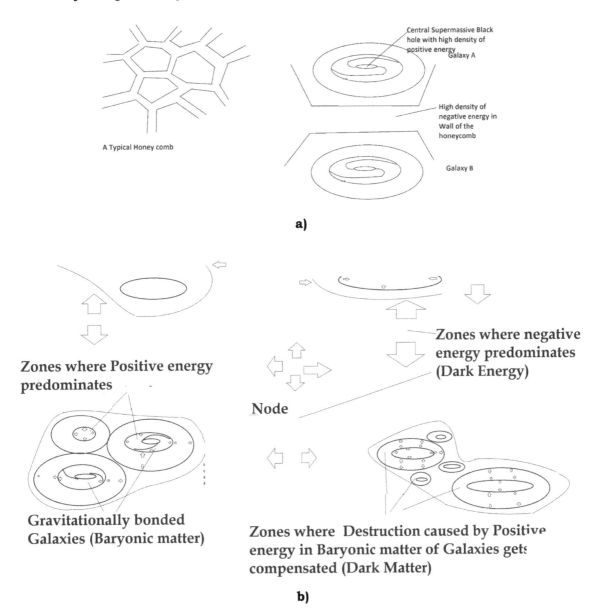

Figure ii/23.2: a) shows the Honeycomb model of the Universe as predicted by DGR with zones dominated by positive energy, where time dilation spatial contraction predominates, forming the chambers of the honeycomb and the ones dominated by negative energy where time contraction spatial expansion predominates. Note that this is just a diagrammatic representation. In reality, the walls of the honeycomb may be multiple times the size of the chamber, more appropriately called "Islands of positive energy". Figure b) shows how the gravitationally bound Galaxies and Galaxy clusters lie in positive energy zones and intergalactic regions where negative energy predominates and spatial expansion occurs. It also shows a node and zones in the periphery of the Galaxy where Dark matter (ENEA particles) lie compensating the spatial contraction caused by baryonic matter in the centre of the Galaxy.

The PCs forming the edge of such a zone would be constantly pulled inwards and a sphere that they form has a PC vacuum on the inside and thus cannot contribute to the compensation of Expansion of the Universe which happens in the intergalactic region between non-gravitationally bound Galaxies.

Such a positive energy-predominant zone is like a chamber of the honeycomb. The walls of the honeycomb constitute the zones where negative energy predominates and here space is in the constant state of expansion.

When the PCs at the edge move out, individual islands or positive energy predominant zones move outwards away from each other. In essence, as time passes, the walls of the honeycomb increase in thickness while the chambers remain the same size.

What lies at the centre of the Universe?

DGR, like the Big Bang theory, would predict that there is a good likelihood that there is a point which we can call the centre of the Universe. However, like the Cosmological principle which suggests that there is no decipherable centre and that the Universe is expanding uniformly everywhere irrespective of the location of measurement, the DGR centre of the Universe would behave no differently.

At the centre of the Universe, all the nearby positive energy zones would be moving away from each other rapidly. We know that this is the case with every other point in the Universe wherein all other points in the Universe would seem to move away irrespective of where we measure it. But there is a significant difference between the centre of the Universe and any other point. This is the negative energy predominated zone where Creation predominates.

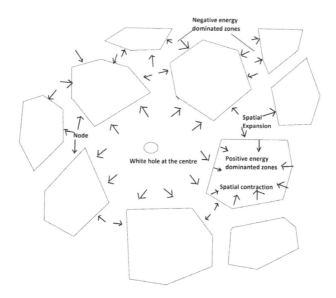

Figure ii/23.3: Shows what lies at the centre of the Universe - A white hole or a region predominated by negative energy so that everything else is moving away from it. The geometric centre of the Universe is probably an unimportant location as most of the nodes and negative energy zones and for that matter, any point in the Universe would tend to behave in a similar manner with all the surrounding non-gravitationally bound entities moving away from it.

The negative energy predominated zone is akin to a White hole. Thus DGR predicts that the walls of the honeycomb or intergalactic zones between non-gravitationally bound structures constitute the White holes. In essence, white holes aren't points in space like their counterparts, the Black holes. They are, in fact, zones or regions where varying but a high density of ENEA particles exists that is responsible for the active expansion of space happening here.

In short, the weird conclusion that DGR is hinting towards is that at the edge of the Universe lies infinite amounts of positive energy and at the centre along with the other nodes and negative energy zones, lies an infinite amount of negative energy. Infinite, because whatever may be the extent to which the edge PCs move outwards, they would not reach the end of the PC vacuum and to whatever extent does the Universe expand, there will be enough negative energy at the centre, nodes and the intergalactic zones to compensate for it by compensatory Creation.

Negative-energy dominant alternative to this model

In this alternative model, there exists a genuine edge of the Universe with no PC vacuum beyond but there is a predominance of Creation within. This is like blowing air in a balloon. The newly created PCs at the centre and the negative-energy dominant zones are not compensatory but are drivers of the expansion of the Universe. The edge of the Universe will expand due to pressure from the inside. (This is in line with the currently believed expansion model wherein fresh space is filled in between the Galaxies that causes the expansion)

This is likely to lead to a positively curved Universe.

This would then be called the "Negative energy dominating at the centre" (NEDC) model.

Both these models indicate an imbalance in PE (positive energy) to NE (negative energy).

A third model exists, with PE excess at the edge and NE excess at the centre, both of which balancing each other so that the entire Universe is balanced and has no overall predominance. One can call this the combined PEDE-NEDC model.

Can we see the centre of the Universe?

What we can see is just the observable Universe. The edge of the Universe is surely beyond our observation and cannot be directly observed.

We do not know if the centre is also beyond our observability. Even the centre is likely beyond our observable limit. But the centre would look like a huge void with no observable Baryonic matter. (note that a White hole, unlike the common misconceptions, isn't emitting light and isn't white. What it is constantly emitting out is negative energy EPCAs which contain PCs created within it.) Also note that it is unlikely that the White hole, like a Blackhole, has $T=T_{max}$ clumped or aggregated together in a small region. Instead, a white hole is a huge region of space with a high density of ENEA particles The Galaxies lying beyond the centre are moving away from us at a rate much faster than the speed of light and are thus unobservable even if they fall within the limit of observable Universe.

Can there be more than one Centre?

Like the walls of a honeycomb, multiple negative energy zones might coalesce at a single point forming a nodal point with a predominance of negative energy. Thus the unexplained voids that we observe like Booty's void might represent one such node.

Potential Explanations to the Cosmological crisis

There are multiple possible explanations for the difference in the measured rates of expansion at the recombination epoch and today.

We will assume that the observations and calculations of the teams of scientists are error-free and show what they claim to show.

Some potential reasons, as per DGR, for the cosmological crisis

1. There was an inwardly directed pull at the time of the recombination epoch that led to a reduced rate of expansion, which the scientists haven't yet accounted for.
2. There is a real difference in the negative energy/positive energy ratio between today compared to the ratio that existed at the time of the recombination epoch.

Inward pull unaccounted by the Scientists

This is the more likely explanation offered.

At the Big Bang, i.e. at a few moments following the creation, there is a likelihood that multiple energy density fluctuations were SMBH or UMBH. The question is what happened to them at the time of the recombination epoch?

Did they remain as SMBH or a part of them disintegrated and formed the baryonic matter visible at the time of the recombination epoch? If this process of disintegration of a Blackhole was common, we would have observed it by now.

In short, DGR would predict that just after the Big Bang, there were points that represented the SMBH and UMBH and represented the density fluctuations in which PE predominated. These points were surrounded all around by points where NE predominated. These NE points represent the White hole counterparts which are NE predominant zones today.

None of the baryonic matter we see today was likely formed at the time of the Big bang and only the PE and NE predominated zones formed. As the Universe expanded, the edge between these zones created all the Baryonic matter, a process that keeps going on even today.

The increased density zones seen in the CMB would then represent the SMBHs of today which are the current Galactic centres. The question of how these SMBH at the galactic centre are formed cannot be solved by modern Cosmology.

The difference, however, was that the distance between these SMBHs was much less than it is today so that most of them were lying within the Einsteinian gravity of each other. Because Gravity in DGR travels faster than light, their individual gravitational influences and the linear EPCA forming time resetting waves formed due to their presence would move out to a significant extent beyond the radius of each BAO. This would create an inward pull that is proportional to the entire mass of the Universe at the time of the recombination epoch, which is likely to be not accounted for till now by modern cosmology.

This gravity would prevent the expansion of the Universe as much as today. Today, these SMBHs lie at such distances that their gravitational influences are weaker now.

Note that, as discussed in the section on variable G, DGR predicts that at the recombinant epoch, G was smaller than it is today. This means that the outwards pull between two gravitating objects was lesser at that time than it is today. This means that the Gravitational attraction would be much stronger at this time compared to today, simply because of the difference in the total mass of the Universe.

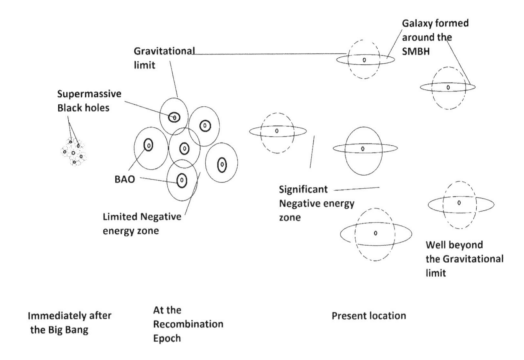

Figure ii/23.4: Shows the location of many SMBH immediately after the Big Bang, at the time of recombination epoch and present location. Given that DGR predicts significantly faster than the light transfer of gravitational interactions and linear EPCA based MONDian gravity beyond the Gravitational limit, there could be an effective extra inwards directed force leading to slowing down of the expansion rate during the recombination epoch.

There is a real difference in the negative energy/positive energy ratio

Some things are beyond explanation in DGR like there are in almost all theories of physics like the Big Bang theory.

The question of whether the PE/NE ratio is balanced or unbalanced and the reason why it is the way it is is beyond DGR. If we presume that there is indeed an imbalance, the extent of it can vary as much as is needed. However, this won't form a good theory as you cannot derive any prediction from it and can change the assumptions of the theory as much as needed to explain the discrepancies. Another enigmatic question that arises with this would be "who or what decides this ratio, is there a conscious mind deciding it?"

CHAPTER 24

INFORMATION LOSS WITHIN A BLACKHOLE OR SINGULARITY IN DGR

Another troubling problem in physics is information loss that occurs at the centre of the Blackhole at the singularity. *(Hossenfelder, Sabine – 54)*

A particle, in DGR, has a positive energy core in the centre where destruction is going on depending on the mass. More is the mass; more is the destruction happening within it per PT. So the number of events e or "з" happening within it per PT is proportional to mass. More is the value of "e" or "з", more is the velocity of outward propagation of the time resetting wave that attempts to compensate for the destruction happening at the centre.

Any particle has certain properties. The most basic of them being mass and velocity which includes direction of propagation.

Let's imagine a particle with mass m (i.e. number of events at the centre e), velocity v, moving from left to right and passing across a point P in its path. Depending on its velocity v, it will spend a small time δt at P. At P and also at every point on its path, it leaves a unique signature in the form of the time resetting wave that propagates from P outwards and keeps travelling indefinitely towards infinity. This wave will have velocity dependant on "e". As the particle moves from P to the next point, the starting point of this wave changes. Essentially, the velocity of the particle v determines how much disturbance is created at a point.

The particle thus leaves a footprint or a trail of its whereabouts within the Universe.

If one has the resources, one can trace these time resetting waves and find out at a later date, when the particle was at P, how much time it spent and also what is the mass of the particle, i.e. its composition.

Note that this resetting wave travels much faster than the speed of light.

If the particle enters an event horizon of the Blackhole, beyond the horizon, EPCAs propagate inwards faster than the speed of light, i.e. these EPCA forming TRWs keep informing the Universe about the whereabouts of the particle even within the event horizon. These TRWs come out of the particle and escape the Event horizon without much effort. If the particle is made of multiple particles bonded together like a rock, the distance between each of these constituent particles abruptly starts reducing with the speed of light and the electromagnetic bonds between them, which according to DGR are due to push-pull bands, are significantly affected. Thus each of these constituent particles (each of them possesses a positive energy core) would collapse into a single positive energy core.

The Negative energy particles forming the part of the particle would probably be ripped apart and due to extreme repulsion, expelled out of the event horizon. (note that this is one prediction of DGR) while all the Positive energy particles would form a single aggregate of Positive energy.

Essentially, any in-falling matter will be converted to energy. Any in-falling matter in a Blackhole is disintegrated into its positive and negative energies and thus the Blackhole keeps emitting out ENEA particles ripped apart from the infalling matter). This is akin to what is described in the AMPS firewall hypothesis given by Almheiri et al. *(A Almheiri et al – 5).* Further, the course inwards would be with extreme speeds possibly much faster than the speed of light along with the faster than light inwards travelling EPCAs until this positive energy core merges with the positive energy core at the centre of the Blackhole. During each point of this journey outside and inside the event horizon of the Blackhole, it keeps leaving its trail of time resetting waves travelling outwards faster than light.

Essentially, till the end of its journey, the particle keeps communicating information about its whereabouts to the Universe irrespective of its location within the event horizon. Even as it merges with the Core of the Blackhole, it keeps manifesting as an additional mass of the Blackhole.

DGR thus predicts that there is no information loss happening when a particle falls into an event horizon. Information about trajectory and individual constituent particles of the particle was communicated in the form of time resetting waves. If "time" is reversed, these waves travel backwards and this information can potentially become available again for recreation of the particle. Thus, Blackhole doesn't lead to time-reversal asymmetry.

Also note that the more time the particle spends travelling, the more information is provided outwards as time resetting waves by creating a disturbance in the Planck Compartments. Thus total "information" of the Universe keeps rising and there is no point in time when the amount of information reduces or is destroyed.

This explains the constant upward trend of entropy of a system.

CHAPTER 25

THE TROUBLE WITH HAWKING RADIATION

Hawking radiation is the probable radiation emitted by a Blackhole. It was first described by Stephen Hawking. *(Hawking, S. W-57)*

This is one of the first attempts at defining the phenomenological sequence of quantum events to derive a prediction. In this, Hawking argued that at the event horizon of a Blackhole, quantum phenomena are still happening with particle-antiparticle pairs coming in and out of existence. If, after the formation of one such pair, one of the two remains within the event horizon, it will be pulled inside the Blackhole and cannot escape as it cannot have a velocity higher than the velocity of light. The other particle being outside the event horizon can escape the clutches of the Blackhole. In this manner, can radiate out its mass and eventually evaporate if given enough time.

DGR suggests otherwise.

The event horizon of a Blackhole is just a point where EPCAs start moving inwards faster than the speed of light so that inward pull is so high that escape velocity is faster than the speed of light. The EPCA at the event horizon keeps moving inwards and is replaced by the EPCA just outside it. Thus, the space-time just outside the event horizon is also moving inwards, and to escape its inward pull, the second partner particle has to have a certain escape velocity and a certain amount of energy within it. If this particle does not possess enough velocity or energy to overcome this, even it will be sucked inwards and will eventually cross the event horizon.

Furthermore, the pair can possess both positive energy or negative energy. If the positive energy one goes inside, it adds to the mass of the Blackhole, and the negative energy of the overall Universe increases. The two pairs will still be able to remain entangled. The opposite is also true that when the negative energy particle enters the event horizon, it presumably loses positive energy and thus loses mass - if we assume that what current physics teaches us is right and the two can annihilate each other. (This is because as per DGR, their annihilation will result in a PC at Universal time which amounts to the destruction of energy – contrary to contemporary Physics which says that their annihilation will release energy. Annihilation of the two or annihilation of Positive energy with Negative energy creating Zero energy or PC with Universal time is the reverse of the reaction "the spontaneous creation of Positive and Negative energy particles from nothing" which presumably does happen as per DGR eventually leading to Creation of new matter particles).

If we assume that 50% times positive energy particle enters and 50% times negative energy particle enters the event horizon, both these will cancel each other out and there would be no change in mass of the Blackhole.

Even if the second i.e. escaping particle has enough velocity to enter into the orbit of such a Blackhole instead of getting sucked inside it, it can encounter other such opposite energy particles in orbit and thus cancel each other out.

Effectively, DGR predicts that the effect of HR on the mass of the Blackhole would be negligible. Only smaller Black holes can have significantly low escape velocity at EPCAs just outside the event horizon to have a significant loss of mass through this mechanism.

Note that whether a Positive energy particle can cancel out a Negative energy particle is also unclear. If not, "what happens to the Negative energy particles entering the Event horizon of the Blackhole?" is a question DGR is unable to answer yet. It is unclear if a Negative energy particle (which cannot coalesce) can travel faster than the speed of light due to its inherent property of Creation happening within, thus easily escaping the clutches of the Blackhole.

If the Negative energy particle indeed is repelled away from the Blackhole every time it forms, due to its negative mass, then every time a particle-antiparticle pair forms at the edge of the event horizon, the Negative energy particle will be pushed away and the Positive energy particle will be pulled within. This implies that the mass of the Blackhole keeps increasing like the overall mass of the Universe and the overall Negative energy of the Universe also keeps increasing with it. It is unclear if such a particle-antiparticle pair creation does happen at the edge of every Blackhole. In any case, complete evaporation of a Blackhole seems unlikely as per DGR.

CHAPTER 26

DGR AND THE BIG BANG THEORY

The Schwarzschild radius of a Blackhole is a critical radius around the centre (where the singularity resides) within which the Gravitational attraction is so high that even light cannot escape the Gravity. In DGR terms, it is the point within which the inwards moving EPCAs start moving faster than the speed of light so that one needs to have a velocity faster than the speed of light to escape the clutches of the Gravity of the said Blackhole.

The Schwarzschild radius of a Blackhole is given by a formula $2GM/c^2$.

With this formula, if the mass M of a heavenly object is known, its Schwarzschild radius can be easily calculated.

The Schwarzschild radius of a lot of objects (or the Black holes with equivalent mass) is known. The Schwarzschild radius of a Blackhole equal to the mass of the Sun is about 3 km. This essentially means this Blackhole will have an event horizon up to this radius around the Singularity.

Supermassive Black holes (SMBH) at the centre of the Galaxies may have solar masses of 0.1 million to several million solar masses. Some Black holes may have masses as much as 1 billion solar masses or more, aptly called the Ultra-massive black holes (UMBH).

It is indeed clear that these Black holes will have a much higher Schwarzschild radius than the Sun. The SMBH at the centre of the Milky Way has a mass of 8.2×10^{36} kg (4.1 million solar masses) and has a Schwarzschild radius of 17 light hours.

The Observable Universe has a mass of 8.8×10^{52} kg and the corresponding Schwarzschild radius of a Blackhole with a mass equal to the current mass of the Universe is about 13.7 billion light-years.

The Big Bang Theory is a theoretical framework describing events happening from the start of the Universe. What happens at the time less than 10^{-35} seconds is not clear and what happens exactly at or before the Big Bang happened is equally unclear.

But the basic premise of the theory is that the entire energy contained in the present observable Universe including all the observable photons and electromagnetic radiation and all the energy contained in Baryonic matter was concentrated at an extremely small point in space.

It is clear from the discussion of mass and Schwarzschild radius of SMBH that the Universe at the early epochs was filled with energy equivalent to much higher solar masses and was almost like an SMBH with the mass equivalent to the mass of the Universe.

This means that whatever events happened at this early age of the Universe happened within the event horizon of this SMBH. Until the size of the Universe exceeded the size of the Schwarzschild radius of an SMBH with mass equal to the mass of the entire current Universe - 13.7 billion light-years (which arguably would be several Astronomical units or Mega parsecs higher than the Schwarzschild radius of the known biggest UMBH), current theories of physics cannot be used to predict any event within them. (As all our theories break down or are not applicable within the event horizon of a Black hole especially at the singularity.)

DGR says that not only all the positive energy in the Universe but the Negative energy in the Universe was also concentrated in a small point at the Big bang. There was, probably, also an inevitable imbalance towards Negative energy which prevented everything from collapsing into itself. This means that the Positive energy wasn't concentrated into a single point but formed multiple points which were surrounded by highly concentrated negative energy all around. The number of events happening per PT within each of these points would be astronomically high. Thus EPCAs would be travelling at insane velocities at such chaotic times. The high velocities of spatial contraction and expansion imply that these events happen faster than the speed of light like those happening within a Black Hole event horizon. These event horizons of positive energy concentrated zones would bend the path of and suck in any photon, thus not allowing any photon to escape.

The CMB probably started becoming visible after Universe expanded enough to have space for the photons to escape without getting scattered or sucked in.

DGR however permits spontaneous creation of Positive and Negative energy particles from nothing. There are physically possible ways by which these pairs can stay together without ever running into each other and in fact form extremely stable elements from it like Protons and Electrons and essentially all the Baryonic matter. This essentially means that all the Baryonic matter particles which we observe today might not be present at the time of the Big Bang and at least a proportion of it was probably created from nothingness later. This has the potential to completely change the equation.

Enough is not known about this process of spontaneous creation of particles from nothingness. This essentially means that important aspects of knowledge of all the processes happening in the Universe that is absolutely essential for playing the video of an ever-expanding cosmos backwards are lacking with us. This creates serious doubts about the various predictions of the Big Bang theory in the rare event that DGR is true.

But this process of spontaneous creation of matter and energy from nothingness provides a good explanation for the daunting problem of a constant energy density in an expanding Universe.

CHAPTER 27

IS STRING THEORY COMPLETELY WRONG?

One of the basic premises of String theory is that particles are two-dimensional strings of energy vibrating in 11-dimensional hyperspace.

So a question emerges, if DGR is right, is String theory wrong?

Does 11-dimensional hyperspace exist? Do additional dimensions exist?

To try and answer this question, we have to come to how the dimensions are defined geometrically.

A point is said to be zero-dimensional with no dimensions involved to describe it. However, in a 3-dimensional space, we have to provide the three spatial coordinates for defining the location of the point. In the case of DGR, which assumes a three-dimensional space allowing definition of a point up to a PL of coordinates, with a single universally applicable fourth dimension of Time with active spatial contraction or expansion allowable, thus allowing the active change in the location of the said point, defining a point at one point in time is not enough and how it's position varies with time is relevant.

Geometrically, a higher dimension to this zeroth dimension can be obtained by doubling the number of points. Once we define two points, we can define the length and thus reach a one-dimensional entity of "the straight line".

Doubling these two points to get four points gives us the second dimension which when joined together forms a square. To get a three-dimensional entity, we have to further double the number of points to get eight points and thus define a cube.

Most of us have no problem imagining these up to this point.

Higher dimensions like the fourth, fifth, etc. can be obtained by this same process of doubling the number of points. So when we have a three-dimensional space with eight points and we double them and get sixteen points, we have a four-dimensional object, when we have sixteen points and we double them, we get a five-dimensional entity with 32 points. This process can continue as many times as you want. Thus theoretically, any number of dimensions can exist in geometry.

However, actually drawing these higher dimensional entities in a two-dimensional paper or even a three-dimensional space is difficult, and here comes the trouble with higher dimensions.

Most people cannot fathom or imagine anything beyond three dimensions with reasonable accuracy.

DGR has given a clue to what these higher dimensions might mean. For this, we have to completely redefine what we think about higher dimensions.

In DGR, a cube of space, which is a three-dimensional entity, can evolve by active spatial expansion or contraction. With time, the location of the eight vertices can evolve. At one point in time, the four points form a cube, as the next fragment of time arrives, the active spatial contraction at one end might distort the cube, and the points now represent some other three-dimensional entity, with a surface area of one space contracted compared to the other faces. As the next fragment of time arrives, the eight points further change their locations and further distort the cube.

Thus, although we are talking of the same eight points, their evolution with the additional dimension of time can be variable and need precise definition. This way, understanding the fourth dimension is also not difficult.

Now imagine what if the cube is not under the influence of one but two different processes, possibly two positive energy particles at a distance "d" on either side of the centre of the cube, sending time resetting waves and causing spatial contraction. Now, the time resetting waves from both particles have their independent influence on the cube and the ultimate location of the eight points and the curvature of the faces depends on the resultant of the two effects. Thus we have to derive eight new locations for the effect of the first wave, and then a further eight locations for the next eight locations. The shape of the cube would change due to the influence of two different processes. This can be redefined as a fifth dimension.

Now further imagine that the cube is within a charged particle so that it is under influence of a high Positive energy core and multiple negative energy particles forming a lattice. The evolution of the cube due to the influence of these two different processes, one with destruction and the other with creation, can be said to be in a higher fifth dimension.

Now imagine that this same cube is also under the influence of the rotation of the lattice which also influences their ultimate position. Indeed, the final location of these eight points will have to be calculated only after taking into account all these processes, creation at each negative energy particle, destruction at positive energy core, and rotation of the core.

Now further imagine that the particle is in motion and thus has velocity, momentum, and kinetic energy. Each of these will influence the ultimate resultant location of these eight points. Thus as the next small fragment of time arrives, each of these forms a dimension. Thus the space-time here is in the sixth (due to rotation of lattice) and seventh dimensions (due to velocity/momentum/inertia).

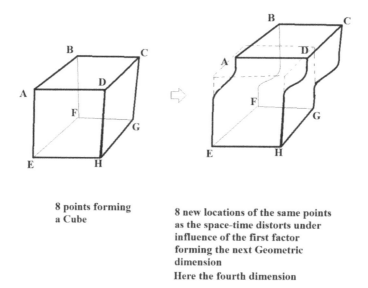

Figure ii/27.1: Showing a three-dimensional cube defined by 8 points namely A, B, C, D, E, F, G, and H. After this is defined, the next location of the same points can be defined by another eight points after considering the influence of a new yet unrecognised factor, defining the fourth dimension.

Further external influences like time resetting waves from an external particle or Gravitational effects of nearby objects or far away objects like linear EPCAs arriving here due to mass of the Andromeda galaxy or mass of the Entire Universe can influence the ultimate location of the eight points after the same fragment of time.

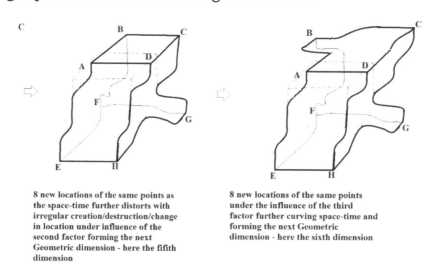

Figure ii/27.2: Shows the next location of the same 8 points by using 8 new points after considering the effect of yet another new factor on the location of these points, defining the fifth dimension. Similarly adding the influence of another third factor leads us to 8 new locations of the same point, thus defining the sixth Geometric dimension.

Another minor influence can be due to the yoyo-like movement of the negative energy particles in the lattice as described in the Superposition of spin section, thus changing the poles of the particle periodically and thus changing the direction of handedness of the particle.

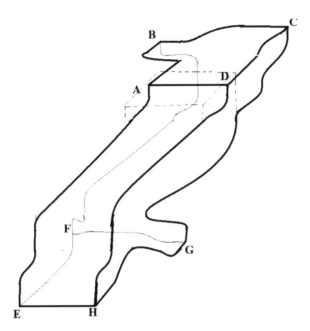

Ultimate position of all the 8 points as determined by multiple factors possibly 8 or 9 or possibly even 11 dimensions

Figure ii/27.3: Shows the final position or the ultimate resultant location of the given 8 points after considering the influence of multiple factors on their location, thus defining the 8th or 9th dimension.

Thus an accurate description of the internal geometry of space within a charged particle depends on multiple influences and the coordinates evolve at higher dimensions geometrically.

Thus, the mathematics of String theory may be of tremendous benefit and the only thing that is needed is to redefine what we mean by higher dimensions.

CHAPTER 28

TAKING A CLOSER LOOK AT G, NEWTONIAN UNIVERSAL CONSTANT AND C – DGR AND VARIABLE SPEED OF LIGHT

DGR hints at the fact that the attraction between two objects with mass should be directly proportional to their individual masses and inversely proportional to the square of the distance.

Thus roughly, it predicts

The force of attraction $\approx m_1 m_2 / R^2$

But Newtonian gravity is given by

$F = G (m_1 m_2 / R^2)$

One can see that the two equations differ by the existence of G

We know that the Newtonian equation works remarkably well at scales up to the solar system and thus is valid.

What is this G and why should it appear in this equation?

G is Newton's Universal constant of Gravitation. It is an extremely small number and has a value of 6.67×10^{-11} $Nkg^{-2} m^2$.

Why should this extremely small number exist as a proportionality constant?

This question has never occurred to mainstream physics and is accepted without further questioning.

Although mainstream physics doesn't question it, some physicists had the curiosity to question why such an unexpected proportionality constant should exist.

One of them was Dicke *(Dicke, R-26)*, who in his 1957 paper suggested that Einstein's variable speed of light (VSL) *(Einstein, A -39)* derivation of General Relativity could hold the key and that G could be much more fundamental and could be related to the Machian principle.

The Machian principle

Einstein further took forward the true meaning of the principle first stated by Ernst Mach.

In this principle, an observer who is within a rotating framework with respect to all the other massive objects in the Universe feels outward centrifugal force which could be due to the gravitational attraction of all the other massive objects in the Universe combined. Thus Mach noted that there is a significant difference between the rotating wheel on which the observer is standing forming his frame of reference with a static Universe and a static observer in a static wheel forming a static local frame of reference with the whole Universe rotating in the opposite direction.

This Gravitational influence of the mass distribution of the rest of the Universe on the gyrating observer was thought to be the explanation for inertia.

In this principle, a rotating bucket with water inside is given as an example. If the bucket with water in it were really a completely independent frame of reference, the masses in the rest of the Universe should not influence it whatsoever. However, what is observed is that the outer water level rises in the rotating bucket suggesting that the centrifugal force does develop whose value is minimum at the centre and increases as one goes towards the edge.

Further Mathematical analysis of G

G is an extremely small number. This means that the numerator in G is much smaller than the denominator. This means that the existence of G in the equation reduces the overall force between the two bodies.

In effect, the fact that G exists in the Gravitational Force equation means that there is an existence of a counterforce that keeps pulling the two bodies apart thus reducing the resultant force. If G was not there in the gravitational force equation, the resultant force due to multiplication of masses and inverse square of the radius would be higher.

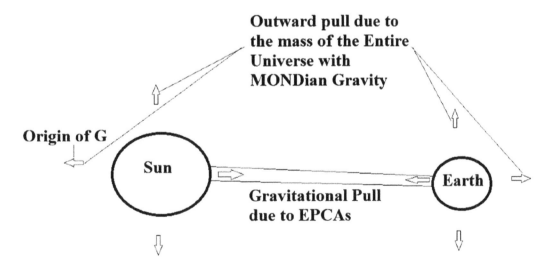

Figure ii/28.1 shows the probable origin of G which reduces the overall Gravitational pull of the Gravitating objects like the Sun and the Earth. This could possibly be due to the mass of the entire rest of the Universe exerting a MONDian force on the two objects effectively opposing their Gravitation.

This outward pull, according to Newton's theory is a universal constant, not changing with time and not varying with the place.

DGR gives a clue to what G could be

In DGR, due to the absolute frame of reference, the distribution of masses around a point of focus can be inferred. Regardless of the distance between this point and any other point in the Universe, all masses in the Universe should have a Gravitational influence on this point. Since the beginning, linear EPCAs from this point directed outwards towards each of the masses in the Universe should exist.

Within the Gravitational limit, circular EPCAs form leading to both horizontal and vertical spatial contraction. Beyond this limit, the horizontal spatial contraction tends towards zero and only a very low but constant vertical spatial contraction denoted by a_0 would happen. This a_0 due to all the constituent stars of the Galaxy can add up and thus lead to MOND forces.

```
Force of MONDian Gravity ≈ a₀ x mass of the Galaxy
```

Instead of using the mass of the Galaxy, a_0 can be multiplied with the mass of the entire Universe to give a term for the overall outwards force existing at a point due to all the masses forming a part of the Universe.

```
Thus MONDian force for the mass of the entire Universe ≈ a₀ x mass of the
Universe

Or Force of MONDian Gravity in one direction ≈ a₀ x ½ mass of the Universe
assuming that the gravitating objects are in the centre so that masses are
distributed equally in the two quadrants.
```

The relation between this outward MONDian force due to mass of the entire Universe at a point in space at a given time and G at that point in space and at that point is of relevance.

```
f(G) -› a0 x mass of the Universe
```

Is G constant?

It is assumed in Newtonian and Einsteinian Gravity that G is a Universal constant i.e. its value does not vary with time and at any point in Space irrespective of coordinate location.

In DGR, G is possibly related to the MONDian force of the mass of the Universe. And thus, being related to the mass of the Universe which is constant at a point in time, looking superficially, it should be constant in magnitude everywhere in the Universe irrespective of the coordinate location at a given point in time.

This isn't entirely true, as noted below, as the direction of G can probably vary with location and so does the magnitude (as predicted by DGR)

DGR allows for the spontaneous creation of Baryonic Matter from nothing which suggests that the mass of the Universe can keep increasing with time.

This suggests that DGR, like Dirac's Large Number Hypothesis, is hinting towards a gradually declining G which essentially means a gradually weakening force of Gravity. (*Dirac, P.-30*)

Can G vary with spatial location within the Universe?

We know that the force of MONDian gravity acting on the peripheral stars of a Galaxy, due to the mass of the entire Galaxy, is much more inwards than outwards. This same MONDian force is distributed in all directions almost equally in a centrally located star. Explaining this is not difficult, given that for the peripheral star, the majority of the mass of the Galaxy lies in one direction while in a central star, the mass of the Galaxy is distributed in all directions almost equally. This excessive inwards directed pull of MONDian gravity can explain the higher than expected rotation velocities of the peripheral stars and the flat rotation curves.

If this is true for the Galaxy, can the same logic be applied to the entire Universe?

The answer isn't clear yet. But can potentially give us ways to confirm DGR i.e. offer predictions of DGR that can be potentially testable.

The question worth pondering is

"like MONDian gravity within the Galaxy has a variable distribution within the Galaxy depending on the distribution of masses, can the MONDian gravity within the Universe (which we call G) have a variable distribution depending upon the distribution of masses in the Universe at different locations?".

A Galaxy that is located in the peripheral part of the Universe will have masses within the Universe predominantly distributed in one direction than the other, which means that in this direction, the inward force of G would be greater. A Galaxy which is located in the central region of the Universe

Another question worth pondering is

"Is the Sphere of the whole Universe rotating or not and if so, is there an outwards directed centrifugal force acting at the Galaxies at the edge of the Universe?"

The stars forming the Galaxy revolve around the centre with such velocities that the peripheral stars revolve almost with the same velocities. The Galaxy thus behaves like a disc with all the peripheral stars bonded together like molecules of the disc, all moving around the centre with the same velocity. The stars in the central region of the Galaxy however revolve like Newtonian Gravity predicts with velocities reducing with distance.

Even if the Universe has a rotation, it is impossible to fathom it for us here on the Earth, unless we go outside the Universe and have a reference frame outside the Universe.

However, we can ask the following questions.

1. Whether there are layers of Galaxies around the centre of the Universe revolving around the centre of the Universe with a variable velocity like the Galactic central stars followed by MONDian type of bonding with no relative change in position of the various layers in which the Galaxies lie

    OR

2. there are no layers of Galaxies revolving at different velocities and all of them are bonded in a MONDian manner and thus the whole Universe behaves like a Solid metallic sphere or a marble (rotating or otherwise) with constituent particles bonded together like covalent bonds.

What is apparent till now, is that there is only outward expansion but no evidence of a revolution of the Galaxies around a centre. Even if the Galaxies revolve, we cannot find it unless the revolution is variable and thus the layers change their relative position with time.

If the Universe does have a rotation, then variable G will be the reality, with more inwards directed force leading to a much higher revolution velocity of the peripheral Galaxies and a block Universe with all the constituents moving with a constant velocity of rotation.

This variability of G with location can also have predictable effects on Galaxy-Galaxy bonding thus paving the way for us to distinguish which Galaxies are more peripherally located than the Milky way and which ones are more centrally located. For example, we may be able to predict a more prevalent pattern of distribution of Satellite Galaxies in the direction where G is located whereas fewer satellite Galaxies are in the opposite direction.

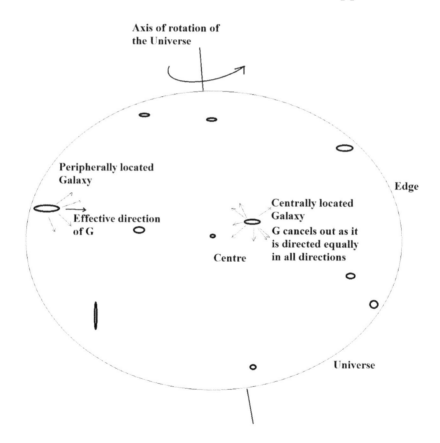

Figure ii/28.2: Shows How MONDian force due to mass of the Entire Universe which we call G can vary according to the location of the object within the Universe, as predicted by DGR. The centrally located single objects say a Galaxy will have no effective direction as G cancels out in all directions. When considering two or more objects gravitation towards each other, G would still come into the picture reducing the amount of Gravitational attraction between the two.

Peripherally placed Galaxies will have most of the rest of the mass of the Universe directed only in one direction and thus will have an inwards directed G. If the Universe does have a rotation, this inward pull and a much higher G can compensate for much higher rotation velocities (if there exists a centrifugal force acting outwards on these peripheral Galaxies), thus enabling a Block type of Universe with the peripheral Galaxies maintaining their relative relations irrespective of location.

Is c a constant?

The entire physics of Special Relativity and Cosmology today is based on this primary assumption that the speed of light is constant in a vacuum (*M. N. Roy – 85*). This constancy is irrespective of coordinate location in the current Universe or any of the states in the Universe in the past.

Many models including the Lambda CDM model of cosmology are completely based on this understanding.

Einstein, in his paper in 1911 (*Einstein, A. – 39*) and further refined by Dicke (*Dicke, R -26*) in his paper in 1957 suggested that c may not be that constant at all and that by applying Huygens principles, the bending of light can be successfully derived with variable speed of light in flat space-time instead of the fixed speed of light in curved space-time.

This concept was however not developed further due to universal acceptance of General Relativity and with it the acceptance of curved space-time. Some renewed interest in the same idea of Variable speed of light has emerged with multiple papers by Alexander Unzicker on the same. (*Unzicker, Alexander – 95*)

"c" as per DGR

Imagine a photon which is a T=0 compartment travelling tangentially in the Earth's atmosphere.

We can imagine many concentric spheres all around the Earth at 1PL distance from each other. These spherical EPCAs will keep shifting inwards due to Earth's gravity.

Let's imagine that the photon enters one such EPCA. If we put a photon clock within this EPCA, we know that the photon clock slows down due to the horizontal spatial contraction due resulting in the Gravitational time dilation.

If we put a similar photon clock at many inner and outer EPCAs like these, they will also show progressively increasing time dilation and thus resultant slowing of the ticking of the photon clock as we move towards the Earth.

The T=0 photon (or for that matter any photon) would be slowed down as it enters the EPCA in focus. It figures that how much the speed of the photon slows down depends on the spatial contraction which in turn depends on the Gravitational time dilation. In fact, the gravitational time dilation, in DGR is an illusion created due to the existence of active horizontal spatial contraction in these EPCAs.

Thus light will have a variable velocity at each of these EPCAs.

Thus DGR is indeed a VSL type of theory and suggests a Variable speed of light.

Not only this, if there is active spatial expansion, as is present in the intergalactic region of non-gravitationally bound Galaxies, here the photons of light would be pushed forwards due to active spatial expansion at their backs thus increasing their speed.

Thus c, in DGR depends on the status of Local time.

Note that the photon is indeed moving 1PL per 1PT even at these places. Thus the speed of causality is not changing. The variation in velocity is solely due to the presence of active spatial expansion or contraction just before the photon.

Variation of c with respect to time

This is indeed a difficult question to answer.

At earlier stages of the Universe, the masses that constituted the Universe at that time were closer so that there was much more gravitational time dilation in between them. The light which left at that time would thus have moved much slower than it is moving today.

There is a slight confounding variable here. The spontaneous creation of Baryonic matter and thus increase in mass that DGR predicts, if true, would mean that there is more mass now than before. The spontaneously created matter would, however, constitute a very small proportion of the total mass of the Universe.

Origin of Inertia

Mach's Principle suggests that Inertia at rest, i.e. the tendency of a body at rest to stay at rest is because of this Gravitational pull of "mass of the rest of the Universe" on the body keeping it in its co-ordinate location. This goes well with DGR, wherein, a body at rest would have a pull of MONDian gravity on all sides due to the mass of the Universe and these linear EPCAs would oppose any movement of the body and would have to be opposed or overcome to set the body in motion.

CHAPTER 29

DGR AND DIRAC'S LARGE NUMBER HYPOTHESIS

Large number Hypothesis

(Dirac, P.-28, Unzicker, Alexander, 95,96,97,98,99,100)

Paul Dirac was an excellent mathematician who believed that his equations were telling him about some hidden reality. With this approach, he had predicted the "Antimatter" particles while deriving the relativistic version of Schrodinger's equation. He had displayed this ingenuity again when he published the "Large number Hypothesis". *(Dirac, P. -28,29,30, Unzicker, Alexander – 97,)*

To understand this hypothesis, we have to first understand the significance of "Dimensionless constants" in nature. These are constants whose value comes the same irrespective of the system of units used for measurement. One of the best examples of such a constant is pi which is the ratio of a circle's circumference to its diameter. This number comes as a constant irrespective of the units used in the measurement of the circumference or the radius. Other examples include Reynold's constant in fluid mechanics, the fine structure constant (also called alpha), and the ratio of the mass of the Proton to Electron.

These dimensionless physical constants are considered to be highly significant and the theory that could successfully explain them would be more likely to be a credible theory of Everything in the opinion of many physicists.

According to Paul Dirac, there are some peculiar dimensionless constants whose values are very large and their values co-incidentally match too close not to notice.

For example, one such constant is the ratio strength of Electromagnetism and Gravity. The value of this is $\approx 10^{40}$.

This is closely connected to the age of the Universe which also comes out to be about 10^{40} if expressed in atomic units of time.

Another constant of interest is the ratio of the radius of the Universe to the radius of the Proton. The value of this ratio is also found to be 10^{40}.

He also noted that the total number of baryonic matter particles in the Universe including Protons and Electrons is the square of 10^{80}, which is the square of 10^{40}. Thus the total number of baryons in the Universe is also closely linked to the Age of the Universe and is directly proportional to the square of the age

Dirac contented that there must be a deep connection between these large numbers which is yet ill-understood. These large dimensionless numbers must be closely related to the age of the Universe. Dirac agreed with a rather unpalatable claim previously given by Edward Arthur Milne, that the Gravitational Constant G is actually not a constant but keeps varying with time. (*Unzicker, Alexander – 97*)

Not only this, Dirac claimed that the strength of electromagnetism compared to gravity is closely related to the Age of the Universe. This can be obtained if G, which determines to a significant extent, the extent of gravitational interactions, is not constant but constantly varies with time.

His Large Number Hypothesis states that

1. *The strength of Gravity, as represented by the Gravitational Constant G is inversely proportional to the age of the Universe*
2. *The mass of the Universe is directly proportional to the age of the Universe*
3. *Physical constants are not constant but their value depends on the age of the Universe.*

His theory indicated that there is some yet unknown way by which active creation of matter happens so that as the Universe's age increases, its mass keeps growing and newer particles with mass keep getting added to it. The Gravitational Constant G has a value that is closely linked to the total mass of the Universe and thus as the mass of the Universe increases the value of G should also increase with time as the Universe increases in Age. This is. in a way, consistent with the Machian principle given by Einstein which says that the information about the distribution of mass in the entire Universe is somehow communicated to a rotating object. (*Unzicker, Alexander, 95, Sciama - 91*)

DGR and Dirac's Large Number Hypothesis (LNH)

It is evident that DGR predicts that every particle with mass produces Time Resetting Waves forming EPCAs which are curved within a specific distance called the Gravitational limit and beyond these the EPCAs produced by it are linear and create a universally constant acceleration called a_0. These linear EPCAs can have an effect irrespective of distance.

This indicates that any two particles under the influence of each other's Gravitational influence are also at a Gravitational influence of all the other massive particles of the Universe irrespective of the distance between them. the Linear EPCAs produced by the mass of the rest of the Universe produce a net outward attraction which should result in a net force that pulls the two bodies apart. Thus the resultant gravitational attraction between the two objects with mass in focus would be significantly reduced. G which is the Gravitational constant could be the resultant outward pull of the mass of the entire Universe on the two objects.

Furthermore, DGR allows for spontaneous Creation of matter particles from Energy which is the deviation of Local time from Universal time as explained in the section of Genesis.

There is a claim in DGR, that at the time of the Big Bang, none of the Baryonic matter that we see today probably existed. As the Universe expanded, newer fluctuations in the density of energy or in other words newer deviations in Local time were created which produced matter particles. This process of spontaneous creation of "Matter Particles" continues even today and thus the mass of the Universe is not a constant term but it keeps increasing with the Age of the Universe.

The ratio of the strength of Electromagnetism to the Force of Gravity would also depend on G which appears at the denominator of this ratio. As the age of the Universe increases, the value of the mass of the Universe and with it the number of Baryonic particles in the Universe increase. Each of these Baryonic particles adds to the counter-pull which leads to a reduction in the Strength of Gravity since the force opposing it in the form of G is increasing steadily with the age of the Universe. Thus the force of Gravity would keep decreasing in strength with increasing age.

As discussed separately, DGR predicts that the velocity of light c at every EPCA around the Earth would vary and would depend on the presence and extent of active spatial contraction or expansion of space just before the Photon.

All these suggest that DGR is perfectly consistent with Dirac's Large Number Hypothesis, the Machian principle, and also Einstein's VSL theory of light. (*Unzicker, Alexander, 95, Einstein, A- 39*)

Chapter 30

THE RATIO OF ELECTROMAGNETIC FORCE TO GRAVITY

This ratio is about 10^{40}.

Let's assume that this constant is indeed related to the age of the Universe.

Another assumption here would be that like the outwards pull between two Gravitating objects that we call G when two charged particles are bonding together, they do not have any similar outward pull. This probably happens due to the extremely small fragments of time and extreme speeds at which the electromagnetic bonding is accomplished.

It is peculiar to note what will happen to this ratio between Electromagnetic force and Gravity as we reverse our video of reality down until the moment of Creation.

As we look at an earlier time in the history of the Universe, according to DGR, the mass of the Universe was much lesser back then than it is now.

This means that the overall outwards pull was much lesser at that time than it is now. Further backwards, and we come to a point where the mass of the Universe was zero at the moment of Creation. It figures out that as time reverses and approaches zero, the Mass of the Universe tends towards zero and thus the value of G tends to zero. At the moment of Creation, G is zero and Gravity is extremely strong or as strong as the Electromagnetic force. Literally, all the forces that are present i.e. even strong and weak nuclear forces would be unified at the moment of Creation as there would be no separate particles to create the Push-Pull bands and there would be no gradient with corridors of deviated Local time.

What will happen to the G if we fast forward our video and see what would happen in the future if this is true?

As we go far into the future, the mass of the entire Universe keeps increasing and the value of G keeps increasing after which there will come a point when G becomes so high that the outwards attractive force of MONDian Gravity becomes greater than the attractive i.e. gravity becomes a repulsive force. This would happen billions of years into the future, so nothing to worry about for us. However, the force of Gravity would indeed get weaker with time.

"What will happen after this?" is still a mystery.

However, as the mass of the Universe rises, a point may come in the future when the force of MONDian Gravity at the edge might increase enough to overpower the outwards drift of the PCs at the edge. This means that the Universe might start collapsing and end up in a big crunch. If the MONDian Gravity cannot take over and the outward drift continues, the positive energy or the PC Vacuum beyond the edge being much more powerful, would mean that the Universe would end up in a big Rip and everything we see will be shred into pieces. There might come a point when the outwards pull of MONDian Gravity equals the Electromagnetic attraction and thus every atom would be ripped apart.

Which of these two is the fate of our Universe is unclear.

What is clear is that we are indeed living in a privileged time since the ratio of Electromagnetic force to Gravity is indeed favourable for life formation at present.

Chapter 31

MANIFEST AND UN-MANIFEST FORM

This chapter discusses a completely different way of imagining the Universe.

At the most fundamental level, DGR suggests that the entire Universe is made up of two aspects.

1. A "Manifest" form which is the Planck Compartment which obeys the exclusion principle and cannot have an identical location to other PCs and thus is capable of exerting a push on neighbouring PCS and is responsible for forces that push particles away from each other.

2. An "Un-manifest" form which we call the "PC Vacuum" that exists in between the PCs which is responsible for the entanglement and which exerts force and pulls the PCs together.

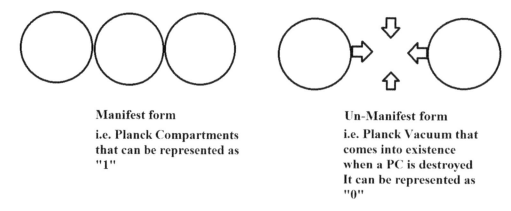

Manifest form
i.e. Planck Compartments
that can be represented as
"1"

Un-Manifest form
i.e. Planck Vacuum that
comes into existence
when a PC is destroyed
It can be represented as
"0"

Figure ii/31.1: All the phenomena happening throughout the Universe can be explained based on just these two aspects of the Universe, the Manifest form and the Unmanifest form.

There are three different Properties which every phenomenon can have,

- Creation of the Manifest
- Destruction of the Manifest
- Compensation of the above two processes i.e. entanglement

Each of these processes leads to the interplay between the Manifest and the Un-manifest.

The Un-manifest is the main means by which attraction force is exerted. The Manifest can push each other and is primarily responsible for the push force or the repulsion force.

The rate of Interplay between these three phenomena happening between the Manifest and the Un-manifest, determine the Local Time and also determine the state of Space I.e. Determines if the space will expand or contract.

This is almost similar to the 1 and 0 of digital code.

It is thus not unrealistic to say that the Universe is like a Hologram with Information akin to Digital information stored in computers.

Creation is the process of conversion from Un-manifest to Manifest i.e. conversion from 0 to 1.

Destruction is the process of conversion from Manifest to Un-manifest i.e. conversion from 1 to 0.

Even the process of Entanglement involves a change in co-ordinate location of the Manifest due to the Un-manifest pulling on to it.

The process of Entanglement involves temporary Positive energy particles (i.e. destruction) or negative energy particles (I.e. creation).

The process of Destruction happening incessantly at a point is the Positive Energy Particle. The Process of Creation happening incessantly at a point is the Negative Energy Particle. These particles could be permanent or temporary appearing only temporarily during the process of Entanglement.

All the Baryonic matter that we see and all the Energy that we see including electricity, heat, light, matter particles and even Humans, is made of the Un-Manifest component.

The Manifest component, due to the invisible Un-manifest component in between is like an extremely thick elastic jelly (magic jelly) that resists the creation of any Gaps in between. The only reason why we don't feel it as an extremely viscous jelly is that each of our constituent particles having the property to constantly "Eat it up".

As more Universal time passes forwards, the spontaneous creation of Baryonic matter, that involves the Creation of entangled pairs of permanent Positive Energy Particles and permanent Negative Energy Particles.

Additional information is needed to encode the coordinate location of every permanent particle that forms and the cycles of entanglement that each of it creates with each passing Planck time.

As more Planck time passes post the moment of Creation I.e. the Big Bang, the Diameter of the Universe keeps increasing and so does the surface area of the Universe.

Thus contrary to the previous opinion, much more information is encoded in the Universe.

Earlier it was believed that the following information is encoded.

1. Number of Baryonic matter particles
2. Locations of each particle at every instant in time
3. Their velocities at every instant and thus their momenta
4. Pace of time at every co-ordinate location
5. Information regarding Tensors or forces acting at every point in space deforming its shape and the exact shape of Space-time at every instant at every point in space.
6. Information regarding what interactions are happening at every instant of time at every corner of the Universe.
7. Information about forces acting on each of these Baryonic matter particles and their other quantum properties like spin, magnetic moment, charge, strangeness, attraction, and repulsion forces acting on each of these particles.

Now, with DGR, the amount of information encoded in the Universe seems to have increased significantly.

With DGR, the new Universe has encode (in addition to the above)

1. The number of Manifest, number of Un-Manifest, their coordinate locations, the timing of their appearance and disappearance, their coordinate location and change in their coordinate location with time,
2. Information about the formation of aggregates of PCs formed I e. Entangled to each other i.e. the information regarding EPCAs forming and their velocity of movement and the forces that their existence exerts, and the effect of interference of such EPCA forming Time resetting waves from every corner of the Universe.
3. The changes happening between Manifest and Un-manifest i.e. Creation/Destruction
4. Temporary and permanent Positive Energy Particles and Negative Energy Particles, their location, direction of propagation, their effects on the surrounding, the TRWs emanating from them, their Momenta,
5. Information about the interference of TRWs emanating from each of them

The constant increase in entropy and the arrow of time

With every tick of the Planck clock, the outermost PCs at the edge of the Universe would move outwards and form newer Manifests, honouring the outward pull of the positive energy-rich infinite PC Vacuum present just outside the Universe. Every outward hop of these Edge PCs per Planck length would determine 1 Universal Planck time. Walls of the Honeycomb chambers, in the negative energy-rich zones, would also create more Manifests thus adding to the number of Manifests.

It is thus inevitable that the total information contained within the Universe keeps increasing with time.

Thermodynamics

Thermodynamics would then be the rate at which there is a change in the coordinate location of the Manifest component (basically how fast are the PC's shifting their position or what is the average velocity of EPCAs in a given region of space). And how stable the Manifest component is, i.e. how fast the PC's are being modified. This is the ratio between total T=UT PCs which are relatively stable and the T=0 or T=Tmax PCs which are highly unstable and add to the instability by invoking movement of the neighbouring PCs for compensation. A T=0 PC is a PC where the process of shifting from Manifest to the Unmanifest is happening and the T= Tmax is a PC where the process of shifting from Unmanifest to Manifest is happening. In short, a place where more T=0 PCs clump together is hot, a place where more density of T=Tmax components exists is also hot, while a place where the majority of PCs are T= UT and are stable in position is cold.

CHAPTER 32

AN ELEGANT WAY TO TEST THE THEORY OF DGR

We know that GR says that the Earth due to its Gravity produces Gravitational time dilation. This has been tested to be true experimentally multiple times. For this, the most precise method of testing time that we have -the atomic clocks were used.

What was essentially done was to keep the two atomic clocks at different time zones and start them, stop them after a specified period simultaneously and compare their reading.

Theoretically, as per GR, Gravitational time dilation would produce multiple time zones and a time gradient. These changes in time happen with the speed of light. Theoretically, these are symmetrical on all sides as per GR.

The Earth is moving in Space with a speed of almost 30 Km per second. This would indeed create some asymmetry as per GR but because this speed is insignificant compared to the speed of light, the asymmetry would be small.

Contrary to this, DGR predicts that at the core of every particle that constitutes the Earth, there is the destruction of PCs going on and due to the movement of the Earth, much more compensation would come from the direction of motion of the Earth, i.e. the preceding border rather than the receding border. This means that the time gradient should be different at the preceding border compared to the receding border.

The experiment

One has to construct an exact model of how the Earth is moving with respect to the Sun in the Solar system and also how it is moving with respect to the rest of the Galaxy with a co-ordinate system beyond the Galaxy.

One has to locate the preceding and the receding borders of the Earth.

The most formidable challenge of this experiment is going to be finding this accurately. Their location keeps changing and thus timing is crucial.

Once the two borders are known, multiple weather balloons or drones with atomic clocks on board, kept at different heights from the surface would start measuring time and the Gravitational time dilation at each level is determined.

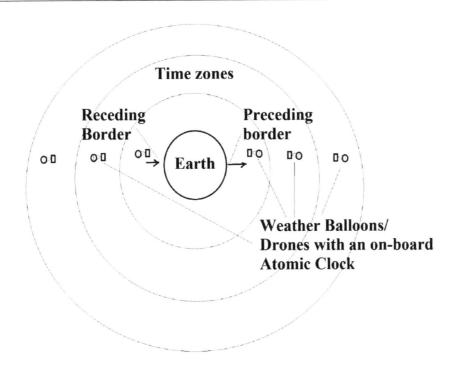

Figure ii/32.1: Shows an experiment to confirm DGR. The exact location of the Preceding and Receding borders of the Earth is located at a specified time and different Weather Balloons or Drones with an Atomic Clock on board are used to measure Gravitational Time dilation at each height. Any difference in decline of GTD with height is noted.

The Gravitational time dilation at each level is noted and compared to that at the surface of the Earth and the gradient of Gravitational time dilation between receding and preceding borders is measured.

If the difference in time gradients is detectable, it could be regarded as proof of DGR. DGR predicts that the EPCAs at the Preceding border have to move inwards faster just to compensate for the Destruction happening at the core of each particle of the Earth which could not be compensated due to the movement of the Earth. This time changes constitute the kinetic energy of the Earth due to its movement. The faster-moving EPCAs at the Preceding border will definitely affect Local time at each EPCA and the time gradient. Exactly at the Preceding border, at a given height h, the GTD will be much more than predicted by General Relativity. The GTD, exactly at the receding border will be much less than predicted by GR at a height h. (It remains to be seen if the difference is measurable)

Caution

The maximum difference in time gradient would happen exactly at the preceding and receding borders. Where these lie at any particular time is difficult to predict. They may be on the ocean or land. The real challenge is to decipher the exact movement of the Earth with respect to the Solar system and also with respect to a coordinate location outside the Galaxy and identify the Preceding and Receding borders. The drones or balloons should be having exact locations. GPS can be utilised to help confirm their locations.

Alternative Experiment

Logistically, measuring the difference in GTD gradient in the preceding/receding border of the Earth is easier. However, the same is true for any object moving through space including the Sun. The preceding and receding borders of the Sun are easier to find simply because we know that the Sun is moving within the Milky Way Galaxy in a plane that is perpendicular to the plane of the Solar system and all the planets are in reality moving around the moving Sun in complex spiral paths. Thus the preceding and receding borders of the Sun lie perpendicular to the plane of the Solar system and Satellites launched in each of these borders can measure the difference in this GTD predicted by DGR. The challenge would be to launch such satellites adequately close to the Sun at equal distances at both its preceding and receding borders and measure the GTD. Also, we have to take into account the movement of the Milky Way Galaxy towards the Andromeda Galaxy which has the potential to change the location of the borders.

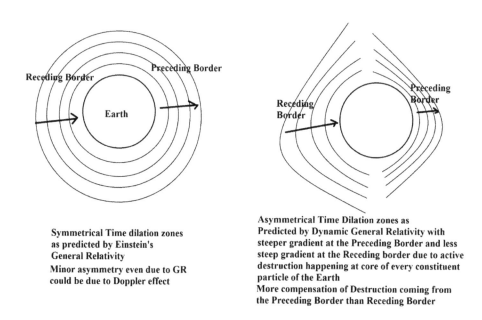

Symmetrical Time dilation zones as predicted by Einstein's General Relativity
Minor asymmetry even due to GR could be due to Doppler effect

Asymmetrical Time Dilation zones as Predicted by Dynamic General Relativity with steeper gradient at the Preceding Border and less steep gradient at the Receding border due to active destruction happening at core of every constituent particle of the Earth
More compensation of Destruction coming from the Preceding Border than Receding Border

Figure ii/32.2: Shows how to test DGR. General relativity predicts perfectly symmetrical time zones all around the Earth and has no concept of Destruction happening at the core of particles forming the Earth. DGR predicts that the Time zones at the Preceding borders have to have a steeper time gradient than the Receding border which happens because much more compensation comes from the Preceding border compared to the Receding border. (Steeper means more change in time dilation per unit distance.) Minor asymmetries that may be explained predicted by GR are those due to the Doppler effect. Note that this is a diagrammatic representation and that the changes are shown to be overly exaggerated. It remains to be seen if the changes are really large enough to be measured by current technology.

The Earth-Sun zone

Figure ii/32.3 shows another way of confirming GR/DGR and in a way deriving Newtonian dynamics.

Time dilation zones created due to the Sun and the Earth

Zone on the Earth-Sun line where GTD of the Sun and the Earth add up and form a zone where fall in Gravitational time dilation is milder as we move from Surface of the Earth outwards due to cancelling effect of the GTD of the Sun. This forms a zone with active spatial contraction akin to a string pulling the two the two towards each other

Figure ii/32.3: Shows how DGR and even GR predicts that at the Earth-Sun line, the decline in GTD caused by the Earth is negated by the increasing GTD caused by the Sun as one moves from the Earth's surface towards the Sun. This zone lying between the Sun and the Earth will have milder fall in the Gravitational time dilation gradient of the Earth as we move away from the Earth due to the cancelling effect of the GTD of the Sun. This according to DGR would be the zone where constant active contraction of Space is happening and keeps the Earth bonded to the Sun.

General relativity states that the Sun will produce Gravitational time dilation zones around it and so will the Earth. A line joining the Earth and the Sun would have a milder drop in the GTD as we move away from the surface of the Earth since the drop in GTD is cancelled out by the increasing GTD of the Sun. This means that there should be a zone extending from the Sun to the Earth, where the two are facing each other, where a milder drop in GTD gradient should exist. If the mass of the Sun and the Earth were the same, there would be a constant time dilation on this line. Because the mass of the Sun and the Earth are not the same, the fall in GTD of the Sun as we go from a point near the Earth on the Sun-Earth line towards the surface of the Earth would be lesser than the gain in the GTD due to increasing proximity to the Earth. This gradient would not be as much as typically predicted by GR.

This, in DGR, represents a zone extending from the Sun to the Earth, where a relatively constant (i.e. un-varying) and relatively higher active spatial contraction exists and acts as a string that applies force and keeps the Earth bonded to the Sun. In reality, this is not the only zone contracting and in a way, the entire space around the Sun and the Earth and the relatively peripheral zones to the Sun-Earth line are also contracting as shown in the figure. The peripheral layers, with active spatial contraction shifting inwards and taking the place of the inner layers. The active volume loss and the active spatial contraction happens in all the zones seen below and the effect of all this spatial contraction combine to bring the two bodies together. This is true for all bodies made up of positive energy particles.

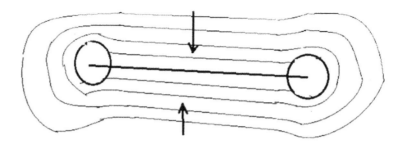

Figure ii/32.4: Shows the various time dilation spatial contraction zones created by the distribution of local time around and between two Positive energy particles (like the Earth and the Sun). The maximum time dilation and spatial contraction would happen at the line joining the two particles with gradually reducing time dilation as one moves from the line outwards. But active volume loss and spatial contraction are happening to some extent depending on the Local time in all the above zones, the effect of all of which combine to pull the two bodies together.

The Experiment

A tower or skyscraper or a set of Weather Balloons or drones are kept in position, with multiple Atomic clocks at different levels, at a location on the Earth where the Sun rays will fall exactly perpendicular to the Earth's surface. Because of the constant rotation of the Earth and its movement, the location of this zone would keep changing and thus time interval of taking time measurements will be extremely small. Thus it is easier to set up the Clocks and keep measuring the Time dilation Gradients. As soon as the zone coincides with the tower, DGR predicts that there will be a much lesser decline in Time dilation as one goes away from the Earth due to cancelling effect of the GTD of the Sun. This effect stays for a very short period and immediately as the Earth rotates, another point will come into focus, this relatively constant GTD will shift away and the GTD gradient will reappear.

Alternatively, multiple drones which are designed to stay at different heights from the surface but follow the Overhead Sun can be used.

Miraculous complete disappearance/reduction of the GTD gradient, a rather drastic event is the key finding.

Caution:

The major source of error would be the wrong location, completely missing the spot where sunlight was perpendicular.

It is unclear yet what would be the size of this zone, however, given the astronomical size and masses of both the Sun and the Earth, it is a sheer hope that the size of the zone is detectably large.

Another important drawback of this experiment is that it just confirms General relativity and doesn't prove DGR. The predictions of GR and DGR would match perfectly here and thus this cannot be taken as absolute proof of DGR.

CHAPTER 33

DISCUSSION

In DGR, we have a phenomenological description of a potential theory of Quantum Gravity which discusses the Quantum events unfolding at Planck realms of length and time.

To reach the theory, multiple believes or "walls of the box" created by modern physics have to be broken and the physicists have to think out of the box. One has to give up the urge to think in the usual lines and think in a different direction. This could be a possible turn-off for many physicists and is the most important test for the theory.

These include giving up, for a while, the ideas of absolute relativity of time and space given by GR and beliefs like information exchange can only occur at the speed of light or that Gravity travels at the speed of light. Relativity of time and space emerges in the theory at appropriate places and DGR explains how despite time and space being nonrelative, become relative. That is, the relativity of space and time is itself relative.

The theory which we have is a non-local theory allowing information exchange faster than the speed of light. The theory assumes an absolute spatial coordinate system of the entire Universe up to a single PL and also assumes that time is absolute and there is absolute simultaneity of time with respect to Universal time everywhere in the Universe. These are difficult concepts to accept by physicists who have accepted SR and GR for over a century.

Note that DGR has a dynamically changing space and time like the description of SR and GR, but the spatial contraction or expansion is different compared to SR in that it evolves every PT instead of remaining static with time and evolving only with distance. Even the description of time is different in that there is a Universal time and a local time in DGR wherein local time is the apparent or illusionary local time that dilates or contracts based on the status of spatial contraction or dilation.

The theory is absolutely deterministic in that if enough knowledge becomes available to the tune of 1PL and 1PT, the path of any particle can be determined accurately even in sub-Planck length distances and velocities. It is indeed a "hidden variable theory" in that the relative indeterminism which is apparent in Quantum mechanics is basically due to our ignorance of multiple hidden variables. However, the determinism of this theory, is itself of no use to us Humans, since at Human scales, the extent of hidden variables is so great that it becomes extremely in-deterministic. It is practically impossible for us Humans at our scales to know every time resetting wave arriving at a point and direction and relative velocity and type of EPCA acting on a Quantum particle. Quantum particles are also having velocities and momenta which change the relative structure of time resetting waves and EPCAs around it.

The description of absolutely symmetrical spherical EPCA formation is highly theoretical and applies only to a static particle with no momentum. For a particle with constantly changing velocity, momentum, and kinetic energy and whose velocity is constantly under the influence of unknowable influences of EPCAs produced by elements at extremely far-away distances like far-away Galaxies or supermassive Black holes, the number of these hidden variables becomes too great and is practically unknowable. The indeterminism is also due to events happening at extremely small fragments of time i.e. At Planck time scales. It is also practically impossible to keep a track of every event happening at Planck scales and the number of events that can potentially happen at a fragment of time is extremely large, contributing to indeterminism. Thus, although it is a highly deterministic theory, it retains and even explains the indeterminism of Quantum mechanics.

At appropriate levels, the theory can completely reproduce Newtonian mechanics, MOND and GR and thus is in good agreement with each of them. It is also in good agreement with most aspects of SR and GR except for a few minor disagreements like the limit of velocity of transfer of information. Although it is in agreement with the fact that any particle with mass cannot travel faster than light. The velocity with which gravitational influences travel also varies in it compared to what is assumed in GR. Because GR agrees that there is no real limit to the velocity with which space can expand or contract, the disagreement is not significant.

The theory derives some assumptions from Loop Quantum Gravity (LQG). However, it differs from it significantly in that there are no spin networks and that there do exist smaller distances than PL and smaller moments of time than PT. It is apparent that most disagreements arise from the inability of LQG to give up the relativity of time.

DGR has some parallels with String theory in that even it has extremely small curled-up regions of space-time. Significant differences arise in the fact that String theory relies heavily on the geometry of these curled-up spaces at Planck lengths to define properties of substances while completely failing to describe what causes these curls and in fact assumes that these curls or extra dimensions are an integral part of the structure within which the Strings vibrate and give properties to the particles. It thus predicts a Multiverse with multiple possible curvatures or Calabi-Yau manifolds possible. The manifolds are not dynamically changing and have no space-time contraction or expansion. If looked closely, the charged particle predicted by DGR having a negative energy lattice of many negative energy particles around a dense positive energy core has an actively changing curled-up space-time around it.

One of the foundational assumptions of DGR, the existence of Quantum foam comes from Quantum Electrodynamics or QED. That said, DGR comes at loggerheads with certain aspects of QED in that the electromagnetic forces in QED are caused by the exchange of photons which are localized quantum virtual particles. DGR would predict that Electromagnetic forces arise because of zones of active spatial contraction or expansion between the particles, instead.

The Photons drawn as lines in a typical Feynman diagram, in DGR would become spherical changes in time resetting waves arising from positive energy core and negative energy lattice, that can get positively and negatively reinforced and can also form an interference pattern.

Much work is needed in this regard before firm conclusions of agreement or disagreement within these can be made.

DGR can potentially recreate the Standard Model of particle physics. Recreating all the properties of all the particles described in it will need much more work. The only two variables at hand that are available with DGR are "e" i.e. the number of events happening at the core of the particle which represents mass and types of particles interacting or types of EPCAs produced. With these two variables, a wide range of particles with varying properties would emerge without effort.

There are two approaches to theory development. The first one being observing the phenomena, gathering experimental description, and then formulating a theory and describing the Mathematics of the theory. With this description, one can then derive predictions and test if the predictions can get experimental confirmation. This approach was used by Galileo, Newton, and also Einstein. This approach requires observing phenomena. With the description of Quantum mechanics and superposition, Heisenberg's uncertainty principle, and wave-particle duality, this method of theory development has become extremely difficult. This is because if we don't know where the particle is with certainty, we don't know if the particle, at the time of measurement, is a wave or a particle and we have the uncertainty of its position, momentum, and even spin, it is difficult to proceed with actually observing phenomena for conceptualizing the theories.

The other method of theory formation utilized by Paul Dirac, Higgs, Feynman and Pauli, etc. (for describing Antimatter particles, Higgs boson, Quarks, and Neutrinos) is to observe the experimental data, derive a mathematical description of the same, and on this basis derive predictions of how the theory would be. With this, Pauli after observing Beta-decay predicted the existence of the Neutrino, Paul Dirac after his famous description of the relativistic equation of Quantum mechanics predicted the existence of antimatter.

This second method, however, involves a lot of complicated mathematical operations, and the fact that some experts believe that current physics is "lost in Maths" is because most physicists are trying to derive their theories based on this second method.

The current theory is derived purely by the first method and requires only basic Mathematics including summation and basic geometry.

The likelihood that just by employing this methodology, one can get an accurate description of reality was indeed small and this will contribute to the scepticism regarding DGR.

The basic assumptions of the theory are novel and no theory in my knowledge has such a description of phenomena of Creation/Destruction and Compensation. However, as the new theory was extrapolated to previously known experimental results and theories, I realized that many of these concepts are not new and at some point in history, something similar was already described.

For example, Gravity being an emergent phenomenon, emerging out of multiple events at the microscopic level have been described by multiple authors before (*Guillerminet, Bernard – 52, Padmanabhan T - 75, Sakharov - 81*). One of the particular theories worth mentioning is "Entropic gravity" by Eric Verlinde (*Verlinde, Eric - 101,102*).

The existence of miniature black holes has been first proposed by Stephen Hawking (*Hawking, Stephen – 56*) and since then multiple authors have published various articles including some saying that miniature Black holes might be a good dark matter candidate (*Scardigli, Fabio – 90*).

Louis de Broglie had suggested that reality is highly nonlocal and that particles (Electrons) are real and have a real and localised position (Realism) regardless of the presence of an observer (and are just riding on independent waves like a surfer) rather than the probabilistic nature of Quantum mechanics with Electrons being wave and particle at the same time and in Quantum Superposition of multiple states at the same time. (*de Broglie, Louis 21, Dewdney, C – 25*). This was further developed by Bohm as Bohmian mechanics (*Bohm, David – 16*)

The concept of Variable speed of light (VSL) was first given by Einstein as early as 1911 (*Einstein, A – 39, Rosen, N - 82*) but went into oblivion and wasn't given any attention by the physics community. The concept of Inertia being due to Gravitational influence exerted by the mass of the rest of the Universe (Mach's principle) was given by multiple authors before (*Dirac, Paul – 30, Sciama, D. W – 91, Unzicker, Alexander – 95*) most notably Paul Dirac in his Large Number Hypothesis. Gong et al attempted to describe how this Einstein Dicke Cosmology with variable speed of light might look (*Yungui, Gong – 51*). Unfortunately, due to the ugly implications of these rather revolutionary ideas on the then accepted beliefs, (most of the physics would have had to be revised if this was true), it wasn't given much importance. Quite literally, the physics community couldn't give up their prejudices and accept their mistakes. This lead to Physics remaining lost for almost a century.

The fact that Newtonian dynamics and General Relativity are not entirely correct was established multiple times by multiple studies which established flat rotation curves of peripheral stars of most Galaxies and the extreme speeds of Galaxy clusters which meant that there is an additional yet unknown or unexplained (Dark) source of Gravity (*Babcock, H. – 13, Rubin, Vera - 86,87*). It became clear that Gravity worked slightly differently at large distances when the Baryonic Tully Fischer relationship was established. (*McGaugh, S 67,68*) which meant that the inwards gravitational pull on peripheral stars of a Galaxy was dependant on the mass of the whole Galaxy and independent of the distance, unlike what General Relativity would predict. But instead of correcting the theories, the Physics community chose to invent "Dark matter" as undetectable particles which interact only through Gravity. Milgrom's Modified

Newtonian Dynamics (MOND) (*Milgrom Mordhei – 69,70,71,72,73*) which was an attempt to correct the mistakes in Newtonian dynamics at great distances, met with significant resistance. Despite the resistance from mainstream physics, MOND now cannot be completely ruled out.

Negative mass is a conceptual idea that is consistent with Relativity and has been described in the past by many authors and their properties were already known for a long time (*Bondi, H – 12*). Because particles with negative mass do not fit the preconceived notions of the Physics community, not much attention was given to the concept. However, some physicists like Farnes have described how small particles with negative mass dispersed throughout the intergalactic space could explain the expansion of the Universe and had already described how such a cosmology might look. (*Farnes J.S – 42*)

Hubble described the redshift of Galaxies and established that the Universe is expanding (*Hubble, E – 60*). Friedman had described an expanding type of cosmos from Einstein's field equations much before this (*Friedman, A – 43*) for which Einstein had to add a cosmological constant to his equations. Lemaître, G described a uniformly expanding Universe with a stable mass (*Lemaître, G. - 63*). Penzias and Wilson discovered the Cosmic microwave background in 1965 (*Penzias and Wilson – 77*). Since then many teams of physicists have been involved in obtaining detailed maps of the CMB. (*Abbott, B et al – 4, The Planck Collaboration 2020 -78, The Planck Collaboration 2014 - 79, Smoot et al – 94*)

To explain the uniformity of the Cosmic microwave background and also the uniform appearance of the Universe, the theory of cosmic inflation (inflationary Cosmology) at the early age of the universe i.e. 10^{-32} seconds after the big bang and just lasted 10^{-36} seconds was put forward. It was contended in it that the Universe expanded exponentially due to some yet unknown entity of Dark energy. (*Guth, A H – 53, Cirigliano, D - 19*) The uniformity of the CMB and the Universe at large is a problem only because in General relativity, information cannot travel faster than the speed of light and thus different points in the early Universe would have no way to communicate their Energy content or temperature to other distant points moving away from it faster than the speed of light. With DGR having information travelling faster than the speed of light, all this confusion of uniformity of cosmos and CMB can be explained without additional assumptions of the inflationary epoch.

Soon evidence mounted that not only the Universe is expanding but its expansion is accelerating. Thus not only at an early age, but even today the Universe is accelerating faster than it was expanding at the time of the release of the Cosmic microwave background. The riddle of Dark energy responsible for this accelerated expansion is not solved yet. (*Peebles, P. – 76*) The spontaneous creation of Baryonic matter particles from Energy, in DGR, means that the value of the mass of the Universe keeps increasing with the age of the Universe and that G is variable and thus the outwards pull on the two Gravitating objects, is much lesser at the time of the release of the CMB means that the Gravitational pull was much stronger at that period and this does explain why the Universe was expanding slower at that time than it is today. (*Unzicker, A – 97*)

Very recently Khoury et al described that Dark matter could be like a superfluid with very little viscosity *(Khoury, Justin – 62)*. A superfluid is something fluid like i.e. it can take the shape of the container and that the individual particles are hardly interacting with each other or not mutually bonded to each other. Negative mass particles described in DGR i.e. ENEA particles would have similar properties with individual particles mutually repelling each other. Thus DGR can explain multiple observations without the need for additional assumptions as far as Cosmology is concerned.

There are enough unsolved riddles on the other side of the spectrum i.e. the extremely small. The interpretation of Young's double-slit experiment created the weird world of Quantum mechanics with the unintuitive Quantum Superposition and Entanglement *(Young, Thomas – 105)*. Dirac's prediction of antimatter particle *(Dirac, Paul – 28)* which was confirmed by Carl Anderson *(Anderson, Carl D – 6,7)* started the "Matter antimatter asymmetry problem" with the question "Where is all the antimatter?" *(Rogers, P – 83)*. Also remains unsolved is the wave-particle duality of matter and wave-particle duality of light. Dirac's equations predicting the negative energy of the Electrons, to explain which, he proposed the Dirac's sea, still lives on *(Dirac, Paul – 29)*. After the description of Black holes, there came more enigmas as to what is there within a singularity and what is the entropy of a Blackhole *(Bekenstein, A – 14)*. Whether information is lost within a Blackhole is still unclear *(Hossenfelder, Sabine 54)*. Each of these riddles remains unsolved. DGR solves all of these elegantly without any additional assumptions.

The currently accepted theories of Quantum Electrodynamics *(Dirac, Paul – 27, E. Fermi – 44, Feynman, Richard - 45)* and Quantum Chromodynamics *(Gell-Mann, M – 47,48,49)* and the Weak nuclear force *(Landau, L -64)* have many questions unanswered including multiple constants that need to be assumed and an almost endless list of assumptions that the theories demand including new exotic particles and exotic charges. Gravity as one of the fundamental forces remains separate and there is no sign of inclusion of Gravity in the Standard Model. The unification of terrestrial and cosmological Gravity by Newton, unification of Magnetism and Electricity by Maxwell and unification of Electromagnetic force and Weak nuclear force by Steven Weinberg and others *(Weinberg, S - 103,104, Glashow, S – 50, Salam, A - 89)* had created hope of unification of all the fundamental forces of Nature into a single force which is still not achievable by current theories. The anomalous behaviour of the Weak nuclear force with the demonstration of CP, CT and CPT symmetry violations by various particles like kaons, mesons etc. remains a mystery. *(Aaij, R – 1,2, Abe, K – 3, Lee, T. D. – 65)*

The question of what is the most fundamental constituent of the matter remains and attempts to theorise these remain in full swing *(Sabine Hossenfelder – 55)*. Loop Quantum Gravity says that space-time is fundamentally made up of loop or spin networks *(Rovelli, Carlo – 84)* String Theory says that it is the strings of energy vibrating in 11-dimensional hyperspace.

DGR can prove to be an overarching theory that bridges all these theories and explains much more with a minimum number of assumptions.

The success of DGR is in the fact that with a handful of basic assumptions, it can successfully explain or solve many hurdles that current physics faces. It can successfully recreate Einsteinian GR and Newtonian Gravity. At great distances, it can explain MOND, flat rotation curves, and a0 and even offer a potential explanation for the Tully Fisher relationship. Thus it can be a potential candidate theory, that over a period of time forms the universally accepted theory of Quantum Gravity.

It can offer a potential explanation about abstract concepts like Inertia, Momentum, and Kinetic energy. It can offer a satisfactory explanation of the weird findings of the double-slit experiment without the real need for a superposition i.e. the electron travelling through both slits at the same time. It can also explain the wave-particle duality of light as well as the wave-particle duality of matter. It can solve the Blackhole information paradox and explain how information loss does not occur at the centre of a Blackhole and also offers clues to the events happening at a singularity. It offers a potential explanation for Dark matter and Dark energy. It offers clues to solve the problem of Matter-Antimatter asymmetry problem. It gives an excellent description of energy and divides it into positive and negative energy. It also provides an explanation about most findings of GR including gravitational time dilation, Gravitational lensing, and Gravitational redshift. It has the potential to unify all the fundamental forces of nature and thus gives a potential window towards a description of a "Theory of Everything".

A lot of work extrapolating DGR to get predictions explaining thermodynamics, Quantum electrodynamics, electricity, and magnetism, Electromagnetic forces, and chemistry, weak and strong nuclear forces, and Quantum Chromodynamics (QCD) (just to name a few) needs to be done before one can extract a Theory of Everything from DGR. However, DGR's full description isn't completely ready yet and many aspects of its application to the above-mentioned fields are pliable enough to be adjusted to fit experimental observations, in case of disagreement with the observations.

Performing the calculations that DGR demands and deriving the complete mathematical description of DGR is also going to be a challenging task.

Although the phenomena that are described in DGR are in the Quantum realm and are such that no potential mechanism exists for them to be observed or confirmed, the theory has the potential to offer multiple testable predictions. This, however, will be possible only after detailed computer simulation models are ready which will be hopefully available in the future.

PART III

EVOLUTION OF THE THEORY AFTER THE CONCEPTUAL IDEA OF EPCAS CAME

Chapter 1

UNDERSTANDING LOCAL TIME

One of the more controversial and conflicting assumptions of our theory is that there are two different types of time, i.e. the Universal Time and Local Time.

This is in contrast to the theory of Quantum mechanics in which there is a single universal independent entity of time that is constantly present in the background and all the events happen with the flow of time happening at a universally constant forward progression.

This is also in contrast to the General theory of relativity in which there is a highly variable time intricately linked to space so that every point in space can have a different measure of time and any change in the measure of time is intricately linked to changes in the curvature of space.

Because of this conflict between General Relativity and Quantum mechanics of background independence of time, it was in a way obvious that either of the two theories is wrong or incomplete or both are right and there are two independent "time" entities.

In our theory, the Local Time is the time akin to what time is in General Relativity and Universal Time is the time akin to the time in Quantum mechanics. That is, in our theory, Universal Time is a universally forward progressive universally constant background entity and Local Time is the highly variable local entity intricately linked to space.

To understand Local Time, we can use the analogy of a pendulum.

Let's consider a spherical pendulum, its centre corresponds to a point C. We plan to use this pendulum to measure time.

The way we would measure time with this pendulum is by setting it off to motion. As the pendulum swings, it moves from C to a point R on the right and a point L to the left.

Because of Gravity, the pendulum is constantly pulled downwards and at both R and L, it has a resultant pull towards C which causes it to swing. Depending upon the length of the thread and the Gravity of the planet it is on, the periodicity of the swings is decided.

If we already know how much is 1 second (say with another atomic clock), we can count the number of times our pendulum swings in one second.

Now if we hide the atomic clock and we have no other means of measuring time but the pendulum, we can calculate how much time elapsed by counting the swings of the pendulum. For this, we have to keep the pendulum from stopping by giving it a slight extra push.

That's how a pendulum is used to measure time.

The first step which we did to compare the periodicity with "a second" measured by an atomic clock, is called the process of calibration. Every measuring instrument needs calibration.

Once calibrated, our pendulum can keep swinging and if its periodicity is maintained, it can be used to calculate subsequent time.

One might ask, what time is it measuring?

Is it Local Time or Universal Time?

The answer to this question will be given soon.

But one has to note that all the other types of clocks utilize a similar method.

Almost all clocks have an event that happens periodically. We measure the event and compare it with a standard to calibrate it or know its periodicity with respect to the standard and then start using it.

This is true for a sundial or a quartz clock. In the sundial, the periodic event is the position of the Sun. In a quartz clock, the vibrations of a quartz crystal.

In an atomic clock, the periodic event is the vibrations of a Caesium atom.

(Note that in all of these, the standard measure is needed for calibration.)

Now let's further imagine that our pendulum is in a box, and we make this apparatus rigged by keeping an electromagnet close to both L and R points. We will also assume that the pendulum ball is made of ferromagnetic material like iron and thus would be attracted to the electromagnets.

Let us set the small hidden electromagnet to be arranged in such a manner that it gets activated only during the pendulum's swing towards point C i.e. when it starts moving towards the centre after reaching the ends.

One can realize that the electromagnet will pull the ball and slow it down. This will reduce the periodicity of the pendulum swing.

Let's give this rigged apparatus to our friend scientist X.

During the initial experiments of Mr X, the electromagnet is off, and so the error in measurement of time would be negligible.

As Mr X starts trusting the time measured by the apparatus, we start the electromagnets so that the swings slow down.

What will effectively happen is that the pendulum seems to swing slower and thus the periodicity seems reduced. Every second, the pendulum will make a smaller number of swings. But X is unaware of the rigged apparatus. So he presumes that the apparatus is showing true time.

However, soon when he corroborates his findings of measurement of time with the pendulum with the measurements done by the atomic clock, he realizes the error. He concludes that the pendulum is slowing down or the time measured by the pendulum is dilated one, i.e. one second measured by the pendulum is longer than one second measured by the atomic clock.

Note here that, every time the pendulum reaches L or R, an extra force starts which pulls the ball towards the adjacent electromagnet and slows it down.

This is true even if both the electromagnets are constantly left on so that they keep pulling the ball towards themselves.

Now let's imagine a Caesium nucleus/atom within the atomic clock. Let's assume that C is the centre of the atom.

We know that the Caesium atom vibrates rapidly with an extremely precise periodicity and these vibrations are the basis of the measurement of time in an atomic clock. That is how an atomic clock measures time. A Caesium atom vibrates 9,192, 631,770 number of times in a single second.

This means that a single oscillation is completed by the atom in $1/9,192,631,770^{th}$ part of a second.

Let us consider one such oscillation when the atom starts at the centre C and goes onto the right until it reaches point R and comes back to C and overshoots to reach point L and back. I.e. CR-RC-CL-LC.

What one should focus on at every instant during its journey is the distance between the instantaneous location of the atomic centre and the two corner points R and L.

Now like our pendulum, what if Nature rigs the entire surrounding of this Caesium atom. What if, due to some reason, the space in between the instantaneous location of the centre and the corner points is constantly in a state of contraction? What if space here has more destruction happening than creation leading to active spatial contraction.

What will effectively happen is the slowing down of the periodicity of the Caesium atom and thus slowing down of the time measured by our atomic clock.

Our theory says that the mass of the Earth means that there is constant destruction going on at every particle that constitutes the planet. <u>To compensate for this rapid destruction, EPCAs start forming around the Earth.</u>[1*]

> 1*- Refer to the Chapter on "Destruction as a compensation to avoid a void" for explanation of formation of an EPCA (Eddy Planck Compartment Aggregate)

If the Caesium atom is kept in one such EPCA, the constant destruction happening around it would act like the rigged pendulum.

This inevitable apparent slowing of time varies depending on where the Caesium atom is. If we take the Caesium atom from the surface of the Earth onto a satellite at height H where the active spatial contraction within the EPCA is lesser, the time dilation caused due to EPCAs here is much less and thus the apparent time dilation would be measured as lesser.

Note that the EPCAS are ever-present and almost never go away. So until the Caesium atom is on the Earth, it will be slowed down by the EPCAs.

To minimize this effect and to get a true measure of time, we may have to measure the Caesium vibrations after taking it at a place beyond the Galaxy in the intergalactic region. Unless we do that, we cannot calculate the Universal Time. Even here in the intergalactic regions, there is no guarantee of a complete absence of EPCAs especially the EPCAs starting from the ENEAs present here which give an opposite push to the Caesium atom and actually makes its vibrations quicker.

This EPCA-led slowing down of all the periodic events means that we cannot get a standard for calibration as the standard itself is also rigged.

Local Time or the time our time measuring instruments measure here on Earth is an illusion. Although it is an illusion, it is an absolute and persistent irrevocable illusion.

Note that the C/D ratio and with it the tendency of space to contract reduces as we go away from the Surface of the Earth. This means that the rigging is more effective nearer to the surface and Local Time would be more dilated on the surface than at a distance.

If we take the same Caesium atom near a supermassive Black hole with a mass equal to million solar masses, near this massive object, the EPCAS forming to compensate for the massive destruction taking place at the centre need to have a much higher deviation towards destruction which means that the EPCAS here would rig the Caesium atomic clock apparatus even more. At such places, the atomic clock will show an even more dilated time.

Now let's look at another time-keeping apparatus described in Einstein's thought experiment - the photon clock.

A photon clock is two mirrors facing each other at a fixed distance (say x metres) and a single photon bouncing in between them. Let this photon travel in the X-axis from a point L on the left mirror to R, a point on the right mirror. Let the instantaneous location of the photon be P. So at every instant in its journey, the LP and PR distances are critical.

We know that a photon can get affected by changes in space. (note that here, there is a rather unpardonable assumption that a photon is a point particle and that it does not occupy space).

It is a bit unclear how the photon will behave in a situation similar to the rigging of the Caesium atom by EPCAs described above. What is likely is that the Photon would be slowed down by the contracting space just behind it. That is when the photon is travelling towards

R, the space LP contracts due to EPCAs and slows the photon. When the photon is travelling towards L, PR contracts and slows it down.

If this is true, even the photon clock would be capable of measuring just the Local Time and it won't measure Universal Time.

Explaining Local Time dilation due to travelling with high speeds like travelling close to light speed:

Let's keep the above Caesium atomic clock apparatus or the photon clock apparatus in a spaceship that is theoretically capable of light-speed travel. Let the spaceship travel along the X axis from left to right with the speed of light.

During this theoretical travel, our spaceship and each of its constituent particles have to eat up or destroy the immediately adjacent PCs towards the right and there is a void formation immediately to the left. This void will have to be compensated with EPCA formation with dilated Local Time on the receding or left edge. Each of these EPCAS, unlike the gravity-related EPCAs, will pull the individual particles to the left, in a way preventing the spaceship from moving at light speed.

"What happens on the proceeding border?" is unclear. Possibly, due to the destruction happening at every particle level, EPCAs form even here. However, there appears to be an asymmetry in the formation and time dilation gradients of EPCAs on the proceeding and the receding border. Thus there is a constant effective pull towards the receding border. This excessive pull on the left is how our theory explains relativistic mass or mass due to velocity.

It is interesting to see what will happen to this photon clock during our spaceship's light-speed journey. While going from left to right, the photon is traveling in the direction of the spaceship and while going right to left it goes exactly opposite. While going from right to left, the periodicity is hastened but while going back from left to right, the photon has to cope up with the increasing speed of the spaceship and as soon as the velocity of the spaceship touches the speed of light, the left to the right journey of our photon can no longer reach its right mirror and the photon clock would stop ticking. This stopping of the photon clock or for that matter vibrations of the caesium clock during the Light speed journey, is, however, a phenomenon of relevance or can be experienced by the astronaut within the space-ship alone. At best, it can be described as an illusion which the astronaut experiences.

It is only the Local Time that has slowed down or stopped ticking. The entire spaceship has become like a T=0 compartment and has stopped experiencing Local Time.

Universal Time would however keep running in the same manner within, just like elsewhere in the Universe, just that there is no way for us or the Astronaut on the ship to measure it.

CHAPTER 2

IMPLICATIONS OF OUR THEORY ON EINSTEIN'S SPECIAL RELATIVITY

Our theory is based on (or is an extension of) both the Special and General theory of relativity.

Thus it is compatible with most postulates or consequences of Special Relativity.

Like Special Relativity, our theory agrees with the relative invariance of laws of physics at inertial frames of reference and the relative constancy of the speed of light.[1#]

(*Miller, Arthur - 74, Jammer, Max – 61*)

> 1# - Much later, it was realized that the speed of light which is 1 PC per 1 PT is in fact constant but this constancy is relative and changeable. In fact, DGR goes in line with Einstein's VSL theory (Variable speed of Light). The Local Time dilation that happens is essentially a result of the active spatial contraction just behind the photon which leads to a reduction of the speed of light and a much slower propagation of the photon leading to the recording of a dilated time by the Planck clock.

The unequivocal consequence of the Special Relativity of space and time merger into space-time is also maintained. The theory manages to explain why space and time are wedded and should be seen as one. In reality, although space and time are connected, they are still separate entities and aren't really the same entity called space-time unlike what Einstein stated in Special Relativity.

It agrees well with the various other consequences of Special Relativity like mass-energy equivalence, the speed of causality and relativistic mass.

Other consequences like length contraction need further discussion which will be done at length later.

The theory differs from Special Relativity on several fronts.

The relativity of time, a fundamental postulate of Special Relativity, is different in Special Relativity and our theory. In Special Relativity, this relativity is absolute, and in our theory, this relativity isn't absolute.

This means that there is no concept of an absolute Universal Time in Special Relativity so that there is only one time and thus it is indeed going to be variable everywhere. This elasticity of time is the fundamental feature of Special Relativity.

In our theory, on the other hand, the relativity of time is itself relative and not absolute. This means that it applies only to one entity called Local Time while the other integral entity of Universal Time is still absolute and thus does not vary anywhere. This means that "a second" (or for that matter any other unit of time) of Universal Time is absolute everywhere in the Universe. And Universal Time would be expected to progress relentlessly everywhere regardless of the Local Time.

Our theory adds its own consequence of equivalence of "deviation of time" and "energy" which is distinctly different from Special Relativity.

In Special Relativity, there is no universally applicable coordinate system. This absolutely relative space which is married to time to produce space-time of Special Relativity is different from the dynamic space and time in our theory.

Movement or position in Special Relativity is relative. This means that we cannot describe the exact location of a particle unless we have some landmarks. This means that there is no absolute up direction or down direction. And up and down change depending on the coordinate system you chose i.e. it is relative. This does seem logical.

Relativity of movement is comparatively less intuitive but proven fact.

To understand this, imagine that you are an observer O on a planet. You are watching a spaceship that is floating at a distance h from the planet surface. To you, the spaceship seems static and thus you conclude that it isn't moving. However, it could be possible that both the observer (with the planet he is on) are also moving with the spaceship at the same time. To know this, we will have to change the frame of reference and examine a bigger frame of reference, i.e. maybe look for a bigger landmark like the galactic centre or the solar system. With these as reference points, we can say that they are both moving.

Now let's come back to the reference frame of the observer O.

Now imagine that the spaceship starts moving spontaneously away from the planet. The observer sees this and concludes that the spaceship is in motion. This could be a fallacious conclusion as it may be possible that the planet and the observer with it are moving in the opposite direction. Thus "movement" is relative and highly dependent on the frame of reference. Special Relativity says that there is no absolute reference frame.

However, our theory says that although all the above is true, there is an absolute frame of reference. The frame of reference is having extremely small individual units i.e. 1 Planck length. Thus the position of an object can be determined to as much accuracy as 1 Planck length. Due to this small unit, the space is highly invariant and the universal coordinate system is highly elastic and can be rotated or moved without any consequence to the tune of 1 Planck length.

This means that it would theoretically be possible to know the absolute motion the Earth is currently engaged in. This includes all its "motions" one can make out, in all the reference frames possible.

This is because as the Earth moves, it moves with respect to the static Planck Compartments of the space and it would recruit fresh PC₁S.

The inertia of the Earth or for that matter any "body" or particle can be explained on the basis of this. Even the kinetic energy of a moving body depends on the velocity of the object with respect to this absolute co-ordinate framework given by the Planck Compartments.

This doesn't go well with Relativity.

Length contraction and time dilation due to motion

Special Relativity states that when a body moves at a high speed, it experiences time in a dilated manner and when its speeds are close to the speed of light, its length contracts. Length contraction happens at very high velocities that are close to the speed of light and thus is, to a great extent, theoretical as Special Relativity itself forbids speeds equal to or beyond the speed of light due to an increase in relativistic mass and infinite energy required to continue its acceleration. However, time dilation due to speed is real and accurately measurable and needs to be taken into account routinely especially in GPS and other satellites.

Our theory suggests that it is the Local Time that shows time dilation while the Universal Time goes ahead relentlessly with the same pace.

The photon clock is useful to understand what happens in a spaceship travelling close to the speed of light.

As explained elsewhere, a photon clock is a theoretical arrangement of two mirrors with a photon bouncing between the two. The distance between the mirrors could be set as anything. Let's arbitrarily assume that it is 10 Planck lengths. It would then take the photon 10 Planck times to reach from one mirror to the other (assuming of course that our theoretical photon is single Compartment one i.e. it occupies a single Planck Compartment.

When such a clock is kept in a spaceship that is accelerating close to the speed of light, the photon has to take a longer route with multiple Planck lengths instead of the 10 Planck lengths as it has to take an angulated path.

Exactly at the speed of light, the photon would stop moving and remain stand-still which means that the ship is travelling at speed the same as the photon. To the astronaut travelling within it, all light would cease to flow and Local Time would cease to move forward, i.e. To the astronaut time would stop. To the observer on Earth, the time would continue. Let's say that now the astronaut slows down the ship, the photon clock starts ticking again and the photon starts bouncing again.

The twin paradox

Special Relativity has a single time. It also has no absolute spatial coordinate system. Due to these two problems, some paradoxes result. One of them being the twin paradox. *(Debs – 20)*

In this, imagine that two twin brothers A and B, are respectively present on the surface of the Earth at his house and in a spaceship floating around the Earth. Both are presently stationary with respect to one another.

Let's imagine that this is an imaginary Earth-like planet and that the surrounding space contains no other landmark like the Sun or the Moon or the Galaxy.

In this case, neither of them can compare their position with respect to anything else other than each other.

Let's assume that the spaceship (without the knowledge of the twin B) starts moving fast away from the planet. Both A and B will see each other moving away from each other.

Logic says that as the twin B is moving away from the planet on a space ship, he would experience time dilation and would age slower. Let's say he stays for a year in his journey and then returns. By this time, logic says that the twin A has remained stationary and thus should have continued ageing with the original or faster pace. So A would have aged 5years instead of 1 year.

However, Special Relativity differs from this logic.

Since there is no universal coordinate system in Special Relativity, both can say that "I am stationary and the other twin is moving". Because both are moving with the same velocity away from each other, both should have time dilation due to movement. Then which twin is ageing slower and which one is ageing faster?

The paradox doesn't arise in our theory as the planet can theoretically remain stationary with respect to the universal coordinate system around it in the form of Planck Compartments in which case there wouldn't be any recruitment of fresh PC'S. Although practically impossible, in theory this is possible. The spaceship, in contrast, is recruiting fresh PC'S with a high speed and thus would have its on board photon clock slowed down. According to our theory, only the Local Time would stop and Universal Time would continue. This means that both the twins would age at the same rate as Universal Time would pass at the same pace for both the twins.

Length contraction

Talking of length contraction at speeds close to the speed of light by a space ship is just theoretical crap as traveling with speed of light can never be achieved.

Our theory suggests that every particle which makes up the space ship, when travelling with the speed of light would have fresh formation of T=0 Compartments at the receding border of the particle which will add to the mass of the particle. Thus every particle will become more massive.

More importantly, the particles that constitute the space ship would be linked together with forces like Electromagnetic force between Electrons and Protons. <u>Each of these is made possible by resetting waves which start from the particle and which presumably travel at</u>

speeds of light.²# It would imply that at speeds close to speed of light, not only the energy content of individual particles and thus their masses change, but the forces which are keeping them together would no longer be possible as every particle bonding to the next one would travel with the speed equal to the speed of the resetting wave emanating from it to maintain the bond. This indicates that at speeds close to or at speeds of light, bonding of particles together to form a space ship would be impossible and the space ship will be ripped apart from its constituents. As the spaceship acquires the speed of light, its mass increases tremendously and tends towards infinity. At the same time, local time in it is turned to T=0. Essentially, the spaceship will be turned into a Black hole with extremely high mass (tending towards infinity). The individual particles might start collapsing into a point.

> 2# - Later it was realized that the Push Pull bands responsible for electromagnetic, weak nuclear and strong nuclear forces have to necessarily travel significantly faster than speed of light while forming. This would mean that this assumption that faster than light speed would disrupt all the bonds seems unlikely. Whether the push-pull bands survive this or not remains a mystery.

Gravitational time dilation

A special feature of the Theory of General Relativity is gravitational time dilation wherein time runs slower nearer a gravitationally large body.

This is described as one of the means of time travel.

Imagining the same twins, A and B, wherein A stays at home and B goes for a voyage in space.

General Relativity predicts that the astronaut B, on his journey close to a supermassive black hole, would age slower due to gravitational time dilation. As even General Relativity has a single concept of time, when the twin B returns back, A would have aged much more than B and effectively B has travelled faster in time or travelled to the future.

Our theory would predict that at the region close to the supermassive black hole, the Local Time would run slower which would lead to the active spatial contraction leading to the high gravitation of the black hole. However, the Universal Time would run in the same manner and our twin B would age the same manner as the twin A. From Universal Time point of view, their age would remain same.

This essentially rules out time travel.

Travelling on a space ship beyond our Galaxy limits into a different galaxy

Both the General and Special Relativity do not forbid travelling beyond the limits of the Galaxy except for the universal speed limit.

However, our theory predicts that beyond the limits of our Galaxy and in fact well beyond the local cluster or other Galaxies to which our Galaxy is gravitationally bound, there exists regions of space which are having Local Time running very fast. These regions which are responsible for the expansion of space probably have very high rates of appearance of new Planck Compartments leading to spatial expansion. <u>A spaceship entering such a region would be ripped apart of all its individual constituents as the individual molecules making the space ship cannot resist the rapid spatial expansion happening between them. The typical bonds between particles due to Electromagnetic force are impossible to form here and the particles cannot form clusters here. This can possibly set a limit to the distance humans can travel through space.</u>[2#]

> 2# - Again to note here that the strong nuclear, weak nuclear and electromagnetic forces, being carried by push-pull bands which form faster than speed of light, as noted later in the development of the theory, will survive the intergalactic travel as was realized much later. However, our fragile bodies and Electronic circuitry of our ships and many things we use like fluids, solvents like water, DNA depend on much weaker forces like hydrogen bonds which may not survive the constantly expanding space in our intergalactic journey. Thus, such a journey would inevitably be lethal for us Humans as predicted by our theory.

These regions however are well beyond the universal speed limit which itself is difficult to break by humans. Even if we somehow travel with the speed of light, to reach the edges of our Galaxy would take thousands of years.

Indeed, for humans and sci-fi movie writers, it is a bad news.

Our theory predicts that like the centre of the Galaxy (where there is a singularity which can rip us apart), there is another no-go zone beyond the galaxy in the intergalactic region. This is a zone where molecular bonds cannot form and is the contemporary of a singularity. It can be considered like a white-hole band where expansion of universe happens.

Is Special Relativity wrong?

There are many experiments that prove that relativistic effects are true. This means that gravitational time dilation and time dilation due to speed were noted by atomic clocks. Are all these experiments or their inferences wrong?

Indeed, not.

All their inferences are true and are valid.

But in all these experiments we measure Local Time which is elastic and can change at places.

Thus relativity is correct but there is more to reality than just relativity.

Physical processes and Local Time

All physical processes including bonding between atoms is inextricably linked to deviation of Local Time and the associated spatial contraction/expansion.

Thus deviation of Local Time at a place would affect every physical processes taking place there.

Most accurate experiments performed to test Special Relativity are done with atomic clocks which tick due to vibration of an atom of Caesium. A system where an atom of Caesium is vibrating, when taken to a Zone of space where time is extremely dilated, this dilated time will affect the vibration of the Caesium atom (which is a physical process).

We do not have any machine or mechanism of measuring the forward progression of Universal Time yet. The Local Time here on Earth may be significantly dilated compared to Universal Time since we are under constant gravitational influence of the Earth, the Sun and all the other stars and Central Black hole of the Milky way galaxy. Thus Universal Time propagation forwards may be significantly faster than our intuition.

Possible method to test the Hypothesis

This brings us to a probable means of testing our theory and confirming its predictions.

General Relativity taught us that Gravitational influences are not instantaneous and travel with the speed of light.

Our theory goes one step further.

It predicts that even the routine Electromagnetic force and nuclear forces are not instantaneous and happen due to similar time resetting waves which travel with the speed of light and which modify Local Time and thus cause local active changes on space that result in formation of bonds.

This means that formation of molecules and inter molecular bonding happens because of such waves travelling with the speed of light and is not instantaneous.[3#]

> 3# - Note that because we realized in later part of the Theory development that these electromagnetic bonds happen much faster than the speed of light unlike what I believed while writing this, **this entire section of "possible method to test the hypothesis" is rendered wrong and possibly useless**.
>
> The discussion, however, is useful to understand the inevitability of faster than light information transfer. It is quite an exercise for our biased mind which fails to accept that faster than light transfer of information can be possible, to imagine how easily the hypothetical EM bonds that form with waves that travel at the speed of light would fall apart at extreme conditions.

Presently, we have particle accelerators that can accelerate smaller charged particles like Proton or Electron to near light speeds. At these high speeds, majority of their properties remain unchanged.

What we need to do is to increase the energy involved in these particle accelerators and accelerate higher molecules like organic molecules probably made of multiple carbon chains or rings with high charge to near light speeds or speeds that are comparable to light. We should have the means to compare higher properties of these molecules pre and post near light speed travel.

It is a prediction of our theory that at speed of light, the bonds just break or are impossible to be maintained. Even at smaller velocities like velocity half the velocity of light, the molecules can show a significant change in their overall orientation or arrangement of atoms. Changes in weaker bonds like hydrogen bonds pre and post light-comparable velocities travel in these higher molecules and resultant changes in molecular geometry may be large enough to be detected by us.

The basic assumption is that at velocities near the speed of light, the bonds which are weaker may not be possible or may breakdown.

This would be especially true for complex composite molecules in which some part of the molecule is electrically charged and can be accelerated while the other part of the molecule is electrically neutral and wouldn't experience force in a strong magnetic field.

To understand this further, let's consider a simple system with two particles A and B which are bound together with a bond.

(Note that the entire discussion is theoretical and is impractical)

This bond, according to our theory happens because a resetting wave that resets the Local Time and makes local space contraction, starts from A and a similar wave starts from B. Let's assume that distance between A and B is 100 Planck lengths. Then the two resetting waves that transmit the Local Time dilation-spatial contraction wave in between them would meet somewhere at 50 Planck lengths i.e. after 50 Planck times.

Now let's assume that B instantly starts moving at speed of light in a direction away from A. (This instantaneous acceleration from absolute rest to speed of light is practically impossible in reality)

The resetting waves starting from B would be highly red shifted as everyone Planck length the wave propagates towards A, B propagates in the opposite direction.

This will have a serious negative effect on the bond between A and B. The bond may break or become weaker leading to changes in orientation of the molecule. Which may be predictable depending on the geometry of the molecule.

This effect may start appearing at much lower velocities than actually achieving the speed of light. That is, a protein chain may have changes in configuration or bond strength on accelerating to speeds as large as half the speed of light.

Imagining a spaceship traveling with light speed

According to our theory, the constituent particles forming the spaceship like Protons and Electrons lose their bonding ability at speed of light.[4#] As all of them are moving at high speeds, due to their inertia, they might maintain their relative positions which they had at the instant the space ship started moving at light speeds.

> 4# - Although the above statement might seem incorrect, given that electromagnetic and other bonds, as discussed later in the theory, are being carried by faster than light travelling push-pull bands, the extreme spatial contraction that happens would probably be similar to the one happening within the event horizon of a black hole. What would happen isn't completely clear. The space in between the bonded particles would start contracting faster than the speed of light and the bonded particles would collapse and form their constituents – i.e. release the trapped energy within, which is similar to the firewall concept, anything crossing the boundary instantly turning into energy and collapsing into itself.
>
> This won't be good news for anybody alive on that spaceship.

But the relative balance of forces between them might get disturbed so that the movement of Electron around the Proton within the nucleus might get disturbed and the Electrons motion may get distorted within every atom that constitutes the space ship or the astronaut. At light speeds, no external light would be capable of being reflected off the Electrons and the space ship would thus become invisible to an outsider. (Note that here we are assuming the theoretical possibility of a spaceship moving with the speed of light, the argument discussed later about the increasing mass of a spaceship like this might change/nullify this whole argument.)

For the astronaut, the time will stop along with the ticking of the on board photon clock. But the astronauts brain would disintegrate into separate constituents and would not be in a position to comprehend that this happened. The astronauts heart wouldn't be able to pump blood, oxygen wouldn't be able to bind to haemoglobin to supply oxygen to his brain and there wouldn't be a finger remaining with the astronaut for his brain to give an order for it to push the button to make the spaceship to slow down. In effect, the inertia of the particles would make the constituent particles to keep moving with the light speed with Local Time running zero. The on board computers would disintegrate and no Electronic equipment would work at these speeds. So the on board computer cannot stop the space ship as well. In effect the astronaut would be trapped in T=0 and the journey would be his penultimate journey. There would be no turning back.

This is if we somehow achieve light speed. For this, one absolute requirement is to prevent each of the constituent particles with mass from giving off resetting waves that prevent its movement and cause inertia at rest, the so called Higgs mechanisms.

Each of the T=0 Compartment within each of the particles of the space ship would be moving at one Planck length per one Planck time. This means that at the receding edge of each of the particles, a void of fresh T=0 Compartments, much like the vacuum of Planck Compartments would be created. Thus at every Planck length translation of each of the particles making the space ship, the T=0 compartments within the constituents of the space ship increase which is the relativistic mass.

<u>At light speeds, the resetting waves starting at the preceding edge would be unable to catch up with each of the particles. But the resetting waves starting from the receding edge would keep moving. For a particle at rest, these time resetting waves emerge from all sides and cause spatial contraction around, which leads to inertia. At light speeds, this spatial contraction would be only at the preceding edge i.e. in the direction opposite to the direction of motion of each of the particles. All these resetting waves and the resultant spatial contraction leads to exponential increase in inertia i.e. all this process would prevent the further acceleration of each of the particles.</u>[5#]

> 5# - Again some faulty line of thinking caused due to the presumption that the Time resetting waves would travel outwards with the speed of light due to the SR bias of "speed limit" set by it.
>
> The TRWs would in fact travel faster than speed of light. Also, the opposition to acceleration that the body would display is because of the void or PC vacuum created just before each of the constituent particle.

At near light/light speeds, not only the integrity of the bonds is destroyed, but each of the now free/separated un-bonded particles becomes exceedingly difficult to accelerate further. That is there inertia and mass keep increasing exponentially. It figures that our theory, thus, agrees well with the theoretical speed limit set by Relativity that nothing can travel faster than light.

Is Alcubierre drive possible?

Theoretical physicist Miguel Alcubierre described engines that utilise spatial expansion and contraction in immediately adjacent spaces as a way to travel faster than light. *(Hiscock – 58)*

Our theory, does not set any limit to the expansion or contraction of space. By giving away time resetting waves, every particle unintentionally keeps doing similar things by modifying the Local Time of its surroundings.

The Alcubierre drive thus isn't a theoretical impossibility, if our theory is correct.

However, for this, we need spatial contraction in the front and spatial expansion at the back while there should be no unbalanced spatial contraction or expansion at the sides or up down direction. This asymmetrical space contraction/expansion is difficult to achieve. For this, probably we may have to find out processes which can separate the PEM and ENEA particles and throw the ENEA particles at the back and PEM particles at the front. However, the massive particles thrown in front would cause a backward push opposing the motion of the space craft.

There are multiple other hurdles including the constant mass attrition of the spaceship as we have to keep extracting particles and losing them.

Another possibility is to extract the positive and negative energy from these PEM and ENEA particles and utilise these. The negative energy involves expansion of space/contracted Local Time and positive energy involves contraction of space or dilated Local Time.

All of this is theoretical, as we don't even know of the PEM or ENEA particles actually exist.

Simultaneity in our Theory vs that in Special Relativity.

Relativity states that there are no simultaneous events and that two observers may differ in what they measure as simultaneous or non-simultaneous events *(Jammer, Max – 61)*. This is because information about whether an event has occurred or not travels with the speed of light which is limited. Speed of light is constant independent of all the observers.[6#]

> 6# - Later it is realized that DGR is in fact hinting towards variable speed of light instead of constant speed of light. And that this limit to transfer of information isn't true.

As there is no concept of separate Local and Universal Time in Special Relativity, simultaneous events need not be simultaneous for every observer and simultaneity is relative and varies according to frame of reference.

In our theory, two events can be simultaneous, occurring at a single instance at two corners of the Universe. Their perception of being simultaneous or not simultaneous is relative and can vary depending upon who is observing and from where. The perception is highly dependent on light and thus limited by the velocity of light.

Let us consider two events A and B.

In A, there is a supernova exploding at the core of a far-away Galaxy millions of light years away.

Let's assume that A happened $10^{10^{10^x}}$ (i.e. 10 to the power 10 to the power 10 to the power x) Planck times of Universal Time after the big bang.

This instance i.e. "$10^{10^{10^x}}$ Planck times of universal Planck time" following the big bang happens all throughout the Universe at the same instance.

At the same instance, on the Earth B, a scientist is sitting in his office and he drops his coffee mug which breaks instantly.

In Special Relativity, time nearer the Black hole near the centre of a Galaxy where the supernova explosion took place is running slower. On the Earth, time is running faster. From God's Eye view or with the entire Universe as reference frame (which is impossible experimentally), one can make out that these are simultaneous events even in Special Relativity, but with a smaller reference frame, say that of the scientist, it is impossible to comprehend the simultaneity. The information that a supernova explosion happened at A will take millions of years to reach to the scientist.

This is in contrast to our theory which says that the non-simultaneity of events is only due to limitation of speed of light and also perceptual glitch.

There does exist, a set of events that are indeed truly simultaneous. But they are simultaneous only because they happen at an identical Planck time following start of Creation.

Where C/D ratio is deviated towards creation, destruction runs according to Universal Time, when C/D ratio is deviated towards destruction then creation runs according to Universal Time. When there is no change in spatial status I.e. C/D ratio is one or Local Time runs equal to Universal Time, both creation and destruction happen according to the Universal Time.[7#]

> 7# - Note that at the start of the theory, the assumption was that destruction is linked to Universal Time and Creation is linked to Local time. The assumption evolved.
>
> The current thinking is that there is no link. That is, the physical processes of Creation or Destruction and the spatial changes they create are fundamental and are the primary events. The time changes are just secondary. That there is only one time and that is the Universal Time and the Local time that we see or measure is just an illusion.

Einstein's equivalence principle

Einstein's equivalence principle states that it is locally impossible to differentiate between acceleration and Gravity.

This means that when an astronaut is in a closed lift, there is no way for the astronaut to know if the downward pull his body is experiencing is due to the Earth's Gravity or is it due to the lift accelerating upwards.

Although practically this is true for our theory as well, i.e. in practical sense, our theory would agree, at the particle level analysis, the two are not equivalent.

In our theory, Gravity results due to a downward pull which is due to the contracting space under the lift towards the centre of the Earth which in turn results from the time resetting waves that start from the Earth. In essence, each particle which constitutes the astronaut and the lift move downwards towards the centre of the Earth and here i.e. locally at the particle level, Planck Compartments will be recruited towards the downward direction of the lift and the astronaut. In contrast, when the lift is accelerating upwards, the Planck Compartments that are being recruited lie at the top direction of each particle forming part of the Astronaut and the lift.

They are thus completely different and not equivalent.

Experimentally testing if the Upper Planck Compartments are getting recruited or lower ones, is impossible and thus practically they can be called equivalent. It may as well happen that the freely falling particle changes its position solely because of the volume changes happening around it, i.e. relatively more vertical spatial contraction towards the Gravitationally active body than the opposite side and that there is no fresh recruitment of PCs happening at the leading edge of the particle. If this is true, the freely falling body and floating body are truly equivalent and Equivalence principle holds.

CHAPTER 3

MATTER ANTIMATTER ASYMMETRY

Matter particles and anti-matter particles

A matter particle, in our theory, would be a particle which has a core made up of positive energy matter, i.e. T=0 Compartments collapsed into Sub-Planck length prisons. Such a particle, would have small negative energy antimatter particles arranged as a lattice around it.[1**]

> 1** - How I reached this model of a charged particle is explained in the subsequent chapters.

An antimatter particle would have it reversed. That means that an antimatter particle would have a dense core of negative energy antimatter which has T=Tmax Compartments collapsed and trapped in a Sub-Planck length prison, like a white hole. Such an antiparticle has a lattice of small positive energy matter particles around them.[2**]

> 2** - Note that later it became clear that T =Tmax compartments do not form aggregates but they stay separate and in-fact repel each other and everything else.
>
> Thus, what we call matter particles in conventional physics correspond to a composite particle made of Positive Energy core and Negative energy lattice.

When two such matter and anti-matter particles come close to each other, the positive and negative energy lattices at the edges that face the two particles come together and annihilate to form energy first which is later followed by the cores coming together to form even more energy, finally the entire structure gets converted to energy.

When positive energy matter disintegrates, it forms positive energy which is nothing but extremely time dilated i.e. T=0 Planck Compartments.

When negative energy antimatter disintegrates, it forms negative energy which is nothing but extremely time contracted i.e. T=Tmax Compartments.

Modern physics tells us that when two matter antimatter particles annihilate they form two photons which look identical as photons are antiparticle of themselves.

However, our theory says that the photons look different. One photon would be a positive energy photon made up of T=0 Compartments that formed a part of the matter particle clubbed together. The other would be a negative energy photon made up of negative energy i.e. T=T max Compartments.³**

> 3** - It still remains unclear what events happen when the matter and antimatter particles, that we know to be composite particles according to DGR, annihilate. Now it is clearer that the energy liberated is positive energy. The miniature ENEA particles liberated are not yet detected or detectable (if detected can be confirmation of DGR). What happens to these is unclear. Possibly the Positron keeps falling towards the Electron due to non-existence of the repulsion by the Push bands around forming a particle with bigger mass and twice as many ENEA particles, which is unstable and splits apart releasing the two T=0 aggregates (energy) as gamma ray photons.

A positive energy photon is an aggregation of a huge number of T=0 Planck Compartments concentrated at a small space, probably Sub-Planck length, a number that is not big enough to collapse into a particle with mass.

A positive energy photon would then be massless and would propagate with the speed of light.

Light or other Electromagnetic waves would then be a series of these positive energy photons propagating forward. The distance between two photons in space would depend on the number of T=0 Compartments (i.e. the extent or rate of destruction happening) in it which would then determine the frequency.

A negative energy photon in that sense would have similar negative energy i.e. T=Tmax Compartments aggregated together.

These aggregations would also propagate at speed of light.

This indicates that like the routine positive energy waves, there must be the negative energy photons and their EM wave counterparts i.e. negative energy waves which are nothing but negative energy photons travelling together with the speed of light

Detecting these exotic negative energy photons and negative energy waves can be a confirmation of our theory.⁴**

> 4** - It was much later that I realized that these negative energy particles or what I called them ENEA particles would not attract each other but repel each other and thus they would not form PC aggregates. Given that they do modify their surrounding PCs and thus create negative energy time contraction type time resetting waves, they can produce something similar to an EM wave which would be just like a train of T=Tmax PCs. However, as they cannot form PC aggregates, the formation of an entire spectrum of EM waves made of negative energy particles, which I thought of here seems unlikely now.

The positive energy photons collapse due to the higher rate of disappearance of Planck Compartments than appearance. Here, the neighbouring Planck Compartments are literally sucked inside. The photon would then be surrounded by Planck Compartments with progressively less time dilation finally having the most peripheral ones merging with the Local Time of the surrounding.

The negative energy photons would have Tmax Compartments which means that these aggregates wouldn't disappear. On the contrary, these Planck Compartments would have a higher rate of appearance of new Planck Compartments than the surrounding so that they will keep pushing newer Planck Compartments into the surrounding and would be surrounded by Planck Compartments that show a progressively smaller time contraction until it merges with the surrounding.

The size of these photons would probably be much bigger than T=0 photons and their properties would also be different.

However, they would also interact with the neighbouring Planck Compartments and thus when emerging out would get entangled to the neighbouring Planck Compartments and with each other and would form wave-fronts in a manner similar to what the positive energy photons do.

<u>Their frequency, like the positive energy photons would be probably determined by the source and possibly has a similar mathematical relation with energy contained, within so that higher the frequency, higher is the content of the positive energy or lower is the content of negative energy contained within.</u>[5**]

> 5** - This seems unlikely now as ENEAs cannot aggregate together.

Matter-Antimatter asymmetry and its counterpart the positive photon-negative photon asymmetry.

As discussed multiple times, the Matter-Antimatter asymmetry is a difficult riddle which physicists are struggling to explain.

Does a similar, difficult to explain riddle of positive energy-negative energy asymmetry exist?

This is another worthwhile scientific question.

Whatever energy we see, is it positive energy i.e. dilated time with active spatial contraction or there also exists active spatial dilation/negative energy?

Is there an abundance of T=0 photons? Are the T=Tmax photons, which cause active spatial expansion equally common? Are they detectable and if so are they more difficult to detect? Have we detected them yet? Is there a way to detect them?

If there exists an equal abundance of T=Tmax photons, why we haven't detected or described them yet.

If there exists an asymmetry, why there is an asymmetry?

These are perplexing questions and their answers could have possible mechanisms of proving our theory right.

To answer some of these questions, let's delve a bit deeper into what we routinely see as energy.

Most common source of energy in the universe

Logic says that the most abundant sources of energy that our minds can see or our eyes can detect are the light or heat given out by stars. These should, by no means be the most abundant form of energy in the Universe. But let's consider them.

The light and heat which are the infra-red radiation are positive energy waves that arise from the constant nuclear fusion taking place in the core of the stars. What is essentially happening in the stars is that hydrogen is fusing into higher elements like helium, nitrogen, carbon etc., and in the process, a small bit of mass is being lost. This lost mass is nothing but unlocking the energy locked in the mass which is then being released. This is due to the weak nuclear force in modern physics.

We know that here in this region of the universe, we find predominantly matter particles that are made up of positive energy. It thus figures that the positive energy locked in the particles is released as positive energy waves.

Even in our theory, the matter particles are more stable and the ENEA particles surrounding the matter core are extremely small, so the more commonly found antimatter is extremely small. When they are lost or annihilate, the resultant negative energy photons would be small and this can explain the apparent positive energy-negative energy asymmetry.

In short, the explanation for apparent positive energy abundance stems from the explanation of the Matter-Antimatter asymmetry itself. Positive energy seems more as there is more Positive Energy Matter present to give off the positive energy.

Explaining the Matter-Antimatter asymmetry

At the time of the early universe, there must be the creation of both matter and anti-matter particles in equal proportions.

Why do we see more matter particles than antimatter particles then?

Where are all the antimatter particles gone?

This is the crux of the matter anti-matter asymmetry problem.

The solution given by our theory is simple.

In the early universe, modern physics says that a high amount of energy was converted to matter and antimatter particles. Many of them annihilated reforming the energy. More antimatter particles were annihilated than matter thus more matter particles remained. But how is this possible if matter and anti-matter particles annihilate in 1:1 proportion? This means if 100 antiparticles annihilate, they must do that with an equal 100 matter particles.

Our theory says that the particles we call matter or anti-matter particles are in reality not fundamental but are composite particles.

There exists no Matter-Antimatter asymmetry.

The missing antiparticles are present within the matter particles as lattices. <u>Individually, their masses are negative, extremely tiny and thus not detectable by us yet.</u>[6**]

> 6** - In fact, these particles, as noted by Bondi et al, have a negative mass.

But they have many detectable effects which we can accurately detect but which we have attributed as fundamental properties of the particles rather than considering a possibility of them being a result of the presence of these lattices.

The particles wherein positive energy forms the core and negative energy forms the lattices are more stable. During the early times of the Universe, when particles formed like this, with a positive energy core and negative energy lattices, they survived.

<u>While the opposite kind of particles, with negative energy core of constantly expanding space with a positive energy lattice made of contracting space fail to form stable bonds and are highly unstable.</u>

<u>The particles annihilate with an equivalent amount of mass rather than their number. This means that an energy-dense core with energy -E can annihilate with n number of particles if all the positive energy contained in them equals E.</u>

<u>Even the converse is true.</u>

<u>That is, an energy-dense core of Positive Energy Matter with energy E can potentially annihilate with n number of negative energy antimatter particles if the sum of all their negative energies equals -E.</u>[7**]

> 7** - This is wrong line of thinking as we now know that the PEP and ENEAs cannot reach each other let alone annihilate each other. Thus the true Matter Antimatter particles do not annihilate each other. It is the composites with positive energy core and opposite charges that annihilate. The whole basis of the matter antimatter model is based on the fact that they don't annihilate but form stable lattice around the PE core.

This, however, doesn't happen in reality and particles with Positive Energy Matter at its core are stable for reasons discussed in detail elsewhere.

The charged particles have a small asymmetry so that negatively charged particles have a slight excess of antimatter lattice particles and thus have a small extra expansion tendency while the positively charged particles have a slight excess amount of matter mass. Even in this scenario, there exists an almost equal number of positively and negatively charged particles in our Universe which explains that the amount of negative energy forming antimatter particles does balance out the amount of positive energy forming matter particles. This means that the deficit of Positive Energy Matter within the negatively charged particles balances out the presence of an excess of the same in positively charged particles.

The Positron which is a positively charged counterpart of the Electron, due to the excess of collapsing tendency due to extra matter compared to antimatter, would arguably fail to remain stable as explained elsewhere in detail. This matches well with the reality, we do know that Positrons are unstable and have a smaller half-life.

The Neutron being electrically neutral has no deficit or excess of either i.e. has no asymmetry of Matter and Antimatter.

The Proton being electrically positive, would overall have an excess of matter than antimatter. The comparable number of Protons and Electrons means that the excess anti-matter in the Electron is balanced by the excess amount of matter in Quarks in the Proton.

Within the Proton and the Neutron, apart from the Quarks, a high quantity of positive energy exists which means that a high density of T=0 Compartments exists within them. We know that both Proton and Neutron have a much higher effective mass than the constituent Quarks can account for. In fact, the Quarks account for just about 1% of their mass. The origin of this extra mass is mysterious. However, the possibility of multiple short-lived matter anti-matter particles which form and re-annihilate within a few Planck seconds is likely. Here there is a likelihood of an overall asymmetry in the form of excess matter than anti-matter which accounts for most of the mass of the Proton and Neutron.

This excess matter within the Proton and Neutron can account for a significant overall imbalance between matter-antimatter. How exactly this process unravels isn't clear but probable mechanisms is given subsequently?

What is clear is that a lot of uncompensated positive energy i.e. a lot of T=0 Compartments are present within the small volume of a Proton.

The Positive Energy- Negative Energy symmetry problem

In modern physics, one of the puzzling questions is whether there exists symmetry between how much positive energy and negative energy or there exists an excess of either.

If there is an excess of positive energy in the universe, the universe would eventually re-collapse into itself in a big crunch while if there is an excess of negative energy, the fate of the Universe would inevitably be a big rip that is the universe would continue to expand forever. It is unclear with current theories to decide which kind of Universe we live in.

Even in our Theory, this is true.

The definition of this problem in our theory is slightly different.

Negative energy is contracted or positively deviated time. Contracted time causes the expansion of space. Positive energy is negatively deviated time. Positive energy leads to the contraction of space. Expansion of space is literally adding more Planck Compartments, while contraction of space causes a relative loss of the Planck Compartments.

Thus the consequences of imbalance are pretty clear.

There are three possibilities. Either the total amount of positive energy and negative energy added together in the universe is equal or there is an excess of either energy.

To understand this further, imagine an intermediate state of the Universe or let's say imagine a hypothetical universe.

Let this Universe be made up of X number of Planck Compartments. We would understand that X is a variable and depends on whether this Universe is expanding or contracting, i.e. more space is being added or being subtracted.

Let Y be the number of Planck Compartments being added due to spatial expansion and Z be the number of Planck Compartments being removed at a given time say every Planck time.

If at any given point in time, Y is equal to Z, this hypothetical Universe of ours would neither contract nor expand.

For having an expanding universe, the Y should be constantly greater than Z.

This means that for a Universe to be expanding, there should be a slight excess of negative energy than positive energy.

Now Imagine the instant of the Big Bang. The first Planck Time after the moment of the Big Bang. (Modern physics cannot tell us what happened before 10^{-35} seconds.) At this instant, there are several possibilities.

We are not sure how many Planck Compartments appeared from nowhere or were created at the first moment of Creation.

We know almost for sure that it would not be a single Planck Compartment as if this was a single Planck Compartment, and if it was a T=0 Compartment, for it to be extremely hot, time would never progress as creation is slower than destruction at a T=0 Compartment which cannot happen.

We also know that there weren't two Planck Compartments as in this case, one T=0 would make the other one T = Tmax which means that there is a balance between creation and destruction, this would mean the number of Planck Compartments created next moment would be destroyed as well.

Let's assume for argument sake that 100 Planck Compartments were formed on this first Planck second.

<u>We know that the temperatures were insanely high. This means that at least some of them were T=0.</u>[8**]

> 8** - It was realized later that these would not be just T=0 compartments but compartments with a hyper-dilated time, i.e. a compartment with a huge number of T=0 compartments aggregated together which probably exist even today as Supra-Massive Black holes at the centre of the Galaxies. In between these compartments with hyper-dilated time probably were aggregates of Hyper contracted time (i.e. aggregates of negative energy particles – the only place where they can possibly be found if they are real.

We know that there were more T=Tmax than T=0, just to make creation slightly more than destruction so that more creation can keep happening. So we can imagine that there were 20 T=0 Compartments and 30 T=Tmax Compartments, the rest 50 are "non-nonzero" "non-Tmax" Compartments. At the instant itself, all the T=0 and T=T max would start moving and would influence the neighbouring Compartments.

<u>Some of them can run into each other. When they run into each other, they can potentially destroy each other and would cause destruction of energy to create a Compartment where T runs the same as Universal Time.</u>[9**]

> 9** - This was taken as a direct extrapolation of common belief that matter particles annihilate with antimatter particles to form energy. It was later realized that the T=0 compartments and T=Tmax compartments do not annihilate each other. They can never come close to each other as T=Tmax compartments would keep pushing the T=0 compartment away and T=0 compartment would keep pulling its partner towards it. This causes the perpetual motion which was first described by Bondi et al for positive and negative mass particles. The anti-matter particles which are described in physics textbooks have a positive mass and opposite charge. These can keep falling into its partner oppositely charged particle as the repulsive effect that exists between Electron and Proton due to Protons mass doesn't exist with them. This causes them to fall into each other until their positive energy core merge and the lattice gets disrupted thus releasing two photons of positive energy and individual ENEA particles which are hard to detect.

If many of them cancel each other out, that would mean, the universe cooled too quickly, which is insane. So, we have to assume that there was space for them to move so that they don't cancel each other out too fast. The next, Planck time, all the T=0 Compartments vanish but all the T=T max Compartments start expanding. The effects of the vanished Compartments still persist.

The gist of the whole discussion is to emphasize the fact that we would indeed need a slight preponderance of Negative energy over Positive energy for the creation to dominate over destruction. If the inverse was true, the newly formed Planck Compartments would be destroyed even before being created.

At any given instant of Universal Time since the Creation, there has to be more Negative energy than Positive energy. This asymmetry is inevitable at least in theory.

This is paradoxical given the fact that what we can see is more often positive energy than negative energy.

Explaining the apparent paradoxical imbalance towards positive energy when it should be the other way round is a complicated process.

Not everything is known about what keeps this constant imbalance towards negative energy constantly. Whatever process keeps this imbalance towards negative energy, need not be permanently functional.

So our theory, in principle, doesn't necessitate either the Big rip or the Big crunch. Both are likely and more knowledge would be needed before we can conclude which of the two is predicted by our theory. But if we assume that the apparent imbalance towards creation persists forever, the theory does rule out just Gravity to be strong enough to eventually be capable of reversing the expansion of the universe and make it contract again.

Cubicle model vs Honeycomb model of expansion-contraction symmetry

Imagine a huge cubical space that contains multiple galaxies or Galaxy clusters that are all gravitationally linked together.

One can say for sure that the expansion of the universe isn't happening in between these gravitationally bound Galaxies.

Our theory would say that each of these Galaxies is bound to the other with Extended MOND bonds due to the constructive interference of the resetting waves that emerged out from each of the stars and reached the space in between them in a state of a Sub-Planck time time-dilation. These time dilation-spatial contraction waves cannot cause any active spatial contraction at Sub-Planck time dilations but when they get added up together with other such waves from other stars, they cause enough time dilation together to cause spatial contraction.[10**]

> 10** - Now we know that these time dilation spatial contraction waves with time dilation at Sub-Planck length realms are nothing but linear EPCAs. Now we know that in these, horizontal volume contraction is negligible or non-existent but their location keeps shifting which means that there is a small constant vertical spatial contraction that goes till infinity which we call a_0. This lead to the insight that all masses that constitute the Universe can exert MOND kind of forces on each gravitating object or a pair of gravitating objects, which we call G. This gives further support to the assertion that the Universe is finite, has a finite edge and has a finite mass.

All this discussion isn't relevant here except the fact that we are interested here in the distribution of time dilation in this region of space. The Central region where a majority of the stars and the Supra-massive black hole exist would indeed have time in a dilated state. Even the periphery of the Galaxy, due to the above discussed constructive interference, would have time dilation. The space between two gravitationally bound Galaxies would also have similar Extended MOND bonds and would thus have time dilation.

The space just outside all this would probably have a progressively smaller dilation until time dilation ceases and time here would run close to Universal Time. This region would have no Local Time deviation related spatial change. But the constant suction coming from the Central masses of constituents of the Galaxy would keep the centrally directed escalator of Planck Compartments sliding. This is the reason why gravitational lensing would cause the bending of light here.

This region would keep getting depleted of Planck Compartments i.e. reduced in volume if there was no constant supply of extra Planck Compartments from elsewhere. Thus beyond this zone of Local Time equal to Universal Time, there would be progressive time contraction and progressive spatial expansion until a zone comes at the extreme periphery where time T=T max i.e. maximally possible time contraction is present.

Depending on how much is the size of this maximally time dilated zone, the following two possibilities exist.

Either the zone of maximum time dilation is a narrow strip so that the width is several light-years. Or the zone is significantly bigger.

The cubicle model[11**]

> 11** - We know that the model is wrong. But it is still retained here to be aware that a thin zone of Hypercontracted time forming the faces of a cube, as envisioned in this model, isn't really possible. What is more likely is a vast expanse of small particles of contracted time repelling each other (ENEA particles) with varying densities in between many PC's running at UT.

So in our cubicle region of space discussed above, the Central region has maximal time dilation due to a Supermassive black hole. The periphery i.e. the eight corners of the cube along with the eight sides and the six faces would have a narrow zone of space have a maximally expanding time contracted region. This is akin to having a white hole at the periphery of the cube (although in reality, a white hole is never so dispersed and a white hole would give out light but light can never enter it and pass through it. As we can see that light from distant Galaxies does reach us, which means this region has to be transparent, this region isn't exactly a white hole.

Now we have to imagine multiple such cubes arranged side to side. The current Universe (the cubicle model) is made of multiple such approximately Cubical zones completely separated from one another. Note here that the "cubical" shape is not mandatory and what's more relevant is the comparatively thin localised T=T max zones.

Note that any matter within the cube has to have a velocity more than the speed of light to actually cross this T=T max zone and thus no matter can pass through it.

Thus the region of space bounded on all sides by these zones are independent universes in themselves as nothing beyond these zones except light can cross these boundaries.

Light can pass through these boundaries unhindered i.e. they are transparent to light but due to its property of spatial expansion, it would get redshifted when it emerges out from the other side. The contracted Local Time would indeed mean that the light travels faster than its usual speed in this region of space or communication between neighbouring PC'S becomes very fast for maintaining the apparent constancy of the speed of light. All the massless particles travelling with the speed of light can travel through it.

This model, however, is wrong. This is clear if you look at the figure. Here, the distance between Galaxy A and B is similar to Galaxy C and D, however, because of the tangential direction of the line CD, the spatial expansion between them would be much higher. This is not what is observed. The expansion of the universe is found to be symmetrical and increases proportional to the distance irrespective of direction.

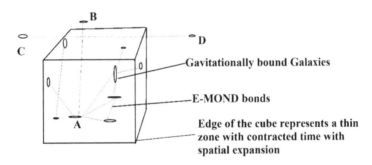

Figure iii/3.1 Shows the Cubicle model where the surface of the Cube represents a thin zone of contracted time where active spatial expansion is happening. The three outside galaxies C, B and D would appear to move away with light from them red-shifted because of the spatial expansion happening at the edge. (Note that C and D, being located tangentially to the edge would apparently move away from each other faster than A and B)

Alternative to this is the honeycomb model which seems more appropriate.

Honeycomb model

The space, in this model, can be divided into multiple compartments like a honeycomb. The zones where galaxies form and matter exists and where relative spatial contraction exists are surrounded from all sides by empty hexagonal honeycomb-like Compartments where time exists in a highly contracted state and the spatial expansion happens in these zones.

Each hexagonal Compartment with matter exists is thus like an island in vast expanses of empty space some of which may have time in various extents of contraction with actively expanding space.

This model explains how, although we see a positive energy abundance in the region of space we are in, the overall balance is tilted towards negative energy or expansion of space.

In today's age of the Universe, all the individual islands of positive energy are extremely far and huge expanses of empty space lie in between. This region is what we call intergalactic space. Thus there is enough cushioning available and it is probably highly unlikely that two neighbouring positive energy islands meet or collide. An island, to overcome the pressure of constantly expanding space in between the second, has to collectively have a velocity higher than the speed of light.

Even at earlier stages of the Universe, when these intermediate distances between the positive energy islands was not much, the likelihood of these positive energy islands colliding would be remote as the expansion of space in between them would act like air being constantly pumped in between them.

As the expansion of space is equal in amount and on all sides of an island, there is less likelihood of a resultant unbalanced force being created which pushes one island closer to the other.

There may indeed be zones in these huge voids where the Local Time is running with Universal Time and expansion of space is limited.

Matter particles can form in these zones and the intergalactic gas formed here can be slowly sucked in gravitationally by the neighbouring positive energy island and its constituent Galaxies.

The direction of movement of T=T max and T=0 Compartments

T=T max particles that form the predominant negative energy source in these negative energy zones cannot remain stationary, them being massless and have to keep moving with the speed of light.

How then can the time here be constantly maintained at a contracted state?

This question is still unanswered by modern physics.

Modern physics tries to answer it by saying that it is a fundamental property of the Universe that space-time keeps expanding unless acted upon by some other force like gravity which is said to oppose the inherent tendency of expansion.

This is in a way similar to what our theory says.

<u>There is some yet unknown mechanism that keeps the negative energy predominating in these inter-island regions.</u>[12**]

> 12** - One of the possibilities, as explained in the section on Crisis in cosmology is that there is predominance of Positive Energy at the edge of the Universe with an unlimited PC Vacuum all around and beyond the edge of the Universe (if the Universe has an edge, that is!) along with an unlimited Negative energy in the centre and at these so called walls of the honey comb or nodes. Essentially, the Universe is being pulled outwards on all sides by an unknown force which is endless and potentially infinite.

As positive energy emitted in the form of light photons arising from the constituents of the islands enters these zones, their direction is divergent going outwards away from the centre of the islands. Due to the actively expanding space in the region, if we consider a spherical shape all around the positive energy island, the Geodesic here is extremely curved. This means that a light beam going tangentially would curve around the island and can cause gravitational lensing.

Are T=T max Compartments really there?

We, at the moment, have to understand that these are just hypothetical and postulated by theory to be present. We don't know for sure if they do exist.

Even if such Compartments do exist, their properties may differ significantly than what is given, they may remain independent miniature entities remaining apart from each other never to coalesce and thus remain as a background entity never to be detected.

CHAPTER 4

WAVE-PARTICLE DUALITY OF MATTER AND LIGHT

History of Wave-particle duality of matter

At the start, it was believed that matter is made up of small indivisible balls called atoms. This philosophical belief was given by Democritus. The balls were theorized to have specks of different sizes which was used to explain the various states of matter like liquids, solids and gases. Unfortunately, nothing more was known about atoms.

This changed when J. J. Thomson discovered the negatively charged Electron in 1897 in his famous Cathode-ray tube experiment followed by the discovery of the "Neutron" by James Chadwick in 1932.

It was theorized that negatively charged Electrons are embedded in a positively charged matrix. This was called the "Plum-pudding model" of the atom, the Electron being the plums embedded in the positively charged pudding.

This changed after Rutherford did his experiment in which he bombarded alpha particles i.e. positively charged helium nuclei onto gold foil. He concluded from his experiment that the atom is mostly empty since most of the alpha particles went through unimpeded. It must also have a positively charged dense core which would repel and deflect at least some of the alpha particles by various angles (*E. Rutherford – 80*). With the conclusion of this experiment, the "Plum-pudding model" was abandoned and the "Planetary model" of the atom was proposed. In this model, it was proposed that the atoms have a positively charged dense nucleus at the centre with multiple Electrons orbiting around in random circular orbits.

Physicists soon realized that even this model could not be right. The theory of Electromagnetism says that any charged particle which is accelerating must radiate energy. This meant that the Electron in a circular motion should keep radiating energy and would keep losing its radius of circular motion and eventually within milliseconds would fall into the nucleus. This meant that the atom, unlike what we find in reality, would be highly unstable. Also, the planetary model of the atoms predicts that the absorption or emission spectra of all elements should be continuous as the Electrons can be placed at any distance from the nucleus and thus should be able to emit or absorb lights of all wavelengths. This also goes against observations.

Niels Bohr revised this model and proposed his orbital model of the atom (*Bohr, Niels – 10,11*). In this, he proposed that

1. The Electron would be allowed to occupy only specific energy levels which he called orbitals. Only those orbits were allowed in which the angular momentum of the Electron was an integral multiple of the reduced Planck's constant h/2π or simply ℏ (h-bar). He said that only in these orbits, the Electron will not radiate electromagnetic waves and thus lose energy and spiral into the nucleus.
2. The Electrons can absorb energy from very highly specific frequencies and when they do so, they would vanish from the lower orbital and reappear in the higher orbital.

This explained the emission and absorption spectra of hydrogen very well and thus was immediately accepted. But problems remained.

Is light a wave or a particle?

There were two distinct and opposing views in the eighteenth-century physicists on what light was. Many supported Newton's view and believed that light was a particle (Newton's corpuscular theory proposed to explain the rectilinear propagation of light). While some supported the Huygens-Fresnel wave theory of light which successfully explained phenomena like diffraction, refraction, interference, optics. The theory worked so well and there was so much evidence to "light being a wave" that there was little doubt.

James Clarke Maxwell had given his equations to explain Electromagnetism in 1865 and stumbled upon the fact that light is an electromagnetic wave and that a wave of electric and magnetic fields would indeed travel with the speed of light. Maxwell's view furthered the wave theory of light.

Edison had invented the light bulb in 1879 and it became possible to choose what frequency of light one wants to experiment with. Armed with the knowledge that light is an electromagnetic wave, many physicists started experimenting on what happens when you focus a beam of light of a specific frequency on various circuits.

The wave theory of light was soon called into question due to the phenomenon of the photoelectric effect. In this phenomenon, a beam of light when focussed on an atom, if energetic enough, is capable of ripping off an Electron from its outer shell and thus results in the creation of a current.

It puzzled physicists when they realized that the ability of light to rip off an Electron from an atom, which was the basic measure of the amount of energy trapped within it, depended on the frequency of the light rather than the intensity of the light.

When an intense beam of red light was focussed on a metal foil, no Electrons were emitted out but when a relatively less intense beam of ultraviolet light was focussed on it, this resulted in a knocking off of Electrons resulting in a detectable current. This was not wave-like behaviour. When we see waves on the sea, a high-intensity wave causes more damage than a low-intensity high-frequency wave.

Before this, another problem which the physics community faced was that of Black body radiation called as the Ultraviolet catastrophe or the Rayleigh-Jeans catastrophe. If an ideal black body radiated energy in a continuous manner i.e. radiated energy in all possible frequencies, the body would lose an exponentially increasing amount of energy with increasing frequency and it would radiate out its entire energy in no time. This meant that no black bodies should remain which meant that the Sun or stars should not exist. This problem was solved by Max Planck who took help of what he thought was a mathematical trick. He theorized that the black body can emit energy, not continuously, but in small packets of energy which he called "energy quanta".

Einstein took this concept further and theorized that light is made up of small packets or quanta of energy, which he called as photons. This solved the problem of the photoelectric effect nicely and won him his Nobel prize. The amount of energy trapped within a photon depends on the frequency of the light which means that a single photon of ultraviolet light has a much higher amount of energy compared to a similar photon of Red light.

Arthur Holly Compton, described the Compton effect in which a high frequency and thus high energy photon of X rays can be deflected by a charged particle, usually an Electron, and in the process transfers a part of its energy and results in a lower frequency/lower energy photon deflected out in a particular angle *(Compton, Arthur H – 18)*. This experiment almost conclusively proved that light cannot be considered a wave (despite the phenomena like diffraction and interference) and proved the photon theory of light, i.e. light should be considered as a stream of particles or packets of energy called photons. 1927 Nobel prize was awarded to Compton for the same.

Both the photoelectric effect and Compton scattering are scientific phenomena that are applied extensively and thus the above-mentioned experiments keep getting repeated countless times. Every time we use solar electricity or treat a cancer patient with radiotherapy, we are proving that light is a particle. Every time we use our eyes or use a camera & click a selfie or we use optics, we know that light is a wave.

Thus it is accepted without doubt today, that light behaves like waves in certain situations and particles in other situations that is the wave-particle duality of light is an accepted and extremely well-tested phenomenon.

The advent of Quantum mechanics and the wave-particle duality of matter (Quantum particles)

Young's double-slit experiment showing interference of light was important evidence in favour of the wave nature of light *(Young, Thomas – 105)*. After it was published, it was repeated at multiple locations with many variations.

Scientists performed the same experiment with particles like Electrons *(Rosa, R – 81)*, Neutrons and atoms, and were surprised to find that an interference pattern emerged even with these quantum particles.

Louis de Broglie in his PhD thesis described how considering the wave nature of the Electron or the Electron-wave can solve the riddle of Bohr's orbitals *(de Broglie, L – 24)*. The Electrons are allowed around the nucleus only in those orbitals where a standing Electron-wave can be created. So the allowed orbits are those in which the circumference of revolution is an integral multiple of the wavelength of de Broglie waves of the Electron.

Wave-particle duality of Light in our Theory

Wave-particle duality of light comes out effortlessly in our theory and is woven in it right from the start. The real seat of energy in our theory is the "T=0 Compartment".

The particle of light or the photon is the Planck Compartment where multiple T=0 Compartments have vanished and their effects still persist. If "10 T=0 Compartments" vanish at a location, the ratio of Creation to Destruction gets skewed to that extent here. This means that at this location 10 Planck Compartments are sucked in every Planck time.

The presence of a highly deviated time here leads to the neighbouring Planck Compartments modifying Local Time within them to accommodate the extra destruction happening.

When two such photons or T=0 Compartment aggregates are released from the source, they instantaneously get entangled with each other. This happens as the neighbouring Planck Compartments react to their presence and accordingly modify Local Time within them.

These reacting Planck Compartments form a wave-front which travels with the speed of light along with the photons/"T =0 Compartment aggregates"[1*].

> **1*** Note that it was realized later that number of events that can potentially happen within a particle could be more than one per Planck time. Thus the TRW or time resetting waves that start from a particle move outwards much faster than the speed of light. Thus while forming the bridging or connecting wave-front or connecting the properties of the two T=0 compartment aggregates, this Wave-front can travel much faster than the speed of light. But forward progression along with the T=0 compartment aggregates would happen at the speed of light.

The intervening space can be divided into Compartments with equal Local Time. Each of them has a slightly different dilated Local Time so that with every Planck time, active contraction of space takes place in all of them[2*].

> **2*** Initially I had thought that every Planck compartment can have its own individual measure of time. Later it was realized that the deviated local time was an emergent phenomenon which depended only on the status of spatial contraction or expansion at a point in space. Thus an individual Planck compartment can have only three states, Either a T=0 state, T = Tmax state or a T=UT state. The local time dilation is decided based on the ratio of T=0 compartments to Total number of PCs and local time contraction is decided based on ratio of T = Tmax compartments to Total number of PCs.

In a way, the two Planck Compartment aggregates which left the source at the same time are kept pulled together and propagate together. The Local Time within the intermediate Compartments is such that closer to the photons they are, more time dilation or spatial contraction they have while as one goes towards the centre, time dilation reduces.

As the photons travel in a divergent manner, the distance between the two entangled photons increases. After they have travelled for a light-year or so, the distance between them is insanely high. But the entanglement between them persists. The time dilation as one goes from one photon to the second keeps reducing for a distance. After this, the time dilation continues in a "Sub-Planck time" manner. This "Sub-Planck time" time dilation causes no spatial contraction but it can still modify the trajectory of the photons.

This wavefront connecting the two photons is not information travelling from one photon to another but the information travelling from the source to the present location of the wavefront. Like the Huygens-Fresnel theory, the wavefront is due to information about Local Time being passed from the intervening Planck Compartments forming the bridge between the newly originated photons towards the Planck Compartments that lie in the direction of propagation of the wavefront.

In effect, the electromagnetic wave is a wave of the passage of information, and this information is about how much C/D ratio there should be, i.e. how much Local Time to Universal Time ratio there should be or what should the Local Time be. The wave is information about how the space over that region should behave, whether it should expand or contract or stay as it is.[3*]

> 3* - Later realized to be wrong, as it is the changes in space i.e. active spatial contractions that is primary event and the time changes are secondary. Thus it is the movement of PCs as EPCAs that is primary event and the time changes happen due to reduced velocity of light due to spatial contraction.

Note that the packet of energy is only the T=0 compartment aggregates and the wave-front that is forming in between is only reactionary. It does however have deviated Local Time which means that it has energy.

This energy is energy derived from space itself and is not energy that the source lost.

Note that there is a limit to the number of T=0 Compartments that can aggregate together like this to form a photon. After a certain amount, the energy density becomes too much and after this, the T=0 Compartments collapse in a so-called Sub-Planck length prison to form a singularity or a Positive Energy Matter particle.[5*] A photon or T=0 Compartments aggregated together can travel with the speed of light. But as soon as they aggregate enough and collapse to form a singularity, they can no longer propagate with light speed.

> 5* - This highly theoretical "baggage" of unnecessary assumptions taken initially from string theory was later removed for good as it was not only not helping but adding more problems

Note that the frequency of light depends only on the frequency of these photons being released from the source. Whether the amount of energy i.e. T=0 Compartments contained within the photon has any effect on the frequency is unclear although it may also have an effect so that the Local Time in the intermediate Compartments between two successive photons determine the distance between them and this determine the frequency of the wave. The role of this however is unclear. Simply because if this is the case, any change in the wavelength that happened during the intergalactic travel of light from distant Galaxies would not manifest itself and would get autocorrected instead. This meant that we would never have detected the Doppler shifts due to expanding Universe. The possible explanation is that the increase in wavelength due to space in between the photons expanding due to Expanding Universe can be and does get corrected when the light wave reaches us but the change in frequency happening due to a farther starting point for each subsequent wave (the originally described cause of the Doppler effect) does get manifested and does not get autocorrected completely and can be measured.

On the contrary, the energy contained in light depends on the number of T=0 Compartments contained in a single photon. The lower energy waves like Radio waves may not have any T=0 Compartment aggregates while the higher energy Electromagnetic waves have an increasing quantity of T=0 aggregates within them.

<u>When these T=0 aggregates are absorbed by an Electron, they coalesce with the inner positive energy core and increase the mass which forms the relativistic mass.</u>[6*]

> 6*This is just an assumption. What exactly happens when a photon is absorbed by a particle is not completely clear yet. The asymmetry in time changes which essentially constitute the momentum of the Photon is transferred to the surroundings of the particle absorbing the Photon.

This means that an Electron which has absorbed a high energy photon would possess higher energy and momentum, which may be enough for it to elevate it to a higher orbital. If the photon does not have enough energy, enough momentum is not gained and the photon is given off without any change to Electron orbit.

A higher energy photon like an ultraviolet or x-ray or gamma-ray photon, which contains a relatively high quantity of T=0 Compartments, when absorbed by outer shell Electrons, have enough energy to cause expulsion of outer Electrons. These radiations are thus called ionising radiation. In this manner, the photoelectric effect can be explained in our theory.

The Compton effect with a change in frequency and wavelength of the light wave after it is absorbed by an Electron and scattered in a different direction can also easily be explained as the entire photon with all its energy is absorbed by the Electron and after going to a higher orbit, the rest of the extra T=0 Compartments are released as an aggregate with a slightly lower frequency. The angle of scattering may also have an explanation when we have a detailed model as to how the space in the neighbouring Planck Compartments reacts.

Wave-particle duality of Quantum particles in our Theory

Even the wave-particle duality of particles emerges just out of the assumptions of the theory without any effort. Quantum particles are aggregates of T=0 Compartments (Positive Energy Matter - PEM particle) or T=Tmax Compartments (Negative Energy Antimatter particle-ENEA). Due to the existence of a highly dense highly deviated time in such a small volume, the neighbouring Planck Compartments react and all such particles give off a time resetting wave. The PEM particles give off a spatial contraction time dilation wave while ENEA give off time contraction spatial dilation waves.

These waves are intrinsically associated with the particle and would be present wherever the particle is or will be or was. However, they are separate from the particle and are not the particle itself.

The Electron, according to our theory is a composite particle made of a dense core of PEM in the centre and a Buckyball like spinning lattice around it. The wave which starts from such a complex structure is a mixture of time dilation and time contraction zones. The complexity of the wave that originates is responsible for the unique properties of the charge of an Electron.

The fact is that all particles are singularities that trap T=0 or T=Tmax Compartments[7*], all Quantum particles are associated with either time dilation or time contraction waves.

> 7* - It was realized much later that T=Tmax compartments would actually repel each other and would not form aggregates or singularities.

Explaining the findings of a single Electron double-slit experiment

If one fires an Electron towards a double slit, the spinning particle, at every Planck time till it approaches the slit, gives off these complex waves (*Rosa, R – 81*). These spherical waves zoom past the Electron and are thus expected to enter just like any other wave through both the slits at the same time. The waves emerging out from the other side of the slits would diffract and can interfere with the counterpart emerging from the other slit in the region immediately past the slits. The Electron would, just like any other particle, pass through a single slit. However, as the Electron is constantly giving out the waves while it emerges from the slit and the waves which come out of the Electron interfere with other waves emerging from the second slit, the interference of these waves results in deflection of the path of the Electron and the Electron forms an interference pattern on the screen despite passing through a single slit.

De Broglie-Bohm "Pilot-wave" theory and our theory

Our theory, at least in principle, does agree with the pilot wave theory first given by Louis De Broglie and further improved by David Bohm. Both these agree on the fact that the waves determine the location of the particle at any given point. There are some subtle differences between the two. *(Dewdney, C – 25, Bohm, David – 16)*

According to de Broglie-Bohm pilot-wave theory, the particle is just taking a free ride on the wavefronts just like a surfer rides on the ocean waves. The particle's location at any point in space at any given time depends on where the particle lies on the wave and also depends on what is the situation of the interference. In it, there is no concept of variable time, Local Time and spatial expansion/contraction due to deviated time.

In our theory, the difference comes here. The interfering waves determine the pattern of interference which in turn determines the Local Time in the space around the particle. The distribution of spatial expansion or contraction determines the location of the Electron on the screen.

Non-locality and our Theory

Like the de Broglie-Bohm pilot-wave theory, even our theory needs non-locality.[8*]

> 8*Our theory needs a limited form of non-locality in that it necessitates faster than light communication of information. There however may exist a maximum speed of communication via entanglement. The theory does not necessitate instantaneous transfer of information and in that sense although the TRW travels faster than speed of light, information travels from one point to the next and involves no jump in the information. Thus in strictest sense, the theory is relatively local.

The information about whether the wave-front is intact and all the photons within the wave-front are still entangled or are "no longer entangled" can, like in de Broglie's theory, travel faster than speed of light i.e. non-locally.

This means that two photons, after formation and after getting entangled, in the entangled state, after travelling for a very long distance- say a light year and after the distance between them has already increased to a light-year or more, can still respond to each other as they are still entangled due to the time changes in the Planck Compartments that lie in between them that are passing on the information about C/D ratio together. If the entanglement gets lost at such a point, the information about the lost entanglement can potentially travel faster than speed of light.

Young's double-slit experiment and its variations

(Young, Thomas – 105)

This is an absolutely critical experiment whose results and variations have a significant role which helped physicists determine the rules that governed the Quantum particles.

At first, we would discuss the experiment and all its typical variations and how physicists explain the findings based on current understanding.

Only after this entire discussion is complete, we will discuss how our theory provides an easy explanation for almost all the weird findings without the need for weird assumptions like Quantum Superposition.

The double-slit experiment classically described by Young was repeated with many modifications. The first one of them was to insert a detector just before the slits so that they know from which slit the particle went. This path of the photon or particle is called the "which way" experiment.

The same experiment can also be done by letting the light beam pass through a beam splitter. Even in such a set-up, the exact path of the photon can be made out.

Whenever an attempt is made to know the pathway taken by the photon, the interference pattern disappears and two slits are seen on the screen. Whenever the "which way" information is not known, a proper interference pattern results.

There are two potential locations where this "which way" information can be gained, the zone before the slits and the zone after the slits.

If the measurement of "which way" information is done after the slits, the experiment is called the "Delayed choice" experiment.

Another addition to the experiment is an addition of a "Quantum eraser". This equipment or array of equipment can completely erase the "which way" information so that the knowledge of "which slit the Quantum Particle or the photon went through?" is completely removed. This can be done by merging the beams together or passing them through partially silvered mirrors which reflect 50% of the photons.

The result of this modification is that whenever the "which way" information is completely erased, the interference pattern reappears and whenever it isn't completely erased, the interference pattern disappears.

All these experiments have a puzzling but simple explanation.

Light behaves like a particle when someone measures it, it behaves like a wave when nobody is measuring.

This insane looking experimental observation and result has been confirmed to be true countless times.

Quantum physics gives an explanation of this phenomenon in an equally controversial manner. It says that the location of a photon, it being a quantum particle, is determined by Schrödinger's equation and is given by probabilities. The wave function, when not being measured is fussy and is in a Quantum superposition of all the possible states possible. The location of the photon before the measurement is impossible to comprehend and is just given by probabilities. However, when the measurement is made, the wave function collapses and the photon's location becomes known. This is called the Copenhagen interpretation of Quantum mechanics. An Electron passing through the slits, being in a superposition of multiple possible possibilities, passes through both the slits at the same time and thus forms an interference pattern. When the measurement is done with a "which way" information measurement device, the wave function collapses from a probability wave to a single location. Thus the Electron or photon, instead of showing an interference pattern forms just two slits on the screen as if behaving like classical particles like tennis balls.

Note here that what constitutes "a measurement" here is ambiguous and ill-defined. It may be a conscious mind noting down the findings or just a non-conscious measuring device. This ambiguity or inability to zero down on what constitutes the process of measurement is what is called the "measurement problem" of Quantum Mechanics.

The weirdness of the Quantum Superposition of Schrödinger's cat being both alive and dead when not being measured puzzled scientists for a long time.

Even today, scientists have not understood why the experiment results are so weird and what exactly happens. The accuracy of results shown by the mathematics of Quantum mechanics is well known so that it gives accurate predictions for physicists. Physicists often say "shut up and calculate" to those who ask too many questions and are unable to grasp the weirdness of Quantum mechanics.

Another set of physicists, who do not buy the Quantum Superposition argument for being too weird, give another weird explanation for the same phenomenon. This second way of explaining the weird findings of the double-slit experiment is called "the Many-World's interpretation". In this, every time the photon or the Electron is trying to pass through the slits, the world sort of diverges or divides (more appropriately multiplies) into a separate world so that every likely possibility does occur. In one such world, the photon or Quantum Particle goes through one slit while in others it goes through the second slit. According to this, every moment the world is branching into infinite worlds.

Our theory explains the findings of the double-slit experiments without any trouble and without the weirdness of Copenhagen or "Many world's" interpretation. The experimental results, in a way, come naturally in it.

<u>The photons when formed and diverge from the source are connected by a zone of Planck Compartments with a variable time, the zones which have a decreasing time dilation and thus spatial contraction within it.</u>[9*]

> 9* - At this time of the Theory development, the concept of EPCA formation was not completely understood by me. This came only after a detailed analysis of Entanglement process was done as described in the Entanglement chapter.

It is just like the photons are like small spheres which are held together by elastic strings. The zones with different dilated Local Times have active spatial contraction happening within them. This means that they are more like balloons of air from where the air is being actively pulled out.

For understanding purposes, an analogy of "two footballs joined together by 20 Balloons" is appropriate. The footballs here are the photons and the balloons pasted together side to side and partially inflated represent the wave-front. The balls are at a distance apart and all the balloons that lie in between them are under some pressure. The amount of air within all the balloons combined together determines the distance between the footballs. The balloons are still attached individually to a separate pump so that air can be pumped in or out. As the footballs and the balloons travel in one direction, more balloons are added and the air pressure within them changes.

Here the balloons are pasted side by side which is there to indicate entanglement of the adjacent Planck Compartments or more appropriately a collection of Planck Compartments.

Note here that this is just an analogy. The balloons in the real wavefront don't have air but Planck Compartments. And it is not the air pump but the active Creation or Destruction of PCs from nowhere or into nowhere in the Quantum foam that causes the deflation/inflation. Here an arbitrary number of 20 Balloons is taken but in reality, there would be an insanely high but not infinite number of such balloons.

This analogy is explained in more detail in a different section.

What is important here, is that by some unknown mechanism, the process of measurement of the location of the photon or knowing from which hole the photon is travelling leads to an instantaneous collapse of the wave-front or dissociation of the entanglement of all the intermediate zones which, by means of the Local Time changes within them, were binding the two photons together and we're determining their trajectories.

If tennis balls or other non-quantum objects were thrown through such slits, they would follow a straight trajectory and would not receive any deviation. They would form two strips.

The photons or other quantum particles, on the contrary, get deflected due to the status of space within the intermediate wave-front. The wave-front of the two waves, when interfering, affect each other and thus affect the final position of the photons. The resultant wave-front formed after the interference of the two wave-fronts leads to outward or inward deflection of the photons within the wavefront. This deflection no longer occurs if the wave-front collapses and the entanglement of the Planck Compartments in between breaks. In such a case, the photon continues un-deflected and forms two slits instead of an interference pattern. Whenever the wave-front remains intact, the quantum Particle behaves like a wave and an interference pattern forms.

Delayed choice quantum eraser experiment

This is a more complex array of equipment first described by Kim et al but since then repeated by many researchers. Its findings when viewed from a conventional physics point of view are enigmatic to explain.

The experimental set-up is shown in figure iii/4.1.

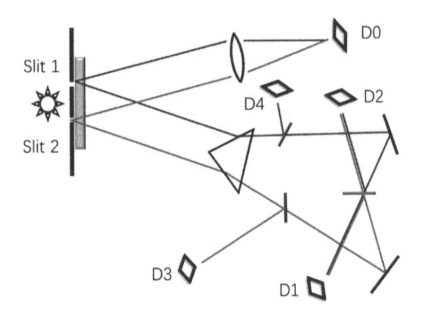

Figure iii/4.1 Experimental set up of the Delayed Choice Quantum Eraser experiment.

(Image Source: BOOK: Shan, Liang 2018/06/24 Consciousness Is an Entity with Entangled States: Correlating the Measurement Problem with Non-Local Consciousness)

Photons from a laser beam are passed through a double slit experimental setup. Immediately after the slits, even before interference can form, the beams which emerge out of the slits pass through barium borate crystals which split each of the entering photons into two polarized entangled pairs. These diverge out through a prism. One of them, called the detector photon passes towards the left towards the detector set up which contains a convex focusing lens and detector D0. The other photon, the entangled twin of the first one, passes towards the right towards the Quantum eraser setup. This photon is called the idler photon.

In this complex setup, the idler Photon passes through a beam splitter which is a 50% silvered mirror that allows passage of 50% of the photons and reflects 50% of the photons. There are two such Beam splitters BS1 and BS2 each for photons coming from the two slits A and B respectively. The reflected light from BS1 goes to detector D1 and one can confidently say that the photon that passes into D1 has entered through slit A i.e. which-way information is not lost here. Similarly, the reflected light from BS2 goes to detector D2 and even in this, the which-way information is preserved.

The photons that pass straight without reflection from BS1 and BS2 enter in the quantum eraser set up where a beam splitter BS3 and 2 mirrors are so arranged that the beams merge and pass to detectors D3 and D4 so that the which-way information is lost and if a photon is detected in D3 and D4, one cannot say from where it came from.

Experimental results and retrocausality

The photons are split up into two entangled photons by the crystal. The detector photon can go to the detector and can either show an interference pattern or two slits pattern showing loss of interference and thus loss of entanglement.

The results clearly show that whenever photons are detected by D3 and D4, the which-way information is lost and there is an interference pattern seen in the detector photon. But whenever the idler photons are detected in D1 and D2, which way information persists and the interference pattern is lost and the detector photon shows a two slits pattern instead of an interference pattern.

There is however a glitch here. The results of the experiment are unchanged even if the "which way set up" is much longer than "the detector setup" so that the detection of detector photon happens much earlier than the detection of the idler photon in any detector.

This just doesn't make sense, the main reason being that the detection of idler photon, an event that occurs much later than detection at the detector photon, can influence the interference or no interference at the detector photon much earlier. How an event that happened in the future can determine, so precisely, an event in the past is unclear.

This experiment, in the opinion of some experts, suggests the possibility of retrocausality or the possibility that the information about the idler photons passed backwards in time.

How does our theory explain the DCQE experiment?

The exact explanation is not known even after applying our theory and some questions remain.

However, our Theory supports absolute time and thus passage of information forward in time or backwards in time is not supported by our Theory.

Our Theory can offer some explanation of the weird phenomenon.

When the photons pass through the crystal, there are two possibilities. Some photons emerge with an intact entanglement and some without any entanglement. This loss of entanglement happens at the crystal level itself.

The entangled photons when emerging out have a wave-front of Planck Compartments with altered time connecting them. When non-entangled pair emerges out, there is no such wave-front present or it might get lost in their path. So that the non-entangled pairs have no wave-front and behave like particles.

When an entangled detector photon along with an intact wave-front passes through the detector apparatus, an interference pattern is noted. There is no enigma in this. The entangled idler pair with intact wave-front, while passing through the second beam splitter will have an intact wavefront along with it. This will affect its path. The intact wave-front will be an additional factor in determining where the photon might go. By some yet unknown means, the wave-front makes the photon go towards the detector D3 and D4 instead of detector D1 or D2. The collapsed wave-front in disengaged photons, on the contrary, cause the opposite effect so that they would make it more probable for the photon to enter the detector D1 or D2 paths.

A detailed analysis of exact structural drawings will be needed to understand why this happens.

In this explanation, the 50% probability of emergent photons going into either the "known which way" apparatus or the "quantum eraser or unknown which way" apparatus is altered by the intact wave-front so that an intact wave-front somehow favours the photons entry into the "unknown which way" or "Quantum eraser apparatus" while a disentangled or collapsed wavefront favours the idler photons entry into the known which way apparatus.

Whether this is the correct explanation of the phenomenon or something more needs to be added remains unclear and indeed this explanation seems incomplete even to me.

Only time will tell if this explanation is good enough and there is a high likelihood that some other explanation may need to be added to this at a later date.

"How exactly does detection of which-way path collapses the wavefront or the entanglement?" is also unclear yet. Possibility of another wave-front coming from the detector reacting to the parent wave-front and modifying it or collapsing it cannot be ruled out.

The enigma of the DCQE experiment thus remain.

Answers are likely to come in the future as progress is made in our Theory.

What is needed is a detailed analysis of all the photons and the wavefronts involved.

Just before the Crystal, there are two slits. When the photon enters one of the slits, the wave-front travelling along with the photon enters the other slit. The wave-front with the photon then interferes with the wave-front without the photon. This interference happening within the crystal along with the splitting of the photon as an emergent property of the crystal means that there is not just the photon but multiple other wavefronts or entangled Planck Compartment groups whose effect needs to be analysed. Complex computing and analysis may be needed with every passage of photon, both the idler and detector photon before a sound explanation would emerge.

Chapter 5

A Particle

The standard model of particle physics divides all particles into fundamental particles and composite particles i.e. particles made by joining fundamental particles together. The fundamental particles are further divided into fermions, leptons and bosons, each of them having varying properties including masses.

Of these, the bosons are the force carrier particles and are responsible for carrying the fundamental forces of nature.

The photon carries the electromagnetic force, the gluon carries the Strong Nuclear force and the W and Z bosons carry the Weak Nuclear force. The Higgs Boson is responsible for giving mass. While "the Graviton" is a hypothetical carrier Boson for gravity. Evidence of Graviton existence of Graviton is not yet confirmed.

Bosons like "the photon" are massless particles and they move with the speed of light. The W and Z bosons on the other hand possess properties like mass, electric and colour charge.

The leptons and fermions, on the contrary, have a definite mass and charge, and cannot move with the speed of light.

Every Quantum particle possesses various properties like mass, electric charge, colour charge, spin, magnetic moment, Quantum number etc.

Furthermore, all particles have lifespans, many of the particles described in the standard model decaying rather quickly. The more commonly occurring particles have very high life spans with Proton decays almost unheard of.

Every particle possesses momentum and energy and thus can transfer momentum and energy to other particles. Every particle has its own antiparticle.

The equation given by Einstein means that each particle has lots of energy trapped within. The exact state of this trapped energy and the transferable part of the energy is not known.

Quantum mechanics, in fact, says that every quantum particle is in fact a dual of particles and waves.

Quantum mechanics further says that every quantum particle is in a state of superposition, wherein it is present in reality in all the quantum states that are possible, the probability of which is given by the square of wave function given by Schrodinger's equation.

Heisenberg's uncertainty principle states that the product of uncertainty in the knowledge of the exact position and the exact momentum of a Quantum particle is constant and thus there is a limit to which either of these can be known. If we want to know the exact position by reducing the uncertainty in its location, the process increases uncertainty in the exact measurement of its momentum. The reverse is also true.

In short, a Quantum particle is in a superposition of multiple states at once, is a particle and wave at once and has a limit to knowledgeability of its position and its momentum.

Thus any theoretical framework, which assumes an accurately knowable location of a Quantum particle -say an Electron, is wrong in its fundamental level as it would violate Quantum mechanics.

Note that almost every theory that described the formation of atoms, molecules and higher compounds in chemistry and biology need Quantum particles to have a definitive position and predictable properties to explain the phenomena and reactions happening at a larger scale.

The wave function describing the quantum particle collapses on making a measurement which then enables us to have a high probability location of the particle.

The standard model suggests that the fundamental particles are indivisible into anything smaller i.e. they are fundamental ingredients of matter. What they are further made of is not described.

String theory, however, describes every particle as a string of energy vibrating in an 11-dimensional hyperspace, with its frequency determining the properties of the particle. According to this theory (which at the present stage is theoretical) all particles at the most fundamental levels, are sub-Planck length in dimensions. The "strings" are made of energy, and thus are equally abstract in nature.

Presently, the standard model along with Quantum mechanics is the accepted version of the science of particles. Multiple questions still remain unanswered.

Quantum mechanics says that Quantum particles can pop in and out of existence. Specifically, a packet of energy like a photon can get converted into a particle-antiparticle pair which can remain for a short period of time followed by annihilating with each other to reform the packet of energy in the form of a photon, more likely two photons. In this reaction, momentum is conserved.

Particles in our theory

A detailed definition of "a particle" is crucial given the central role played by various particles as building blocks in physics.

Massless particles or virtual particles, although described to be present in standard models, are difficult to describe further in terms of what they are made of.

In our theory, massless particles are a bunch of Planck Compartments with a deviated Local Time, thus constituting a packet of energy. As the massless particles move with the speed of light, their interaction with the adjacent PCs is minimal and lasts for a short period of time.

The photons of light or the particle of light are basically an **aggregate of PC'S that have T=0.** When these T=0 Compartments move, they recruit fresh PCs at the preceding edge and at the receding edge, some T=0 Compartments become T=non-nonzero Compartments.

Thus they have actively contracting space all around. As these PCs do not provide any resistance to movement, they move with the speed of light.

There can be two types of such virtual particles, the ones with positively deviated i.e. contracted time and expanding space, or the ones with negatively deviated or dilated time with actively contracting space.

The maximal negative deviation possible is T=0 and the maximum positive deviation possible is T=Tmax.

The region with T=0 would be called photons. Multiple photons aggregated together would have compensatory changes in the space between them.

The Gluons, unlike photons, although massless, do possess colour charge and anti-colour charge as discussed in detail elsewhere. They occur only inside the nucleus where the conditions are difficult to fathom. It is thus difficult to determine as yet whether Gluons are also, like photons, just a bunch of PCs with deviated Local Times that occur in an actively contracting vibrant space or are Composite Particles that are made of smaller exotic or non-exotic particles that possess mass.

In contrast to massless particles, there are particles possessing the property which we call Mass.

What exactly leads to particles obtaining their mass is not known. But in Modern physics, what is accepted is the Higgs mechanisms. With the discovery of Higgs Boson, the Higgs mechanism for obtaining mass has significant evidence.

In this, there exists a field called as Higgs field all throughout the universe. Any particle which interacts with this field possesses mass and any particle which doesn't interact with it has zero mass.

Mass itself is the property of matter by way of which it resists change of its present state, I.e. it possesses the property of inertia. This means that

"Mass is the property of a particle due to which

- it offers resistance to being moved from resting position to uniform motion by an unbalanced force or

- it offers resistance to acceleration or deceleration i.e. any change in its velocity of the direction of motion by an unbalanced force while in uniform motion."

We will not go into details about how the Higgs mechanism works. Just one thing is worth mentioning, that a particle that interacts more with the Higgs field has more mass.

Types of mass

Usually, three types of mass are described.

1. Inertial mass,
2. Active gravitational mass,
3. Passive gravitational mass.

Inertial mass is the one due to which any change in velocity is opposed by a particle with mass. This type of mass is measured by simply applying a known amount of force/energy to a particle and accurately measuring the acceleration it achieves.

The Newtonian equation F=ma then gives us the mass.

Active gravitational mass is the one that enables the particle to create its gravitational field and thus attracts other objects with mass towards it.

The passive gravitational mass is the one due to which a particle possesses the property to react to or be affected by a gravitational field created by some other particle.

Einstein's equivalence principle suggests that the inertial mass is the same as gravitational mass and even the active and passive gravitational masses are one and the same i.e. all these three things are achieved by a single entity.

Experimental data supports the same view and time and again proves all these masses to be one and the same.

Consequences of Mass

One of the major consequences of mass in relativity is that a particle with mass cannot travel with velocities faster than the speed of light as this requires infinite energy. As the velocity of a particle approaches the speed of light, its relativistic mass increases and as a consequence, its inertia increases i.e. more energy is needed to get the same acceleration.

In a way, inertia at rest behaves like friction. Inertia at rest, like friction, causes the particle to stick to its previous coordinates and resists any movement. Inertia in uniform motion is different, it makes the particle stick to the uniform velocity it is travelling with and resists any change in it.

This means that if we try to move a particle to the right by applying an unbalanced force, another force appears whose direction is towards the left which tries to prevent the particle at rest from starting to move. Similarly, a particle is moving from left to right with uniform velocity and we apply an unbalanced force to decelerate it, a force directed towards the direction of motion appears that tends to maintain the velocity or prevent it from changing.

Higgs Boson and Higgs Mechanism:

As discussed before, the Higgs field is a field that is thought to permeate the entire universe. Any particle that has the property to interact with it has more mass and any particle with less interaction with it has less mass. This interaction of a particle with the Higgs field is what produces the Higgs Boson.

The Higgs mechanism is explained to laymen with the example of "Paparazzi in the crowded party". For this, we have to imagine a big hall which has many people in it and a party is in progress. The people are staying separate with minimal tendency to form clumps. A famous celebrity enters the hall. When that happens, everyone instantly recognizes him/her and the Paparazzi form a crowd or a clump around him/her and this would make it difficult for the celebrity to move in the crowd and would slow down his/her movement. On the contrary, when a random guy enters the crowd, nobody knows him and so nobody clumps around him or interacts with him to take a photograph. So his movement within the crowd is unimpeded.

Like the Paparazzi clumps around the celebrity, the Higgs field by the Higgs mechanism is supposed to clump around the more massive particle hindering its movement. While less clumping of the Higgs field causes less hindrance to the movement of a relatively less massive particle.

Negative mass

Exotic particles with negative mass have been described.

These particles behave differently from the usual particles in that when an unbalanced force tries to move them, instead of resisting the motion, their inertia accelerates the particle in a direction opposite to the direction of the unbalanced force.

These particles are theoretical and not experimentally verified and may just be imaginary products of Mathematics. However, they come up very naturally in our theory and are thus discussed in detail elsewhere.

Mass in our theory

A particle in our theory is an aggregation of T=0 compartments trapped in a Sub-Planck length prison.

We know from classical physics that when enough energy is concentrated at a single point or a small volume I.e. when the energy density goes beyond a certain point, a black hole is created.

A Blackhole can be considered a prison for light as light cannot escape it. Even though what happens inside a Blackhole is beyond our understanding as to the laws of physics break down at the singularity, the presence of a black hole can be made out due to its interactions with things outside it.

A particle with mass in our theory is a specific amount of energy given by the equation "$E = mc^2$", concentrated in a small sub-Planck length region thus creating a miniature black hole, the T=0 compartments contracted to sub-Planck length size, moving still with a speed of light in higher dimensions which lie within the Planck length but which are curled up. Thus these T=0 compartments moving with the speed of light in higher dimensions are equivalent to the strings described in the string theory.[1#]

> 1# - Note that this concept of a Sub-Planck length prison, originally taken from String Theory, was later given up as unnecessary baggage of unhelpful assumptions.

This miniature black hole remains in a specific coordinate location accurate to a single Planck length and from here it can interact with the neighbouring Planck Compartments.

We know that energy has to keep flowing from a region of higher energy to a region of lower energy. This region of space-time which is sub-Planck length in dimensions but extremely energy-dense, and where there is an incessant presence of a Local Time T=0 and thus there is constant disappearance of Planck Compartments but no appearance of fresh compartments acts like an open drain in the centre of a Bathtub full of water. It keeps sucking Planck Compartments from the surrounding, thus there exists a constant unbalanced force or an escalator directed inwards.

These T=0 compartments, because they stay at a particular coordinate for a long time and they represent energy, interact with the neighbouring Planck Compartments and modify or reset time in them. These neighbouring Planck Compartments whose time has been reset, further interact with their neighbours and reset their time in turn. Thus starts a resetting wave that travels in all directions around the particle. These waves travel with the speed of light outwards. These are not typical waves and can be considered as standing waves as the time reset will remain until it is changed by the influence of some other such wave travelling towards it. This is in contrast to the EM waves where the time reset remains for a very minute period of time, as it travels one Planck length per Planck time.[2#]

> 2# - Note that the concept of EPCA and C/D ratio had not been understood yet. The Time resetting waves do not move with the speed of light but much faster, as was understood after concept of EPCAs became clearer.

They are in a way interference waves I.e. they arise due to interference of multiple miniature spherical waves as in Huygens's theory.

Let's imagine that the particle is static.

The resetting waves arising from it, (which have the active component in the form of a region with dilated time and thus are basically waves with spatial contraction) depending upon its size and the formula for lambda, after every distance equal to lambda that it travels, is associated with a reactionary time contraction- space dilation zone which represents the passive component of the wave. This is a passive reactionary component because it is needed just to keep an absolute coordinate system a reality. This passive component does not get added up when multiple spatial contraction waves get added up although the higher spatial contraction that will result from the merger of multiple spatial contraction waves will result in an increase in contraction of space which leads to a larger compensatory component.[3#]

> 3# - It was later realized that this compensatory expansion cannot possibly compensate for the destruction happening at the centre. Thus it is clear now that destruction happening at the centre of a particle is compensated by active volume contraction all around and no expansion. This compensatory spatial expansion introduced just to introduce the shape of a sine wave to these time resetting waves, is now considered as wrong line of thinking and this concept is given up in present version of the theory.

As the spatial contraction waves spread all around the particle, each of them pulls the particle towards the direction they are going. Thus the particle is being pulled from all sides by something akin to rubber strings all around it.

Now if a force wishes to move this particle towards the right, the spatial contractions towards the left oppose it. These standing space contraction waves in a way bind the particle to its coordinates. The force has to also cause the T=nonzero Compartments immediately adjacent to the particle in direction of its future path to be reset to T=0 and thus need outside energy.

These waves keep emerging irrespective of the state of motion of the particle and thus can explain the inertia of motion as well as angular momentum.

If the particle is moving with uniform motion, the velocity will necessarily be less than speed of light, which means both the receding and the preceding borders will have these waves emerging from them. The waves emerging at the preceding border will be more crowded due to the Blueshift due to the Doppler effect while waves emerging from the receding end will emerge at distances higher than the wavelength of the particle. The crowded waves lead to higher net traction at the preceding edge and this would keep the particle from changing its velocity.

Comparing Higgs mechanisms to our theory

The crowding of the Higgs field hypothesized is similar to the crowding that occurs in our theory due to the resetting waves that modify Local Time around the particle. These waves constantly keep coming out of the particle and thus the constant pull from all sides remains. In a way, the Higgs field is the same as our dynamically reactive space-time.

Types of masses in our theory and the Equivalence principal

A particle sends time resetting waves in the surrounding space and causes active spatial contraction. At the same time resetting waves form the gravitational field of the particle and does interact with similar waves starting from other particles and leads to mutual attraction between particles with mass. This same mutual attraction keeps multiple particles together close to each other in space, to form meteorites, planets and even stars. Waves originating from one particle add up with the ones originating with others, so as to produce concentric waves around bigger aggregates of such particles that form bigger structures like the Earth.

This basically indicates that active and passive gravitational masses are equivalent even in our theory.

The inertial mass, however, creates a problem. If the more commonly occurring particles like Protons/Quarks and Electrons are made solely of matter particles without any exotic negative mass particles, then there is no contradiction. However, if there is even a small truth in the Matter-Antimatter model and the repulsion caused between charges are due to the presence of exotic negative mass particles which give of time contraction spatial expansion waves as resetting waves instead of the typical spatial contraction waves given off by matter particles, then the effective mass is an addition of the positive mass of the matter particles and the negative mass of the Antimatter exotic particles.

The mass we get from Einstein's mass-energy equation would also vary. That is energy associated with the charged particles would be both positive and negative. How this impacts the experimental results and whether this is compatible with findings of experiments performed till now remains to be seen. It remains to be seen if we can detect this negative mass associated with the particles.

Other properties of particles

Particles in the standard model have multiple other properties like charge, spin, magnetic moment, colour-charge just to name a few. All these properties are fundamental, i.e. modern physics offers no explanation for any of these. All these are rather abstract, although all of them have been tested extensively and thus accurate methodologies are available to calculate them. Experimental evidence for their existence is strong.

Let's discuss them one at a time.

Charge

A particle with a charge has a property by way of which it can get affected by or exert electromagnetic force. A charge is one of the extremely important properties, as it enables the formation of bonds between various particles to form atoms, molecules and elements.

It is not wrong to say that according to the classical theory of Electromagnetics, a charged particle has an intrinsic property by way of which it can throw/disseminate in its surroundings, electrical flux which is represented by lines and arrows but represent lines of forces. In other words, a charged particle can generate or modify the electrical field around it.

Any change in the electrical field due to the movement of the charged particle creates another field called the magnetic field.

Any change in location of the charged particle causes a change in the fluxes i.e. location of flux lines. This change is represented by a kink as shown in figure iii/5.1. This kink in the flux lines travels outwards with a speed of light and thus any movement of charges produces its effects with a speed of light and not instantaneously.

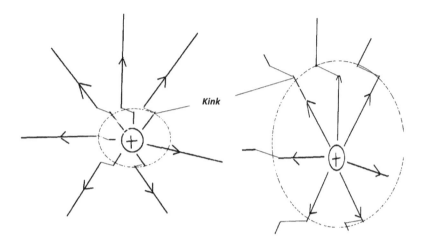

Figure iii/5.1: Shows a diagrammatic view of a positively charged particle with electrical flux lines emanating from it, also showing how a kink appears when the charge is moved. The kink in the electrical field moves outwards with the speed of light.

Quantum electrodynamics, in fact, says that the charged particle generates "a virtual photon" which is nothing but a packet of energy. This photon travels with the speed of light and interacts with other charged particles and transfers its energy to it. For example, an Electron produces a photon that leaves the Electron and reaches another Electron, transfers its energy to it and in the process, scatters or deviates the second Electron. This process is represented for ease of presentation by Feynman's diagrams shown below.

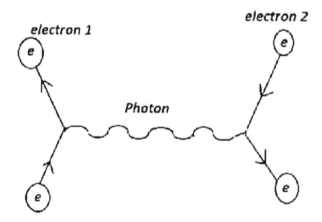

Figure iii/5.2: Shows a typical Feynman diagram showing an exchange of a virtual photon between two Electrons as per QED.

"What is the exact difference between photons generated by the Electron that repels another Electron from the photon that attracts a Proton?" is not described in detail. However, this does imply that during the formation of an atom, the Proton in the nucleus and Electron in the orbit should also communicate their affinity towards each other through the force carrier particle of a photon.

If two opposite charges exist close to each other (like in atoms) and the distance between them oscillates between two fixed points, this causes oscillation of the electric field in the midpoint between them. This oscillating electric field disturbance travels forwards with the speed of light. As the electric field here is constantly changing, it is associated with a constantly changing magnetic field which is usually at right angles to the electric field disturbance. This travelling disturbance in the electrical and magnetic fields is what we call an Electromagnetic wave.

When the Proton within the nucleus binds with the Electron in the outer orbitals of an atom, and these outer Electrons absorb a photon, the Electron jumps to a higher orbit. When it re-emits the photon, it jumps back to the lower orbit, and this oscillation gives rise to another EM wave. This is how atoms keep emitting visible light and enable us to see things.

The charge is thus an important fundamental property.

However, modern physics assumes that charge exists as a property rather than attempting to explain "how charges operate or how a charged particle manages to do all these miracles?"

Charged particles in our theory

If we want to attempt to explain charges on the basis of our theory, a charged particle manages to modify time around it in such a manner that there are zones of actively contracting space and/or actively expanding space.

The exact location of the distribution of time is indeed difficult to predict. However, certain models can be described.

The basic properties of charged particles need to be explained. These include

1. Like charges repel
2. Opposite charges attract.

This means that there need to be both types of zones around the charges, time contracted and time dilated zones.

So that when like charges come together, the space-expanding time contracted zones of both charges align and form bands extending from one charge to the other. This causes repulsion between them.

When opposite charges come together, the space-contracting time dilated zones align together and form bands that extend between charges and lead to attraction.

These concepts look convincing but a lot needs to be explained and things aren't going to be that easy.

1) Like, what happens to the attraction zones when like charges face, why they do not align, why their attraction doesn't work as well.
2) What about the photon of Quantum mechanics? What about the non-instantaneous speed of light interactions?
3) What gives rise to these time contraction/time dilation zones? How do they change with time? How does space contraction/expansion change?
4) How do we explain the strength of the electromagnetic force?
5) How do we explain other properties like spin/colour charge and how does the property of charge interact with other properties like colour charge or spin?

To answer these questions isn't easy and possibly some of them might remain partially unanswered.

However, our theory has the potential to answer each of them eventually.

Defining Matter and Antimatter

The definition of matter and antimatter is significantly different as per our theory compared to the accepted definition in physics. Therefore, what is needed here is to first define matter and antimatter as per our theory.

As described above,

Matter or matter-particle is formed when high amounts of positive energy i.e. T=0 compartments are concentrated in a small volume of space. What happens by this is the creation of a miniature black hole or a sub-Planck length prison for the T=0 compartments.

An antimatter particle on the contrary would then be a similar sub-Planck length prison created due to the concentration of T=Tmax Compartments in a small volume of space.

Matter particles possess positive mass. Antimatter particles would then possess negative mass.

These definitions are different from what contemporary physics defines antimatter as. As per contemporary physics, each type of matter particle has a counterpart that possesses negative energy, they have similar mass but opposite charge.

For example, the anti-particle of an Electron is a Positron with the opposite charge and the same mass.

Anti-Proton is composed of antiparticles of its constituents i.e. it is made of anti-Quarks of various types.

Matter and antimatter, according to conventional physics possess the property to annihilate each other to produce energy.

(Significance of the need to have such a different definition will be clearer soon, although nomenclature is not really important and if needed, the old one can be restored.)

To avoid confusion antimatter as defined by our theory would henceforth be referred to as exotic antimatter.

Exotic antimatter

An exotic antimatter particle, like a matter particle, is a localised concentration of energy, just that it is negative energy i.e. T=Tmax. These T=Tmax Planck compartments have an extremely contracted time and thus expand space.

Like matter particles, the energy from them also tends to flow from the high energy region to the surrounding low energy region. This means that like matter particles, the T=Tmax compartments affect the time in Planck Compartments immediately adjacent to it and reset time in them. The neighbouring Planck Compartments in turn reset time in the Planck Compartments that lie beyond. This means that all around the exotic antimatter particle, a resetting wave starts and travels outwards as a concentric expanding sphere. <u>This wave is a space-expanding wave and also has a passive contraction component.</u>[4#]

> 4# - This concept of an associated passive spatial contraction component, introduced initially to give a sine wave like waxing and waning to the associated wave was later found to be difficult to explain after insight about EPCAs emerged, and was given up. Note that a wave does have two components each, the "Manifest" one in the form of the PC aggregates and the "Unmanifest" component in the form of the PC vacuum.

The Exotic Antimatter particle will thus behave more like a white hole and it will have actively expanding space all around it.

These space-expanding time resetting waves can interfere with other space-expanding waves and their space expansion would get added together. Similarly, these time-contracted waves can interfere with time-dilation waves emitted by an adjoining matter particle and its space expanding nature can get completely or partially cancelled.

Like matter has mass, these particles would have negative mass.

There is a correlation between how much is the mass and how much is energy involved in the creation of the matter particle. This energy would indeed have a correlation with the number of Planck Compartments collapsing into a sub-Planck length prison to form the particle. The more the mass, the more the energy is involved and more is this number of PCs collapsing. More number of PC'S with T=0, more are the PC's disappearing every Planck time, and thus more is the suction effect or faster is the inward-directed escalator of Planck Compartments.

Geodesics

A Geodesic is the shape of space-time at a particular region of space-time.

In General Relativity, with space-time being completely relative, the pace of time and status of contraction or expansion of space varies with location or point in question. It is extremely difficult to predict what is the present state of space-time curvature at a point at a given instant of time.

There is a single way of knowing about this shape or status of expansion/contraction of space-time.

What we need to do is pass a ray of light and observe how it passes through the point. A ray of light is also a rather unscientific description. Thus it is better to use a particle of light i.e. a photon for this purpose.

If we want to know the status of space-time curvature at point P, pass a photon of light in any direction so that its future path passes through point P.

In relativity, the space-time curvature is rather static and doesn't change unless the particle causing it moves.

An apt example of a Geodesic is the region of space-time just outside the event horizon of a Blackhole. In this region, the space-time is such that a photon of light travelling inwards tangentially keeps curving around the edge of the event horizon and it keeps revolving around the black hole in circular orbits and eventually, the concretion disc of the black hole is formed.

Significance of Geodesics

If two particles with conflicting effects on time around them co-exist at a small distance between them, both sending time-resetting waves to the surrounding areas, the overall effective space-time curvature formed in their immediate vicinity and beyond has relevance.

The space-time curvature created by one particle can have a significant effect on how the resetting waves starting from the other particles propagate.

The figure iii/5.3 shows one such particle with waves emanating from it going all around unhindered when there is no other particle in the vicinity, compared to another situation where the same concentric waves are propagating at different directions and get distorted due to a presence of a strong inward pull like an escalator of Planck Compartments moving towards another point.

The velocity with which the Planck Compartments are sucked in due to the presence of P, specifically how much it is compared to the velocity of light is important. If the Planck Compartments are being sucked with the speed of light (which happens in the centre of a Blackhole beyond the event horizon) none of the waves would escape. This means that the part of the wave in the PQ direction would also be pulled inwards and can't escape. All the other directions would curve downwards towards P.

However, we know that P, even if it is like a Blackhole, its event horizon doesn't extend till Q. The velocity of suction of Planck Compartments would then necessarily be less than c. Also, higher is the mass in P, higher is the amount of energy trapped within, i.e. higher are the number of T=0 Compartments trapped within P.

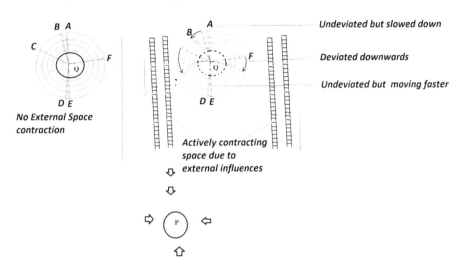

Figure iii/5.3: Shows significance of Geodesics. Note how A, B, C, D, E and F move out from the centre Q without any deviation when there is no external space contraction as shown in the left part of the figure. When active spatial contraction due to the presence of Positive energy core at P is introduced, the paths taken by each of these deviate. The equatorial ones have the maximum deviation of path while the polar ones have less deviation of the path but their velocity is significantly affected.

(Such a situation commonly exists near the nucleus of an atom or even within the charged particles like Electron as per the matter-antimatter model.)

We can assume that the point P in the figure is the centre of a nucleus. And Q is at a distance from P. The particle at P due to the energy contained within causes a high degree of suction effect to the Planck Compartments all around due to the spherical concentric waves propagating from it. The spherical waves starting from Q will get distorted. In the direction towards QP. The waves propagate with a speed of light outwards.[5#]

> 5# - Later it became clear that these TRWs or time resetting waves, in fact, travel much faster than the speed of light. The linear velocity of formation of the Push-Pull bands is much faster than the speed of light, but drift velocity of these bands is equal to the speed of light. While formation, being formed faster than the speed of light, these bands could potentially go into the Black hole event horizon (of the Nucleon) and come out from the other side with nothing but a changed trajectory.

The suction would indeed be directly proportional to the mass of P and inversely proportional to distance PQ.

If we analyse lines starting from Q in all directions, the line going directly towards P i.e. QP will be minimally affected in direction but maximally affected in quantity. While the lines perpendicular to that will deviate maximally.

Here the ratio of the velocity of suction of Planck Compartments to speed of light i.e. v/c is important.

At some ratio v/c and at some distance PQ, it is possible that the lines perpendicular to PQ will deviate in such a manner that they form a circle around P (let us name them QA and QB).

If there is an exotic antiparticle made up of negative energy at Q, i.e. There exists T=Tmax Compartments collapsed into a Sub-Planck length prison at Q, the resetting waves arising from Q will be a time contraction wave or a spatial expansion wave.

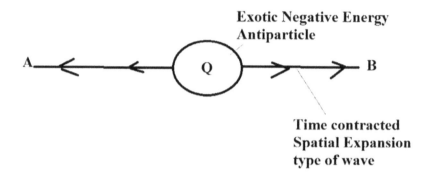

Figure iii/5.4: Shows an ENEA Q with spatial expansion waves emanating from its equator, namely A and B when there is no spatial curvature.

The spatial expansion wave arising in direction QA and QB at the equator can meet each other at a point diametrically opposite to Q as the geodesic causes them to curve round P.

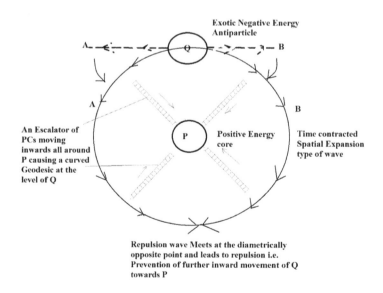

Figure iii/5.5: Shows the same ENEA in Figure iii/5.4 named Q with the spatial expansion waves, now under the influence of a Positive energy core at P leading to a curved geodesic at the level of their paths. Due to the curved paths, the spatial expansion waves A and B curve around the PEP and meet at the diametrically opposite end.

If the suction effect is the same as the expansion effect, the expansion waves can provide the much-needed repulsion that can counter the attractive forces. For this to happen, the masses at P and Q should be equivalent.

But if Q has very little negative energy compared to P, the repulsion won't be enough and Q would keep falling towards P.

Let's say that mass at P be equivalent to energy E. The negative energy at Q is much smaller than E, indicated by e1. Let's presume that E is so high that

$E = ne1$

This means that if we have n number of negative energy exotic antimatter particles with mass equivalent to e1 arranged in a spherical manner around P, the resultant repulsion due to curved geodesic would be sufficient to counteract the attraction caused by the suction caused by energy E present at P.

This takes us to the Matter- Exotic Negative Energy Antimatter model of a charged particle.

Chapter 6

MATTER-ANTIMATTER (OR MATTER-EXOTIC NEGATIVE ENERGY ANTIMATTER) MODEL

This model is a probable model of a charged particle based on concepts described in this book.

It could be completely wrong or it could have some errors which might need corrections. It is by no means a complete explanation and can be considered a direction of thinking rather than a complete model.

A charged particle, as explained earlier needs to have both actively contracting and actively expanding regions of space-time around it. The electromagnetic force is almost 10^{30} times stronger than gravity itself. This means that the model should offer an explanation of why the force is so strong compared to gravity (which happens due to the resetting waves in their original form).

The model is able to provide some answers. But some are still yet to come.

Imagine a matter particle with positive energy (i.e. T=0 Compartments) E within it. Due to this high amount of energy, it will have a region of actively contracting space around it. Let this particle be at co-ordinate location P.

At a distance r, let there be n number of comparatively smaller negative energy exotic antimatter particles, each with energy e1 arranged in a hexagonal or triangular lattice,

They are at such a distance from the centre P that the expansion waves starting from them are bent forming a circle. Each individual wave of expansion is insignificant as E is much more than e1. But when waves from all the particles are put together, that is ne1 is considered, the spatial expansion effect caused by these, as they all add up would cancel or exceed the inward suction effect due to P.

Due to the cancelling of suction effect due to P and expansion effect due to each of the waves arising from each antimatter particle, the Antimatter particles lattice will remain in a stable orbit and will not collapse or expand.

It is as if there are balloons in between each of these particles pushing them apart and there is a vacuum-like suction sucking each of them towards the centre.

The ratio ne1/E is relevant as well.

If ne1=E, then the negative and positive mass energies involved are equal, which means the expanding and contracting tendency would cancel out. This probably would make up neutral or uncharged particles. If ne1 is bigger, negative energy becomes more and the tendency of expansion is higher than contraction. While if E is bigger, positive energy is bigger and thus contracting tendency prevails.

When E<ne1 probably it is a negatively charged particle

When E>ne1 probably it is a positively charged particle

When E=ne1 probably means a neutral particle.

Note that all these are possibilities and the actual model would need significant computer modelling following the application of mathematics

The lattices can also have spins which would explain properties like the spins of various particles, the magnetic moment of the Electron etc.

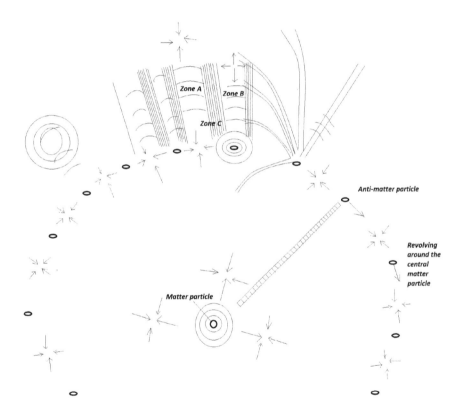

Figure iii/6.1: Shows how a positive energy core will give off Time dilation spatial contraction type time resetting waves in the centre and would cause an escalator of PCs being sucked inwards. The lattice is made up of much lesser negative energy each can remain stable at a distance. The zone immediately around the charged particle would be divided into various time zones depending on the Local Time changes. Zone A probably having excessive time dilation and active spatial contraction, zone B having excessive time contraction and thus active spatial expansion and zone C, i.e. intervening zone with no spatial expansion or contraction due to Local time running with Universal Time.

What explains the strength of the electromagnetic force?

As explained earlier, the electromagnetic force is 10^{30} times more powerful compared to gravity. That is why a small magnet can completely overcome the gravitational force of the humongous object called the Earth and pull the metallic pin upwards.

This needs to be explained why the Electromagnetic force is so strong.

There are inherently two things that need explaining,

Why the attractive EM force is so strong?

And

Why the repulsive EM force is present and is also equally strong?

The detailed mathematical explanation will need to wait until detailed mathematical equations of the theory are derived. But the probable explanation is described below.

<u>The phenomenon of interference holds the key. In the typical double-slit experiment, when two waves of light interfere, we get alternate light and dark bands of light on the screen. Essentially what is happening here is that at the places where it is dark, the crests and troughs are meeting, i.e. the waves are meeting out of phase and thus the crests cancel the troughs leading to a dark band.</u>[1**] In the place where light can be seen, the intensity of light is found to be more than the intensity of the individual incident waves. This happens because here the waves are meeting in phase so that troughs of the first add to the troughs of the second wave leading to an adding of the intensity of light. There is no creation of energy from nowhere taking place. There is just a redistribution of energy. The energy which was supposed to reach the dark spots is being redirected towards the brighter band.

> 1** - We know a part of the probable solution to this riddle now. MONDian type of Gravity caused by mass of the entire Universe leads to outwards pull that reduces the overall gravitational pull by a factor of 10^{10}. This is what we call G. indeed, unanswered questions persist as to "why this outwards pull doesn't act between two interacting charged particles thus reducing the Electromagnetic force between them?"

This positive or constructive interference has to happen 10^{30} times at each band.

In the case of light waves, there is a crest and a trough. So both constructive and obstructive interference is possible. But in case of the resetting waves produced by the energy I e. T=0 Compartments within the particle, the only active component is the spatial contraction component and thus in these, only constructive interference can happen, i.e. they can add up to form higher time dilation and a higher rate of spatial contraction.

In the case of an exotic negative energy antimatter particle (ENEA particle), the resetting waves will be a spatial expansion waves and even in it, the only possible interference will be constructive interference.

Each of the ENEAs forming the lattice will keep giving off such expansion waves. Originally spherical, the ultimate wave would be significantly distorted due to the inward suction of Planck Compartments towards the central matter particle.

Each of the spatial expansion time contraction resetting waves emanating from the ENEAs can and will interfere constructively with a similar wave originating from other ENEAs and also interfere destructively with similar time dilation spatial contraction waves emanating from Positive Energy Matter particle in the centre.

There is another very unique way of constructive interference possible.

To understand this, one can imagine two adjacent ENEA particles forming a part of the lattice of a charged particle namely A and B. Let the centre be P.

As spatial contraction waves start from P and move outwards towards A and B, they cannot pass right through the ENEAs but can pass in the space in between them. While passing in the space between A and B, the waves are modified by the spatial expansion waves emanating from A and B.

For this, analysing two geodesics R and S starting from P going outwards in between A and B is useful. The geodesic R is closer to A and S is closer to B. As space expands between A and R and also between B and S, the Geodesics are pushed away from A and B respectively. This means that the spatial contraction waves passing in between A and B will be pushed towards the central band-shaped region in between A and B where all of them Interfere constructively and increase in strength.

As these waves emerge out between A and B, the strengthened waves diffract, that is they spill over on both sides. Here they can interfere with the waves from other sources including the spatial contraction from other holes and spatial expansion waves from other ENEAs.

The result is a complex pattern of distribution of time.

The ultimate result is mostly related to the ratio $ne1/E$.

If an adequate number of ENEAs is present everything cancels each other out and it results in an un-charged or electrically neutral particle.

If there is an excess of ENEAs then the tendency to spatial expansion would be more and spatial contraction would be less. If there are fewer ENEAs, the contraction waves dominate.

One of these situations is a negatively charged particle and the other is a positively charged particle.

Most likely, the excess ENEA situation with a $ne1/E$ ratio of more than 1 is a negatively charged particle.

The deficient ENEA situation with a ne1/E ratio less than 1 would be a positively charged particle.

It is unclear how many ENEA particles will be present around the Positive Energy Matter (PEM) particle but the number n could be as big as a million.

This means that the individual negative energies of the ENEAs could be as little as a millionth of the positive energy at the centre. The negative energy of an individual ENEA could even be as small as $1/10^{30}$ th of the central positive energy. This means that the positive mass at the centre is huge as compared to the individual negative masses of individual ENEAs forming the lattice.

Note here that each of these particles is sub-Planck length, be it the Positive Energy Matter particle at the centre or the negative energy Antimatter particles forming the lattice. However, the composite particle formed due to the lattices gains a definitive but small diameter of 10^{-18} m in the case of an Electron.

This way we can explain the charge of charged particles like electrons or positrons.

The small difference in mass is probably beyond our measurement capacity

Another possible alternative arrangement could be that negatively charged particle is the one with ENEA particles arranged in the periphery as lattice while positively charged particle with the same mass, i.e. the Positron has a dense ENEA at the centre and an equal number of small PEM particles arranged as a lattice around it.

This way one can explain the stable configuration and longer half-life of an Electron while a comparatively unstable nature and early decay of the Positron.[2**]

> 2** - It was later realized that ENEAs cannot form aggregates and thus central negative energy core is not possible.

What is needed here is detailed computerized modelling to find out how the distribution of time will look with various arrangements and whether these arrangements do in fact result in composite particles whose properties resemble that of the known charged particle.

One of the extremely important implications of these is that the inertial mass of a charged particle which we calculate by measuring its resistance to motion when a known amount of energy is given to it is the composite mass of both the type of particles involved. The behaviour of a PEM particle when energy is applied is known. The behaviour of an ENEA particle is unclear. However, we should remember that whatever experiments have been carried to date to measure the mass of an Electron have been done in its exact form and thus if an Electron does compose of more fundamental particles, then whatever mass we have measured till now is the composite mass and not the real masses of all the particles involved.

Explaining the spin and angular momentum

We know that Electrons have a property called spin. This is an integral fundamental property of the Electron or for that matter any Quantum particle.

It is said that the spin of the Electron gives it intrinsic angular momentum.

Linear momentum or moment of inertia of a body is a property due to inertia by way of which, the body continues to move in its state of motion in a given direction and that additional work or forces will be needed to counteract this momentum for the body to be brought back to rest.

In that context, the angular momentum of a body or angular moment of inertia would be the property of a body to maintain its rotational motion due to its inertia. This force will have opposed by an opposing force to bring the body to a stop. Angular momentum is thus a force that actively rotates a body and keeps it in rotation.

Intrinsic angular momentum would then indeed, by logic, also be an intrinsic property of a particle that keeps the particle in its rotational motion. That is, it involves forces that are constantly present and are directed tangentially. In our theory, the application of force is due to either active spatial contraction or expansion. This means that the spinning charged particles must have zones of active unopposed tangentially directed or possibly spiralling outwards zones of spatial expansion or contraction.

We know for sure that the spin is a property fundamental to the make-up of the Electron (or for that matter other particles). Physicists often explain that it is not simply a spherical object spinning around an axis, what is pictured by our mind when reading this word.

This is because for achieving the angular momentum needed ($=1/2*\hbar$) at its size of 10^{-18} m, the various points on the surface of the sphere will have to move a million times faster than the speed of light.

Alternatively, applying the formulas for angular momentum in classical mechanics to the Electron sphere predicts that an Electron has to have a diameter that is bigger than the diameter of the atom itself. This erroneous prediction indicates that an Electron cannot be like a classical mechanical spherical object spinning around an axis.

The unique arrangement of negative energy particles like ENEA particles in lattice causes the emanating spatial expansion waves or the spatial contraction waves coming in between them to spread spirally as they move out due to the spin. These spatial changes will be tangentially directed as needed and will keep pushing the particle into the rotation and are akin to angular momentum, just that this angular momentum is due to intrinsic waves and not due to inertia of the mass of Electron.

This would indeed give a much higher value of angular momentum at much lower velocities or masses/diameters. What we are talking of are quantum particles, the formula for angular momentum in classical mechanics cannot be applied here. A single ENEA particle would produce one Planck Compartment per Planck time extra which amounts to 10^{43} Planck compartments per second. This is a significant extra volume of space. Each cycle of creation would push the Buckyball lattice. Thus the lattice isn't moving just because of its mass but is moving around because of the push caused by this extensive volume Creation.

The spin leads to the Electron behaving like a miniature dipole magnet and this imparts a miniature magnetic potential to it. How this happens is explained in the section on magnetism.

An Electron can be either spin up or spin down. Like Electron almost all particles have this spin. Electrons are spin 1/2 particles. When the Electron is in an external magnetic field, it encounters an angular deviation called a magnetic moment.

<u>The lattice has to have extreme symmetry in the position of individual particles and their arrangement so that multiple particles can be equidistant to other particles.</u>[3**] This will ensure that waves emanating from each particle may constructively interfere at a line diverging from the centre of the particle outward.

> [3**] - Contrary to this belief, it was later realized that the poles of a charged particle have to have a deficiency of ENEAs or excess of ENEAs. This causes a pendulum like movement of the ENEA lattice which is used to explain superposition of Spin.

The region of the hole would have constructive interference of spatial contraction waves while the region of ENEA would have a diverging band where spatial expansion waves constructively interfere and it is here that the spatial expansion band of the charged particle will form.

But for the lattice to have extreme symmetry of arrangement, the individual particles should be bonded to each other. The adjacent particles push the other apart at an angle of 60 degrees. Any particle trying to move faster or slower would be pushed firmer and thus pushed back into position. The lattice would indeed remain stable like a Buckyball arrangement of carbon atoms. This structure would be extremely rigid and would spin around the central core with a high but constant velocity. Note that the individual particles are not in a circular orbit around the centre like the planets in a solar system.

If we consider four ENEA particles, they cannot be in a single plane. This means that there will have to be an outward push and as such this structure cannot remain stable and would keep expanding and fall apart unless there is a high-density central core pulling it inwards together.

If we consider the space in between two ENEA particles, the individual spatial expansion waves are too weak to completely counter the spatial contraction waves coming from inside. This would mean that the spatial contraction happening here would collapse the structure unless the structure is rotating rapidly so that the spatial contraction helps the rotation of the structure instead of collapsing it.

Here, one can note that the inward pull is counterbalanced by the outward push along with the centripetal force.

This centripetal force differs depending on the radius of curvature of the path of revolution. For the ENEA particles near the equator of the Buckyball like lattice, the centripetal force will be much higher as the radius of curvature of the revolution of the individual ENEAs here is much higher. The radii of orbits of the pole based ENEA particles would be comparatively smaller and here the centripetal force much lesser. The inward pull will still remain the same here.

This can be explained on the basis of individual rearrangement of pushes. This means that the central core pulls the lattice as a single structure instead of pulling each ENEA particle separately. The lattice also rotates around as a single structure instead of individual particles orbiting at varying orbits.

The perpetual motion of Bondi et al

This is probably the best place to describe what was described by Bondi et al.

Bondi et al, in their paper, described how negative mass particles are fully compatible with relativity and what will be the possible properties.

They suggested that two particles with positive masses will attract each other as we know very well and what we call gravity.

Two particles with negative mass will repel each other.

Two particles, one having positive mass and one with negative mass, according to them will have a curious motion which they called "the perpetual motion" in which the negative mass particle will keep running away and the positive mass particle will keep chasing it. In short, the negative mass particle will repel and the positive mass particle would attract.

This mysterious motion was one of the reasons why many rejected the idea of the possibility of existence of a particle with a negative mass.

Can this perpetual motion explain the relentless perpetual rotation of the Buckyball of ENEA particles or the high angular momentum of the Buckyball lattice?

The model looks very attractive to me despite its current deficiencies and I do feel that the spin of the Buckyball-like lattice of equal but negative mass could be explained by this perpetual motion of Bondi.

The real answer however will have to wait till complex mathematics can be included in our theory.

Why imagine so much complexity?

One can see that the model of charged particles proposed here is a bit shaky. Its foundations at the moment are a bit weak and even a small change in its structure or unexplained behaviour of some forces or changes in Geodesics at places can make the entire structure unstable and can hit the credibility of the model.

The model is particularly of interest given the fact that it can help us solve one of the equally perplexing problems of matter-antimatter asymmetry. Laws of physics say that Matter and Antimatter particles were produced in equal numbers at the start during the early stages of the Universe. As explained elsewhere how the formation of a charged particle from nothing can be explained relatively easily if this model is correct. The Matter antimatter asymmetry problem will be solved without much effort i.e. we would find the lost antimatter.

Structure of Antimatter I.e. Positron.

One can see that the PEM particles and ENEA particles can exchange places. If a high-density Negative Energy Matter core is present at the centre and is pushing the lattice of PEM particles, the structure can still work. The expanding space beneath providing the push force and the contracting space in between providing the counterbalancing pull. However, in this case, there is a likelihood that the pull may prove to be inadequate given the fact that the rotating of the lattice structure cannot produce an inward-directed force and its direction is always directed outwards.

Thus, the matter version of this particle would be stable but the antimatter version would be very unstable and would have a smaller half-life.

CHAPTER 7

THE CHARGED PARTICLE

The emerging description of the charged particle (Electron)

The typical charged particle, the negatively charged Electron, according to conventional notation is a spherical region of space or more intuitively a spherical particle, into which electrical flux lines coming from all directions and starting from infinity, sink in.

Note that this was the old version of an Electron described in classical Electromagnetism and we know it to be flawed or wrong. In today's world of physics, the quantum description of the Electron is significantly different and involves Quantum superposition and wave-particle duality.

But this classic description of the charged particle is helpful in explaining many phenomena routinely encountered like the flow of electric current. Here the electric flux is lines of forces that a unit charge would feel when in the vicinity of the classical Electron. Classic description only talks of flux directed inwards. The flux lines represent the ability of a charged particle to create an electric field around it. When the charged particle moves, another field appears called the magnetic field.

Both these unique abilities, to generate electric and magnetic fields are considered fundamental properties of all charged particles and classical or Quantum physics do not feel the need to further explain why this happens.

The above description of a charged particle in our theory leads us to a slightly different description.

In it, a charged particle would have two types of lines of force. The sphere would be about 10^{-18} m and the centre of it has a matter particle with a high positive mass equivalent to a singularity where T=0 Compartments are trapped. The periphery of the sphere has a Buckyball-like sphere of ENEA particles with a small negative mass spinning around at high speeds. Both the central singularity and each of these peripheral ENEA particles give off time resetting waves. The region of the gap between the ENEAs lets out dilated time and radiates out, the zone of active spatial contraction. The region beyond each ENEA particle gives rise to the spatial expansion zones.

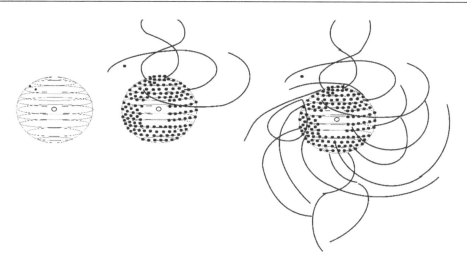

Figure iii/7.1: Shows a diagrammatic representation of a charged particle with a central positive energy core and a peripheral ENEA lattice. Note the zones of excessive time dilation/contraction forming bands emanating around the particle. Note that the density of these bands will be highest at the poles and least at the equator.

A spherical wave-front of time-wave starts every Planck time and determines the spatial configuration and Local Time at this sphere. The sphere emerges out and moves outward. After a few Planck seconds, the Buckyball lattice rotates and so the location of zones of maximum time dilation/contraction and the intermediate zones in the emerging spherical wave-front also rotates with it with respect to the outer sphere.

If we analyse each such spherical time wave, each point in it would be entangled to each of the other points on the wave-front. Also, each point would also be entangled with the outer spherical wave-front. Until the next rotation of the Buckyball which may happen in a few Planck times, the location of the maximum spatial expansion zones and contraction zones coincides with the first such sphere emanating out.

On analysing many adjacent waves together, the zones of contraction or expansion lie in a spiral. The distance between them keeps increasing as they move outwards.

If we focus on the Electron (the charged particle) one thing is clear, as the spin can be up or down, the Electron isn't a symmetrical particle which means that it has an up and a down. If one cannot make out the upper pole from the lower pole, rotating the same by 180 degrees will reverse the direction of spin. In this case, the up spinning and down spinning Electrons would be identical which is absurd.

The upper pole of the Electron probably has one or more extra ENEA which explains the extra expansion potential and which determines it to be negatively charged. <u>On the contrary, a positively charged counterpart i.e. the positron would have a deficient ENEA at the opposite pole (or an extra PEM particle the antimatter Electron) (note here that according to our theory there are two ways of producing a positively charged particle which has very similar mass as that of Electron. The anti-Electron with central ENEA core and peripheral PEM lattice and another matter version in which there is a central PEM core and an ENEA lattice just like</u>

our Electron but with a deficient ENEA rather than extra ENEA which can be called Proton counterpart of the Electron. Note that a composite particle with identical PEM mass and ENEA mass/negative mass would be neutral electrically and would be a Neutron counterpart of an Electron.)[1*]

Note that in our theory, in negatively charged particles, there is more ENEA mass rather than PEM particle while in a positively charged particle, PEM mass is more than ENEA mass. (that's an assumption and one must keep in mind that, in reality, this may be exactly the opposite with positively charged particles having an extra ENEA and a negatively charged particle having a deficiency of ENEAs.)

> 1* - These particles with an ENEA core at the centre were later found to be impossible as we noted that ENEA particles cannot form a large aggregate. The antimatter as defined by conventional physics is a particle with equal mass but opposite charge. Thus the Electron-antiparticle also has positive charge. What is more likely is that antimatter like Positron is a particle with positive energy core with mass similar to the Electron but a slightly deficient ENEA lattice so that there is an imbalance towards positive energy

Interaction between two charged particles and its comparison with the Quantum electrodynamics

The charged particle, in our theory literally is giving out time resetting waves which go out in all directions and form a situation where every point in space around it would have a deviated time and that another charged particle present at any point around it would experience a force, either attractive or repulsive.

This, on superficial analysis, contradicts significantly with the accepted theory of Quantum Electrodynamics (although on deeper analysis, it may very well gel well with it).

Quantum Electrodynamics states that an Electron when coming near another Electron gives off a virtual photon (a packet of energy) which emerges out from the first Electron and in the process changes its direction, carries a part of its momentum forward and subsequently collides with the other Electron to transfer its momentum to the second Electron while thrusting the second Electron in a different direction in the process.

This is well illustrated by the famous Feynman's diagrams. These diagrams (as Feynman's also pointed out occasionally) are just for understanding and may not necessarily be what happens in reality.

Many philosophical questions remain for a questioning mind like

1. ***How does the first Electron know that the second Electron does exist?***
2. ***How does it know in which direction it is, how can it calculate its velocity so precisely for it to send the photon in exactly the same direction? One should***

> ***realise that the first Electron has no mechanism by which it can know the exact location or coordinates of the second Electron.***
>
> 3. ***The Uncertainty Principle and Quantum Superposition would make it almost impossible even for nature itself, let alone the first Electron to know where the second Electron is located at a moment or in which direction and at what velocity it is moving.***

An Electron which is just 10^{-18} m in diameter and is moving rapidly and is a particle and wave at the same time and whose location is unknowable would indeed be a difficult target to aim at (and throw its photon) for the first Electron. The conventional Electromagnetism would conversely make more sense as it would convey that both the Electrons modify space around them to create the electric field and the interaction between these fields causes the two Electrons to move apart.

It is known that the virtual photons of Quantum Electrodynamics do not obey conservation of energy and the energy in them (although they are a packet of energy) isn't taken away from the Electron itself. It is energy that is taken up from the surroundings and given back to the surrounding. Only the momentum of the Electron is taken up by the photon and transferred to the second Electron, conserving the momentum at each step.

Note that the spin of the Buckyball like lattice and thus the spin of the spiralling zones of contraction or expansion are an intrinsic property of the particle, they keep pushing the Buckyball into the rotation and thus Electron or for that matter, any charged particle will keep spinning, whether it is isolated from other particles, in process of bonding with another particle or interacting with another Electron.

The second Electron will also be spinning. As the time resetting waves starting from both reach each other, both the spatial expansion and contraction of the first will interact with the similar ones of the second one. What will happen to the Electrons will depend on which zones come and remain in contact. At a Planck second when maximum expansion zones of the two overlap, there is a net expansion of the space and they repel. Wherever the spatial contraction zones are interacting, the space in between them contracts and they would seem to attract each other. The net effect depends on what interaction dominates.

Understandably, there will be confusion here as there is a significant deviation from Quantum Electrodynamics. There is no mention of spontaneous particle anti-particle pair of virtual particles produced and other such relatively rare events which remain a hallmark of Quantum Electrodynamics. Particle and anti-particles here in our theory could be akin to Push-Pull bands simply because they possess opposite energy.

The spherical time resetting wave could be considered as the virtual photon or packet of energy. The difference of illustration (a single photon in QED vs spherical wave in our theory) could be regarded as an insignificant philosophical difference and thus at least the foundations of the theory match with QED. The math and the details will have to wait till our theory takes shape.

Explaining the positive or negative charges of Quarks

We know that Up Quarks have a 2/3rd positive charge and the Down Quarks have a 1/3rd negative charge.

We know that in Beta plus decay a Proton within a nucleus (with too many Neutrons and Protons) decays and produces a Neutron (which stays within the nucleus), a Positron and an Electron-Neutrino (which are emitted or ejected out). In beta minus decay, a Neutron in a crowded nucleus decay into a Proton, an Electron and an Electron Anti-neutrino.

In these decays, one can see that the electric charges are conserved along with the conservation of other properties.

We also know that within the nucleus, according to the Meson theory, the Neutrons and Protons keep exchanging small particles with mass called Mesons and in the process, a Proton is converted into a Neutron and vice versa.

We also know that the Neutron has a slightly higher mass than Proton.

We also know that a high density of energy is present in the nucleus which means that the density of T=0 Compartments here is very high.

The nucleus may also have regions of actively expanding space which keeps the Protons and Neutrons within the nucleus from colliding.

We know that the Quarks possess colour charge and that they constantly exchange particles called Gluons which constitute particle-antiparticle pairs like red, anti-blue etc.

All in all, the nucleus is a messy place where space is contracting rapidly due to extremely dilated Local Time at places.

Enough energy density exists here for the spontaneous formation of particles from energy. Many of these particles may be extremely short-lived and spontaneously annihilate back into energy. The particles, when present would behave like any other particle giving off resetting waves. These waves would however propagate in a difficult to predict manner because of the chaotic space-time here, twisting and curving as the space around them contracts in an uneven manner.

We know that the mass of Quarks constitutes less than 1% of the mass of the Proton and thus the majority of the mass of the Proton would be due to such resetting waves emanating from such short-lived particles which last only for a few Planck times before annihilating back to energy.

We know that Quarks are bound together within the nucleus by gluon exchanges with what is called the colour charge. Even Quarks are bouncing within with speeds close to the speed of light and possess electrical charge which means that waves of space-time might start from them. The path these waves take would indeed be curvy as they curve around the intermittently forming and annihilating particles. These particles may indeed be both Positive Energy Matter particles or Negative Energy Antimatter particles.

The structure of individual Quarks and their respective anti-Quarks would indeed be similar to the above with variation just in the masses. However, here the chaotic surroundings and the strong pull of colour charges would provide additional forces to counterbalance some forces making certain versions unstable elsewhere to remain stable here.

It is unclear whether colour-charge constitutes a separate charge from an electrical charge. However, there is the likelihood that it is a completely separate charge which means that the structure of Quarks might differ significantly from the above to accommodate Quark confinement and colour charges. The forces of attraction between neighbouring Neutron-Proton pairs are described to be due to these colour charges. The Strong Nuclear Force within the nucleons and even within the Proton or Neutron is extremely high and easily surpass the columbic repulsion between closely placed "like" electrical charges.

There are short-lived particles that are probably responsible for this colour charge. These have their antimatter versions as well. The red, green and blue ones having antimatter versions called anti-red, anti-green and anti-blue. These keep forming and annihilating within the nucleus and are exchanged by Quarks. The Gluons are made up of a combination of a matter particle and another antimatter particle of a different colour.

This is the first indication where modern physics suggests that matter and antimatter particles can stay close to each other for a brief period of time.

Even within the nucleus, matter and antimatter may be present as clumps that remain for some time and then annihilate.

The spin of the Protons or Neutrons

Experimental evidence has baffled scientists. Physics is all about the accounting of conserved quantities. When scientists tried to measure the spin of individual Quarks and compared it with the spins of Protons or Neutrons, they realized that the spins don't add up. This was a real problem. The individual spins of Quarks account for a very small fraction of the total spin of Protons. It is still an enigma as to where does the rest of the spin of the nucleons comes from.

A possible explanation of charge and spin of the Protons

We know that the Quarks are not static within the Proton (or Neutron). They are constantly moving. More appropriately, they are bouncing off the edge of the nucleon and keep revolving in complex paths at velocities close to the speed of light. They are held together by Gluons within the nucleon.

In our Theory, we can imagine the individual Quark to be of similar morphology to the spinning Electron described above.

Due to the central core of the positive matter and peripheral lattice of ENEA particles, each of the Quarks would be expected to give off complex waves like an Electron would give. The complex waves which have zones or <u>bands of high spatial contraction and expansion spiralling out of the surface of the particle move outwards with the speed of light.</u>[2*]

> 2* - Later termed as Push-pull bands, these zones, when forming travel much faster than the speed of light. But when they are formed and are moving outwards, they move with the speed of light. In short, their propagation along their length is much faster than the speed of light but once formed, they move outwards with the speed of light.

At very short distances of travelling outwards, these waves and the bands would encounter other similar waves from their neighbours. These would act as Eddy's so that the wave travelling outwards is bend as an eddy and the individual bands are bent towards each other. The bands are likely to make a complete circular turn if they lie at the edge of the newly formed singularity i.e. its zone of interaction. In a few Planck times, the singularity vanishes or is sucked within the primary singularity and the turning influence on the band is no longer present and the band is released from its influence albeit in a completely different direction than before. This gives rise to the complex turbulent-like flow of these time resetting bands.

We can understand that whenever the Quarks are at the edge of the Nucleon, the outgoing time wave will not encounter any singularity and the resultant time resetting wave will pass without deviation outwards. The Quarks are revolving very rapidly at close to speeds of light and the distances we are talking about are within a Femtometer which is the approximate diameter of a Proton. The rapid spin of the Quarks added to the rapid revolution of the Quarks around the centre of the nucleon sends Electron-like waves with little deviation all around the Circumference of the nucleon. The spin of a Nucleon is possibly given by the revolution of the Quarks around the core. The complex paths taken by the rotating and revolving Quarks give off waves with corridors of excessively dilated or contracted time. Each of these waves moves outwards from the rapidly spinning and revolving Quarks. The charge of the Nucleon is the probable end result of all this process, with these corridors of active spatial expansion and contraction emanating outwards.

Why did we not detect ENEA particles yet in the particle accelerators?

There was a time when the Electron was considered the tiniest of all the fundamental particles. The Electron has a diameter that ranges around 10^{-18}m.

This is 1000 times smaller than that of a Proton which ranges from 10^{-15}m.

With current energies, we can probe only up to this limit of about 10^{-18} metres. Probing anything smaller needs higher energies which we do not have access to, yet.

Can Neutrinos give us a clue?

Now we know that Electron is not the tiniest particle. Neutrinos, which are electrically neutral, are extremely small particles that interact very little in their original independent form. Neutrinos were initially considered massless particles. But now we know that Neutrinos have a very tiny mass which is about a millionth of that of an Electron.

There are three flavours of Neutrinos. In-flight, a Neutrino keeps changing its flavour.

The Neutrinos are so tiny that their detection is extremely difficult. Another interesting property about a Neutrino is that there is a likelihood that Neutrinos are antimatter particles of themselves.

Neutrinos have the property of weak interaction and gravitational interaction. They are not affected by the strong interaction. This means that in their parent form, they are not attracted by the nuclear force and are incapable of forming a part of a nucleus.

The weak interaction has a very short range. By this, they can collide with a nucleus and cause changes in it. This was the first means of detecting them. This is however very rare. The gravitational interaction is also very small. Thus majority travel with velocities close to the speed of light and pass right through matter particles without much interaction.

They are one of the most abundant particles in the standard model. Neutrinos are produced in many reactions.

On the Earth, the majority of the Neutrinos reaching the surface arrive from nuclear reactions taking place on the Sun. Every square centimetre of the Earth's surface is bombarded by 6.5×10^{10} i.e. 65 billion solar Neutrinos every second.

However other sources include artificial particle accelerators, solar flares, cosmic rays, Neutron stars etc.

They are a spin 1/2 particle. This is an important clue.

Although they are considered fundamental particles, it is not impossible if they are made up of a much smaller number of PEM and ENEA particles clumped together arrangements like the ones given above.

Proving that would be extremely difficult. However, their existence proved that there can be things smaller than the Electron.

All is not known yet about Neutrinos and experimental research is in full swing. Scientists have detected that Neutrinos change their flavour or oscillate between flavours. The three flavours have slightly different masses.

These might open up the possibility that they aren't fundamental and are also made up of a similar combination of matter and antimatter particles.

Chapter 8

A BOND

Almost all scientific disciplines describe various types of bonds. Particle physics involves description of multiple bonds between particles of various sizes.

An Electron forms a bond with the nucleus to form an atom. A proton forms a bond with the Neutron to form a hydrogen nucleus. Two hydrogen atoms form a bond in between each other to form a hydrogen molecule.

Classical physics explains these bonds based on different fundamental forces. The bond between two nucleons, for example, is explained to be due to existence of strong nuclear force, a force between two nucleons. The bond between Electron and Proton (or positively charged nucleus) is due to the electromagnetic force.

A bond, as defined by plain logic, is a tendency of the particles to maintain distance between each other and more importantly resist any change in distance between them.

The two particles forming a bond exhibit two important necessities.

- The first one is rather Logical. They should have a force constantly directed towards each other that prevents any increase in distance between them.
- The second one, often ignored by our intuition is a force directed away from each other. This other force prevents the two particles from collapsing into each other thus annihilating each other and forming energy.

The bond between Earth and the Sun is a good example. The Newtonian gravitational pull was supposed to be cancelling the centripetal force generated due to the revolution of the Earth around the Sun. The existence of an inward pull is counterbalanced by an outward pull.

In our theory, a force is exerted by either actively contracting space or actively expanding space due to a deviation in Local Time.

Thus, in our theory, a bond would be the existence of zones of both actively contracting and actively expanding space in the region between the particles or the region around them.

Let's consider two particles A and B that are bonded together by some force.

The distance AB remains constant or at best there is a play or degree of freedom offered by the bond which allows some change in the distance AB (say d).

If the distance increases more than this, the inward force increases and pulls A and B together again. If the distance AB reduces, the outward force increases and pushes the two particles apart again.

Let's now consider another particle C bonded to B, thus forming a complex bonded arrangement ABC. The particles can be arranged in a linear fashion or maybe at an angle. This angle is called a bond - angle. There is a limited play or degree of freedom that most such bonds provide, i.e. the angle might increase or decrease to a limited extent. However, as the angle increases or decreases beyond a point, another force often starts acting, thus limiting the amount of deformation of the angular bond that exists.

Thus, in a complex system with three particles with angular bonds, additional forces might exist between the non - neighbouring particles.

Here, A and C might have forces between them that help determine the angle along with the space-contracting and space-expanding forces described above.

These additional forces determine the degree of freedom afforded by the bond.

In the space in between A and B would exist different zones which are like bands extending from one particle to the other.

The typical toy model of bonds between particles shows the particles as spheres and the bond as a single cylindrical link between the two. The bond, however, need not be that simple in reality and might consist of multiple such links/bands where active contraction of space is happening.

The bond between Protons and Electrons forming atoms:

The electromagnetic attraction between the positively charged Proton and negatively charged Electron explains the inward acting force without any trouble.

The problem comes when one has to explain the reason why Electrons don't continue falling into the Proton and annihilate with them but instead remain in well-defined orbits.

Modern physics can tell us with certainty that Electrons behave in this manner.

The "why they behave that way?" is a question that should not be asked. They behave that way because of their intrinsic properties.

"Protons are not antimatter of Electrons and are a completely different animal, which explains why they are not meant to annihilate with Electrons" is a probable explanation.

This however fails to explain why, at distances smaller than the first orbit of Electrons, the columbic attraction no longer matters or even if it does exist is completely neutralized or overtaken by an unexplained force that prevents the Electrons from further descending towards the nucleus.

Our theory can potentially offer an explanation as to why this happens.

The repulsive force which counterbalances the columbic attraction between Electrons and Protons is the expansion wave that starts at the equator of the Electron and due to the active contraction of space present around the nucleus, curves around the nucleus and forms a standing wave around of constantly expanding space around the nucleus. [1#]

(see the figure iii/8.1)

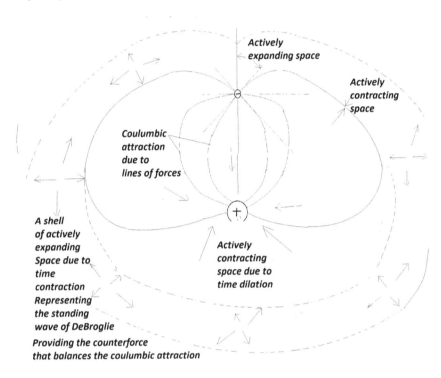

Figure iii/8.1: Shows how the space-time around a positively charged particle bonded with a negatively charged particle might look according to DGR. Especially note the actively contracting space pulling the two particles together and the wave of actively expanding space going all around the positively charged particle providing the repulsive counterforce preventing the particles from coming closer than a limit.

> 1# - There are two other less likely possibilities that I could only figure out much later. First being that the Repulsion-wave exists in the centre and the Attraction-wave goes all around the Proton. Second being that they are in a Superposition of Repulsion wave in the centre – Attraction wave going around the Proton and Attraction wave in the centre and Repulsion wave going around the particle with every pendulum movements of the lattice around the charged particles. It is extremely difficult to pin point which of these three mechanisms is actually happening and is responsible for the bond.

This, however, does not counter balance the contraction and expansion waves emanating from the distant pole of both Protons and Electrons. To balance these, they need to bond with other Protons or Electrons in the same nucleus or the nucleus of other atoms. This can offer an explanation of the various orbitals spdf and covalent bond formation of different atoms. (As explained in detail later)

Charged particles, as explained elsewhere in detail, possess a spin and bands of highly contracted or highly dilated Local Time spiral around it. At the poles, the density of such bands is much higher and there is an asymmetry in the poles which probably is responsible for the North and South poles. The charged particle is thus a miniature magnet with one pole acting as North and the other acting as South.

If bonding of two oppositely charged particles does take place, it would take in a manner that respects their magnetic properties and so the bond is likely to bind two poles. They would have spins in a direction perpendicular to the direction of the maximal attractive forces. The spinning positively charged particle also acts like a miniature black hole causing significant distortion of the space around it due to actively contracting surrounding space.

The geodesic around the Proton is such that the space is almost spherical here so that a tangentially directed ray of light here would enter into a circular orbit around the Proton and keep going around.

As the poles of our oppositely charged particles interact, they attract each other. They keep coming closer to each other until a situation comes wherein the bands emanating from the equatorial region of the Electron lie in such a tangential orbit and turn around the Proton and come back. At such an orbit when the expansion bands arising from the equatorial region form a standing wave so that the circumference of their path is an integer multiple of the wavelength of the Electron wave, the attraction between the Proton and Electron is counterbalanced by the repulsion caused by the returning expansion standing wave.

Note that both the attraction and repulsion are momentary. As both attraction and repulsion type bands emanate from both particles, at any instant which predominate determines the resultant effect.

Also, the location of the Protons within the nucleus varies and after a period of time, the Neutrons become Protons and old Protons cease to be Protons and are converted to Neutrons due to the exchange of Pions or pi Mesons. So the bond is no longer present at the brief time when the pi meson is in flight and there are two Neutrons instead of one Proton and one Neutron. At this brief period of time, the Electron Proton bond no longer exists. But within a fraction of time, the Proton reappears at a different location and the attraction and repulsion bonds are re-established for another brief period. This cycle of forming and disappearance of bonds keeps going on.

Note here that only some of the bands emanating from the Electron and Proton are interacting, mainly the ones emanating from poles facing each other.

The ones emanating from the opposite poles or the other half are free to interact with other surrounding particles like other Electrons.

Other Bonds of importance

Other bonds including bonds between Proton and Proton or Proton and Neutron within the Nucleus are dealt with in the section of Strong nuclear force.

CHAPTER 9

MAGNETISM AND ELECTRICITY

Faraday was the first one to describe an invisible field that permeates space and is the reason why Magnets have the property to attract or repel. He also gave significant insight into the close connection between electricity and magnetism.

James Clark Maxwell gave his Maxwell's equations and conclusively showed that light was an electromagnetic wave.

Modern physics has understood magnetism and Electronics to a great extent and has made a significant impact on human lives.

Unlike Quantum mechanics, classical explanations of charges, electrical flux, magnetic flux, which are philosophical methods used, can answer most of the questions. Very few unanswered questions exist in this regard which includes explanations for superconductivity and Quantum tunnelling.

But how all these compare with our new theory is what interests us.

This will necessitate at least a basic discussion about what philosophical methods are used to describe electricity and magnetism.

Classical physics says that a charge modifies its surroundings and produces an electric field. Within this field, lie electrical flux lines. The density of flux lines determines the strength of the field.

A field in classical physics is some quantity that has a value at every point. A vector field is a field that has a value at every point and a varying direction at every point. An electric field is a vector field. In an electric field, every point has what is called electrical potential. This is nothing but the ability to (or force exerted to) repel a positively charged particle.

When there exists a potential difference between two points, an electric current starts flowing which is nothing but the flow of charges from one point to another.

With these fields, the charges can affect other charges so that like charges attract and opposite charges repel. Thus where a potential difference propels a charge depends on the charge.

When charges move, they create a magnetic field. The faster is the change in electrical potential, the stronger is the strength of the magnetic fields.

Another very interesting fact which classical physics and Maxwell's equations teach us is that divergence of magnetic flux is always zero (Gauss's law). This means that there exists no point where magnetic flux is entering but not leaving or leaving but not entering. This is in contrast to the electrical field which has no such law. An important consequence of this law is that, unlike electric field which can have monopoles (which we call positive or negative charges from where electric flux only exits out or into which electric flux only enters but doesn't exist), magnetism cannot have any monopoles so that there will always be a North pole and a South pole joined together, there can never be an isolated North pole.

When a magnet is moved towards or away from a wire, it causes a change in flux travelling across the wire and this changing magnetic flux leads to the generation of an electric current.

When two charges oscillate between two points, the electric and thus magnetic fields in between them oscillate as a sine wave in between them. When they move apart against their pull, energy is spent which gets stored in the system as potential energy. When they come back close towards each other, this potential energy propagates forwards as a wave. they would soon stop at their closest distance and repeat the cycles. This oscillatory movement of charges most commonly is noted in an atom, where positively charged particles stay in the nucleus while Electrons might jump between orbits. This gives rise to electric and magnetic field oscillations which travel as electromagnetic waves which we call light. Visible light given off by everything that we see is due to this phenomenon.

However, this oscillatory motion isn't limited to atoms. When it happens in antennae, it gives rise to waves of a higher wavelength and smaller frequency which we call Radio waves. Most of our telecommunication is based on this. Every particle which is having some temperature has particles that are vibrating. These vibrations also result in the movement of such charges which causes the formation of waves that we call infrared waves which are electromagnetic waves of a higher wavelength than visible light but smaller than Radio waves and microwaves.

Entire molecular bonding and chemistry depend on the electromagnetic forces between charges. Thus how charges interact with the environment and other charges or particles within it and how this affects their surroundings is a crucial part of the study of physics.

Every particle described in the Standard Model of Particle physics which is the currently accepted model for Particle physics has a magnetic moment and can be considered as a miniature magnet.

How this happens is ill-understood, but it is explained on the basis of an intrinsic property of particles which is called a spin. Spin is also called the intrinsic angular momentum of the particles.

However, as the name suggests, physicists often advise not to take the analogy too far as the particle isn't actually spinning here. The momentum and velocity of a particle are extrinsic properties that can vary from time to time. But spin is an intrinsic property and it remains constant irrespective of the surroundings.

For a charged particle, say an Electron, the spin causes a magnetic moment, because a magnetic field is caused by movement of charges and here the Electron is having intrinsic spin (again caution is needed as contemporary Physics says that an Electron is probably a point particle and thus cannot spin).

The movement of an electron around the nucleus in its orbit can cause a magnetic moment called orbital magnetic moment. Accordingly, an atom can have a magnetic moment which is obtained by a vector sum of all the magnetic moments of particles within the atom i.e. nuclear magnetic moment plus a sum of both spin and orbital magnetic moments of all the particles.

Multiple such atomic magnetic moments often cancel each other out and thus most solid materials have no resultant magnetic moment. This isn't, however, true for permanent magnets in which all the small magnetic moments align to form small magnetically active islands. Stronger magnets may even have magnetic moments from many or all these magnetically active islands aligned properly so as to produce permanent bar magnets.

Magnetism in our theory

Like positive or negative charges, even magnets have both attractive and repulsive forces around them. Thus like charges, zones with active spatial contraction and expansion must be present as bands around them.

Description of a typical charged particle is described elsewhere as per our theory.

According to it, it is made of the central core of Positive Energy Matter where the C/D ratio would be highly deviated towards destruction. The outer shell is formed by a Buckyball lattice made of ENEA particles which revolves rapidly. The central core is like a singularity or a rotating Blackhole. Thus Planck Compartments are sucked within it rapidly. This distorts the space around it and the geodesic around it is a curved one.

The lattice has multiple notable properties. It is made up of negative mass particles or Negative energy antimatter particles (ENEA) where the C/D ratio has extremely deviated towards creation so that Planck Compartments are formed rapidly and are pushed outwards. They give out space expanding waves around. The strength is much smaller than the central core as each ENEA particle has a negative mass, which is probably (approximately) a million times smaller than the negative mass which can completely neutralize the positive core.

The sum of negative masses is comparable to the central positive energy core but not equal.

In particles in which these two masses are equal would have a balance between space expanding and space contracting tendency and thus would result in an electrically neutral particle like Neutrons.

It is unclear if such particles which have masses similar to an Electron but which have slightly less negative mass ENEA particles in the Buckyball lattice so that the negative and positive masses exactly the balance and electrical charge us negative do exist in nature or not.

A corollary to the above discussion is that the sum of all the positive energy masses and negative energy masses in the two down Quarks and one up Quark within a Neutron is exactly equal which leads to cancelling out of space contracting and expanding tendency which makes the Neutron electrically neutral. The Buckyball lattice around it still moves and thus although it is electrically neutral, it still possesses magnetic moment and spin.

If the negative mass is in excess in the form of an additional ENEA particle, the spatial expansion tendency is more and it forms a negatively charged particle. If the positive mass in the core is excess or the ENEA lattice is deficient, it forms a positively charged particle and has more tendency of spatial contraction.

The time contraction waves emanating from each of the ENEA particles interfere in the space just beyond the lattice. The time dilation waves emanating from the central core comes out of the gaps in between the rapidly rotating lattice and also interfere in the space beyond. The time contraction waves from all the ENEA particles must constructively interfere and add up. This addition is maximum at a very narrow strip or band starting from an ENEA and diverging out. The time dilation wave emerging out from the gaps between lattice structure constructively interfere with others and destructively interfere with other such waves emanating from other gaps. The constructive interference would be maximum at a strip starting from the centre of the gap and diverging out.[1**]

> 1** - This is a rather vague description of the Push-Pull bands which were not completely defined by the time I wrote this.

As the Buckyball rotates close to the speed of light, these maximum constructive interference zones also rotate. Thus the zones of maximal spatial contraction or expansion spiral out from this spherical lattice.

The spin of the particle is due to the perpetual runaway motion of the ENEA lattice.

Spin can be spin up or spin down. For the direction of the spin to be discernible, the lattice has to have an upside and a downside i.e. It has to be asymmetrical.

One of the poles must have an extra ENEA particle and the other must have a hole or gap. Note this is one way of making the lattice asymmetrical. Others could be differences in relative numbers or relative distances between ENEA particles. Which of these is the one that makes it asymmetrical is unclear yet.

Another possible and more likely reason for this asymmetry is the slightly deformed shape of the lattice. We know that the Earth is not a perfect sphere but is a geoid due to its rotation. Planets revolve around the sun, not in perfectly circular but elliptical orbits. It is likely that some similar mechanism is in action which causes the lattice to deform a bit in some yet unknown manner. This deformation is solely due to the rapid rotation of the lattice. When the lattice rotates in the exact opposite direction, the deformation is exactly reversed.[2**] The deformation leads to a slight breaking of the symmetry of interference between the waves and leads to a slight preponderance of time dilation spatial contraction at one pole while time contraction spatial expansion at the other pole, one of which becomes the South pole and the other becomes the North pole.

> 2** - A much more complete description of this deformation process is described in the Part III Chapter 28 in the Superposition of spins of charged particles section. The Pendulum like motion of the ENEA particles, which leads to automatic switching of the spins of the charged particles, was also not known at the point of writing this.

Every resultant wave that emerges out and forms have points where time is extremely contracted and thus space expanding, and other points where time is dilated and space contracting, while still other points where time and space are in different states of contracting and dilation. Every such point is in a state of entanglement with other points in the wave. Every wave is in a state of entanglement with the outer or previous wave and in turn, will get entangled with the next or inner wave as it moves out.

As the lattice rotates, each subsequent wave has a different relative location of the maximally expanding spatial bands and maximally contracting spatial bands.

If we see a cross-section of the space at the equator, the spatial expansion bands or spatial contraction bands lying here spiral outwards with the least density. As we analyse such sections higher or lower part of the rotating Buckyball lattice, here the bands spiral out similar to that near the equator but their density keeps reducing as one goes from poles towards the equator. At the poles lies the axis of rotation of the lattice. Here lies a single static ENEA and a static hole lying on the axis and the holes or ENEA particles around it rotate with much smaller diameters and thus the bands emanating out from it spiral around the axis like a spiral staircase. Here their density is maximal and varies much less as one moves away from the poles.

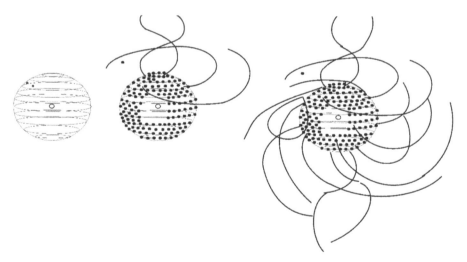

Figure iii/9.1: Shows a typical charged particle with a central positive energy core and negative energy lattice. Note the spiral staircase like the arrangement at the poles. The high time dilation (spatial contraction) bands and high time contraction (spatial dilation bands) are aptly called the Push-Pull bands emanating out of the particle and moving outwards with the speed of light. Only a few bands are shown for clarity and push and pull bands are not shown differently.

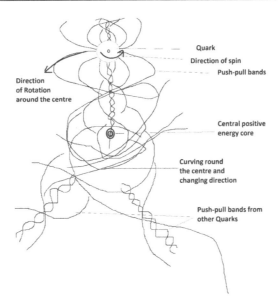

Figure iii/9.2: Shows a charged Quark that is bonded to the central positive energy core of the Nucleon. Note the spiral staircase like the arrangement and more density of the high time contraction/time dilation bands i.e. push-pull bands at the poles. Also, note that the other pole of the Quark and the Equatorial and non-polar push-pull bands can fly away and keep affecting the neighbouring nucleons. Their linear velocity being faster than the speed of light, they can curve around the centre of the Nucleon and affect their fellow Nucleons.

These polar regions where the spatial expansion and contraction bands spiral around the axis of rotation of the lattice form the magnetic poles that give rise to the magnetic moment of the charged particle.

All this description is when the charged particle is static. When it is moving, every subsequent resultant spherical wave formed following complex interference of multiple emerging and interfering waves starts with a different centre. This changes the direction of the bands of expansion or contraction so that now the bands are not only spiralling outwards but curving with convexity towards the direction of motion of the charge.

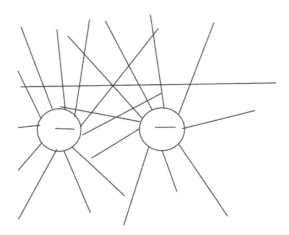

Figure iii/9.3: Shows two Electrons remaining static in a conducting wire. Note that the spin is ignored just for understanding purposes. Because they are static, the Flux lines move out in varying directions and thus cancel each other out. Thus there is no magnetism.

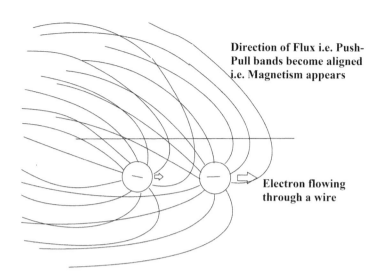

Figure iii/9.4: Shows the same two Electrons shown in Figure iii/9.3, but now moving from left to right due to the electric current flowing through the conductor from right to left. Note that the Flux lines now curve back so that their direction is aligned with others and thus their density increases. This is when effects of Magnetism start appearing around the conductor.

To note that the same spatial expansion or contraction bands that emerge out of the charged particle responsible for interaction between charges is also responsible for the magnetic field interactions, just that their direction and density changes due to movement of the charges.

The spiral staircase like arrangement of the polar bands also might bend and curve depending on the movement of the charged particle. In an external magnetic field, the particle will arrange itself to align this maximum density spiral bands zone with the direction of the magnetic field.

If we analyse the zone of space just inside the poles of the rotating Buckyball lattice of ENEA particles, here also similar bands of interference exist and these spiral inwards, so that even at this point, the bands are in a way moving from the interior of the charged particle towards the pole and then further outwards. Thus this point cannot be considered a monopole.

Which of the two poles, i.e. the pole with a hole or the pole with an extra ENEA particle constitutes the North or the South pole of the miniature magnet is unclear presently but with detailed computer modelling, it will become clear subsequently.

When the charged particle is revolving in a circular orbit, the direction of these spiralling expansion/contraction bands becomes increasingly more complex and given that even while revolving, the intrinsic spin of the lattice continues and thus the deviation of the bands due to revolution would be on top of the deviation occurring due to the spin. When multiple charged particles are present in the vicinity like what happens in an atom, the interaction of such contraction/expansion bands of other charges with those belonging to the charged particle in question and the resultant changes in orientation of the charged particles has to be taken into consideration.

A spinning sphere has an angular momentum that is directed perpendicular to the plane of rotation of the sphere. For a charged particle, this intrinsic angular momentum would be directed perpendicular to the plane of rotation i.e. in the direction of one of the two spiral staircase-like bands at the poles.

In the atom, one of the poles of the Electron would be directed towards the nucleus towards the Proton which it is bonded to. The other one would be directed away from it. These phenomena could offer potential explanations on why within an atom, Electrons have quantised angular momenta, I.e. Electrons can have only specified values for angular momentum, both direction and value. It also potentially explains why in the same orbit, two Electrons with opposite spin remain more stable than two Electrons with spin in the same direction.

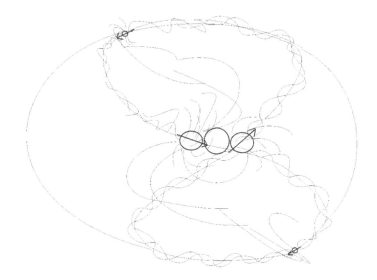

Figure iii/9.3: Shows a diagrammatic representation of two magnets (Protons) within the nucleus, whose one pole is binding to another similar pole of the Electron magnet. The opposite poles of these bonded partners and the equatorial and non-polar push-pull bands (only some of which are shown) bond with the other one forming a stable Helium atom. Note the Spiral staircase-like arrangement of the Push-Pull bands at the poles. Also, note the curved geodesics due to the mass (extremely high positive energy) at the centre of the nucleus.

The same arrangement of space contracting and expanding bands with widely spread spiralling equator and closely packed spiral staircase like arrangement near poles, can be extrapolated to the Quarks bonded together and probably revolving around a central positive energy core within a Proton. However, the individual strengths or trajectories of the contracting or expanding zones or the distances would vary. The mechanism of formation of these bands would be similar except that these interferences take place around the individual Quarks. How these interact with the chaotic space-time within the nucleus and how they emerge out without much difference is a matter of enigma.

The Proton would also have a spin and a magnetic moment. The Spin of the Proton is likely to be independent of the spin of individual Quarks and its spin is likely the effect of the rotation of Quarks around the centre.

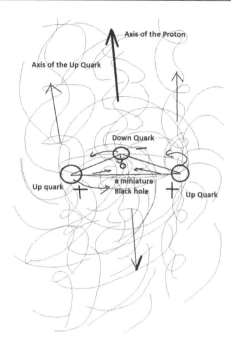

Figure iii/9.4: Shows a possible model of a Proton with the Quarks moving around a central massive Positive energy core like a miniature Black Hole actively destroying PCs. The Quarks stay apart and the push-pull bands emanating from them interact with their partner Quarks and also the neighbouring nucleons. Note that the direction of Spin of the three Quarks is more likely to be horizontally directed towards and away from the central Positive energy core instead of vertical or random as shown, given that the miniature Quark magnet would align with the Spatial changes taking place due to the Positive energy core.

Figure iii/9.5: Shows a still from a freely available animation of a spinor which is one of the possible models of the charged particles given by LQG. Note the similarity of the spinor with the charged particle in DGR. (Image Courtesy: Wikipedia)

Electric current flowing through a wire forming a closed circuit

A closed-circuit formed by a good conducting element like copper, let's say a wire, has multiple loosely bound Electrons in the outer shells which form the sea of Electrons. These are bound to the Protons within the nucleus. But these good conducting elements are invariably metals.

They have a high number of nucleons within their nucleus and many orbitals. The valence electrons are said to be loosely bound and keep shifting from one valence shell to another.

When an electrical potential difference is applied by a battery, one end becomes electrically negative and the other becomes electrically positive. Here, the electrically negative end of the wire has a higher density of electrically negative charges and the positive end has a deficiency of electrically negative charges. Let's name the electrically positive end as P and the negative end as N.

At P, the battery takes up Electrons from the wire and creates a deficiency. This creates a net positive charge here. The immediately neighbouring zone of the wire towards N where no positivity exists yet has more Electron density. Electrons from here immediately move towards the positively charged zone. This movement of Electron is intended to neutralise positivity. But this movement creates another zone of positivity which can be called an "Electron hole" at the zone which donates these Electrons. If the battery is still connected, the movement of Electrons cannot completely neutralize the positivity created and it keeps reforming at P.

The Electron-hole created leads to positivity here which causes the immediate neighbouring zone in direction of N to respond by donating Electrons to this positively charged zone but creates another zone of positivity at this new location. This cycle continues and the Electron hole keeps travelling towards N.

Thus it is not just a convention that we say the electric current travels from positive to negative. It is the Electron-holes that are actually travelling in this direction.

All these events are extremely rapid and thus what happens effectively is that Electrons keep hopping from N towards P and Electron holes keep travelling from P to N.

Note that negatively charged particles move but positively charged particles stay where they are.

Any increase or decrease in this potential difference applied leads to an increase or decrease in the flow of current.

We know that any change in the passage of current leads to the development of a varying magnetic field around the wire which can then act on the Electrons within a nearby conductor and induce new currents within it.

Explaining the Direction of magnetic field I.e. the right hand thumb rule

We know that a magnetic field is formed around the wire in the direction which follows the right-hand thumb rule. Thus rule states that

"If the thumb of our right hand closed fist points to towards the direction of current in a wire, the fingers point towards the direction of the magnetic field."

Although this is known, the reason for this is explained on the basis of the magnetic moments of individual charged particles.

When the electric potential is applied along the length of a wire, when the electric current is flowing, charged particles get aligned within the wire according to their individual magnetic moments.

Let's assume for now that a charged particle, say an Electron, has spin-up if it has a clockwise spinning lattice. It would have an anticlockwise spinning lattice if it has spin-down. We can apply the same right-hand thumb rule to charged particles in our theory. If done so, the particle with a clockwise spinning lattice would have a magnetic moment directed inwards towards the paper and a particle with an anticlockwise spinning lattice would have the magnetic moment directed outwards from the paper.

It would be evident that in event of passage of current through a wire PN, each charged particle within its influence would reorient itself according to this electrical flux. If the current is flowing in the wire such that all spin up Electrons orient themselves so that their magnetic moments point to the right, then all the spin down Electrons will also reorient themselves so that their magnetic moments point to the right in which case the rotations of the spins of both spin up and spin down Electrons will orient in phase or would be parallel and in the same direction, not in opposite directions.

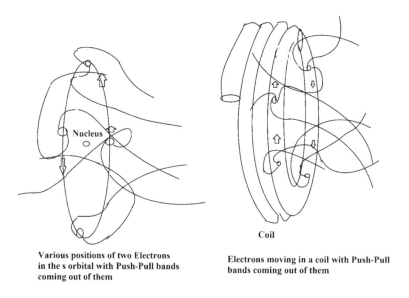

Various positions of two Electrons in the s orbital with Push-Pull bands coming out of them

Electrons moving in a coil with Push-Pull bands coming out of them

Figure iii/9.6: Shows a diagrammatic representation of Magnetic flux in the form of Push-Pull bands emanating out of the moving Electrons around a Nucleus (orbital magnetic moment) and also within a coil with electric current. (only few bands have been shown for clarity). Note that the arrows show the direction of movement of the Electron so that the more distant locations of the bands have originated when the Electrons were at an earlier position.

In the same orbital, say "s" orbital, if two Electrons with spin of the same direction come up, they will have to bond with two Protons in such a manner that the magnetic moments of Protons are in the same direction and the Protons would have a higher repulsion due to their like poles coming closer. When opposite spin Electrons come inside an s orbital, their magnetic moments face in different direction i.e. their like poles might face each other so that their binding Protons have their individual magnetic moments facing in opposite direction.

Now imagine an Electron revolving around a circular orbit, say around an atom or going around a circular coil. When revolving around an atomic nucleus, one of its poles constantly points towards the centre towards the Proton it has bonded with. While it revolves, the Spatial expansion-contraction bands emanating from it at various locations around the orbit meet at the centre of its orbit. Here, the curved bands have a much higher density compared to the zone around the orbit where they are spread apart. This is akin to the higher density seen at the poles and is the orbital magnetic moment. While rotating around a circular coil as a part of electric current, similar Spatial expansion-contraction bands emanating from different Electrons forming the part of the flowing electric current will meet at the centre while curving around the coil so that their density here in the centre of the coil will be much higher. This represents the magnetic field created by the passage of current through the coil.

Electromagnetic induction

Current moving through a coil can be considered as the movement of spinning charged particles oriented according to their spin magnetic moments. Higher is the potential difference applied, more is the number of Electrons or Electron holes travelling and more is the current.

As explained above, the negatively charged Electrons or positively charged Electron-holes travelling along the coil leads to the generation of the spiralling bands of highly contracted or highly dilated time which represent the bands that cause attraction or repulsion. The density of these spiralling bands is much higher at the centre of the coil which represents the magnetic field produced around the coil. If another coil is kept in the vicinity of the first coil, the lines while spiralling out at light speed move through the second coil and can potentially alter charges within it. However, to produce a current within the second coil, there should be a steadily changing current within the first coil so that the density of these flux lines fluctuates and leads to the flow of Electrons within the second coil. If the current passing through the first coil is steady, the magnetic field would be unchanging and there would be no current in the second coil. When the current is changing, especially in alternating current, the magnetic field is ever-changing. This creates variation in density of these field lines and causes a significant induction of current in the second coil whose frequency corresponds accurately to the frequency of the alternating current.

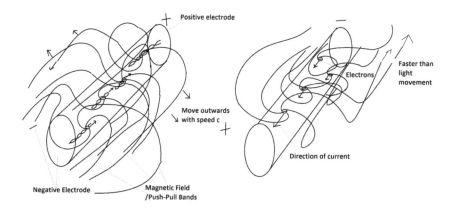

Figure iii/9.6: Shows a wire with an alternating current flowing within. Note that the individual spins orient themselves along the electric field so that the Spiral staircase like polar push-pull bands orient towards the electric field within the wire. As the charges move, the non-polar and equatorial push-pull bands move out spiralling around the wire forming the magnetic flux.

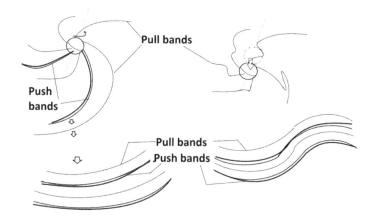

Figure iii/9.7: Shows how the spinning charge with emanating push-pull bands form EM waves/ Electric and Magnetic flux.

Summary

The electric and magnetic flux lines which we routinely hear in Electromagnetism are the bands of highly contracted or dilated time emanating from charged particles. A lot of behaviours of charges can be explained on the basis of these. Many of the phenomena known in contemporary physics remain as it is and the possibility of discrepancy between predictions of our theory and conventional Electromagnetism is less likely.

CHAPTER 10

ENTANGLEMENT**

> **Note that this section was written at a very early part of the theory development and it is literally loud thinking done by me to dissect the potential meaning of the word "Entanglement". Many things mentioned or discussed in this section were later given up or subsequently proved to be wrong. The discussion might look naïve but the best part is that it led my thinking to what I believe is the "Idea of the century" which is EPCA formation.

Entanglement is the weirdest part of Physics that has conclusive proof.

Entanglement is also one of the basic assumptions of our theory.

The exact cause of Quantum entanglement is not described in Quantum mechanics. It is defined for Quantum particles.

Definition of entanglement in Quantum mechanics is in a way closely related to and somewhat made unclear by Quantum superposition.

Two Quantum particles are said to be in a Quantum entanglement state when they have the ability to communicate their quantum state to each other and affect each other's Quantum state regardless of the distance between them.

Thus in theory, if particle A is in a state of entanglement with particle B, regardless of the distance between them, particle A can communicate with particle B about its Quantum state possibly faster than light speed. Whether this faster than the speed of light communication is instantaneous or just faster is not known. The mechanism by which this communication happens is also not known.

Quantum superposition complicates the understanding of entanglement a bit.

Why is this weird and non-intuitive?

Imagine that particles A and B form at the same instance from energy E. In this case, laws of physics state that multiple things must balance out or must be conserved. Energy, Linear Momentum, Charge, Spin, Angular Momentum must all be conserved when the two particles form. This means that after the formation of the two particles, all these quantities in the preformation Quantum state should be equal to all these in the post-formation Quantum state.

If all the energies of both particles (including all types of energies i.e. energies locked within the particles, kinetic and potential energies) are added together, the sum should be equal to the energy of the system when formation hadn't taken place yet. If the charge and spin in the preformation state were zero, the sum of charges and spins should also be zero in post-formation states. This means that if particle A is spin up, particle B should be spin down. If particle A is neutral, particle B should also be neutral but if particle A is positively charged, particle B has to have a negative charge to cancel it out. The angular and linear momenta of the two particles after formation, after they are summed up with a vector sum, should be equal to the vector sum done in the preformation state. This means that before their formation, whatever linear and angular momenta were possessed by the energy-dense photons that formed the particles, would be conserved even after the photons formed the two particles.

After formation, the particles would move apart but despite the increasing distance between them, their properties would still be mathematically and in some manner physically related so that laws of physics are not violated and all these quantities are conserved.

The two particles being in Quantum state are supposed to be in Quantum superposition of all possible states until a measurement is made. Let's consider just the spin.

After formation, Quantum mechanics would say that both A and B would be in Quantum superposition of all possible spins at the same time so that both would have a superposition of both up and down spins. After they move apart and their distance crosses one light-year, if a scientist named Alice makes a measurement on the spin of particle A and finds it to be spin up, the Copenhagen interpretation says that the particle collapsed into this spin "Up" state just now from its earlier state of superposition of both spin up and spin down states. This information would then travel instantaneously or at least faster than the speed of light and would be available to another scientist Bob who would measure the spin of Particle B and find it to be spin down.

If general relativity holds true, as the distance between the two is 1 light-year away, this information will take one year to reach Bob as it cannot travel faster than light speed. This, however, is not what is observed in experiments that test entanglement.

Note that above was a slightly modified version of the EPR paradox (Einstein Podolski and Rosen paradox) *(Einstein, A et al – 38)*.

Einstein's "spooky action at a distance", although he didn't like it as it proved him wrong subsequently, is tested to be a part of reality.

Not only particles with mass but massless particles like photons and virtual particles can also form an entangled pair. And entangled pair may or may not annihilate with each other subsequently.

Bell's theorem proves that quantum physics is incompatible with local hidden-variable theories. It was introduced by physicist John Stewart Bell in a 1964 paper titled "On the Einstein Podolsky Rosen Paradox." He proved that "local hidden variables" cannot fully explain the transfer of information between entangled particles. *(Bell, J. S. – 8,9)*

Quantum entanglement is still a spooky misunderstood and enigmatic part of physics and little is known about why it happens although it is known without a doubt that it does.

Entanglement in our theory

Entanglement is a fundamental assumption in our theory and thus it makes our theory inherently non-local.

Our theory is able to discuss the exact cause of entanglement in stark detail.

Two particles become entangled when they come into existence simultaneously due to the intervening Planck Compartments that develop local time changes in response to the presence of these newly formed particles.

Two photons, which in our theory are an aggregate of T=0 Compartments, formed simultaneously affect the PCs around them and especially in between them develop time changes simultaneously with them and as the photons diverge away, the wave-front formed by these connecting PCs also moves away from the source.

But Quantum entanglement doesn't appear just here. In fact, the core of our theory is based on entanglement.

Every Planck Compartment is supposed to be in Quantum entanglement with the surrounding PCS. Any change in local time and resultant spatial changes that happen in one PC affect the neighbouring Planck Compartments due to this quantum entanglement.

Universally fixed coordinate system, no possibility of a void in PCs and such high turnover of PCS i.e. destruction or creation rates are made possible only due to the presence of entanglement.

To understand this further, let's imagine a Planck Compartment called C. We can imagine it to be cubical or spherical. In either case, we need to imagine the centre of the cube and let's call this point P. Let's draw our x, y and z axes taking this point P as the centre so that the three planes which are defined by these axes divide the PC into eight different portions. The coordinates of P would be (0,0,0).

Now let's assume that the C/D ratio in our compartment C is 1/11 or 0/10 i.e. 1 new Planck Compartment forms and 11 Compartments vanish or are destroyed. This means that the space at C sucks in Volume of space equivalent to 11 PCS from the periphery and it adds a volume of 1 Planck Compartment back. Or it sucks in 10 PCS and no PC is created. All these events have to happen in a time of 1Planck time.

We can divide the Planck time into 11 equal parts. So the first event out of these eleven events happens in the first 1/11th of the Planck time.

Another assumption of our theory is that there can be no Planck Compartment Vacuum or Void or "space where there exists no PC". This means that if at any place, excessive destruction has taken place, the volume of space from the neighbouring PCs is sucked in to accommodate for the loss.

In our PC "C", 10 PCs are destroyed effectively which means every 1/11th of a Planck time, some volume is sucked from the neighbouring PCs.

Now the question whose answer remains unanswered to us is how many PCs are there in direct contact with our PC "C". If its shape is cubical or even a sphere, it can accommodate 8+18 i.e. 26 PCs around it as shown. (This arrangement and this number 26 is arbitrary and could be different in reality).

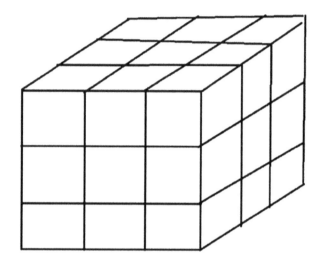

Figure iii/10.1: shows that there can be 8+9+9=26 cubes around the central cube within this 3x3x3 cube. If the central cube is the PC in question, there will be 26 cubes i.e. 26 PCs around it

One must understand here that there is no wall around every PC. This means that the boundaries between one PC to the next is arbitrary and is for our understanding.

Thus, when a volume of one PC vanishes from C, the surrounding PCs move diametrically inwards to accommodate the void. This happens at a high speed of 1/11th of a Planck time and the reaction time would indeed be faster than the speed of light.

After this process of partial shifting of the neighbouring PCs has occurred after 1/11th of the Planck time, boundaries of the original PC "C" and coordinate locations of its corners would be restored and the boundaries of the neighbouring PCs of C would lie in a different location of Space. The destruction of the next 1 Planck Compartment volume would start which will take another 1/11th of the Planck time.

Note here that if there are 26 neighbouring PCs, then only 1/26th portion of each of the neighbouring PCs would be needed to accommodate for the loss of volume that has occurred. Thus it is not exactly shifting of blocks like a blocks game.

Also worth noting is that as the PCs shift, the outer edge of the PCs would have a void created. But these are also entangled with their outer neighbours. Their neighbours shift to fill the void which creates a void outside. This shifting void would eventually be filled up by a change in local time in the zone around C where there occurs a compensatory passive time contraction and additional formation of new PCs occurs.

All this complex rapid compensatory shifting process takes place within a Planck time and is thus due to the entanglement of neighbouring PCs.

This process, which I have described rather slowly, in reality, would occur at insanely fast speeds.

For our PC "C", 10 PCs are lost every Planck time. This compensatory shifting of PCs would happen for some distance from C, all around.

In this hypothetical example, our PC C is not moving. But photons would be moving One PC per one Planck time. As the Photon or the T=0 PC aggregate moves around, this sub-Planck time compensatory PC shift also moves.

In a T=0 Compartment, one PC is lost and no PCs are produced or more appropriately One PC is produced and two PCs are lost i.e. C/D ratio would be 1/2.

As more PCs with T=0 aggregate together, the C/D ratio shifts more towards destruction and more destruction per PC formed keeps happening.

As more destruction happens every Planck time, more of the compensatory PC shift and compensatory time contraction spreading around within a Planck time happens.

<u>When this PC aggregate is moving its coordinate location, what is happening is the change in location of this C/D ratio. The information being transferred one PC per one Planck length is thus this "C/D ratio".</u>[1*]

> 1* - Initially this was the conceptual thinking that "information about how much is the flow of Time" was the key. However later it was realized that what is possibly more fundamental is just movement of PCs and the apparent deviation of time is just secondary.
>
> Even then this concept of transfer of "information on extent of deviation of time" is useful and can be utilized as a tool as the theory progresses.

As T=0 Compartment is considered as positive energy, more T=0 Compartments get aggregated at a point, more is the energy density. After energy density goes beyond one point, the T=0 Compartments collapse into a Sub-Planck length prison.

What this means at Sub-Planck length isn't important and is just theoretical jargon taken directly from String Theory.

However, one thing is clear, the C/D ratio exceeds a point so that so much destruction per creation is happening that a particle with mass forms. This particle will start resetting the time in the surrounding PCs with such high speeds velocity that the inwards movement of PCs exceeds the speed of light,

Effectively, the destruction happening is so much that a time resetting wave starts forming and moves all around the particle. This wave is a time dilation wave and thus prevents the particle from travelling with the speed of light as it slows it down.

In reality, there doesn't seem to be a limit to how much the C/D ratio can sway towards Destruction (or for that matter creation). The Supermassive Black Holes at the centre of Galaxies have masses a million times the mass of the Sun and the Sun has a very high number of such particles, each having a very high number of T=0 PCs locked in. Thus these supermassive black holes would have an insanely high extent of the shift in the C/D ratio and would have an insanely high amount of destruction happening with every Planck time within a small volume. As discussed separately in greater detail, the resetting waves emerging out of such regions of highly swayed C/D ratio would be so time dilated as to have mass and would form secondary and tertiary singularities.[2*]

When this limit (beyond which a "particle with mass" is formed) is reached, just entanglement cannot compensate for the swayed C/D ratio. So along with the compensatory shift of PCs and the compensatory time contraction spatial dilation, local time resetting i.e. dilation wave starts as a compensatory mechanism to avoid the void from being formed. The high amount of destruction happening in the centre is carried to the periphery with a relatively slower more spread out filling of the void created.[2*]

[2*] - Again this jargon of formation of primary, secondary and tertiary singularities was thought of initially, to conform with Quantum Chromodynamics in which spontaneous formation and destruction of Quarks and anti-Quarks is said to happen routinely within the Nucleus.

However, what is clear now is that none of these un-necessary assumptions are needed and what essentially happens is faster than light inwards movement of the EPCAs and faster than light vertical as well as horizontal contraction of space. That said, for faster than light speed inward movement, the EPCA has to develop volume loss at insanely high velocities. In usual EPCA's, this volume loss happens due to temporary T=0 events dispersed uniformly throughout the EPCA. For the EPCA to travel faster than c, the extremely high volume loss needed would require not only T=0 events but multiple temporary "T=0 aggregate events" (multiple T=0 events aggregated together) dispersed all throughout the EPCA. These T=0 aggregates are nothing but singularities. Once formed, as the EPCA moves inwards, more volume loss is needed as faster inwards drift is needed and thus this T=0 aggregate might keep increasing and moving inwards along with the EPCA or newer ones might emerge.

Thus, the concept of emergence of multiple temporary singularities in this zone isn't without merit.

> Even the idea of compensatory time contraction spatial dilation for a highly deviated local time was given up later as it was realized that spatial expansion cannot compensate for the volume loss as spatial expansion would lead to increase in the surface area of the involved sphere and thus cannot compensate for volume loss at a fixed point.

To conclude, entanglement is an integral interaction between neighbouring Planck Compartments and that the whole theory breaks down without it in assumptions.

The EPR paradox in our theory.

In our theory, there is no quantum Superposition as every particle has an absolute location or coordinate values. Superposition is not needed to explain the weird findings of the Double slit experiment.

Every particle also has a definite spin which is the direction in which the Buckyball of ENEA particles rotates. Each particle possesses a well-defined spin which is well known to it and nature. For obvious reasons, we may remain ignorant of the particles spin before we perform an experiment and measure the spin. <u>There is no collapse of the particles spin in a particular direction, instead, the spin was always the one which we measured. It is not the particles spin that changed post measurement. Just our ignorance or lack of knowledge of the same.</u>[3*]

> 3* - This was written much before the insight of meaning of Superposition of spin of a particle emerged. Now we know that Spin direction and handedness of a charged particle keep changing due to pendulum like movements of the lattice. Also, now we know that the information of this handedness can potentially travel much faster than speed of light along with an EPCA forming Time Resetting Wave.

This is somewhat like Einstein had said. The particles are like two parts of a pair of gloves, one right and one left, right before when they formed. And the "which spin does A have" information did not travel with velocities higher than the speed of light to Bob before he performed his measurement on B.

However, the information does travel faster than the speed of light due to the collapse of the wave-front.

In the case of a pair of photons that became entangled together when formed, behave like a particle producing no interference if the wave-front collapses and the entanglement between them ceases to exist. This information of "collapse of wave-front" or "cessation of entanglement" does travel faster than the speed of light. This is tested countless times during the various variations of the double-slit experiment especially which way experiments, delayed-choice experiments and those using quantum erasers. These are discussed separately.

This, however, creates a problem of how to explain the 85% opposite spins measured in Bells inequalities.

The real answer to this is unclear to me and may become clearer only after a detailed study of the experiment and its methodology.

The 10^6 Planck Compartments sphere

Because every Planck Compartment is in close entanglement with its neighbour and every neighbour is in turn in the entanglement of its neighbours, one can imagine small spherical imaginary compartments probably made up of 10^6 PCs (a cube of 100 x100 x 100 PCs), which are working as an individual unit working together. This number is also taken randomly and subsequently one can zero into a different much bigger or smaller number as well.

These 10^6 Compartments can be considered to have equivalent local time and their volumes can increase or decrease as local time varies within them. These are just meant for understanding purposes.

This conceptual idea of a 10^6 Compartment can be useful in explaining gravity, inertia, momentum etc. as explained ahead.

Concept of the C/D ratio and zone of interaction

A particle with mass is a Planck Compartment where a high amount of energy is locked within.

In our theory, a particle with positive mass (Positive Energy Matter particle) would then be a Planck Compartment where a huge number of T=0 Compartments are concentrated or locked within. The theoretical jargon taken directly from String theory is T=0 Compartments or energy is trapped in Sub-Planck length prison of higher dimensional loops. The reason why this is purely theoretical is that there is no way of confirming the existence of these higher dimensional loops. Even if they exist, they won't be exactly like what String theory predicts as most of the properties of particles like spin and charge are not determined by the shape of these loops. In our Theory, the shape of these loops is of no relevance for explaining the properties of particles. This is in contrast to String theory where the various possible shapes lead to the prediction of the existence of a Multiverse.

The same thing in our theory can be explained by the C/D ratio. What is essentially happening in this specialised PC is a highly deviated local time towards destruction equivalent to a high number of Planck Compartments disappearing per Planck time. In a T=0 compartment, no creation of fresh PC occurs. This means that in a single T=0 Compartment, in one Planck time, a volume of 1PC is lost but no creation happens and the neighbouring PCs shift.

When more T=0 Compartments aggregate, as explained earlier, the C/D ratio further deviates towards destruction, so that at the single PC, every Planck time, more than one Planck volume is lost.

In a particle with mass, the amount of energy trapped is huge. This means that the number of T=0 Compartments trapped or the extent of deviation of the C/D ratio is also huge.

<u>The destruction happening at this PC, every Plank time, is compensated by two mechanisms, a zone of entangled PCs which instantly (faster than the speed of light) contracts local time within them to generate extra volume to compensate for the loss as happens around a</u>

photon. The destruction happening here, however, is too much for this process to completely compensate for it. The second process to compensate for the rapid destruction happening at our PC with positive mass is time resetting i.e. time dilation waves emanating from the particle. This process gives mass to the particle and is responsible for its properties like inertia. It also slows the particle down so that it cannot travel faster than the speed of Light.[4*]

> 4* - This compensatory expansion of space was later understood to be incapable of compensating the volume loss happening at the centre and is no longer included in the assumptions. Thus the only way Volume loss can be compensated is by development of volume loss in the surrounding PC aggregates and thus forming EPCAs. This compensatory expansion of space or Creation of new PCs can still happen when two regions of Space are moving apart. This happens when two gravitationally bound regions of space with Galaxy clusters move away due to expansion of the Universe as noted in the "Why the Universe Expands and what happens at the edge of the Universe?" in Crisis in Cosmology section.

It is difficult, at present, to guess the amount of destruction happening at the PC. There probably exists an equation that equates the number of T=0 PCs to energy E which would then make it possible to equate the number of PCs destroyed with mass with the help of Einstein's equation $E=mc^2$ One of the aims of applying Mathematics in our theory is to find this equation.

However, the number could be huge to the tune of 10^{1000} or $10^{1000000}$ PCs destroyed per Planck time.

Note here that PCs in which local time is dilated but not so much that it becomes T=0 would also have a sub-Planck volume loss wherein a sum of volume-loss of multiple such PCs with less than T= 0 time-dilation would add to produce a significant (more than a Planck volume) volume loss.[5*]

> 5* - This thinking was later replaced by the Volume loss ratios. Note that this sub-Planck volume loss of volume is contradictory to the basic assumptions of our theory. Thus regions where Local time is dilated but not as dilated as to have T=0, will have much higher number of PCs in the UT state than T=0 state. This means that all the T=UT PCs will have no change in volume while T=0 PCs will lose volume.

Movement of a particle with mass and Momentum

A particle with mass or in other words a PC where the C/D ratio has highly deviated towards destruction, can be static or moving. A particle with mass would have time dilation-spatial contraction waves all around it so even if it moves, it cannot move faster than one PC per one Planck time, i.e. it will move a Planck length in multiple Planck times or sub-Planck length per Planck time.

This highly time deviated PC will have a coordinate location that is specific, accurate to one Planck length in all three axes.

It will also possess a highly accurate direction. Let's say it is moving along the x-axis from left to right with a speed close to the speed of light, let's say it is moving one Planck length every three Planck times.

When it moves, it converts a non T=0 PC into a PC with a highly deviated local time. When the particle is moving, what is essentially happening is "the information" of the highly deviated C/D ratio is being transferred from one PC to the next PC.

An alternate way of imagining is that the PC with highly deviated time remains at the same place but space beyond it i.e. the space lying to the right of it on the X-axis, contracts and space lying to the left of it on the x-axis expands.

Thus every particle that moves, depending on the direction it moves has a zone that has spatial contraction/expansion around it. The changes happening here are inevitable and occur faster than the speed of light.

This movement in a way distorts the zone of interaction of the particle or in other words, makes the compensatory mechanisms taking place for destruction happening in it distorted.

Momentum

Every physical object has mass. Classical physics says that when an object is moving, it possesses a force that will have to be opposed to stop its movement. This force is directly proportional to the mass and also directly proportional to the velocity.

Newtonian physics states that the momentum of an object with mass m and moving with velocity v is **P = mv**

Where both P and v are vectors and have both a direction and a magnitude.

Imagine a small toy train made of plastic and a real train made of metal. When both of them are moving with a velocity v, it would be relatively easier to stop the plastic train as it has negligible mass. But the inertia of the real train, i.e. its tendency to keep moving forward, would make it difficult to stop and one would need a much higher force to stop it from moving. The reverse is also true so that a toy train that is at rest is easier to move than a real train at rest and it would take significant effort and much higher force for the real train to start moving. In short, the inertia of the real train will be much higher for both, changing from rest to motion or changing from motion to rest.

Why does this happen?

No one knows the answer to this question and as usual, the answer given is that inertia is the property of everything that has mass and all this happens because of inertia. (Continuing the tradition of contemporary physics - whatever you can't explain with your theories, it is easier to make it into fundamental property or law of nature so that you have no reason left to explain it further. This is like saying God or Creator made it that way - something which should not be typical of science).

But few things are clear.

Inertia at rest is a constant force existing all-around a particle with a mass that increases in the direction opposite to the movement of the particle in an attempt to maintain the particles co-ordinate location or prevent it from moving.

The inertia of motion or moment of Inertia or just "momentum" would then be a force that is constantly present in the direction of motion which keeps trying to maintain the body in its state of motion.

Momentum can be linear momentum for linear motion or angular momentum for non-linear curved motion or circular motion or for an object which is rotating around an axis passing through itself.

For our particle above, moving from left to right starting from coordinate location (0,0,0) along the x-axis, linear momentum is directed towards the right.

For a particle with a complex curvilinear or Frank circular motion, each point on its path can be considered to have three different directions and magnitudes of momenta, each direction parallel to each of the three axes.

Explaining momentum in our Theory

The particle at its core has a grossly deviated C/D ratio and has high destruction happening within the small Plank Compartment volume of its coordinate location which causes a lot of compensatory changes in the zone of interaction due to entanglement.

When this particle is moving with velocity v (which is much smaller than the speed of light) in direction left to right as stated above, there will be a difference in the compensatory changes at the preceding border and the receding border. The Zone of interaction (ZoI) at the preceding border immediately right to the particle will have a much higher disappearance of Planck Compartments than the receding border immediately to the left of the particle. In fact, there will be a much higher spatial expansion on the left side of the particle in the receding ZoI. This Spatial expansion is happening as a reaction to the motion of the particle and is independent of the particle and is directly proportional to the mass of the particle and the velocity of the particle. This spatial expansion at the receding ZoI is equivalent to the momentum of the particle. If an object is made of multiple particles, say x number of particles, the total momentum would be the vector sum of momenta of all the particles put together. Since the direction of all these momenta will be the same as the direction of propagation of the object, it won't change for the resultant momentum and thus the scalar sum of momenta of all particles with the same direction as each of these individual particles gives the total resultant momentum.

A particle that is at rest will keep sending time resetting waves, all of which will move in all the directions around the current coordinate location of the particle. Any attempt to move the particle will be opposed by the spatial contraction caused by all these spherical time dilation waves. A force that tries to move the particle has to oppose the spatial contraction caused all around by these waves. When already in motion, the force moving the particle will have to constantly oppose these waves which attempt to stop the movement of the particle. These time dilation waves would prevent the particle from moving faster than the speed of light as well regardless of the momentum. These can thus explain the inertia at rest and reluctance to motion well.[6*]

> 6* - These were some of the early attempts to make sense of the enigmatic concept of Inertia. Due to the ignorance of the idea of EPCA formation at these stages along with some faulty concepts like compensatory spatial expansion at the back, I was being steered away from logical explanations of Inertia at motion/momentum. Although still enigmatic, now we know that PC vacuum formation at the back prevents forward progression (inertia at rest) and excessive compensation from the front leads to Inertia in motion along with a yet unclear contribution from MONDian forces due to mass of the entire Universe.

If we imagine a spherical zone of interaction around the Planck Compartment which represents the coordinate location of the particle where the deviated C/D ratio exists, the momentum can be visualised as a compensatory & additional (i.e. in addition to the one caused to compensate for the rapid destruction happening at the Planck Compartment that represents the particle) spatial expansion happening at receding ZoI. The additional spatial contraction happening at the preceding ZoI is the reason why the coordinate location of the particle changes i.e. the particle moves.

For a massless particle, there is no mass, so there would be no Spatial contraction waves emanating around and any movement would make it move with the speed of light. The momentum would then depend on its frequency or the amount of destruction happening in it or the extent of deviation of the C/D ratio within it.

Even a massless particle would have a zone of interaction around it, and destruction happening within it would lead to compensatory changes in the surrounding entangled Planck Compartments.[7*]

> 7* - Idea of Zone of Interaction was found to be too vague and later given up. Even the spatial expansion at the receding border was found to be incorrect later once EPCA formation was understood. The basis of Inertia is still an enigma and some questions persist, especially the role of mass of the entire Universe and its MONDian Gravity in explaining Inertia at rest and to some extent Inertia in motion.

Understanding recruitment of PCs further

Let's imagine a point P where a Planck Compartment sized particle exists I.e. P lies at the centre of the Planck Compartment which has the C/D ratio characteristic of the particle. Also, let's assume that this particle has a C/D ratio such that it eats up X number of Planck Compartments every Planck time. Let's draw another circle (more appropriately a sphere) around P in such a manner that the spherical volume it incorporates is equal to X Planck volumes. (Alternatively, one can draw a sphere with a point P as the centre and X as the radius. At the end of 1 Planck time, this entire spherical region will end up being destroyed by the destruction under progress at the PC.) From this, we understand that in the next Planck time, all the PCS existing within this circle around the P will be eaten up or sucked up within P. To compensate for the void formed, surrounding PCs will have compensatory time dilation. Let's draw a second circle around P well beyond the first one which represents the limit of Zone of Interaction. The compensatory changes due to the presence of the particle at P lie within this circle.

Let's presume that our particle is moving from left to right with a speed one fourth the speed of light. This means that the particle finishes 1/4th Planck length per Planck time or it covers 1 Planck length in 4 Planck times.

Let's now imagine four other points Q, R and S along the left to the right axis which are 1/4 th Planck length away from P. The particles centre coincides with P at the start.

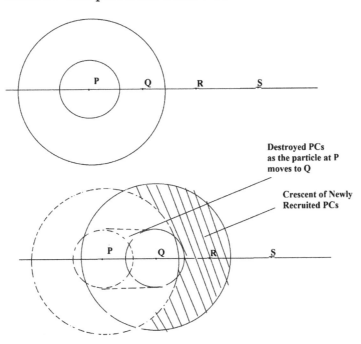

Figure ii/10.2 Shows Recruitment of PCs.

By the end of the first Planck time, the Planck Compartments in the inner circle are gone and compensatory changes in the outer circle have occurred. However, at the end of this Planck time, the centre of the particle has shifted by 1/4th of Planck length and now lies at Q.

At this new location of the centre, the PCs that formed part of the old first circle are replaced by newer PCs that came from the periphery. At Q, again a fresh inner and outer circle will need to be drawn around Q and all the PCs that form a part of the inner circle would be destroyed or sucked in Q, while compensatory changes spread until the new outer circle. But again by the end of the second Planck time is over, the particle is again out of its original position and its centre now lies at R. Here a newer inner and outer circle need to be drawn and the fate of newer PCs in the inner circle is now in question. All these cycles keep repeating.

Note in these figures, that the inner circles at each of the points are offset by a small amount. When the particle shifts from P to Q or Q to R, newer PCs, representing a shaded crescentic region in the inner circle fall prey to the destruction of the particle. The PCs falling in this crescent are the ones freshly recruited by the particle.

Note that the location of this crescent would depend on the direction of movement of the particle and if the particle is moving upwards, the crescent would point upwards.

Now imagine a ball being thrown in space in some random direction. When the ball moves, each particle that constitutes the ball and is bonded together to form the ball will behave like the above particle and would go through each of these cycles. Each of the particles constituting the ball would thus be forming a fresh crescent and recruiting fresh PCs.

If one has the magical ability to see all these changes in the PCs and make out which direction the Crescent lies, the motion of the ball can be made out irrespective of any landmark or other frame of reference. The universal frame of reference of the PCs is more than enough to define the exact direction of motion of the ball in space. This, however, is practically impossible and thus in practice, we need a broader frame of reference or landmarks like the Earth or the Solar system or the centre of the Galaxy to decipher the movement of the ball or for that matter any object.

For convenience, we can call this inner circle (or sphere) the zone of destruction. This zone of destruction can be defined for a particle with mass or a massless photon I.e. an aggregate of T=0 Compartments.

Zone of Destruction of a moving particle

In the above example, we examined the inner circles at 3 different points namely P, Q and R. Each of these is about 1/4th of Planck length apart.

This is in a way absurd as we know that Euclidean geometry fails or is no longer applicable at distances smaller than Planck length.

Here, we have to analyse the Sub-Planck length distances, as we take the measure of time as Plank time. At such insanely small units of time, even light can travel by just one Planck length. So any particle moving slower would travel a distance that is much smaller than a Planck length. Using a bigger measure of time may give us something meaningful.

Here, we have analysed three points. But the particle wouldn't just jump from P to Q or Q to R as a digital clock would jump by one second. The particle would have its presence at each of the points in between P and Q. Again, using a larger unit of time would make this statement more meaningful.

The point is, as the particle moves, more PCs of the preceding border fall prey to the destruction taking place within the particle which can be called recruitment.

<u>Another crescent in the receding border would be formed where PCs are being released from the destructive effects of the particle- aptly called "surrendered".</u>[8*]

> 8* - Later it was realized that even these PCs belonging to the receding border are in the state of time dilation spatial contraction to compensate for the void remaining due to movement of the PC and thus are in the state of being sucked in and destroyed by the particle. However, more number of PCs are sucked in from the preceding border than the receding border.

CHAPTER 11

DESTRUCTION AND COMPENSATION TO AVOID A VOID

Imagine a point P where there exists a Planck Compartment.

Let us assume that the C/D ratio here is highly skewed towards destruction so that it is like a singularity or a particle with significant positive mass.

This would mean destruction here is very high and very fast. It is akin to a Blackhole.

Let's presume for argument sake, that number of Planck Compartments destroyed at P per Planck time be 10^{10}.

With every Planck time, so much destruction happening at P means that it will keep sucking Planck Compartments from the surrounding PCs.

The question that arises is how so much destruction is compensated for and why there is no void of PCs created at P.

Let's imagine multiple circles/spheres around P with radii progressively increasing by 1 Planck length. The sphere immediately adjacent to the Planck Compartment P would have more than one Planck Compartment, possibly 26 PCs. As we examine outer spheres and compare them with the immediately adjacent inner spheres, we would realise that there are inevitably more Planck compartments in the outer ones.

When destruction is happening at P, the Planck Compartments from the immediately adjacent sphere would be sucked in. This causes the immediately neighbouring outer Shell sphere with more PCs to be sucked inside taking its place. This can only happen if some of the PCS are destroyed.

All the PCS belonging to volume in between these spheres can be considered in entanglement, each one of them can be considered a single unit.

They are like Eddy's around a localised low-pressure point in a fluid. As they are aggregates of Planck Compartments and they are spherical and have a nonzero volume, we will call them "Eddy Planck Compartment volume aggregates or EPCA".

Each EPCA has a constant measure of local time and C/D ratio.

Note that

1. In such a sphere composed of PC aggregates, if creation is equal to destruction, the volume remains the same I.e. number of PCs aggregated together remains the same and so the surface area of the sphere they create remains constant and thus the diameter of this sphere remains constant.

2. If additional destruction is happening, i.e. C/D ratio has deviated towards more destruction, the number of PCs forming it reduces over time with every Planck time and it loses volume. The sphere they form loses its surface area and thus lose its diameter. This PC aggregate would thus move inwards and take place of the inner sphere or inner PC aggregate.

3. If in this PC aggregate, the C/D ratio has deviated towards creation and thus additional creation is taking place, the number of PCs increases and the sphere gains volume and the PC aggregate sphere gains surface area i.e. its diameter increases.

Local time in it would be expected to be in various stages of time dilation I.e. skewed more towards destruction. As more destruction happens in the EPCA than creation, it loses volume. Effectively what is happening is a reduction of the circumference of the circle/surface area of the sphere involved, which would lead to a reduction of the radius.

Thus as destruction happens at P, EPCAs form and start moving inwards towards P.

The question which arises is that if entanglement can happen at faster than light speed, can EPCA's form faster than light speed as well?

Another question is what is the pace of inwards movement of these EPCAs and is there any connection with the extent of skewed C/D ratio.

At the periphery or boundary of a Blackhole i.e. event horizon (a particle is also akin to a Blackhole), the local time is T=0 which means no creation, only destruction. Well within it, there is higher Destruction per Creation. This means the EPCA'S move inwards faster than the speed of light. A photon of light travelling outwards is in a way climbing up on an escalator which is going down at high speeds, and thus cannot escape.

An EPCA represents a volume of space. There is no limit to how fast space moves, even in Relativity. These eddy volume aggregates will be sucked inside at variable speeds depending on the rate of destruction at the centre.

<u>Let's now put a number to each of the EPCA'S around P to ease identification starting from 1 till n.</u>[1#]

> 1# - This is discussed in much greater detail in the "defining the levels" and creation of a model section in Part II

The EPCA 1 can contribute a very small number of Planck Compartments to the destruction taking place at P. Thus, the faster is the destruction happening at P, the faster the EPCA'S have to move to replenish the PCs lost at EPCA 1.

If we assume that EPCA 1 contributes a volume of 26 PCs, then even EPCA 2 or 3 or n, after moving inwards will ultimately become as small as EPCA 1 and would contribute to just 26 PCs volume loss compensation.

Note that the 26 surrounding PCS will each contribute a volume of 1/26th portion of each PC destroyed.

Alternatively, EPCA 1 loses volume and out of the 26 PCs which make up EPCA1, 25 PC volume is destroyed and only one PC volume remains which takes over the lost volume at P lost due to destruction happening at P.

We have to divide a Planck time into 10^{10} equal parts and in each of these parts, 1/26th volume of each of the 26 PCs in EPCA 1 gets sucked in. This moving inwards creates a small PC Vacuum or PC void in between EPCA 1 and 2.

This void will also be of 1 Planck volume but as it distributes around the 26 PCs, it will reduce in height. Thus the subsequent outer EPCA has to travel inwards much lesser than the inner one.

In the next $1/10^{10th}$ Planck time, another PC is destroyed and another 1/26 th of each of the surrounding PCs in EPCA 1 will be sucked in. In a single Planck time, a destruction of 10^{10} PCs occurs. That is this cycle is repeated 10^{10} times every Planck time. This is extremely rapid.

To compensate for the void coming around EPCA 1, EPCA 2 has to develop time dilation so that it loses volume and its circumference/surface area reduces and this process causes reduction in radius of EPCA 2 which then takes over the place of EPCA1. This shifts the void outwards just outside EPCA2 and to compensate for this void, EPCA 3 has to develop time changes and contract in volume to take place of EPCA 2. This cycle repeats endlessly.

One can understand that if the above "entanglement based time changes propagation" is true, in a single Planck time, 10^{10} PCs are being destroyed here and the destruction compensated by EPCAs whose numbers may be as big as 10^{10}.

The time resetting wave we were talking about so often, instead of travelling with the speed of light one Planck length per one Planck time, would propagate 10^{10} PCs per Planck time if this mechanism of entanglement based spatial changes is true. The movement of space to compensate for central destruction should have no limit.

If EPCA 2 has to replace EPCA 1, it has to destroy some of the PCs within it before. In other words, space has to contract within it. That is, the C/D ratio within EPCA 2 has to get more deviated towards destruction and thus Local time here has to dilate.

An alternate method of imagining this is with the help of 10^6 PC aggregates and a single Planck time.

At the end of 1st Planck time, a void of 10^{10} PCs develops around P. One can imagine multiple 10^6 PC aggregates all around this void forming a spherical Eddy Multiple (10^6) PC aggregate or EMPA.

The EMPAs can also be numbered like above. So the EMPA immediately outside this 10^{10} void can be called EMPA 1, the next EMPA can be called EMPA 2 etc.

The EMPA1, being entangled to the PCs forming the void around P, will develop Local Time changes i.e. time dilation to enable extra destruction within it to lose volume. The EMPA1 loses volume and thus reduces in diameter to compensate for the void created. The EMPA2 in turn develops its time dilation and lose radius to compensate for the void created just outside EMPA1 due to its volume loss.

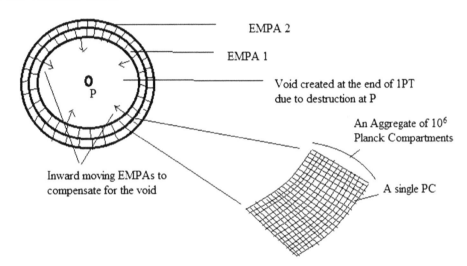

Figure iii/11.1: Shows the PC Vacuum or void created at the end of a Planck Time and the compensation by EMPAs which are an aggregate of multiple 10^6 compartments, getting entangled spontaneously and moving inwards while losing volume and thus compensating for the void.

The second method is easier to imagine as we don't have to divide the Planck time further into smaller units or imagine Lengths or volumes smaller than a Planck length. But it is more difficult to fathom how fast the EMPAs will move.

Thus we will revert to the EPCAs.

One can understand that EPCA1 will be sucked in, 10^{10} times in a single Planck time. Thus EPCA2 has to develop spatial contraction and take the place of EPCA1 10^{10} times in a single Planck time. If we imagine EPCA number 1000, the volume within it will keep getting destroyed as this Planck time progresses and even before the Planck time is over, it will have to keep contracting spatially to compensate for the inward void until finally, it comes into the position of EPCA 2 and finally EPCA1.

One can realise that information transfer is taking place much faster than the speed of light. The velocity of the inward flow of the outer EPCAs may or may not be faster than the speed of light and it all depends on the extent of destruction taking place at P. This means that, if instead of 10^{10}, the destruction of volume at P was much higher, say 10^{100000}, then the velocity of inwards movement of the inner EPCAs can exceed the speed of light.

This may be the distinction between a massive particle and a massless particle.

A massive particle having so much destruction happening at its centre, that at least the innermost EPCAs have to travel faster than the speed of light.

What if the C/D ratio at P was skewed excessively towards creation instead?

Let's imagine a scenario when there is excessive creation at P, possibly 100 Planck Compartments created and one Planck Compartment destroyed every Planck time.

The volume at P increases. So outwards pressure is exerted at the EPCA1. This EPCA being entangled to P would have to develop compensatory time contraction i.e. extra creation within to increase its volume. With extra creation happening at EPCA 1, its diameter increases and it puts pressure on the next EPCA i.e. EPCA2. The EPCA 2 now has to develop compensatory time contraction or excessive creation within to increase in volume and the process increase in Diameter. This cycle goes on outwards indefinitely.

At a single Planck time, this expansion at P happens 100 times. This means we have to divide the Planck time into 101 equal parts. The 1st part of the Planck time will see the creation of a single Planck Compartment, the second would produce another Planck Compartment and so on. The volume expansion of EPCAs and the time contraction wave originating would also propagate faster than the speed of light i.e. information is being passed faster than the speed of light.

This number which we took as 100 randomly can be any number from 10^{10} or even $10^{10,000}$ PCs formed per Planck second.

Imagine two points Q and R, both having a highly contracted local time and thus a creation to destruction ratio deviated towards creation. Imagine that they are immediately adjacent with no distance between them. As more Planck time passes, more PCS form and create EPCA's around both Q and R. This means that distance between both of them keeps rising i.e. the two points will keep repelling each other and they would not tend to fuse together but rather remain apart, unlike points like P where the inward movement of EPCAS would keep sucking any other potential point like P inward. This means that points like P tend to coalesce.

CHAPTER 12

CREATION AS A COMPENSATION TO AVOID A VOID

Imagine two neighbouring PCs that are being pulled apart by some forces. If the forces that are pulling them apart are more than the force which keeps them together (which is 1 PC worth Planck Vacuum) then the PCs will continue being pulled apart. In such a situation, there will remain a persistent void in between the PCs and in the next cycle of Creation, the Creation of new PCs will happen.

Imagine two aggregates of PCs with a boundary marked as in figure iii/12.1.

The two aggregates can constitute any number of PCs. They could be a few meter cube in volume or have a volume equivalent to the volume of an entire cluster of Galaxies and satellite dwarf Galaxies gravitationally bound together.

Let's name them A and B.

Imagine that for some reason, A is moving towards the left and B is moving towards the right. At the edge, the PCs forming the edge of A and B will be under tremendous stress as they are pulled on both sides.

The void that is created in between, thus leading to a formation of a persistent PC vacuum between the edge PCs cannot be filled in by positive energy EPCAs. The only way this can be filled in is by active Creation happening at the edge.

Note how space can be divided, as shown in the figure given below, into the hexagonal zones like honeycomb wherein the gravitationally bound zones can be the chambers and the expanding zones can be the walls. This is the honeycomb model of the Universe.

Note that it is just diagrammatic as the walls can be significantly higher in volume than the individual chambers themselves so that most of the chambers of the honeycomb are empty and only a few are filled with the Baryonic matter of Galaxies.

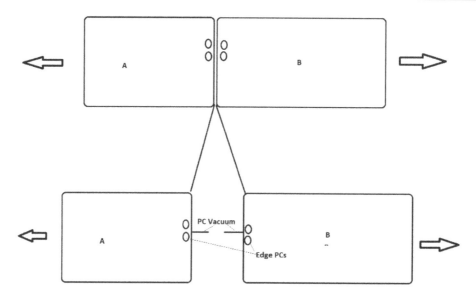

Figure iii/12.1: Shows two aggregates of PCs namely A and B moving away in two opposite directions with no other means of compensation i.e. neighbouring PCs cannot rush in and compensate for the PC Vacuum formed due to this movement. The only way this can be compensated is by the creation of Negative energy particles in between, which can create additional PCs to fill the PC vacuum.

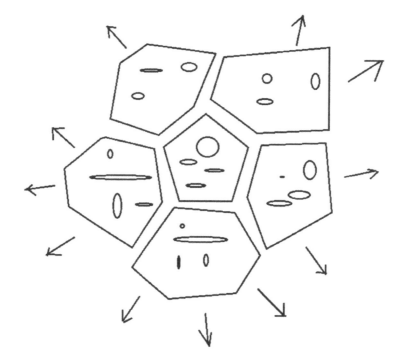

Figure iii/12.2: Shows the various chambers of the Honeycomb Universe moving away and the compensation of void being done by active filling up of new space (i.e. fresh PCs) in between done by freshly formed Negative energy particles.

Chapter 13

HOW DO "EPCA FORMING TRWS" (TIME RESETTING WAVES) EMANATING FROM DIFFERENT PARTICLES INTERACT?

It is relatively easy to imagine a single positive energy particle where more destruction is happening and is being compensated by EPCAs formed all around it resetting time and moving space to compensate for the void created but the destruction happening at the centre.

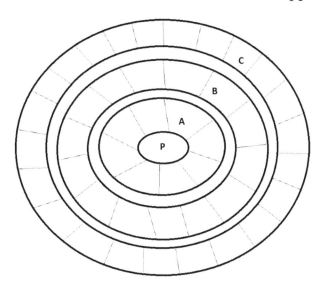

Figure iii/13.1: Shows a single Positive energy particle at P with perfectly symmetrical time resetting waves forming EPCAs around it to compensate the void created by its presence.

But we know that there will be invariably many more numbers of such positive energy particles and their negative energy antiparticle counterparts. The EPCA forming TRWs emanating from many such objects located in specific locations and their interactions with other such particles or antiparticles is highly relevant.

We will try to take it one by one.

First, let's imagine a single Positive energy particle at rest that starts moving with a velocity v.

At rest, the EPCA forming TRWs emanating from it would form a perfect sphere all around it. The time gradients of these TRWs would also be symmetrical and thus there would be a uniform spatial contraction all around it forcing it to maintain its coordinate location.

Let's now consider a small unbalanced force that tends to move our particle towards the right.

For us, an unbalanced force is caused by some phenomenon that leads to a zone of spatial contraction towards the right of the particle.

When this happens, the coordinate location of the particle changes. The particle moves with velocity v. The velocity cannot be faster than the speed of light so that in a single Planck time, the particle cannot move more than one Planck length i.e. one Compartment. Let's consider that it moves a Planck length in 10 Planck times i.e. it is moving one-tenth the speed of light. During these 10 Planck times, destruction happening at the centre continues and EPCAs keep forming.

As the particle moves, the time zones created by the EPCAs that form after movement do not coincide with the previous time zones. The movement of the particle creates a void in the left side and creates a pressure (opposite of void) at the right preceding or front border.

At the preceding border or right border, it is akin to excessive creation. While at the receding border, it is akin to excessive destruction. Both these need to be compensated. The excessive destruction like void being created at the receding left border would create its own independent EPCAs to compensate for the void. The excessive creation like situation created at the preceding right border would create its own independent EPCAs to compensate for. This, however, would be time contraction spatial expansion type of EPCAs like seen around an ENEA particle as these are meant for compensation of excessive creation.

These EPCAs would interfere with the EPCAs produced due to destruction happening at the centre. It figures that the time dilation spatial contraction type EPCAs produced at the receding border would interfere constructively with the EPCAs produced to compensate for destruction at the particle. The time contraction spatial dilation EPCAs would interfere destructively with the time dilation spatial contraction type EPCAs produced due to destruction at the centre. [1#*]

This means that at the preceding border, some cancellation takes place while at receding border constructive interference and thus the addition of time dilation takes place.

All this leads to a creation of distribution of local time around the particle so that there is a much more time dilation at the receding border and much less time dilation at the preceding or right border.

> **1#* - Note that, although this line of thinking looks ok, it is wrong. The asymmetry that we get from this is exactly opposite to the asymmetry that we must have to explain inertia of Motion.** The concept of Inertia derived from our theory is not completely clear yet. But the other approach discussed below is better, in which the asymmetry is produced because of additional compensation from the receding border so that there is a much steeper time gradient along the preceding border. In this, in the receding border, the movement of particle creates a void and fails to contribute to the destruction happening at the particle as much as on the preceding border. The Vacuum cleaner analogy explained in Part II while discussing TRWs is useful.

This asymmetry remains until the particle keeps moving towards the right. [1*#]

Effectively speaking, there is a much higher spatial contraction in the space around the preceding border compared to the space around the receding border. This means the particle is pushed towards the preceding border. Note that this push is solely due to the EPCAs which are solely due to the destruction taking place at the particle centre. This push is independent of the force causing the movement of the particle.

This starts a vicious cycle so that a small nudge to the particle leads to the creation of this asymmetry in spatial contraction which leads to another push that makes the particle move towards the preceding border i.e. the direction of the previous motion. This new push again leads to its own asymmetry in spatial contraction which in turn causes another push.

Newton's first law of motion or inertia of the particle can be effectively explained by this phenomenon.

The extent of asymmetry in the local times in the surrounding space which is directly proportional to the mass of the particle and also directly proportional to the velocity of motion of the particle is what we call "Momentum" in classical physics

If another unbalanced force tries to stop such a moving particle, this push created by spatial contraction asymmetry opposes it.

If multiple such particles are bonded together, the effective momentum of all of them would be essentially the simple addition of all their individual momenta. So the momentum of a billiard ball is the sum of individual momenta of every constituent particle which it is made of. When a small nudge is given to one such billiard ball floating in space, the EPCAs emanating from each of these particles cause an effective spatial contraction asymmetry between the spaces around the preceding and receding border of each of them and the sum of all these asymmetries put together gives the effective spatial contraction which determines the effective push the billiard ball experiences. Each of these particles enters into a similar vicious cycle and the billiard ball would keep moving in the direction of the initial push. (note that here we are talking of a Billiard ball floating in space. The real motion of a Billiard ball is much more complicated with a combination of rolling and linear shifting.)

Also worth noting is that EPCAs at the preceding border are starting in a moving source and would show Blue Doppler shift while the EPCAs at the left or receding border would be red-shifted.

Implications are that the time gradient when one moves from the particle towards the left would reduce slower than inverse square law. Contrary to that, the time gradient would fall faster as we move from the centre of the particle towards the right. In short, the EPCAs move inwards faster in the preceding border.

One of the implications of this process is that when the particle is moving, there is a void in the receding border. The EPCAs here have to compensate for this void as well as the destruction happening at the centre of the particle. Due to the void, the destruction at the centre cannot be compensated from this border and the EPCAs here compensate mainly the void due to movement. The destruction happening at the centre will then have to be compensated by the EPCAs from the rest of the directions namely all directions other than the left.

The left to right movement of the particle facilitates the destruction of the Planck Compartments immediately adjacent to the particle on the Right i.e. Preceding border. This can also be called recruitment of these Planck Compartments. Recruitment will occur only for those Planck Compartments on the Right and won't happen for Planck Compartments on the left.

It is almost like a particle which has a small nudge towards the right, due to its appetite for destroying Planck Compartments, will keep sucking in or recruiting or preferentially destroying more of the Planck Compartments on the right and keep causing Spatial contraction towards the right and this keeps pulling it towards the right.

This explains how the massless particle keeps moving in the direction of its propagation relentlessly with the speed of one Planck Compartment per one Planck time.

Note that destruction happening at the centre is much more than one Planck Compartment per Planck time and compensation is mainly by EPCAs moving from the right quadrant towards the particle and not by the particle moving towards the right and recruiting fresh PCS.

If this was true, the particle would move much faster than one Planck Compartment per one Planck time. For example, if 100 Planck Compartments are being destroyed per Planck time at our particle centre and if every time the particle destroys a Planck Compartment, it recruits its neighbour and gobbles it up, it would move to the right. It would move 100 Planck Compartments per Planck time and the void created at the receding border would also be too big. A 100 Planck Compartment void every Planck second is too much and cannot be part of reality.

With the majority of destruction compensated by EPCAs rather than recruitment or movement of the particle, explaining why the particle moves 1/101 th of Planck length with every destruction and thus the end of one Planck time has moved one Planck length becomes a bit difficult.

Observing the pattern of time gradients or time zones

It is critical to observe how the time gradients or time zones lie around the particle.

For a positive energy particle that is at rest, the maximum destruction would happen at the centre. The EPCAs would cause perfectly spherical concentric time zones. These time zones or EPCAs can be numbered according to their location away from the centre. As time proceeds forward, the EPCAs at number n would lose volume and shift to become EPCA number n-1. Thus the time zones remain the same but the space is moving inwards. Every concentric time zone would have a uniform C/D ratio which means that local time is running at the same pace throughout the EPCA.

Two Positive energy particles

When EPCAs from two Positive energy particles interact, they interfere constructively. Because of the curved shape of each EPCA and because local time in the EPCA would depend on distance from the maximum destruction, the resultant distribution of time zones would be complicated.

The probable distribution is shown in figure iii/13.2.

Consider two positive energy particles A and B. The TRWs forming EPCAs emanating outwards from each one of them and interacting in the space in between is shown.

The time dilation caused by each of the TRW can effectively add up with the time dilation caused by the TRW of the second particle.

We analyse the time dilation caused at various points in between A and B due to the presence of destruction at A and B.

Consider the line AB. As we go from the first particle to the second particle, time dilation caused by one particle reduces but time dilation caused by the second increases. Essentially, the reduction in time dilation caused due to distance is cancelled by the increasing time dilation by the other. Thus line AB will have maximum time dilation due to the TRWs.

The points C and D are at a greater distance than AB and thus will have a greater fall in time dilation i.e. time dilation in these will be lesser than the line AB. Also time dilation at the more peripheral points E and F will be further lesser as compared to C and D.

Essentially, the space in between A and B will have a central zone of maximum time dilation and as one moves away from this line AB on either side, time dilation reduces. Thus the shape of EPCAs would change here and possibly multiple layers of EPCAs parallel to line AB moving inwards from the peripheral areas is likely. The EPCAs not only move inwards towards AB but also contract in length. The maximum time dilation at AB will also correspond to maximum spatial contraction at AB. This is literally like a string of space-time pulling the two particles together.

Note that in all these, the time dilation will have to be calculated using the mass of A and B using Einstein's equation for gravitational time dilation.

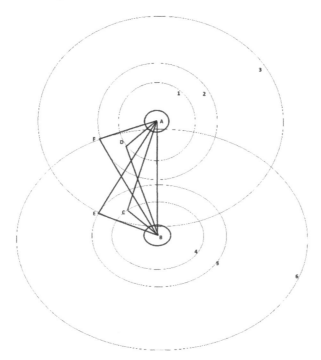

Figure iii/13.2 shows two particles A and B with time resetting waves emanating out from them interacting with each other in the space in between them. Maximum time dilation and thus spatial contraction will be at the line AB. At C and D there will be lesser time dilation compared to the line AB and time dilation at the more peripheral points E and F would be still lower.

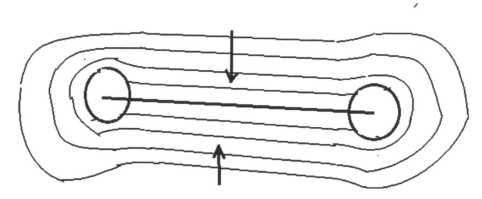

Figure iii/13.3: Shows the same Positive Energy Particles A and B with line AB having maximal time dilation and the space in between them divided into multiple time zones with a time dilation gradient from the line AB outwards. These time zones have an increasing time dilation (and thus active spatial contraction) as one goes from the periphery towards line AB. Thus space in between would form structures akin to elongated EPCAs as shown. The active spatial contraction will pull the two particles together.

Two negative energy particles

The time distribution caused by time resetting waves emanating from two negative energy particles would be similar except that the TRWs in this case would be time contraction spatial expansion type of EPCA forming waves.

The analysis of the figures would be similar but reversed so that the time resetting waves would cause time contraction and spatial expansion and instead of maximal time dilation at the line AB, there will be a maximum time contraction and thus spatial expansion at the line AB which would push the particles apart.

Positive energy and a negative energy particle

Imagine what will happen if, in figure iii/13.2, A is a positive energy particle and B is a negative energy particle (with equal but negative mass). The time resetting waves emanating from A would be time dilation type waves while those from B would be time contraction waves. The space in between them would be divided in a rather complex manner with some places where time dilation caused by TRW from A completely cancelling time contraction caused due to TRW from B leading to zones where there would be Local time running along Universal time and there would be neither spatial expansion nor contraction. There will be still other zones in which time dilation caused by TRW from A predominates and would lead to spatial contraction. There would also be still other zones where time contraction caused by TRW from B predominates and there will be spatial expansion. At line AB, half of the line will have active spatial contraction thus pulling the particles together while half of it would be pushing the particles apart.

Detailed maps will have to be drawn with the help of a computer to determine how these particles would behave. But the Positive energy particle would pull the Negative Energy particle while the Negative energy particle would keep pushing the other away leading to a peculiar perpetual run-away motion first described by Bondi et al.

CHAPTER 14

ATTRACTION AND REPULSION

Physics as we know it is full of particles attracting or repelling each other. Gravity is one such attractive force.

Whenever attraction or repulsion is happening between two particles, it is an effect that displays action at a distance.

The puzzle about this action at a distance is how exactly does this exchange of force take place at a distance with nothing happening in between.

As per classical physics as well as the "Standard Model" description, these "action at a distance" forces are transmitted at the speed of light by something called the exchange of virtual particles. Thus an Electron exchanges a virtual particle with the other Electron it is interacting with and this virtual particle causes the repulsion between them. The virtual particle as the name implies is virtual and thus there is no real particle there. What exactly happens thus is left to the imagination of the observer.

The attraction would then be explained on the basis of the exchange of particles with negative momentum. Here, the negative sign just shows the direction. So that the momentum of the exchanged virtual particles is directed towards the centre of the two interacting particles instead of opposite direction during repulsion.

Action at a distance in our theory

In our theory, space in between two particles is made of Planck Compartments which are having fixed volume and are non-collapsible.

The space in between the particles can be considered made of an array or row of these Planck Compartments.

In our theory, Planck Compartments cannot lose or gain volume but they can be created or destroyed and this process can cause expansion or contraction of the space they contain.

With this simple arrangement, it becomes clear that

- ***if a process happens that causes a creation of a band of Planck Compartments where creation supersedes destruction, the space in the band expands. This spatial expansion pushes the two particles apart leading to repulsion.***

- *if one such process causes the creation of a band of Planck Compartments wherein destruction outruns creation, space loses volume, the band loses length and this leads to attraction.*

Note that this band stretching from one particle to the other, effecting repulsion or attraction, can stay for a minute period of time before disappearing.

The exact geometry of the band is also irrelevant. The exact point or pole of the first particle from which the band starts and the exact pathway it takes in between the particles and the exact part of the second particle it ends in are all irrelevant.

The band can start from the far pole, spiral around the first particle, cross the space in between and spiral around the second particle to finally reach the other pole of the second particle. Any such complex path can be taken. What is important is where it starts, where it ends and whether it is losing or gaining spatial volume.

There is also no limit as to how many such bands should exist and a pair of particles may have multiple such bands traversing the space in between them.

The state of the two particles is also irrelevant. The particles can be in the process of rest or linear motion or they may be in the process of spinning or all at the same time.

There is no size limit as to the breadth or length of these bands. The bands can be a single Planck Compartment wide or several hundred or several million Planck Compartments wide. Given the miniature size of each particle that is equivalent to the size of a Planck Compartment, the bands are also likely to be not very wide.

The formation of these bands, as explained elsewhere, happens due to time resetting due to Entanglement-EPCA formation. This process can be extremely rapid so that the bands can form or propagate at speeds much higher than the speed of light.

How the bands propagate in space and how they curve around depends heavily on the local space-time curvature which depends on many factors.

It is also likely that such bands keep emanating from both particles during their spin and as they traverse the space in between, they instantaneously touch each other and affect each other's stability.

For example, a spatial contraction band instantaneously interacting with another spatial contraction band emanating from the second particle can cause an instantaneous attraction that lasts only till the band's touch or interact. As the band's move in a different direction due to the spin of the particles, their direction changes and they might lose contact and lose their action. At this same instant, some other bands might touch and effect the attraction. In this manner, the interaction between different bands might remain instantaneous but the interaction of multiple bands in sequence leads to a more prolonged and sustained attraction.

Similarly, if the bands that interact are spatial expansion time contracted bands, the instant they touch, they effect repulsion. Subsequent instances might make them change direction and lose contact thus losing the repulsive effect.

Low Energy state or complex formed by the particles

It is interesting what would happen if a repulsive spatial expansion band touches an attractive spatial contraction band.

The repulsion and attraction will get counterbalanced.

The excessive creation happening at the spatial expansion band would be counterbalanced by excessive destruction happening at the spatial contraction band.

Although this leads to no effect like repulsion or attraction, the balance achieved would essentially mean much less overall deviation of local time. Any state in which there is an excessive deviation of local time is a high energy state. Thus it figures that such a counterbalanced state would be a lower energy state.

We know that particles prefer to bond in such a manner that they assume a lower energy state.

Thus if the particles can assume spatial coordinates and positions in 3 dimensions in such a manner that the spatial expansion bands of one counterbalance the spatial contraction bands of the other, this will be a low energy state and it would be preferred by the particles. In fact, additional energy will be needed to make the particles leave such a state.

Because both particles are inevitable to have a spin, it is apparent that they would align in such a manner that their spins also balance out and their axes of spinning are also matching.

It is clear, however, that if a pair of particles bond in this manner, they have to be of opposite charge so that they have mutual attraction between them to keep them bonded. Particles with the same charge would repel each other away and would be unlikely to form such pairs.

Even after such a stable pair is formed, the spatial expansion and contraction bands emanating from the opposing poles (or outer poles or far sided poles) would still remain uncompensated and would remain reactive.

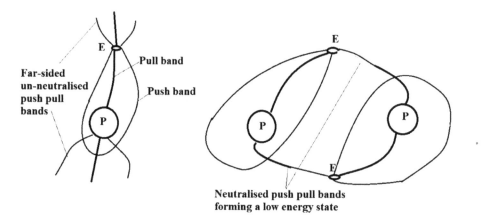

Figure iii/14.1: Shows Un-neutralised push-pull bands of an Electron-Proton pair making it unstable even after a bond is formed between them. Stability is obtained only after their far sided push-pull bands are neutralised by bonding with another such Electron-Proton pair.

However, if two such bonded pairs come close to each other so that each pair neutralize each other's remaining uncompensated outer pole Spatial contraction and expansion bands, the stability achieved can be profound.

What we have described above is how a pair of Protons bind with a pair of Electrons forming the first stable element, the noble gas molecule of Helium.

Here, each Electron bonds with its Proton. Not only that, but the Electron also binds with the other Proton to neutralize the expansion/contraction bands emanating from its outer pole.

Boundaries of a particle

Particles could be positive or negative energy particles or composite matter particles made of a positive energy core and negative energy lattice.

One might ask what's the boundary of a particle then.

In principle, a particle is energy or extremely deviated local time in a single Planck Compartment. Thus a particle should be considered limited to the Planck Compartment where the destruction or creation is happening (for a PEP or ENEA respectively).

But at the end of a Planck time, due to destruction (or creation) at the Planck Compartment at the centre of the particle, a large spherical void would be created. As the Planck time passes, this void is compensated by EPCA formation even before it can form. Thus extreme destruction at the Planck Compartment leads to the creation of a disturbance around.

This zone can be called the "zone of disturbance". This zone could be considered an alternative limit of the particle.

As the EPCAs form and move inwards to compensate for the void, the innermost EPCAs may have to move inwards faster than the speed of light if the destruction being caused is significantly rapid. This is akin to the formation of a Blackhole.

There exists a radius, within which the EPCAs have to move inwards faster than the speed of light to compensate for the void. This radius is determined by the extent of destruction happening in the centre which in turn is analogous to how much energy is trapped within the particle. This can be considered analogous to the Schwarzschild radius of the Blackhole.

This radius could be considered as another potential boundary of the particle, alternatively called the "ZOI" or zone of interaction i.e. events happening within this zone happen so rapidly that they and their effects are transferred faster than the speed of light.

For a composite matter particle made of positive energy core and negative energy lattice, the boundary would be constituted by the lattice itself.

Wave emanating from a composite matter particle

As soon as the positive and negative energy particles form, they get bonded with each other. The positive energy core keeps sucking the peripheral lattice of ENEAs inwards but the lattice particles keep spinning around in a way running away from the Positive energy core and also running away from each other. This is analogous to the perpetual motion of Bondi.

Thus this spin becomes an intrinsic property of this composite particle.

To note here is that each of these ENEAs, according to typical SR logic, have to move within the speed limit of the speed of light and cannot exceed it. But with active spatial expansion happening within it, it is unclear whether this speed limit applies to them since it is likely that this speed limit will not apply even to the photon of light in the presence of actively expanding space just before it

As destruction happens at the positive energy core, compensatory EPCAs form around it and the time resetting wave moves outwards.

This wave interferes with other time resetting waves emanating from individual ENEA particles. As it permeates the perforated lattice structure and comes out, it gets broken down into many waves and shows diffraction as well as interference.

This complex interference pattern leads to redistribution of time around the particle.

This possibly creates bands of highly contracted and highly dilated time described above.

However, these bands along with the rest of the time resetting wave moves outwards spherically. It is almost like a disturbance travelling outwards. With every bit of distance travelled, the radius of the sphere increases. The velocity with which this disturbance wave travels depends on the extent of destruction happening at the positive energy core. Even the wavelength of this disturbance wave depends on the extent of destruction and thus the mass of the particle.

Linear velocity vs lateral drifting velocity of the Push-Pull bands

When the Push or Pull bands form, a cylinder of multiple Planck Compartments (which could be a significant number) are displaced in a single Planck time. Thus linear velocity as shown in the figure iii/14.2 can be several times faster than light.

Once formed, the surrounding PCs react to the actively contracted space within this cylindrical region leading to the shift of the deviated Local time outwards. This can be called the Drift velocity. This occurs at the speed of light at 1 PC per 1PT.

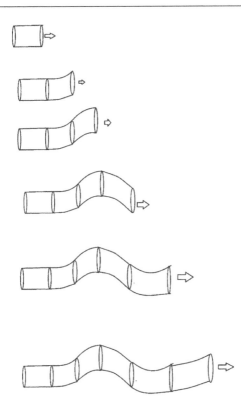

Figure iii/14.2: Shows Linear velocity of the Push/Pull band with a new cylindrical volume with several hundred to thousand PCs wide cylinder of space being displaced every Planck time. This linear velocity is likely to be significantly higher than the speed of light.

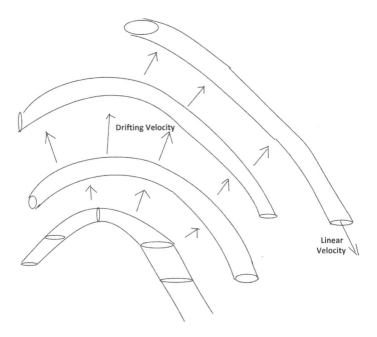

Figure iii/14.3: Shows the drift velocity of the Push/Pull bands once formed. The surrounding PCs react to the presence of the highly deviated Local time and their response leads to the deviated Local time to drift outwards with the speed of light.

Levels in which the Push-pull bands lie

Figure iii/14.4 given below shows a charged particle with a Positive Energy Core at the centre and a Negative energy lattice of ENEA particles. Just two bands emanating from it are shown as they move away.

The Push bands start from the particle at a different level than the Pull bands. As they spiral out, they would remain separate from the Pull bands. As the lattice rotates, the origin of the bands changes at the surface of the Charged particle and their location within the space also changes as they drift outwards.

Because the positively charged and negatively charged particles have an asymmetry between Positive and Negative Energy, the levels at which these bands come out also does not match. Even the two poles have a different arrangement of the ENEA lattice with one pole having a relative deficiency and the other having a relative excess of an ENEA. Thus even here the levels and densities would vary. As the particle is spinning, the densities would also vary at poles and the equator with poles having a significantly higher density.

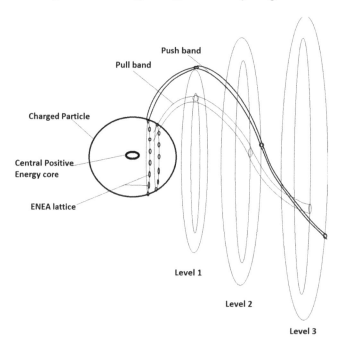

Figure iii/14.4: Shows a charged particle with a single Push band and a single Push band illustrated diagrammatically. Note that the Push bands start at a different level than the pull bands and as they spiral out, they remain separate.

Cones of Attraction and Repulsion

Study the figure iii/14.5.

The level on the Charged particle where the Push bands and Pull band lie, would be different. As the bands move outwards they would also lie in a different level so that starting from the surface of the charged particle moving outwards, the space-time would be divided into multiple cones. In some of these cones, only the Push band lie, while in others only Pull bands would lie as shown in the figure iii/14.5 where the Cone of Attraction where the Pull bands lie is marked by oblique shading while the Cone of Repulsion where the Push bands would lie is left unshaded. Note that the Bands lie in these cones in a spiral manner. Also worth noting is that one pole of the particle being Positive Energy excess would have a cone of Attraction while the other would have a Negative Energy excess and would have a cone of Repulsion at the axis. At both the S and N poles, the size of these cones will vary. In an isolated charged particle without its partner, the sizes keep on varying as the poles switch and as the ENEA particles move in a pendulum-like manner.

When two charged particles are in proximity and are in proper spatial alignment with the N pole of one facing the S pole of the other, the attraction cones possibly align properly and the repulsion cones are cancelled out more leading to a net attraction as explained below.

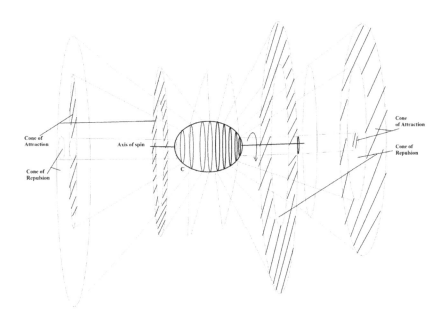

Figure iii/14.5: Shows various levels of cones around the particle where these Push or Pull bands lie.

Attraction and Repulsion

When two charges are together while in the state of spinning around their axes, it is likely that the levels of push bands or the pull bands momentarily match each other. When this happens, the particle experiences a rotational torque and changes its alignment.[1**] At this new alignment, other sets of Push/pull interactions might occur. All these events add up to make the two charged particles face each other's poles so that the North pole of one coincides with the South pole of the other. If more Push interactions happen between push bands of the two, the particles push each other. If more pull interactions happen, there is a resultant pull between the two particles.

1** - After superposition of spin of a charged particle was understood, these methods were slightly revisited. The effects of change in the configuration of the ENEA lattice under space-time deformation caused by coming in vicinity of the oppositely charged particle were noted down.

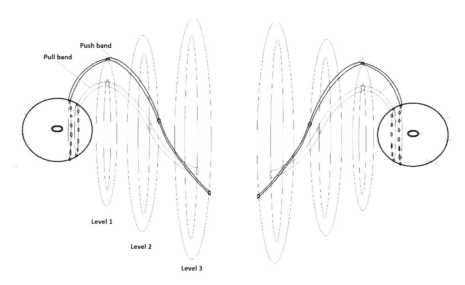

Figure iii/14.6: Shows how two identical charged particles interacting together leads to the interaction of the Push-Pull bands

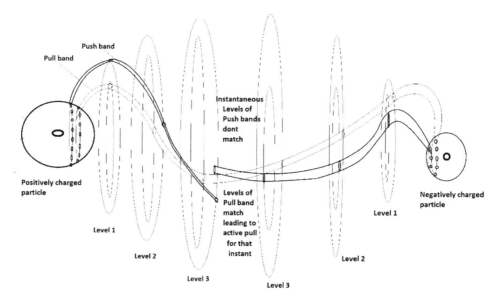

Figure iii/14.7: Shows two slightly different charged particles with only a few Push-Pull bands. Note how the Pull bands from both particles as shown in the figure are coinciding, thus attraction would result.

Chapter 15

THE MATTER-WAVE AS PREDICTED BY DGR

Every Quantum particle is moving and thus has a velocity and a momentum. Every Quantum particle also has destruction going on at its core. Thus every Quantum particle possesses a "preceding border" and a "receding border". Every Quantum particle also has a spin due to its composite nature with the Negative energy lattice particles producing the "Push bands" that constantly push the particle into a spin.

Thus the preceding and the receding borders keep shifting their orientation. At the preceding border, the velocity of EPCAs moving inwards is much more to compensate for the destruction at the centre. This direction can be called the D_{max} (Direction of maximum inwards velocity of the compensatory EPCAs). The other diametrically opposite direction would then be D_{min} (Direction of minimum inwards velocity of compensatory EPCAs). As the particle spins around, the location of D_{max} and D_{min} keep shifting. The inward movement of EPCAs (both at D_{max} and D_{min}) at a point in the trajectory of the particle starts a cycle and forms Time resetting waves that move outwards. Due to this, the positive energy core of the Quantum particle creates a disturbance in its surroundings which are characterized by alternating zones of higher velocity EPCAs and lower velocity EPCAs akin to a crest and a trough of the matter-wave. Note that all this will happen in addition to the emanating push-pull bands that spiral around the particle.

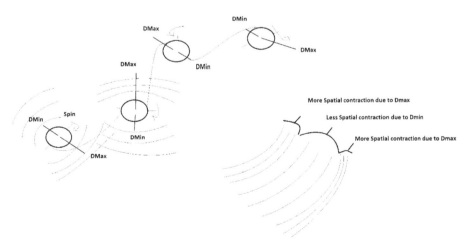

Figure ii/15.1: Matter wave as predicted by DGR

Chapter 16

STATIC GENERAL RELATIVITY VS DYNAMIC GENERAL RELATIVITY

Space-time is an entity that is supposed to be intricately linked in General Relativity or is, in fact, a single entity. Although this is true, space-time is said to be continuous at the smallest level. This means that there is no concept of a Planck Compartment in General Relativity.

In Dynamic General Relativity or our theory, space and time are two separate entities but the intricate link between the two is defined in detail. At Planck scales, the space-time in Dynamic General Relativity is not continuous but made up of small granularity of individual Planck Compartments.

These granular Planck Compartments can be destroyed or created leading to spatial contraction or expansion.

The spatial contraction is akin to a low-pressure region of a sound wave. The spatial expansion, on the contrary, is akin to a high-pressure region of the sound wave.

Einstein's General relativity doesn't speak about the underlying mechanisms which cause the mass within the massive objects to cause their effect on the fabric of space-time. It just says that this deformation travels outwards at the speed of light.

Dynamic General Relativity describes an in-depth mechanism as to how this deformation is achieved.

However, for this to be accomplished, some assumptions of the General Relativity of Einstein have to be given up. The Eddy PC aggregates formation is the process described to compensate for destruction happening at the centre of a particle. These Eddy Planck Compartment aggregates form at extremely high speeds in line with the fact that destruction at the centre can happen at extremely high speeds. So that when destruction can happen so fast that 10^{10} Planck Compartments can get destroyed at the "massive (- i.e. the one having mass) particle every Planck time. To compensate for this destruction, the Eddy Planck Compartment aggregates have to form at similar rates, which means that Planck Compartments that form a part of every Planck Compartment aggregate gets entangled at extreme speeds. Each of these Eddies has a particular C/D ratio and thus have local time running at an equal pace. As the Eddy compartment aggregate moves inwards, its local time becomes increasingly more dilated and destruction hastens in it. The pace of inward movement of Eddy's can be faster than the speed of light at places in the interior of the Blackhole beyond the event horizon or extremely close to the positive energy core of the nucleons like Protons or Neutrons.

The formation of these Eddies can potentially move outwards at very rapid rates like 10^{10} PCs per Planck time and thus these processes spread outwards multiple times the speed of light. This counters the claim of Einstein's General Relativity that information cannot travel faster than the speed of Light.

Einstein's General Relativity, although it speaks about variability in time happening all around the massive object so that close to a massive object there occurs Gravitational time dilation and as we go further apart the Gravitational time dilation effect reduces significantly, does not talk of the exact mechanism due to which this variability of time occurs.

Dynamic General Relativity explains in minute detail how this variability in time in General Relativity happens. The Gravitational time dilation is explained by the presence of Eddy Planck Compartment aggregates all around the massive object meant for compensating the destruction caused at the centre of the massive object.

Balloon models

Simple Balloon models can be imagined to understand how Dynamic General Relativity explains the various complex phenomena like gravity.

Described below are basic models described just for understanding purposes.

A 10^6 compartment aggregate – a single Balloon

Arbitrarily, we can keep a cube or sphere made of 10^6 PCs as a single balloon. Thus our balloon has a fixed volume. The balloon has an inlet and an outlet so that air can be pumped in or out of it. In a similar manner, by a yet unknown process, Planck compartments can be removed or extra PCs can be inserted into our balloon.

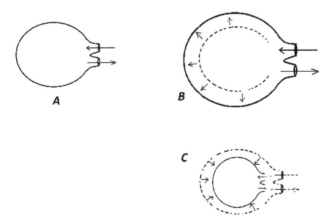

Figure iii/16.1: "A" Shows a Balloon with an inlet and outlet which represents an aggregate of 10^6 PCs. "B" shows an inflating balloon which indicates excessive creation within it. "C" shows excessive destruction leading to active loss of volume or active contraction.

Any excessive Creation happening within our 10^6 PC sphere will increase its volume like an inflating balloon and any excessive destruction will cause it to lose volume.

Explaining Gravity

The space in between a freely falling apple and the Earth can be divided into multiple compartments of size 10^6 PCs or 1 mm3 or 1 m3 etc. These represent our balloons. Each of them has an inlet and an outlet.

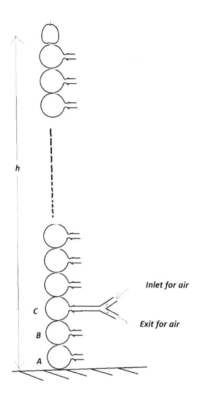

Figure iii/16.2: Shows an apple at a height h freely falling towards the Earth and the multiple Balloons which represent 10^6 PC cubes or spheres or for that matter PC aggregates of any volume like 1 mm³ or 1 m³ etc. Each of them keeps losing volume thus pulling the apple down.

However, if the incoming air is much less than the outgoing air, the apple will keep falling downwards. Note that as we move downwards starting from the apple, the rate of pumping air out of the balloons increases so that the lowermost balloon is losing its volume fastest. The rate of removal of air from the Balloons reduces as we move from the Earth towards the apple. Thus a simple model can explain the change in the shape of space-time as described by GR.

Explaining a Charged particle

The same logic can be applied for a charged particle so that the centre of a charged particle has a much bigger rigid sphere with a lot of air within and is connected to a pump so that air from it is being actively pumped out (shown as a central tube). The balloon has multiple sieve-like holes so that the extremely strong suction of the air moving out of it produces a suction effect at each of these holes. In between the holes are smaller balloons attached to a pump in such a manner that the air from outside the apparatus is being blown inside these smaller balloons but because of the limited ability of these balloons to inflate or rigid size, the air is escaping out causing a push. In this manner, the particle will have zones of high and low pressure, the high-pressure areas pushing outwards and low-pressure areas pulling inwards.

Alternatively, one can imagine a rigid sphere immersed in water as explained below.

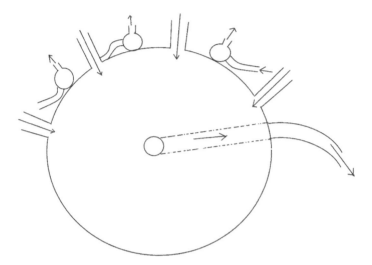

Figure iii/16.3: Showing a simple balloon model of a Charged particle as described in DGR

Chapter 17

EXPLAINING PUSH-PULL BANDS

Imagine a spherical hollow ball with two diametrically opposite holes. Imagine that it is made of a non-collapsible material like steel. Imagine that the sphere is fully immersed in water so that water fills it completely.

If we attach a pipe to one hole and attach a vacuum apparatus that sucks out all the water in it, what will happen?

As the water is sucked out, a void or vacuum is formed within which reduces the pressure within the sphere. The difference in pressure inside and outside the sphere creates a suction-like effect near the second hole and the water molecules outside the sphere start entering it. This flow of water can be seen, characteristically with the creation of turbulence here.

Now imagine that instead of a single hole, the whole sphere is sieve-like with multiple holes all around the surface of the sphere. Let the pipe attached to the vacuum apparatus be fixed at the centre of the sphere. The vacuum or void with no water will now be created first at the centre. Water starts flowing from the peripheral part to the centre within the sphere. Usually, the water spirals around and thus forms a whirlpool. There is a pressure gradient along the length of these spiral lines. As water exits, the pressure difference inside and outside the sphere, water starts flowing through these holes from the region immediately around the sphere, forming a miniature spiral here as well.

Now further imagine that the metallic walls of the sphere between holes have smaller pipes fixed to them and water is being pumped out from there.

Further, imagine that the outflow of water in the sphere from the central pipe is the same as the inflow of water from all these smaller pipes combined.

Now further imagine if this whole assembly starts rotating rapidly.

The sphere is rotating. The whirlpools starting from every hole also spiral around with an axis same as the axis of rotation of the sphere. The inflowing water from the miniature pipes also creates their own whirlpool equivalents.

The whirlpools starting peripherally and entering the holes are like the pull bands.

The whirlpool equivalents starting from the mini pipes from where freshwater is rushing in are equivalent to push bands.

This example is not exactly equivalent and there are significant differences in this and a charged particle in our theory.

CHAPTER 18

ENEA PARTICLES, THEIR PROPERTIES, INTERACTIONS AND IMPLICATIONS

ENEA or exotic negative energy antiparticles are Planck Compartments where the local time has deviated towards creation. That is, in them, Creation is happening faster than Destruction. In them, the time is contracted.

A PC will have a volume equal to one Planck volume. After every Planck time, an extra Planck volume of space is created at these particles, if the C/D ratio in one such ENEA is 1/0 or 2/1.

As the C/D ratio increases, the number of extra Planck volumes of space created per Planck time keeps increasing.

A T=0 Compartment has a C/D ratio of 0/1 or 1/2. So the counterpart of the T=0 Compartment would be a T=Tmax compartment having a C/D ratio of 1/0 or 2/1.

We know that a T=0 Compartment constantly sucks volume from the surrounding space. This causes it to create Eddy Planck Compartment Aggregates around it. These EPCAS compensate for the volume loss. This compensation can potentially happen for an infinite length away from the T=0 Compartment in focus.

When two T=0 Compartments approach each other, the EPCAs forming around them would also interact and are likely to interfere constructively, so that the zone between them ends up having a much more time dilation within it and lose volume. In effect, the two T=0 Compartments attract each other until they fall into each other and are likely to fuse. Thus multiple T=0 Compartments will tend to form aggregates. There is no limit to how many such T=0 Compartments can fuse.

With entanglement as the cause of the formation of these EPCAS, the formation can happen faster than the speed of light.

Analogously, when excessive creation is happening around a PC, EPCAS would be expected to form around it which bring about C/D or local time changes and push these EPCAS outwards.

Two such T=Tmax PCs, when approaching each other would have their respective EPCAS interfering with each other. This interference will also be constructive or additive interference and thus the two negative energy particles would tend to repel each other.

This is exactly what Bondi et al noted in their paper.

When a T=0 Compartment approaches a T=Tmax Compartment, a weird thing happens. The T=0 Compartment keeps sucking or attracting the other while the T=Tmax Compartment keeps repelling its partner.

In a way, one of them will chase the other and the other will keep running away from it.

This perpetual runaway motion of Bondi spooked many physicists and was the reason why many physicists still don't believe that negative mass particles can exist.

Implications

1. The ENEAs can't form aggregates.

The ENEA particles with. C/D ratio 2/1 i.e. the Tmax Compartments repel each other and are unlikely to form aggregates.

This may possibly have serious implications.

Like we described a hyper-dilated local time with multiple T=0 Compartments fused together so that much more than one PC volume of space is lost per Planck time, a hyper contracted time with the formation of aggregates of T=Tmax Compartments seems unlikely.

Does this mean that the formation of an actual white hole with an extremely highly deviated C/D ratio towards creation is unlikely to form?

Does this mean that there is a limit to deviated local time towards creation?

Like a supermassive Black hole at the centre of the Galaxies having an extremely deviated local time with destruction happening significantly faster than creation, does this mean that a supermassive white hole with a C/D ratio deviated extremely towards creation does not exist or cannot remain stable?

We know that there are trillions of Galaxies and each of them has such a supermassive Blackhole.

We also know that laws of physics require that the positive and negative energies be formed as a pair.

This means that when a T=0 Compartment forms, another T=Tmax Compartment also forms simultaneously. <u>Contemporary physics says that being matter-antimatter counter-part they should annihilate each other reforming the energy. In our theory they are the energy and that they cannot reach each other let alone annihilate each other.</u>[1*]

> 1* - Note that the matter- antimatter definitions are different in contemporary physics and our theory. In our theory, the T=0 compartments pull the T=Tmax compartments while the T=Tmax compartments push the T=0 compartments. So they cannot come near each other or annihilate each other. But the matter antimatter that contemporary physics define can and do annihilate.

So if X number of T=0 Compartments formed till now in the universe, it is likely that X number of T=Tmax compartments also formed.

In this kind of Universe, there is a perfect balance between Creation and Destruction. In such a Universe, all the Creation would be destroyed somewhere else and such a Universe would essentially be a static universe. Alternatively, such a Universe is unlikely to have formed from a High-density beginning as such a Universe would remain the same size.

This doesn't apply to our Universe. <u>In the Universe that we live in, right from the moment of formation, there is a small imbalance between Creation and Destruction. This means that more Creation is happening than Destruction.</u> [2*]

> 2* - The Positive energy dominance at the edge- Negative energy dominance at the centre model and other possible models of the Universe are described in Crisis in cosmology section in Part ii. In this section, the various possibilities about PE and NE distribution are discussed. If we can accept that there is infinite PC vacuum beyond the edge of the Universe i.e. there is infinite Positive energy at the edge of the Universe, then there is infinite PE at the edge and infinite NE at the centre which means that the Universe has a maintained balance of PE and NE.

This imbalance is critical as without it, not only the Universe wouldn't expand, but there would be no existence.

"What is the mechanism, why this imbalance is?" is not clear.

The implication is that there must be much more ENEAs i.e. T=Tmax Compartments throughout the Universe than the total number of T=0 Compartments.

2. The intergalactic space is filled with ENEAs with a C/D ratio of 2/1 and expansion of the universe happens due to them.

The ENEAs that represent the counterparts of the T=0 Compartments that merged to form the central supermassive Black holes of the Galaxies remain separate in the intergalactic region. These would create space equal to the space destroyed by the Blackhole.

The ENEAs corresponding to the mass of protons and neutrons also roam free.

Note that ENEAs corresponding to the mass of the Quarks and the electrons, get attached to them as soon as the pairs form. The positive energy particles merge to form the core and negative energy particles form the lattice and get trapped in the perpetual runaway motion around them, forming their electrical charges.

3. ENEA particles have actively expanding space just behind and in front of them. When they are moving, the compensation of excessive Creation happening within them, like in a Negative energy particle, would become asymmetrical. The compensation of excessive creation, leading

to much more contracted time and thus actively expanding space just behind a particle would have serious implications. We know that even a Positive energy particle like a photon, when having an actively expanding space just behind it, would travel faster than the speed of light and would reach the mirror of the photon clock faster, thus measuring a contracted time. Thus an ENEA particle with unequal actively expanding space just behind it is likely to move faster than the speed of light despite itself not breaking the Einsteinian speed limit. The ENEA is just simply riding on an "Escalator" which is moving in the same direction.

4. A white hole would be bright and would be opaque to the passage of light and if they existed, would not be as hard to find.

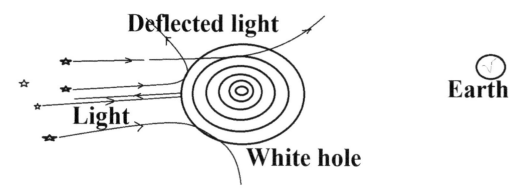

Figure iii/18.1: Shows how a White hole that is an extremely dense aggregation of T=Tmax compartments might look. Note that most of the light behind a true white hole will get deflected or reflected and thus it would be completely black with a boundary with a significant gravitational deflection of the light which can be detectable. It would not cause any Gravitational lensing like a Galaxy.

A white hole, that is an aggregate of T=Tmax Compartments having a highly deviated C/D ratio towards Creation i.e. a region of space-time where entering is impossible unless we travel faster than the speed of light so that even light cannot enter inside, however, light or for that matter EPCAs keep moving out of it at a fast rate so that it has a negative or repulsive gravity.

Such an entity, if existed, would be indeed opaque to light and would not allow passage of light from Galaxies hiding behind it. Note that this opacity is a slightly different type of opacity. The light trying to enter into the white hole would either be deviated to one side or would be completely stopped or bounced back instead of going through such a white hole.

Such an object would have created a significant curvature of space-time in its vicinity and would indeed have caused at least some gravitational deflection which is inverse of the typical Gravitational lensing.

It would have caused at least some detectable effect on the neighbouring stars. Thus if it was a reality, it would have been detected by now. In short, such regions with T=Tmax compartments packed together within a single Planck Compartment are unlikely to be real.

Separate freely floating ENEAs, however, would be difficult to detect being minimally reactive to any other entity and would be dark and transparent to light.

A huge void of space filled with such freely floating independently individually repulsive ENEAs would indeed be transparent to light passing by.

A positive energy photon is an aggregate of T=0 Planck Compartments and when it enters such space, would pass through a zone full of such ENEAs without much interaction. <u>The ENEAs present in between two adjacent photons would successfully explain the redshift i.e. increase in wavelength or decreased frequency of the resultant light.</u>[3*] This is because the ENEAs would create additional space in between photons and thus increase the space between two photons.

> 3* - This happens primarily due to the source (from which every subsequent photon comes out) receding away, what we call as the Doppler effect but a part of the stretching and also the expansion of the space in between the source and the Earth happens because of the addition of PCs by the ENEA particles in the path.
>
> Another alternative explanation is that the stretching of light happens primarily due to receding away of the source which happens due to the infinite positive energy (PC Vacuum) at the edge of the Universe and the ENEAs are primarily involved in compensation of the void by filling it with their new creation.

ENEA's and Dark matter

We know that a central supermassive Black hole exists at the centre of each Galaxy and it keeps sucking in Planck Compartments. At the smallest, all the particles which form all the detectable and undetectable heavenly bodies have a Blackhole-like aggregate of T=0 Compartments and only a part of this destruction is compensated by the peripheral ENEA lattice.

Thus space around the Galaxy would be continually sucked inwards.

There is a likelihood that all this extra destruction (especially accounting for the unbalanced destruction happening at the core of all Neutrons and Protons that form the parts of the Galaxy) needs to be compensated by something.

There is the likelihood that all the T=Tmax Compartments which are counterparts of the mass of Protons and Neutrons in all the matter in the Galaxy are still in entanglement with their parent T=0 Compartments. So that there exists a mass of these ENEA particles which surround the Galaxy and are bonded to the central region of the Galaxy.

These particles would compensate for the loss of volume of space caused due to the mass in the matter present in the Galaxies. If this compensation is absolute, then it indicates that there is a limit beyond which mass within Galaxy or Galaxy clusters can cause Gravitational lensing. Beyond this limit, the mass contained in the Galaxy cannot cause any deviation of space or form an EPCA. The EPCAs formed at the Galaxy's mass, after travelling for millions of light-years might get cancelled by the EPCAs starting from this layer of ENEAs.

To note that the only way by which EPCAs emanating from ENEAs (which are supposed to be just a Planck Compartment in size) can compensate EPCAs emanating from massive stars and Central supermassive cores of the Galaxy is due to numbers.

Even then, there has to be a limit to the density of ENEAs in the intergalactic space. This means that at least some of the EPCAS emanating from the centre would extend beyond the Galaxies into the intergalactic space towards the other Galaxies or Galaxy clusters. These EPCAS could explain the extended MOND bonds and intergalactic bonding and also explain the purpose of G in Newton's gravitation equation.

CHAPTER 19

WHAT'S THERE IN THE EMPTY SPACE?

Contemporary physics says that everything is made up of atoms that bond together to form molecules.

The atom is composed of a centrally located positively charged nucleus and a negatively charged cloud of Electrons present around it. The total diameter of an atom is much more than that of the nucleus so the nucleus is measured in a femtometer while atoms are measured in Armstrong units. The Atom, it figures is mostly empty space.

In solids and liquids, the particles are bonded together by electromagnetic forces. However, the space in between individual atoms bonding to form molecules is also mostly empty and has no particles. Even in gases, most of the space in between the particles is empty.

The question which one might ask is "what is empty space or what is there in the empty space or what are the events that happen in this empty space or what is the significance of empty space?

These questions are troubling and there is no answer to them in contemporary physics. In short, for contemporary physics, the monotonous answer would be, there is nothing in this empty space or no events are happening here and that empty space is genuinely empty and that it is of little significance.

Although Quantum Electrodynamics says that at Planck lengths, there exists Quantum foam which is minute fluctuations of energy wherein virtual particles come in and out of existence and that's the starting point that empty space is not really empty.

Empty Space according to DGR

Space, according to our theory is made up of an array of Planck Compartments that stay entangled to their neighbours. The "particles" could be Negative or Positive Energy Particles and could be with mass or massless.

Whatever a particle is, it is an aggregate of Planck Compartments with a highly deviated local time so that either Creation or Destruction taking place at that point in space is dominant. Thus both massless and with mass particles lead to a formation of a time resetting wave in the form of EPCAs that emanate at high speeds all around the particle.

The typical charged particles like Electrons and Quarks are composite particles and have a much more complex pattern of EPCAs around them.

When a particle is present at a co-ordinate location, all the surrounding volume of empty space gets filled with EPCA forming TRWs emanating from it. If a single Positive Energy particle is present, the maximum time dilation i.e. maximum destruction would be present at the coordinate location of the particle. As we move away, due to entanglement, the surrounding space would show local time showing progressively lesser time dilation. This time dilation ideally should follow the inverse square law and thus time dilation due to EPCAs would also be expected to follow the inverse square law and reduce in proportion to the square of the distance.

In Solids, the density of such particles is highest. This density keeps on reducing as we go from solid to liquid to gaseous. This means that in solids, the particles are closer and there is lesser empty space while in liquids or gases, there is more empty space. The EPCA forming TRWs emanating from each particle would have to travel much more to interact with others and thus the average time dilation would be much less in gases or liquids and much more in solids. What that means is that the TRWs interact (mostly constructively) when they are closer and have a much higher time dilation in solids. When the temperature of a gas or liquid is more, the particle density rises as more Positive Energy Particles are now present. These cause a much higher density of particles and a much higher density of presence of EPCAs.

The temperature in the case of liquids and gases is basically based on individual velocities of the particles which depend on their momenta. The presence of a higher time dilation and hence a higher spatial contraction around a particle in an irregular manner i.e. more in one direction and less in another direction leads to the creation of unbalanced forces that lead to movement of the particles, i.e. induce momenta in them. Thus if the empty space has more Positive Energy mass-less particles which are photons or packets of energy, they will add to the density of the emanating EPCAs and increase the velocity of the with-mass particles or matter particles.

Thus temperature can be the density of presence of these photons or packets of energy or aggregates of Positive Energy mass-less particles in the empty space between matter particles within the liquids or gases

Solids have particles arranged either in a systematic and orderly manner or a haphazard manner. When they are orderly, the solids are said to be in a crystalline phase. When they are disorderly arranged, it is called the amorphous phase.

Metals are usually crystalline while non-metals can be crystalline or amorphous depending on the constituents. Like NaCl is crystalline and so is Silicon oxide or glass. Carbon can be arranged in a crystalline manner as in diamond or Buckyballs or Graphite or an irregular manner in burnt ash.

In crystals, especially metals, the particles are bound tightly and are thus having limited freedom of movement unlike in liquids or gases. The only freedom they have is to vibrate or alternate between higher and lower energy states absorbing and emitting energy packets i.e. photons or Positive Energy particle aggregates.

The temperature of solids especially metals is thus determined by a number of these Positive Energy Particles within it which inadvertently determine the density and the time dilation of the EPCAs interacting in the empty space in between.

If an atom of metal has an excess of time dilation spatial contraction on one side and then a similar spatial contraction on the other diametrically opposite side alternating at regular intervals, this leads to a vibrating atom. Thus vibration of the metallic atoms can be explained based on the interaction of EPCAs in the empty space between the atoms.

If the Positive Energy Particles don't have enough energy, they can only achieve the elevation of an Electron from a lower-energy orbit to a higher orbit. But if the energy is good enough, it can exert enough force on the atom and transfer momentum to the with-mass particle.

A particle with mass and massless particles are basically similar. The difference in them is just the amount of destruction happening with the amount of destruction being much more in a particle with mass.

Thus a massless particle can be absorbed in a with-mass particle. When this happens, the consequence would be just an addition of the amount of destruction happening at the particle if we are talking of two Positive Energy Particles with excessive destruction interacting.

The with-mass particle after acquiring the additional destruction of the massless particle increases in energy content. The additional destruction causes an additional mass to the with-mass particle which Relativity calls relativistic mass.

What is interesting is that although the particle merges, the place where it did merge in three dimensions onto the destruction sphere of the with-mass particle is remembered i.e. this information is not lost. The mass-less particle transfers its momentum to the massive particle. Essentially, when the massless particle comes close to or interacts or is absorbed in the sphere of the destruction of the with-mass particle, the additional destruction created in that quadrant where it enters remains and creates EPCAs at the quadrant of impact differently than the diametrically opposite side.

For example, let's assume that the Positive Energy mass-less particle A hits the with-mass particle B while moving from left to right on the x-axis when the central point of the particle B where destruction is happening has coordinates (0,0,0,)

When this happens, the destruction within particle A gets added to the destruction happening within particle B and the empty space in the quadrants of space left to the XZ plane have more time dilated EPCAs than the quadrants lying to the right of the XZ plane. This difference in time dilation EPCAs on left and right or asymmetry in time dilation zones from left to right, caused because of the particle, makes the particle move in one direction and acquire momentum.

More the number of such Positive Energy mass-less particles or quanta of energy the "with-mass" particle absorbs, more is its relativistic mass and more is its velocity and momentum as more will be this asymmetry. Mass-less Positive Energy Particles or quanta of energy can enter the with-mass particle from anywhere. If the second particle enters from a different direction, the resultant momentum would be a vector sum of the two momenta.

In effect, the empty space can also have a temperature. And the temperature of vacuum or empty space in between gas particles or liquid particles is determined by the EPCAs present in it and their interactions.

There are significant implications of this statement.

Let's consider water in two containers. The first container has water at 30 degrees Celsius and the second one has water at 99 degrees Celsius. The second container has water close to the boiling point. When a person puts a finger in both these, the 99-degree water feels different. What is the difference between these two glasses of water? Chemically, no difference exists. Both contain H2O. But individual velocities of molecules in the second container would be much higher than the first one. Is that the only difference?

If our theory is true, the second container has a lot more mass-less Positive Energy Particles which are nothing but quanta or packets of energy. These particles produce EPCAs as well and thus the second container will have many more particles, although the number of with-mass particles may be the same. The density of EPCAs and the resultant time dilation would also be much more.

Explaining refraction of light

We know that light travels slower in water as compared to vacuum. Quantum mechanics gives a complex explanation of why this happens which involves the interaction of the particles of light with Electrons in the molecules of water.

Whether refraction of light happens or not can very easily be proved and demonstrated by putting a pen or a pencil in a glass made of glass and filled with water. The light coming out of the pen or pencil, when seen from various angles clearly shows refraction.

Our theory would say that the EPCAs interacting in empty space create an average time dilation within the empty space. When particles of light travel through this empty space, this time dilation causing secondary spatial contraction leads to disturbance in the velocity of the particles of light which are basically Positive Energy Particles of comparatively less energy content.

Contemporary physics says that as the temperature of water rises, the water (or for that matter any other fluid) loses density and this leads to an increased velocity of light within it and thus a reduced refractive index.

I.e. Temperature of fluid is directly proportional to the velocity of light within it which in turn is inversely proportional to the refractive index.

Light is made up of Positive Energy Particles of predefined energy content and thus having a predetermined C/D ratio. There is a predetermined amount of destruction taking place at each of these particles and the EPCA forming TRWs emanating from them cause entanglement of the neighbouring particles. The frequency and thus wavelength of each ray of light is determined by the distance between the adjacent particles. Each of these particles forms a pattern of EPCAs around them and thus reset time around them and are having a contracting space around them. All these spatial contraction time dilation zones formed due to the EPCAs represent the electromagnetic field variations characteristic of the EM waves.

Each of these particles are having small energy, i.e. not enough to cause so much destruction to create mass. Thus each of these is massless particles and each of these waves travel one Planck Compartment per one Planck time i.e. with the speed of light in the vacuum.

In a vacuum or empty space between gas molecules or air, this train of particles, there are no EPCAs strong enough to significantly alter the time dilation zones formed by the EPCAs.

In transparent solids like glass or transparent liquids like water, when this train of entangled Positive Energy Particles enter, the EPCA forming TRWs emanating from the particles of the solid or liquid interact with the TRWs of the particles of light. Thus the time zones around it are affected. In short, the Photon may get transformed into something different. This is called in contemporary physics as a Polariton which is a quasi-particle formed by the interaction of a photon of light with a magnetic dipole like an electron-electron hole or an electron-proton in an atom

As explained in the propagation of a massless particle section, the forward propagation of the Photon gets reduced significantly due to the time dilation spatial contraction produced by EPCA forming TRWs emanating from the particles of the transparent material where the light has entered. This is just like the particle trying to climb an escalator moving downwards. The Photon will keep on getting pulled backwards and thus its propagation within the medium would be slower. The slower propagation of one Photon causes bending or change in direction of the wave-front due to change in direction of neighbouring Photons. These phenomena explain the laboratory observed Refraction.

The temperature of the material determines the individual velocities of the particles. With the rise in temperature, the velocities increase. The density of the material reduces as individual particles acquire more energy to get out of the bondages. As the average distances between particles increases, the average time dilation caused due to the EPCAs in the empty space in between the particles reduces and the spatial contraction due to it reduces. This means that in such a medium, the velocity of propagation of the same photon would be higher. The velocity of propagation closer to the velocity of light in a vacuum means the deviation of photons while passing in the transparent medium reduces and refraction reduces.

It is thus clear that the refractive index would reduce with an increase in temperature.

Events happening at the transition

This is another question that conventional physics is unable to answer or has a limited answer.

What changes happen exactly at the transition of water and air on the surface of a pond?

We know that the density of particles is much higher within the water and much lesser in the air. This means that the amount of destruction happening would be much more within the Water compared to within the air. Each of these particles of water, i.e. molecules of water, would give off EPCAs which are meant to compensate for the destruction happening at the particle's centre. The average intermolecular distance would be much more in the air than within the water.

Each molecule of water also has multiple particles bonded together. The oxygen atoms have multiple particles within the protons and electrons bonded together, each of them not only give off EPCAs but also give off the Push-pull bands which are responsible for the electromagnetic bonds.

The push-pull bands emanating from hydrogen atoms (which are basically protons) that face the oxygen atoms are responsible for the hydrogen-oxygen electromagnetic bond. The ones not facing the oxygen atom forming the H2O molecule can escape and keep moving outwards at the speed of light. These can interact with similar push-pull bands emanating from other neighbouring oxygen molecules within the interiors of water giving rise to the hydrogen bonds which are responsible for various properties of water.

At the surface of the water, the transition is from more EPCAS and thus more time dilation and spatial contraction along with much higher density of these push-pull bands to a much lesser density of EPCAS and lesser time dilation spatial contraction.

CHAPTER 20

WHAT'S THE SHAPE OF THE UNIVERSE?

There are three possibilities. Either the Universe is absolutely flat so that two parallel rays of light keep remaining parallel even after travelling for thousands of light-years. The other two possibilities are the Universe having a positive curvature or a Universe having a negative curvature. In the first, the two parallel lines keep drifting away so that after some distance, they start diverging. A negative curvature, in contrast, means that the lines converge eventually meeting or crossing each other.

All evidence currently points towards a flat Universe.

A ray of light is a very ill-defined term. Alternatively, we can use a particle of light or a photon of light.

A photon of light is itself an ill-defined entity in contemporary physics given the wave-particle duality of light. The exact dimensions of a photon are not known although assumptions based on frequency can be used. At least one thing is certain that when contemporary physics talks of a photon, it is indeed not a point sized thing.

When photons of light are used, their paths will have to be compared at different portions of their journey to check for parallel direction or divergence or convergence. Often a single crest and trough of the electromagnetic wave forming the light are loosely termed as a photon.

In our theory, a photon is restricted to a single Planck Compartment although the disturbance created by it moves outward as time redistribution waves in the form of EPCAs.

What's the shape of Universe in our theory?

The space-time in our theory is not static or statically changing (like in General Relativity) but is dynamically changing.

Note that a source usually gives off photons all around it and not in a few directions.

Let X number of photons start simultaneously from the surface of this star. Let's name them A, B, C..... Etc. in order.

As each of these photons is a Positive Energy particle and has a hyper-dilated time within and has excessive destruction happening within each of them. This destruction would create EPCAs which act as a disturbance around them. When the EPCA forming TRWs moving outwards interact with the similar EPCA forming TRWs of its neighbour, the result is attraction. It is these EPCAs that keep all these photons entangled with each other.

Note that these EPCA forming TRWs happen due to the high quantity of destruction taking place per Planck time within the photon and thus would move at extremely high speeds much faster than the speed of light.

Also note that in reality, there would indeed be a sphere of photons entangled together along with their wave-fronts which are formed by their interacting EPCAs.

But let's consider just a circular cross-section of this sphere passing across the photons A, B and C.

Each of these photons would experience two forces perpendicular to their direction of motion due to the suction effect of the neighbour. These forces are likely to cancel each other out.

Thus, at this point in their journeys, they would diverge outward.

Also, note that other points on the star's surface would also give off photons in other directions. Let's consider one such point from which A', B' and C' photons are starting, all of which are parallel to their respective counterparts.

We can consider two photons A and A' which started their journey simultaneously and in a parallel direction.

Let the distance between them be "d".

This distance can be further broken down or divided or expressed in terms of Planck Compartments. So let there be "d" number of Planck Compartments in between the two photons when they start their journey.

As these photons start simultaneously, they would indeed be in entanglement with each other. They would also be entangled with the photons which started just before and just after they started due to the interaction of EPCAs starting from them with EPCAs starting from these other photons.

As these A and A' move away, let's imagine the Planck Compartments extending in the distance A to A'.

If the space in between these photons increases during their journey, the distance AA' increases and they would eventually diverge.

If the space in between these photons reduces due to the destruction of some Planck Compartments in between them, the distance AA' decreases and the photons converge.

Thus, whether the photons will eventually converge or diverge depends heavily on what's in between them which in turn depends on their paths.

If a large Galaxy lies on the line AA', i.e. the two photons are moving on either side of a Galaxy, the effect of mass within it including the central supermassive Blackhole would make additional destruction on the AA' line more likely and this additional destruction would pull the photons towards each other effecting Gravitational lensing effect of the Galaxy.

Honeycomb model of the Universe

Every Galaxy is made up of matter particles which in our theory are a combination of central Positive Energy core with a peripheral lattice of Negative Energy particles. These form the Quarks and Electrons. Here the Positive and Negative Energy particles are effectively in balance.

The Proton and Neutron, however, in our theory are a central high Positive Energy core around which the three Quarks are bonded. This Positive Energy constitutes 99% mass of the Proton and the Neutron. There is no Negative Energy counterpart of this visible to us here.

What is theorized is that each of these has their counterparts around the Galaxy which are still in a state of entanglement with this unbalanced Positive Energy mass.

Thus there exist zones of Positive energy excess (where all the Baryonic matter abounds) which are surrounded by these entangled Negative energy particles followed by extremely dense areas of non-entangled free ENEAs.

This entangled Negative Energy counterpart would be in the form of freely floating mutually repulsive particles having negative mass, i.e. the ENEA particles. These represent the dark matter in our theory.

Concept of the density of Negative Energy particles

It is theorized that the Negative Energy particles we are talking about are the neutrinos.[1#] This essentially means that our theory makes a gross unpalatable claim that one of the essential compositions of the Electron or Quarks is the Negative Energy lattice of the neutrinos around its central core.

> 1# - Alternatively, and more appropriately, Neutrinos may be a small positive energy core with a few ENEA particles attached to them so as to account for the small positive mass that the particles display.

At any given volume in space, "how much is the density of Negative Energy particles and how much is the proportion of Positive Energy vs Negative Energy?" has important implications.

At the core of every matter particle, there is a high density of Positive Energy particles.

Around the heavenly bodies, the compensatory EPCAs making the space move inwards towards the body, thus leading to the Gravitational attraction, means that there is a dilated time and thus comparatively higher destruction. This indicates that even here Positive Energy predominates. The effect of the mass of the stars and central supermassive Black holes in a Galaxy means that the resultant compensatory EPCAs would be Positive Energy ones.

The zone in between Galaxies that are not gravitationally bound to form clusters of Galaxies often called the intergalactic zone is the region where Negative Energy would be expected to predominate in the form of a higher concentration of ENEA particles.

Here, a unit volume of space, say every light-year square, would have a much higher number of these ENEA particles floating within.

In fact, the number of Negative Energy particles here would be uncompensated with Positive Energy particles. This is how despite extreme destruction taking place at the centre of every Heavenly body with a Gravitational attraction, the Universe is expanding.

This extra uncompensated Negative Energy is equivalent to Dark Energy described in contemporary physics.

The space can effectively be divided into zones where there is predominantly Negative Energy vs zones with predominantly Positive Energy.

The zones with Positive Energy would indeed have huge central supermassive Black holes at the centre and the space here is invariably contracting and moving inwards.

On the contrary, the zones with Negative Energy predominance would invariably have space expanding and extra creation happening within.

Possibly, these zones are arranged like a honeycomb, so that the hexagonal (the hexagonal shape is just exemplary and not compulsive) empty space in the honeycomb corresponding to Positive Energy zones and the walls of this hexagonal compartment being made of Negative Energy excess.

The zones which represent the walls of this hypothetical honeycomb would have a very high density of Negative Energy particles and thus would have expanding space. Galaxies lie in the centre of each of these hexagonal Compartments. Multiple compartments joined together form the Galaxy clusters. The inter Galaxy bonds in Galaxy clusters also have a contracting space and thus would have Positive Energy excess i.e. contracting space.

CHAPTER 21

PHOTON AND TYPES OF EM RADIATION

From contemporary physics, we are aware that photons are packets or quanta of energy. They are particles of light.

A photon with a higher frequency has more energy in it and a photon with a smaller frequency or higher wavelength has a lower content of energy.

The description of the photon almost ends there. Photon has no mass. But what it's made of or how it looks spatially or how does it affect space around it or time within or around it is not discussed.

The density of photons is higher in higher intensity of light in a cross-section of a wave-front of light. The intensity of light keeps reducing as it goes away from the source obeying the inverse square law.

It is also clear that as photons travel, their frequency doesn't change as long as they are travelling in non-expanding space-time. When light travels through an expanding space-time, the space in between photons expands and the wavelength increases which essentially means its frequency reduces. This is what we call redshift.

If we imagine photons are spherical ball-like (non-quantum) particles i.e. they have a perfect location in space, then it is clear that the distance between two spheres forming a part of a single wave-front is independent of frequency or amount of energy within the photon but is a function of the distance the wave-front has travelled from the source. The distance between every subsequent wave-front depends on the periodicity which in turn is dependent on the frequency.

We know that photons have momentum and that the momentum is directly related to its frequency. So that a higher frequency photon like an x-ray or ultraviolet photon has much higher energy and a much higher momentum and thus is in itself capable of knocking off electrons from outer orbits of atoms and thus are called ionising radiation. But smaller energy photons with smaller frequencies have much lesser momentum and thus may not have what it takes to cause ionisation.

Photons of low energy red light are like thin foam balls or marbles, incapable of knocking off a pyramid made of heavy plastic blocks. The photons of higher energy light like ultraviolet light are like large steel or lead balls which have enough power to knock the blocks down.

The photons, however, are quantum particles and thus this analogy shouldn't be taken too far.

One of the most important properties of a photon that we missed till now is that the photons travel forward with a speed equivalent to the speed of light in vacuum. And they are massless and hence according to Special Relativity do not experience time.

Photons in our theory

A T=0 compartment is like a photon. It has a C/D ratio such that one Planck Compartment is destroyed per Planck time.

Many T=0 compartments can coalesce and thus the much higher quantity of Planck Compartments can be destroyed per Planck time. Higher is the destruction, higher is the energy in it. Every T=0 compartment constitutes positive energy.

EPCA forming TRWs emanating from every photon move outwards and interact with the TRWs from other neighbouring photons.

The photons also possess a unique direction of motion. This is determined by the difference in time changes or difference in inward velocity of EPCAs at the preceding and the receding borders.

The figure iii/21.1 (and 21.2) shows a diagrammatic representation of how a typical photon would look as per our theory.

A static photon would have uniformly spherical TRWs emanating from it forming EPCAs moving inwards with a uniform velocity and thus the resultant Local time dilation and Spatial contraction caused would be uniform and uniformly reducing as one goes away from the centre.

More commonly, a massless positive energy particle or a photon would move in a direction with the velocity of 1PL per 1 PT.

In such a photon, shown in the second figure iii/21.2, the volume loss compensation will come predominantly from the Preceding border and the TRW would be asymmetric so that the EPCAs coming inwards from the Preceding border in the direction of propagation of the photon would move inwards much faster and the resultant time dilation would change much faster. Contrary to that, in the Receding border, the volume loss compensation will be much smaller, the inwards movement of EPCAs would be slower and thus resultant time dilation would be much lesser. The spatial contraction would be much more and faster at the Preceding border as compared to the Receding border.

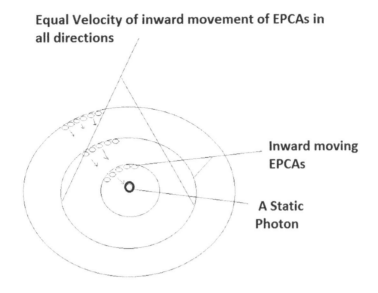

Figure iii/21.1: Shows a static photon with EPCAs forming around it.

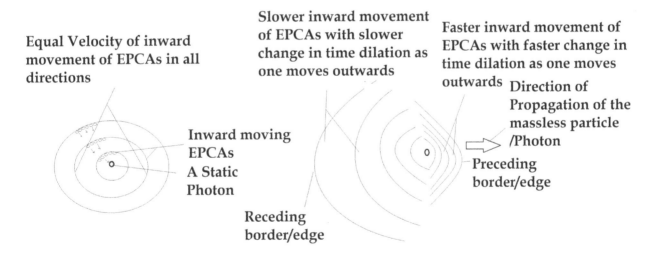

Figure iii/21.2 A photon

A train of photons in an EM wave would look like the figure iii/21.3.

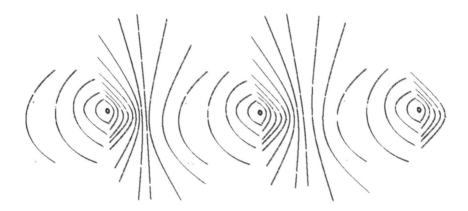

Figure iii/21.3: Shows a train of photons

Radio waves in DGR

Radio waves that form when charges oscillate between two points in an alternating current are their Push-Pull bands drifting away with the speed of light, thus forming zones of highly deviated local time. Note that unlike the above description of EM waves with photons entangled to each other, where there is no zone of negative energy or contracted time, in the Radio waves, there are zones of highly dilated time representing the Pull bands and zones of highly contracted time which represents push bands.

Charged particles can interact with these highly time deviated zones in the radio waves through their push-pull bands and acquire momentum or kinetic energy or deflection thus explaining the creation of an electric current in a coil or antenna by a passing Radio wave, a phenomenon which is extensively used in telecommunication.

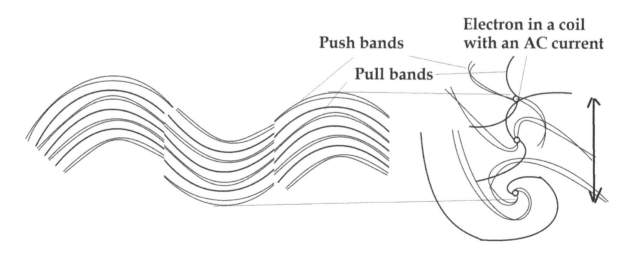

Figure iii/21.4: A Radio wave as predicted by DGR with alternating push-pull bands as electromagnetic flux lines drifting away with the speed of light from the Electron oscillating in an Alternating Current.

CHAPTER 22

THERMODYNAMICS - NEWER INSIGHTS

Presently Temperature of a gas is the measure of the average kinetic energy of the constituent matter particles that the gas contains. Essentially, every particle contained in it would possess some velocity, momentum and kinetic energy. As the temperature of this gas increases, all of these increase so that their average would keep increasing.

DGR gives a slightly different model.

Contemporary physics has no concept of a Planck Compartment. It also has no concept of movement of these Planck Compartments. It also has no concept of formation of PC aggregates formation or movement of PC aggregates.

In contemporary physics, it is difficult to define the temperature at a place where there is no particulate matter and thus there is no kinetic energy. If the question is asked to a physicist, what's the temperature of a Blackhole, the instant answer would be extremely cold almost near absolute zero, simply because a Black hole doesn't radiate anything or it radiates very little Hawking radiation. It is indeed unimaginable to describe the temperature of the Vacuum or space within the event horizon of a Blackhole, thus to ask what the temperature within the event horizon of a Blackhole is almost forbidden.

This paradox becomes a problem when one realises that the time just after the Big Bang was also like a region within the event horizon of a Blackhole. How is it that we call it very hot when there were no matter particles there?

However, with DGR, this changes.

In DGR, the temperature of a space doesn't need particulate matter.

The temperature would be defined based on the average velocity of movement of EPCAs, or how fast the space within is moving. If the space within a region is moving rapidly, automatically all the matter particles would acquire more energy and acquire more kinetic energy.

Within a Blackhole event horizon, the space is moving inwards with extreme velocities much faster than the speed of light and thus temperatures within the event horizon should be extremely high according to DGR. Even the epoch immediately after Creation when the entire Universe was akin to be within the event horizon of a large Blackhole, the space was moving extremely rapidly, much faster than the speed of light. Thus even though (temperatures were too hot and hence) the particles couldn't clump together and form matter particles, and thus at that time there were no particles, it becomes objectively possible to say that temperatures at that time were extremely high.

Absolute zero temperature would then be the temperature of a place where there is absolutely no movement of PCs, an almost impossible task to achieve.

Flow of Energy

It is a universally accepted law of thermodynamics that heat-energy flows from a place of higher density to a place of lower density. Simply speaking, heat energy flows from gas at a higher temperature to gas at a lower temperature.

The above would simply mean that any place where the average velocity of movement of PCs is higher, the PCs will tend to move from such areas towards areas with a relatively smaller average velocity of EPCAs,

The temperature of a region of space, thus, is not just the extent of Heat energy present within, which in Electromagnetic terms is the number of photons having a frequency falling in the infrared range.

Chapter 23

THE ELECTROMAGNETIC BOND

A lot is known about how a few particles namely a positively charged Proton, a negatively charged Electron and an electrically neutral particle, the Neutron, can create the complexity that we see in this world.

We know that just with these three particles, 118 different elements can be made as described in the periodic table, each one of them differing by just the number of Protons present in their nucleus. Each of these elements can be further grouped based on their chemical properties and nuclear properties.

The charges are having a property called spin, due to which they are a miniature magnet. They have the property of charge which means that they modify their surrounding space and give out electric flux, whose density determines the electrical field created by the charge. Initially, it was thought that the attraction between the positively charged Protons and the negatively charged Electrons is enough to explain the structure of atoms.

The Planetary model of an atom meant that the Protons and Neutrons are clumped together in the centre and the Electrons revolved around the Nucleus with the Electromagnetic attraction between the Proton and Electron giving rise to the centripetal force. This however meant that there is no constrain on the Electron and that the Electrons can be at any distance from the Proton. Another problem was the constant radiation of energy that an accelerated particle must do, according to the Classical Theory of Electromagnetism, which meant that the Electron should keep radiating energy and keep falling inwards towards the Proton ultimately falling within the Nucleus in less than a millisecond.

This changed with the advent of the Bohr model of an Atom wherein Niels Bohr proposed that the Electron can orbit around the Nucleus only in orbits in which the angular momentum of the Electron is an integer multiple of h-bar. This meant that the Electrons can remain in only specified orbitals and stay only in these allowable orbitals. de Broglie noted that the Electron could be like a wave and that the Bohr's allowable orbits are actually those orbits in which the wave associated with the Electron or the "Electron-wave" can form a standing wave.

With the advent of Schrödinger's equation and Quantum superposition, it became clearer that the Electron doesn't actually revolve around the Nucleus but is present as an Electron cloud such that we can actually just define the probability of finding the electron at a particular location.

The s, p, d and f orbitals were described and the way electrons are potentially arranged in the inner filled orbitals became better understood and the understanding of Chemical bonding of different elements became better with the knowledge of the half-filled outermost orbitals.

Each of these orbitals and orbits could accommodate a limited number of paired electrons. Every Bohr orbit was believed to have a different number of these allowable orbitals with the first orbit having only s orbital, the second orbit having only s and p, while the third having s, p & d and the final fourth orbit having all four s, p d and f orbitals. Each of the s orbitals can accommodate 2 Electrons with an opposite spin, p orbital can accommodate 6, d can accommodate 10 and f can accommodate 14 Electrons.

The elements which have an extra electron in their outermost shell can easily donate this "valence" electron while those elements who have a deficiency of this electron can take up the extra electron from another element that has an excess of it.

The concept of "formation sp/sp2 hybrid orbitals" was proposed to explain a single bond or a double bond between various elements like C, N, O, H etc.

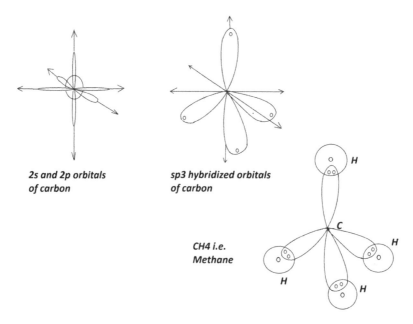

Figure iii/23.1: Shows the 2s and 2p orbitals of Carbon and formation of sp3 hybrid orbitals each having one of the four valence Electrons of a Carbon atom. A Hydrogen atom can bond with each of these 4 sp3 hybrid orbitals producing Methane molecules.

Electromagnetic bonds in DGR

It is not difficult to incorporate or explain all of the above in DGR. Many of the things fall in place nicely in DGR. We can explain the charge and the electric flux emanating from the charged particles, the spin and thus their magnetic moment, we can explain the quantization of the angular momentum in the Bohr orbit and the Electron wave at least phenomenologically. The Push-Pull bands can explain the attraction between opposite charges and the repulsion between P and E as E falls below the allowable Bohr orbit as well as repulsion between two Protons keeping the two bonding nuclei separate.

The binding energy of hydrogen atoms

Quantum mechanics says that an isolated H atom is unstable and that as two isolated Hydrogen atoms come closer to each other, the Proton P1 of one atom starts attracting the Electron E2 of the other and the Proton P2 of the second one attracts the Electron E1 of the second. They come close to each other and reach a sweet spot where their energies hit a nadir. Further reduction of distance between the two H atoms increases the Proton to Proton repulsion increasing the energy of the system. The atoms thus favour remaining in this sweet spot where their energies are the least. In this state, they are said to share the Electron pair as shown in the Figure iii/23.2 given below.

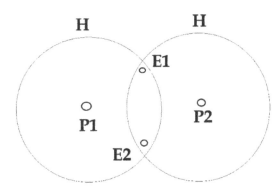

Figure iii/23.2: Shows the conventional notation of a bond between two hydrogen atoms wherein two Electrons are being shared, thus forming a hydrogen molecule i.e. H2.

An isolated hydrogen atom in DGR

Figure iii/22.3 shows how an isolated Hydrogen atom not in the vicinity of another element would look.

The Proton in the nucleus would keep emanating Push-Pull band forming Time resetting waves and these would interact with similar such waves from the Electron. The Push-pull bands from both would keep interacting as both of them spin. The sides which face each other would show a significant interaction with Pull bands pulling the two together. This pulls the Proton towards the Electron and vice versa. This keeps going until the Push bands (shown as darker lines) go all around the Proton and due to the spatial changes here curve all around the Proton to meet on the opposite side and thus provide the repulsion that is needed to keep the two attractive forces from falling into each other.

One can see that only the Push-Pull bands that are facing the other charge are being utilized. The other quadrants would be minimally affected by the presence of their partner charge and even after the bonding is complete and the two have come together and attained minimum possible energy, these unused Push-Pull bands would keep moving outwards and would have no way of being neutralized. These would add to the Potential energy of the system and thus make the whole construct unstable. Thus the next thing these newly formed H atoms would attempt to do is to somehow neutralize these unused Push-Pull bands from their non-facing hemisection.

When these Push-Pull bands do come in the vicinity of another H atom or for that matter any other element which can neutralize their pull, they get attracted to it. As shown in the next figure iii/23.3, the unused Push-Pull bands of the Charges interact with similar ones from a different element and attract or repel each other and thus transition into a state where P1 partially bonds with E2 and P2 partially bonds with E1 thus reducing the overall Potential energy of the system.

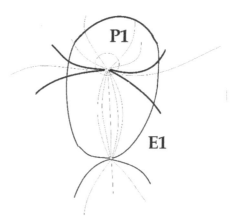

Figure iii/23.3: Shows how the Pull bands emanating from the Proton P1 and Electron E1 interact and thus form an Electromagnetic bond between the two which keeps pulling the two together until the dark Push bands from the E1 reach the Bohr radius where the Push wave emanating from the Equatorial region of the Electron curve around the Proton and provide the push force that prevents the two charges from falling into each other.

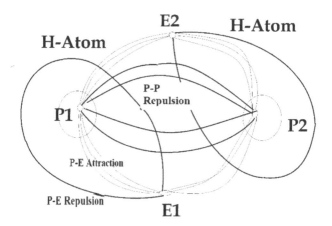

Figure iii/23.4: Shows how DGR explains the sharing of Electrons by two Hydrogen atoms forming a hydrogen nucleus. Note the attraction between P1-E2 and P2-E1 and also repulsion between P1P2 due to the push-pull bands. Why only these interactions happen and not the others, i.e. why the pull bands of P1 do not align with the pull of P2 and so on is unclear in detail yet although it may become clearer once detailed computer models for the same are created.

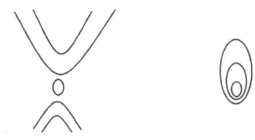

A charge with asymmetric time contraction/spatial expansion (tentatively Positive charge)

A charge with asymmetric time dilation/spatial contraction (tentatively Negative charge)

Figure iii/23.5: Shows how at the start of the Theory, I had thought of indicating the charged particles with the help of asymmetrical space-time changes induced by the same. Although what changes each of the charges might induce isn't accurate, the method of notation seemed good as it successfully noted how the charges interacted with the Push-Pull bands emanating from the two sides. This can make the notation of complex molecules simpler. (Note that the choice of these notations is just random and that one could have very easily switched the two and use them for the opposite charges than they are being used currently)

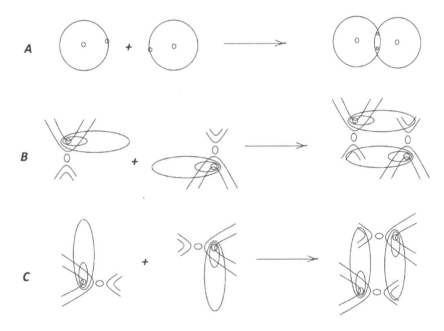

Figure iii/23.6: Shows how a couple of Hydrogen atoms interacting with each other and thus forming an H2 molecule can be denoted.

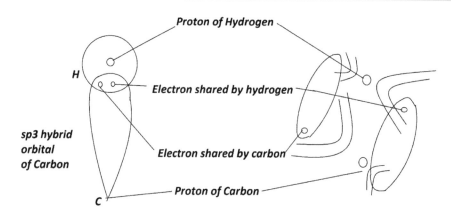

Conventional denotation   Denotation as predicted by theory

Figure iii/23.7: Shows how a Proton- Electron pair from a Carbon nucleus interacts with a bonded Proton Electron pair from a hydrogen atom to form a C-H bond, how it is conventionally denoted and how it is denoted as per the above method.

Chapter 24

THE CHEMICAL BONDING AND THE ENIGMA OF CHEMICAL POTENTIAL ENERGY

We do not have the detailed Mathematical equations of binding energies in DGR for chemical bonding of various elements. Details of the arrangement of various Nucleons within the Nucleus, although described in Lattice Quantum Chromodynamics, isn't completely understood.

We know for sure, that certain elements are more chemically active than others. This can be explained well with the extra electrons in their outermost shells or the deficiency of electrons in their outermost shell. It is known that the atoms of certain elements prefer to bind with atoms of other elements so that their individual deficiency or excesses balance out and they can remain stably with minimal potential energy.

It is also clear that certain "element pairs" form more stable compounds with much less chemical potential energy than others. For example, Carbon can form more stable bonds with lesser chemical potential energy with Oxygen forming CO_2 and Hydrogen forms more stable bonds with Oxygen leading to the formation of H_2O. However, C bonding with Hydrogen in hydrocarbons or gases like Methane I e. CH_4 have a much higher chemical potential energy.

When methane gas or for that matter any fossil fuel is burnt, bonds between Carbon and Hydrogen are broken down and newer bonds are made between C and O and H and O. In this reaction, we know that a lot of energy Is released. This includes heat and light. When a bomb explodes, not only heat and light but sound and shock waves are also generated.

Molecular bonds are thus an excellent battery. Lots of energy is potentially stored in them.

Details of how DGR explains chemical bonds

It is in a way wrong to say that an element has an excess Electron in its outer shell or a deficient Electron in its outer shell. The right way of saying it is that an element has an excess Proton-Electron pair or a deficiency of the same.

Molecular bonds in DGR are explained based on push-pull bands extending from an extra Electron in an atom to a Proton in an atom with a deficient Electron. Essentially, the Electron which is bonded on one side to a Nucleon in one atom also bonds with another Nucleon in another atom. In chemistry terms, this is called as sharing of the Electron.

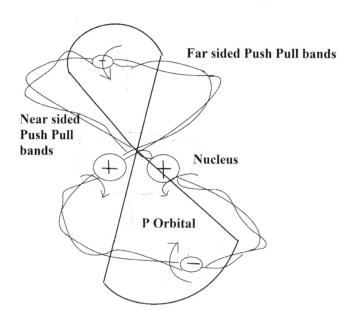

Figure ii/24.1: Shows a pair of Electrons forming a part of the p orbitals of an atom of an element showing both the Far Sided and Near Sided Push-Pull bands interacting and thus bonding with two Protons within the Nucleus. Note the spins of particles. Note that this is a diagrammatic representation and thus not all Push-Pull bands are shown.

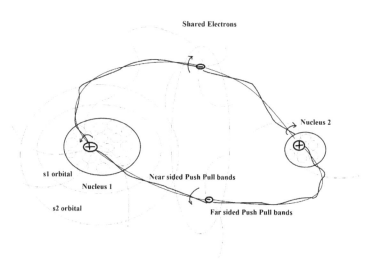

Figure iii/24.2: Shows two Protons belonging to two separate Nuclei, Nucleus 1 and Nucleus 2, bonding with two Electrons. Note the spins of each of the Charged particles. Note that this is a diagrammatic representation and thus not all the interacting Push-Pull bands are shown. Also, only certain portions of the two atoms are shown.

We know that in an s orbital two Protons bond with two Electrons. However, DGR says that the Push-Pull bands on the other far side of the Electron (and for that matter the far side of the Proton as well) would remain unused and make the construct unstable. This is prevented by each of the two Protons bonding with both the Electrons in the s orbital in question. What DGR says is that the far side push-pull bands of the Electron bond with the far side of the partner Proton of the other partner Electron (with an opposite spin) within the s orbital. Essentially, a single Proton bonded to a single Electron is not stable but two Protons bonded to two Electrons becomes more stable simply because both far and near sides of all the four charges are bonded i.e. are neutralised. This two Protons bonding to two Electrons configuration is inevitable if one has to get stability in a molecule.

In every s orbital, 2 Electrons with an opposite spin lie, that bond each with two Protons in the Nucleus.

In the three p orbitals, three pairs of such Electrons lie, bonding with three pairs of Protons within the nucleus. (one electron pair in one p orbital)

The d has 5 orbitals and each of these orbitals can house a pair of Electrons bonding with a pair of Protons within the Nucleus. Thus d orbitals totally can have 10 Electrons (a total of 10 Electrons bonding with 10 Protons).

The f orbital has 7 orbitals, each orbital housing a pair of Electrons. Thus f orbital can house 7 pairs of (i.e. total fourteen) Electrons within it bonded with 7 pairs of Protons within the Nucleus.

In elements with an extra Electron, like Oxygen which has two extra Electrons, two unstable Electrons with un-bonded un-neutralised Proton- Electron pair exists. When two Hydrogen atoms (bonded to other atoms or free atoms or bonded to each other as a Hydrogen molecule) approach such an unstable Oxygen atom, these extra Proton-Electron pairs with un-neutralised far sided push-pull bands bond with a similar pair within each of these Hydrogen atoms and thus attain stability.

Chemical bonding is thus an ultimate example of un-faithfulness with each of these charged particles cheating its partner and having a bond with two partners.

Chemical bonds as a Battery

The Sun gives off light of all frequencies with a continuous Solar spectrum. In DGR terms, the sunlight consists of Photons with multiple values of "e". Each coloured Photon would have a varying amount of destruction happening within it.

Plants take up these Photons and lock the energy to form sugars which are then used to build their bodies. These sugars are also utilised as a source of energy by entire ecosystems.

Essentially what reaction is happening here within the plants is to utilise the Sun's energy in the form of these Solar Photons and break the more stable bonds between CO_2 and H_2O and convert them into bonds that are less stable in the form of molecules with long carbon and hydrogen chains. Each of these molecules is essentially storing the energy within the Solar photons. This energy is released back when the fossil fuels or oils derived from plants are burnt.

Chemical potential energy conundrum in DGR

When a TNT bomb explodes or for that matter, fossil fuel burns, heat and light are released. This means that while the relocation of bonds in the above exothermic reaction is happening, a lot of Photons of varying sizes are released. These include Photons lying in the visible range and also photons lying in the infrared range which have lesser energy and lesser frequency.

We know that the T=0 compartments that are aggregated together in these Photons were the same ones that entered the molecule when the Solar energy was stored within the molecular bonds by photosynthesis.

If we assume that X amounts of T=0 compartments were utilised to break the stable bonds and form the relatively unstable bonds of the carbon chains, the law of conservation of energy suggests that all these X T=0 Compartments have to be recovered back.

The conundrum with DGR is the question

"Where do these T=0 compartments go"

"Where are they stored in the large carbon chains of the hydrocarbon molecule?"

Just the reorientation of the mismatch of push-pull bands cannot explain the generation of these T=0 compartments when the fuel burns.

This question of where the Chemical potential energy lies within the molecule isn't known even to contemporary Physics.

There is one vague explanation in DGR.

A stable system is the one in which there is relatively less movement. Contrary to that, an unstable system is the one in which there is much more movement.

Wherever there is movement, there is more velocity and wherever there is more velocity, there is more relativistic mass.

These T=0 compartments stay within the positive energy core of the bonding particles as relativistic mass. When the reorientation of bonds happens during the oxidation process, the bonding particles become more stable or less mobile and their relativistic mass changes and the excess mass is given off as energy.

The ugly implication of this is that molecules with a relatively unstable bond like CH bonds in methane have much higher relative velocities and relativistic masses than more stable molecules like CO_2 and H_2O, i.e. in these molecules, the particles forming the molecules are moving much more rapidly.

The exact reason why or how this happens is not clear.

The Cogwheel model

One can imagine each Charged particle within the Nucleon as a Cogwheel while the Push-Pull bands emanating from each of them are like the pegs or the cogs of the Cogwheel. Thus every Nucleon is like three Cogwheels bonded together and rotating around an axis but at the same time interacting with other Cogwheels of the neighbouring Nucleons.

Every interaction that the Proton has with the Electron in the Orbitals, even if it is for a very short period, would have a consequence of the transfer of energy and momentum on the constituent cogwheels. Thus if the overall system is unstable and the Electrons at the edge are not stably bonded but are moving rapidly, each of these movements has the likelihood of getting transmitted within these Cogwheels. The Push-Pull bands are like belts or chains in complex machines transmitting torque or push or pull from one cogwheel to the other. The Electron, staying outside the Nucleus but bonded with the Proton within, would act like a Cantilever mechanism. Any angular force acting onto the Electron would cause a much higher force within the Nucleus thus changing the orientation of the Proton. The neighbouring Nucleons might transmit the additional momentum gained by these interactions, to their partner Electrons, with which they bond. Thus even if a single member is unstable or unhappy, all the members are likely to acquire higher energy and velocity.

This however does not explain why sharing of Electrons between two Protons belonging to C and H leads to unstable Electrons and higher overall energy within the system and how this energy gets released. It also does not explain how sharing of Electrons between Protons within Carbon and Oxygen lead to comparatively more stable sharing of Electrons.

(Note that these are just assumptions and extrapolations and that it is unclear if this or any other mechanisms are involved in the storage of energy within a bond)

There are some possibilities one should keep in mind.

Let's consider two such Proton-Electron pairs. Let's call the P1, P2 and E1, E2.

Let's further look at the simplest scenario of a helium atom with only these four particles bonding. (see Figure iii/24.1)

Each of them has a spin. The two Electrons would occupy the s orbital of the first Bohr orbit. We know that Pauli's exclusion principle states that the Spins of E1 and E2 have to be opposite.

Let's assume that E1 has an Up spin and E2 has a Down spin.

Let's further presume that E1 binds with P1 and E2 binds with P2. It is more likely then, that P1 will have an Up spin and P2 will have a Down spin.

Thus when the far side of E1 is trying to bond with the far side of P2, their spins are opposite. It is thus more likely that the far side interaction is slightly more unstable. Thus minor variations in spin directions and bond angles might determine why some combinations are more stable than others.

Water

The physical properties of water are pretty characteristic and diverse. It is called a Universal Solvent since a lot of elements (water-soluble ones) can dissolve in it. The liquid state, the anomalous behaviour while turning from liquid to solid, transparency to light, high resistance to the passage of an electric current in pure form but low resistance when impurities are dissolved within, relatively stable bonds with less chemical potential energy are some of the many characteristics that need to be explained.

The current theories depend heavily on ionization and Hydrogen bonding.

The model of the interior of water, as predicted by DGR, is an interesting one.

Oxygen has two Proton-Electron pairs with un-neutralized far-sided Push-Pull bands. The other three Proton-Electron pairs (atomic number of Oxygen being 8) are neutralized by bonding within. More importantly, the remaining two Proton-Electron pairs within the last shell bond with the Protons within the same atom. However, these Electrons, if coming close to an active freely floating Proton, can open up and bond with it momentarily. But in this process, as more than three Hydrogen atoms cannot be bonded to a single molecule of Oxygen, one of the other bonded partner Hydrogen of the Oxygen atom is released.

Each Hydrogen atom is made of one such pair with one Proton bonded to one Electron.

Thus it figures that the two un-neutralized pairs in Oxygen bond with two such pairs in two Hydrogen atoms forming the H2O molecule. When in the liquid state, the Protons in one of the Hydrogen, instead of curving inwards and bonding with far sided Push-pull bands of Electrons of Oxygen, can just as well bond with similar such bands of Electrons from neighbouring atoms. The Proton or more commonly Hydrogen ion would then be the ultimate definition of unfaithfulness. It can keep getting attracted to neighbouring atoms and thus the bonds keep forming and breaking. Momentarily, the H_2O molecule is broken into H+ and OH-, when the neighbouring Proton would reform the H_2O molecule. This cycle of Protons hopping from one Oxygen atom to the next would keep going. In one cycle, one of the partner Protons cheats with the Oxygen, while in the next cycle the second might cheat. The Oxygen has two Proton-Electron pairs neutralised by infolding, i.e. by Proton-Electron pairs within the atom itself. Thus out of the six Electrons in its outermost shell, two pairs (i.e. four Electrons) are already neutralised and here no far sided Push-Pull bands exist to share. Here the molecules repel any potential neighbouring Oxygen molecules. This explains the angular shape of the Water molecule. This keeps exerting torque to the molecule and it keeps revolving before the above cycles restart. Instead of Hydrogen ions (i.e. Proton), Proton Electron pairs from other molecules like Sodium and Chloride can play this ultimate game of unfaithfulness with momentary bonding and then attraction to the other neighbour leading to breakage and reformation of bonds.

It is essentially an aggregate of Oxygen molecules mutually repelling each other and exchanging Protons. This explains the Universal solvent property of Water, well. Other compounds which can donate or accept electrons like NaCl can form Na+ and Cl- and be an active participant in the cycles of formation and breakage of bonds with momentary neutralization of the Far-sided Push-Pull bands.

The overall message here is that which atom or Nucleon bonds with whom is immaterial. What is important is the overall system should attain stability for a maximum period of time. If this can be obtained at the cost of a series of cyclical interactions, so be it.

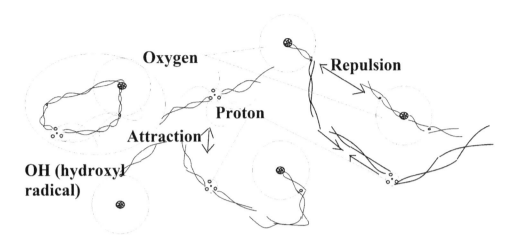

Figure iii/24.3: Shows the model of water as predicted by DGR. The Oxygen Nuclei, within which lie the Protons bonding with their own Electrons, remain unstable due to the un-neutralized far sided Push-Pull bands unless they bond with other Electron- Proton pairs from the neighbouring molecules. Here, the exchange of Protons between neighbouring molecules due to instantaneous interaction of their Push or Pull bands with neighbours can lead to the momentary formation of H^+ and OH^-.

Carbon

Carbon has four Electrons in its outermost i.e.2nd shell. This can be arranged in s orbital as s-2 and the rest of the two Electrons would stay in the p-orbitals (2s1,2s2,2p2). But this is a relatively unstable position. For the formation of four C-H bonds in CH_4 i.e. methane, as per accepted models, the s-orbitals merge with the p-orbitals and for four sp3 hybrid orbitals. Each of these sp3 hybrid orbitals formed by the process of sp3 Hybridization stays at maximum distance from each other and have an Electron within them. The far-sided Push-Pull bands of these Electrons and their partner Protons remain un-neutralized until four H atoms (which are also nothing but a pair of Proton-Electron pair with similar un-neutralized far-sided Push-Pull bands, bond with them, neutralize them and in turn get neutralized, forming methane. Instead of an H, another Proton-Electron pair from an sp3 hybrid orbital from another C atom can bond with one of the sp3 hybrid orbitals of our C atom thus forming Ethane or for that matter explaining the property of catenation (formation of long chains) of Carbon.

The two electrons within the s1 orbital i.e. innermost orbital of the Carbon atoms are neutralized by Protons within the same Nucleus.

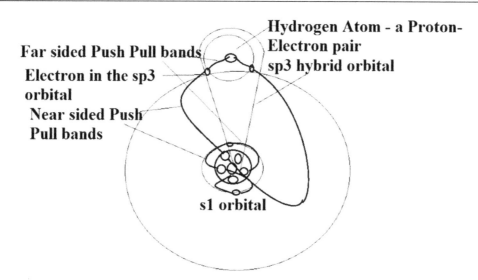

Figure iii/24.4: Shows a Carbon Hydrogen bond with one of the four Electrons in the 2nd orbital forming an sp3 hybrid orbital. Note the oblique orientation of the Proton within the Carbon Nucleus that determines the location of shared Electrons. (Note that this is a diagrammatic image as other sp3 hybrid orbitals are not shown and the bonding is shown as a straight line while in reality, it would be with a spiral staircase like oriented Push-Pull bands at the pole and many other more non-polar Push-Pull bands all interacting with the same ones of the bonding partner.)

Carbon-dioxide

We know that CO2 is a linear molecule with a Carbon atom at the centre and two Oxygen atoms bonded on each side with a double bond. This isn't difficult to be explained on the basis of DGR as Carbon has four pairs of un-neutralised Proton Electron pairs and each Oxygen atom has two pairs. The Oxygen atoms are said to be in an sp2 Hybrid state and the Carbon has two sp hybrid orbitals. Each of the C and O atoms have one sigma bond which is a stronger bond and another one is the pi bond which is the second weaker bond. Thus the middle C atom forms two sigma and two pi bonds. The figure shows how sharing of two pairs of Electrons happens in CO2.

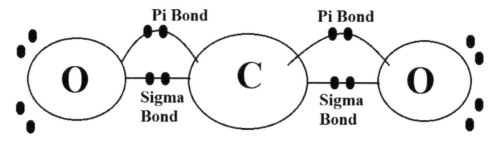

Figure iii/24.5: Shows how the double bond between Carbon and Hydrogen is formed by the formation of one stronger sigma bond and a relatively weaker pi bond.

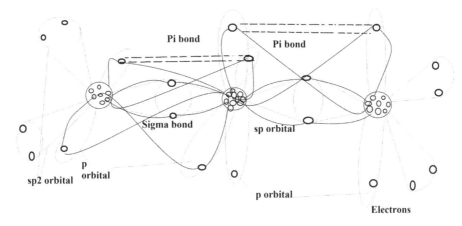

Figure iii/24.6: Shows how DGR would predict how the Proton-Electron pairs within C and O bond to form the sigma and pi bonds. Note that this is a diagrammatic image as not all push-pull bands are shown and the bonds are just shown as a linear line. Note that O atoms are in a state of sp2 hybridization and C orbitals are in a state of formation of two sp hybrid orbitals.

Benzene ring

We know that 6 carbon atoms form a hexagonal ring-like structure called a benzene ring. In this, there are three single C-C bonds and three double C-C bonds. Out of the two double C-C bonds, one is a sigma bond and the other one is a pi bond. It is hypothesized that the double and single bonds switch places constantly and that the benzene ring is thus often designated as a hexagon with a ring inside, suggesting that the bonds are interchangeable.

The Benzene ring is extremely stable chemically and significant energy is needed to break the C-C bonds in it and thus break the ring. The groups or moieties attached to any of the C atoms determine the chemical reactivity of the resultant compound. A diverse group of compounds with varying physicochemical properties can be obtained by attaching various groups to the benzene ring. All these compounds are called aromatic hydrocarbons.

The figures iii/24.7 and iii/24.8 given below show how is the arrangement of bonds and atoms in the benzene ring. The figure shows how the bonds within the Benzene ring would look like as predicted by DGR. We know that each C atom forming the ring has the 2s orbital and two of the three 2p orbitals fused to form three sp2 hybrid orbitals that lie in a single plane and are at 120 degrees from each other. The third p orbital remains the way it is and bonds with a similar p orbital from the adjacent C atoms. The sp2 hybrid orbitals between two adjacent carbon atoms overlap and here forms a sigma type of bond. This is nothing but a pair of Electrons neutralizing the push-pull bands of a pair of Protons within the Nucleus of the two bonding Carbon atoms thus forming a sigma bond. The alternate pair of Carbons having a second weaker pi type of bond is nothing but an electron in the p orbital of a C atom (1st atom) neutralizing the far side push-pull bands of a Proton in the adjacent C atom (2nd atom) and in turn the partner Proton of the same Electron (belonging to the 1st atom) neutralizes the far sided Push-pull bands of the partner Electron of the bonding Proton in the p orbital of the second C atom.

The Electrons in the p orbital, as shown in the figure can hop from one C atom to the next and thus the C-C single and double bonds can keep changing in position.

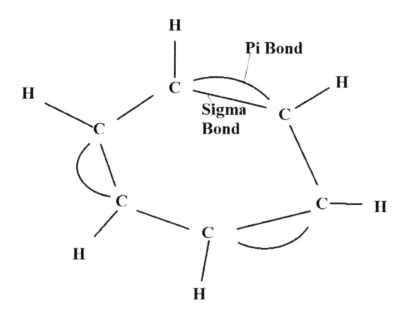

Figure iii/24.7: Shows arrangement of atoms and bonds in a Benzene ring

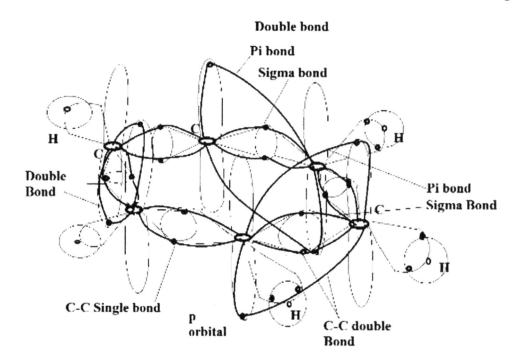

Figure iii/24.8: Shows how DGR predicts the bonds in the Benzene ring. Note that a sigma bond between every neighbouring C-C atom is a pair of Electrons arranged in the middle (in the region we call sp2 hybrid orbitals) neutralizing two Protons in the adjacent C atoms. The pi bond which happens between three of the 6 pairs of C atoms is simply an Electron in the p orbital of one atom neutralizing the far sided push-pull bands of a Proton within the Nucleus of an adjacent Carbon atom and the partner Proton of the first Electron neutralizing the far sided push-pull bands of the partner Electron of the first Proton, thus forming a bonded electron-proton pair.

CHAPTER 25

METALLIC BONDS PREDICTED BY DGR

Conventionally, the metallic bonds are explained on the basis of the "sea of Electrons" model. This model can explain the electrical conductivity of metals well due to the free availability of the Electrons in the outermost shells that overlap and form the sea. This also explains the sheen that is characteristic of metals. However, this does not explain the torsional and bending strength, malleability and the high boiling points of metals satisfactorily.

DGR can offer an elegant model of the metallic bonds that can explain the high flexural strength of metals and also their malleability. This model like the model of water described above is also based on the un-neutralized Electron-Proton pairs. The atomic number of metals is generally high. Thus the number of Electrons in their outermost shell is also high.

We will take an example of Fe i.e. Iron. Iron has an atomic number of 26 And has a total of 14 Electrons in its second last i.e. 3rd shell and two Electrons in its last i.e. 4th shell. (as shown in the figure below)

The second last shell with 14 Electrons are in the third Bohr orbit and are significantly far from the Nucleus. Each of these has a partner Proton with which they are bonded with their near-sided Push-Pull bands. The far sided Push-Pull bands of each of these 14 Electrons probably remain un-neutralised by themselves and thus each of the 14, bonds with similar Push-Pull bands from the Protons from the neighbouring Nucleus. Thus every atom of Fe remains bonded to its neighbours although these are not Covalent bonds and thus they can break and reform without much effort explaining the malleability and ductility of metals. The donation of outermost Electrons to the sea of Electrons, and free availability of un-neutralized far sided Push-Pull bands of neighbouring Protons means that the Electron can hop from one atom to the next forming and breaking bonds momentarily and thus move along an electrical field.

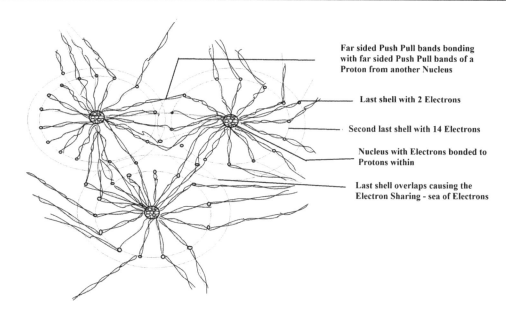

Figure iii/25.1: Shows metallic bonds of Fe, as predicted by DGR with the Far-sided push-pull bands of all of the 14 outer Electrons (and their partner Protons) in the second last orbital and two Electrons in the last orbital, bonded and being neutralized by similar far sided push-pull bands from other neighbouring atoms thus explaining many properties of metals like strength, malleability, ductility, sheen and electrical conductivity

One can consider other examples.

The metals with small atomic numbers like Lithium, Beryllium and Boron are good examples. Lithium has an atomic number of 3, Beryllium has atomic number 4 and Boron has atomic number 5. We know that Hydrogen has one electron in its 1s orbital and Helium has two Electrons, both within the 1s orbital.

The third Electron in Lithium occupies the 2s orbital. One can imagine that the inner i.e. 1s orbital Electrons bond with the partner Proton of the second Electron by infolded far sided push-pull bands. When they are in the pure metallic form of Lithium, these 1s electrons can also open up and instead of infolded push-pull bands, bond with the neighbouring Lithium atoms. Thus each atom of Lithium binds to other atoms with three bonds. This explains the extremely high strength of Lithium metal. The freely available 2s electron explains the good electrical conductivity. Lithium like Na+ can donate its 2s electron and form salts.

Similarly, Beryllium and Boron have multiple Proton-Electron pairs which are neutralized by infolding and can become available for interatomic bonding forming metallic bonds. Beryllium, with 2 Electrons in 2s orbital has excellent flexural stability, thermal stability and low density all of which make it an excellent choice in aerospace applications like components of aircraft, missiles, satellites. Similarly, Boron with atomic number 5 and with 3 Electrons in the second shell (2s1,2s2,2p1) again have multiple Electrons which can unfold and form metal-like bonds. However, naturally occurring boron is an amorphous brittle solid and crystalline Boron is metalloid solid. Clearly, one cannot overdo the same process for higher-order elements forming metallic-like bonds for all elements.

Recently even Hydrogen present within the surface of the Sun is being described to be in a hexagonal metallic Hydrogen state.

Alloys

One of the well-known properties of metals is the ability to be in alloys. This unique bonding of Electron-Proton pairs from neighbouring atoms can happen irrespective of the presence of atoms of other metals as impurities within the crystal. This explains the ability to participate in alloy formation. The un-neutralized Electron Proton pairs that remain lead to variation in the properties of the alloys compared to the pure metals. Alloys may have varying physical properties like strength, appearance and electrical conductivity.

Chapter 26

THE STRONG NUCLEAR FORCE

What's inside a Nucleus?

The nucleus of every atom is a pretty messy place. Little is known about it for sure and a lot is theoretical.

We know for sure that the nucleus has protons and neutrons packed together in a very small space. The dimensions of the nucleus are in the range of a few femtometers. (10^{-15} m)

The protons and neutrons are pulled together by what we call the strong nuclear force.

The nucleus is positively charged due to many protons packed in a small space. How the mutual repulsion between them is completely overtaken and how the strong nuclear forces keep the nucleus together, at least for the smaller nuclei is a mystery.

We are aware that the strong nuclear force is 10^{30} times more powerful than Gravity and is 10^4 times stronger than the electromagnetic force which would expect to repel the protons apart.

"Why this specific proportion?" is not clear.

The individual masses of protons, neutrons and electrons need to be taken as assumptions as yet and none of the theories gives predictions or reasons why these take the specified values.

We are quite certain that Quarks exist and that there are mainly Up Quarks and Down Quarks. The Proton has two up and one down Quark while a Neutron has one up and two down Quarks. Why this exists is not known. We also know that Quarks possess fraction charge. Up Quarks are positively charged and down Quarks are negatively charged.

The question one might ask is, why do the Quarks remain in the nucleus while the electrons remain outside of it. What if we make a nucleus out of two Up Quarks and one Electron instead? Would it remain stable or would it decay rapidly?

The true nature of the Colour charge possessed by the Quarks is also not clear. Nature of the Gluons and the reason for Quark confinement, the interaction of various nucleons and interconversion of Proton to Neutron within the nucleus is not known for sure although theoretical concepts like Quantum Chromodynamics are described as discussed elsewhere.

The spin of Quarks contributes little to the spin of the Proton or the Neutrons. Also, their masses are pretty insignificant compared to masses of the Proton and the Neutron.

The stability of the various particles in the bonded state as well as in the free state cannot be explained yet by any theory.

We don't know why the Electrons and Protons don't decay when kept without their partner while a free Neutron has a half-life of just about 10 minutes and decays unless it is bonded in the nucleus.

The nucleons in our theory

The Proton and the Neutron have much more mass compared to the mass of Quarks combined. Where this mass emerges is not known and is speculated that this mass comes from the energy stored in the Gluon based strong nuclear force.

In our theory, the Proton and the Neutron both have a very high centrally located positive energy particle where a high extent of destruction is happening. This is akin to a Blackhole. This leads to the creation of a zone around it where even light cannot exit out.

This extreme energy which is densely packed indicates a significant number of T=0 Compartments within and a significant deviation of the C/D ratio towards destruction. To compensate for this destruction, EPCA forming TRWs start from the centre and move out. At the initial zone immediately around the particle, the EPCAs move faster than the speed of light.

The destruction being done leading to the EPCAs causes a significant curvature of space around it. This curvature however is just akin to Gravity and the force with which it can pull some other particle would be small and insignificant and not single-handedly capable of opposing the electromagnetic repulsion.

Just beyond the event horizon would lie the Quarks.

Like the Electron, each Up and Down Quark is a composite particle with a positive energy particle core with a given mass at the centre and a lattice of negative energy particles forming a sphere around it.

The EPCAs emanating from each of these interact with the EPCA forming TRWs emanating from the positive energy core. The result is a complex redistribution of local time around the particle.

Essentially, very narrow zones of extreme time dilation or time contraction are produced which when formed move outwards with very high velocities faster than the speed of light. While in this process of moving outwards, these zones are in entanglement with the next zone created near the particle.

These could be called time dilation spatial contraction bands and time contraction spatial dilation bands. Alternatively, they can be called "push bands" or "pull bands", simply because the Spatial contraction or expansion they produce enable them to push or pull.

As the negative energy particles i.e. ENEA lattice spiral around the core, these push-pull bands spiral around. The spin will have an axis of rotation and that there will indeed be two poles that represent the North and South poles of this magnetic dipole. At the poles, there is necessarily an asymmetry between the two poles, one having an extra ENEA or one having an extra unbalanced positive energy.

Quarks spin around their own rotational axis. The Quarks also revolve around the central core of the particle with speeds close to the speed of light. This means that the push-pull bands emanating out of them representing the electric charge of the Quark in question, move out faster than the speed of light taking a complex path.

Given the proximity of the positive energy core, the space here is curved and accordingly, all the push-pull bands emerging in the quadrant facing the core would be curved inward along the spatial curvature caused by the destruction taking place in the core.

The two Up Quarks would be expected to repel each other. However, we know that their location is much closer to the first orbital of the electron. This means that even the negatively charged Down Quark and Positively charged Up Quarks would be expected to repel each other and thus would stay maximally far at vertices of an equilateral triangle.

How exactly, push-pull bands emanating from them in the curved space lie is unclear. However, it seems likely all are attracted to the central core.

The EPCAs emanating from the opposite quadrants not facing the Positive energy particle core would also get affected to a much lesser extent by the curvature of space. They can still fly away, being faster than the speed of light. These can interact with the push-pull bands emanating from the neighbouring nucleons. If enough curvature of space is present, these push-pull bands can bend enough so that they can interact with other push-pull bands emanating from other Quarks.

Depending on the proximity of these Quarks to other Quarks from neighbouring nucleons, the attraction and repulsion forces may line up in such a manner as to rip the Quark off from a more massive Up Quark to form a less massive Down Quark and in the process forming a smaller particle which is the Meson.

Meson in our theory is again a combination of matter particles and antimatter particles, essentially a positive energy core of a much smaller mass and a smaller lattice of ENEAs. These mesons might have a fraction of the charge being able to emanate push-pull bands. However, being in the curved space-time in the proximity of the positive energy particle core, it would be called the colour charge rather than the routine electromagnetic charge.

Exchanging these mesons between nucleons would transform one type of Quark to another leading to a transformation of Protons to Neutrons and vice versa.

It is worth noting that the Quarks are held apart so that any tendency of the Quarks coming closer to other Quarks creates an excessive repulsion and any tendency of them pulling away from each other aggravates the centrally directed pull.

It is also likely that the location of the ENEA particles around the Positive energy core in the Quarks is also affected due to the curved space-time due to the central destruction of the nucleon. This can modify their location so that the push-pull bands emanating from them are slightly different in density and location than if they were spinning out in a space-time with no spatial curvature. This is discussed in detail in the Superposition of spin section.

Each Quark is thus under the influence of the pull of the central destruction in the central positive energy core in the nucleon. At the same time, the push-pull bands emanating from it directly facing the centre of the nucleon add to the centrally directed pull. The push-pull bands that emanate from each Quark lying in the hemisphere facing the positive energy core, travelling faster than the speed of light, would curve around the Positive energy core and thus are able to interact with the push-pull bands of its other partner Quarks. The push-pull bands emanating from the hemisphere facing opposite to the positive energy core can also curve around due to central destruction induced by curved space and thus can interact with other similar push-pull bands from other partner Quarks. These interactions can explain the attraction-repulsion forces which keep the Quarks apart as well as bonded to the central Positive energy core and thus can explain the strong nuclear force.

The Quarks are probably in the process of revolving around the positive energy core at very high speeds close to the speed of light. The Quarks are also rotating around an axis which is the revolution of the ENEA lattice revolving around the Positive energy core of the Quarks. This spin of each Quark means that the Quark is a miniature magnet. It also means that every Quark has an asymmetry at the two poles so that one of the poles is a South pole and the other is a North pole. This asymmetry probably means that the density of push-pull bands starting from one pole differs from the other. This asymmetry would mean that the Quarks, in the process of revolving around the central core would orient themselves in such a manner that the axis of rotation aligns along with one of the radii and one of the axes pass through the centre of the positive energy core.

The Colour-charge would then be a combination of the curved space-time created by a positive energy core and the push-pull bands that get affected and curved due to spatial curvature.

Gluons would be the push-pull bands turning around the central PEP core and interacting with other push-pull bands in addition to the central pull of the PEP core.

The nucleons would then be held together by the push-pull bands lying in other quadrants not directly facing the PEP core, the push-pull bands flying away and interacting with other push-pull bands from Quarks from the neighbouring nucleons.

Because the Quarks are rotating as well as revolving at high speeds, each of these push-pull bands takes complex paths and thus interacts with other such push-pull bands from other Quarks for an extremely short time.

What interaction happens between them depends entirely on what bands interact. When a push band interacts with another push band from a neighbouring Quark, repulsion happens while when a pull band interacts with another pull band of a neighbouring Quark attraction happens. When a push band interacts with a pull band of a neighbouring Quark, they just cancel each other out for that instant and no push or pull happens.

Note that Pauli's exclusion principle can be effectively explained as two Quarks cannot attain the same co-ordinate location (in Planck lengths) because any attempt of two Quarks to move towards each other would increase the repulsion and decrease the attraction.

The exact paths of the push-pull bands around the Quarks can only be guessed at present and will need detailed computer modelling to be known precisely.

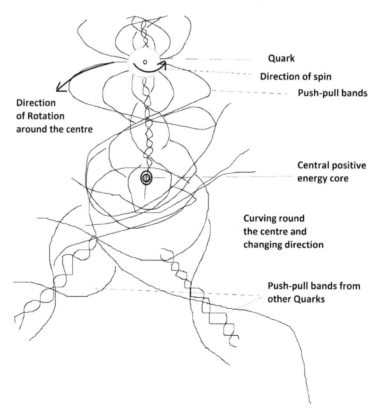

Figure iii/26.1: Shows a diagrammatic representation of a Quark bound to a central Positive Energy Core. Also shown is the spiral staircase like arrangement of the push-pull bands of the other two partner Quarks. Also, note the push-pull bands turning around the central core to affect the other partner Quarks.

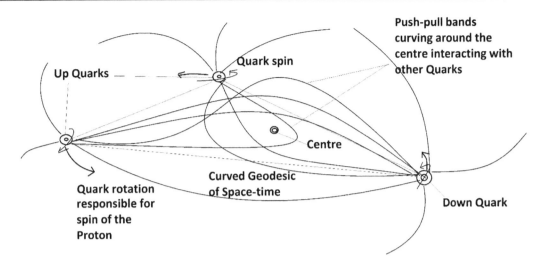

Figure iii/26.2: Shows a diagrammatic representation of three Quarks arranged as an equilateral triangle, bonded with a centra PEP and to each other by push-pull bands curving around the central core. Note that only a few push-pull bands are shown.

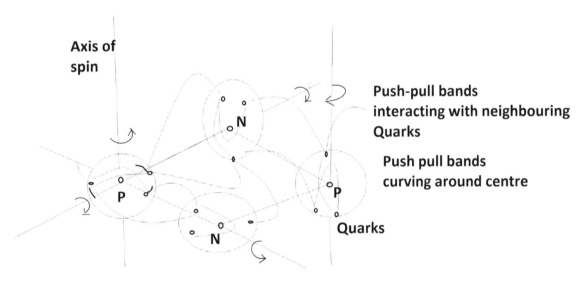

Figure iii/26.3: Shows how neighbouring nucleons interact with each other within the Nucleus. The Quarks forming a part of each Nucleon give off push-pull bands which curve around and take complex paths due to the curved geodesic here due to proximity to the PEP core. Note that each of the Quarks shown is in the process of moving i.e. revolving around the central PEP core at the same time rotating around its axis due to the ENEA lattice.

CHAPTER 27

WEAK NUCLEAR FORCE

The weak nuclear force is the fourth of the fundamental forces of Nature.

It is said to be carried by the W and the Z bosons.

It is one of the highly significant forces in theoretical physics due to its anomalous behaviour.

Although it is weak and is 1000 times weaker than the electromagnetic force, it is highly significant as it leads to the formation of newer lighter elements from heavier ones (flavour change) or the other way round and is used to explain the beta decays.

In beta minus decay, a Neutron is converted to a Proton within the nucleus and an Electron (also called the Beta particle) and an Anti-Neutrino are released. This happens in heavy and unstable nuclei where there are too many Neutrons.

In beta plus decay, a Proton in a heavier nucleus is converted to a Neutron and a Positron (positively charged cousin of an Electron) and a Neutrino are given off. The nucleus thus loses one Proton and forms a lighter element. This kind of decay happens in heavy elements with too many Protons.

In each of these reactions, the main fundamental reaction happening is the conversion of an Up Quark to a Down Quark or the other way round.

Another closely related type of decay is the "Electron capture" in which inner energy-shell Electron is captured by an unstable heavy nucleus which combines with the Nucleus and acts like a beta plus decay with a Proton getting converted to a Neutron.

Other reactions including Kaon decays, Neutron decays are also described.

The weak nuclear force also makes nuclear fusion possible that helps make the stars generate a huge amount of energy and shine brightly. In this two atoms of hydrogen combine to form an atom of helium and the mass difference between the two is released as energy given out by the star.

The Weak force is said to be a short-range force acting at extremely short distances and at extremely small time intervals. In 1960, Sheldon Glashow, Steven Weinberg et al unified the electromagnetic force with the weak interaction by showing them to be two aspects of a single force, now termed the "electroweak force" and were awarded the Nobel prize in physics in 1979. (*Sheldon Glashow 40, Steven Weinberg –103,104, Abdus Salam-89*)

It is shown that weak nuclear force is the force that breaks various symmetries of nature and acts differently on the right or left-handed particles. It also breaks P and CP symmetries and thus people believe that it holds the key in explaining multiple mysteries of nature like why there is something rather than nothing and also why there is more matter than antimatter. *(Aaij, R-1,2, Abe K. -3, Carbone, A-17)*

This means that the weak force acts differently on right-handed and left-handed particles (violation of parity symmetry). It also acts differently for mirror-reversed particles.

Meson creation

Mesons or Pions or pi Mesons are continually exchanged between Protons and Neutrons according to Yukawa's theory. *(Yukawa, Hideki-106)*

The following reactions happen and are said to be responsible for the strong interaction which keeps the nucleons together.

A Down Quark-AntiDown Quark pair is produced and interacts with an Up Quark Within a Proton---> a Down Quark stays in the Nucleon and the anti-Down Quark binds with the Up Quark taken from the Nucleon to be thrown out to be exchanged with another Nucleon as a positively charged Pion. Here this Anti-Down Quark- Up Quark pair react with a Down Quark of a neighbouring Neutron and the Anti Down Quark in it annihilates with the Down Quark of the Neutron while the remaining Up Quark gets incorporated or remains within the Nucleon forming a Proton.

Thus the first particle which was a Proton with UUD gets converted to a Neutron with UDD and the second particle which is the Neutron with UDD Quarks gets converted into a Proton with UUD.

The particle exchanged here is the Down Quark- Anti Up Quark pair which is a positive pi meson.

In this manner, positive pi Mesons keep getting exchanged and Protons form Neutrons while Neutrons get converted into fresh Protons.

Other interactions like the exchange of negative pi Mesons or neutral pi Mesons are also described.

Things worth noting...

Note that there are two important insights which these strong interactions give.

1. Spontaneous creation of particle and antiparticle pairs (here Quarks and Anti Quarks), from nowhere from energy is already described in Physics for the first time here.
2. A particle bonded with a different antiparticle to form a particle-antiparticle pair is first described here (without getting annihilated)

Here Quark Anti-Quark pairs of any type can get generated. In the description of QCD, it is a much more complex process where not only the type of Quark is important but the colour charge is also important.

Quark Confinement

It is said that the Quarks are bound to the central nucleus by extremely high forces. Thus forces needed to separate a Quark from its Nucleon are so high that at the edge, the energy required for the process is sufficient for the creation of another Quark-Anti Quark pair. Thus the Quarks remain confined to their Nucleons and forces that attempt to remove them lead to the creation of an additional pair thus leading to the creation of a Pi meson instead.

Nuclear force has a peculiar characteristic. At distances of 0.8 Femtometer to About 2 Femtometer, it is attractive. As the distance increases, the strength of this force fades away rapidly. When the distance between the two Nucleons reduces lesser than 0.8 Femtometer, the Nuclear force becomes strongly repulsive. In this manner, the Nucleons are kept at a distance of about one Femtometer with an increase in repulsive force with any attempt to reduce this distance.

Bigger Nuclei may have as many as 200 Nucleons.

The Columbic repulsion between two Protons at such small distances (of about 1 Femtometer) is about 230 Newton which for the size of particles is a huge force of Repulsion. The attraction due to Strong Nuclear force, however, is several magnitudes higher in the range of 25,000 N Thus in small Nuclei, the Columbic repulsion is well compensated and the Protons remain stable.

But in larger Nuclei, at least at some places within the Nucleus, it becomes difficult to keep Columbic repulsion in check as the range of Columbic repulsion is higher than Strong Nuclear Force and with an increase in Nuclear diameter, the SNF can no longer hold all the Nucleons stably.

This is the probable reason why there is an upper limit for the size of the Nucleus i.e. we can find elements with $z = 200$ or smaller but cannot find heavier elements.

As the number of Nucleons rises, this Columbic repulsion rises and the nucleus becomes unstable, and the likelihood of a beta decay process that changes the identity of the nucleus increases.

Understanding the basics of weak nuclear force

Imagine that there is a rod or cylinder-shaped dough of flour that is kept for making bread. All the particles within it are bonded together well and thus we can consider the whole dough as a single particle. We know that the dough is not hard before it is put in the oven for baking and the surface is pliable and changes shape if an external force is applied to it.

Now imagine that the two kids of the Baker come in the kitchen where the dough is kept, and start playing with it. Both of them hold the dough on either side and pull apart.

We know that the bonds within are not strong enough and that the dough will be pulled apart into two pieces so that some particles remain in one portion while others remain with the other portion.

This breaking of the particle and thus the release of new particles (here the dough and its portions) is called the beta decay and the force which is pulling it apart (the hands of the two kids) is the weak nuclear force.

Weak nuclear force in DGR

A pair of forces (or accumulation of multiple forces) in two different directions causing a PEP or a composite charged particle to be ripped apart so that there is a creation of two new particles is the weak nuclear force in DGR.

Imagine a charged particle with a central positive energy core and multiple ENEA lattices with push-pull bands emanating out from it, within a nucleus, say an Up Quark. The Up Quark will be spinning at its axis and also rotating around the central positive energy core of the Nucleon, of which it forms a part. It is being acted upon by various push-pull bands from various sources. If two sides of it are acted upon by multiple pull bands from multiple neighbouring particles, the forces may add up and become enough to rip the Nucleon apart into two pieces. If this happens, a part of the positive energy core and some ENEA particles from the ENEA Lattice get pulled to the right and the rest to the left thus forming two completely different particles.

Figure iii/27.1 shows the diagrammatic representation of possible events that may happen as per DGR. Note that the Up Quark in the upper Nucleon, under some push/pull bands or under the influence of a newly formed particle (particle anti-particle pair or Quark Anti-Quark pairs as per QCD) gets ripped apart of some of its positive energy along with some of its ENEA lattices thus forming the smaller Down Quark. The newly formed particle is called the pi plus meson. This is a positively charged particle with a charge of +1 and a positive mass. It is not unreasonable to theorise that even this particle will be having push-pull bands of its own which will enable it to interact with similar bands from the neighbouring charged particles within the neighbouring nucleons. This particle is exchanged between the neighbouring nucleons. As it enters the neighbouring nucleon, it interacts with one of the negatively charged Down Quarks and combines with it, so that the opposite of the above reaction happens wherein a Down Quark and a pi meson interact and give rise to an Up Quark.

The first Nucleon having an extra Up Quark and having lost the extra Up Quark is a Proton and after the reaction, the Up Quark is converted to a Down quark so that what remains in it is UDD i.e. the reaction turns a Proton into a Neutron.

The second Nucleon is the one that has an extra Down Quark and after completion of the whole reaction, forms an extra Up Quark. Thus this Nucleon has to be a Neutron and after the completion of this reaction forms a Proton.

Essentially, A Proton is giving off a Pi plus Meson and getting converted to a Neutron while the Pi plus Meson is absorbed by a neighbouring Neutron and converts it into a Proton.

Note that in the figure iii/27.1, a spontaneously formed particle-antiparticle pair as theorised by QCD is shown where the belief is that a Quark- antiquark pair is formed. The Quarks and Anti- Quarks in these have a positive mass. It is not necessary for DGR and even push-pull bands from multiple neighbouring Nucleons added together and acting in unison can also rip a particle apart without the need for a formation of a particle-antiparticle pair with mass. In this second case, the particle-antiparticle pairs would be represented by the "Push and Pull bands" which are effectively massless and antiparticles of each other.

a) & b)

Figure iii/27.1: Shows how DGR explains the Weak nuclear force a) Up Quark converted to Down Quark thus converting a Proton to Neutron and ejecting a pi Meson or Pion. b) Down Quark converted back to an Up Quark after absorbing a pi Meson thus forming a Proton

The EPCA model vs EMPA model of the positive energy core

Here it is imperative to review these two as the latter seems to be more appropriate for explaining the weak nuclear force. It is also better in explaining the spin of a photon and doesn't need sub-Planck time events.

The two are actually descriptive forms of a single process and differ to a very limited extent.

The EMPA model

In this, the T=0 compartments coming together and forming the positive energy core stay as separate PCs and do not fuse into a single PC. In this case, at the end of each PT, a PC vacuum of the "e" number of PCs is formed at the same time. The PCs at the edge then have to react by moving inwards by developing time dilation within. The large PC Vacuum created at the end of a PT, once compensated by inward moving EPCAs is again recreated at the end of the PT thus restarting the cycle.

The PCs surrounding the PC vacuum can alternatively be grouped into EMPAs I e. Eddy multiple Planck compartment aggregates. An EMPA can be thought to be made of about 10^6 PCs. Thus multiple layers of EMPAs can be defined around the PC vacuum. When, at the end of the PT, the PC Vacuum is formed, these EMPAs form time dilation within and thus lose volume and move inwards.

The T=0 compartments aggregated together but still staying apart can explain the variation in forces at the two ends that rips the particle apart.

The EPCA model

Contrary to that all the T=0 PCs forming the PEP in the EPCA model fuse to form a single Planck Compartment where the volume loss happens. Thus, here it becomes inevitable to divide a Planck time into smaller parts where each event happens. In a particle with a huge mass like 10^{40} events per PT, a PT will have to be divided into that many small equal parts when each event would happen. This problem was known to me at the start. However, the ease of understanding which this model provides was the reason why it was retained.

It is more difficult to imagine the ripping apart of a part of the positive energy from such a PEP.

CHAPTER 28

SUPERPOSITION OF SPIN OF A CHARGED PARTICLE

Quantum mechanics says that a charged particle, say an electron, exists in a superposition of all possible spin directions. This means that at any given point in time before the measurement is made, the electron would be in both possible states i.e. the right-handed and the left-handed state

Every charged particle also is a miniature magnet with a minute magnetic potential. This indicates that every charged particle and every Electron possesses a North and South pole.

It is also clear that every particle possesses a direction of motion and momentum.

The Matter-antimatter model of a charged particle can explain a lot of these findings.

The positive energy core will keep pulling the negative energy particles in the lattice. Each particle forming the lattice would in turn push each other away. The spin of the lattice is the spin of the particle. The lattice is spinning not because of inertia but because the lattice is made of negative energy particles which actively create Planck Compartments.

The inwards pull would be counterbalanced by the outward push.

But the model says that there is a small discrepancy in Positive and negative energy within a charged particle which determines the charge. A positively charged particle tentatively can be considered the one with a slightly excessive amount of positive energy and thus a slightly excessive inward pull. The negatively charged particle would then have an extra negative energy i.e. one or a few extra ENEA particles.

There also should exist, an asymmetry between two poles. One pole having a deficiency of the ENEA particles and another pole having an excess of ENEA particles.

This asymmetrical arrangement would however lead to instability. The resultant particle should ideally be highly unstable and should logically disintegrate rapidly.

If we focus on a pair or a triplet of ENEA particles forming the lattice, they might tend to move towards each other or away from each other. If they move towards each other, the repulsive force increases and if they move away, the attractive force directed inwards towards the central core increases. Thus they might keep moving like a pendulum from a point closer to each other to a faraway point. These oscillations when combined with oscillations of the entire lattice might have other potential oscillatory movements.

The pole where there exists a deficiency of ENEA particle would have a hole in the lattice so that the TRW from inside the lattice can exit from here easily as it does in triangular zones between ENEA particles. This TRW causes spatial contraction here. It means that there exists additional time dilation spatial contraction here.

Contrary to this, at the other pole where an excess of ENEA particles exists, there will be excessive spatial expansion. Thus the ENEAs push each other more here.

The deficiency pole would thus pull ENEAs together and the excess pole would push them away. This leads to ENEAs oscillating in such a manner that the deficiency pole and the ENEA excess pole keep alternating. These pendulum movements take place at extremely high speeds and thus an electron remains right-handed for a short period and left-handed for a short period.

To note that the direction of spin of the particle would remain the same but the poles are changing and thus handedness can change.

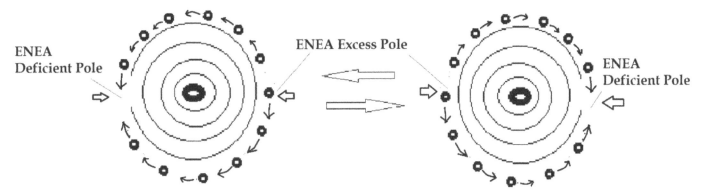

Figure iii/28.1: **Shows how the ENEA particles forming the lattice in a charged particle in DGR can swing or move like a pendulum so that the ENEA deficient pole and the ENEA excess pole shift places constantly. With the same direction of spin, any shifting of poles leads to a reversal of the handedness of the particle. Thus the charged particle is sometimes right-handed and sometimes left-handed. This shift keeps happening very rapidly, much faster than Human scales and thus the particles appear to be in a superposition of all possible spins.**

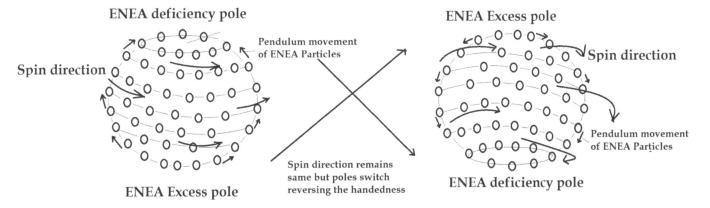

Figure iii/28.2: **Shows an ENEA lattice around a charged particle with the two poles, one with ENEA excess and one with ENEA deficiency. The ENEA deficiency creates spatial contraction**

which pulls **ENEA particles upwards and the poles switch. The direction of spin doesn't change but the poles switch and thus handedness changes. The Particle can keep switching from one handedness to the other and thus can be called to be in superposition of handedness until a measurement is made.**

When the spin is measured or when an electron comes in an electric or magnetic field, it leads to stabilization of the spin and reduction of this pendulum movement of the lattice. This also happens when the Electron comes within the electric field generated by a Proton within an atom. In other words, this happens when the Electron comes to interact with the push-pull bands of the Proton.

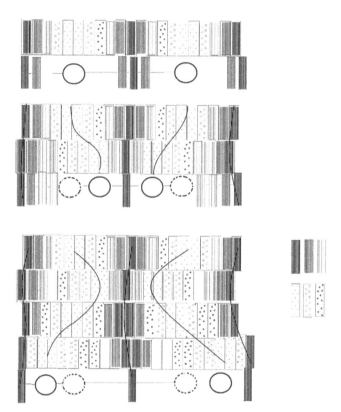

Figure iii/28.3: Shows two adjacent ENEA particles represented by circles and the various time changes caused by them represented by various shades. The vertical stripes represent spatial contraction and circular shading represents spatial expansion.

How the resultant wave would look?

Figure iii/28.3 shows two ENEA particles forming a part of the ENEA lattice around a charged particle in a relatively detailed manner.

At an instant, the space-time just beyond the particles would be divided into sections or bands where time is maximally contracted and at other places where time is maximally dilated. The zones which lie in between them would have a time gradient where time lies in between these two extremes.

At the next instant, this wave moves outwards and the position of the ENEAs changes and so does the location of the resultant zones as described above. The zones of maximal time dilation and maximal time contraction would lie slightly offset to each other but still be in contact i.e. entangled to each other.

At the next instant, these two offset locations further move outwards and the ENEAs move slightly again creating another different position of the zones.

This cycle repeats till the extreme end of the position of the ENEA oscillation comes. After this, the ENEA start moving in the opposite direction. So if they were moving left to right first, now they would start moving right to left. In each of these positions, an offset location of maximally time-dilated and time-contracted zones forms and move outwards.

Thus the resultant wave that is generated due to this oscillation of ENEA particles would have alternating time dilation and time contraction zones as shown in figure iii/28.3.

Note that the maximally time contracted zones would have active spatial dilation and the zone of maximal time dilation will have a spatial contraction.

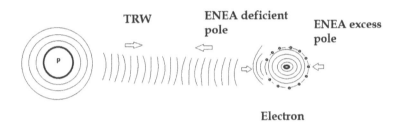

Figure iii/28.4: Shows an Electron and a Proton within a nucleus bonded together with a time resetting wave. In this location, the ENEA pendulum motion gets stabilised and so the spin no longer switches direction/handedness

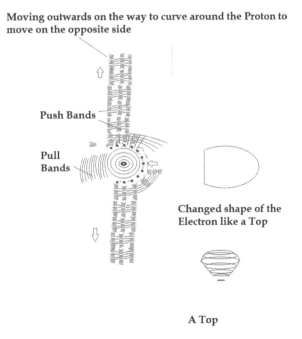

Figure iii/28.5: Shows the top model of a Charged particle. Due to the extreme inward pull of the TRWs from the nucleus, the ENEA lattice would be repelled outwards and the ENEA lattice will

undergo a change in configuration. It may take the form similar to a top with a relatively large hole in the lattice towards the central positive charge. (Note that these are just diagrammatic images and not all time changes or bands are shown for clarity.

Top Model of the Electron

There is another likely possibility.

When the Electron is present in an extremely strong electric field, like close to the Proton, the changes in the shape of the Geodesic here become important. The Superposition of spin gets stabilised so that now there is no pendulum movement of the ENEA lattice and there is no shifting of poles. The repulsion force on the ENEA particles and attraction force on the inner core might make the Electron change its configuration to something similar to a top wherein the ENEA lattice has a hole in the direction of the Proton where the attraction of the inner core with the Positive Energy core of the Proton takes place and the ENEA lattice arranges itself in a crescent in the rest of the directions.

These changes will be more exaggerated extremely near the Positive energy core as inside a Proton or a Neutron, where the Quarks lie near the Positive energy core of the nucleon. This can partly explain a higher strength of the Strong Nuclear force compared to the Electromagnetic force.

Chapter 29

WHY THERE IS SOMETHING AND NOT NOTHING? - THE PROCESS OF GENESIS

A profound question that is often asked physicists which they just have no answer to is

"why there is something and not nothing".

Although intuitively the answer to this question seems simple, it is not.

If everything is mutually attractive due to Gravity, why everything doesn't collapse into everything else instead of forming matter particles, stars and planets. Another possibility is if matter and antimatter particles are made in pairs when they are formed, why did all the antimatter not annihilate all the matter?

The answer to this question is given elsewhere in this book and involves the ENEA particles or ENEA lattices that occupy the surrounding region of the positive energy particles which provide the necessary repulsion.

It all boils down to the busting of a pre-existing firm prejudice in the physicist's mind. The fact is that

Matter particles (or Positive energy particles) and Antimatter particles (or Negative energy particles) that form in pairs can exist beside each other and don't necessarily annihilate each other.

Alternatively, we can stick to the routine definition of Matter and Antimatter in conventional physics, wherein "Matter particle" is a composite particle and so is an Antimatter particle.

In which case the myth is that "Matter and antimatter particles are formed in pairs" while the reality is the PEP and ENEA particles form in pairs.

Another fundamental and equally profound question would then be "how is this process of formation of something out of nothing get accomplished".

In short, ***"how matter can get generated out of nothing"***. (At the edge of the Universe and also at every Negative energy particle, Planck compartments are formed constantly out of nothing which is an equally daunting task to explain in DGR. Once you assume that, assuming that a PC where Destruction predominates i.e. a T=0 compartment forming spontaneously is less of a problem)

Here "nothing" does not literally mean nothing. It just means "Planck Compartments" which are a definitive component of things being created. The Planck Compartments can be said to be made up of the elusive "ether" which is the medium pervading the entire Universe. The Creation process thus creates this Ether and the Destruction process destroys this Ether.

In physics, we have learnt a rule. If a reaction can happen in one direction, it can likely happen in the reverse direction as well. In short, if "Mass→Energy" can happen within the Sun and all the stars and also in nuclear reactors, then why not the inverse reaction i.e. "Energy→ Mass"? DGR suggests that, at the Edge of the Universe, as the edge PC aggregates are pulled outwards forming negative energy EPCAs, a fresh PC Vacuum is created within this EPCA which is essentially assimilation of new Positive energy within the Universe. It is literally like new Positive energy is being absorbed all around the surface area of the Universe at the edge. As the age of the Universe increases, its surface area increases and so does the infusion of fresh Positive energy in proportion to the surface area.

In short, the total Positive and Negative energy content within the Universe keeps increasing with the age of the Universe.

If the total energy of the Universe keeps rising, and mass can get converted into energy and the other way round, it is not difficult to imagine that Energy can be utilized, at certain places in the Universe, to spontaneously create Baryonic matter particles from nothing.

Our theory can give some insight into how this Genesis happens. That is how Planck Compartments get converted into positive or negative energy particles and form matter.

Although the question of why this process happens, what or who makes this process happen is equally elusive as the basic assumptions of our theory and these questions will not be dealt with.

Certain things are clearly known.

1. The creation process or Genesis started right from the instant of the Big Bang. The process is incessantly going on since then and has evolved.
2. At that time, an extreme amount of energy was concentrated at a very small region of space-time. The temperature of space-time at the instant of creation was insanely high. For our theory, this means that insanely high fluctuations of Local time from Universal time were present so that Creation/Destruction ratios at points were insanely deviated and rates of unchecked Creation at places or unchecked Destruction at other places was extremely high.
3. Cosmic microwave background shows evidence of an overall uniform temperature with miniature fluctuations. These miniature fluctuations would be profound at the time of genesis.

4. Our current theories of physics cannot give any insight as to what happened before or exactly at the instant of Creation i.e. 10^{-35} seconds after the creation.

5. We know from conservation laws that whenever there is a fluctuation, it comes with pairs. This means that if a point with an extremely high C/D ratio appears, to compensate for that, another point with an extremely low C/D ratio also has to appear. They cannot appear alone but always in pairs.

There are exceptions to this rule, which are ill-understood presently. The exception is the slight excess of Negative Energy that our Universe displays which has led to an expanding universe. This means that there are at least some fluctuations that are stand-alone negative energy fluctuations so that they are not compensated by any positive energy fluctuations.

Imagining Genesis

For this, we have to imagine portions of the early Universe and analyse how the fluctuations can potentially lie.

Imagine a spherical region of the early Universe. Let's assume that at the start, the Local time is running along Universal time so that no uncompensated creation or destruction is happening. It figures that the sphere will retain its volume and there won't be any expansion or contraction of space in it.

Now let's imagine that an energy (i.e. time) fluctuation develops in this sphere. Let's imagine another concentric sphere within this sphere which has a diameter such that it divides the volume of the sphere into two equal parts.

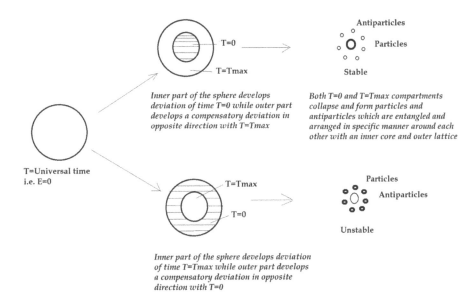

Figure iii/29.1: Shows One of the possible ways by which Baryonic matter particles are spontaneously formed from nothing due to a fluctuation in energy that develops at one spherical region. Note that the second part with a T=0 lattice and an antiparticle core shown here can also form but is unstable and probably disintegrates quickly.

This inner sphere represents a spherical plane of demarcation between deviated local times. Let's imagine that Local time in the outer region deviates in such a way that it forms negative energy and the equivalent volume of space of the inner sphere has deviated local time towards destruction i.e. positive energy.

All this negative energy in the outer portion of the sphere would tend to repel each other and thus get divided into multiple small ENEAs. Each of these starts moving or spinning along lines tangential to the sphere. The inner positive energy starts causing destruction and thus attract itself and collapse into a positive energy core that now starts destroying the Planck Compartments according to the total positive energy lying within. Thus destruction curves the space-time around it and thus the geodesic near the lattice is curved so that an ENEA particle travelling tangential would tend to spin around the core instead of flying away.

Although these ENEAs are moving separately to the positive energy core, they are in the process of Entanglement with them.

The scale of this process can vary from several million Planck Compartments to huge numbers like 10 to the power 10 to the power 10 to the power thousand. Such insanely high volumes of space can also get converted into such positive and negative energy pairs.

If it is a smaller scale, it forms composite matter particles, mainly Electrons or Quarks.

If at supra-massive levels, this process can create a central supermassive Blackhole of the centre of a Galaxy and its surrounding Dark matter consisting of ENEA particles.

The emergence of Charged particles

Note that like the parent fluctuations, minor fluctuations in the process which cause slight excess positive energy or slightly excess negative energy can lead to a discrepancy between the amount of negative and positive energy within a composite particle.

When such a composite matter particle has a slight excess of negative energy, that is the mass of positive energy in the core is slightly less than the mass of negative energy (negative mass) of all the ENEAs added together, this creates a composite particle which is highly reactive and with unchecked potential for expansion. This is essentially the negatively charged particle or the Electron or down Quark. Here there is an excess of ENEAs.

On the contrary, if in such a composite matter particle, the positive energy core has more mass than the ENEAs combined together, i.e. there is deficient ENEAs, what forms is a positively charged particle like a Positron or an Up-Quark.

It is likely that the charged particles are formed in pairs so that there is an equal number of particles with excessive ENEAs as there are particles with excessive Positive energy particles, maintaining the Positive Energy/Negative Energy balance. Being paired, the overall charge of the Universe will also balance out with an equal number of positive to negative particles.

Although such a process of "Spontaneous Baryogenesis" seems a far-fetched idea, there is nothing new in it and has been described by many before. In fact, Dirac said that as the age of the Universe increases, more Baryonic matter forms and causes the total mass of the Universe to rise. Dirac hypothesised that as G is connected to the overall mass of the Universe and the process of spontaneous Baryogenesis means that the mass of the Universe is not constant but is a constantly growing number, the value of G should also be not a constant but a constantly reducing number.

We know that in Uranium reactors, the energy locked in the baryonic matter is converted to free energy which is then used to generate electricity. If this reaction of the release of energy from Baryonic matter is possible, why is it that the mind feels that the inverse reaction won't be possible?

Thus laws of physics should have no problem with the same reaction being reversed i.e. a significant amount of energy being focussed at one point to create enough energy fluctuation to trap it and form Baryonic matter. We know that as the Universe expands, the vacuum energy of the Universe, doesn't get diluted but its density remains constant. This means that as the age of the Universe expands, the total energy of the Universe keeps increasing. There is thus no real scarcity of energy with the Universe for carrying out this inverse reaction.

The figure iii/29.2 shows one of the probable hypothetical reactions which can form Baryonic matter, although there is no reason to believe that more complex reactions with many more steps may be involved before the emergence of stable baryonic matter from focussed energy fluctuations.

The hypothetical reaction

A spontaneous Positive Energy/Negative Energy fluctuation develops due to the interaction of various large forces. This fluctuation (essentially a large deviation in Local Time) can collapse into a PE/NE balanced particle with a PE core and NE lattice. Because it is not reactive to any other particle, it may be unstable. This particle may come under further forces that rip it apart and form a NE excess intermediary complex and a NE deficient particle. This NE deficient particle would effectively form a positively charged particle like a UP Quark. NE excess intermediary complex may be unstable (alternatively – neutrinos which are extremely stable particles could be an example of one such Negative Energy excess intermediary complex). This NE excess intermediary complex merges with the PE/NE balanced particle to form a NE excess particle i.e. a particle that has more ENEA particles than PE particle in the core. That is the particle has more Negative mass in its lattice than Positive mass in its core. Such a particle forms a negatively charged particle like a Down Quark or an Electron.

Three Quarks once formed (in the extremely hot plasma of the early Universe) or the much colder sections of the present Universe, would merge with the neighbouring PE cores to produce a Proton or a Neutron depending on which Quarks are taking part in the reaction.

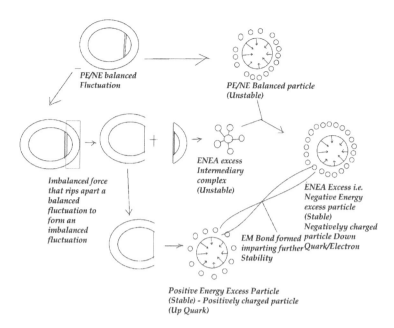

Figure iii/29.2: Shows a diagrammatic representation of spontaneous Baryogenesis. A Positive Energy Negative energy balanced fluctuation originates and is ripped apart of some ENEA elements and forming an intermediary element and what remains is PE excess particle which is akin to a positively charged particle. This element is absorbed by another such PE/NE balanced particle and creates a NE excess particle which is akin to the negatively charged particles.

The emergence of paired particles of opposite polarity

Imagine a small spherical region of space. Imagine that a fluctuation develops within it which divides it into two equal volumes by a vertical plane. The left side of the spherical volume develops Local time dilation and the right side develops Local time contraction.

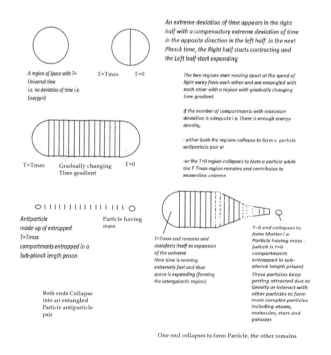

Figure iii/29.3: Shows Spontaneous particle and antiparticle (PEP and ENEA) pair creation.

Imagine that the two separate for a moment and the positive energy particle formed on the right side comes under influence of some other force, possibly attracting it towards the right. The pair separate away but are still connected or in a way are entangled with each other by a region of space-time which probably has a time gradient. The negative energy particle has excessive creation which causes the formation of a time contraction type of EPCA. A similar EPCA time resetting wave starts from the Positive energy particle as well. When they reach each other, the pair stay entangled.

Perpetual motion

Bondi et al suggested that such a pair of equal but opposite mass particles come near each other, one keeps pushing the other away while the other keeps pulling the other. Thus the two will enter a perpetual motion of one particle chasing the other particle. The ENEA pushes away and tries to fly away while the PEP keeps pulling at it and thus follows it everywhere.

Clumping

Once such a pair of Permanent positive and permanent negative mass particles form, the PEP might come under the influence of another PEP and get attracted to it while still in bondage with it. Thus, the pair starts moving towards the second PEP, which in itself would also be in bondage with its partner ENEA particle. The two PEPs fuse and form a complex with 2 PEPs fused together and two ENEAs in bondage at a distance from the clump. In this manner, the PEP clump or core can keep attracting PEPs from surrounding regions and increase in mass. The ENEAs in the bondage of each of these will keep pushing each other and form a lattice around the core.

All these processes happen spontaneously due to the TRWs emanating from each of the particles without any additional input of energy from outside.

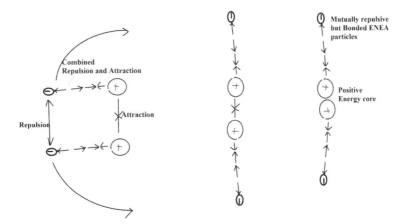

Figure iii/29.4: Shows two Positive Energy Particles paired with two Negative Energy Particles (ENEAs). Note that the PEPs attract each other, the two ENEAs repel each other while the PEPs attract the ENEA but get repelled by them literally chasing them or entering a perpetual motion with them. Eventually, the PEPs will clump together and the bonded ENEAs will remain separate.

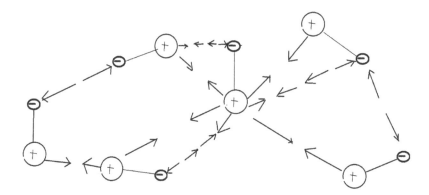

Figure iii/29.5: Shows a zone of the Universe with multiple such PEP-ENEA pairs interacting with each other. PEPs attracting each other while ENEAs repelling each other

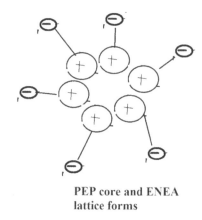

PEP core and ENEA lattice forms

Figure iii/29.6: The particles from figure iii/29.5, eventually form a dense core of PEPs in the centre with a ENEA lattice at the periphery. (Note that these are diagrammatic images and the positive and negative masses shown by different sized circles are supposed to be equal masses with opposite signs.)

Neutrinos

The Neutrinos are likely made of a few PEPs clumped in the centre with a few ENEA particles attached to them. As the particles move through their flight, it is likely that they acquire more such pairs or lose some of the bonded pairs. There may be loss of a pair or just ENEA particles leading to the particles acquiring a miniature charge. These events can possibly explain the fluctuation of Neutrino mass and flavour during their flight. Note that these are speculations and not yet confirmed.

Does the Entanglement remain or does it vanish?

The positive energy particle formed in the above reactions might get attracted towards a particle with a high mass like a supermassive Black hole in the Galaxy centre. Once trapped in it, the Positive energy particle cannot escape its clutches. However, it can add to the mass and the energy trapped in the Blackhole thus increasing its Schwarzschild radius.

The trouble with this is what would happen to the Entanglement between the negative energy particle and its counterpart that just merged with the Blackhole?

What is likely is that it stays and the negative energy particle remains independent orbiting the Blackhole at a distance attached to its core by Entanglement.

Note that this is a classical description of Hawkins radiation.

The spontaneously formed negative particle keeps causing creation which counteracts the destruction done by the Positive energy particle lost within the Blackhole.

If the Entanglement persists and the negative energy particle persists as a signature of the Positive energy particle which was just lost, it figures that information about the particle is not entirely lost and information loss has not occurred.

In fact, the ENEA particles repelled away and escaping the clutches of the Blackhole can be potentially detected by us. This is one of the predictions of DGR, i.e. a Blackhole emits only Hawking radiation in the form of ENEA particles which may clump with PEPs on their way out thus forming heavier with mass particles like Neutrinos.

In situ formation of hydrogen gas

A hydrogen atom is nothing but a Proton and an Electron bonded with each other. The Proton, in turn, is made of two Up Quarks and one Down Quark. The two Up Quarks are positively charged and the Down Quark is negatively charged.

Note that majority of the positive and negative energy present in these can be considered to be balanced. Only the energy corresponding to the mass of the nucleus is contributed by the so-called "Gluon field" which in our theory is nothing but the positive energy core akin to a miniature Black hole at the centre of the Proton. This Positive energy particle must have a counterpart that remains entangled to it at a distance but cannot come close or interact with it directly. The negative energy counterpart once created flies away from this Positive energy particle and probably contributes to dark matter.

Around this Positive energy core or miniature Blackhole, further paired fluctuations must develop. The fact that an atom of hydrogen is electrically neutral and has no charge means the positive charges of two Up Quarks neutralizes negative charges of an electron and a Down Quark.

Each of these particles to form would need the above-mentioned process of paired formation of positive and negative energy concentric spheres, in which the central positive energy core collapses and forms the central positive energy core where significant destruction keeps happening. The peripheral negative energy fluctuation forms the ENEA particles forming the lattice. One extra ENEA particle within the Down Quark is compensated by the extra Positive Energy Matter particle in the Up Quark. In the same manner, the extra ENEA particle in the Electron is compensated by the extra positive energy particle within the core of the Up Quark.

The three Quarks form around the Positive energy core and start rotating and get entangled with each other. The effective charge which is the Push-pull bands emanating from them, while in the process of their complex motion around the core, interact with their partner Quarks and can be held together with the combination of inward pull of the Positive energy core and the attraction of the opposite charges.

Although the theory allows such a spontaneous creation of each of these particles including ENEAs, positive energy core thus forming Quarks and Electrons and Protons, the theory currently has no way of explaining what is the sequence of events and also no way of explaining why only these events happen and not the countless other combinations possible. Currently, there is no way of explaining the Proton to Electron mass ratio of 1836.152:1. There is no way of explaining why only selected energy fluctuations contribute to this genesis process and lead to a Proton or an Electron with the known mass.

One possible explanation is that fluctuations of all sorts happen and keep happening and all the combinations of particles will all sort of masses form. But only the particles having these combinations of masses remain stable while others fall apart due to some underlying instability. This is probably because only the particles with these masses create space-time changes around them that are compatible with the formation of electromagnetic bonds. Many potential zones in the Universe exist, especially in the intergalactic region at the edge of Positive energy predominant and Negative energy predominant zones, which might provide such fluctuations in Local time needed for the spontaneous formation of charged particles. These might provide fertile grounds where the spontaneous genesis of Baryonic matter can take place.

This forms the basis of the rather revolutionary 'claim or prediction" of DGR that

The baryonic matter that we see may not be present at the time of the Big bang and that it formed and is still forming spontaneously from energy fluctuations.

Until detailed computer models of the theory of DGR become available and detailed computer simulations of processes of genesis with various combinations of masses become available, progress in this regard is difficult.

Chapter 30

STRING THEORY AND OUR THEORY

The basic intention why "the String theory" was developed was to derive the theory of Quantum gravity and the theory of everything.

It is impossible to discuss all the details of String theory including the complex mathematics which constitutes the backbone of the theory. But String theory attempts to explain all the physical phenomena based on one-dimensional entities called strings.

Strings could be open strings or closed strings. What these things are made of, is not discussed. But each of the particles of the Standard model, at the Planck scale, are made up of different strings and the properties of each of the particles are determined by the vibrations of these strings.

These strings are supposed to be 11 dimensional for the Math to work.

Thus every particle has a unique way in which its String vibrates and this gives rise to the properties of the particle.

The strings are said to vibrate in 11-dimensional complex spaces which are called as Calabi-Yau manifolds.

The vibrations of these strings create properties of these particles just like vibrations of a taut guitar string generates sound waves. The details of how this is achieved or which particle corresponds to which shape is not known for sure and cannot be discussed here as well.

The basic assumption of this space-time curvature in these manifolds is that they are static unchanging i.e. they are not dynamic or they do not dynamically change shape.

In our theory, in contrast to General Relativity, the space-time curvature is dynamic. The curvature of space is in fact due to either excessive "Creation" or excessive "Destruction" of Planck Compartments and that the curvature can change rapidly around a particle

The manifolds described above can be compared to the space-time changes occurring around a charged particle.

In our theory, a charged particle is a positive energy core with a negative energy lattice rotating around it. One can thus understand that the destruction or creation happening at these particles leads to a predictable change in the shape of space-time. This change in shape persists for a very short time and the changes are dynamic. The changes are also not static in location but lead to progressive changes in the surrounding space which in turn lead to space beyond them. These dynamic changes in space-time around the particle thus ultimately lead to a progressively spreading disturbance in space surrounding it along with the local time in it. All this process is described in a slightly different manner elsewhere as spreading of EPCAs. The EPCAs lead to the formation of push-pull bands which emanate outwards and their interaction with other push-pull bands from other charges is the reason why electromagnetic interactions become possible.

Thus the complex manifolds described in String theory could be compared to space-time curvature happening in surrounding space due to the presence of the positive energy core and negative energy lattice.

The harmonious oscillations and vibrations of the Strings in String theory would then represent the combination of the rotation of lattice and the creation/destruction processes happening at the particles bonded together.

Thus although not identical, our theory can be comparable to what is described in String theory. Where the two theories deviate need a much deeper understanding of both and would need to wait.

CHAPTER 31

CONTEMPORARY PHYSICS AND DGR

Contemporary Physics	**DGR**
Gluons carrying the Strong Nuclear Force	Push-pull bands with Pull of central positive energy core
Virtual Photons carrying the Electromagnetic Force	Push-Pull Bands between oppositely charged particles
W/Z bosons leading to Weak Nuclear Force	Push-Pull bands from neighbouring Nucleons adding up to cause enough Pull or Push to cause ripping off of a part of the positive energy core
Graviton	EPCAs
Electric Flux	Push-Pull Bands emanating from Charged particle
Magnetic Flux	Push-Pull Bands when the charged Particle starts moving or spinning
Dark Matter	ENEA particles
Dark Matter or MOND	Linear EPCAs
Neutrinos	A single ENEA particle or possibly a small number of them with a small positive energy core leading to its small positive mass

	Contemporary Physics	**DGR**
Gravity	Change in the curvature of Spacetime moving with the speed of light - Einstein's General relativity + Dark Matter/MOND	Dynamic change in curvature evolving every Planck time consistent with Einstein's VSL theory, Entropic Gravity, MOND and Large number Hypothesis

	Contemporary Physics	**DGR**
Light	Electromagnetic Wave + Photon - Wave-particle duality	Positive Energy Particles with Destruction happening at their core with compensatory EPCAs around, being entangled together to form a wave-front
Mass	Higg's mechanism- Higg's Boson – Higg's field	Destruction at the core of the Particle with compensatory EPCAs around it
Energy	Capacity to do work	Any Deviation in Local time, Dilated Local time is Positive energy and Contracted Local time is Negative Energy
Time	Variable in GR Single unchanging background entity in QM	Two types Variable time named Local time due to active spatial expansion or contraction Absolute time namely Universal time
Electromagnetic interactions between Charged particles	QED – exchange of virtual photons	Exchange of Push-pull bands
Electromagnetic bonds	Shared electrons	Acquiring Low energy state by neutralizing far sided Push-pull bands of Proton-Electron pairs
Speed of Causality	Follows the Universal speed limit i.e. information cannot travel faster than light except in Entanglement	Information can be transferred much faster than the speed of light but a particle with mass cannot break this barrier.
Double slit experiment – Explanation of the Weird findings	Copenhagen Interpretation with Quantum Superposition and collapse of wave-function/Many world's interpretation	In line with Realism, In line with de Broglie-Bohm mechanics,
Wave-particle duality of matter	Superposition of wave and particle	A separate Positive energy particle with Destruction happening within causing compensatory wave formation around it.

	Contemporary Physics	**DGR**
Baryonic Matter	Made up of Fermions like Electrons and Quarks	Consisting of a Positive energy core with a negative energy lattice
Charged particle	Fundamentally a point particle (Possibly made of strings vibrating in 11-dimensional hyperspace as per String Theory)	Composite particle made up of a Positive Energy core with a lattice of Negative Energy particles or ENEAs with a slight imbalance in PE/NE ratio

CHAPTER 32

FINAL ASSUMPTIONS OF THE THEORY

1. Space is made of Planck Compartments which are one Planck volume with sides one Planck length.
2. Space is made of an array of these Planck Compartments stacked onto each other and in direct contact with neighbouring ones.
3. Each Planck Compartment is in entanglement with its immediate neighbour so that anything that affects one Planck Compartment can affect its neighbour.
4. Planck Compartments are free to move. There is no limit to how fast they can move or change their coordinate location. They can move slower or faster than the speed of light.
5. There is a Universal coordinate system accurate to a Planck length. Every Planck Compartment has an accurate coordinate location which can change. This essentially means that the relativity of space given by Special and General relativity is in itself not absolute. The relativity of space is relative. That is although space can change, the universal coordinate system doesn't change.
6. Two Planck Compartments cannot occupy the same co-ordinate location. When the Creation of a new PC happens in between two PCs, it pushes the two PCs apart so that it changes their co-ordinate location.
7. There cannot exist a void in between Planck Compartments. If in a process, a void is created in the vicinity of a Planck Compartment, the Planck Compartment is sucked towards it or moves in coordinate location to compensate for the void. This movement itself creates another void which is then in turn filled by the movement of its neighbouring Planck Compartment. Thus there is a constant movement of Planck Compartments into their neighbourhood. An array of such movements creates a wave.
8. Process of Creation and Destruction

    Planck Compartments can appear from nothingness. This process is called Creation. Planck compartments can be destroyed into nothingness. This process is called Destruction. At the Planck level, Quantum foam constituting of spontaneously appearing and disappearing PCs is present.
9. Time scale

    The cycle of Creation and Destruction continues incessantly every Planck Time.

In short, there is spontaneous Creation and Destruction of Planck Compartments everywhere. The ugly implication of this is that at the Planck level, the Universe is being "Destroyed" and "Recreated" constantly.

10. The process of Creation or Destruction need not occur simultaneously or in pairs. Three possibilities exist. At some locations in the Universe, Creation can balance Destruction. At other locations, Creation can predominate and at still others, Destruction can predominate. The Creation/Destruction ratio at a point is the most important information about any point in Space. This ratio determines the Local time running at every point.

11. Number of times Creation or Destruction can happen per Planck time

    When balanced, the process of Creation and Destruction happens once every Planck time. However, there is no restriction as to how fast these processes can happen. The number of events of Creation or Destruction happening at a point per Planck Time is denoted by "e" or "ꜫ". This is especially true for Destruction so that the process of Destruction can happen millions or billions of times per Planck time.

12. Active Expansion of space

    At places where Creation predominates over Destruction, space can expand actively. When this happens, neighbouring PCs are pushed apart.

13. Active Contraction of space

    In other places, Destruction predominates over Creation. Here, space actively contracts and pull the neighbouring PCs together.

14. What causes the Creation of new Planck Compartments or the Destruction of Planck Compartments is not known. It could be actual Creation or Destruction or it could be teleportation to a different point in the Universe. How exactly these processes happen is beyond the theory and need to be assumed as it is as integral assumptions of the theory.

15. Entanglement/Compensation

    Entanglement is an integral assumption of the theory. Every Planck Compartment is in entanglement with its neighbour. Entanglement can be considered as the phenomenon happening due to the "No void allowed" assumption. It is a process meant to compensate for a void formed during any process.

    In this process, any creation of void or any tendency of Space to expand due to excessive Creation taking place at a point is compensated by the reaction that is shown by the neighbouring entangled Planck Compartments.

    As explained in detail, a void created at any point leads to immediate entanglement of all the neighbouring PCS forming an "Eddy PC aggregate – (EPCA)". Each PC forming this aggregate is pulled towards the central PC vacuum. The Local time in this aggregate of space, auto-adjusts itself so that there is enough uncompensated Destruction happening in it to cause spatial contraction. The EPCA loses volume and thus fills up the void.

Similar EPCA develops even with excessive Creation. Excessive Creation expands space which pushes the neighbouring PCs apart causing a void formation in between them. This newly formed void around the pushed PCs is compensated by additional uncompensated Creation. This PC aggregate has Local time in a contracted state due to excessive Creation. This further pushes the outer PCs and forms another contracted time type of EPCA around it.

There is no lower time limit in which these processes can happen. In short, the entanglement process can happen millions or billions of times in a single Planck time.

16. Time resetting wave formation and propagation

The entangled Eddy PC Aggregate (EPCA) formed following the Creation of void (at a place where PCs were destroyed) gets pulled inwards to compensate for the void. This, however, forms another void around these inward moving PCs. This void pulls the outer PCs inward and forms a void just outside them. Thus the void keeps travelling outwards. Every EPCA that forms in the process due to the inward pull and additional uncompensated Destruction has a dilated Local time. Thus this wave that starts at the point of Destruction and moves outwards is a time resetting wave (TRW).

It moves outwards much faster than the speed of light and its speed depends on the number of events of Destruction happening at the centre per Planck time.

This is the basis of the Non-Locality of our theory.

17. Universal time

There are two concepts of time in our theory. There is one Universal time and there is one Local time. Universal time is a time that is common to the entire Universe and at every corner of the Universe, it runs at the same pace. This can also be considered Absolute time. This concept, violating Special relativity, upholds the presence of universal simultaneity. There is a simultaneous moment for all points in space, irrespective of the distance between them.

18. Local time

In contrast to that, there is another concept of time called Local time. Universal time is independent of space. However, Local time is intricately linked to space. It is intricately linked to the process of Creation and Destruction and thus intricately linked to the C/D ratio.

There are two possibilities. The spatial changes occurring are primary and secondarily lead to change in the Local time. This is more likely. In this case, the difference in Local time is a mere illusion. In this, the active spatial contraction just before the photon causes a slowing down of its forward propagation i.e. the speed of light reduces leading to an illusion of slowing down of time. An active spatial expansion just before the photon hastens the forward propagation of the photon leading to the illusion of a faster ticking of the Photon clock causing the illusion of faster progressing time.

The second possibility is that the time which is determined by how much Creation and Destruction is happening at a place changes primarily and the changes in time secondarily create the spatial changes.

19. Relationship between space and Local time

    Local time running at every point varies. If at any point, Creation is more than Destruction, time is contracted or running faster than Universal time. At any point, if the process of Destruction is faster than Creation, the Local time here is running slower than Universal time and thus Local time here is dilated.

    Wherever Local time is dilated, space is actively contracting due to excessive uncompensated Destruction.

    Wherever Local time is contracted, space is actively expanding due to excessive uncompensated Creation.

20. Energy is nothing but a deviation of Local time away from Universal time.

    In essence, when Local time is running equal to Universal time, there is no uncompensated Creation or Destruction and the C/D ratio is one. This constitutes zero energy.

    Any deviation of the C/D ratio towards Destruction constitutes Positive Energy.

    Any deviation of the C/D ratio towards Creation constitutes Negative Energy.

    It figures, that positive energy causes excessive Destruction and spatial contraction.

    If energy can be negative or positive, Mass which is equivalent to the high amount of energy trapped in a particle can also be negative and positive.

21. Gravity

    Gravity is explained based on this Entanglement-EPCA based Local time readjustment. The Sun or the Earth, due to their mass, have high Positive Energy within their substance, at each of these locations, extreme deviation of C/D ratio happens and there is constant Destruction and void formation taking place. The EPCAs that form to compensate lead to Local time readjustment which moves outwards as a time resetting wave (TRW) which leads to a gradient of gradually declining gravitational time dilation (as one moves outwards), dilated Local time and active spatial contraction. This active spatial contraction or excessive Destruction happening at every point around these heavenly bodies leads to active curvature of space around them and an active force that pulls these towards each other.

    In a way, Gravity is an emergent phenomenon arising from spatial contraction happening at the quantum level, with these contractions adding up to show the macro-level phenomenon of Gravity.

    Gravity in our theory, in a way, aligns well with Entropic Gravity described by Eric Verlinde.

22. MOND

    At extremely high distances like intergalactic distances, the curved EPCA's are replaced by linear EPCAs in which there is just shifting of EPCAs without horizontal volume contraction.

The shifting of these linear EPCAs happens at a universally constant acceleration called a_0. Due to this, the mass of the entire Galaxy can have a gravitational effect at each of the constituent stars and thus the resultant force is directly related to the mass of the Galaxy. Gravity at high distances in DGR thus goes in line with MOND.

23. PEP and ENEA

    For explaining Electromagnetic forces and Strong Nuclear forces, another set of assumptions need to be added.

    There are two types of particles.

    Positive energy particles and negative energy particles.

    Positive energy particles are those in which there is a highly dilated Local time i.e. hyper-dilated Local time. These particles cause a fixed Destruction of the space per Planck time and thus lead to EPCAs around it with dilated Local time. These EPCAs cause spatial contraction within them and travel inwards towards the particle. Thus such positive energy particles are mutually attractive. They will keep attracting each other until they fall into each other. As they come close, the Destruction happening within them gets added so that their attractive power also increases.

    There is no upper limit to which positive energy particles can form aggregates.

    After a limit, the amount of Destruction happening within them increases so much that the EPCAs that form around them move faster than the speed of light and light cannot escape them. When this happens, they are said to gain the property of mass. So smaller positive energy particle aggregates which do not form EPCAs travelling inwards faster than the speed of light, form massless particles.

    In contrast, there are negative energy particles, also called Exotic negative energy particles or ENEAs. These are particles that have negative energy within. This means there is excessive Creation happening within them so that they cause EPCAs around them with contracted time.

    These EPCAs with contracted time travel outwards, i.e. push outwards and cause the formation of more contracted time EPCAs outwards.

    These ENEAs are thus mutually repulsive and they also repel other positive energy particles.

    They are thus, unlikely to form aggregates and form bigger particles with higher deviations of C/D ratios. This means that hyper contracted Local time seems unlikely.

    These remain separate.

24. Spontaneous Creation of PEP and ENEA

    Positive energy particles and ENEAs appear in pairs. So that when 100 PEPs have appeared, there has to be 100 ENEAs appearing as well.

    The PEPs and ENEAs after their formation are likely to stay in a state of entanglement.

There are certain corners in the Universe, where the spontaneous Creation of PEP and ENEAs is happening. These can further combine and create Baryonic matter i.e. Matter particles.

25. Asymmetry between PEP and ENEA

    There is a slight imbalance between the PEP/ENEA ratio in favour of ENEA.

    This means that there is a slight excess of ENEAs.

    In short, there is higher negative energy than positive energy in this whole Universe.

    This excess negative energy is used to explain the expansion of the Universe.

    The ENEAs are dispersed in the entire intergalactic space as a superfluid, i.e. a fluid with extremely low internal friction.

    (Whether this is true or not is unclear and all three possibilities still can occur)

26. Charged particle – the Matter-antimatter model

    A charged particle is a high Positive Energy core with a Negative energy lattice made of almost equal amounts of Negative Energy. The slight difference in Positive or Negative Energy determines whether the charged particle will have a positive or a negative charge.

    The Time resetting EPCA forming waves (TRW) from the Positive Energy core and similar TRWs emanating from each of the ENEA particles forming the Negative energy lattice interfere and lead to a complex interference pattern in the space immediately outside the charged particle with the formation of bands of excessive time dilation i.e. spatial contraction called as Pull bands and bands of excessive time contraction spatial dilation called as Push bands. The property of charge of a particle depends on the ability to form these bands. These push-pull bands and their interaction with other such push-pull bands from other neighbouring particles determines the complex array of interactions like electromagnetic bonding, strong nuclear force etc.

    These Push-Pull bands form much faster than the speed of light (linear velocity) but drift outwards with the speed of light.

27. The spin of a particle

    The ENEA lattice is made up of multiple particles with Negative Energy having active Creation happening within, which rotates around the positive energy core with high speeds. This rotation is not due to the inertia of the particles but is because of the intrinsic property of the ENEA particles to produce an extra PC per PT. Due to this spinning ENEA lattice, all the matter particles and charged particles will have a miniature magnetic moment.

28. Antimatter and Matter asymmetry

    ENEA particles are not the same as the antimatter particles described in physics.

    Matter particles as described in contemporary physics are made of PEP at the core and ENEA revolving around as a lattice. The antimatter particles described in physics are also made of PEP core and ENEA lattice with equivalent positive mass but a differing PEP/ENEA ratio so as to invert the charge. These antimatter particles are not the ones forming in pairs with matter particles. What are formed in pairs are the PEP and the ENEA particles.

That is, in our theory, the PEP and ENEA are the equivalents of matter and antimatter. At least a part of the missing antimatter thus lies around the charged particles as a lattice.

The antimatter particles (ENEAs) corresponding to the Positive energy corresponding to the mass of the Protons or Neutrons are not present in the vicinity and are probably lying around the Galaxy as Dark Matter.

29. The Bonds and chemistry

The Electromagnetic bonds and Nuclear bonds are explained based on these spatial contraction and expansion bands (Push-Pull bands) emanating and spiralling out of all particles. These bands move much faster than the speed of light while forming, i.e. while moving along their length. But once formed, they keep moving outwards at the speed of light.

While bonding with the Protons in the nucleus, the Electrons also give out the Push bands along their equatorial region. These Push bands move outwards and curve around the Nucleus and meet at the other side. The bonding Electrons do not fall further within the nucleus beyond the first orbital or for that matter the outermost unfilled orbital because this repulsion from the Push bands counterbalances the attraction. This can also potentially explain the quantization of angular momentum.

30. The Strong Nuclear Force

The Quarks revolving around the Positive energy core, within each nucleon, irrespective of whether the nucleon is charged or neutral, give off these Push-Pull bands. These push-pull bands are modified because of the curved geodesic due to the presence of a Positive energy core at the centre of the Nucleon. These modified Push-Pull bands interacting with similar bands from neighbouring Nucleons causes the Nucleons to stay in bondage. The constantly moving Quarks means that the location of these bands keeps varying. When multiple bands act on an Up Quark, the push or pull force may be strong enough to rip it off to form a Pion which is then pulled towards the neighbouring Nucleon. The Nucleon thus keeps exchanging Pions and the location of the positively charged Nucleon keeps changing.

31. The Weak Nuclear Force

The Quarks from neighbouring Nucleons keep interacting with their Push-Pull bands. If multiple such bands are acting on a single Quark, it can be ripped apart thus reducing its mass and changing it to a different smaller mass variety i.e. Up Quark is converted to a Down Quark. In Nuclei with a high number of Nucleons, this process can change the number of Protons within the Nucleus and thus transform the elements.

32. The wave-particle duality of light and electromagnetic radiation

    The higher frequency electromagnetic radiation is explained based on positive energy aggregates forming small massless particles that form particles of light. These cause EPCA forming TRWs which interact with each other and this process entangles these particles together when they are in flight with other such particles.

    The particles with the EPCAs form the wave-front of the light.

    The wave-front creates the waves of the light while the PEP aggregates form the particles of light thus explaining the wave-particle duality of light

33. Wave-particle duality of matter

    Every particle like an Electron due to the positive energy core and negative energy lattice, due to the complex interference pattern that forms around it, gives off a time resetting wave all around it.

    This explains the wave-particle duality of particles and the Single Electron Double Slit Experiment.

34. Superposition

    The Superposition described in Quantum Mechanics is not the real one but an apparent one happening solely due to events happening at an extremely rapid pace at Planck scales of time.

    Thus the probabilistic nature of predictions of Quantum Mechanics is basically due to this inability of our instruments or methods to decipher these individual events happening at Planck scales of time, something which Einstein called hidden variables theory.

35. Dark matter

    This is explained based on modified Gravity due to MOND and the presence of entangled ENEA particles. ENEA particles which constitute the antiparticles that formed along with the PEPs that form the mass of the Nucleons, are probably still entangled with these and are bonded to the mass of the Galaxy and exist as a layer all around the Galaxy.

36. Dark energy

    This is explained by the excess ENEA particles i.e. asymmetry between negative energy and positive energy.

37. Magnetic field and electric field or current

    This is explained due to the dissemination of push-pull bands from charged particles which are positive energy particle aggregates at the core and negative energy particles forming a lattice structure.

    The magnetic moment of charged particle and also the spin is explained by the same push-pull bands spiralling around the particle due to the spinning of the lattice structure due to its inherent property of excessive Creation.

In short, there are no fields in our theory.

The Push-pull bands represent the counterparts of Electric and Magnetic Flux.

38. Higgs mechanism

    Higgs field is analogous to extreme positive energy particle aggregates induced Destruction happening at the centre of a particle leading to EPCA formation that moves inwards, faster than the speed of light.

39. Speed of light

    Unlike special relativity, the speed of light is not the limit with which information can travel in our theory. Instead, the time redistribution can take place much faster.

    The Universal speed limit, however, does apply to any Positive Energy particle to move across space. This limit however does not apply to the actual movement of space.

    Also, note that the Speed of light isn't as constant as it is considered in Special relativity and can vary depending on the presence of active spatial contraction of expansion just before the photon.

40. Electromagnetic waves

    All the Electromagnetic waves (except Radio waves) are aggregates of $T=0$ particles or Positive energy particles with various amounts of destruction happening within, entangled by EPCA forming TRWs which together form the wavefront. The Wavelength and the frequency are determined by the extent of positive energy within each of the PEPs. These waves do not have any zones of negative energy.

41. Radio waves

    Radio waves are the push-pull bands emanating from a charged particle emanating out and drifting away with the speed of light. Thus these waves have alternating zones of high time dilation and high time contraction. The Wavelength and Frequency are determined by the frequency and amplitude of oscillation of the charged particles creating the Radio wave.

42. Gravitational waves

    Gravitational waves detected by LIGO are different from the TRWs responsible for Gravitation. These are waves that are generated as a disturbance or as turbulence in the smooth laminar-like flow of EPCAs when EPCA creation and compensation of volume loss is hindered. For example, a large machine is lifted up and held by a Robotic arm and thus prevented from falling.

PART IV

VERY EARLY IMMATURE STAGES OF THE THEORY

CHAPTER 1

EVOLUTION OF DGR IN MY MIND (1**-6**)

Part IV contains my notes that I wrote while thinking about the very early versions of my theory. Those who are interested in how my thinking process was directed and how the theory of DGR took shape in my mind starting from the current physics can read the entire section keeping in mind that some of the written stuff is the wrong line of thinking. Those in a hurry can just read this "Evolution of DGR in my mind" section to get a fair idea of how the theory evolved and took shape at these early stages.

> **1** Evolution of DGR in my mind**
>
> Most people would be interested in what DGR is and how it can explain phenomena we observe. But some may also be interested in how I reached the present state of the theory from what is already known. Given that the theory of DGR requires a paradigm shift in the thinking process that the physicists are required to do, this process of evolution may be important and it may be important to monitor the direction in which the theory is going so that if things aren't working in the future, the direction of thinking can be corrected.
>
> One should realize that at the start, the theory was significantly different and possibly in a bizarre direction with many mistakes. Thus publishing these wrong lines of thinking amounts to letting the world know how foolishly I thought at the start.
>
> *After pondering for a while about whether to publish the complete journey or whether to publish only the present status of the theory, I took the rather ugly decision of publishing my notes while thinking about DGR at the very start.*
>
> These are minimally edited and meant only to bridge the gap between the present understanding of physics and the present status of DGR.
>
> One can note that there is a significant difference between what the fundamental assumptions were in the earlier versions of the theory and the present versions of the theory.
>
> *(Contd.)*

Sometimes, the later versions of the theory contradicted the earlier versions. Instead of deleting or editing these notes, I have added text boxes or notes (which were added at the end of the writing journey, during editing). These text boxes were chosen to point out the mistakes in the thinking in the earlier versions, instead of making corrections or choosing to delete the earlier versions.

So note that each text box including this and all the other text boxes seen hereinafter in this book are added later to point out the mistakes in the thinking in earlier versions or giving further clarifications or insights that emerged later.

In the earlier version of the theory, I started with General Relativity, LQG, and QED. Thus, deviation of time in this version was seen by me as the fundamental phenomenon occurring and the changes in space happening were seen as secondary. The Link between creation and local time and destruction with Universal Time, which is seen in the earlier versions was later replaced by the concept of the C/D ratio. In the initial version, there was no EPCA formation and there were some rather poorly defined concepts like Zone of Interaction which were omitted in the later versions. In the early versions of the theory, the rather unrealistic theoretical jargon of "sub-Planck length prisons of space-time curvature which trap T=0 compartments was used after taking it directly from String theory. In later versions, it was found to be unhelpful and thus was omitted in the present version. It was pretty late that the true meaning of the higher dimensional entities became clearer. The Matter antimatter model came much later and thus in this earlier version the charged particle was a rather ugly ill-defined one.

Even in this premature form, it was more attractive than I expected and could explain (although rather vaguely) a lot of concepts like Newtonian gravity, MOND, inertia, momentum, and wave-particle duality of light and particles.

2**

These successes of the earlier versions prompted me to keep on thinking about it, acknowledging that. this could be the right direction of thinking.

There are some vague generalizations and vague diagrammatic representations in these notes which I had a chance to edit or delete if I could have. But I kept them solely so that the bridge of thinking from current understanding of physics to present DGR becomes clearer.

(Contd.)

An earlier version of DGR

How it all started?

Since my childhood days, I have had an eerie feeling about Einstein's theory of Relativity. In those days, my ignorance was probably key to my feeling. As I acquired more knowledge, I expected that the concepts would become clearer. However, even after significant effort, the theory of GR looked unconvincing or looked incomplete. Initially, I used to accept it as an inevitable consequence of my ignorance. But when I stumbled upon Quantum Mechanics and its universally accepted phenomena like superposition and wave-particle duality and also heard of the conflict between GR and QM, I smelt an opportunity.

When I had acquired enough working knowledge of General Relativity, QED, and LQG, it all started.

I stumbled upon the idea of Quantum foam in that there are small fluctuations of energy at Planck scales which are extremely minute levels.

With the newfound knowledge of GR, I started imagining concentric spheres around the Earth with a variably running time. Whatever Einstein would say, I knew that the observation that the apple actively falls down cannot be completely ignored.

When we imagine an apple at a height h in the process of falling down, space in between it and the Earth undoubtedly reduces. I knew that space can expand. In fact, this spatial expansion is the reason why Galaxies are drifting away from us. If space can expand, why is it difficult to imagine an actively contracting space?

But for the space between the apple and the Earth to contract actively, the only thing available was a variably running time provided by GR. How can a variably running time with a Gravitational time dilation gradient provided by GR cause active spatial contraction? Reading the description of the working of an Alcubierre drive in which space expansion behind and contraction in front can potentially drive a spaceship faster than the speed of light, further strengthened my thinking that the space between the apple and the Earth can undergo active contraction.

The only thing needed was to provide an assumption that makes this active contraction of space with deviated time possible.

3**

Understanding LQG and the description of Quantum foam was the point where I felt that I had almost everything I needed to complete this new theory. The realization that events in nature happened at the Planck scales was there before. The description of Planck Compartments taken straight from LQG was enough to start the process of imagining this new theory. This array of PCs stacked on top of each other which make up the space-time made a good background on which events would happen. Quantum foam meant that these PCs would possibly spontaneously appear and disappear as the virtual particles do in Quantum Mechanics. What if this appearance of new PCs or disappearance of PCs is linked to time, I thought!

However, I soon realized that if both "Appearance" and "Disappearance" are linked to a single entity of time, their rate of appearance and disappearance would not vary. Thus active expansion or contraction won't happen. For active expansion or contraction to happen, what was inevitable was two different types of time running at speeds different from each other. Einstein's description of the stationary time or the time of a distant observer to compare the dilated time of a clock kept in a Gravitational field while describing Gravitational time dilation gave a clue.

From here, heralded the insight that the other time I am looking for is the Universal Time or stationary time. This could as well be the stationary background entity of time in QM.

If the appearance of PCs is linked to this variable dilated time due to the gravitational effect of the Earth and the disappearance of PCs in Quantum foam was linked to the Universal Time which is running relatively faster here, it can explain the active contraction of space. This formed a part of the very first assumptions of the theory. It became clear later, however, that this insight was wrong and that appearance and disappearance are not linked to time but time is linked to them.

The insight that Energy was indeed the appearance or the disappearance of PCs took a bit longer time. The discovery of Gravitational waves by LIGO which were detected as miniature fluctuation in the size of space-time gave the first clue to this. A significant proportion of the mass of the two black holes that merged was converted into a fluctuation in time that also led to fluctuations in space. The energy released in the process was travelling as a time fluctuation.

The second clue came from Special Relativity which stated that a person in motion would experience time dilation due to motion and that a hypothetical spaceship travelling with the speed of light would have time running at zero pace. That is all clocks on this spaceship would stop ticking.

It also stated that a photon of light experienced no time due to the fact that it travels with the speed of light.

4**

In other words, for the photon, there is no concept of time. This rather self-conflicting statement given by Special Relativity that a photon travels with the speed of light i.e. one PL per one PT but doesn't feel even a single Planck time and doesn't feel the Planck clock move forward was a bit counterintuitive. But it made sense if our original hypothesis of two separate types of times was true. There was even another interpretation of the same concept in which it is said that everyone or everything moves forward with the speed of light. If a thing is static with no movement in space, it starts moving forward in time with maximum pace and thus starts experiencing un-dilated time. Contrary to that, if a thing starts moving, time dilates for it until its velocity reaches the speed of light when it stops.

Combining all these, the thought emerged, as to what if the photon is a Planck Compartment where this variable non-stationary time is 0. This PC where T has stopped running was later called a T=0 compartment for ease of communication and understanding. One Planck compartment would appear and two PCs would disappear in a T=0 compartment causing a net loss of one Planck volume every Planck time, I thought!

It was soon realized that these T=0 compartments could be an ideal packet of energy and could form the nodes in the EM waves. If active spatial contraction is happening at these Compartments, what would happen to the surrounding PCs? Could the surrounding PCs contain a Time gradient with a variable time dilation or time contraction like the one present around a gravitationally active body? A chain of these T=0 compartments running with the speed of light along with the compensatory changes around it in the form of some sort of time gradient emerged as a good working model for the EM waves and light.

How to get the entire Electromagnetic spectrum from just one T=0 compartment? was the next question. I did not have to go far for the answers. What if multiple T=0 compartments come close to each other?

If two T=0 compartments cause spatial contraction, they would attract each other and form aggregates. How many T=0 compartments clump together, would then determine the frequency of the EM radiation.

5** What about the particle with mass?

To explain this, a little help was taken from String theory. It was theorized that as more energy gets focussed at a single point, i.e. the quantity of T=0 compartments getting aggregated increases, a point comes when the extent of additive spatial contraction happening due to it is so high that these T=0 compartments would collapse into a sub-Planck length prison characterized by additional higher dimensions as theorized in String theory. These assumptions later turned out to be unnecessary and were given up in a later part of the theory.

The emergence of the concept of C/D ratio

Instead of imagining time changes as primary changes and spatial changes happening secondary to the time changes due to the appearance or disappearance of new PCs, insight emerged as to use concepts of Creation and Destruction. In this, the spatial changes became primary events happening and the time changes became secondary. C/D ratio became central to the development of the theory and Creation, Destruction and Entanglement or Compensation became the only three assumptions needed to construct an almost complete complex theory that could potentially explain a wide variety of phenomena.

6**The emergence of the concept of EPCAs

The concept of the formation of EPCAs was a turning point in the theory development.

Many things became clearer after this idea emerged. The concept that a compensatory wave emerged out from every particle, went very well with the wave-particle duality. After this was applied to the matter-antimatter problem, further insights came as to what a charged particle might look like.

The Matter- Antimatter model of a charged particle emerged. The time contraction spatial expansion type waves emanating out from a particle with excessive creation, when interacting with a time dilation spatial contraction type waves emanating from a particle with excessive destruction, would indeed form complex patterns. What if these patterns form interference bands like the usual waves, with narrow zones of spatial contraction explaining attraction and narrow bands of spatial expansion explaining repulsion? These narrow zones could represent the electrical and magnetic fluxes.

Chapter 2

GRAVITY

Gravity is still an enigma for physicists.

We still don't know how the most drastic and the most important physical phenomena happen.

The currently accepted theory of Gravity is The General Theory of Relativity. But for all practical purposes like satellite launch or space exploration (going to the moon), Newtonian mechanics hold.

In extreme situations like the nearest orbit of the Sun, the changes caused by General Relativity can no longer be ignored and thus it has significant effects on the orbit of Mercury.

However, at places where the curvature of space-time becomes infinite, like near a singularity or at the start of the Universe, General Relativity breaks down.

Newton proposed that Gravity is a force. This force acts between any two objects with mass. The "action at a distance" which his theory of Gravity necessitated, bothered him. He could not explain how an object can exert force without continuity. Although troubling, he accepted it as an assumption of the Theory.

Thus Newtonian Gravity says that Gravity acts instantaneously. This means that distance was not an issue. This means that an object can gravitationally interact with another object, a thousand light-years away without any limit. This also meant that the range of Gravity was infinite, that is, for example, the Sun can interact gravitationally with a star at extremely high distances close to infinity.

The Newtonian Gravity had to however obey inverse square law which meant that the longer is the distance, the weaker is the interaction. Thus after some distance, the forces related to Gravity would be expected to become negligible in strength.

Any change in location of an object, say the location of a Star will have an instantaneous change in its gravitational interaction with another object bound gravitationally to it. This meant that simultaneity is implied in Newtonian theory. This is in contrast to General Relativity and Special relativity which presume that the concept of time is variable at variable places which means that simultaneity doesn't exist for the two gravitationally interacting objects. (*Miller, Arthur – 74, Jammer, Max – 64*)

The attractive force of Gravity, in Newtonian dynamics, counterbalances the centrifugal force during circular I.e. elliptical motions of the planets. This centrifugal force happens due to Newtonian laws of motion which necessitate a body to continue its state of motion or rest in a straight line due to inertia. This "Inertia" is thus an integral part of Newtonian dynamics.

In General Relativity, the objects with mass are supposed to curve space-time (*Einstein, A -33,34,35, H. A. Lorentz – 66*). This curvature of space-time is just followed by the planets and thus there is no centrifugal force in GR. The inertia makes the Planets continue in their straight paths but the mass of the Sun curves their straight paths into elliptical orbits.

Here, Gravity is said to act, not instantaneously like Newtonian Gravity, but through a gravitational wave which is a wave of gravitational interaction which starts at the Sun and travels with the speed of light curving space-time. When this reaches the planets, it leads to the curvature of the paths and thus keeps them in orbit.

Gravity thus acts, not instantaneously, but honouring the "speed of light limit" set by GR. This has gross implications.

This means that the Gravitational influence of central Supermassive Black holes of the centre of a Galaxy start as Gravitational waves and travel at the speed of light, travelling for millions of years and covering millions of light-years before they exert their influence on the peripheral stars of the Galaxy thus keeping them in orbit.

Einsteinian Gravity is also based on the principle of Equivalence. In short, Einstein noted that there is no difference between a person under Gravitational influence and a person on an accelerating spaceship. This is almost akin to saying that when we are standing on the Earth, the Earth is accelerating upwards towards the sky like standing on an accelerating spaceship which is the cause of the feeling of Gravitational force which we feel.

Chapter 3

EINSTEIN'S GENERAL THEORY OF RELATIVITY, IS IT RIGHT?

General Relativity has made a lot of important predictions (including Gravitational lensing, Gravitational redshift) which have been tested and found to be true multiple times. Thus it cannot be said to be completely wrong. However, it has consistently come in conflict with Quantum Mechanics and its inability to go with Quantum Mechanics suggests that there is a lacuna in either of these theories.

The main premise on which this theory is based is called the Equivalence principle.

This is explained in brief as follows.

Consider the figure iv/3.1.

It shows a closed box or an elevator in various states in A, B, C and D.

The box has a person entrapped within who has an apple. The person can do experiments within the box but has no means to communicate outside the box.

Einstein argued that A and B are equivalent and so are C and D.

"A" shows a state wherein the box is freely floating in space without any gravity acting on it. In this, the box itself, the person within it and the apple all three are under no force and are thus freely floating. The person would thus see a freely floating apple.

In "B", the box is falling freely towards the Earth. In this, all the three structures in question, i.e. the box, the person and the apple are under the influence of the same force of gravity which is pointed downwards towards the centre of the Earth and are accelerating towards the Earth with an equal acceleration. Here, none of them can exert any force on each other and are thus going to feel weightlessness and like A, the person within the lift will see an apple (and himself) floating within the box. Since the box is completely closed and the person cannot see him selves accelerating down towards the Earth, there is no way he can make out if he is in state A or B.

In "C", the box is being accelerated in an upwards direction, that is its velocity is constantly increasing. Just before the instant when the rocket started, it figures that all three objects were floating freely with an equivalent velocity. When the rocket starts, the person and the apple continue their previous velocities while the box acquires the new accelerating velocity. Since there is no direct physical contact of the box with the person and the apple at this point, the rocket cannot exert any force on the person or the apple as yet. The y coordinates of the person and apple will not change much but that of the box will start increasing rapidly. The person would thus note that both himself and the apple rapidly move towards the floor of the box as if a gravitational force has been created. This force persists till acceleration persists (which needs a constant source of energy to burn the rocket). As soon as the rocket stops burning and the acceleration stops, that is change in velocity stops, all three would acquire the same velocity and would start floating together due to their inertia. That is, they will adopt state A.

One thing to note here is that the space in between the apple and the floor of the box would remain relatively unchanged in A. But when the box starts accelerating upwards and the apple stays in the same y coordinates, this space starts shrinking.

A hypothetical skyscraper of Planck Compartments in between the apple and the floor would keep losing floors. When the velocity of the box becomes a constant, all three acquire the state A and whenever the acceleration restarts, this difference in velocities manifests and the gravity-like downward force and the shrinking space manifests.

Einstein explained gravity in this manner in his theory. In Newtonian gravity, the apple is acted upon by a force of gravity that pulls the apple downwards. However, in Einsteinian gravity, like the box travelling upwards and coming close to the apple, the Earth travels upwards close to the apple.

The ridiculousness of this becomes apparent in D. When the same falling box reaches the Earth and is touching the ground, the Earth, as well as the box, travels upwards reaching the apple as per Einstein's gravity as Einstein's equivalence principle says that C and D are equivalent. That is like the floor of the box travelled upwards to reach the apple, the floor of the box and the Earth travelled upwards to reach the apple even in D.

Einstein argued that there is no way that the person within the box know that the box is accelerating towards the Earth in B or is freely floating in space as in A. And thus gravity in B could as well be like the illusion of force created as in C. This is the most controversial and the most difficult part of General Relativity to understand and thus a place where a majority of the lay people are either unable to understand relativity or object to what General Relativity says.

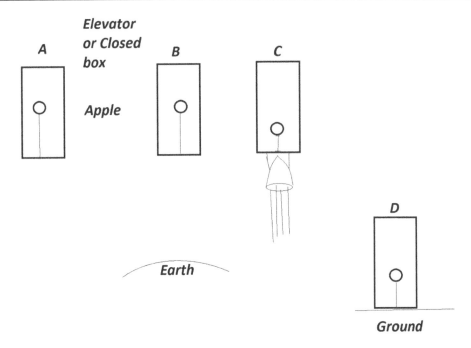

Figure iv/3.1 Shows Four closed boxes A, B, C and D at various states to explain the Equivalence principle. A is floating in space, B is freely falling towards the Earth, C is on a spaceship accelerating upwards with 9.8 m/sec2 and D is at rest on the Earth's surface.

To understand it further, let us consider a thought experiment of our own. Let's have four such boxes falling freely at four different corners of the Earth as shown in the figure from a height of h.

All these boxes have 4 different individuals who possess apples which they would release simultaneously.

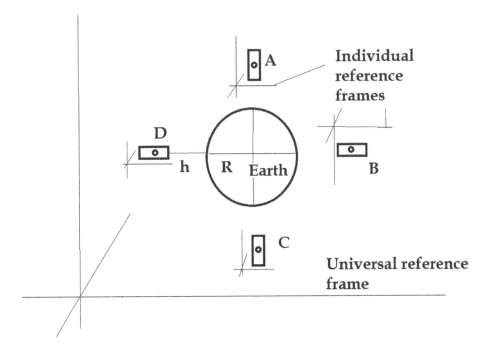

Figure iv/3.2: Shows a thought experiment to re-analyse the Equivalence principle.

In this, it would be clear that all of them can have their individual completely different frames of reference as shown. Although in the figure, it appears that they are at diametrically opposite ends, no such compulsion exists and that they can be anywhere on the surface of the Earth. It thus figures that their individual frames of references and thus their individual axes (X, Y and Z axes) are not parallel)

If all the four people release the apple at the same instant, what would happen? Logic says that the Earth with its frame of reference cannot go in all four directions simultaneously.

The above clearly shows that to know what happened at this instance, we need to consider a more universal frame of reference which encompasses all the references together. Logic says that the coordinates of the Earth in this universal frame of reference (shown separately) would remain unchanged and the coordinates of the boxes and their apples would keep changing so that all the four boxes with the apples within them would fall freely with an equal acceleration towards the Earth.

All the four boxes would be in state B, and in all of them, the apple and the person would remain freely floating without any apparent force acting on them (as viewed by the person in the closed box).

This universal frame of reference, although logical is inconsistent with General Relativity which says that there can be no fixed universal frame of reference as the space-time can bend themselves and thus the frames of reference would also bend. This is one of the primary places where General Relativity comes into conflict with Quantum Mechanics which assumes a flat space-time.

There are many other apparent paradoxes in General Relativity.

General Relativity points out that time itself is an illusion. This means there is no consistent past, present and future. This is because time itself is not absolute and is relative to different observers.

This means that what comes in past of one observer might come in "the present" of another and the same can be in future of the third. This is in contrast to the absolute time which others believed before Einstein. Because of this weird paradox, Relativity points to a universe with a Block time wherein everything that has happened and everything that has to happen is predetermined.

This is precisely the reason why General Relativity does not preclude time travel.

General Relativity vs Newtonian Gravity

There are significant differences between these two.

In Newtonian gravity, there exists a force that a body with mass exerts on another body with mass. This force acts at infinite distances and travels with infinite velocity. This means, for example, if the Sun moves in one direction, it instantaneously exerts a force on all the other planets to make necessary corrections in their orbits.

In Einstein's gravity, nothing can travel faster than light. Thus even gravitational influences travel at the speed of light. In Einsteinian gravity, gravity is just an illusionary force created due to the bending of space-time created due to the mass of the Sun (*Einstein, A -33,34,35, H. A. Lorentz – 66*).

This bend in space-time travels with the speed of light. So any change in this bend that may occur due to a movement of the Sun in a particular direction, would have to travel with the speed of light towards all the planets and the innermost planet will make the corrections in its orbits earlier than the farthermost planets.

Newtonian gravity does not mention any effect on time. Einsteinian gravity affects time, such that closer to the Earth, according to Einstein, time is dilated. This time dilation reduces as one travels away from an object with mass.

Chapter 4

BACKGROUND THINKING

General Relativity says that Gravity causes Gravitational time dilation (GTD). Beginning with this, we can start imagining multiple spheres with varying diameters around the spherical Earth. Each of these will have a variable GTD and thus in the sphere closest to the surface, the time dilation effect will be maximum.

If we consider a skyscraper, each floor will lie at a different time dilation zone. The height of each floor of this skyscraper is immaterial. Even if the skyscraper has each floor of the height 1 km or 1 m or 1 cm or 1 mm, there will be some difference in time dilation caused because of the mass of the Earth.

We know for sure that the "Newtonian apple" falls down. Thus the fact that the space in between the apple and the Earth is losing volume due to some yet unknown process cannot be denied.

General Relativity says that the fabric of space-time is curved and contracted nearer the surface and as we move away, the contraction caused by Gravity reduces in line with the reduction in Gravitational time dilation. It is often difficult to visualize how exactly does this change in the shape of space-time achieve this "falling apple" trait.

The statically contracted space in GR doesn't help in our imagination process. The analogy of a trampoline is good to visualize planetary motion but it fails miserably to help us imagine the shape of space-time while imagining a falling apple.

When I realized that at the most fundamental level, i.e. at Planck level, there exists Planck foam wherein spontaneous creation and destruction of virtual particles or tiny fluctuations of energy occur at every nook and corner of the Universe (*Feynman, Richard – 45, E. Fermi – 44*), it got me thinking.

What if gradually decreasing time dilation provided by GR as we move apart (which forms a Time Dilation Gradient) is intricately related to the active volume loss we see in between the apple and the surface of the Earth? What if, at the Planck scales, these actively created and destroyed virtual particles forming the Quantum foam are intricately related to the flow of time?

Two assumptions are mandatory here.
  1. The virtual particles have to be more real than what is presumed that is their appearance should add to the volume and their disappearance should lose volume.

2. Even if the rate of appearance and disappearance of these virtual particles vary with Gravitational time dilation, there will not be any change in the volume at each level unless the appearance and disappearance are dissociated. This means that it is vital to have two types of time, a constant universally progressive time and a variable time. The extent of time dilation at any zone would then be compared to this universally progressive time.

Let's now focus on our skyscraper. Let's presume that each storey of this skyscraper is 1mm thick.

Let's look at two adjacent floors x and x-1.

Both will have a different Gravitational time dilation provided by GR and thus time would run differently in both of them.

But the universally progressive time would still run with the same speed on both these floors and only the variable time will run differently.

If we presume that the process of appearance of new virtual particles in the Quantum foam in both these floors is linked to the variable time and the process of disappearance of new particles is linked to the universally progressive time, this can give us what we need.

Because the rate of disappearance is linked to a universally constant entity, it will be constant in both these floors. Because the rate of appearance is linked to a variable entity, the rate of appearance of new virtual particles will be variable in both these floors.

Not only is the time variable, but also dilated in both these floors. This means that a second in the universally progressive time is smaller than a second on the x floor and a second in the x-1 floor is even longer.

Let's presume 100 virtual particles appear in one second and disappear in one second.

But the appearance is linked to the variable time and disappearance is linked to the universally progressive time.

This means that from every corner of the Universe, 100 virtual particles will disappear. The appearance of new particles however will be variable depending on the variable time.

In our x floor, at the end of one universally progressive time "second", the variable "second" is not yet over and thus at the end of universally progressive time second the number of newly appearing virtual particles would be much lower than 100. In our x-1 floor, the variable time is even more dilated and thus the disappearance is constant 100 virtual particles but the appearance is even smaller

In short, there will be much more disappearance of the virtual particles in x and x-1 floors than appearance. This indicates that the floors will keep losing volume and contract.

Extrapolating this in each of the floors starting from the ground floor up until the top floor of our hypothetical skyscraper beyond which the Gravitational time dilation ceases to exist i.e. becomes insignificant or goes below 1 Planck time, we would realize that each of these floors would keep losing volume with every Planck time, more volume loss occurring in floors with more gravitational time dilation i.e. the floors closer to the Earth's surface.

Thus with these basic assumptions, the active volume loss beneath "the Newtonian apple" can be explained and Gravity can be explained.

In this, Gravity literally emerges out of multiple Quantum phenomena distributed throughout the skyscraper determined by the Gravitational time dilation i.e. Gravity is an emergent phenomenon.

The universally progressive time can be labelled Universal Time. The variably progressive time, on the contrary, can be called the Local Time as it is the property of the location and is dependent on the location.

As I studied the basic assumptions of Loop Quantum Gravity (LQG) and came across the terms Planck length, Planck volume and Planck time, I had almost everything that I needed to define this new theory of Gravity.

Some assumptions from LQG

(Rovelli, Carlo – 84)

Planck units are units derived by using all the fundamental constants of nature like h, c and G.

One Planck length is equal to 1.616255×10^{-35} meters. This is an extremely small unit of measure of length and is considered to be the smallest possible length of space that can exist.

It is presumed that smaller than this length cannot be measured. If one attempts to measure anything smaller than this length, we will need a light wave (EM wave) having a wavelength smaller than this. This kind of light wave will have such a high frequency that the concentration of energy at that place will be so much as to form a Blackhole. Thus this can be considered the smallest measurable length possible. A square with a side with 1 Planck length has a surface area of $10^{-70} m^2$ which is the Planck surface area. A cube with each side as 1 Planck length has a volume of $10^{-105} m^3$ which is called the Planck volume.

LQG thus assumes that the most fundamental unit of space-time that can exist is a Planck Compartment which is a cube (or possibly sphere) with the length of one Planck length and a volume of one Planck volume.

Another important measure in our theory would be the Planck time. One Planck time is 10^{-43} seconds. This is an extremely small measure of time and is almost incomprehensible to Humans. Thus, most of the events happening at these scales would remain in the realms of theory.

Most of the events that would happen in our theory would happen at these time and length scales and thus these can be considered as the most fundamental scales or "Scales of Nature".

Light travels with the speed of 1 Planck length per 1 Planck time.

Other Planck units include Planck temperature (1.41×10^{32} K), Planck energy (1.96×10^9 J) and Planck mass (2.18×10^{-8} kg). Presently these do not have any importance in our theory. However, as the theory advances, these might become important at a later date.

CHAPTER 5

BASIC ASSUMPTIONS

> Note that these are given in an unedited manner to understand how the theory looked when it started. However, some of the assumptions are wrong or unnecessary and in the current version of the theory, some of them are given up while others significantly modified.

The theory intends to utilize certain known characteristics already described in Quantum Mechanics, Loop Quantum Gravity and String Theory and with a single additional assumption of - decoupling of "disappearance of Planck Compartments (PCs)" in the Quantum foam with Local Time (which is coupled to the Universal Time instead of the Local Time) while the coupling of the "appearance of PCs" in Quantum foam thus creating a link between space and time and creating a discrepancy in the rate of appearance and disappearance of PCs with changes in Local Time like time dilation and time contraction.

1. Space-time is made up of quanta called Planck compartments (PCs) which are 1 Planck volume in size and for practical purpose, can be considered cubical although the reality is that they are not.

2. <u>Every single Planck Compartment has an individual measure of time. Throughout the volume of the Planck's compartment, time flows at the same rate.</u>[1*]

> 1* - In later part of the Theory development, it was realized that time changes are in fact secondary, and spatial changes happening at a point are the primary events. Thus Creation or Destruction of PCs lead to active changes in volume and the time changes that happen are secondary to these active spatial changes.

3. Planck length is the minimum possible length of space without disruption of Euclidian geometry. (note that this is not an entirely accurate statement as explained separately later, i.e. Planck length is not the minimum possible length and smaller lengths are possible). In line with that, 1 Planck surface area is the minimum possible surface area and 1 Planck volume is the minimum possible volume in space. The time comes in a quantum of 1 Planck time i.e. 10^{-43} sec. Time thus flows in ticks which are 10^{-43} sec long and is not smooth and continuously advancing (even this is not entirely true and sub-

Planck-time time-dilations are possible as in the description of MOND) (note that all these are in accordance with Loop Quantum Gravity)

4. Each of these compartments is non-expandable or non-compressible so that expansion or contraction of one PC is not possible. If space expands, it has to expand by adding additional Planck compartments. Space in space-time cannot expand at fractions of PCs

5. Quantum foam made up of these compartments is not static, and in fact, there is a constant fluctuation of space-time so that at places, these compartments keep appearing and disappearing, these quantum fluctuations happening at Planck times.

6. The appearance of new PCs is either synthesis of a new PC (i.e. not teleported from anywhere in our universe) or Quantum Teleportation of the PC from some other area of the Universe.

7. The disappearance of an existing PC is either destruction (i.e. not teleported to anywhere in our universe) or Quantum Teleportation of the PC in question from its original position being observed by the observer to another location in space.

8. Note that the latter is more likely in both 6 and 7. That is the appearance of one PC is more likely to be associated with the disappearance of a PC at a different location in the universe. Thus PCs can be quantum teleported in the immediate neighbourhood or at great distances from their original location.[2*]

> 2* - As the theory progressed, it was realized that this unnecessary baggage of assumptions of teleportation of PCs from one place of the Universe to the other, just to avoid the word spontaneous Creation from nothingness or spontaneous Destruction which might suggest the presence of a Creator, was neither provable nor worthy of continuation and was later given up.

9. There is a time-dependent discrepancy between the appearance and disappearance of these PCs. The appearance of new PCs in this quantum foam is a time-dependent process. That is, the rate of appearance of new PCs is linked to Local Time and thus processes that dilate Local Time or contract Local Time lead to changes in the rate of appearance of PCs.

10. The disappearance of the PCs however is a time-independent process and occurs throughout the universe at a constant rate. That is, there is a Universal Time, to which the disappearance of PCs is linked.

11. The Special Theory of Relativity holds for individual PCs so that each compartment can have an independent measure of time and just adjacent compartments can have a significantly different pace of flow of time. [3*]

> 3* - Later found to be the wrong line of thinking, as discussed earlier.

12. If the difference between the pace of time between two different PCs is less than 1 Planck time, it is assumed that the difference in the flow of time between them is insignificant and that both the PCs would have the same pace of flow of time.

13. Space and time are thus intricately related. Any zone where time is contracted (i.e. runs faster than the observer's measure of time), would have more appearance of new PCs but the same constant rate of disappearance of PCs, which means that the number of PCs in this zone would be more, i.e. space here will expand. If time runs slower in this zone compared to the observer's time, for a unit of the observer's time, the number of appeared PCs would be lesser and disappearance would again be independent of the flow of time. In short, any place where time runs faster dilates. In short, wherever time runs slower space contracts, while wherever time runs faster space expands.

14. Neighbouring PCs are entangled with each other. The pace of flow of time in one PC has the potential to influence the flow of time in its neighbouring PC as well (as explained in detail in communication between PCs)

15. A particle and it's Mass:

    <u>A particle is nothing but energy trapped in an extremely small space by extreme curvature of space-time akin to a sub-Planck length prison.</u>[4*]

    This curvature of space-time around every particle with mass leaves a distinct pattern of spatial contraction/expansion and with it, a distinct pattern of time gradient around the particle, with some Planck compartments around the particle having time dilated and in others in a contracted state. This curvature of space-time is integral to the structure of every particle with mass and it travels with it as the particle moves. This curvature of space-time is akin to creases that appear in a bedsheet when a part of it is squeezed between fingers.

    > 4* - This concept of sub-Planck length prisons for T=0 Planck compartments/ energy, initially introduced in the assumptions from String Theory was later found to be of no particular use and was removed.
    >
    > However, the time changes and secondary curvature of space happening around a charged particle do appear to be similar to this concept.

    <u>This zone of space-time creases around a particle with mass can be called the zone of interaction (ZOI). This is the zone where the information about time changes that happen around the particle arrive much faster than the speed of light, i.e. almost simultaneously.</u>[5*]

> 5* - This concept of Zone of Interaction was not found particularly useful and was given up to be replaced by r_{cric} which is the critical radius around a particle where space around the particle contracts faster than speed of light akin to an event horizon around a Black Hole.
>
> Note that with the Time Resetting Waves for every particle caused due to entanglement travelling millions or billions of PCs, these ZOIs would be extremely large and impractical for every particle, let alone for the Earth or Sun which are individually made of large number of such particles.

The routine practice of identifying a particle just by the central portion where the density is located may have to be changed and a particle should be identified not only by the zone where the energy is trapped but with the space-time creases which it creates

These creases could be symmetrical or asymmetrical, with the extent of time gradient more in one direction than in the other.

The trapping of energy in the form of strings could be due to the extreme curling of space-time into multiple spatial dimensions which are smaller than Planck length as advocated by string theory.

In the figure iv/5.1, a central zone of extreme curling of space-time akin to what is described in String theory is shown with the accompanying space-time distortion or creases. The cubical compartments shown here would probably have time running at differing speeds.

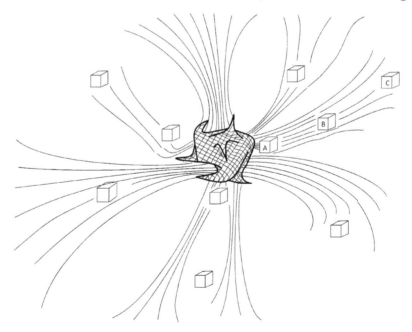

Figure iv/5.1 Space time creases produced by a particle with mass. Just a diagrammatic representation. Note that all the cubes, especially A, B and C would have Local time running differently.

16. Charge

    A charged particle is a particle with mass with asymmetrical space-time curvature. Such a particle will have an asymmetrical zone of interaction wherein there exists a direction around the particle where individual Planck compartments have time running either slower or faster than the rest. Alternatively, a charge may mean an overly exaggerated time dilation or time contraction (much more than a particle without any charge)

17. Zone of interaction

    Every particle with mass would have a zone in which there exist space-time creases i.e. contraction or expansion, the exact size shape and extent of which is determined by the geometry of curvature of the central space-time prison where energy is trapped. The time gradient present in the zone of interaction determines the physical and chemical properties of a particle to a large extent. The zone of Interaction flows like a wave around a moving particle which could potentially explain the wave-particle duality and findings of the double-slit experiment.

18. Fields

    There are no fields. All the fields, according to this theory can be successfully explained by space-time curvatures and time gradients. In other words, all fields can be represented by space-time variations.

19. Force carrying particles or Bosons

    All the naturally occurring forces including the Electromagnetic, Strong Nuclear, Weak Nuclear and Gravity can be explained based on time gradients. The force-carrying particles described in the Standard Model of particle physics would boil down to communications between neighbouring PCs which takes place with the speed of light.

    Force carrying particles could be explained as aggregates of PCs with varying time gradients.

    Out of these, the structure of photons can be deciphered more readily than others.

20. Energy

    Energy would be the entity that induces a difference in the pace of time in a particular PC. An alternative view would be any deviation in Local Time from Universal Time constitutes energy.

    The most concentrated form of energy would be a PC that has T=0.

21. Transmission of energy

    Energy transmission from one point to another happens due to communication between adjacent PCs which is described in detail separately. Most often, the transmission of energy happens in the form of a wave and obeys the inverse square law. However, it can also be transmitted as quanta. The transmission of energy takes place at the speed of 1 Planck distance at 1 Planck time, i.e. at the speed of light. The propagation of photons or electromagnetic waves is the best example and would be dealt with in detail subsequently.

22. Entropy

    Any change in time within one Planck Compartment affects the adjacent Planck compartments as well. As explained in detail later, when time in a PC becomes zero, it would induce variations in time in the surrounding region which would travel with a speed of light and propagate. This would continue until all the adjacent compartments have time flowing at the same rate. this constant flow of energy from a region with higher energy to lower energy indicates that the disorder or entropy of the universe is always towards the increasing side.

23. Temperature

    The ratio of PCs where time has stopped flowing i.e. T=0 to the number of PCs with flowing non-zero time. Alternatively, absolute zero temperature is a state where there is not a single PC with T=0 and not a single PC with a time deviated from Universal Time in a region.

    Planck's temperature on the other hand probably is the maximum possible temperature. If all the PCs (i.e. 100% of the PCs) in a given volume are T=0. This is almost impossible to obtain and maintain.

    A body with non-zero temperature has a proportion of PCs where T=0. These PCs disappear instantly with 1 Planck time. However, they leave their effects in place in the form of T=0 in adjacent PCs or time gradient in adjacent PCs or both (explained later in detail). The T=0 compartments propagating forwards to adjacent PC with every Planck time, with the associated time gradient is nothing but propagating EM radiation. In short, a body with a non-zero temperature would keep emitting EM radiation.

24. Virtual particles

    All virtual particles are a collection of PCs with varying time running within each of them with varying time gradients, two identical virtual particles having an identical arrangement of PCs and identical pace of time running in each of its constituent PCs.

25. Bonds

    Nuclear bonds between various particles within the nucleus or the atom and for that matter chemical bonds between two atoms are formed due to overlapping of zones of interaction around these particles which leads to maximum stability. Any attempt to remove the bonded particle would lead to dynamic contraction and expansion of space caused by time changes in PCs around the particle preventing the breakage of these bonds. These bonds can be broken by preventing these time changes from happening or cancelling their effects, by addition of T=0 compartments (in short supplying more energy)

26. Electromagnetic waves

    EM waves are T=0 compartments travelling around with the speed of light. These compartments would carry along with them, a series of PCs with a time gradient either with time in a progressively dilated state compared to Universal Time or in a progressively contracted state. The two usually follow each other in an EM wave. The intensity of an EM wave is the density of T=0 Compartments/photons.

27. Photons

    These are nothing but aggregates of T=0 Planck Compartments.

CHAPTER 6

HOW TIME GRADIENT LEADS TO DILATION OF SPACE?

Let us consider a cube with a side of 1 meter. This cube would have $10^{35} \times 10^{35} \times 10^{35}$ I.e. 10^{105} Planck compartments.

Let's imagine that this hypothetical cube is at an infinite distance from any gravitational body and thus it has no gravitational time dilation.

We know that with every 1 Planck time, some of these Planck compartments would disappear and some would reappear constituting the Quantum foam.

Let x be the number of Planck compartments appearing in every Planck time in this cube and Y be the number of Planck compartments disappearing with every Planck time.

According to our theory, x (small x) should be variable depending on the pace of time and Y should be a universally constant entity at all places in the Universe i.e. x is coupled to Local Time but Y would be coupled to the Universal Time.

At the first Planck time, the number of PCs would become $10^{105} + x_1 - Y$.

At the second Planck time, the number of PCs would become $10^{105} + (x_1+x_2) - 2Y$....

At nth Planck time, the number of PCs would become $10^{105} + (x_1+x_2+x_3 ..... +x_n) - nY$

If we assume that at each of these Planck times, the value of x is >Y then the entity $(x_1+x_2+x_3 ..... +x_n)$ would be greater than nY and the entity $[(x_1+x_2+x_3 ..... +x_n)-nY]$ will be positive. This means that as time passes, the number of PCs in our cube would keep growing which means the space keeps expanding. This probably happens at the space in between galaxies where there is no Gravitational time dilation. This can explain the expansion of the Universe.

If on the other hand, the value of Y is greater than x, then the entity $(x_1+x_2+x_3 ..... +x_n)$ would be smaller than nY and the entity $[(x_1+x_2+x_3 ..... +x_n)-nY]$ will be negative and the number of PCs in our cube keep reducing. This happens when the time is dilated due to gravitational time dilation near a massive object like the Sun or the Earth. The time dilation reduces x but has no effect on Y.

Now consider a time gradient as shown.

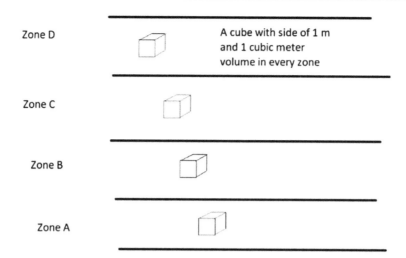

Figure iv/6.1: Shows different zones namely A, B, C and D with variable Local time.

Zone A is closer to land and thus has more time dilation than higher time zones.

Let the observers time correspond to the time zone D.

This means that a distant observer has time running at the same pace as zone D.

In short distant observers 1 Planck time (Universal Time) = 1 Planck time in zone D.

However, time is dilated in zone A. As we go towards higher time zones, time dilation reduces.

Let

1 Planck time at time zone C = 2 Planck times of observer

1 Planck time at time zone B = 3 Planck times of observer

1 Planck time at time zone A = 4 Planck times of observer

And so on.

This has important implications.

At 1 Planck time of the observer, due to time running slower in zone C, B and A, the appearance on the new PC will be 1/2, 1/3 and 1/4 times as much.

Now consider a cube with one cubic meter and thus 10^{105} Planck compartments in each of these zones.

At the end of observers 1 Planck time, each of these 4 cubes would have 1(for D), 1/2 (for C), 1/3 (for B) and 1/4 (for A) Planck times elapsed.

Let X (capital X) be the number of PCs that disappear and re-appear in one Planck time in a zone where Time = Universal time, so as to keep the Volume unchanged.

The number of PCs appearing is coupled to Local Time.

At 1 Planck time of the Observer, each of these cubes will lose X PCs.

At the end of 1 Planck time of observer i.e. at 1 Universal Planck time,

- ❖ the cube in zone A would have a new appearance of X/4 PCs
- ❖ the cube in zone B would have a new appearance of X/3 PCs
- ❖ the cube in zone C would have a new appearance of X/2 PCs
- ❖ the cube in zone D would have a new appearance of X PCs

In short, if assumptions of our theory are right, the cube with 1 metre cube volume in each of the above four zones, at the end of 1 observer's Planck time, would have different volumes as they will lose an equal number of PCs but will regain a progressively smaller number of PCs depending upon the extent of Local time dilation within them.

At the second observer, Planck time and every subsequent tick of the observer's clock, further loss of PCs occurs at zones which are time dilated compared to those which have time running faster (i.e. which have a lesser time dilation).

Essentially, the 1 cubic meter volume cube contracts from all sides as shown in the figure iv/6.2, much more in zone A and almost none in zone D. The void created around it is instantly filled by the neighbouring compartments taking their place. In this manner, time dilation leads to space contraction.

As shown in the figure iv/6.2, the neighbouring compartments shift to take the place of the contracted cube, thus shifting some of the PCs from one Time-zone to another.

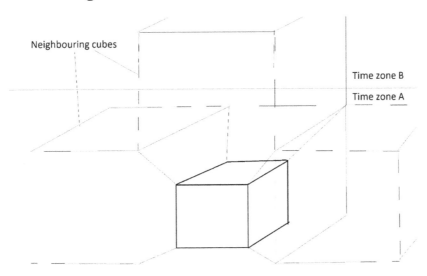

Figure iv/6.2: Shows active spatial contraction of a cubical region of space and the resultant shifting of neighbouring cubes.

The mechanism of gravitational time dilation (which is mostly the mass of the planet) remains the same and so does the location of time zones.

With every tick of the Planck clock of an observer, this cosmic game of shifting of blocks in the form of Planck compartments continues incessantly with a part of the shifting of compartments compensated by new appearing Planck compartments at the higher time zones.

In this manner, there is literally an **eternal "Shift the block puzzle game"** going on. In other words, whenever there is an object possessing enough gravity, there is a constant flow of Planck compartments from the periphery of the zone of time dilation towards the object like

an "eternally moving cosmic escalator".

This is unlike Einstein's view, in which space contracts and remains static, in DGR, this contraction of space with gravitational time dilation is a dynamic process happening with every tick of the Planck clock.

Note that the pace of time can vary from the observer's time by a factor of Planck time, i.e. if a zone has time flowing at a pace slower than 1 Planck time, it is insignificant.

Like in the above example where every subsequent zone had time in a dilated manner, there can be a system where, with respect to the observer, time is contracted compared to the observer's time, wherein

2 Local zonal Planck time = 1 Observer's Planck time

3 Local zonal Planck time = 1 Observer's Planck time

4 Local zonal Planck time = 1 Observer's Planck time

5 Local zonal Planck time = 1 Observer's Planck time

Etc.

We can denote this as PT i.e. Pace of time compared to observer's time and can be shown as in the following table

Pace of time in the zone	Comparison to Observers time	State of the time
PT_5	1 Local Time here = 5 Planck times of the observer	Contracted
PT_4	1 Local Time here = 4 Planck times of the observer	Contracted
PT_3	1 Local Time here = 3 Planck times of the observer	Contracted
PT_2	1 Local Time here = 2 Planck times of the observer	Contracted
PT_1	1 Local Time here = 1 Planck times of the observer	Running with equivalent pace
PT_{-2}	1 Planck time of the observer time = 2 Planck times locally	Dilated

Pace of time in the zone	Comparison to Observers time	State of the time
PT_{-3}	1 Planck time of the observer time = 3 Planck times locally	Dilated
PT_{-4}	1 Planck time of the observer time = 4 Planck times locally	Dilated
PT_{-5}	1 Planck time of the observer time = 5 Planck times locally	Dilated

Alternatively, one can compare the pace of time with Y, i.e. Time of Disappearance which is universal and the same everywhere according to our assumption.

Comparing with the universal Planck times, Local Times could be re-written as follows

Pace of time in the zone	Comparison with Universal Time	Status of time	Status of space
PT_{U5}	1 Universal Planck time = 5 Local Planck times	Contracted	Actively Expanding
PT_{U4}	1 Universal Planck time = 4 Local Planck times	Contracted	Actively Expanding
PT_{U3}	1 Universal Planck time = 3 Local Planck times	Contracted	Actively Expanding
PT_{U2}	1 Universal Planck time = 2 Local Planck times	Contracted	Actively Expanding
PT_{U1}	1 Universal Planck time = 1 Local Planck times	Equal to Universal Time	Neither Actively Expanding nor Actively Contracting
PT_{U-2}	2 Universal Planck time = 1 Local Planck times	Dilated	Actively Contracting
PT_{U-3}	3 Universal Planck time = 1 Local Planck times	Dilated	Actively Contracting
PT_{U-4}	5 Universal Planck time = 1 Local Planck times	Dilated	Actively Contracting
PT_{U-5}	5 Universal Planck time = 1 Local Planck times	Dilated	Actively Contracting

How time gradient can lead to application of force or movement of a particle

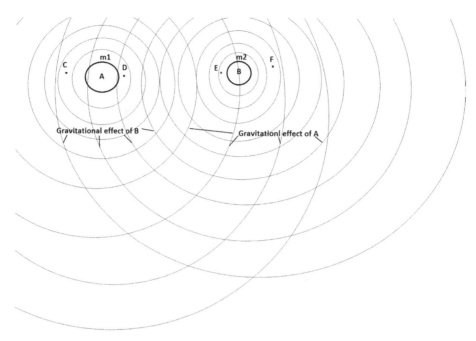

Figure iv/6.3: Shows two Gravitationally active bodies A and B, with masses m1 and m2, under each other's Gravitational influence. Each of them modifies time in their surroundings in such a manner as to create concentric spheres of Gravitational time dilation gradient.

The figure iv/6.3 shows two particles/celestial bodies A and B, with masses m1 and m2, suspended in space that comes in proximity to each other and their gravitational time dilation zones overlap. The concentric circles (in reality spheres) in the figure represent the time zones with the region immediately close to the bodies having maximum time dilation so that at points D near A and at point E near B, time dilation is maximum. In addition, in the region in between the bodies, and especially at D and E, the time dilation caused by the two gets added up so that in the entire figure, maximum time dilation would exist at D and A. As noted earlier, wherever time dilation is present, space contracts. This means that the space in between the two bodies would contract and they would start moving towards each other.

With every movement towards each other, they would end up deeper into the time dilation zones of each other and would enter into a cascade of time dilation leading to space contraction leading to further time dilation. This continues until they come close to each other and there is no space left in between, i.e. they collide.

Figure iv/6.4 shows a Particle (say a meteorite) suspended at a height above the Earth's surface. Also shown is the time zones due to Gravitational time dilation caused by the Earth.

A hypothetical skyscraper column of cubical compartments is shown above and below the particle, each cube being 1 cubic meter volume as discussed above. Some of these compartments lie in different time zones. As discussed above, the compartments lying in the lowermost zone have maximally dilated time and time dilation reduces as we consider subsequent concentric outer zones.

As our Planck Clock ticks, at the end of the first universal Planck time, i.e. disappearance Planck time, due to time dilation, the compartments in the innermost zones would have more disappearance of PCs than appearance. This causes space here to contract. At the same 1 universal Planck time, at higher compartments, there is comparatively faster local time, and more appearance of PCs compared to disappearances leading to a lesser overall contraction.

However, throughout the skyscraper, the compartments lose volume and pull the particle towards the Earth.

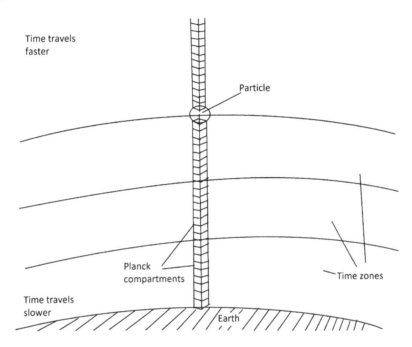

Figure iv/6.4: Shows the string of Gravity. Note the Skyscraper with cubical compartments stacked on top of each other. Each compartment will have a varying Gravitational time dilation and thus will have a varying appearance of PCs in Quantum foam. However, the disappearance of PCs will remain constant, being coupled to the Universal time. This leads to active loss of volume at each of these compartments/stories of the skyscraper puling the particle P downwards as a string.

Two things are absolutely needed for any kind of movement to occur due to a time dilation gradient like this. They are

1. Significant time gradient (i.e. pace of time differing by more than 1 Planck time)
2. The particle with its own field of dilated or expanded time in its zone of Interaction directly interacting with the time gradient.

It is worth noting that in the above case, the particle has a gravitational time dilation around it. This is why the particle kept on moving towards the zone of time dilation away from the zone of contraction.

On the contrary, if the same particle due to its internal properties, had concentric zones with time contracted with respect to Universal Time, so that time contraction (lesser time dilation) in higher zones gets added to the time contraction surrounding it, the contracted time would cause the particle to move towards the upper time zones with relatively contracted or less time dilated zones, i.e. against the Gravity, which is counterintuitive.

Acceleration due to gravity

Note that if we consider 2 particles with different masses m1 and m2 with negligible masses (and thus gravitational pulls) compared to the Earth, stationed at the same height, the skyscraper beneath both of them would be identical. This means that the two masses would be pulled towards the Earth by identical time zones of the Earth and would accelerate towards the Earth equally fast, i.e. their velocities would change identically irrespective of the masses.

This however is only true if their gravitational field is insignificant and cannot significantly add its own time dilation. For objects with a mass comparable to the mass of Earth, the resultant acceleration due to gravity might be slightly higher.

Even then, their gravitational pulls would indeed determine the forces acting on them which are dependent on their masses as well.

Chapter 7

THE POSITIVE FEED-BACK LOOPS OR ROLLER COASTER

Any time gradient leads to formation of a Positive feedback loop which leads to movement of the particle.[1#]

> 1# - There is a minor mistake in this. The gradient isn't absolutely necessary and just time dilation is enough for spatial contraction. However, development of a gradient is almost inevitable. What gave rise to the time gradient was not known to me at this point. Now we know that the EPCAs getting sucked in and while moving inward developing time dilation within to compensate for the destruction at the centre leads to a time gradient. Thus the spatial changes are primary and development of time gradient is secondary.

If the particle is kept in a time gradient with its zone of interaction which has time dilation in it, interacting with the gradient, the time dilation gets added up all around it. In the direction of the time dilation of the gradient, the space contracts and pushes the particle towards the direction of the time-dilated zone. This pushes the particle further into higher time dilation. The time dilation adds up again and this positive feedback loop keeps pushing the particle towards the time-dilated zone.

Chapter 8

DYNAMICITY OF THE APPARENTLY STATIC SPACE TIME CHANGES IN GENERAL RELATIVITY AND HOW IT IS DIFFERENT THAN CURRENT THEORY

Imagine a point Q in space, on the Earth's orbit which is being approached by the Earth revolving around the Sun.

General Relativity says that due to the Earth's gravity, point Q will have space in a contracted state while time will be dilated. The spatial contraction and time dilation will keep on increasing in extent as the Earth approaches closer to Q.

Thus, if we focus only on point Q, with time, due to Earth's motion, space keeps contracting more and time keeps dilating more. The space-time of relativity isn't as static as it appears superficially, but the change is directly related to the motion of the Earth (or for that matter any other heavenly body in question).

We know that the Earth revolves around the Sun with a velocity V that is nothing compared to the velocity of light.

Let's also consider another point P on the surface of the Earth that touches the Earth's orbit. As the Earth moves, there will come an instant wherein P will coincide with Q.

For the sake of discussion, let's ignore the Earth's rotation for the moment.

Also consider multiple points A, B, C, D, E, F etc. that are at one Planck length distance from one another and all of them are lying on the Earth's orbit and are potential points that will coincide with P at some point in time in the past, present or future.

What we are interested in is the changes that take place in space-time as per General Relativity, at the point Q due to the Earth's motion towards it.

Let's imagine that at this moment, the Earth is such that P coincides with B while at some point in the past, P coincided with A.

At this moment in the past when P coincided with A, a gravitational interaction in the form of the gravitational wave started and travelled with the speed of light to change the status of space-time at Q.

We know that since this gravitational interaction travels with the speed of light, it would travel one Planck length every 1 Planck time.

The Earth will take a comparatively large number of Planck times, to move by a distance of 1 Planck length, since its velocity is negligible compared to the velocity of light. Let the Earth take X Planck times to move 1 Planck length.

The number X is bound to be large.

(Earth's velocity of revolution is 30 km/sec compared to the speed of light which is 300,000 km/sec. In one Planck time, the Earth would move just 1/30,000th of Planck length or it would take 30,000 Planck times to complete one Planck length.)

Let's assume that the distance between A and Q is small and thus it takes Y Planck times (wherein Y is negligible I.e. much lesser than X) for the gravitational wave interaction to reach Q from A. Once the space-time change occurs, the next space-time curvature change is destined to occur when the Earth moves further and the gravitational interaction that starts from it reaches point Q again.

Now P coincides with B. This will remain for a time of X Planck times. After X, the Earth will move again and P will then coincide with C. Again a gravitational interaction starts and changes the space-time curvature at Q after a time of Y Planck times.

Subsequently, P coincides with D and space-time curvature changes after X+Y Planck times.

It is clear that, during this entire period of X+Y Planck times, no change in curvature happens as per General Relativity.

So according to General Relativity, the next change in the extent of space-time curvature would occur when the Earth moves again and for the entire period of X Planck times, there is no change in space-time curvature at Q. That is space-time curvatures are relatively static in General Relativity.

This is in contrast to our theory.

In the theory of dynamic General Relativity, the space-time curvature changes with every Planck time, how much it would change depends on the distance.

Chapter 9

WHY PLANCK LENGTHS ARE NOT MINIMUM POSSIBLE LENGTHS AND PLANCK TIMES ARE NOT MINIMUM POSSIBLE TIMES?

Let's do a thought experiment.

Consider a particle Q and a photon P. Q is moving in a straight line with half the speed of light. The photon P is also moving in the same direction besides the particle, again in the straight line.

We know that the photon would move with the speed of light.

At the first tick of the Planck clock, at the end of 1 Planck time, the photon has travelled one Planck length. At this same time, the particle is also moving. However, it has moved just half that distance. This means that at Planck scales of time, almost everything that moves would experience distances less than Planck lengths and thus Planck lengths cannot be the smallest possible measure of length.

Note that for the photon itself, there is no measure of time and within it T=0. However, we know that an outside observer can experience a photon experiencing time. A photon from a star 1 light year away from Earth (for example) would take 1 year to reach us after being emitted.

We know that pace of T=0 is possible in a Planck compartment. Thus Planck time, although an important landmark for understanding, is not the smallest measure of time.

Now consider a zone where a time gradient exists. Let's consider 2 zones wherein when the time of 1 Planck time passes in the first one, 4 Planck times pass in the second zones. Clearly, at the end of the first Planck time of the second zone, the first zone had experienced just 1/4th of the Planck time, and thus Planck time could not be the smallest possible measure of time.

However, for the observer entirely within the first compartment, there is probably no means of experiencing time less than 1 Planck time.

Chapter 10

TIME

A proper definition of "Time" is still an enigma.

Defining the arrow of time and the exact reason for it is one of the unsolved mysteries of Physics.

When we see "time", we remember the subjective experience of time, the relentless march of the second hand which keeps pulling the minute and the hour hand forward without any halt. The clock then reminds us of the calendar which says to us the inescapable looking fact that there was a yesterday, there is a today and there will come a tomorrow. Even today, what we have is just the present moment and there are more moments in the past and future than the present at any given instance. With such an experience happening to us every moment of our life, the subjective belief of humans would undoubtedly be that there is only one arrow of time, which points forward. This however could just be an illusion.

Why the arrow of time is pointed forwards and not backwards? Why the past is in the past and the future in the future? is not known. Although this seems ridiculous, it is a reality in physics.

One of the explanations to define the arrow of time is "by the second law of thermodynamics" wherein the increasing entropy of the universe is regarded as the arrow of time. This definition of the arrow of time is evidently flawed simply because there are many processes (especially biological processes within a living cell) where entropy is reduced by utilizing energy without any external help. In these systems, time is progressing forwards, without doubt, however, entropy is apparently reducing. One has to remember that this apparent reduction of entropy is only apparent as we cannot have a system with 100% efficiency and thus while the creation or utilization of the energy sources used here, some energy gets wasted or is lost to the system. This means that if one includes all the processes involved, the entropy of the system still increases despite the illusion of a decreased entropy.

CPT Symmetry

All laws of physics display a property which is called time-invariance.

This Time-Invariance is accompanied by other symmetries namely charge conjugation invariance and Parity invariance.

The real meaning of these statements is difficult to fathom for many, but it is simple. The laws of physics are independent of time, coordinates and parity. This means that if I measure the gravitational potential energy in an apple 5 feet high on a tree today, tomorrow or yesterday it would give the same result irrespective of time. If I apply a voltage of V across a circuit with resistance R, the same current C will be generated across it, irrespective of the time of the day or day or year. The gravitational potential energy of the above apple will remain the same irrespective of whether we measure the height of the tree from the ground or the ocean floor (i.e. we introduce a minor change in the coordinates). The above circuit will generate the same current irrespective of its exact position. This means that the left end of the circuit can have any X coordinate, which means that we can move the left end of the circuit by any amount along the x-axis without any change in the amount of current. The parity invariance means that the laws of physics are invariant to the right or left change and would look the same in a mirror-inverted universe. It means that if we switch the positive terminal from right to left in the above circuit, the direction of current would be also reversed but the result would still be valid.

These three are called CPT-symmetries and they are displayed by the three fundamental forces of nature i.e. Gravitation, Electromagnetic force and Strong nuclear force. The Weak nuclear force is an exception as it does not show parity symmetry and behaves differently with left-handed electrons than right-handed electrons (for example).

However, all of the forces display Time-invariance. This means that they behave the same irrespective of the arrow of time and that they will behave identically even if the arrow of time was running backwards.

This is puzzling for many, simply because the laws of physics do not help us in defining the arrow of time or answering the question "why time relentlessly moves forward?" or "Does it?".

Before Einstein, time was considered absolute. This meant that when a clock ticks one second at one point in space, all the other clocks located at any other point in space irrespective of the distance between the two points should tick the same one second. This was indeed proven to be true almost at all the points on the surface of the Earth. Einstein, however, left no doubt of the reality that a clock ticking a second at the surface of the Earth would tick slower than a clock far above on a satellite. This is called gravitational time dilation. A stationary Clock would click faster than a clock that is moving at great speeds.

In short, a second (or for that matter any unit of time being used) at one point could be different than the same at a different point. This relativity of time was a significant deviation in our understanding of time.

The concept of universal simultaneity was literally demolished by the Special theory of relativity (SR) (*Miller, Arthur – 74, Jammer, Max – 64*). This means that the pace of forward movement of time may be different for different people with a change of frame of reference.

This means that two events which seemed to happen simultaneously for one observer in one frame of reference would be non-simultaneous to another observer in another frame of reference. This was illustrated well by Einstein by "the train thought experiment" in which a train is moving at a high speed. An observer S standing is standing on the platform and another observer R is standing in the middle of the train. While in motion, at a point in time, two lightning sparks appear at the two ends of the train. To the observer on the platform, the two sparks happened simultaneously as light from them reached S simultaneously. However, as the train is in motion with high speeds, light from the lightning reaches the observer R at different times and thus observer R would see the lightning happening in front of the train earlier than the second one i.e. the events would not be simultaneous.

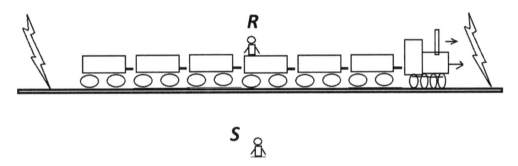

Figure iv/10.1: Simultaneity in Special relativity

Other interesting concepts like space contraction and time dilation due to high speeds were described by Einstein.

Time has never been the same again after the description of Special and General Relativity theories.

How we Measure time

In ancient times, men used to measure "time" by taking into account a number of recurrently occurring events. Most notable was the setting and rising of the Sun dividing time into day and night. The most primitive clocks utilised these events to define time. These clocks were called Sun-dials and utilized the position of the Sun i.e. shadow of a pole to define time. As humans advanced further, a day was divided further into hours and minutes. The year was deduced by other recurrently occurring events like the arrival of summer or winter which happened once a year. These helped these primitive humans in planning important events like sowing or harvesting. Even the location of the Moon was utilized to divide time and is known as the lunar calendar in contrast to the solar calendar where the Sun is utilized.

With further progress came the pendulum clocks where further accurate timekeeping of smaller units of time like hours and minutes became possible using the recurrent movements of a pendulum. When further accuracy was needed, we started utilizing finer recurrently happening events like the vibrations of a Quartz crystal or oscillations of a Caesium atom. Soon the central timekeeping was started with highly accurate atomic clocks and other clocks could be adjusted to match the time kept. Time zones were created around the world as timekeeping became more important to plan events around the world across time zones to be done simultaneously.

Einstein described his "light clock or photon clock" in which there is a light photon and two mirrors kept at a fixed distance. Here the recurrent event of the light photon travelling from one mirror to the other and then getting reflected back is utilized to measure time. If time slows down, the photon will take a longer time to reach the mirrors and would appear to slow down while if time runs faster, the photon would appear to move faster.

This "Photon Clock" is of particular interest to us and would be further discussed in detail.

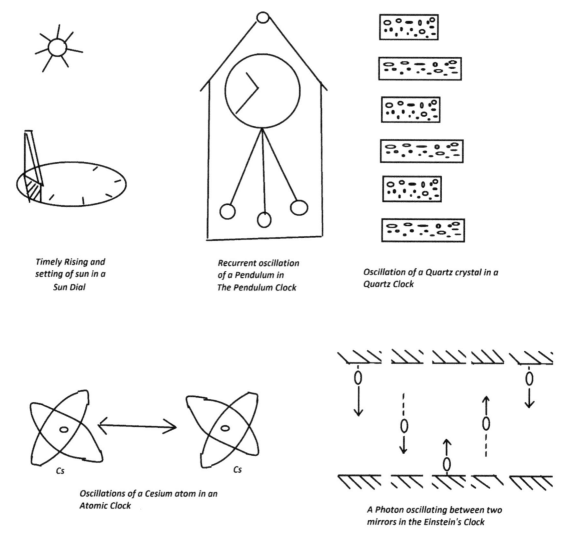

Figure iv/10.2: Shows how we measure time. In each of these types of clocks, there is an event that is repeated periodically. We measure time by calibrating the periodicity or number of repetitions of the event with a known standard.

In all of these, one can note that nowhere is there an accurate definition of what we are measuring? Although today we can measure time with extraordinary accuracy, we are still unable to define "What progresses forward or what changes take place which we should equate to time progressing forward?" or answer "Can time travel backwards?" Or "is travelling backwards in time possible?"

What we are doing in all these is to accurately divide "whatever is progressing forward" into smaller and smaller divisions and keep counting these.

Although the theory of relativity – especially General Relativity did elucidate the fundamental interplay between time and space and in the process married the two together forming Space-time, the definition of "what constitutes time" doesn't come in it and some interpretations of relativity steer into the controversial idea of "Block-time" or "Illusion of time" i.e. absolute determinism, the idea that past, present and future all exist simultaneously and that everything is like a video cassette being played forward. This means that the past and the future are already pre-set and nothing can be changed in any of them.

The following things, however, are adequately certain

1. The Universe had a beginning that happened at a definable point in the past at the Big Bang. Thus,
2. time has a beginning. This means that the arrow of time (although it possibly has no direction) does have a beginning. That is the universe as a whole has a finite age. (whether the universe has a finite end or is destined to exist eternally is still unknown. Some theories suggest an eternally expanding universe while others suggest that there will be a definite point in the future when the universe will stop expanding and will re-collapse into a "Big-Crunch" or a "Big Bounce" and thus time has an end as well)
3. If the universe has a well-defined starting point, then it has a well-defined age, which means that there has to be a Universal Time that is definable and universally constant. This Universal Time, however, goes against relativity simply because according to relativity different sections of the universe age at different rates and thus this universal age of the universe has no consequence.
4. The interplay between space and time has good observational evidence or agreement with experiments, with space being contracted at the place where time is dilated (like near a massive star or a black hole) and space being dilated where there is lesser dilation of time (i.e. relative time-contraction – e.g. as we go away from the same star or black hole). Although we know this happens, it is considered as a law of nature and no mechanism of how this may happen is described.

Redefining time

At the time when Einstein described his theories of relativity, Quantum Mechanics was not fully described. Concepts like antimatter and quantum foam were also not adequately described.

With our theory, we can have a universally applicable definition of time and can define a universally applicable reason for the direction of the arrow of time.

As per our Theory

"Time" has two components.

1. Universal Time
2. Local Time

Universal Time

Universal Time is universal and thus constant for the entire universe. This conceptual idea would mean that there is a universal benchmark of time-based on which time elsewhere can be defined or with which time elsewhere can be compared with. It would be better to measure this time in Planck-times. This universal clock started ticking at the Big Bang and has kept on ticking once every 1 Planck time. Thus every event that happened in the Universe can actually be earmarked or identified with a unique measure of time. The event "X" happened at the 10th tick of the Planck clock. Note that other events "Y", "Z" and "A" would also have happened at the same 10th tick of the Planck clock and thus would be genuinely simultaneous although they are occurring at different regions of the universe with a completely different Local Time.

__The recurring event that defines a tick of the Universal Planck Clock, (as explained in greater detail in arrow of time), is the destruction (or removal from the universe) of a universally constant number of Planck compartments from a given number of PCs.__

Let's presume that this universal constant number of PCs removed in one Universal Planck Time be 10^x PCs removed from 10^y PCs.

__That is, the time taken to remove 10^x PCs from 10^y PCs is the Universal Planck Time.__

In this, 10^y can be taken arbitrarily as 10^6 PCs and value of x would need to be calculated subsequently.[1#]

> 1# - All this erroneous line of thinking to define time, was simply due to the mental glitch that time changes are the primary event. Now we know that time changes are just an illusion and happen secondary to spatial changes causing a variation in velocity of Light.

Significance of Universal Time

1. Universal Time, if true, means that everyone ages with the same velocity forward with respect to Universal Time and thus time travel to a point at a past of Universal Time that has already happened is impossible. In the Twin paradox, the two twins age at a similar pace as far as the Universal Time is concerned and only their "Local Time" ages would differ as one of them travelled. A photon thus has a definite age with respect to Universal Time and does experience Universal Time but has Local Time = 0.

Local Time

Local Time is the time that is linked to space and this is the only part of the time that varies from place to place as described in Einstein's relativity.

This Local Time is the one which we experience. (this is because the conscious experience of time by a conscious human being depends primarily on the light that we perceive.)

It is defined by the amount of Planck compartments added by the time 1 Universal Planck time happens.

This Local Time is in reality just an illusion created due to the fact that it controls the velocity of light.[2#]

> 2# - Although I got this insight pretty early in the development of the theory, the real significance of this statement was understood by me much later. The insight that the active spatial contraction is primary thing occurring and that the active spatial contraction leads to just an illusion of delay in time, came much later, i.e. every PC cannot have its own measure of time. Here, I had believed that change in the Time occurs primarily and the spatial effects are secondarily happening determined by the appearance or disappearance of the PCs.
>
> Possibly, even now the real significance of this statement remains a mystery but one thing is clear, DGR comes close to Einstein's theory of Variable Speed of light (VSL). This might potentially have significant consequences on the Cosmos and acceptability of DGR as a theory among physicists.

To understand it, we need to understand certain ratios.

Local Time can be equivalent to the Universal Planck time i.e. LT=UT or UPT

Or

Local Time can run slower than Universal Planck time i.e. LT<UPT

Or

Local Time can run faster than Universal Planck time i.e. LT>UPT

When Local Time is equal to Universal Planck time, this means that the number of Planck compartments appearing by the time Universal PT happens is the same as that being removed i.e. 10^x

In this case, the number of PCs appearing is equal to those disappearing and thus space will neither contract actively nor expand actively.

Wherever Local Time runs slower, less number of PCs are appearing and thus space would contract. With every tick of UPT, more and more PCs would be lost and thus a persistent dilated i.e. slower Local Time would mean an actively contracting space.

Wherever Local Time runs faster, more number of PCs are appearing and thus space would expand. With every tick of UPT, more and more PCs would be gained and thus a persistent contracted Local Time i.e. faster Local Time would mean an actively expanding space.

Thus the ratio

LPT/UPT = 1 indicates a Local Time running equal to Universal Time,

LPT/UPT > 1 indicates a contracted Local Time

LPT/UPT < 1 indicates a dilated Local Time

Significance of Local Time

1. It determines if space at the location would be expanding or contracting or static.
2. Light travels at 1Planck Compartment per 1 Local Planck time. This means that the transfer of information depends on Local Time. If Local Time is contracted, compared to the observer on the Earth, light there would appear to travel faster than it travels at or near the Earth. On the contrary, if Local Time is dilated or running slower, light appears to travel slower. This essentially means that Local Time determines the transfer of information between Planck adjacent Planck compartments.
3. Information regarding the length of Universal Time remains constant throughout the universe and thus any information transfer that occurs from one point in space to another is information regarding the length (or pace) of Local Time.
4. Adjacent compartments are entangled with each other and thus Local Time in one affects the other.
5. Any deviation of Local Time from Universal Time has the potential to move stuff which is nothing but doing work. The ability to do work is called energy. Thus any deviation of the Local Time from Universal Time would be equivalent to energy.

Arrow of time

For this, we need to imagine the first click of the Planck Clock, i.e. the first moment of creation of the universe at the Big Bang. (this is discussed in great detail elsewhere as well in a separate section)

For discussion sake, let's imagine that at the time of creation, i.e. by the end of the first tick of the Universal Planck clock, 10^9 Planck compartments were created (i.e. appeared from nowhere) and 10^3 Planck compartments were destroyed. What thus remained would be 10^6 Planck compartments.

Note that in this purely hypothetical example (with no connection with reality), x=3 and y=6.

(It is indeed clear, that at this juncture, the number of PCs created cannot be less than the number of PCs destroyed. If this happens, no universe would form.)

Here we can say that a well-defined arrow of time can be identified which is the destruction of 10^x PCs in a volume of 10^y PCs.

Is the Speed of light Constant?

Imagine 5 regions or zones in the space A, B, C, D and E such that C has Local Time = Universal Time whereas A and B have a dilated time while C and D have a contracted time (as shown in figure iv/10.3)

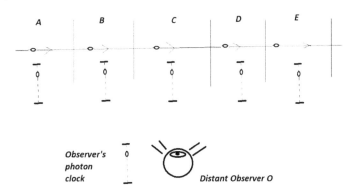

Figure iv/10.3: Five Photon clocks in 5 different time zones

Imagine a light particle (or a photon) travelling from left to right, i.e. from A to E.

Imagine a distant observer O, having a hypothetical photon clock is observing the particle travel from one end to the other. What we are interested in is the perceived velocity of the particle of light by the observer O.

For this, let's imagine that the photon clock which the observer has, is made of 2 hypothetical mirrors and a photon travelling back and forth with the distance between the two mirrors fixed and remaining constant irrespective of the expansion or contraction of the space. Let's further assume the distance between the mirrors be 10 Planck lengths and so every journey from the up mirror to down mirror or the other way round measures 10 Planck time.

Let's also imagine that a photon clock is being placed in every zone so that the Observer can compare the pace of photon clocks at the various zones with his photon clock.

Further, let's assume that the observer lies in a zone where there is no dilated or contracted Local Time i.e. the photon clock of the observer would tick in unison with the photon clock in zone C.

We know that the space in A and B, just before the path of the Photon, would be actively contracting and halting the forward progression of the photon, thus slowing its journey down. Thus at A and B, the photon will take much more time to reach the mirror and the photon clock would seem to tick slower.

The space just before the photon, contrary to this, would be expanding in D and E and thus pushing the photon forward and thus literally carrying the photon towards its destination, thus making the photon reach faster towards every tick and giving the illusion that the photon clock is ticking faster.

This is similar to walking on a long treadmill. If one is moving along the direction of movement of the treadmill, the resultant velocity is an addition of the velocity of the treadmill and the velocity of the individual moving on it. And if the person is walking in a direction opposite to the direction of the treadmill, one of the two velocities in question would acquire a negative sign and the resultant velocity would depend on who is moving faster, the individual or the treadmill.

Chapter 11

COMMUNICATION BETWEEN ADJACENT PLANCK COMPARTMENTS

What determines the Local Time of a PC?

Every Planck compartment can have a Local Time which can significantly differ from its neighbouring one.[1#]

> 1# - At the time of writing this, the concept of time was central and time deviation happening secondary to spatial changes was realised much later. Later, it was realised that every PC having its own local time is wrong since local time itself is an illusionary time and is apparent only due to active contraction or expansion of space due to creation or destruction of PCs. The right thing here is that Local time is due to presence of active spatial expansion or contraction just before the measurement device like the photon of the photon clock or the Cs atom of the atomic clock. It is thus just an illusion.
>
> Thus the "information of time" transferred from one PC to the next turned out to be a wrong concept. However, it is a useful concept.

Various influences can determine or influence the Local Time.

Imagine a Planck compartment named A.

As discussed earlier in the section of Inertia, any movement of a celestial body, far away from the sun can create a resetting wave. When this resetting wave reaches our PC, it can reset the time in it.

The time in the neighbouring PCs can influence the time in A.

Resetting waves from multiple celestial bodies can have an additive effect. If the PC is close to a massive particle, in the zone of interaction[2#], it would cause time dilation in it.

> 2# - This concept of Zone of interaction was dropped in later part of the Theory as it was not seen useful

If the PC is close to a charged particle, it would reset its time according to the charge it has.

If an Electromagnetic wave passes right through compartment A or passes close to it, it can potentially influence the time in it, albeit for a very short period i.e. for 1 Planck time.

We know that energy is the deviation in time according to our theory and we know that energy tends to flow from a point of higher energy to a point of lower energy.

So, it figures that if a PC has a highly deviated time nearby, this event can influence the time in our PC as well.

Conversely, if our PC has a highly deviated time, either time dilation or time contraction, it can influence the same.

Also, we have assumed that space and time are intricately associated. This means that if there is any spatial change occurring in the neighbouring PCs, that can influence the time in our PC.

Note that the Planck compartment cannot expand or contract itself. And thus if time becomes zero in a PC, it just disappears.

Also, irrespective of the Local Time, some PCs will mandatorily disappear. This disappearance could be quantum teleportation to some other region of the universe (more likely) or just destruction.[3#]

> 3# - I later realised that this quantum teleportation of disappearing PCs is theoretical crap and is not absolutely needed for this theory.

A PC that is in the centre of a particle (charged or otherwise) would have time in an extremely contracted state with sub Planck length loops of Planck compartment like compartments of sub Planck length size with T=0 compartments trapped in them.

Neighbouring PCs communicate and influence each other such that time variation in one probably can influence the neighbour. It is almost as if the neighbouring compartments are entangled with each other

For this, we can imagine a cubical space with 100 PCs on one side and thus 10^6 PCs.

What would happen if time suddenly becomes zero in this 10^6 PCs cube?

The appearance of PCs is coupled to Local Time. Disappearance is however coupled to Universal Time and will be unaffected by this local T=0.

After 1 Planck time of Universal Time, wherein Local Time remains zero in these 10^6 PCs, disappearance rate takes over and it loses the fixed amount of PCs. As Local Time is zero and remains zero, no appearance of PCs would occur and with time, the "T=0 10^6 PC" cube keeps losing PCs until it vanishes. As our cube reduces in size, the neighbouring 10^6 PC cubes would have to expand to accommodate the contraction which in turn modifies time in them.[4#]

This modification in space-time in the immediate neighbouring 10^6 PCs cube might affect the subsequent neighbour etc. This possibly creates a gradient in time on all the 10^6 PC cubes, all around the original cube as a sphere.

> 4# - I realised later that this compensatory expansion of surrounding PCs is wrong and the only way the surrounding PCs compensate the volume loss is by compensatory volume contraction. It was deleted in later version after the insight of formation of EPCA's became understood.

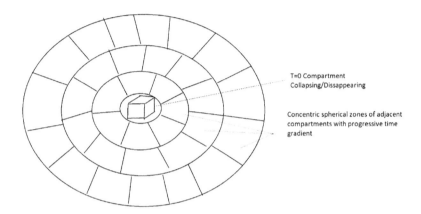

Figure iv/11.1: Shows how neighbouring PCs react to a presence of a T=0 compartment, with formation of a time gradient.

Note that figure iv/11.1 shows a T=0 PC that is static at one place.

What would happen if the T=0 Planck compartments are not static but they are moving?

2nd law of thermodynamics states that the entropy of a system keeps on increasing. We also know that energy tends to flow from a region of space-time with higher energy to a region of lower energy state. This happens in a vacuum in the form of giving off EM waves which are nothing but packets of energy.

Thus, if at any corner of the universe, there appears a time deviation from the surrounding Local Time, this time deviation (which represents energy in our theory) would also show a tendency to flow. The mechanism by which this time deviation moves in space-time is what we call EM waves.

<u>As all EM waves travel at the speed of light which is 1 Planck length per 1 Planck time, this deviation from surrounding Local Time would also travel around with the speed of light.</u>5#

> 5# - In the initial thinking, the Einsteinian limit that transfer of information can happen only at the speed of light was in mind. Later came the realization that if multiple events can happen in the centre at the core of a particle, the EPCA formation would occur much faster than the speed of light and thus the TRW or time resetting wave would travel much faster than the speed of light

Types of communication theoretically possible

Newtonian mechanics says that gravitational influences happen instantaneously, i.e. the Sun pulls the Earth with a force of gravity without a delay. However, Einsteinian theory says that nothing can travel faster than light and that gravitational influences travel with the speed of light in the form of gravitational waves.

Einstein explained this with a thought experiment (which is considered relatively impossible), in which he asked the question

"what would happen if the Sun disappeared suddenly?

According to Newtonian gravity, the Earth would instantaneously lose all gravitational influence on it and would be thrown into a tangential orbit along with all the other planets. In Einsteinian General Relativity, however, the sudden disappearance of the Sun would lead to a formation of a gravitational wave that would travel along space-time at the speed of light. In this manner, the Earth would continue in orbit for about 8 min before the gravitational wave along with the last light emitted by the Sun reaches the Earth. It would be thrown out of orbit only after a lag of 8 minutes.

Our theory relies heavily on the time dilation zones that the celestial bodies cause as a result of the mass present in their constituent particles. These time zones were an integral part of and are an important prediction of General Relativity.

The question which will come to an inquisitive mind is do these time zones happen instantaneously or do these time zones also form around the sun with the speed of light.

This is not an easy question to answer.

To understand this, let's first understand the two theoretical types of communications between PCs that are possible.

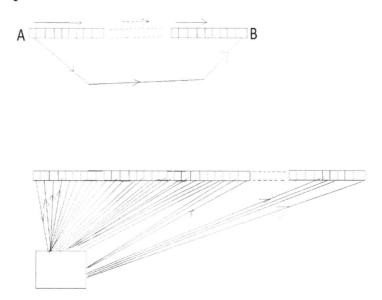

Figure iv/11.2: Shows types of communication theoretically possible between two PCs A and B at the corner of the stack shown.

1. Communication between the neighbouring compartments (direct communication):

   For example, as shown in figure iv/11.2, consider the two compartments A and B which are multiple Planck compartments away. If A communicates with its neighbouring compartment which in turn communicates with the next one, and this wave of communication continues one PC per one Planck time, i.e. with the speed of light. This kind of communication can be called direct communication. Einsteinian gravity communicates in this manner.

2. Communication between A and B without communicating with all the other PCs i.e. indirect communication:

   In this, there is a yet unknown mechanism that exists by way of which A communicates with B without the need for involvement or communicating with the intervening compartments. We know that this kind of communication (which Einstein believed to be impossible and called it spooky action at a distance) does happen in entangled particles. The exact mechanism with which this happens is not yet known.

   This might be explained with the help of higher dimensions. That is, the Planck compartment communicates to another compartment in higher dimensions which then communicates to B without the need for communication with the intervening compartments.

   Although it seems a bit spooky, it has been proven time and again in experiments with entangled photons, to be not just theory.

   The first mechanism of the two looks a bit closer to reality. However, today how exactly a neighbouring Planck compartment reacts to the difference in time of its neighbour and resets its own time is equally mysterious.

   Consider that a particle with mass "M" arrives at PC just before A. We know that due to its mass, it tends to cause time dilation at all the PCs starting from Planck compartment A and progressively decreasing till B. This means that every progressive Planck compartment would have time running faster than A, some of them may have differences that are below Planck time and thus insignificant.

   The question remains whether the news of the arrival of the particle with mass M would travel along the chain of PCs like a train or that its arrival would be known to all the PCs instantly. It is as if, A is entangled to, and communicates with a compartment in a higher dimension which then communicates with all the other PCs from A to B instantly without a delay.

3. Both means of communication exist

   This is a more likely scenario. In this, some PCs around the mass M are entangled and receive the news of the arrival of A instantly (faster than light can reach) while others receive the information with the speed of light.

   Implications of these and the reason why this is needed are explained in detail later while explaining its implications on inertia

Note that presently it is unclear which of the three types of communications exists and which does not. The second type of communication cannot be completely ruled out and would have drastically different predictions and thus potentially can be confirmed or ruled out.

Thought experiment to understand the significance

The Einstein's thought experiment of disappearance of the Sun is in reality not as uncommon as it seems.

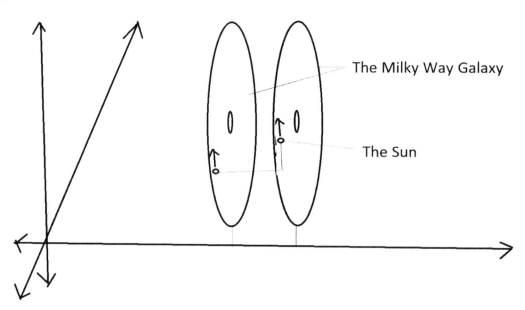

Figure iv/11.3: Shows movement of the Sun.

Consider the following figure iv/11.3.

We know that the Milky Way Galaxy is moving towards Andromeda Galaxy due to its gravitational pull almost with the speed of 110 km/sec. This movement is represented diagrammatically in the figure along the X-axis of a hypothetical coordinate system existing well outside the Galaxy. We also know that the Sun is a peripheral star in the Milky Way galaxy which is a spiral galaxy and thus the Sun is revolving around the centre of the Milky Way at a high speed. This motion is denoted with a slight upward movement of the sun as shown in the second location of the Milky Way.

It is clear that not only the X co-ordinate of the Sun changed but also the Y co-ordinate and even possibly the z co-ordinate.

Indeed, as the position of the Sun changes with time, a situation similar to the above ensues in which, the Sun is no longer in the position where it was some time ago.

Let's assume that this figure shows the position of the Sun a few minutes apart.

The question would then be,

What happens to the time gradients around the Sun due to this movement?

Do they change their positions instantly or do they start adjusting as a gravitational wave that starts from the new location of the Sun and flows around with the speed of light and gives the message to the Earth that its parent star is no longer in its old location and that it is time to change its location to confirm with the new location of the Sun?

(Note that this is a hypothetical question, simply because when the Sun and the solar system were forming, both the Sun and the solar system were revolving around the centre of the Milky way in which case, the Earth would also continue to follow the same path due to its inertia irrespective of the fact that a message has been received from the Sun or not.)

Although our opinion would be more in favour of the "no indirect or instantaneous transfer of information" i.e. "the information travels as a wave with the speed of light", it is difficult to conclude

If an indirect exchange of information is the reality, the transfer of the Sun to its new location would result in instantaneous resetting of the time according to the time dilation zones map.

This indeed would have to happen till infinity every time the Sun (or for that matter any celestial body moves). This instantaneous transfer of Information will happen till infinity, i.e. well beyond the Sun's gravitational field limit, after which the time dilation gradients caused by the sun would go to insignificantly small i.e. below Planck times. The significance of this is given in the discussion of MOND wherein although individually the time dilation gradients lose their significance as they have gone beneath Planck times, they can potentially add up with similar Sub Planck-time time dilation gradients of other stars to form MOND bonds.

The time dilation gradients would, in that case, be considered an integral part of the celestial body itself and that all the Planck Compartments surrounding it who bear the time dilation gradient caused by the mass of the particle would be in a way entangled to the central $T=0$ compartments imprisoned in the centre of each of the constituent particles made by the celestial body.

Although this directly conflicts with the general theory of relativity, and thus seems unlikely, the alternative is equally ugly. The alternative would be that with every movement of the celestial body, a resetting wave starts which travels with the speed of light, while every compartment communicates with the next Planck compartment to reset its time according to its own time. This resetting wave would travel for astronomical distances and would influence PCs which are millions of light-years away. These waves are different from the typical gravitational waves detected by LIGO as these do not drain/carry energy from the celestial body like the Sun but utilize the intrinsic energy within space-time.

The reason why this looks ugly is that, if this is true, MOND would result due to many such waves emitted thousands of years in the past by the stars in the central part of the Galaxy wherein the interference of these waves with similar such waves from other stars would add up to form the MOND bonds. The Sun being a peripheral star, its stability in the Milky Way Galaxy would also then depend on events that started thousands of years in the past. Note that here there is only constructive interference, no destructive interference. The time dilation caused by these resetting waves from different stars adds up to become significant despite being individually sub-Planck time to start with.

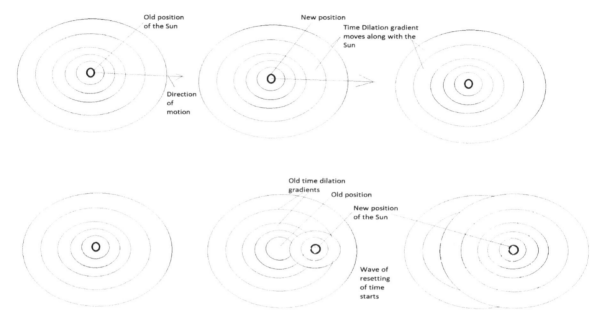

Figure iv/11.3: Shows different possible methods of time resetting around the Sun. Lower part of the figure shows the situation if the time resetting happened as a time resetting wave that starts at the surface and moves outwards with the speed of light. The Upper part of the figure shows instantaneous resetting of time.

CHAPTER 12

AN ELECTROMAGNETIC WAVE

PCs in between 2 T=0 compartments

Consider the following figure iv/12.1.

It shows a train (a row) of 10^6 PC cubes.

Imagine that the first and the last have the pace of time turned to T= 0.

What would happen to the PC cubes in between?

Indeed, the space would contract rapidly where T=0 compartments are present. A compensatory change in time appears in the neighbouring PC cubes so that the time would run slower at the ends and faster as we approach the centre. With faster time, the space would also expand as shown.[1**]

> 1** - This compensatory expansion was found to be wrong in higher energy EM waves like light or UV light. But Radio waves do have alternating zones of highly contracted and highly dilated time. However, they would not have T=0 compartment aggregates.

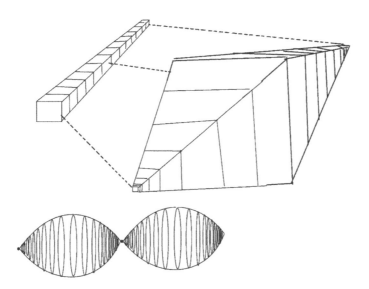

Figure iv/12.1: Shows spatial changes that might happen if the two corner PCs in the train of 10^6 PC cubes become T=0 PCs so that active volume contraction due to resetting time creates a

time gradient. (Diagrammatic image –note the central PCs expanding due to compensatory time contraction- spatial expansion)

Now consider the figure iv/12.2 below which shows similar rows of 10^6 cubes lined up in a wedge shaped section of a sphere. Every sphere would have multiple such PC rows. One can imagine what would happen if each of the ends of these rows now has a T=0 compartment.

The time gradient created in between the two T=0 compartments creates complex spatial changes.

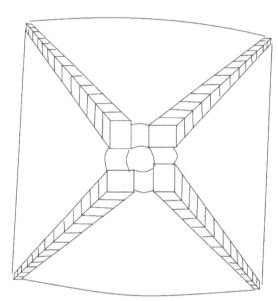

Figure iv/12.2: Shows a similar train or row of 10^6 PC cubes as in figure iv/12.1 which are a part of a spherical area of space (a wedge shaped slice of a sphere with four such trains is shown). If the corner PC cubes of each of these have their time turned to T=0 creating active spatial contraction, the entire PC train would develop time and spatial changes similar to those shown in figure iv/12.1

A single Photon

It is difficult to imagine what a single photon would look like.

It probably looks like a T= 0 compartment followed by a time gradient.

Alternatively, it would be like a T= 0 compartment with all the neighbouring PCs reacting to its presence forming a time gradient all around, for a fixed length, forming a complex arrangement of expansion and contraction of space.

Now the question which comes to mind is

"If these neighbouring compartments are entangled ones, do these changes happen one PC per one Planck length or faster than light as suggested by experiments."

Indeed, the answer is ugly.

The changes have to most likely occur significantly faster than the speed of light.

So two questions remain.

1. Whether a photon looks like a T=0 compartment running one PC per one Planck time followed by a time gradient for a fixed length (or until the next T=0 compartment)? In this case, faster-than-light transfer of information won't be necessary.
2. Or it looks like a T= 0 compartment surrounded on all sides to a fixed length of PCs with a time gradient in which case the time gradient should necessarily happen faster than the speed of light.

Energy and momentum of a single photon

We know that every photon has energy.

Quantum Mechanics tells us that the energy within a photon and the momentum of the photon are related to wavelength and frequency.

$E = h\nu$

Or

$E = h/\lambda$

(where E is the energy of the photon, h is the Planck's constant, λ is the wavelength of the EM wave in question and ν is the frequency of the EM wave in question)

These equations suggest that every photon belonging to a different colour of light would possess a different amount of energy packed within itself and thus in white light, all these photons should be distinguishable.

We know that a T=0 compartment would disappear and thus the dimensions would be zero. However, its presence can be felt due to its effects on the time in the neighbouring PCs.

There is no rule as to how many PCs can have time turned zero in them at a moment.

One can imagine one PC with T=0, 4 PCs having T=0, 16 PCs turning T=0 etc.

In this manner, any number of PCs can be turned into T=0 PCs.[2**]

> 2** - This was probably the point where realization occurred that T=0 compartments can aggregate together and aggregates of T=0 PCs can be the potential light particle.

Note here that the above numbers are used for understanding purposes and the actual number of PCs that correspond to what energy of a photon would need further research.

An even larger number of PCs like a 10^6 PC cube with 100 PCs as one side or, multiples of these forming a cubicle shape might have T=0 in them. The cubical shape is not mandatory and all the PCs included in a sphere of x diameter might be converted to a T=0 compartment.

Larger the number of compartments with T=0 forming a single photon, the larger the energy contained in it.

The larger the number of PCs disappearing, the larger is going to be the distortion in space and the larger would be the compensatory spatial distortion around it.

The higher the frequency of the wave, the lesser is the wavelength which is the distance between two T= 0 compartments in the photon. Thus higher energy photons need a higher extent of compensatory time deviations and associated spatial changes at the smaller available space between the photons. The higher spatial contraction i.e. higher mutual attraction between neighbouring photons due to higher time changes could explain the smaller wavelengths and higher frequencies in more energetic photons.

See figure iv/12.3 below.

Higher frequency electromagnetic radiation like high energy cosmic rays or x rays would have a photon with a higher number of PCs disappearing within, thus the compensatory changes occurring in the neighbouring PCs would be larger and extending into a greater number of surrounding PCs. The lower frequency EM rays would have individual photons with lesser energies individually and thus they have a smaller extent of compensatory changes around the central T=0 compartment.

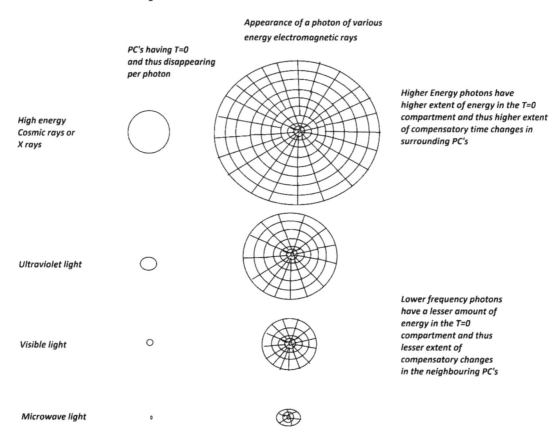

Figure iv/12.3: Shows how a photon might look as per DGR with multiple T=0 compartments aggregated at the centre with highly deviated time at the centre and a time gradient all around up to a particular length all around, possibly happening faster than the speed of light.

Communication between entangled photons

Consider the following figure

S is a source of multiple photons (like any celestial body or star)

At a given instant, multiple photons A, B, C, D, E, F and G have started from the source in different directions. <u>These photons represent the (aggregate of) T=0 compartments.</u> [3**] P, Q and R represent their positions at three different moments in their journey forward. The circles represent the wavefronts.

> [3**] Note that these notes represented a preliminary analysis of what would happen in terms of time changes happening around T=0 compartments which were the new found quanta of energy. At the time, the focus of thinking was solely how time changes would be distributed and little focus on spatial changes. As such, the mind wasn't aware of what would be the implications of multiple T=0 PCs aggregated together on surrounding space. As it later became clear, the process of entanglement is basically spatial changes taking primarily and time changes happening as a secondary consequence. The concept of EPCA formation had not occurred to me at this stage of the theory.

At P, just after being emitted, the circumference of P or the arcs AB, BC, CD etc. represent the zones where the neighbouring PCs would react to the presence of these T=0 PCs and would have a time gradient in such a manner that space progressively contracts nearer to the (aggregates of) T=0 compartment and expands as one goes away from the two T=0 compartments in the middle of the arc.

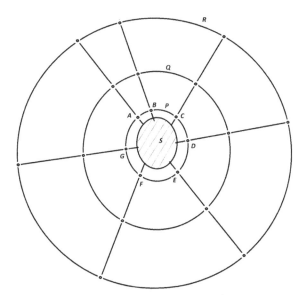

Figure iv/12.4: Shows a source of photons S with multiple photons A, B, C, D, E, F and G moving outwards from its source. To note is that each of these photons is entangled to others and the line AS, BS, CS etc. and the arcs AB, BC etc. will have time changes forming a time gradient.

Entangled Photons

As shown in the figure below, the presence of T=0 compartment aggregates would create a wavefront of T=0 compartment aggregates connected by PCs which are entangled to each of these T=0 compartments such that there develops a time gradient between them. This time gradient is just a compensatory response of these PCs to the presence of these PCs with extremely contracted space and dilated time. Thus the time deviation caused in between the T=0 PCs is compensatory and utilizes the intrinsic energy of the space-time and is not energy lost by S. The only loss of energy that happens from S is in the form of the T=0 compartments.

As these T=0 compartments aggregated together to travel in divergent directions, the length of arcs between them increases. When the photons have travelled for years together, the distances between these T=0 compartment aggregates (lengths of these arcs AB, BC etc.) would run into several light-years or more. The time gradient in them would continue, however possibly at such great distances the differences between the time between the neighbouring PCs in these arcs would be extremely low, going below Planck times.

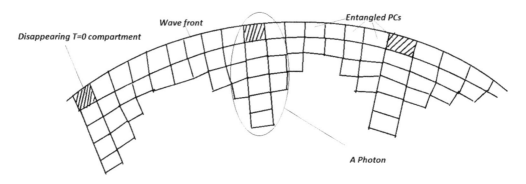

Figure iv/12.5: Shows one probable configuration of how entangled photons might look in case faster than light resetting of time forming the time gradients in between is not possible.

Entangled PCs in a wave front

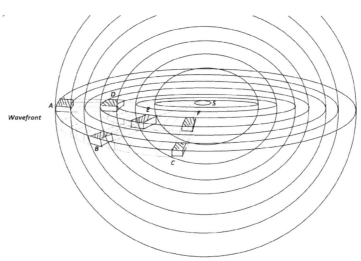

Figure iv/12.6: Shows a source S with multiple photons moving away from it named A, B, C and D, E, F. Note that A will be entangled to B and C and it will also be entangled to D. the arc ABC

represents a single wave front which is composed of photons at the same level with the time changes in between leading to secondary spatial changes.

Examine the figure iv/12.6.

The concentric circles denote the various wavefronts emerging out of the source S.

A, B, C denote three entangled photons or T=0 compartment aggregates with the intervening waveform made up of multiple PCs with time gradient. After some time has elapsed from the start of these photons, (say 2-3 Planck times) another set of 3 photons or T=0 compartment aggregates emerges from the source. One can see that the difference between A and D or for that matter B and E or C and F constitutes the wavelength, while the number of T=0 compartment aggregates being released per second would be the frequency of the wave. One can note that the space between A and D would have a time gradient due to the compensatory expansion/contraction of the neighbouring PCs.

The intensity would indeed be the number of photons being emitted per unit surface area. The photons i.e. the individual T=0 compartment aggregates represent the only energy lost by the source, all the other time changes represent just reactions to the presence of the T=0 compartment.

Note that all these events described happen at extremely rapid speeds i.e. they last for one Planck time.

Thus, as soon as the T=0 compartment aggregates appear, possibly even before the first tick of the Planck clock, the time and spatial changes that are reactionary to its presence appear, and the T=0 aggregate starts moving at one Planck compartment per one Planck time.

The following figures show the probable distribution of 10^6 PC cubes around a T=0 compartment aggregate in a wavefront.

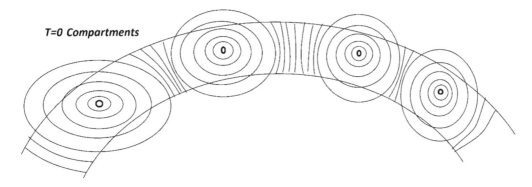

Figure iv/12.7: Shows another possible appearance of a wave front with multiple T=0 compartment aggregates along with compensatory time changes forming a time gradient in between. (if faster than light transfer of information is possible)

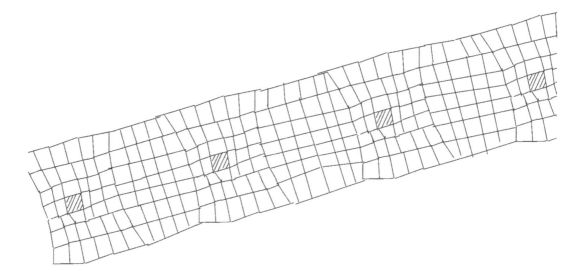

Figure iv/12.8: Shows another possible configuration of multiple photons, here shown as an array or row of 10^6 PC cubes with the central zone having the T=0 compartment aggregates and the 10^6 PC cubes in between developing the compensatory time changes and spatial changes.

Note that in both the above figures, only the 10^6 PC cubes being affected by the presence of the T=0 compartment aggregate is shown.

What happens to space in these 10^6 PC cubes:

We know that time is running zero exactly at the T=0 compartments aggregated at the centre. The space here would indeed be zero, I.e. in a state of active contraction losing volume with every Planck time. As we move outwards from them, there would be a time gradient so that time running at PCs immediately adjacent to the T=0 compartment aggregates would be in a state of high time dilation and the space here would be in a state of active contraction i.e. actively losing volume, although slightly lesser than the centre.

As we move outwards away from the T=0 aggregates, there would be regions in between where there is time in a state of contraction and where space is actively expanding and gaining volume. In between these zones of highly contracted and highly dilated time, there may exist zones where time runs along Universal Time and there is no spatial expansion or contraction.

Note that we can go in multiple directions.

As we go in the direction of propagation of the wave, the status of the space-time depends on whether faster than light travel of information is allowed or not. Without faster than light travel of information, the space-time at the preceding edge of the wave would remain unaffected and there would be no time gradient or spatial change in this region.

In the direction opposite to the propagation of the wave, i.e. the receding end, the T=0 compartment aggregates would be arranged at a distance corresponding to the frequency of the wave and the space in between would have a sinusoidal i.e. fluctuating time gradient and thus space would be maximally contracted nearer the T=0 compartment aggregates and maximally dilated at the midpoint between the two nodal T=0 compartments aggregates.

In the side to side directions all around the direction of propagation of the wave, parallel to the wavefront, the T=0 compartment aggregates would be arranged at distances dependant on the intensity of the EM radiation and thus at higher intensities, the distances between T=0 compartment aggregates forming a part of a wavefront would have lesser distances than at lower intensities.[4**] As we move along the wavefront from one T=0 compartment aggregate to another, time would gradually vary from highly dilated at the two ends to maximally contracted at the centre. Space in between the T=0 compartment aggregates forming a wavefront would also be in a state of active contraction nearer to the T=0 compartments and would be in a state of active expansion at the midpoint between the two.

> [4**] - Here again, one can see that the sole focus was on how the time gradient around the T=0 compartment is distributed. By this time, the concept of EPCA formation had not occurred to me. Even then, the distribution of time gradient and the logic of faster than light transfer of information came into the realization without effort

As the two T=0 compartment aggregates travel in different directions, the intensity reduces and thus the distances between individual T=0 compartment aggregates increases and so the rate of change of the time gradient in between them would also reduce. This means that the spatial changes in between the adjacent T=0 compartment aggregates forming a wavefront would keep getting smaller and smaller with more distance travelled by the wavefront. At great distances from its source, although the compartments in between the T=0 aggregates are entangled, the time gradients go below Planck time and thus the spatial changes also become sub-Planck length or negligible.

Although the space-time gradients have become smaller than significant, the entanglement does remain and is detectable.

What if Faster than light information travel is impossible

Imagine a PC named A as shown in figure iv/12.9.

The question we are trying to answer is that if the time is turned zero at A at a moment, without faster than light information transfer, how would time changes evolve and what are the possible repercussions.

If the faster-than-light transfer of information is possible, all the surrounding PCs can react to the presence of disappearing PC instantaneously without any gap. Without faster than light information transfer, for one Planck time, there will be a void at A as the neighbouring PCs cannot react to its disappearance until the next tick of the Planck clock.

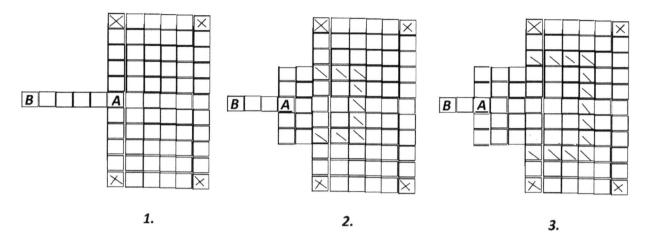

Figure iv/12.9: Shows how information about time changes would evolve if faster than light transfer of information is not possible.

If the neighbouring PCs react to the presence of T=0 at the speed of light, then by the time changes reach the corner PCs marked x, A would have reached B. during its journey from its original location to B, the T=0 compartment would induce time changes not only in the PCs marked with an oblique line, but also its neighbouring PCs as shown in 2nd and 3rd part of the figure.

This would indeed mean that a photon can induce no PC-time changes at the PCs to which it is yet to come and the time changes would be restricted behind the PCs and the changes would not necessarily be spherical.

The time changes connecting the neighbouring photons, as shown above, in a wavefront would also have to travel just at speeds of light, which would be too slow. The theory would fall apart if faster than light information transfer (which is shown to be present in multiple experiments on entangled photons) is not allowed.

Does information travel instantly from one corner of the universe to the next along a wavefront?

As the wavefront is forming at the source, the T=0 compartment aggregates get entangled to each other by the 10^6 PC cubes which lie between them. Here distances could be pretty small, although even here there is an implication of faster than light travel of information. However, as the wavefront propagates, the information need not propagate from one T=0 compartment to the other. What is happening here is the transfer of information from PCs lining one position of the wavefront to the PCs lining the subsequent position of the wavefront not only at the level of the T=0 compartments but throughout the wavefront. These could be the "local hidden variables" Einstein was talking about. Thus, although the theory possibly points to faster than light communication of information within the PCs which are reacting to the presence of a T=0 PC, there can still be a significant restriction in the velocity of flow of information.

At high distances from the source, how would the Electromagnetic waves look like?

At high distances from the source, the side to side time and spatial changes would be negligible along the wavefront. However, the time and spatial changes would persist in between the T=0 compartment aggregates generated one frequency-time later or those positioned in the receding direction opposite to the propagation of the EM wave, at a distance of one wavelength from the T=0 compartment aggregate in question.

This distance between the T=0 compartment aggregates emitted in a row arranged at a wavelengths distance away from each other gradually increases with the expansion of the universe as space in between expands. This accounts for the cosmological redshift.

Possible mechanism/explanation of faster-than-light transfer of information

As per the law of thermodynamics, energy keeps flowing from a higher concentration to a lower concentration. A T=0 compartment would mean a highly concentrated form of energy. As soon as it appears, all the surrounding PCs (i.e. 10^6 PC cubes or higher) would react to its presence and time in them gets highly contracted to compensate for the extreme time dilation in the T=0 compartment.

A highly contracted time would mean that every tick of the Universal Time would correspond to multiple ticks of the Local Time. As Local Time is running faster, even if the information travels with the velocity of on PC per one Planck time, a higher number of Planck times can happen in Local Time during a single tick of the universal Planck clock, thus despite not violating the velocity limit of one PC per one Planck time, information can travel locally faster than the speed of light.[5**]

> 5** - An interesting although completely wrong direction of thinking. As was later realized, there is no zone of highly contracted time (which indicates negative energy) around the photon which is positive energy and thus has only dilated time and active spatial contraction. (This is the reason that this philosophical phenomenon-based thinking is a difficult method to develop a theory as it is easy to get lost.)
>
> That said, one should note that the Radio-waves, as predicted by DGR much later, have no aggregates of T=0 PCs and are basically drifting Push-Pull bands from the charged particles and thus have both regions with dilated and contracted time.

Electromagnetic waves [6**]

> 6** - In such an early stage in development of theory, the EM wave predicted by the theory, although not completely accurate, is pretty close to what the finally predicted wave looks like, with aggregates of PCs with EPCAs moving inwards due to the time resetting wave which move out faster than the speed of light causing secondary time changes that form a time gradient.

The Electromagnetic waves that we know of including light are probably made of many such photons (T=0 compartment aggregates) with entangled neighbouring PC cubes with various time gradients coming together.

We know that white light is made up of light of all frequencies. This means that a single wave is a complicated mixture of multiple waves of different wavelengths superimposed onto each other to form a complex pattern, all of which would indeed travel with the speed of light but all of which would have a differing wavelength and thus would have an effect on space.

A comparatively less accurate representation of an EM wave of a single frequency is shown below, akin to a radio wave being received by the radio antenna. One can see that the wave has alternating zones of contracted and expanded space, the contracted space being the one where the T=0 compartment aggregates are and the expanded ones are the region in between the T=0 compartment aggregates.

The EM waves predicted by our theory would have multiple T=0 photons of the same frequency connected together by a wavefront of entangled PCs with time changes in them, multiple such photon types with different energies superimposed onto each other or at least travelling close by.

A radio wave and the electric current generated in the antenna

The Radio works on the principle of resonance. With the knob of the radio, the resonance of the circuit is modified to match the frequency of the radio wave. When the resonance matches, as the radio wave comes in contact with the antenna, the T=0 compartment aggregates along with the time gradients following (and probably preceding) them would create a time gradient within the substance of the antenna.[7**] This alternate spatial contraction and expansion would indeed create a similar time gradient within the substance of the metal which causes the charges in it to move. With alternating contraction and expansion zones, the time gradients would fluctuate from contracted time to dilated time, which cause charges to move once in one direction and then in the other. This process thus produces an alternating current in the circuit. This miniature current flowing within the antenna can be magnified with a transistor and thus decoded. Note that the working of Radio or for that matter other applications of this technology is not as simple as is shown above and intricate technical details cannot be discussed here, and further research into this would be needed.

7** - This relatively naïve explanation at the start of the theory is later replaced by the presence of the push-pull bands emanating out from the charged particles which represent the Electrical and Magnetic flux in our theory. Although the basic logic remains constant in that the time changes induced by the radio-wave within the antenna induces a current of the same frequency.

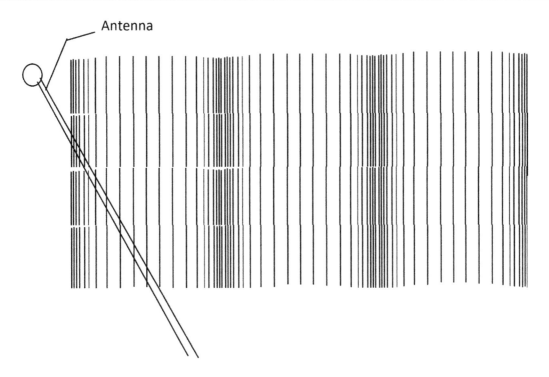

Figure iv/12.10: Shows a diagrammatic representation of a Radio-wave represented by alternate zone of time dilation (active spatial contraction) and time contraction (spatial expansion)

CHAPTER 13

GRAVITATIONAL LENSING EXPLAINED[1#]

> 1# - The fact that even Gravitational lensing is explained phenomenologically with reasonable correctness at such early stage of development of the theory is commendable. Although the current explanation is much simpler and is based on constantly inward movement of the EPCAs that cause active spatial contraction and lead to inward bending of the path of the photon.

Now imagine a celestial body, say a star or a black hole, as shown below with its gravitational time dilation gradient. 6 photons from another star travelling from right to left are shown. A. B, C, D and E represent the paths these photons would take if the star was not present.

Note that all these photons are entangled and thus connected.

Let's presume for a moment that these photons are the photons of light as represented in classical physics, being disturbances in the electromagnetic field. Einsteinian theory predicts that there is spatial contraction nearer a body with a strong gravitational field and as we go outwards towards the outer circles, the space is relatively expanded.

We know that light travels with the speed of light in a vacuum, i.e. light travels at one PC per one Planck time.

These photons indeed would also travel with the same speed. Each of these six photons would travel at different time zones at different times.

If we assume Einsteinian static spatial expansion is true, then there would be more Planck compartments in their path when the path goes through the region where space is expanded and lesser Planck compartments in the region where the space is relatively contracted.

The upper photon would travel at relatively expanded space (slowly contracting space to be more precise) and would have to cross more Planck compartments and thus would lag behind. The lower photons would have to come across a lesser number of Planck compartments and would travel faster.

If the photons behave like individual particles and are disconnected from each other, they would just travel in a straight line with variable speed, which does not match the observations. Thus indeed, the photons have to be connected and they have to affect each other's path.

The lower photon travelling faster and upper travelling slower would mean that the ultimate resultant wavefront would be directed upwards and the light would be deviated upwards, which is also contrary to what is observed.

What our theory would say is that as the photons enter the time dilated zones, at B, the lower 4 photons are in the outermost dilation zone. In this zone, the space is dynamically contracting. The space just below the lower photon is also contracting towards the centre of the celestial body and thus the photon would tend to move down.

As they move towards C, the space between each of them is contracting and thus all of them would have a tendency towards downwards deviation. The time dilation gradients created by the gravitational time dilation zones add up with and modify the time gradients in the waveforms in between the photons and thus reduce the distance between the individual photons or change their direction. (Note that D and E, in the figure, just represent the potential locations of the same photons if the Gravitational field/Gravitational time dilation gradient did not exist so and there was no resultant bending of the path of the photons.)

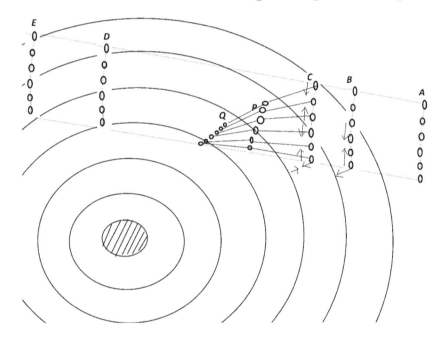

Figure iv/13.1: Shows 6 entangled photons of light entering the Gravitational time dilation zones created by an object with significant gravitation, say the Earth. The inwards bending of their path is due to active inwards contraction due to the Gravitational time dilation, which is more in the inner zones and also due to a difference in velocity of the inner photons compared to the outer photons due to active spatial contraction.

CHAPTER 14

ENERGY

Energy is defined as an entity that has the ability to do work.

Energy is central to the study of almost all subjects. However, it is at the core of physics.

Energy has many physical properties that form an integral part of the universal laws of physics.

The "law of conservation of energy" states that energy can never be created or destroyed but can be converted from one form to another.

Energy can be in various forms namely – potential energy, kinetic energy, elastic energy, chemical energy, nuclear energy, electromagnetic energy etc. One of the most important forms of energy is heat energy which is in itself a part of electromagnetic energy. Heat also determines the kinetic energies of molecules in a solid or liquid or gases which determines its temperature. Thus temperature is also a determinant of the amount of energy in a solid/liquid or gas.

Although energy is so important in physics, it is still an abstract entity. The assumption is that it just exists as a fundamental entity with which the universe is made and that there is no need to further characterize energy as to what it contains or what it is made of. The number of such abstract entities which need to be assumed without further characterization being derived from energy like potential energy, kinetic energy, temperature etc. just keeps getting added with an increase in knowledge of the complexity of systems.

No theory has attempted to actually characterize energy or define it. It is a firm belief that this is not relevant or needed.

Recent observations which suggest that the universe is expanding and that the expansion is in fact accelerating has implied the existence of another such abstract entity called dark energy. This dark energy along with another abstract entity called dark matter probably constitutes 95% of the Universe, which literally means that the knowledge we have about 95% of what the Universe is made of, is pretty limited.

Indeed, this is the right time for a theory that helps characterize what energy really is.

The first clue came with the discovery of gravitational waves that were once predicted by Einstein's general theory of relativity, by LIGO (laser interferometer gravitational field observatory). These gravitational waves were generated when two black holes, the first one 35 solar masses and the second one 30 solar masses spiralled into each other and collided forming a single black hole. This relatively dramatic event led to the disappearance of 3 solar masses worth of mass of the black holes. Literally, this mass was converted into energy and this energy was converted as a space-time fluctuation which travelled for about 1.4 billion light-years before it was detected by the laser interferometer. This indicates that energy can propagate in the universe as a space-time fluctuation.

The second clue came a century ago when Einstein removed the bias of the scientific community of absoluteness of time and proposed that the pre-Einsteinian scientific community made a blunder by considering time as a separate entity and that in reality, time is relative and that space and time are integrally connected. That is, in most pre-Einsteinian theories, time and space were considered separate variables and the pace of time was considered equal for all Individuals or observers irrespective of the frame of reference. This means that often scientific thinking gets biased about some frequently repeated opinions (which could be wrong). These biases lead to the formation of thick walls in the minds of the scientific community due to which their minds get locked into certain presumptions and they are unable to think beyond these prejudices. The question which arises is, is modern physics riddled with such prejudices or biases? Do we need to think completely out of the box to arrive at a completely new way of thinking about reality?

Einstein also proposed that mass is in a way equivalent to energy, that a particle with mass has high amounts of energy stored within it, which was proven conclusively in Hiroshima and Nagasaki atomic bomb explosions. His famous equation $E = m_0 c^2$ (where m_0 is the rest mass) clearly shows the mass-energy equivalence.

The final clue came while working at the present theory wherein, in an effort to quantize gravity and in an effort to develop a theory which could derive Newtonian mechanics, Einsteinian General Relativity and successfully unite Quantum Mechanics and relativity, I came across the fact that the time dilation caused by the mass of the Earth could potentially explain the movement of the apple i.e. a time gradient can successfully get work done.

If a time gradient can get work done, and if energy is an entity that gets work done or has the potential to exert force, can there be any equivalence between time gradient (time deviation) and energy?

Is it possible that due to biases of thinking in terms of energy and fields, the scientific community was/is/will be unable to comprehend this equivalence?

If this is true, the implication is that every fundamental force of nature is caused due to some kind of time deviation or time gradient and that in every place where energy is said to be stored, there is a deviation of time. One can perceive a "Theory of everything" from this. But for this, the outlook towards energy that one has should change dramatically.

The only way forward is to compile this into a theory and see if it can explain all the findings or observations or that it conflicts with what is already known.

This is attempted here. Indeed, it is naïve to expect that answer to every question will be known at the end of this exercise, one can definitely promise a new direction of thinking for modern physics.

Redefining energy

Energy according to our theory is any part of the space-time where time is running at a pace different than Universal Time.

For this, the three most important fundamental things need to be defined.

1. There is a single universal frame of reference which is made of 3 dimensions. This is flat. Although the absolute coordinates are flat, as needed for General Relativity, locally space-time can curve.
2. There is a Universal Time that is absolute and which flows at a single pace all over the universe.
3. Space-time is relative i.e. time running at any point in space may vary compared to Universal Time. As space and time are intricately connected, which is implied from the theory of Dynamic General Relativity, any change in the pace of time locally would have a direct effect on the space there. These entities could be called Local Time and local space respectively.

As space-time is divided into multiple compartments which are roughly cubical and have a side of Planck length. And each Planck compartment could potentially have a different pace of time.[1#]

> 1# - This was later found to be the wrong line of thinking and was replaced by the fact that the changes in space i.e. destruction or creation are fundamental and changes in Local Time happening are secondary to slowing of the velocity of time due to the spatial changes.

Thus the smallest unit of energy would be a Planck compartment with a deviated Local Time (any deviation from the Universal Time, above or below is counted – thus both contracted time or expanded time both constitute energy)

Energy density

More the deviation in a PC from Universal Time (UT), more is the energy density.

If a PC has time running exactly with the pace of Universal Time, there is zero energy density in it.

The maximum possible energy density would be in a PC where Time has stopped i.e. T=0.

Note that in this case, time running faster than Universal Time should constitute negative energy.

When we see a bigger volume, the energy density would be = number of PCs with deviated time/total number of PCs

Note that even with this there could be variation, some of the above PCs with deviated time could be T=0 PCs, and still others would be having dilated or contracted time compared to UT.

Another way of putting it is

$$Energy\ density = \frac{(Total\ number\ of\ T=0\ PCs + Total\ number\ of\ PCs\ with\ time\ dilated + Total\ number\ of\ PCs\ with\ time\ contracted\ with\ respect\ to\ UT)}{Total\ number\ of\ PCs}$$

Note that the individual PCs communicate with each other and thus influence each other in such a manner that Law of conservation of energy is followed.

The law for this modified out-look of energy would be – the sum of all the time deviations introduced in a system by a source would remain embedded in it and cannot be destroyed, time deviations cannot be created nor can they be destroyed by the individual PCs. The PCs can only transfer their time deviations.

Even the second law of thermodynamics will have to be followed, which says that the entropy of a system keeps increasing. We know that there are places in the universe where energy is concentrated. For example, at the surface of the Sun, where temperatures are extremely high. At these places, the energy density would be very high i.e. the number of time-deviated or T=0 PCs would be very high. We also know that energy tends to radiate from such areas in the form of electromagnetic waves in the vacuum or in the form of conduction or convection where media are available. This tendency indicates that neighbouring PCs must have a tendency to self-adjust the time in them to match with the times of the neighbour. This can happen by two mechanisms as discussed below.

We know that once created, a T=0 compartment cannot be destroyed, i.e. it would keep moving from one place to another. This means that when time in the PC becomes zero, Local Time is zero but Universal Time continues and thus after one Planck time, the PC disappears. To compensate for the loss of space, the neighbouring PCs modify time in them leading to the creation of a time gradient in the neighbouring PCs. With every tick of the Planck Clock, this T=0 compartment shifts to its neighbour so that now the time in it becomes zero and the previous PC forms a part of the time gradient. In this manner, a T=0 compartment would keep transferring its T=0 status to the next compartment and this T=0-time deviation would travel at the speed of one Planck compartment every one Planck time, i.e. it would travel with the speed of light. At the source of the formation of the T=0 compartment, probably it would not be just one but many such compartments that would form. Depending upon how much energy is being generated, a wave of such T=0 compartments followed by a time gradient would form.

As explained in the section on electromagnetic waves, waves with different frequencies and waves with different intensities would be formed. How they would look and what effects they would have on Local Times of individual constituent PCs is discussed elsewhere. But according to the Local Time, they would indeed have space dilated or contracted. Thus EM waves are not drastically different from sound waves which are nothing but waves of pressure difference.

The T=0 energy in short would travel in the form of electromagnetic waves.

Is there any other manner in which deviation of time (energy) can travel?

<u>Gravitational waves, more comprehensively discussed in the "communication between PCs" section and also in the "Inertia" and "MOND" sections, are also waves in which energy can travel from one place to another. In fact, as predicted by Einstein's theory, any movement of any celestial body would indeed create a gravitational wave. This wave would be a resetting wave. We know that every celestial body has time dilation zones all around it, with any movement in the location of the body, the immediately adjacent PCs would reset their time according to the new position and this resetting would keep travelling outwards in all directions.</u>[2#]

> 2# - Note that later it was realized that the Time resetting waves that effect gravity are different from the Gravitational waves detected by LIGO. These TRW travel much faster than the speed of light. But any movement of a heavy body or any opposing of the Gravitational pull leading to potential energy accumulation (like a heavy machinery being held at a height h by a robotic arm) would produce disturbance in the smooth flow of EPCAs in the surrounding space which would travel 1 PC per PT and are the typical Gravitational waves detected by LIGO.

Is T=0 the maximum possible energy density per PC?

We know that according to Einstein, energy and mass are equivalent, i.e. there is energy in the centre of every particle with mass.

String theory presumes that there could be a higher number of dimensions that are smaller than a Planck length and curled up within a small pocket of space-time. The energy associated with the mass is trapped in these curled up dimensions as a string of energy that vibrates in a frequency unique for every particle.

According to our theory, however, energy is equivalent to a deviation in time, mostly a T=0 compartment. Thus, multiple T zero compartments superimposed onto each other, curled up into higher dimensions, all restricted to a small curled up region of space-time which can be considered a prison for these T=0 compartments, is what constitutes mass. In short, there is a miniature black-hole like prison made specifically for these T=0 compartments. These compartments get curled up and thus disappear from sight, but their effect on the neighbouring PCs remains i.e. they keep causing time dilation of their neighbouring PCs thus creating a gravitational time dilation gradient.

Because these T=0 compartments are within their prison, they get restricted to a small space and do not have a tendency to move around like the free T=0 compartments. <u>The time running within them is still zero and thus within the curled up dimensions, they still move around.</u>[3#]

> 3# - This added baggage of assumptions of additional dimensions or prison for T=0 compartments was given up later as it proved to be un-useful.

Another controversial assumption, the need for which will be explained subsequently in the section on Inertia and also the section on Atoms, molecules and covalent bonds and charges is the fact that these T=0 compartments present in higher dimensions get entangled with some of the PCs around the particle and lead to space-time deviations which are responsible for the various properties of the particles like positive charge, negative charge etc. These entanglements would necessitate faster than light –i.e. instantaneous transfer of information.

It is almost as if, these entangled PCs which are well beyond the boundaries of the particle, move or rotate along with the particle instantaneously as if they form a part of the particle.

These assumptions, although seemingly counterintuitive, are mandatory to explain certain observations.

CHAPTER 15

TEMPERATURE

The usual definition of the temperature of a body is "an entity" that depends on the kinetic energies of all the constituent molecules (for a gas or liquids or solids).

In a vacuum, it is the measure of energy density.

In our theory, indeed, the temperature would be a measure of the time changes in the PCs surrounding a particle especially in the zone of interaction. In the section on inertia, how a particle that is moving interacts with the PCs at the edges of the time dilation zones surrounding the particles and how various PCs are recruited and surrendered leading to time changes in the surrounding PCs is explained. These time changes would constitute the kinetic energy of the moving particle. Temperature is basically a measure of the sum of these kinetic energies.

Thus a body with a higher temperature would have particles with higher kinetic energies and individual velocities which would translate into higher time changes. Higher temperatures would often indicate a high presence of fast-moving T=0 compartments, what we call electromagnetic radiation.

In a solid, the kinetic energies represent velocities of vibration while in liquids or gases, it involves random movement of the molecules.

Temperature can also be considered a measure of the number of electromagnetic waves (which are time deviations themselves and include T=0 compartments) roaming around within a solid or liquid or a gas. A photon of an electromagnetic wave gets absorbed by electrons around an atom and gets energized and moves up into a higher orbit. Once the electron comes back to its original orbit, another photon with similar energy is released back. This photon would then get absorbed by another atom and this cycle is repeated here as well.

The number of such events happening would be expected to increase with increasing temperature.

There comes a temperature when the energy of photons (i.e. the time deviations caused by them) and the number of such (absorption/release of the photon) events increases so much that individual bonds between constituents of atoms no longer hold. This gives rise to the fourth type of matter- the plasma. In this, the amount of energy in the form of T=0 compartments, the time deviations and the EM waves are so great that all the particles like electrons, protons and neutrons lose their bonding. This happens in stars routinely.

With further increase in temperature, probably higher than temperatures at the centre of the Sun, even the strong nuclear bonds which hold the quarks together in the protons or neutrons are lost and what is formed is called the quark-gluon plasma. This has been achieved in minute quantities in particle accelerators like CERN's Large Hadron Collider (LHC) and Heavy Ion Collider of Brookhaven National Laboratory. This state is extremely unstable and stays for an extremely short period before stabilizing back again into larger particles.

Absolute zero temperature

If a region of space has no Planck compartment with any deviation from Universal Time, it probably has attained absolute zero temperature. Attaining it is extremely difficult and maintaining it is even more difficult simply because energy (time deviation or T=0 time from other compartments with higher time deviation) would tend to flow into such a region so that isolating it from other non-absolute zero regions becomes difficult.

Planck temperature

In this, almost all the compartments in a region are T=0. For example, a 10^6 cube of PCs in which all the 10^6 PCs are T=0. This probably happened at the moment of the creation of the universe at the moment of the big bang in which all the PCs were T=0 and time was running at a zero pace. Given the tendency of the T=0 compartments to immediately shift off their T=0 status to their neighbouring PCs, it would be impossible to create or maintain this temperature.

Other forms of energy

Other forms of energy like potential energy, kinetic energy, electromagnetic energy, chemical energy (due to the formation of covalent bonds), nuclear energy etc. (how they can be explained in terms of time deviations) are explained individually later.

CHAPTER 16

INERTIA

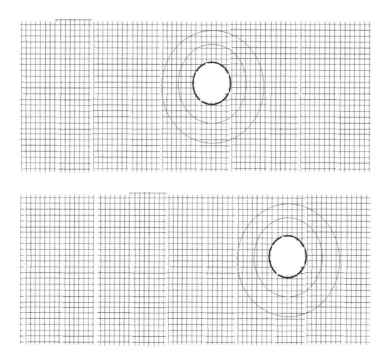

Figure iv/16.1: Shows a spherical body suspended in space time made up of tiny Planck Compartments causing Gravitational time dilation in the surrounding space which can be represented as concentric circles (spheres)

Figure iv/16.1 shows a spherical body E (say the Earth) suspended in space-time. The concentric circles represent some of the zones of time dilation (only some are shown for better presentation)

If we imagine E is at rest and has mass M, the gravitational time dilation caused by its mass can be calculated and the zone of interaction and its limits can be theoretically guessed. <u>If the exact location of the centre of E is known, one can also calculate the edges of various time zones</u>[1*] At these edges, neighbouring PCs are located in different time zones. <u>This includes PCs located immediately next to the surface of the sphere and the PCs located at the outer edge of the zone of Interference</u>[1*] beyond which time dilation caused by the sphere's mass leads to a time dilation which is insignificant, which in the figure is represented by the outermost circle.

> 1* - This concept of Zone of Interaction was a rather ill-defined concept which was the zone around the particle within which the time changes happen instantaneously, i.e. much faster than the speed of light. This was envisioned by me assuming that both types of communication exist between the PCs, namely
>
> -instantaneous communication of time related information between the PCs (as the process of entanglement show) and
>
> -local communication of time changes travelling as a wave, one PC at a time in the form of gravitational waves.
>
> Within the zone of interaction, I imagined, that the time changes happening around the particle due to its mass travelled instantaneously and thus form an integral part of the particle.
>
> Later this concept was given up as it wasn't helpful.
>
> In the current theory, it is replaced by the "r_{cric}" which represents the critical radius around the particle where the destruction happening within the particle leads to EPCAs moving inwards faster than the speed of light. It roughly represents the Event horizon of the Black hole of a Positive Energy Particle.
>
> The zone of interference which lies beyond the zone of interaction starts beyond zone of interaction and lies until the Gravitational limit beyond which limit the time changes happening due to mass of a particle go below Planck time. even this name wasn't found useful although this is the zone where the time resetting waves lie and interact/interfere with similar time resetting waves from other particles, thus determining a lot of properties of the particle including momentum, kinetic energy and gravitation.

<u>Any attempt to move E (acting from left to right, left being the receding edge and right being the preceding edge) would mean that multiple PCs will have to be moved from one time zone to their neighbouring time zones.</u>[2*] As E moves to the right, the PCs at the preceding edge of constituent particles that make up E, would disappear into the centre of the individual particles which has T=0 i.e. PCs with non-zero T would be converted to T=0. PC vacuum created immediately before the particle as it moves will need PCs from the receding borders to move for compensation. New PCs, with lower or no time dilation before, would be in a way recruited at the right edges of each time zones of E. All these changes in the pace of time (resetting of time) need energy (in fact Energy itself is defined as resetting of time i.e. any deviation from Universal Time is the probable moving force in all the natural forces.)

> 2* - Note that this was a significantly wrong line of thinking as initially, due to the effect of GR, time changes induced by a particle with mass were the only means in the mind along with the new found "Active spatial contraction/expansion with deviation of time" according to the assumptions of the theory.
>
> Although relatively flawed, it still can successfully explain a lot of concepts like inertia, kinetic energy, potential energy and even MOND (subsequently).
>
> This line of thinking of time changes occurring primarily and secondarily causing spatial changes was later replaced by spatial changes (destruction) happening at the centre leading to reactionary spatial changes happening around the particle (EPCA formation) which leads to secondary Local time changes due to variable speed of light.
>
> Also note that much later, Inertia at rest is probably easier to explain on the basis of MONDian forces acting on a resting body due to mass of all the bodies in the entire Universe.

Gravitational potential energy

Due to the zones of time dilation around the Earth (or other heavenly bodies), there would be a constant dynamic spatial contraction that pulls every object towards the surface of the Earth. Indeed, if a body with mass (say an apple or a rock) is lifted at a height against this constant tendency of the space in-between the body and the surface of the Earth, at the height h, the gravitational time dilation caused by the body would add up with the local gravitational time dilation caused by the Earth at that height, which would create local changes in time around the body, different than the surrounding. This would be a dynamic process, constantly trying to reset time around the body and in between the body and the Earth's surface, with every tick of the Planck clock. These changes or variations in time constitute the Gravitational Potential energy.

Kinetic energy

If a particle P is moving rapidly, with a velocity v, the kinetic energy possessed by it, in Newtonian mechanics is given by

$KE = 1/2\ mv^2$

According to our theory, with Quantized space-time, as the particle moves (provided it has a constant supply of energy from some source, which keeps providing enough energy to overcome the inertia of particle due to its mass) it has to keep recruiting newer PCs and resetting their time, which constantly needs energy. With every 1 Planck length shift, newer compartments are recruited at the preceding edge. According to the law of conservation of energy, this energy has to be stored in the system as changes in pace of time, till the acceleration of P continues or the energy source of P continues supplying more energy.

At the preceding edge, as the particle P moves from left to right, space immediately on the right contracts even more than the gravitational time dilation would have caused, making time move even slower here. <u>At the receding edge, where the constituent particles of P are running away creating a void, the PCs adjacent to them would have to shift instantly to take their place which means space here would expand and time here would contract or run faster.</u>[3*]

If the particle P is stopped forcibly, the energy stored in these time changes can be released back as space rebounds back and time dilation returns back to gravitational time dilation level.

> 3*This compensatory expansion was later found to be incorrect as the void cannot be compensated by expansion. Instead, the PCs in the receding border have to get entangled, form "an EPCA" and develop volume loss or time dilation to move inwards to compensate for the PC Vacuum.

Note that if the particle is travelling with a speed lesser than the speed of light, at one tick of the Planck clock, it would move less than a Planck length as only light can travel 1 Planck length at 1 Planck time (reaffirming that Planck lengths are not the smallest lengths possible).

At the end of a second, the particle would have travelled multiple PCs. Faster is its velocity, more energy is needed to move forwards to recruit more PCs, more would be the time changes. As the time changes are happening at both preceding and receding edges, one can understand why kinetic energy is proportional to the square of the velocity rather than just velocity. It also figures that if the mass of the particle is increased, more would be the time dilation in various zones, more would be the original energy needed and more would be the time changes

Relativistic mass of a particle travelling with speed of light

When the particle starts travelling with the speed of light, or velocities comparable to that i.e. probably 90% or even 99 % of the speed of light, each constituent particle, it is made of (which are nothing but energy imprisoned in compactified sub-Planck scale prisons by curling of space-time) is shifting 1 Planck length every 1 Planck second, i.e. at its receding edge, a void is being created which is 1 Planck compartment wide. This is a compartment where Time is running extremely slow i.e. time T is tending towards 0. These T=0 compartments forming with every tick of the Planck clock, extremely close to the T=0 compartment-prisons at the centre of the particle, could account for the relativistic mass of the particle as it keeps travelling with the speed of light. The more Planck lengths it travels, the more T=0 compartments are formed and added to its mass, the more energy it needs to propel forward.

Thus the theory allows for a relativistic mass and can explain difficulties encountered to propel particles faster-than-speed of light and also possibly explain the exact basis of relativistic masses as confirmed in various particle accelerators.

Newton's first law – non- accelerated motion in vacuum with no unbalanced force acting on a body

Newton's first law states that

"A body continues to be in the state of rest or of uniform motion until it is acted upon by an unbalanced force."

Although this law has been proven time and again and is one of the foundational laws on which our conventional mechanics depends, the exact mechanism of why it happens is not explained. It is assumed that this is the way nature is without the need for explanation.

But our theory can explain both these. As explained above, it is not difficult to explain why a body at rest offers some resistance to motion (the first part of the law). The second part of the law is a bit more difficult to explain.

To explain this, let's consider a body B which is perfectly spherical and with a mass m and diameter d.

We know that the body, due to its mass, causes multiple time dilation zones arranged in perfect concentric spheres around B. We know that there is a finite distance (depending on the mass) beyond which the time dilation ceases to be significant. Let this diameter be D.

The length D-d is also thus a finite number.

Let's assume that we draw such zones around B, and realize that there is n number of zones and thus the number of edges between individual zones is n. Note that even n will be a finite number and cannot be infinite for a finite mass.

Let's also draw concentric spheres around B every 1 Planck length away. As (D-d) is a finite number, we can draw a finite number of spheres like these. It may be possible that the time running within and beyond at least some of these spheres may have a difference less than Planck time and thus the edges of zones don't need to be equal to the number of these spheres. That is, the breadth of individual zones may be broader than one Planck length.

We know that each of these edges of zones is a sphere with a finite surface area and thus each of these zones is lined both inside and outside by a finite number of PCs.

We know that for this system, as long as we assume that the properties like mass and shape of B do not alter, the distance between these spheres representing the edges of different time zones will remain constant, irrespective of whether the body is at rest or in a state of motion.

Let's draw a plane in the vertical direction dividing B and the time zones into right and left regions.

When B is moved by 1 Planck length, each of these n spherical time zone edges moves by 1 Planck length as well. The time zone on the right region moves towards the right and so the PCs which were lying to the right of each of these edges now has to reset their time to match the higher time dilation of the inner time zone. The time zone edges on the left also move to the right. Thus PCs that were previously lying in a zone with higher time dilation will now lie in lower time dilation and their time would be reset accordingly.[4*]

> 4* - It became clear later that this line of thinking wasn't really consistent with General Relativity, as GR would have suggested that the new position of B would have created a Gravitational wave starting from the new location of the surface of B moving outwards with speed of light resetting time around the particle and also modifying the Contraction status of Spate-time fabric.
>
> The time zones moving spontaneously up until long distances look like fantasy.
>
> However, it was realized later as DGR took shape, that although this shifting of time zones doesn't happen spontaneously all throughout the distance and in fact happen in the form of time resetting waves (TRWs) which travel huge distances (millions of Planck lengths) per Planck time. Thus Gravitational influences travel much faster than speed of light. Essentially, although these influences don't happen spontaneously, from Human perspective, these happen extremely rapidly and thus the following discussion remains valid. However, the inertia, momentum and kinetic energy can better be defined in terms of EPCAs described later, during the formation of which, essentially same things are happening, just spatial changes happening before and time changes following them.

Let X be the number of PCs lying in the right region which had to reset their time to a higher time dilation. Let Y be the number of PCs lying in the left region which had to reset their time to lower time dilation.

What is happening in the right region could be called "recruitment of new PCs to higher time dilation". What is happening in the left region could be called "surrendering of old PCs to a lower time dilation".

Thus with every movement of B by 1 Planck length, X PCs are recruited and Y PCs are surrendered. Note that X and Y just represent the number of PCs. All these PCs had a completely different spectrum of time changes happening within them which are not represented in X and Y.

One can note that, because B is perfectly spherical, the frame of reference can be rotated by as much amount as is needed without changing any of the distances. This means that the numbers X and Y would remain the same even if we move body B up or down or to the left by a Planck length.

Now let's assume that B is moving with velocity v m/sec.

As v is small compared to c, in one Planck time, the distance travelled by B would be lesser than a Planck length and thus we have to consider bigger Planck times to get at least 1 Planck length distance travelled (as Euclidean geometry fails at smaller distances).

Let's thus stick to per second.

B travels v meters in 1 second i.e. it traverses $v \times 10^{35}$ Planck lengths in one second.

Let's assume that B travels in a perfectly straight line, and thus with every 1 Planck length stride, the above process of recruitment and surrender of PCs continues flawlessly for 10^{35} times.

Thus, in 1 second, $(X \times v \times 10^{35})$ new PCs are recruited and $(Y \times v \times 10^{35})$ are surrendered.

All the energy invested in moving the body B by v meters in one second would thus be utilized in bringing about recruitment and surrender of a known number of PCs.

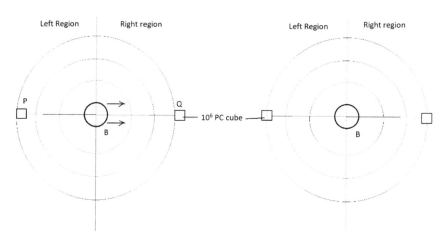

Figure iv/16.2: Shows recruitment and surrender of PCs.

Let us now consider a 10^6 PC cube located just outside the outermost edge on the right side (let's name it Q) and a similar 10^6 PC cube located just inside the outermost edge (let's name it P).

As B moves from left to right with velocity v, after some time, B would progress by 50 Planck lengths. At the same time, the outer edge on the left region would progress well within P and would divide P into two regions, the left one now lying outside the influence of B and thus without any time dilation, i.e. with a contracted time. The PCs lying in this left region of P can be said to be surrendered PCs.

Similarly, even the preceding outermost edge would progress deep into Q and would divide it into right and left zones wherein PCs lying in the left region of Q are newly recruited and thus would start having a slightly dilated time.

Note that like P, the outermost edge would have a high number of 10^6 PC cubes throughout the surface. In fact, every edge of time zones lying in the left region of B would have 10^6 PC cubes like P where PCs are being surrendered to a lower time zone.

Likewise, all the edges lying in the right region of B would have a high number of (but a finite number of) 10^6 PC cubes which would have PCs that are freshly being recruited into a higher time dilation zone.

The left region of P, as time was reset in it, now would have dilated space, i.e. the left region of all the P-like 10^6 PC cubes would expand or contract comparatively slowly.

Likewise, the left region of Q, as time was reset in it, now would have actively contracting space, i.e. the left region of all the Q-like 10^6 PC cubes would contract.

Note that space on the left region would expand and space on the right region would contract, the overall result being that B would be pushed further to the right.

Even if the force which pushed body B to start its movement from left to right vanishes now, every movement of B from left to right would continue this endless process of surrendering and recruitment of PCs and the body would continue its motion until an unbalanced force stops the movement.

In this manner, the current theory can explain the inertia of a body in a non-accelerating frame of reference.

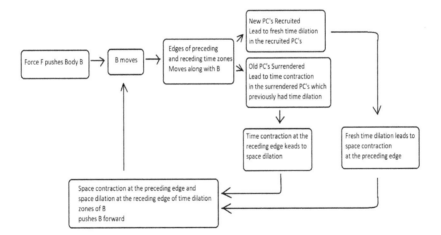

Figure iv/16.3: Shows the cyclical changes in time around a moving particle with mass with recruitment and surrender of PCs leading to active spatial contraction at the preceding edge and spatial expansion at the receding edge that gives another push and keeps it in motion

Newton's second law of motion

It suggests that

"Acceleration of a body produced by a net force is directly proportional to the net force in the same direction and inversely proportional to the mass of the object."

A body that is accelerating is one whose velocity keeps growing with time.

For our body B, if its velocity is v1 in the first second, v2 in the second sec and so on so that

v2>v1 etc.

In this case, the force pushing the body B continues, so that with every second, the number of PCs recruited and surrendered per second would keep increasing.

For example, let's presume that the distance travelled per second doubles.

In short,

1st second - PCs recruited= X, PCs surrendered= Y

2nd second - PCs recruited= 2X, PCs surrendered= 2Y

More is the Force which pushes B, more would be the distance travelled, more is the mass m of B, more is the time dilation per PC, i.e. more time resetting would be needed to overcome the body's inertia.

i.e. acceleration acquired is more if force is more and less if the mass is more, i.e. acceleration of the body is inversely proportional to its mass and directly proportional to the force applied.

Thus $a = F/m$

Or $F = ma$

The theory thus agrees with the second law of Newton.

Newton's third law

It states that

"For every action, there is an equal and opposite reaction."

In space, imagine that an astronaut is floating in zero gravity inside a spaceship or space station. There is another piece of heavy machinery (or a fellow astronaut) floating in one of the passages. Let the masses of the astronaut and machinery be comparable. In short, the zones of interactions and time dilation zones would be comparable. Both the machinery and the astronaut are not acted upon by an unbalanced force.

Now, what will happen if the astronaut tried to push the piece of machinery in one direction?

We know that both the astronaut and the machinery have mass and thus have zones of time dilation around them. When the Astronaut pushes the machine, the energy he puts into the act attempts to recruit and surrender PCs in the vicinity of the machine, these (representing inertia of the machine) time changes oppose the force and this would prevent the machine from moving until the force is enough to overcome these opposing forces. These opposing forces would indeed exert an opposite force on the astronaut's body. When the force applied becomes greater than these opposing forces of inertia, the machine starts moving, and with this nudge, it would keep on moving until it finds an unbalanced force to oppose its strides (as explained above). The inertia of the machine, exerting an opposing force on the astronaut's body would also be an unbalanced force acting on the astronaut, which means that the astronaut would also start moving in an opposite direction to the movement of the machinery.

In short, the current theory does not contradict Newton's third law.

A rock stationed on the surface of the Earth would be pulled downwards by the Earth's gravity. The individual constituent particles from which the rock is made of, along with their respective zone of interactions and time dilation zones, when coming in close proximity with the constituent particles forming the Earth, would, at one distance from each other, be opposed by electromagnetic forces between them, which are explained later and are consistent with the theory and which are the reason why the rock stays separate from the Earth and does not sink in. The Earth exerts an equal and opposite force on the rock as much as the rock exerts on the Earth.

Law of conservation of Momentum

When a body of mass m starts moving with velocity v, it gains momentum which is nothing but an additional unbalanced force needed to stop the body from moving.

The momentum of linear motion is given by

$P = mv$.

Velocity could be constant in which case it will be a non-accelerated frame or it could be increasing or decreasing in which case it will be an accelerating or decelerating frame.

In all these frames, the momentum could be explained based on time changes occurring around the body, which are directly proportional to the mass (as it increases the time dilation) and the velocity as it increases the number of PCs recruited or surrendered per minute.

It is clear that when two bodies collide, there can be a transfer of force, according to this theory, due to two possible mechanisms, the first being direct transfer through the bodies and the second one through the time changes in the zones of interactions of the two objects.

Later it will be shown how the neighbouring PCs are entangled in a manner that the time changes are transferred without any loss or gain of time (which is equivalent to energy) in accordance with the law of conservation of energy.

Most of the energy in the system including the kinetic energy is probably around the particle in the zone of interaction and not the particle itself. When the two bodies collide, the individual constituent particles would not come in direct contact with each other ever and the direct transfer of force is also due to transfer of time changes in PCs immediately around the constituent particles.

Angular momentum

In a rotating body, the influence which the individual constituent's mass has on the zones around the body also rotate and thus the entire zone of interaction rotates. In a perfectly spherical body, the zones of interaction might not appear to change in distance from the centre. However, in an irregular body or non-spherical body, the zones of time dilation vary at different angles and thus rotation would make the zones to vary. At a single point, time may vary according to which time zone it lies at different stages of rotation of the non-spherical body.

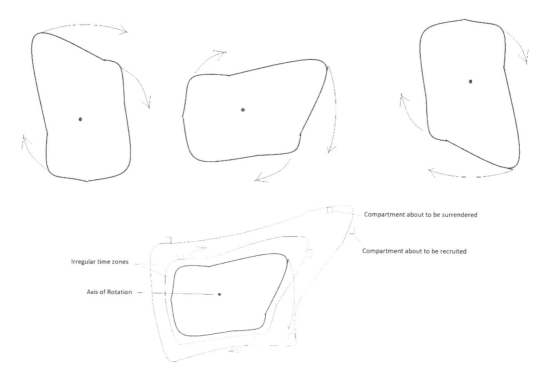

Figure iv/16.4: Shows how PCs are recruited at the preceding edge of a rotating body.

In the above figure, many potential compartments lying at the edges are shown that would get recruited into a higher time dilation zone or surrendered into a lower time dilation zone. As the body moves, more and more PCs change in time which represents the inertia of the body to rotate, which when fulfilled can enable the body to keep rotating in the absence of an unbalanced force (like the Earth or other planets).

Here we come across a contradiction.

Intuition says that in a perfectly spherical body with zero ups and downs of the surface, the time zones would also be perfectly spherical. In such a body, when rotating exactly at the centre, no recruitment and surrender of PCs would result and it indicates that such a sphere would show no resistance to rotation and would have zero angular momentum.

In reality, such a sphere would never exist as most spheres are made of constituent particles which are much higher in diameter than PCs and thus even the smoothest of the spheres would be highly irregular at Planck scales. This invariably means that the time zones would invariably have waviness and there can be no perfectly spherical sphere and there would always be angular momentum.

In addition, the PCs which are present within the substance of the body in between its constituent particles can also undergo time changes. Every point on the rotating body is in fact in a circular motion around a point on the axis of rotation of the body and is under the influence of particles present at other such points.

In short, the perfectly spherical sphere is also in reality, like a porous sieve with significant empty space in between the individual particles.

All these time changes can account for the angular momentum of the body.

The typical example given for the law of conservation of momentum is an athlete doing ice skating. When the athlete is rotating around an axis passing through her body with arms spread out with a certain angular velocity when she brings the arms closer to her body while still in the process of rotating, her angular velocity increases to conserve the angular momentum (to compensate for the change in the distribution of mass).

The energy in the system utilized by the athlete to gain the angular velocity would remain in the system as time changes around her. With the redistribution of mass, the redistribution of time changes in the PCs around her can explain the increase in her angular velocity when she brings her arms closer.

Centrifugal force

This is a force acting on a body that is revolving around the centre of its circular path, which is due to the inertia of the body and is directed away from the centre.

If we consider the above body B and assume that it is travelling in a circular path, the body's time changes (recruitment and surrendering of PCs) responsible for its inertia would force the body into moving in a tangential path. The centrally directed centripetal force would, however, keep pulling it towards the centre. As the body moves rapidly, the radially divergent component of the tangentially directed force generated by inertia adds up from every position along its circular path and forms the "Centrifugal force". Thus the theory can explain how centrifugal force can be created.

Conflict with General Relativity

Here the theory comes in conflict with the General theory of relativity.

General Relativity states that the gravity of the Sun curves space-time and thus Earth which is moving along a straight line seems to turn as the space-time itself is curved.

Our theory however says that irrespective of the curvature in space-time, a body that follows a circular path would inevitably lead a conflict with its inertia which would keep pushing it in a straight line and would experience a centrifugal force which would need a constant centrally directed pull to keep it in orbit.

Fortunately, the theory also provides with the constant centrally directed pull to keep the Earth (and other planets in orbit) due to the dynamic time-dilation mediated spatial contraction.

The following figure iv/16.5 shows the typical example of centrifugal force being in the outward direction and being completely negated by the normal reaction force exerted by the wall on the wheels of the motorcycle or car in the "Well of Death" stunt often performed (in which a motorcycle or a car is driven at very high speeds on the walls of a large well with obliquely placed walls below, but almost vertical walls above.)

Another example often quoted to explain Einsteinian Gravity is the trampoline and the heavy metal balls as shown below.

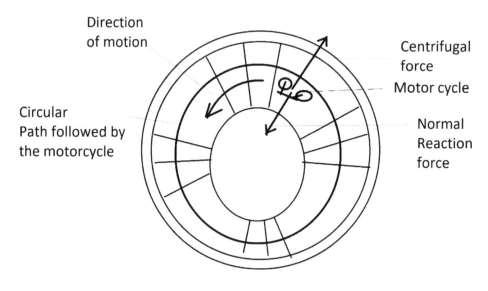

The Well of Death

Figure iv//16.5: Shows "The Well of death" with a motor cycle being driven on oblique walls of a well, with the Centripetal and Centrifugal forces.

Here, the heavier metal ball causes a warp in the elastic surface of the trampoline and the lighter metal ball just follows the curved path of the warp. Even here, despite the curved medium on which it travels, the smaller metal ball does experience a centrifugal force due to its curved path which would be balanced by the elasticity of the material of the Trampoline.

Note that General Relativity argues that the warp caused by the Sun causes the space-time to curve, the Earth is just following a straight line and thus is not experiencing any centrifugal force.

This becomes possible because Relativity insists that there is no rigid reference frame and the reference frame is flexible. However, for our hypothesis, a rigid external reference frame is mandatory despite the flexible space-time.

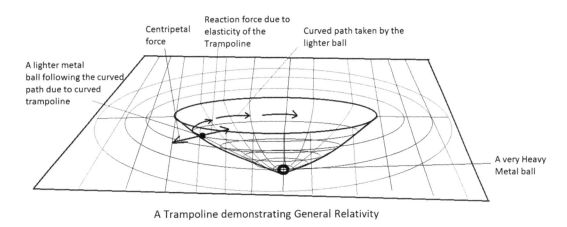

Figure iv/12.6: Shows trampoline analogy to explain GR

Gravitational lensing of the Sun

As predicted by General Relativity, even our theory conforms with the fact that massive objects like Sun, within their gravitational fields, can bend space-time around them due to differential spatial contraction according to the time dilation, its mass causes. This has been experimentally confirmed.

Gravitational lensing caused by Galaxies is a more difficult thing to explain and is explained later following discussion on MONDs.

Gravitational red shift

General Relativity predicts that as any EM wave propagates outwards from the Earth's surface towards the outer space, space just before individual waves is actively contracting thus slowing the innermost waves more than the outer ones. This leads to Red-shifting of the propagating EM wave. This has been confirmed experimentally.

CHAPTER 17

MOND

Range of Gravitational field

The theory predicts, contrary to the usual teaching of infinite Gravitational field, that there exists an upper limit of distance, based on gravitational time dilation which in turn depends on the mass of the celestial object, beyond which the gravitational time dilation caused by the body would become smaller than Planck times and would thus be insignificant.

This means that any additional increase in the distance, beyond the celestial body's gravitational field limit, would make the mass of the body incapable of producing a gravitational time dilation of more than 1 Planck time. Thus there would be no attractive effect on any celestial body at this distance.

This is where the difficulty arises. The Sun or for that matter any star with its finite mass, according to the theory, has a distance called its Gravitational limit beyond which it causes no significant time dilation. How then can the Galaxy have its own gravity, which arguably would probably be at a distance much farther than the gravitational limit of most stars?[1#]

> 1# - Later it was realized that this "Gravitational time dilation" going below Planck time just means that there will be no horizontal spatial contraction i.e. the curved EPCA would be converted to a linear EPCA. This was later called as the Gravitational limit. It was later realized that the vertical volume loss which is essentially the inwards pull due to gravity continues with a constant acceleration even beyond this point and thus the range of Gravitational force is truly infinite. When curved EPCA are converted to linear EPCAs, MOND forces take over. Here the acceleration due to gravity becomes a constant and is called as a_0.
>
> In fact, G which is the proportionality factor in Newtonian gravity equation is probably due presence of this MONDian gravity acting on the gravitating objects with mass from the mass of the entire Universe.

Gravity at higher distances as per Dynamic General Relativity

To understand this, study the figures iv/17.1 "a" and "b" carefully.

The first figure shows multiple stars A, B, C, D, E and F along with a Supermassive black hole G in the centre of a Galaxy. The central circle denotes the star and the outer circle denotes the gravitational field limit. The elliptical region is the gravitational limit of the central supermassive black hole suggesting that even its gravitational limit does not extend beyond the galaxy.

The question would be

"How does the cluster of stars exert forces on the stars like S which are well beyond the Gravitational field limit of most of the stars and also beyond the gravitational field limit of the central supermassive black hole?"

According to our Dynamic General Relativity, all the stars would cause significant time dilation until their gravitational limit.

The question one might ask is

"What happens beyond that?"

The answer is simple. The presence of stars does continue to cause gravitational time dilation well beyond this limit as well. However, it would be below Planck times i.e. it would be smaller than 1 Planck time and thus individually it would be insignificant.

These sub-Planck-time time-dilations, although individually useless, can however still remain relevant.[2#]

> 2# - This concept of Sub-Planck-time "Time dilations" caused by stars beyond their gravitational field limit, although a useful concept may be inadequate in itself to mathematically derive the extent of inward pull that MONDian gravity can provide. The real source of his inward pull, probably is the continued slow inward pull due to linear EPCAs. The real meaning of sub-Planck time time-dilation is that the linear EPCAs have very little horizontal spatial contraction which causes the slowing down of light speed which gives the illusion of time dilation.

Consider the zone P.

The sub-Planck-time time-dilation gradients of the stars A and B would cancel each other out at the zone P and would negate any time gradient caused by each other. They would indeed form a uniform time dilation between each other (despite the fact that it is sub-Planck-time).

The zone P is also under the influence of sub-Planck-time time-dilations from all the other stars, namely C, D, E and F along with the supermassive black hole G. Effects from all these get added up in zone P and form a time dilation gradient of their own which after addition of all individual sub-Planck scale time dilations becomes greater than Planck scale and thus would act just like conventional gravity.

There is a major difference, however. The routine Newtonian gravity and so also gravity caused due to General theory of relativity follow the inverse square law, i.e. the gravitational force caused by it reduces in strength proportional to the square of the distance. Even the strength of gravity proposed in our theory would be dependent on the time dilation gradient which is inversely proportional to the square of the distance.

However, as the time dilation gradient in zone P is caused by the addition of the time dilation gradients of multiple stars, it would follow a slightly different pattern.

The time dilation caused due to G would reduce by inverse square law. But the more obliquely placed stars would prevent its time dilation to reduce so rapidly. The time dilation gradient would probably vary linearly instead of following an inverse square law. In short, the resultant gravity would be a modification of the routine Newtonian Mechanics.

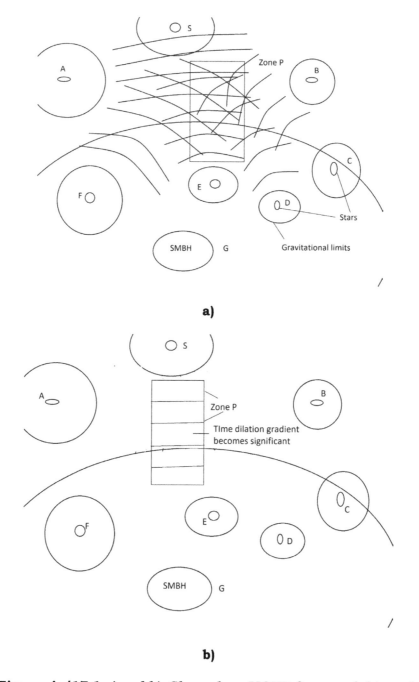

Figures iv/17.1 a) and b): Shows how MOND forces might work

MOND (Modified Newtonian dynamics)

Created by Israeli physicist Mordehai Milgrom, the MOND theory was originally intended to explain the stability of outer stars in galaxies despite their high velocities. The visible mass in the Galaxies was just not enough to explain such high velocities if routine Newtonian mechanics was applied.

Several independent observations suggested this missing mass problem, first described by Fritz Zwicky, subsequently confirmed by the work of Horace Babcock and Vera Rubin.

Newtonian mechanics predicts that the stellar rotation velocities should reduce with distance from the centre of the galaxy. However, these stellar rotation velocities were found to remain constant with distance and the stellar rotation velocities curve remains flat instead of curving downward.

This "missing mass problem" was the reason why the term "dark matter" was introduced to compensate for the extra gravity that is needed to account for these velocities. An alternative to this "dark matter hypothesis" is the MOND theory.

Milgrom figured out that this discrepancy could be solved if the stars present in the outer regions of the Galaxy would experience a gravitational force that is directly proportional to the square of gravitational acceleration (instead of directly proportional to the gravitational acceleration as in Newtonian mechanics.) In other words, if the gravitational force experienced here in the outer regions of the Galaxy varied inversely with the radius instead of the square of the radius as suggested by Newtonian mechanics, the problem would be solved.

MOND and Dynamic General Relativity

Our theory's predictions of the behaviour of Gravity at very large distances goes in line with MOND.

Dynamic General Relativity can predict why gravity behaves differently at such great distances.

It can explain the present rotation velocities without the need for some exotic particle called "Dark matter"

Despite all these claims, detailed mathematical calculations might be needed before concluding that the "dark matter hypothesis" could be wrong.

The same mechanism could potentially explain Galaxy clusters. Even here, due to extremely high distances involved and thus extremely low resultant time dilation gradients (that might be significantly lower than Planck time), detailed mathematical calculations possibly with the help of computers and models may be needed before concluding that this kind of mechanism could explain the gravity acting at these great distances.

Consider figure iv/17.2 given below, which shows an arrangement of stars in a spiral galaxy.

Note here that an average galaxy is about 10,000 to 30,000 light-years in diameter. This means that the individual small circles would be several light-years in size. Thus the small circles are actually gravitational field limits of individual stars and not the stars themselves. Beyond these limits, the gravitational time dilations caused by the stars would be below 1 Planck time, which can potentially add up with similar sub-Planck-time time-dilations of other stars as shown, to form the MOND bonds.

The second figure iv/17.2 b) shows a zoomed-in picture of the same. The four stars A, B, C and D (note that the circles represent gravitational field limits of the said stars and not the stars themselves) are shown to remain bonded by MOND forces. Note that the sub-Planck-time time-dilations of A and C get added up in the zone between them, and create a situation that their time gradients cancel out so that the time dilation closest to the star A gravitational limit is maximum but as we move towards the centre, there will come a point where time dilation will become minimum followed again by further increasing time dilation (again sub-Planck-time time dilation) as one moves from the centre of the two stars (A and C) towards C. <u>The time dilations caused by the other two stars B and D further cancel out the drop in the time dilation as one goes from A to centre and further towards C. The addition of all the time dilations caused by each of these stars in this system causes the sub-Planck-time time-dilations to become greater than a Planck time and thus become significant and form a MOND bond which would constantly keep pulling the two stars together as the Gravity string discussed earlier.</u>[3#]

> 3# - Note the complete focus on time dilations created by the stars with little focus on the spatial changes. A place where there is a sub-Planck time time-dilation was thought to have no spatial contraction and thus explaining MOND was more difficult.
>
> Once the concept of linear EPCAs was understood, the extreme distances no longer limited the vertical spatial contractions which represent just shifting of the linear EPCAs and contribute to the inward MONDian pull that keeps the peripheral stars bonded.

<u>There must exist a stellar density beyond which such MOND bond formation may no longer be possible.</u>[4#]

> 4# - With no limit to MONDian gravity which is essentially due to linear EPCAs, this statement is fortunately not true. In fact, every mass in the Universe is capable of exerting MONDian force on two gravitating objects pulling them apart and acting like an opposition to the attractive force.

Also, note that similar interactions take place with multiple stars, that is A would be bonded to B, C and D due to these interactions. Each of these, i.e. B, C and D would, in turn, be bonded with their neighbouring stars by similar mechanisms.

This is similar to a carbon atom bonded to four carbon atoms around itself in a crystal of diamond.

Thus, MOND creates stars that are bound to each other causing the observed equal velocities of all the peripheral stars in the galaxy. It is as if the galaxy forms a single mega celestial body consisting of all the peripheral stars of the galaxy which is rotating in unison like a mega diamond ring rotating around the centre of the galaxy. Each star that contributes to the mass of the Galaxy would contribute to this force. Thus this force would depend on the overall mass of the Galaxy, the bigger the Galaxy or more massive the Galaxy, the more the MONDian force it can provide and more can be the stellar rotation velocities of the peripheral stars.

Note that these peripheral stars are revolving around the centre of the galaxy with high velocities and thus are subject to tremendous centrifugal forces. The MONDs cancel these out and prevent the stars from flying off in tangential paths.

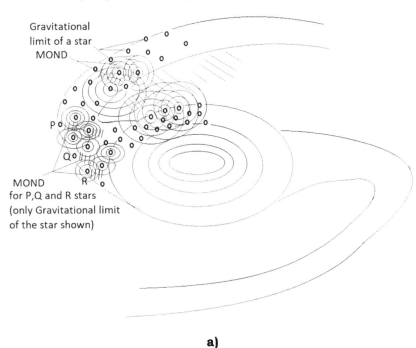

a)

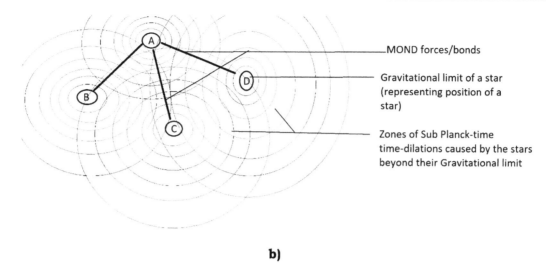

b)

Figure iv/17.2 a) and b): MOND forces in a Galaxy

Dynamic General Relativity and Dark matter

As discussed before, although precise mathematical calculations will be needed to conclude, the theory in principle does not need "Dark matter" to explain the excessive gravity experienced by peripheral stars in the Galaxy.

Instantaneous time dilations or otherwise

As discussed in detail in the "communication between PCs", there is still a controversy regarding whether the information of the location of time dilation zones caused by a star due to its mass is communicated to the individual PCs around the star at the speed of light (as asserted by General Relativity) or it is communicated instantly. Observations suggest that both kinds of communications might be possible.

Indeed, if General Relativity is to be trusted and this information becomes available to the PCs only with the speed of light, the implications are pretty ugly.

Every change in the location of A in the above figure would create a 1 Planck compartment thick resetting gravitational wave that travels across the gravitational limit of the star for years (with the speed of light). The time dilation gradients responsible for these MOND forces would then be the result of a merger of such waves.

<u>The stability of these stars, it implies as per our theory, would be based on waves that left their parent star several years before. Once this resetting wave resets the times in these intervening PCs, it would take another movement of the galaxy and with it, movement of these stars to create another wave that resets time again.</u>[5#] Almost all galaxies are in motion, some approaching other galaxies in a progressing or imminent collision, others bound to other galaxies as clusters. Thus, it is clear that the movement of the galaxies is a rule rather than an exception. This means that the universe would be full of such resetting waves which are formed to reset the gravitational zones around every star or celestial body as it moves.

> 5# - This problem was solved or later was found to be non-existent when it was realized that the TRWs which cause Gravity travel outwards much faster than the speed of light which could be millions or billions of PCs per Planck time.
>
> Also, the TRWs keep evolving every Planck time instead of modifying time with movement. Thus they would arise irrespective of movement of the Galaxies.

EMOND or Extended MOND

We know that Gravity works even for Galaxy clusters. Here we are talking about extreme distances.

As shown in the figure below, in the galaxy clusters, multiple galaxies are observed to be under influence of gravity. Given the distances are astronomically huge, Einsteinian or Newtonian gravity fails to explain their velocities and the missing mass problem persists or in fact, becomes even more pronounced at these levels.

Even MOND may have limitations in explaining what happens in these clusters. A predominant component of these galaxy clusters is intergalactic gas and other matter whose gravity needs to be taken into account. Even after taking that into account, it would take tedious calculations to calculate if the above mechanisms of MOND might be enough to provide with enough gravitational pulls needed to explain their velocities.

Despite these drawbacks, Extended MOND i.e. EMOND can potentially provide enough gravitation needed to explain the motion of the galaxy clusters.

An average galaxy like the Milky Way galaxy contains 100 – 400 billion stars. All of them are arranged in a disc-like configuration in the centre and with multiple spiral arms. Some galaxies may be smaller (dwarf Galaxies) and still others bigger than the Milky way with billions of stars. Each of these stars, on any movement of the galaxy, would generate a resetting gravitational wave as explained above to reset the gravitational time dilations they produce.[6#] This wave can potentially travel outwards in all directions for each of the stars. Each of these waves travels for hundreds or even thousands of years until they meet in the zone between the two galaxies and their sub-Planck-time time-dilations get added up with similar waves which started from other stars forming significant time dilation gradients.

> 6# - This statement is wrong since the TRWs keep forming every Planck time irrespective of movement and have nothing to do with movement as the TRWs are not actually meant for resetting time. they are meant for compensation of volume loss and the time changes happen secondarily. At this point, only focus was on time changes with almost no focus on spatial changes. Thus it was thought that once a time gradient is established, it moves only with movement of the Galaxy.

In figure iv/17.3, are shown two such galaxies. Sub Plank-time time-dilation waves originating from two stars in galaxy 2, namely A and B are shown along with other such waves originating from C, D and E. Only some stars are shown here due to ease of understanding and presentation.

This process would happen for each of the billions of stars forming both these galaxies. The resultant significant time dilation gradients are shown as dark lines. It is indeed clear that these waves interact with multiple other waves at multiple locations and thus there is a possibility of the formation of such E-MOND bonds in multiple directions.

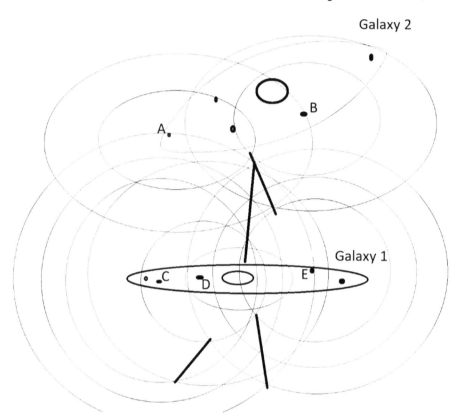

Figure iv/17.3: E-MOND bonds between galaxies and clusters

Gravitational lensing due to Galaxies and Galaxy clusters

A single image of a deep field of Hubble space telescope contains millions of dots, each of them probably a separate Galaxy. The breadth of most of the galaxies observed by highly precise telescopes would be less than a degree. We know that most medium-sized galaxies like the Andromeda galaxy are 3000 to 300,000 light-years in diameter. Most gravitational lensing observed beyond the Galaxies would also be at significant distances from the edge of the Galaxy in question which could be hundreds or even thousands of light-years from the edge. At such high distances, the gravitational time dilations caused by the individual constituents of the galaxies may no longer be significant. Thus even the Gravitational lensing which we see could potentially be caused by zone P-like addition of sub-Planck scale time dilations caused by hundreds of stars.[7#]

Note that unlike the gravitational time dilation caused within the Gravitational limits which would be present as time ones all around the star or the celestial body, the time dilation caused by combined effects of multiple stars would happen only at some regions of the space-time where their sub-Planck scale time dilations add up to a significant level. This means that Gravitational lensing would occur only in certain regions around the galaxy and not all around the galaxy. Even Galaxy clusters can have their insignificant time dilations turned significant and exhibit gravitational lensing.[7#]

> 7# - Again once linear EPCAs were understood, the shifting of linear EPCAs without horizontal spatial contractions, which cause the MONDian gravity as per DGR, can easily explain the spatial curvature and Gravitational lensing done by Galaxies and superclusters without any problem.
>
> Even this claim of irregular gravitational lensing is wrong and is simply due to sole focus on time changes. The Galaxy can keep eating away PCs depending on its mass, wherein every molecule of Hydrogen within it or every Baryonic matter contributing to the mass counts. The linear EPCA forming TRWs from each of the stars just adds up and forms linear EPCA forming TRWs just outside the Galaxy.

Inertia of a Galaxy

As described above, the inertia of an individual particle or a celestial body is caused due to the zones of time dilation that are caused due to the mass of its constituent particles. These zones are arranged concentrically all around the celestial body. Given that the edges of galaxies are at great distances from most of the matter within them which could be well beyond the gravitational limit of most mass in it, the Inertia of the galaxies i.e. its tendency to continue in a state of rest or of uniform motion when not acted upon by an unbalanced force is basically due to individual inertias of the stars which form its structure. Although one might be tempted to think that time dilation zones exist around the Galaxy as a whole due to its mass, the distances are too great and thus no zones would exist at the edges of a Galaxy. The above-mentioned mechanism would mean a zone of highly expanded space or highly contracted (i.e. fast-moving) time might be present around the Galaxy.

Expansion of the Universe

As confirmed by multiple observations, we know that our universe is expanding. More recent evidence suggests that the expansion is in fact accelerating instead of slowing down as what might be expected due to the effect of gravity. Here we are talking about huge distances to the tune of millions or billions of light-years apart. As per our theory of dynamic General Relativity, gravity is caused by gravitational time dilation, which is least expected to act at such huge distances. In fact, at such huge distances from matter, there should be no time dilation, which essentially means that there should be a reverse of time dilation, i.e. time should be contracted.

Contracted time could be relative or absolute. Time running in a time zone 100 km from the Sun is relatively contracted as compared to time running at the surface of the Sun. although even this time is in reality dilated due to the effect of the Sun.

The real absolute contracted time is when Local Time is running faster than Universal Time i.e. the discrepancy between Local Time and Universal Time which is present near the celestial bodies causing the effect of gravity, is reversed so that the appearance of PCs becomes faster than the disappearance of PCs. Here, space would start expanding.

The question one might ask is

"If there exists a mechanism by which presence of a mass in the vicinity has an effect to create a discrepancy between Local Time and Universal Time, what is the default setting in this?"

That is, what would be the setting of this mechanism at places that are at infinite distance from such a body with mass.

It turns out, that the observation of an expanding accelerating universe point to the fact that this default setting, contrary to one might expect, is more inclined towards a more contracted Local Time compared to Universal Time. That is, at great distances from any time dilating influence of bodies with mass, i.e. in the intergalactic space, time is highly contracted i.e. running fast (absolute contraction) and the space is expanding.

The reason for such a default setting is not clear.

This accelerated expansion of the universe, in conventional teaching, is said to be caused by dark energy which constitutes more than 95% of the universe as we know it.

CHAPTER 18

BLACK HOLES

One of the Enigma's of modern physics, black holes are mysterious objects with extremely high density. Although a lot is known about them, a lot of questions remain unanswered.

They are a prediction of Einstein's theory of relativity.

Many sceptical scientists still believe that these objects are not possible practically. Evidence in favour of their existence however is mounting. The first evidence came from the binary system Cygnus X which is supposed to be a system in which a star orbits a black hole.

Astrologers confirmed that the movements of the star imply that it is orbiting something which is not visible, i.e. it is not emitting anything in visible light range, however, x-ray telescopes revealed a source of x rays near it. What is believed to be true is that this is a system of a star and a black hole orbiting each other, in which the black hole is slowly nibbling material out of the star and there is an accretion disc formed around the black hole which leads to the emission of X rays.

Today, a lot of evidence suggests that supermassive black holes exist at the centre of almost all galaxies. Quasars, or quasi-stellar radio sources which are extremely bright objects visible from extremely distant regions of the universe, whose light almost overshadows the light of the entire galaxies to which they belong, are thought to be supermassive black holes that are engulfing material rapidly and have a huge accretion disc formed of orbiting material which generates light of enough intensity to remain detectable despite such vast distances. Stephen Hawking described that black holes, also emit radiation which is called Hawking radiation. *(Hawking, Stephen – 57)*

Black holes have a singularity at the centre, around which there is a region of space-time with an extremely high gravitational field so that even light cannot escape from it once entered.

There exists a boundary around the black hole, called the "Event horizon" which can be considered the "point of no return" beyond which any matter falling into the black hole would need velocities greater than the speed of light to escape from the gravitational pull. It is clear then that what is inside a black hole remains speculative and has no way of being confirmed. However, although we cannot see any event happening inside the event horizon, we can infer the presence of a black hole from its interactions with the surrounding celestial bodies. The black holes possess a mass and more is the mass, larger is the event horizon.

Some black holes are extremely small. These are called primordial black holes.

A spaceship or an astronaut or for that matter any particle with mass crossing the event horizon would never be able to escape from the gravitational pull.

Due to extremely high gravity, there is a significant time dilation (time slows down near a Blackhole) and a resultant space contraction. Beyond a certain point, scientists assert that the difference in space-time between different portions of the spaceship or the astronaut would be so high that they would be forcibly stretched out like spaghetti.

Black holes and DGR

At the centre of the black hole, lies a singularity which is an ill-defined entity where all the laws of physics break down. Here, time would stop i.e. in a singularity T=0.

Black holes have extremely high gravitational fields. So that if we consider concentric spherical rows of Planck compartments around the point of singularity, time would run extremely slowly at them and the time gradient would be extreme.

Thus, by the time any matter/any particles with mass reaching so deep inside a black hole would experience an extreme difference in the flow of time from one end to the other. In short, by the time any particles with mass reach here, all the individual molecules and even atomic constituents would be ripped apart and all the bonds would be broken. Mass would thus be converted into its basic constituent i.e. energy. According to our theory, energy is nothing but T=0 compartments.

Note that according to contemporary physics, as we travel deeper into the black hole, time keeps slowing down and every Planck length which a particle has to travel takes more time so that the particle might take millions of years to reach the region of singularity where it is ripped apart to release the energy contained in the particles. DGR, however, would say otherwise simply because the time which stops at the centre is the Local time, while the Universal time continues relentlessly.

Ultimately, the singularity absorbs these T=0 compartments into itself and this increases the event horizon of the black hole, from where these T=0 compartments can never escape.

In short, the singularity is nothing but a prison of energy i.e. Prison for T=0 compartments. (note that a T=0 compartment vanishes soon, i.e. no compartment remains, it is just its effect that remains and propagates.) The singularity itself also has Local Time T=0 or more appropriately hyper-dilated time with multiple T=0 compartments aggregated together which means time running slower than T=0.

There is another place where we hear this term "Prison for T=0 compartments or Prison for energy" which is inside every particle as described by String theory, wherein the energy is entrapped in additional higher dimensions of space-time due to extreme curvature. Although String theory asserts that it is a small string of energy vibrating within the additional dimensions, which is easier to imagine for our brains as we associate "energy" with "a spark or a lightning", in reality, we see the lightening the way we see as the process gives off light, and in this prison for energy, there is no likelihood of light or any other form of energy escaping. Thus, although there is nothing wrong to imagine it for ease of understanding, it is an inaccurate depiction. More accurately, the centre of every particle could be considered as a miniature black hole where energy is trapped.

There is however a major difference, due to the difference in scales involved. The bigger black holes interact with other celestial bodies with gravity alone. While these miniature black holes interact with other matter surrounding them with other forces as described separately.

A particle with mass can probably absorb more T=0 compartments within this prison to modify its zone of interaction. When an electron absorbs some T=0 compartments from a photon, its zone of interaction probably changes slightly and this leads to time changes around it that make it jump onto a higher orbital level. T=0 compartments can escape from it to form another photon which exits from the electron and the electron comes back to its original lowermost orbital level. This property however is speculative, as it is counter-intuitive to think that a sub-Planck length black hole like a prison can accept or emit further energy.

We know however that a W boson has the property to change a particle from one make to the other. In beta decay, a W boson can get absorbed within a neutron and form a proton. In this manner, an element can get converted into a completely different element. Carbon can create Nitrogen. If W bosons, just like photons are packets of energy, i.e. they are a collection of Planck compartments with a deviation in Local Time from the Universal Time and constitute energy and if it can be absorbed by the nucleus of an atom or one of its constituents, then the above property is likely.

CHAPTER 19

CHARGED PARTICLE

A charged particle, according to classical physics, is a particle that possesses a physical property by way of which it can experience a force when present in an electric field. This is a hypothetical field that forms within a conducting surface when a voltage is applied or more precisely is present throughout the universe. The charge is not described any further and is said to exist as the particle's physical property and is undefinable any further.

When charges are moving, they also generate a magnetic field. Movement of charges is supposed to be mandatory for the formation of a magnetic field or the ripples in it. Although the movement of charges is considered mandatory for the generation of magnetism and thus a static charge cannot show magnetism, nothing more about how the movement actually produces the ripples in the magnetic field and the phenomenon of magnetism is described. "Movement of charges" is an arbitrary term and movement with respect to what is not defined in detail. For example, consider a charged particle which forms a part of a metallic wire through which no current is flowing. Because no current is flowing, there is no magnetism. However, the wire is present on the table which is on the Earth and the Earth is in constant motion. The solar system with the Sun is in constant motion with respect to the centre of the Milky Way Galaxy and even the galaxy is in motion with respect to the Andromeda galaxy. This means that in reality, the wire and the charged particle are in constant motion at a high speed with respect to the cosmos. Does this movement of the charged particle produce any charge? According to classical physics, it should, as any moving charge should. However, we know that no magnetism can be detected if the charged particle and the wire both are static with respect to the Earth and no potential difference exists between the end of the wire.

Electric charges come in quanta, i.e. the minimum possible charge is e. charges could be positive charge or negative charge. Thus protons have a "+e" charge and electrons have a "–e" charge.

Quarks, however, are an exception as they possess a fraction of e as their charge, with Up Quark having a +2/3rd charge and the Down Quark having a -1/3rd charge.

If we have to explain the "charge" as per our theory, it needs to be explained fully by the time running around the particle in various zones and the state of space in these regions. In short, the distribution of time around a charged particle need to be deduced.

These time changes need to explain a lot of observed physicochemical phenomenon like electromagnetism, chemical bonding and physical properties of bonded elements.

The various "time change distributions" possible would be described below and which physical phenomenon they can successfully explain will be discussed. Some of them, we know are grossly wrong and are unlikely. Even then they are discussed to understand how we can possibly arrive at the right time distributions.[1**]

> 1** - The description given below is just a naïve way of imagining charges which I was aware is wrong and would be replaced later by a much more detailed description of the same.
>
> Here again, the excessive focus solely on time changes or time gradient around the charged particle is visible. It was later realized that nature need not be as simple and that the time contraction or time dilation zones need not be as described but will invariably be much smaller. It is critical to realize that although logical, the description was incomplete simply because like charges attract and opposite charges repel. Which means that every charged particle has to have zones of time contraction and time dilation around them unlike what is described with positive charge with one type and negative charge with one type. The slight difference in poles, however, was ingenious and was rightfully retained.
>
> As Matter Antimatter theory came, this naïve way of imagining the charged particles was rightfully given up.
>
> Currently, the description of Charged particle is much more complex.

The simplest time distribution possible around a charged particle is the one in which there is an asymmetrical zone of interaction. In the space-time around the charged particle can be divided into two zones by an imaginary plane passing through its centre. The pace of time significantly differs in the upper zone than the lower zone.

Positive charges and negative charges would indeed have similar but opposite effects.

Two possibilities arise. First, Positive charges are the ones that have time contracted around them and negatively charged particles having time dilated around them. The second is the exact opposite with positive charges with time dilated and negatively charged with time contracted. We know that only one of the two is correct and it will take further research to conclusively prove which of these is true if at all.

The exact geometry, shape, size, extent of time variation and the direction of the gradient is still speculative.

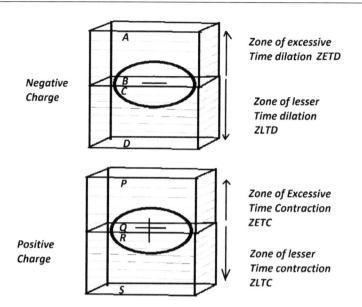

Figure iv/19.1: Shows a diagrammatic representation of zones of time distributions around charged particles, Note the asymmetry between the two poles.

But some assumptions can be randomly made.

We assume here that negatively charged particles like electrons have an asymmetrical time dilation (and thus space contraction) around them while protons (or other positively charged particles) have an asymmetrical time contraction and thus spatial dilation around them.

Multiple possibilities exist

1. The time dilation/contraction zone is uniform and has no time gradient. It forms a cloud on either side of an imaginary plane passing through the charged particle (as shown below). Note here that the two zones exist but the time running in them is uniform and no time gradient exists within the zones. While in both the upper and lower zones, the pace of time deviates from the Universal Time, there also exists significant variation in the pace of time in the upper and lower zones in the same charged particle.

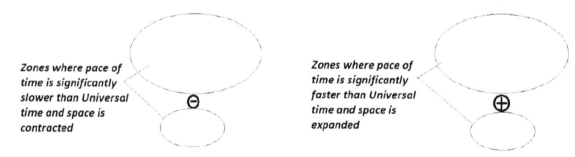

Figure iv/19.2: Shows another possible way by which time changes can be distributed around the charged particles.

1. The time dilation/contraction zones above and below have a time gradient directed towards the centre. This means maximal time dilation or contraction exists closer to the centre and the time deviation reduces as we go away from the centre, in both upper and lower zones.

2. The time dilation/contraction zones above and below have a time gradient directed away from the centre

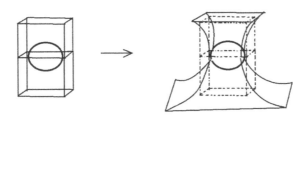

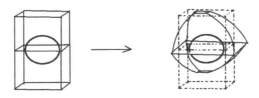

Figure iv/19.3: Shows the status of spatial changes around the above hypothetical charged particles. Note the actively contracting space causing significant spatial curvature.

3. The time dilation/contraction zones above and below have a time gradient arranged in such a manner as to lead to the formation of forces that line up to make the charge as a miniature dipole magnet

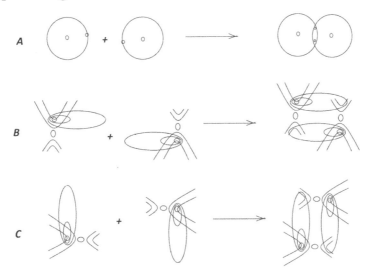

Figure iv/19.4: Shows how the time changes are possibly distributed around a bar magnet. With linear bands of deviated time (either contracted or dilated) starting from one pole to the other.

Each of these would make differing predictions and further research would be needed to elucidate which of the predictions match observations better. Each of these is equally likely presently and deciding conclusively presently is difficult.

For ease of description, we would use the below-mentioned notations. Note that in both the figures, there is eccentricity, and in the first one, greater expansion of space is visible above than below while greater contraction of space is visible below than above in the second figure. The actual geometry may vary significantly than this and these are just used for ease of understanding.

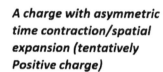

A charge with asymmetric time contraction/spatial expansion (tentatively Positive charge)

A charge with asymmetric time dilation/spatial contraction (tentatively Negative charge)

Figure iv/19.5: Shows how the charged particles can be denoted

Each charged particle necessarily must have a higher time deviation on the upper zone (zone of excessive time dilation –ZETD or zone of excessive time contraction –ZETC) and a comparatively lesser time deviation in the lower zones (zone of lesser time dilation – ZLTD or zone of lesser time contraction-ZLTC)

Charges in absolute zero temperature

Imagine a charged particle is suspended in a region of space.

With absolute zero temperature, there is no availability of electromagnetic waves (which are basically travelling space-time deviations to give energy/velocity to the charged particles.

<u>Even without any external energy source, the time deviations around the particle would persist. This means that asymmetrical spatial contraction due to time dilation around the negatively charged particle (electron) would remain even at such low temperatures. Due to the eccentricity of the pace of time, space would contract faster in one zone than in the other zone. This means that electrons/negatively charged particles will move even in absolute zero. This is true even for positively charged particles as spatial expansion around them is asymmetrical and thus space expands faster in one zone than in the other even at absolute zero.</u>[2##]

> 2## These Zones of excessive time contraction or expansion were later found to be naïve and replaced after the advent of EPCAs by D_{max} and D_{min} which are direction of maximal time dilation and direction of minimum time dilation
>
> It was also realized later that there is a likelihood that the poles keep shifting due to pendulum like movement of the ENEA particles within the lattice which is akin to superposition of spin direction or handedness.
>
> The role of spin was realized after the Matter Antimatter model came to the realization.

See figure iv/19.6 given below.

Note the directions of various zones including those with excessive deviation and those with comparatively lesser time deviations.

For the electron, in the zone on the right, there is excessive spatial contraction compared to the zone on the left which leads to the electron moving towards the right. Even the space in the zone of lesser time contraction does contract. This contraction along with the movement of the particle on the right would create excessive compensatory spatial expansion in the region just beyond the zone of lesser contraction. The rightward movement of the charge would continue incessantly and so will the compensatory spatial expansion on the left side.

The space beyond this compensatory zone would also expand to compensate for the spatial expansion happening in this compensatory expansion zone. This compensatory expansion wave keeps propagating in the direction away from the propagation of this charge.

The Proton or Positron would also have similar zones with excessive time contractions and spatial expansion. The eccentricity of spatial contraction around the proton would keep propagating on the left due to excessive spatial expansion on the right than on the left. Due to excessive spatial expansion on the right, there is also a compensatory spatial contraction just beyond the zone of excessive spatial expansion. This compensatory expansion keeps propagating towards the right as a wave of spatial contraction propagating away from the propagation of the charge.

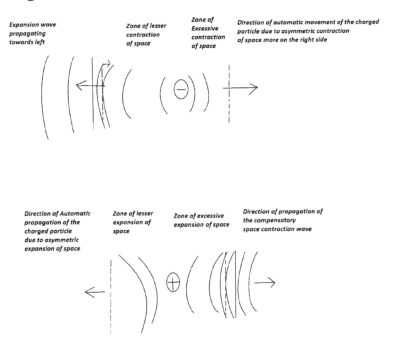

Figure iv/19.6: Shows another possible configuration of the various time zones around charged particles with asymmetry between poles.

Charges at higher temperatures – Pilot wave

At higher temperatures, a higher proportion of Electromagnetic waves would be available in various frequencies.

These electromagnetic waves are spatial contraction/expansion waves with a time gradient propagating through space.[3##]

> 3## - This was indeed a rough approximation of the Electromagnetic waves akin to sound waves. This lodging of a charged particle into an independent EM wave was thought to be an explanation edging towards de Broglie-Bohm's pilot wave theory. This later turned out to be quite different. The vertical time gradients shown turned out to be too vague.
>
> The latest version is an EM wave with nodes being PC aggregates and the surrounding wave being reactionary disturbances including EPCA forming TRW (time resetting waves) propagating outwards from it. Thus the wave is an integral part of the Photon or the Electron in the respective double slit experiments. While the above description matches closely to the current description of Radio-waves which are push-pull bands drifting away and propagating as electric and magnetic flux.

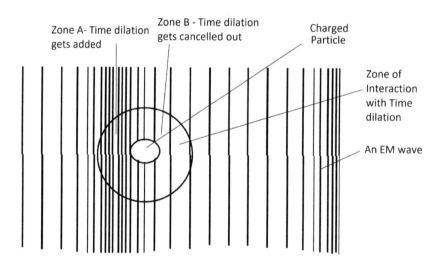

Figure iv/19.7: Shows a charged particle lodged in an Electromagnetic wave as predicted by DGR.

Figure iv/19.7 shows an approximate representation of an electromagnetic wave. This wave is just like a sound wave with a time gradient and a secondary spatial expansion-contraction propagating through space.

When a charged particle comes in contact with such a wave, the particle can get lodged onto the transition zones of these waves.

The figure shows a charged particle with an asymmetrical time dilation around it (a negatively charged particle) lodged in one such wave which is travelling from left to right.

The left zone is the zone of excessive time dilation and the right-sided zone is the zone of lesser time dilation. The left zone of the particle is within the region of the wave with time in a contracted state.[4#]

> 4# - This description was thought of in order to conform with de Broglie's waves wherein an independent wave travelling in space-time where a particle can get lodged within and can be carried around as a surfer is carried around on a wave.

Here the time dilation gets added up. The right zone is in the region where time is in a state of contraction. Here the time dilation gets cancelled out. It is worth noting that the space would start contracting to a higher extent on the left zone and space contraction would be reduced or space might start expanding on the right zone of the particle. In this manner, the particle would start moving in the left direction and would acquire energy from the passing wave.

At room temperatures or comfortable liveable temperatures on the Earth, a high quantity of such electromagnetic waves would be present, more importantly in the infrared spectrum, but in almost all spectra. This explains the mechanism with which particles derive kinetic energy or increase their velocities as the temperature of a gas or liquid increases. This also explains how molecules like H2O (water vapour) with polar covalent bonds with individual atoms retaining some of their charges act as greenhouse gases and can acquire energy from electromagnetic waves.[5#]

> 5# - According to the current description, the charged particles possessing Push pull bands interact with the push pull bands drifting as radio-waves and acquire kinetic energy. Alternatively, the T=0 compartment aggregates in photons can be absorbed by the particles thus transferring their momentum.

Description of a Charged particle based on Classical Electrodynamics

According to Classical electrodynamics (which is the slightly older description and is now replaced by Quantum Electrodynamics), a charged particle modifies its surroundings because of its property of charge. Classically, around a charged particle, lines of forces emerge out which are called "Electric flux". The strength of the electric field is determined by the density of these lines of forces. The electric flux also has a direction indicated by the arrow in such a manner that the flux starts from a positively charged particle and ends in a negatively charged particle. Unlike in Magnetism, there are electric mono-poles. The following figure shows the classic description of a positively charged particle with electric flux emanating from its surface outwards diverging in all directions. A negatively charged particle having radially inwards directed flux coming in from all sides entering its surface.

When the two charged particles interact as shown, the attraction or repulsion happen classically due to the interaction between these electrical fluxes. As shown in the figure, two opposite charges will have the fluxes aligned together so as to have a net attractive force while the like charges (i.e. positive interacting with positive or negative interacting with negative) would have the flux arranged in such a manner that there is a net repulsive force.

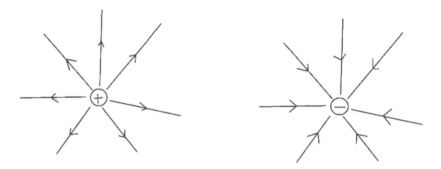

Figure iv/19.8: Shows how charges look as per Classical electrodynamics, with electrical flux lines emanating from the surface.

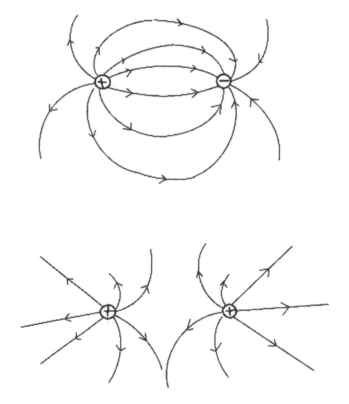

Figure iv/19.9: Shows how like charges repel and opposite charges attract as per Classical electrodynamics.

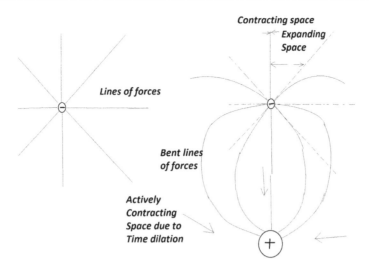

Figure iv/19.9: Shows how the lines of forces or electrical flux could potentially be zones with contracting space with alternating zones of actively expanding space.

Another naïve way of imagining the time changes around charged particles based on this idea would be to consider that these lines of forces or electric flux could possibly be narrow passages with extremely deviated time. Thus two types of such narrow zones or corridors would exist, one with highly contracted time with expanding space and the one with highly dilated time with a contracting space.

The positively charged particle as shown would have narrow zones of time dilation spatial contraction zones coming out radially with interspersed zones with time contraction spatial expansion.

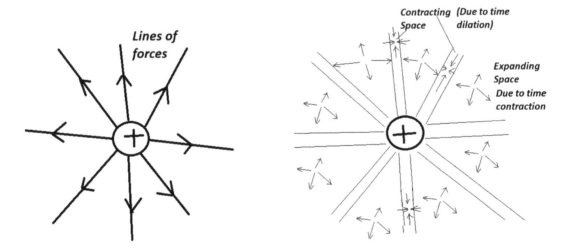

Figure iv/19.10: Shows how lines of forces or electric flux in classical electrodynamics could be narrow passages of highly dilated or highly contracted time with active spatial contraction or expansion respectively.

Pendulum movement of charges

The Electrons staying in orbit will have asymmetrical time changes. It would also possess Momentum including linear and angular momentum. Each of these would add to the asymmetry of time changes around the electron and thus asymmetry in spatial expansion/contraction status around an Electron. Any photon that an Electron in a Bohr orbit can potentially add to these asymmetric time changes and thus add to the Momentum of the Electron. Whether such a Photon can add to or modify these narrow corridors of deviated time is unclear. However, an Electron can acquire some of these asymmetrical time changes from a passing electromagnetic wave and jump up the energy level it is in. The electron can then again release the same time changes with accompanying spatial changes in the form of a photon and regain its earlier lower energy state. This way the Electrons can keep absorbing small quanta of energy and keep moving up down the energy levels in a pendulum-like manner.

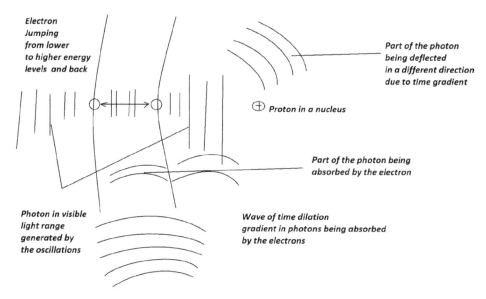

Figure iv/19.11: Shows a naïve representation of how the electron can absorb a photon and jump from one orbit to the orbit with the next energy level.

Lower energy state of oppositely charged particles

Figure iv/19.12 shows how two charged particles can cancel out the asymmetries in time changes around them just by being in the vicinity of another charged particle. In this, the number of time changes would be minimized and this would cause the charges to stay with minimal spatial changes around them, thus remaining stably bonded.

Note that the two charged particles shown to be bonded here have half of their asymmetric time changes negated by facing each other. The asymmetries on the far sides still remain and to negate these and thus attain an even lower energy state, the newly formed bonded particles would try and find another such bonded pair so that even their far sides can bond with and in the process negate some time changes.

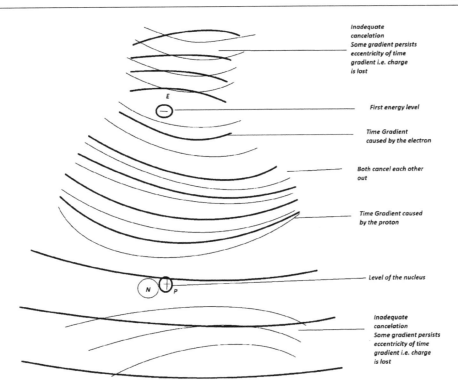

Figure iv/19.12: Shows how two charged particles can cancel out the time changes around each other and thus form a lower energy state while coming closer to each other thus staying bonded. Note that the far sided time changes remain un-neutralized and thus the stability acquired may not be enough.

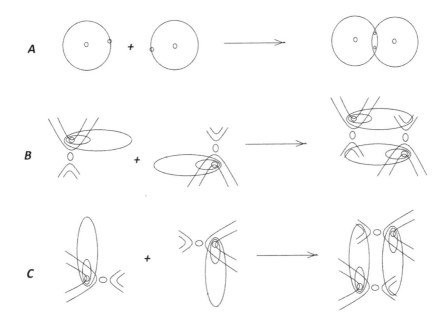

Figure iv/19.13: Shows how a single pair of bonded charged particles have a far-sided time changes un-neutralised and are thus unstable but when bonded to another similar pair so that their far sided time changes are also neutralized, they acquire more stability. A proton-electron pair here is a hydrogen atom while 2 pairs bonded together would be a Hydrogen molecule.

In the above figure iv/19.13 part A, the typical method of depicting a hydrogen atom being bonded with another hydrogen atom to form an H2 molecule is shown. The two Protons bond to their respective Electrons but despite their bond, they remain unstable until they come across another such bonded pair. Part B shows the same two Particles - a Proton and an Electron bonded together with their depicted space-time changes partially being cancelled only on the side facing each other in their "atom" form. When the two bonded pairs come close to each other, the un-negated far-sided space-time changes around them also interact and get partially negated. This minimizes the active spatial changes around them which would tend to disrupt the bonding. Essentially, minimizing time changes around the bonded particles means attaining a low energy state or minimizing disruptive spatial changes thus forming stable bonds. Figure C shows the same thing as figure B, however, the electrons are depicted as Protons and Protons are depicted as electrons just to highlight that these figures are just for understanding and the space-time changes depicted aren't exactly accurate.

CHAPTER 20

ELECTRON

Does an Electron revolve around the nucleus?

The typical Rutherford model of an atom says that Electrons are revolving around the nucleus. (*Geiger, Hans – 46, E. Rutherford – 80*). It figures that the outward centripetal force would then be cancelled by the inward attraction between the positive charged Proton in the nucleus. It was however realized almost instantly that this model is wrong. The Electron is a charged particle. When the Electron revolves, it should keep losing energy and keep radiating energy with every completed revolution. The radius of revolution should keep on reducing with every revolution and the Electron should spiral inwards towards the nucleus until it finally collapses in the nucleus. Theoretically, the above process happens in less than 10^{-10} seconds. Thus, the Rutherford model predicts that every matter should collapse instantly. This isn't what we see in reality which means that this model commonly taught in undergraduate textbooks is wrong.

Niels Bohr suggested that the Electron can stay in only a few well-defined orbits which he called energy levels (*Bohr, Niels – 10,11*) The Electron has to stay in these energy levels and every time it absorbs energy in the form of a photon, it has to jump to a higher energy level. It cannot occupy any intermittent levels. The optical spectrums of various materials constitute solid evidence to support this view.

Quantum Mechanics and specifically de Broglie suggested that Electrons can also behave as waves. This concept is called "the Wave-particle duality" (*de Broglie, L-24*). Although this assumption is being made in Quantum Mechanics, "how exactly is an Electron both a particle and a wave?" is not explained well. However, it is overwhelmingly clear that the Electron is allowed in highly selective orbits.

These orbits, according to the Bohr model, are the ones where the angular momentum is an integer multiple of $\hbar$.

In other words, the angular momentum of an Electron is "Quantised" or can have only certain values, unlike other revolving bodies which can have any value of angular momentum. This weird phenomenon is explained based on de Broglie's standing wave of Electrons. In this, the wave of Electron has a particular wavelength which is determined by the formula

The wave associated with Electron goes all around the nucleus and comes back. Only in those orbits in which the crests of the Electron-wave coincide with crests of the returning wave, the Electron-wave forms a standing wave and the Electron is said to remain stable in these orbits. At any level in which there is a mismatch between the starting wave and the returning wave, there is no formation of a standing wave and thus the Electron wave cannot remain stable.

The diagrammatic depiction of this phenomenon is shown below.

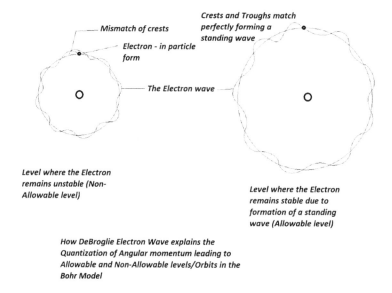

Figure iv/20.1: Shows non-allowable orbit on the left side with the electron wave not able to form a standing wave. Also shown on the right is an allowable orbit where the electron wave has formed a standing wave and can thus remain stable.

Quantum Mechanics suggests that the Electron is in a superposition and its position is given by the Schrödinger equation which just gives the probability of finding the Electron (*Schrödinger, Erwin – 93*)

When no measurement is made, Quantum Mechanics assumes that the Electron is present in all the possible states at the same time, i.e. it is in quantum superposition. The Electron is thus said to be present as an Electron cloud around the nucleus occupying the various orbitals. The exact location of the Electron is impossible to make out.

This is called the Copenhagen interpretation.

An Electron is a quantum particle. Thus it has to follow Heisenberg's uncertainty principle which states that the product of an error in measurement of the exact location of a quantum particle and error in the measurement of its momentum is a constant.

This means that if we are able to measure the Electrons exact location with an acceptable error then the error in the measurement of the momentum increases. Conversely, if we are able to measure the momentum of a quantum particle its location becomes increasingly uncertain. This means that if the Electron can be said to be present at a particular point with reasonable certainty, its momentum becomes almost unmeasurable or the error in measurement of its momentum increases significantly. This theory thus necessitates that an

Electron cannot remain static at one position around the nucleus, if it does its momentum becomes undefinable.

This principle is explained on the basis of wave-particle duality. This is in fact not related to Quantum Mechanics and is a property of all waves. However, the Electron has to also follow another principle called Pauli's Exclusion principle. This states that an Electron cannot take a position that is already taken by another Electron or for that matter another particle with mass. Pauli's exclusion principle does not apply to a wave and thus a wave can superimpose on other waves and cancel each other or strengthen each other. These two principles seem mutually contradictory with Pauli's principle saying that the Electron can possess an exact location in space (say exact x, y and z coordinates) but Heisenberg's principles saying that it cannot.

The more commonly perceived explanation of Heisenberg's uncertainty principle, the measurement problem, is considered an incomplete or erroneous explanation. This explanation is easier to understand and fathom. It says that due to the extremely small size, energy is needed to make a measurement of the position or momentum of the Electron. To know the location of an Electron, energy has to be focussed on a narrow region of space. This means that to measure the small size of the Electron, the EM radiation that is needed has to be low wavelength-high frequency radiation. High-frequency radiation has high energy within as energy contained within any radiation is directly proportional to the frequency.

This energy is big enough to change the momentum of the Electron which is being measured. Focussing of energy to measure the momentum of an Electron accurately leads to a change in the location of the Electron. The majority of measurements of these particles are done by electromagnetic rays which are usually capable of changing both the location and momentum of charged particles.

Wave-Particle Duality and de Broglie Electron Wave emerge effortlessly in our theory

Classical Physics says that the Protons are positively charged and the Electrons are negatively charged and thus both of them attract each other. Quantum electrodynamics suggests that these interactions take place as the two quantum particles exchange massless particles called photons.

The exact nature of what photons are made of or how they make the opposite charges attract (or same charges to repel) is not known or described in physics yet. It is assumed to be the property of charges.

Another important question yet to be answered conclusively is "Why do the Electrons have a ground level?" i.e. Why do Electrons fall into Protons or the nucleus? If Protons and Electrons have an opposite charge, columbic attraction must exist between them even at the lowermost orbits of the atom. Why then does the Electron not fall further inwards? What force prevents it from falling further inwards? Classical Physics has no convincing answer to this apart from the usual "it is the property of Electrons and Protons to form atoms instead of falling inwards towards each other and annihilating with each other (which an Electron does when it encounters a positron – its antiparticle, which also has an equivalent positive charge as the Proton)

The Proton has a much higher mass than a Positron. The nucleus of an atom (which may contain neutrons as well bound together to Protons by the strong nuclear force have significant energy within them. The Strong nuclear force is extremely strong compared to electromagnetism and thus energy is bound to be present in a highly concentrated manner in the nucleus.

Energy is equivalent to T=0 compartments (negative energy is equivalent to T= Tmax)

Thus there is a high concentration of T=0 compartments in the nucleus, especially within the Proton (in the form of what classical physics calls as Gluons that bind the Quarks together) and within the neutron but also in between.

This extreme concentration of T=0 compartments (energy) is akin to a miniature "Blackhole".

This means that there is literally a miniature blackhole at the centre of each atom.

Due to these T=0 compartments, there is bound to be extreme time-dilation in the space immediately around the nucleus.

This extreme time dilation would mean that the space here would be actively contracting (with the appearance of Planck Compartments in the quantum foam remaining much lower than the disappearance)

This is akin to multiple escalators running inwards from the Electron orbits towards the central nucleus.

As discussed earlier in the charges section, charges have a property to modify local time in the space around them.

An Electron and also positively charged particles have linear zones where time is highly dilated or contracted. In the zones where highly dilated time exists, space is actively contracting while wherever highly contracted time exists, space is actively expanding. When like charges come close to each other, the zones where space is expanding come in direct contact and thus there forms an active force of repulsion in the space in between them. In contrast, when two opposite charges come close to each other, the contracting zones come near each other and there occurs a resultant force of attraction between the two.

The exact geometry of these is difficult to imagine here and would need detailed computer simulations.

These zones form due to the standing waves created by interference of waves emanating from the surface of charges.

The Matter-Antimatter model of charges (although hypothetical presently until proven otherwise) can give a pretty apt explanation as to what waves are created, how they lead to what combination of zones as discussed separately.

These waves starting from the surface of an Electron, when it is in the vicinity of a nucleus, with its actively contracting neighbourhood, would be modified accordingly due to the constant inward movement of the Planck compartments which carry the information of these waves.

See figure iv/20.2 below.

A particle of light travelling tangentially near the vicinity of a nucleus where space is contracting actively at a pace comparable to the speed of light is shown. Here, due to the proximity of the nucleus of the atom, with a high concentration of T=0 compartments, we can assume that the space is contracting 1/40th the speed of light.

When the particle of light in question moves with the speed of light, it travels one Planck compartment per one Planck time. After going in a straight line for 39 Planck compartments, the 40th Planck compartment (or a group of PCs) possessing the information being transmitted by the particle move downwards towards the nucleus. Thus, with every 40 PCs, the path of this light particle deviates towards the nucleus.

Faster is the contraction of the space, faster is the deviation.

The same thing happens in the vicinity of a Blackhole. Classical physics describes that immediately beyond the event horizon of a Blackhole, the space-time is curved in such a manner that a ray of light entering tangentially keeps travelling along its circular edge, and thus forms the accretion disc of the Blackhole. This curved space-time which the General Theory of Relativity describes to be the property of matter could be explained due to this active contraction of space.

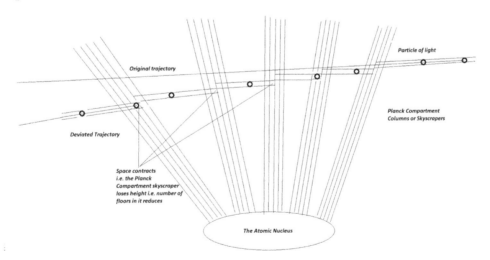

Figure iv/20.2: Shows how the geodesic around the Nucleus curves due to active spatial contraction which deviates the trajectory of a particle of light. Note that the vertical lines are imaginary columns or skyscrapers of PCs and with actively contracting space, they keep losing stories and thus lose height thus pulling the particle of light downwards.

The left part of the figure shows a charged particle P present in a region of space-time where there is no active contraction happening. Here, the waves emanating from the charged particle keep moving all around the charged particle in concentric circles. The exact shape of the standing waves generated due to constructive or destructive interference of the various waves is discussed in detail in the matter-antimatter model of charges.

However, we can consider various conical sections of the spheres A, B, C, D, E and F exiting the charges at different angles as shown. When there is no active contraction of space in the vicinity, their paths go outwards without any deviation and as time passes, their locations form concentric spheres.

If, however, the active spatial contraction is present, the same cones show significant deviation and thus the wave-front they form when seen collectively is no longer a sphere.

Cone A would not see any deviation but would travel forwards slower than the rest. The cones E and to some extent D would appear to move downwards faster than expected. The cones B, C and F would also show deviations in their paths pointing downwards depending on the location of the cones and the velocity of active spatial contraction.

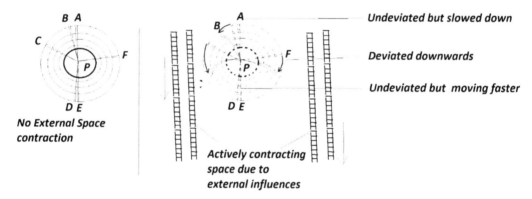

Figure iv/20.3: Shows how conical sections move out un-deviated from P if there is no actively contracting influence in the left part of the figure. In the right part of the figure, actively contracting space with the skyscrapers or stacks of PCs actively losing volume leads to change in geodesic and thus bends or affects each of the conical sections A to F.

Difference between the classical description of a particle-wave and the predicted wave pattern

It is worth noting here that de Broglie described the particle wave associated with the Electron as a two-dimensional wave with crests and troughs like those occurring on a string. These are just descriptive diagrammatic depictions and need not be anywhere close to reality.

Our theory would predict a different type of wave emanating from a particle with mass.

A particle with mass, as it is T=0 compartments condensed in a small region of space, would give off spherical waves which start from the surface of the particle outwards with resetting of time in the immediately adjacent PCs in such a manner that as we move away from the centre of the particle, the reduction in time follows the inverse square law. With every contracted time wave which is the active component of the wave, there is a passive component of the wave which represents the reaction of the surrounding PCs to the presence of the highly dilated time in the resetting wave. The crests of the wave associated with the particles would hen be the time-dilated component and the trough would be the time contracted passive expansion of the space immediately adjacent to it.[1*]

> 1* - This secondary or passive expansion of space, although inserted here so that it matches with the crest-trough type model of typical de Broglie waves, was later found to be difficult to sustain simply because expansion of volume can never compensate for loss of volume as any expansion of volume happening all around the sphere would increase its surface area and thus increase in radius i.e. move outwards instead of moving inwards to compensate for the central destruction.
>
> It became clearer that with spin of the charged particles, the push-pull bands emanating from the surface of the charged particle will spiral outwards with the speed of light without ever meeting each other and this alternate arrangement of highly contracted and highly dilated time can give an ideal wave with crests and troughs.
>
> Even then, the exact description of this predominantly time-contraction spatial expansion type electron wave going all around the nucleon and meeting on the opposite side, deciding the allowable Bohr orbits and also preventing the Electron from falling further inwards below the allowable Bohr orbits, is elusive and will have to be worked out further.

The waves emanating from the Electron, as discussed in the matter-antimatter model, would be even more complicated with standing waves forming zones of highly time dilated and highly time contracted regions of space radiating from the centre.

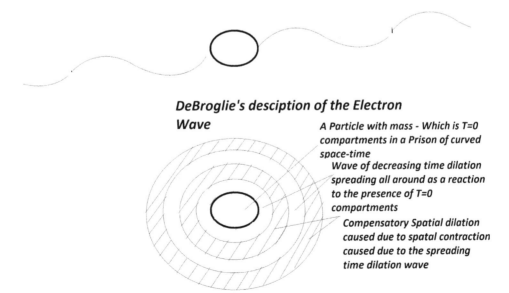

Figure iv/20.4: Shows the difference in waves predicted by de Broglie and the one predicted by DGR.

The waves emanating from the Electron, as discussed in the matter-antimatter model, would be even more complicated with standing waves forming zones of highly time dilated and highly time contracted regions of space radiating from the centre.

The classical description of charges denotes only the lines of forces which probably are corresponding to the highly time dilated zones that are having space that is actively contracting and thus the one responsible for the attraction between opposite charges. The space in between would then indicate the zone which is highly time-contracted and contains a region of space-time that is actively expanding.

The classically denoted negatively charged particle with lines of forces radiating outwards is shown on the left part of the figure.

When such a negative charge comes in close proximity to a conglomeration of T=0 compartments which is the nucleus, the actively contracting space around it would indeed bend the lines of forces downwards or inwards as shown.

This description or depiction is, however, incomplete and in reality, similar lines depicting zones with highly contracted time and expanding space should also be shown. The effect of the neighbouring nucleus with its actively contracting space on these expansion lines will also be similar and that these would also show a tendency towards inward bending like the contraction lines.

The expansion lines that are probably 45 degrees upwards would become horizontal while the horizontally oriented lines would be pointing downwards. Because the nucleus can be considered a miniature black hole and the region where the Electron currently lies can be akin to the region of the accretion disc of the Blackhole, at least some of the inward bending expansion lines would curve around the nucleus as the light travels around the accretion disc of the black hole. This wave of expansion, at specific distances from the centre, would have its crests and troughs match and would form a revolving expansion wave. This circular standing expansion wave can provide the repulsive force needed to hold the Electron stable in specified orbits. However, wherever the crests don't match up with the crests, the revolving expansion wave cannot form a standing wave and would thus not provide the repulsion needed to counterbalance the columbic attraction between Proton and Electron.

<u>The expanding wave would also probably create a shell of expanding space all around the atom thus almost completely obliterating any charge (creating a neutral atom in the case of the Hydrogen atom).</u>[2*]

> 2* - This concept of "an expanding wave all around the atom obliterating the charge" was later removed as it was found to be wrong and unnecessary and not contributing to anything

Note that there would still be some contraction and expansion lines of forces on far sides of the Proton and the Electron which would still make the resultant hydrogen atom unstable unless it reacts with another atom and the remaining lines of forces of the Proton and Electron are negated.

This description has certain implications,

1. The Electron is not the one revolving around the nucleus, the wave emanating from the Electron, more specifically the expansion wave is the one revolving around it.
2. The Electron would thus stay in one position and does have an exact position. This would go against the uncertainty principle, however, that would be the case if Electron was a wave as well, whereas Electron in our theory is a particle that gives off waves.
3. The Electron would still not have a static position simply because of the meson theory, as the location of Proton keeps changing due to the exchange of Pions or pi Mesons (see the description of the Meson theory). The orientation of the new Proton would also vary accordingly and the Electron would thus keep moving with every Proton to Neutron switch to adjust itself to the new Proton position.
4. Superposition i.e. Electron is present in all the possible positions at once is not needed and can be given up.
5. This description doesn't yet explain why only two Electrons, each with an opposite spin to its partner, can be accommodated in a given orbital, which is nothing but Pauli's exclusion principle.

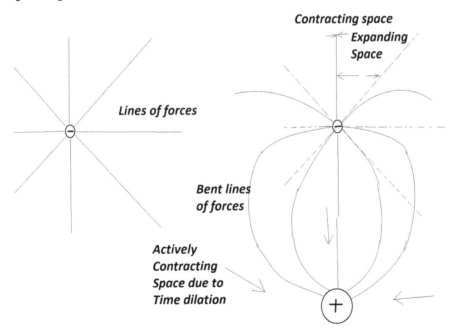

Figure iv/20.5: Shows how actively contracting space near the Nucleus bends the geodesics around it and bends the lines of forces inwards thus facilitating the Proton Electron bond.

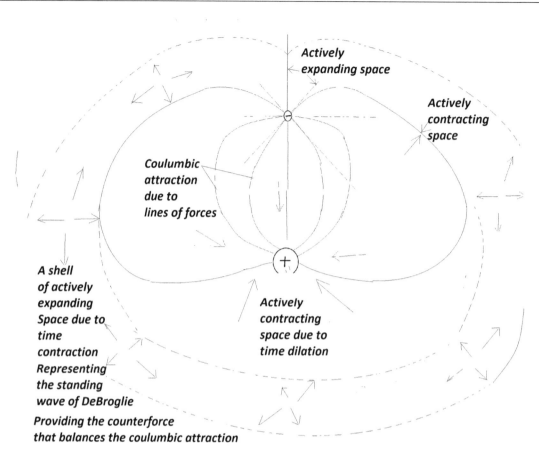

Figure iv/20.6: Shows how a wave of expanding space starting from the Equatorial region of the Electron curves all around the curved geodesic of the Nucleus and forms a standing wave that provides the repulsion required for the Electrons to stop falling inwards and form a stable bond with the Proton

CHAPTER 21

MESON THEORY OF NUCLEUS AND QUANTUM CHROMODYNAMICS

We know that nucleus of the atom is the central region of the atom where the majority of its mass is concentrated. Most of the rest of the space of the atom is free space and the electrons occupy specific orbits around the nucleus.

As discussed separately, the diameter of the nucleus is so small compared to the atom, that if the diameter of the nucleus as compared to the diameter of the sun, the first permitted orbit for the electron wouldn't come until we travel outwards five times the diameter of the solar system. All the intervening space is empty.

The nucleus is where Neutrons and Protons are packed together.

For keeping multiple positively charged Protons together in such close proximity in a stable configuration, there need to be attractive forces between them that exceed the repulsive columbic repulsion. These attractive forces are explained on the basis of the Meson theory which was given by Hideki Yukawa.

The theory suggests that a Neutron needs to be present between two Protons.

The Protons and Neutrons are kept close to each other in the nucleus at a tiny distance of about 2 femtometers. The nuclear force reduces drastically beyond 2 femtometers. Thus the liquid drop model of the nucleus is most favoured which states that the integrity of the nucleus is based on forces exerted on each nucleon on its neighbour.

The Protons emit or exchange a tiny particle called a pi meson or a Pion with the adjacent Neutron.

The Proton, on emitting a Pion becomes a Neutron. The Pion travels towards the adjacent Neutron with velocities nearing the speed of light and on being absorbed, converts the adjacent Neutron into a newly formed Proton.

This means that for a significant period when the Pion is in existence, there are two Neutrons instead of a Proton and a Neutron in the nucleus. Despite this, the nucleons maintain their positions. That is the forces keeping them together remain unchanged while these reactions are happening.

Furthermore, it is clear that while these transformations keep occurring incessantly, the nucleons have no tendency to collapse into each other or touch each other. This points out to a presence of an inwards pointing force that prevents them from moving apart and an outward-pointing force that keeps them apart and prevents their constituents from coming too close to each other.

There are multiple lines of evidence in favour of this hypothesis. The meson hypothesis was described first by Yukawa (*Yukawa, Hideki – 106*)

The pi Meson was first discovered and by English physicist Cecil Frank Powell in 1947.

The hypothesis has serious implications if true. The Neutron and Proton would, not be significantly different in structure and would then differ by the presence or absence or presence of a small particle.

The internal structure of a single nucleon is extremely difficult to study. There is much speculation as to what exactly happens inside a Proton or a Neutron. The exact boundaries, the events happening inside are all speculative at present. However, some things are clear.

It is quite certain that Neutrons have three Quarks within them namely two Down Quarks and one Up Quark. On the contrary, a Proton has two Up Quarks and one Down Quark within it.

Quarks were first described by Murray Gell-Mann and George Zweig in 1964 and Murray Gell-Mann received a Nobel prize in the year 1969 for describing them. (*Gell-Mann, M – 47,48,49*)

Gell-Mann often used to say that it is almost impossible to isolate a Quark independently, if we give adequate energy to extract a Quark from its pair, the energy concentrated will be large enough to produce another pair of Quarks. This might imply that Quarks may represent mathematical tricks rather than being real particles. However, Feynman disagreed and believed that Quarks are real particles and not just born out of mathematics.

The Standard Model of particle physics says that there are six flavours of Quarks. These are further divided into two groups.

The flavours and groups are

Group 1

- Up Quark,
- Charm Quark,
- Top Quark,

Group 2

- Down Quark
- Strange Quark,
- Bottom Quark

As we go down a group, the mass increases so that Top Quark is much more massive than Charm Quark which is, in turn, more massive than the Up Quark. Only the Up Quark and the Down Quark are sable and all the rest are highly unstable and do not survive for a significant time, i.e. decay almost instantaneously after being formed. The other Quarks occur only in high energy particle collisions occurring in particle accelerators. Some of them may be found in cosmic rays which are akin to naturally occurring particle accelerators.

All the members in Group 1 have a positive 2/3 electric charge while all members in Group 2 have a negative 1/3 rd electric charge.

Apart from this electric charge, each Quark also possesses another variety of charges called the colour charge. The strong nuclear force which keeps the nucleons together is said to occur primarily because of this different variety of charges.

The colour charges are of three types i.e. red, blue and green. Thus, the same "Up Quark" can have a red colour charge or blue colour charge or green colour charge. How a Quark acquires a colour charge is not known in detail. However, there are some sub-particles which the Quarks possess which determine their colour. These can thus be named red, blue and green. Each of these can have an antiparticle counterpart which is named anti-red, anti-blue and anti-green.

The exact state of each of the Quarks is unclear within the nucleus. It is likely, that there is spontaneous formation and destruction of multiple Quarks every instant within the nucleus. Thus there may be instances when there are four or five Quarks within a small period.

The Quarks, some say are bouncing within the nucleus at velocities close to the speed of light. One thing is adequately clear. Quarks are bound to the other Quarks. The Quarks can only exist as a doublet or triplet, i.e. it is impossible to isolate a single Quark.

The three Quarks within a Neutron are bound together with bonds that are created due to a constant exchange of "gluons". This means that like the photon is the carrier particle of the electromagnetic force, gluons are carrier particles of the strong nuclear force binding the Quarks together.

The gluons are however significantly different from photons. Photons are massless particles that move with the speed of light. They can pass through each other and thus are incapable of interacting with each other. The Gluons, on the contrary, are capable of interacting with each other and when they cross each other, they form vertices.

The exact nature of gluons (what they are made of) is unclear and they are considered to be fundamental particles i.e. not composed of anything else, i.e. they are indivisible i.e. they are not a composite particle like a Proton. However, every gluon is further made of a particle and antiparticle version of the sub-particles responsible for the creation of the colour charge. Thus gluons can be red-antiblue, green-antired, blue and antigreen etc. The Gluons thus possess colour charge.

The colour charge is completely different from the electric charge and so the Gluons possess no electric charge. A Proton and a Neutron possess three Quarks. However, each Proton and Neutron has to be necessarily colour charge neutral.

This means that every Proton or Neutron must contain one particle each of red, blue and green. This means any reaction taking place within the nucleus has to conserve this colour neutrality of the nucleon.

The anti-colour sub-particles are extremely unstable and remain for a very short period while they are within the Gluons and cannot bind to Quarks. They can, however, bind to anti-Quarks.

The Gluons can change the colour charge of a Quark. However, they do not have the ability to change the flavour of a Quark.

Thus, an Up Quark will remain an Up Quark despite absorbing a gluon. This can change its colour charge. The red colour charged sub-particle in a red colour charged Quark can annihilate with its anti-particle (anti-red sub-particle) in a gluon. The remaining different coloured colour charge remaining with the Quark now becomes its colour charge.

For example, if a red colour charged Up Quark, absorbs a gluon which is composed of a blue and anti-red sub-particle, the Up Quark now becomes a blue Up Quark while the red and anti-red will annihilate to form energy.

From the energy within the nucleus, a pair of blue- anti-blue may form near a red colour charged Up Quark. The red colour charged sub-particle of this Up Quark then binds to the anti-blue sub-particle and form a gluon. The blue sub-particle remains with the originally red Up Quark thus now turning it into a blue Up Quark. The gluon with red and Anti-blue which just formed is emitted out and is absorbed by the adjacent paired Quark. Because of the colour neutrality clause, this adjacent Quark is unlikely to be red colour charged. It would thus lose its original colour determining sub-particle which will annihilate with the incoming anti colour sub-particle and would change its colour into the one which the incoming gluon possesses.

A complex array of such colour-charge changing reactions with the exchange of gluons is constantly supposed to take place within the nucleus.

Within the nucleus in a Proton, a Quark-antiQuark pair with a flavour "Down Quark" may be created with a particular colour charge-anti colour charge.

The Down Quark in it stays within the Proton. The additional Up Quark of the Proton joins the anti-Down Quark.

This "UpQuark-antiDownQuark" pair can leave the Proton to reach the adjacent nucleon.

This Quark-antiQuark pair is called a Pion or pi meson. The exchange of these is responsible for keeping the nucleons close to each other within the nucleus as described above.

When the Pion or pi meson is absorbed by the neighbouring Neutron, the anti-Down Quark within it annihilates with one of the two Down Quarks to form energy and the remaining up Quark binds with the remaining two Quarks, thus transforming the nucleon from a Neutron to a Proton. The Proton that emitted the Pion gets converted to a Neutron.

Effectively what is happening here is that a Proton emits a Pion and gets converted to a Neutron and as the Pion gets absorbed by the adjacent Neutron, it gets converted to a Neutron.

The position of Protons keeps getting shifted from one location to the other within the nucleus.

The Quarks are said to be in a superposition of all the possible reactions possible and the adjacent nucleons would then be in a superposition of being Protons and Neutrons simultaneously.

Each of these reactions has important implications.

Implications of all this information to DGR will be discussed subsequently.

CHAPTER 22

MATTER ANTIMATTER PROBLEM

Einstein's famous equation "E=mc² suggests that energy and mass are interchangeable. A single gram of mass is equivalent to 90 trillion joules (i.e. 9×10^{13} joules) of energy. It figures that every particle with mass has a tremendous amount of energy trapped within it. The atomic bomb explosions of Hiroshima and Nagasaki form a living proof of this fact.

This interchangeability, however, has a limit. Physicist Paul Dirac, while attempting to unite Quantum Mechanics and General Relativity, pondered upon an equation which pointed in the direction of existence of antimatter particles. Antimatter particles are particles which have the same mass as the quantum particle in question, but exactly opposite charge. For the Electron the antimatter particle is called as the positron. For a Proton, the antimatter particle is called as an anti-Proton.

<u>When an Electron comes in close proximity to a positron, physics says that they attract each other due to opposite charges, but unlike a Proton, they do not form stable atoms, but instead keep falling into each other and finally annihilate each other and release all the energy locked within them.</u>[1**]

> 1** In DGR, the matter particles are Positive energy particles and Antimatter particles are the ENEA particles or particles with negative mass which have predominance of creation within them. The ENEA particles, as discussed elsewhere in detail in their properties section, repel both the PEPs and ENEAs. Thus this notion that matter and antimatter attract each other and annihilate is untrue in our theory. Even positrons which have positive charge and same mass as Electron, and represent the true antiparticles of Electron as per physics, would have ENEA particles around them. As the Positron and Electron approach each other, the ENEA particles repel each other and disturb the stable configuration formed and liberate the positive energy locked within the positive energy core, in the form of gamma energy photons.

In the same manner, when a large amount of energy comes in close proximity, it can give rise to a matter-antimatter pair. Note that when energy is converted to mass, it has to necessarily form matter-antimatter pair. However, in the real world, we see everything is made only of matter while in reality antimatter is extremely unstable and it annihilates with matter and is thus extremely short lived. Scientists have prepared antimatter and in fact antimatter has been utilized for certain commercial applications like Positron emissions tomography or PET scan.

There was a time in the early universe when a huge amount of energy was converted into matter and antimatter pairs. The Big Bang theory suggests that a major proportion of the matter and antimatter pairs annihilated and released a large amount of radiation and only a small proportion remained. (*Rogers, P – 83*)

One of the major questions in the modern physics is

- *why there are no antimatter planets or antimatter stars,*
- *i.e. "why our universe has predominance of matter and not antimatter?" and*
- *"where did all the antimatter (which formed while formation of all the matter which we see today) go?"*
- *What makes antimatter so unstable in our universe when scientists suggest that a completely separate universe just like the one visible to us is possible using just antimatter?*

The positron which is antimatter of an Electron also has a very small mass like the Electron but a positive charge like a Proton. The question remains as to why a positron with +1 charge, when interacting with an Electron continues to move towards it unhindered and can annihilate with it but a Proton cannot. What hinders the Proton from falling into an Electron despite the positive charge. What is the difference in the Proton and the positron apart from their mass that make them behave so differently? How come such different particles have exactly equivalent charge but such differing behaviour?

Indeed, we are lucky that an Electron bonds with a Proton to form stable atoms or molecules instead of annihilating with it which if present would be incompatible with formation of the universe.

The unanswered question in a different strongly worded manner is

Why is there something instead of just nothing?

Many theories have tried to give explanation for the matter-antimatter asymmetry. However, none is accepted yet.

After Schrodinger's equation was described, it became clear that it cannot be applied for fast moving objects and is not compatible with relativity.

Paul Dirac described the modified relativistic equation for the same. However, he soon realized that his equation seemed to have 4 answers instead of the 2 expected by applying theories at that time. Paul Dirac realized that the Electron can have positive energy as well as negative energy. To explain the weird nature of negative energy, Dirac had proposed the "Dirac

Hole hypothesis" in which there exists a sea of infinite Electrons and that wherever there is a hole in the sea of Electrons i.e. there is a deficiency of Electron in the sea acts as a positively charged object.

Dirac's relativistic version of the Schrodinger's equation successfully predicted the existence of antimatter which was confirmed later by Carl Anderson in 1932.

Currently, the most accepted explanation of antimatter, according to the Quantum Field Theory (QFT) is that there is a field present all throughout the universe for every known particle in the "Standard Model of Particle Physics" and that each of these particles is nothing but a deviation in the field from its baseline. Thus there is an Electron field and an Electron is a positive deviation in it, wherein the negative deviation in this Electron field is the antimatter of Electron or the Positron.

This however does not explain why our universe seems to be made predominantly of matter and that antimatter is extremely short lived and unstable and decays in no time. Paul Dirac hypothesized that like we have atoms, molecules, planets, stars made of matter, the laws of physics allow for the formation of these same with antimatter particles so that there should be antimatter atoms, molecules, stars etc. An entire universe made up of antimatter is also possible. In fact, he claimed that, formation of the entire living beings just like humans made entirely of antimatter is also possible. However, these antimatter humans would have no way to prove that they are made of antimatter and that they would then call the antimatter they are made from as "matter".

Spontaneous generation of matter and antimatter particles from nothing

Quantum physics says that from "nothing", a pair of matter and antimatter can get generated and the same can indeed annihilate with each other and disappear. For spontaneous creation of particles with mass, enough energy must be present in adequate density. The typical reaction described in nuclear physics is a high energy gamma ray photon (or more appropriately 2 gamma ray photons coming together from different directions) and form a particle-antiparticle pair which when annihilating with each other can reform a photon with same energy as before. Due to law of conservation of linear momentum, this reaction cannot happen with a single photon i.e. when the particle antiparticle annihilates, they produce two photons of equal energy which are moving in exactly opposite directions to conserve linear momentum.

BIBLIOGRAPHY AND REFERENCES

Aaij, R – CP violation

1. Aaij, R.; et al. (LHCb Collaboration) (30 May 2013). "First Observation of CP Violation in the Decays of B0s Mesons". Physical Review Letters. 110 (22): 221601. arXiv:1304.6173. Bibcode:2013PhRvL.110v1601A. doi:10.1103/PhysRevLett.110.221601. PMID 23767711. S2CID 20486226.

2. R. Aaij; et al. (LHCb Collaboration) (2019). "Observation of CP Violation in Charm Decays" (PDF). Physical Review Letters. 122 (21): 211803. Bibcode:2019PhRvL.122u1803A. doi:10.1103/PhysRevLett.122.211803. PMID 31283320. S2CID 84842008.

Abe, K. – CP violation

3. Abe, K.; Akutsu, R.; et al. (T2K Collaboration) (16 April 2020). "Constraint on the matter-antimatter symmetry-violating phase in neutrino oscillations". Nature. 580 (7803): 339–344. arXiv:1910.03887. Bibcode:2020Natur.580..339T. doi:10.1038/s41586-020-2177-0. PMID 32296192. S2CID 203951445.

Abbott, B - WMAP Satellite data CMB

4. Abbott, B. (2007). "Microwave (WMAP) All-Sky Survey". Hayden Planetarium. Archived from the original on 2013-02-13. Retrieved 2008-01-13.

Almheiri's firewall/AMPS firewall

5. by A Almheiri · 2012 · [1207.3123] — Black Holes: Complementarity or Firewalls? Authors: Ahmed Almheiri, Donald Marolf, Joseph Polchinski, James Sully · DOI: 10.1007/JHEP02(2013)062

Anderson, Carl D - Discovery of the Positron

6. Anderson, C. D. (1933). "The Positive Electron". Physical Review. 43 (6): 491–494. Bibcode:1933PhRv...43..491A. doi:10.1103/PhysRev.43.491.

7. (1936). "The Production and Properties of Positrons". Retrieved 10 August 2020.

Bell, J. S. - Bells inequalities

8. Bell, J. S. (1966). "On the problem of hidden variables in quantum mechanics". Rev. Mod. Phys. 38 (3): 447–452. Bibcode:1966RvMP...38..447B. doi:10.1103/revmodphys.38.447. OSTI 1444158.

9. Bell, J. S. (1964). "On the Einstein Podolsky Rosen Paradox" (PDF). Physics Physique Физика. 1 (3): 195–200. doi:10.1103/PhysicsPhysiqueFizika.1.195.

Bohr, Niels - Bohr orbit

10. Niels Bohr (1913). "On the Constitution of Atoms and Molecules, Part I" (PDF). Philosophical Magazine. 26 (151): 1–24. Bibcode:1913PMag...26....1B. doi:10.1080/14786441308634955.

11. Original Proceedings of the 1911 Solvay Conference published 1912. THÉORIE DU RAYONNEMENT ET LES QUANTA. RAPPORTS ET DISCUSSIONS DELA Réunion tenue à Bruxelles, du 30 octobre au 3 novembre 1911, Sous les Auspices dk M. E. SOLVAY. Publiés par MM. P. LANGEVIN et M. de BROGLIE. Translated from the French, P. 114.

Bondi et al - Negative mass

12. Bondi, H. (July 1957). "Negative Mass in General Relativity"". *Reviews of Modern Physics.* **29** (3): 423. doi:10.1103/RevModPhys*Reviews of Modern Physics.* **29** (3): 423. doi:10.1103/RevModPhys.29.423.

Babcock, H. – Flat Rotation curves

13. Babcock, H. (1939). "The rotation of the Andromeda Nebula" (PDF). Lick Observatory Bulletin. 498 (498): 41. Bibcode:1939LicOB..19...41B. doi:10.5479/ADS/bib/1939LicOB.19.41B.

Bekenstein, A – Black hole entropy

14. Bekenstein, A. (1972). "Black holes and the second law". Lettere al Nuovo Cimento. 4 (15): 99–104. doi:10.1007/BF02757029. S2CID 120254309.

Bennett – COBE satellite CMB

15. Bennett, C.L.; et al. (1996). "Four-Year COBE DMR Cosmic Microwave Background Observations: Maps and Basic Results". Astrophysical Journal Letters. 464: L1–L4. arXiv:astro-ph/9601067. Bibcode:1996ApJ...464L...1B. doi:10.1086/310075. S2CID 18144842.

Bohm, David - Bohmian mechanics

16. Bohm, David (1952). "A Suggested Interpretation of the Quantum Theory in Terms of "Hidden Variables" I". Physical Review. 85 (2): 166–179. Bibcode:1952PhRv...85..166B. doi:10.1103/PhysRev.85.166

Carbone, A. – C P Violation

17. Carbone, A. (2012). "A search for time-integrated CP violation in D0→h-h+ decays". arXiv:1210.8257 [hep-ex]. LHCb Collaboration (2014). "Measurement of CP asymmetry in D0→K+K- and D0→π+π- decays". Journal of High Energy Physics. 2014 (7): 41. arXiv:1405.2797. Bibcode:2014JHEP...07..041A. doi:10.1007/JHEP07(2014)041. S2CID 118510475.

Arthur Compton - Compton scattering

18. *Compton, Arthur H. (May 1923). "A Quantum Theory of the Scattering of X-Rays by Light Elements" (PDF). Physical Review. **21** (5): 483–502. Bibcode:1923PhRv...21..483C. doi:10.1103/PhysRev.21.483. Archived from the original (PDF) on 2012-04-15. Retrieved 2011-10-04.* (the original 1923 paper on the APS website)

Cirigliano, D - Inflationary Cosmology

19. Cirigliano, D.; de Vega, H.J.; Sanchez, N. G. (2005). "Clarifying inflation models: The precise inflationary potential from effective field theory and the WMAP data". Physical Review D (Submitted manuscript). 71 (10): 77–115. arXiv:astro-ph/0412634. Bibcode:2005PhRvD..71j3518C. doi:10.1103/PhysRevD.71.103518. S2CID 36572996.

Debs, Talal A - Twin paradox

20. Debs, Talal A.; Redhead, Michael L.G. (1996). "The twin "paradox" and the conventionality of simultaneity". American Journal of Physics. 64 (4): 384–392. Bibcode:1996AmJPh..64..384D. doi:10.1119/1.18252.

De Broglie waves

21. de Broglie, Louis (1970). "The reinterpretation of wave mechanics". Foundations of Physics. 1 (1): 5–15. Bibcode:1970FoPh....1....5D. doi:10.1007/BF00708650. S2CID 122931010.

de Sitter W. on General Relativity

22. W. de Sitter, M. N. Roy. On Einstein's Theory of Gravitation and its Astronomical Consequences: Astron. Soc., lxxvi. p. 699, 1916; lxxvii. p. 155, 1916; lxxviii. p. 3, 1917.

23. W. de Sitter. Space, Time and Gravitation: The Observatory, No. 505, p. 412. Taylor & Francis, Fleet Street, London.

De Broglie's paper on Matter waves

24. de Broglie, L. (1927). "La mécanique ondulatoire et la structure atomique de la matière et du rayonnement". Journal de Physique et le Radium. 8 (5): 225–241. Bibcode:1927JPhRa...8..225D. doi:10.1051/jphysrad:0192700805022500. (de Broglie, Louis: Wave mechanics and the atomic structure of matter and radiation, Journal of Physics and Radium, Volume 8, Number 5, Pages 225-241, Pub Date: May 1927, DOI: 10.1051/jphysrad:0192700805022500, Bibcode: 1927JPhRa... 8..225D)

Dewdney, C. - Pilot waves

25. Dewdney, C.; Horton, G.; Lam, M. M.; Malik, Z.; Schmidt, M. (1992). "Wave–particle dualism and the interpretation of quantum mechanics». Foundations of Physics. 22 (10): 1217–1265. Bibcode:1992FoPh...22.1217D. doi:10.1007/BF01889712.

Dicke, R. H. – Origin of Inertia

26. Dicke, R. H. (1957). Gravitation without a principle of equivalence. Review of modern Physics 129(3), 363–376.

Dirac, Paul - QED

27. P. A. M. Dirac (1927). "The Quantum Theory of the Emission and Absorption of Radiation". Proceedings of the Royal Society of London A. 114 (767): 243–65. Bibcode:1927RSPSA.114..243D. doi:10.1098/rspa.1927.0039

Paul Dirac - Prediction of antimatter

28. Dirac, P. A. M. (1928). "The quantum theory of the electron". Proceedings of the Royal Society A. 117 (778): 610–624. Bibcode:1928RSPSA.117..610D. doi:10.1098/rspa.1928.0023

Paul Dirac's Sea of Electrons

29. Dirac, P. A. M. (1930). "A theory of electrons and protons". Proceedings of the Royal Society A. 126 (801): 360–365. Bibcode:1930RSPSA.126..360D. doi:10.1098/rspa.1930.0013.

Paul Dirac's Large Number Hypothesis

30. Dirac, P. A. M. (1938). A new basis for cosmology, Proc. Roy. Soc. London A 165, 199–208.

Eddington, A.S. - Eddington's paper Confirming General Relativity

31. A. S. Eddington , The Total Eclipse of 29th May 1919, and the Influence of Gravitation on Light, ibid., March, 1919.

Eddington - on THE GENERAL THEORY of Relativity,

32. Report on the Relativity Theory of Gravitation: A. S. Eddington. Fleetway Press Ltd., Fleet Street, London

ALBERT EINSTEIN - Einstein's Special theory of Relativity and General theory of Relativity

33. RELATIVITY, THE SPECIAL AND GENERAL THEORY BY ALBERT EINSTEIN, Ph.D. PROFESSOR OF PHYSICS IN THE UNIVERSITY OF BERLIN TRANSLATED BY ROBERT W. LAWSON, M.Sc., UNIVERSITY OF SHEFFIELD, NEW YORK, HENRY HOLT AND COMPANY, 1920.

34. Das Relativitäts prinzip (The Principle of Relativity) H. A. Lorentz, A. Einstein, H. Minkowski, Fortschritte der mathematischen Wissenschaften (Advances in the Mathematical Sciences),

35. The Principle of Relativity: E. Cunningham. Camb. Univ. Press. Relativity and the Electron Theory: E. Cunningham, Monographs on Physics. Longmans, Green & Co.

36. The Theory of Relativity: L. Silberstein. Macmillan & Co.

37. The Space-Time Manifold of Relativity: E. B. Wilson and G. N. Lewis, Proc. Amer. Soc. Arts & Science, vol. xlviii., No. 11, 1912.

Einstein's EPR paradox

38. Einstein, A; B Podolsky; N Rosen (1935-05-15). "Can Quantum-Mechanical Description of Physical Reality Be Considered Complete?" (PDF). Physical Review. 47 (10): 777–780. Bibcode:1935PhRv...47..777E. doi:10.1103/PhysRev.47.777.

Einstein's VSL theory

39. Einstein, A. (1911). On the influence of gravitation on the propagation of light. Annalen der Physik 35(German org: Uber den Einfluss der "Schwerkraft auf die Ausbreitung des Lichtes), 35

Einstein's photoelectric effect paper

40. MLA style: The Nobel Prize in Physics 1921. NobelPrize.org. Nobel Prize Outreach AB 2021. Fri. 24 Sep 2021. <https://www.nobelprize.org/prizes/physics/1921/summary/>

Einstein's E=mc2 paper

41. A. EINSTEIN, DOES THE INERTIA OF A BODY DEPEND UPON ITS ENERGY-CONTENT? September 27, 1905, this is an English translation of his original 1905 German language paper (published as Ist die Trägheit eines Körpers von seinem Energiegehalt abhängig?, in Annalen der Physik. 18:639, 1905) which appeared in the book The Principle of Relativity, published in 1923 by Methuen and Company, Ltd. of London.

FARNES J.S. - Negative Mass and expansion of the Universe

42. FARNES J.S. A unifying theory of dark energy and dark matter: Negative masses and matter creation within a modified ΛCDM framework. 2018A&A...620A..92F - Astronomy and Astrophysics, volume 620A, 92-92 (2018/12-1)

Friedman, A - Hubble expansion of the Universe

43. Friedman, A. (December 1922). "Über die Krümmung des Raumes". Zeitschrift für Physik. 10 (1): 377–386. Bibcode:1922ZPhy...10..377F. doi:10.1007/BF01332580. S2CID 125190902. (English translation in Friedman, A. (December 1999). "On the Curvature of Space". General Relativity and Gravitation. 31 (12): 1991–2000. Bibcode:1999GReGr.31.1991F. doi:10.1023/A:1026751225741. S2CID 122950995.)

Fermi, E - QED

44. E. Fermi (1932). "Quantum Theory of Radiation". Reviews of Modern Physics. 4 (1): 87–132. Bibcode:1932RvMP....4...87F. doi:10.1103/RevModPhys.4.87.

Feynman, Richard - QED

45. Feynman, Richard (1985). QED: The Strange Theory of Light and Matter. Princeton University Press. ISBN 978-0-691-12575-6.

Geiger, Hans - Rutherford's experiment

46. Geiger, Hans; Marsden, Ernest (1909). "On a Diffuse Reflection of the α-Particles". Proceedings of the Royal Society of London A. 82 (557): 495–500. Bibcode:1909RSPSA..82..495G. doi:10.1098/rspa.1909.0054.

Gell-Mann, M - QCD

47. Gell-Mann, M (1956). "The Interpretation of the New Particles as Displaced Charged Multiplets". Il Nuovo Cimento. 4 (S2): 848–866. Bibcode:1956NCim....4S.848G. doi:10.1007/BF02748000. S2CID 121017243.

48. Gell-Mann, M. (1961). "The Eightfold Way: A Theory of strong interaction symmetry" (No. TID-12608; CTSL-20). California Inst. of Tech., Pasadena. Synchrotron Lab (online).

49. M. Gell-Mann (1964). "A Schematic Model of Baryons and Mesons". Physics Letters. 8 (3): 214–215. Bibcode:1964PhL.....8..214G. doi:10.1016/S0031-9163(64)92001-3.

Glashow, S – Electroweak unification

50. Glashow, S. (1959). "The renormalizability of vector meson interactions." Nucl. Phys. **10**, 107.

Gong, Yungui - Einstein Dicke Cosmology

51. Yungui Gong. Einstein-Brans-Dicke Cosmology, General Relativity and Quantum Cosmology, arXiv:gr-qc/9809015v2 (gr-qc) [Submitted on 3 Sep 1998 (v1), last revised 3 Sep 1999 (this version, v2)]

Guillerminet, Bernard – On Emergent Gravity

52. Bernard Guillerminet, Emergent (Dark) Gravity, 452, chemin du Ventoux, 84120-Pertuis, France

Guth, A H – Inflationary cosmology

53. Guth, A. H. (1998). The Inflationary Universe: The Quest for a New Theory of Cosmic Origins. Basic Books. p. 186. ISBN 978-0201328400. OCLC 35701222.

Hossenfelder, Sabine - Information loss paradox

54. Hossenfelder, Sabine "How do black holes destroy information and why is that a problem?". (23 August 2019 Back ReAction. Retrieved 23 November 2019.

55. Sabine Hossenfelder, Minimal Length Scale Scenarios for Quantum Gravity, Living Rev. Relativity, 16, (2013), 2, http://www.livingreviews.org/lrr-2013-2, doi:10.12942/lrr-2013-2

Hawking, Stephen - Miniature Black holes

56. Hawking, Stephen. (1995). "Virtual Black Holes". Physical Review D. 53 (6): 3099–3107. arXiv:hep-th/9510029. Bibcode:1996PhRvD..53.3099H. doi:10.1103/PhysRevD.53.3099. PMID 10020307. S2CID 14666004.

Hawking, Stephen - Hawking radiation

57. Hawking, S. W. (1975). "Particle creation by black holes". Communications in Mathematical Physics. 43 (3): 199–220. Bibcode:1975CMaPh..43..199H. doi:10.1007/BF02345020. S2CID 55539246.

Hiscock, William - Alcubierre drive

58. Hiscock, William A. (1997). "Quantum effects in the Alcubierre warp drive space-time". Classical and Quantum Gravity. 14 (11): L183–L188. arXiv:gr-qc/9707024. Bibcode:1997CQGra..14L.183H. doi:10.1088/0264-9381/14/11/002. S2CID 1884428.

Hu, W - BAO (Baryon Acoustic Oscillations) in CMB

59. Hu, W.; White, M. (1996). "Acoustic Signatures in the Cosmic Microwave Background". Astrophysical Journal. 471: 30–51. arXiv:astro-ph/9602019. Bibcode:1996ApJ...471...30H. doi:10.1086/177951. S2CID 8791666

Hubble Edwin - Expansion of the Universe

60. Edwin Hubble: A relation between distance and radial velocity among extra-galactic nebulae PNAS March 15, 1929 15 (3) 168-173; https://doi.org/10.1073/pnas.15.3.168 Jammer, Max - Simultaneity and Special Relativity

61. Jammer, Max (2006). Concepts of Simultaneity: From Antiquity to Einstein and Beyond. The Johns Hopkins University Press. p. 165. ISBN 0-8018-8422-5.

Khoury – Superfluid Dark matter

62. Khoury, Justin. Dark Matter Superfluidity, *Submitted on 22 Sep 2021.* https://arxiv.org/abs/2109.10928?s=09

Lemaître, G. - Expansion of the universe

63. Lemaître, Georges (1931), "Expansion of the universe, A homogeneous universe of constant mass and increasing radius accounting for the radial velocity of extra-galactic nebulæ", Monthly Notices of the Royal Astronomical Society, 91 (5): 483–490, Bibcode:1931MNRAS..91..483L, doi:10.1093/mnras/91.5.483 translated from Lemaître, Georges (1927), "Un univers homogène de masse constante et de rayon croissant rendant compte de la vitesse radiale des nébuleuses extra-galactiques", Annales de la Société Scientifique de Bruxelles, A47: 49–56, Bibcode:1927ASSB...47...49L

Landau L – Weak Nuclear Force

64. Landau, L. (1957). "On the conservation laws for weak interactions". Nuclear Physics. 3 (1): 127–131. Bibcode:1957NucPh...3..127L. doi:10.1016/0029-5582(57)90061-5.

Lee, T.D. – C and T violation

65. Lee, T. D.; Oehme, R.; Yang, C. N. (1957). "Remarks on Possible Noninvariance under Time Reversal and Charge Conjugation". Physical Review. 106 (2): 340–345. Bibcode:1957PhRv..106..340L. doi:10.1103/PhysRev.106.340. Archived from the original on 5 August 2012.

Lorentz – On General Relativity

66. H. A. Lorentz: On Einstein's Theory of Gravitation, Proc. Amsterdam Acad., vol. xix. p. 1341, 1917.

McGaugh, S - Tully Fisher relationship

67. McGaugh, S. S.; Schombert, J. M.; Bothun, G. D.; De Blok, W. J. G. (2000). "The Baryonic Tully-Fisher Relation". The Astrophysical Journal. 533 (2): L99–L102. arXiv:astro-ph/0003001. Bibcode:2000ApJ...533L..99M. doi:10.1086/312628. PMID 10770699. S2CID 103865.

68. McGaugh, Stacy S. (2012). "The Baryonic Tully-Fisher Relation of Gas-Rich Galaxies as a Test of Λcdm and Mond". The Astronomical Journal. 143 (2): 40. arXiv:1107.2934. Bibcode:2012AJ....143...40M. doi:10.1088/0004-6256/143/2/40. S2CID 38472632.

Milgrom Mordhei - MONDian Gravity

69. Milgrom, M. (1983). "A modification of the Newtonian dynamics as a possible alternative to the hidden mass hypothesis". Astrophysical Journal. 270: 365–370. Bibcode:1983ApJ...270..365M. doi:10.1086/161130..

70. Milgrom, M. (1983). "A modification of the Newtonian dynamics - Implications for galaxies". Astrophysical Journal. 270: 371–389. Bibcode:1983ApJ...270..371M. doi:10.1086/161131..

71. Milgrom, M. (1983). "A modification of the Newtonian dynamics - Implications for galaxy systems". Astrophysical Journal. 270: 384. Bibcode:1983ApJ...270..384M. doi:10.1086/161132..

72. Milgrom, M.; Sanders, R.H. (2003). "Modified Newtonian Dynamics and the 'Dearth of Dark Matter in Ordinary Elliptical Galaxies'". Astrophys J. 599 (1): 25–28. arXiv:astro-ph/0309617. Bibcode:2003ApJ...599L..25M. doi:10.1086/381138. S2CID 14378227.

73. Kroupa, P.; Pawlowski, M.; Milgrom, M. (2012). "The failures of the standard model of cosmology require a new paradigm". International Journal of Modern Physics D. 21 (14): 1230003. arXiv:1301.3907. Bibcode:2012IJMPD..2130003K. doi:10.1142/S0218271812300030. S2CID 118461811.

Miller, Arthur - Special Theory of Relativity

74. Miller, Arthur I. (1981). Albert Einstein's special theory of relativity. Emergence (1905) and early interpretation (1905–1911). Reading: Addison–Wesley. pp. 257–264. ISBN 0-201-04679-2.

Padmanabhan T. – Emergent gravity

75. T. Padmanabhan, "Dark Energy and Gravity", Gen. Rel. Grav., 40, 529 (2008), arXiv:0705.2533

Peebles, P - Dark energy

76. Peebles, P. J. E.; Ratra, Bharat (22 April 2003). "The cosmological constant and dark energy". Reviews of Modern Physics. 75 (2): 559–606. arXiv:astro-ph/0207347. Bibcode:2003RvMP...75..559P. doi:10.1103/RevModPhys.75.559. S2CID 118961123.

Penzias and Wilson – CMB

77. Penzias, A. A.; Wilson, R. W. (1965). "A Measurement of Excess Antenna Temperature at 4080 Mc/s". The Astrophysical Journal. 142 (1): 419–421. Bibcode:1965ApJ...142..419P. doi:10.1086/148307.

Planck survey of CMB

78. The Planck Collaboration (2020), "Planck 2018 results. I. Overview, and the cosmological legacy of Planck", Astronomy and Astrophysics, 641: A1, arXiv:1807.06205, Bibcode:2020A&A...641A...1P, doi:10.1051/0004-6361/201833880, S2CID 119185252

79. The Planck Collaboration (2014), "Planck 2013 results. XXVII. Doppler boosting of the CMB: Eppur si muove", Astronomy, 571 (27): A27, arXiv:1303.5087, Bibcode:2014A&A...571A..27P, doi:10.1051/0004-6361/201321556, S2CID 5398329

E. Rutherford, F.R.S.* - Rutherford's experiment

80. E. Rutherford: The Scattering of α and β Particles by Matter and the Structure of the Atom Philosophical Magazine, Series 6, vol. 21 May 1911, p. 669-688

Rosa, R - Single electron DSE

81. Rosa, R (2012). "The Merli–Missiroli–Pozzi Two-Slit Electron-Interference Experiment". Physics in Perspective. 14 (2): 178–194. Bibcode:2012PhP....14..178R. doi:10.1007/s00016-011-0079-0. PMC 4617474. PMID 26525832

Rosen, N on Einstein's Variable speed of light theory

82. Rosen, N. (1940). General relativity and flat space. I.+II. Physical Review 57, 147–153.

Rogers, P – Matter Antimatter asymmetry

83. Rodgers, Peter (August 2001). "Where did all the antimatter go?". Physics World. p. 11.

Rovelli, Carlo - Loop Quantum Gravity

84. Rovelli, Carlo. Loop Quantum Gravity, Department of Physics and Astronomy, University of Pittsburgh, Pittsburgh PA 15260, USA. http://www.pitt.edu/~rovelli, Published on 26 January 1998. www.livingreviews.org/Articles/Volume1/1998-1rovelli, Living Reviews in Relativity, Published by the Max-Planck-Institute for Gravitational Physics, Albert Einstein Institute, Potsdam, Germany

M. N. Roy – On Relativity

85. M. N. Roy: Discussion on the Theory of Relativity: M. N. Roy. Astron. Soc., vol., lxxx., No. 2., p. 96, December 1919.

Rubin, Vera - Flat rotation curves

86. Rubin, Vera : "First observational evidence of dark matter". Darkmatterphysics.com. Archived from the original on 25 June 2013. Retrieved 6 August 2013.

87. Rubin, Vera C.; Ford, W. Kent, Jr. (February 1970). "Rotation of the Andromeda Nebula from a Spectroscopic Survey of Emission Regions". The Astrophysical Journal. 159: 379–403. Bibcode:1970ApJ...159..379R. doi:10.1086/150317.

Sakharov – Emergent Gravity

88. A. D. Sakharov, "Vacuum quantum fluctuations in curved space and the theory of gravitation". Soviet Physics Doklady, 12, 1040 (1968)

Salam, A – Electroweak Unification

89. Salam, A.; *Ward, J. C. (1959). "Weak and electromagnetic interactions". Nuovo Cimento. **11** (4): 568–577.* Bibcode:1959NCim...11..568S. doi:10.1007/BF02726525. S2CID 15889731

Scardigli, Fabio - Micro Black holes

90. (1999). "Generalized Uncertainty Principle in Quantum Gravity from Micro-Black Hole Gedanken Experiment". Physics Letters B. 452 (1–2): 39–44. arXiv:hep-th/9904025. Bibcode:1999PhLB..452...39S. doi:10.1016/S0370-2693(99)00167-7. S2CID 14440837.

Sciama, D. W – On origin of Inertia

91. Sciama, D. W. (1953). On the origin of inertia. Monthly Notices of the Royal Astronomical Society 113, 34–42.

Scott, D - LCDM model

92. Scott, D. (2005). "The Standard Cosmological Model". Canadian Journal of Physics. 84 (6–7): 419–435. arXiv:astro-ph/0510731. Bibcode:2006CaJPh..84..419S. CiteSeerX 10.1.1.317.2954. doi:10.1139/P06-066. S2CID 15606491

Schrödinger, Erwin - Schrödinger equation

93. Schrödinger, E. (1926). "An Undulatory Theory of the Mechanics of Atoms and Molecules" (PDF). Physical Review. 28 (6): 1049–1070. Bibcode:1926PhRv...28.1049S. doi:10.1103/PhysRev.28.1049. Archived from the original (PDF) on 17 December 2008.

Smoot et al – COBE satellite CMB

94. Smoot Group (28 March 1996). "The Cosmic Microwave Background Radiation". Lawrence Berkeley Lab. Retrieved 2008-12-11

Unzicker, Alexander - Mach's principle

95. Unzicker, Alexander, Pestalozzi-Gymnasium Munchen, Germany, A Machian Version of Einstein's Variable Speed of Light Theory, Jan Preuss, Technische Universit ̈at M ̈unchen, Germany

Unzicker, Alexander – Origin of Constants of Nature c and h

96. Unzicker, Alexander, On the Origin of the Constants c and h, JOUR, 2015/08/20, http://vixra.org/abs/1508.0170

Unzicker, Alexander – On Large Number Hypothesis

97. Unzicker, A. (2007a). A Look at the Abandoned, Contributions to Cosmology of Dirac, Sciama and Dicke. ArXiv: 0708.3518.

Unzicker, Alexander – On Variable Speed of light theory

98. Unzicker, A. (2007b). The VSL Discussion: What Does Variable Speed of Light Mean and Should we be Allowed to Think About? ArXiv: 0708.2927.

Unzicker, Alexander – Mach's principle

99. Unzicker, A. (2003). Galaxies as rotating buckets - a hypothesis on the gravitational constant based on Mach's principle. arXiv grqc/0308087.

Unzicker, Alexander – On Large Number Hypothesis

100. Unzicker, A. (2007a). A Look at the Abandoned Contributions to Cosmology of Dirac, Sciama and Dicke. ArXiv: 0708.3518.

Verlinde Eric - Entropic Gravity

101. Erik P. Verlinde, Emergent Gravity and the Dark Universe, SciPost Phys. 2, 016 (2017) · published 16 May 2017

102. E.P. Verlinde, "On the Origin of Gravity and the Laws of Newton" (2010), arXiv:1001.0785

Weinberg, S – Electroweak unification

103. *Weinberg, S (1967).* "A Model of Leptons" (PDF). *Phys. Rev. Lett.* **19** *(21): 1264–66.* Bibcode:1967PhRvL..19.1264W. doi:10.1103/PhysRevLett.19.1264. *Archived from* the original (PDF) *on 2012-01-12.*

104. Steven Weinberg, Sheldon Glashow and Abdus Salam: "The Nobel Prize in Physics 1979". The Nobel Foundation. Retrieved 2008-12-16.

Young, Thomas - Young's double slit experiment

105. Young, Thomas (1804). "The Bakerian lecture. Experiments and calculation relative to physical optics". Philosophical Transactions of the Royal Society of London. 94: 1–16. doi:10.1098/rstl.1804.0001. S2CID 110408369. Retrieved 14 July 2021

Yukawa, Hideki - Yukawa's theory

106. Yukawa, Hideki, On the Interaction of Elementary Particles I, PRINT-92-0144, 10.1143/PTPS.1.1, Proc. Phys. Math. Soc. Jap, Volume -17, Page 48—57, 1935, Yukawa:1935xg,

Zwicky, Fritz - Dark matter

107. Zwicky, F. (1937). "On the Masses of Nebulae and of Clusters of Nebulae". The Astrophysical Journal. 86: 217–246. Bibcode:1937ApJ....86..217Z. doi:10.1086/143864.

108. Zwicky, F. (1933). "Die Rotverschiebung von extragalaktischen Nebeln" [The red shift of extragalactic nebulae]. Helvetica Physica Acta. 6: 110–127. Bibcode:1933AcHPh...6..110Z.

ABBREVIATIONS AND EXPLANATIONS

ATP	:	Adenosine Triphosphate: A molecule which acts as a battery for our cells used for storage of energy or utilization of energy
BH	:	Black hole
BAO	:	Baryon acoustic oscillations
COBE satellite	:	A satellite that is a part of the Project to map the cosmic microwave background
CMB	:	Cosmic Microwave Background
C/D ratio	:	Creation to Destruction ratio, a measure of the balance of imbalance between these two processes at any point. A C/D ratio deviated towards Destruction is positive energy and deviated towards Creation is Negative energy
DGR	:	Dynamic General Relativity
DCQE experiment	:	Delayed Choice Quantum Eraser experiment
Dirac's LNH	:	Dirac's Large number hypothesis
Einstein's VSL theory	:	Einstein's Variable speed of light theory
e	:	number of events per Planck time happening at a point, an event can be an event of Creation or an event of Destruction.
ENEA	:	Exotic Negative Energy Anti-particles: a point in space where Creation predominates over Destruction thus leading to active spatial expansion
EPCA	:	Eddy Planck Compartment Aggregate: these are Planck Compartments aggregated and simultaneously under influence of a push or pull force formed around a particle in an attempt to compensate the active Creation or Destruction happening at its centre.
EPCAs	:	More than one EPCA
EPR paradox	:	Einstein Podolski Rosen paradox
GR	:	General Relativity
GUT	:	Grand Unified Theories

LCDM model or Lambda CDM model	:	Lambda Cold Dark Matter model, model of the Universe which is based on lambda i.e. cosmological constant, dark matter and Dark energy - the currently accepted or popular model of Cosmology
LHC	:	Large Hadron collider, the particle accelerator at CERN, Switzerland.
LIGO	:	Laser interferometer Gravitational Wave Observatory
LQG	:	Loop Quantum Gravity
LT	:	Local time - The time in DGR which varies with location and movement. Local time is an absolute illusion of slowing down of time or time running faster due to active spatial contraction or active spatial expansion respectively.
MM experiment	:	Michelson Morley experiment
MOND	:	Modified Newtonian dynamics
MONDian Gravity	:	Gravitational attraction due to MOND
NE	:	Negative Energy
NEDC model	:	Negative Energy dominance at the centre. A model of the Universe where total negative energy exceeds total positive energy
NEP	:	Negative Energy Particle: Equivalent to a T=Tmax Planck Compartment. Here local time is contracted and thus the process of Creation predominates over destruction so that active Spatial expansion results. These do not aggregate like Positive Energy Particles and thus remain as separate entities. In them, one extra PC is created with every tick of Planck clock
PA	:	Planck Area: 10^{-70} Meter square
PC	:	Planck Compartment: The smallest indivisible unit of Space
Planck Clock	:	A clock that is used to measure time in Planck scales. A hypothetical Universal Planck clock ticks with every Planck time of Universal time. It started at the moment of Creation and is relentlessly ticking, maintaining Universal Simultaneity. It's pace is constant everywhere in the Universe. A Planck clock measures Local time and its ticks are delayed or happen faster compared to Universal Planck clock depending on whether Local time runs slower or faster.

PCs	:	More than one Planck Compartment
PE	:	Positive Energy
PEDE model	:	Positive energy dominance at the edge. A model of Universe where the total Positive energy exceeds Negative energy.
PEDE-NEDC model	:	A model of the Universe where there is a perfect balance between Positive and Negative energy with predominance of positive energy at the edge and negative energy at the centre, nodes and walls of the honeycombs.
PEP	:	Positive Energy Particle: An aggregate of multiple T=0 Planck Compartments. This is a point in space where, depending on how many T=0 compartments are aggregated, active destruction of Planck Compartments keeps happening every Planck time. Depending on mass/amount of energy aggregated together, millions or billions of PCs can be destroyed at a point.
PEM particle	:	Positive Energy Matter Particle
QM	:	Quantum mechanics
PL	:	Planck length: 10^{-35} meters
Planck satellite	:	A satellite launched by the European space agency with the intention of giving an accurate map of the Cosmic Microwave Background
PT	:	Planck time: 10^{-43} seconds
Pull band	:	A corridor of space within which extreme Local time dilation and thus active spatial contraction develops and thus pulls the two ends towards each other
Push band	:	A corridor of space within which extreme Local time contraction and active spatial expansion develops and pushes the two ends apart
PV	:	Planck Volume: 10^{-105} meter cube
QED	:	Quantum Electro-dynamics
QCD	:	Quantum Chromo-dynamics
QFT	:	Quantum Field theory
QS	:	Quantum Superposition
SA	:	Surface Area
Single electron DSE	:	Double slit experiment done with single Electrons at a time

SR	:	Special Relativity
SMBH	:	Supermassive Black hole made of several million solar masses
ST	:	String Theory
T=0 compartment	:	A Planck Compartment where creation of one PCs occurs and destruction of 2 PCs occurs i.e. effective destruction of 1 PC. I e. Entire Volume of the PC is lost. Alternatively, in earlier version of the theory, a PC with Local Time zero.
T=Tmax Compartment	:	A PC where creation of 2 PCs happens and destruction of one PC happens. The volume increases from 1PC to 3 PCs and then reduces by one PC. Thus an effective increase in volume from 1PV to 2 PV occurs Alternatively, in earlier version of the theory, a PC with maximum possible time contraction
T=UT Compartments	:	A PC where one PV is created and one PV is destroyed again so that it retains its volume of 1 PV. Alternatively, in earlier version of the theory, a PC with Local Time running equal to Universal time.
TOE	:	Theory of Everything
TRW	:	Time resetting wave
UMBH	:	Ultra-massive Black holes made of up to a billion or more solar masses
UT	:	Universal Time - The time in DGR which is universally progressive at a constant rate at every point in the Universe irrespective of Gravity or movement.
VSL	:	Variable speed of light – relating to Einstein's Variable Speed of light theory
ΔV	:	change in volume, usually loss of volume in a positive energy particle at its core or change in volume in an EPCA
WMAP satellite	:	A satellite that is a part of the Project to map the cosmic microwave background
WW	:	Which way experiment

Milton Keynes UK
Ingram Content Group UK Ltd.
UKHW050923190124
436321UK00012B/482

The Heimskringla...

Snorri Sturluson, Samuel Lang, Rasmus Björn Anderson

Nabu Public Domain Reprints:

You are holding a reproduction of an original work published before 1923 that is in the public domain in the United States of America, and possibly other countries. You may freely copy and distribute this work as no entity (individual or corporate) has a copyright on the body of the work. This book may contain prior copyright references, and library stamps (as most of these works were scanned from library copies). These have been scanned and retained as part of the historical artifact.

This book may have occasional imperfections such as missing or blurred pages, poor pictures, errant marks, etc. that were either part of the original artifact, or were introduced by the scanning process. We believe this work is culturally important, and despite the imperfections, have elected to bring it back into print as part of our continuing commitment to the preservation of printed works worldwide. We appreciate your understanding of the imperfections in the preservation process, and hope you enjoy this valuable book.

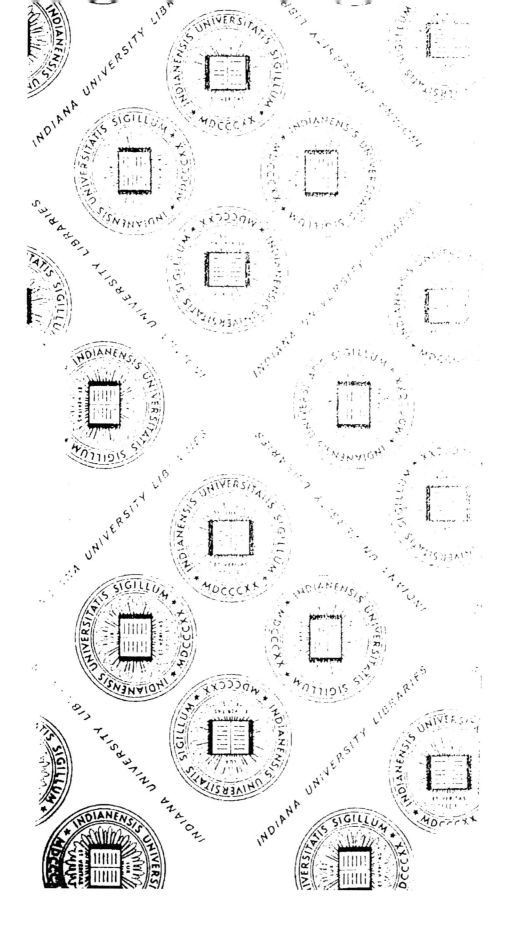

THE HEIMSKRINGLA.

SAMUEL LAING.

VOLUME THE FIRST.

PUBLISHERS' NOTE.

Five hundred and twenty copies in all of this book printed for America and England. Each copy is numbered as issued. Type distributed.

No. 222

THE HEIMSKRINGLA

OR

THE SAGAS OF

THE NORSE KINGS

FROM THE ICELANDIC OF SNORRE STURLASON

BY

SAMUEL LAING, Esq.

SECOND EDITION, REVISED, WITH NOTES

BY

RASMUS B. ANDERSON, LL.D.

UNITED STATES MINISTER TO DENMARK
AUTHOR OF "NORSE MYTHOLOGY," "VIKING TALES OF THE NORTH," AND OTHER WORKS

With Two Maps

IN FOUR VOLUMES

VOLUME THE FIRST

NEW YORK

SCRIBNER & WELFORD

743 & 745 BROADWAY

MDCCCLXXXIX

PT 7277
.E5 L18
v. 1

Copyright, 1889. Scribner & Welford.

TO

WILLIAM F. VILAS,
SECRETARY OF THE INTERIOR,

THIS EDITION OF

SNORRE'S "HEIMSKRINGLA"

IS

Dedicated

BY

HIS GRATEFUL FRIEND,

R. B. ANDERSON.

"In this as in other things let us cry: *England for English! Scandia for Scandians!* THE NORTH FOR THE NORTHMEN!"—Prof. Dr. George Stephens, in RUNIC MONUMENTS, vol. i. p. 20.

"In comparison with his contemporaries, Snorre's broader views and keen statesmanlike tact are certainly remarkable, and every page of his historical works attests his sympathy with the political life and his possession of the peculiar qualities necessary for a ruler of men. Able to value at its real worth the careful truthseeking of Are, he yet takes his own path as an historian; seizing on character and situation with the truest dramatic feeling; letting his heroes speak for themselves; working boldly and vigorously, but with the surest skill; and so creating works which for deep political insight, truth of conception, vividness of colour, and knowledge of mankind, must ever retain their place beside the masterpieces of the greatest historians."—Dr. GUDBRAND VIGFUSSON.

"The crown of Icelandic historiography is SNORRE STURLASON'S HEIMSKRINGLA, which towers above all other Icelandic histories like a splendid tree above the low brushwood."—Dr. FREDERIK WINKEL HORN.

"From whatever point of view, therefore, we consider the relations which exist between England and Iceland, whether from that of primæval affinity and a community of race, religion, and law, or from that of connection by commerce, immigration, or conquest, we shall find the two languages and peoples so closely bound together, that whatever throws light on the beliefs, institutions, and customs of the one, must necessarily illustrate and explain those of the other. Nor should it be forgotten that in the tenth and eleventh centuries the Icelanders were foremost in the history of the time. They were at once the most learned and the boldest and most adventurous of men. From Iceland they pushed on to Greenland and America, and their ships swarmed in commerce or in viking voyages on all the seas. At the courts of kings and earls, whether Norwegian, Danish, Swedish, or Anglo-Saxon, they were welcome guests, for though none were more dreaded as foes, none were more greeted as friends for their gifts of wit and song."—Dr. GEORGE WEBBE DASENT.

INTRODUCTION.

In his interesting little book, "The Early Kings of Norway," the distinguished writer, Thomas Carlyle, says that Snorre Sturlason's HEIMSKRINGLA "deserves, were it once well edited, furnished with accurate maps, chronological summaries, &c., to be reckoned among the great history-books of the world." The Swedish language actually possesses such an edition of this great historical work in the elegant and scholarly translation by Hans Olaf Hildebrand, who is now Riks-Antiquary of Sweden. It is in three volumes, published at Orebro (1869-1871), and furnished with an elaborate introduction, with summaries, commentaries, notes, a chronological table, a map, and a complete index. Hildebrand is a man of profound literary insight and remarkable diligence. He is one of the chief authorities in regard to Scandinavian archæology, and has thus been able to make a careful study of both the antiquities and the topography of the olden North; nor has he failed to examine any collateral record or any foreign chronology that could be of service in elucidating Snorre's Heimskringla. In short, we have in Hildebrand's Swedish translation a monumental edition, which could not have failed to delight and satisfy Carlyle's heart, had he become acquainted with it.

The present editor and reviser of Samuel Laing's translation of Snorre's chronicle lays no claim to that erudition which shines on every page of Hildebrand's work. He does not hesitate to confess that he came to Copenhagen, not to teach others, but to learn himself; he came to the Athens of the North, not as a master but as a disciple, eager to sit at the feet of the great scholars of Scandinavia, in order that he might return to his native country with more knowledge of that weird North, from whose "frozen loins" poured the vikings of the Middle Ages—with more knowledge of that grand old Scandinavia, which was destined to become the mother of England and the grandmother of America. How eminently fitting that the child and the grandchild should listen to the words of wisdom that in times past have fallen from the lips of their mother and grandmother! An acquaintance with the ancient runes, with the Eddas, with the Heimskringla, and with all the old saga-lore, should be the pride of every Englishman and American.

In assuming the revision of this great historical work the editor did not conceive it to be within his province to examine old manuscripts and make original researches, or to criticise the scholars of the North. He deemed his task well done if he should succeed in gathering the ripest fruits and best results of the specialists in this field, and in incorporating and dovetailing them into a new edition of Laing's translation, thus bringing the latter, if possible, up to the standard of modern scholarship. Nor does he flatter himself that his ambitious object has been attained, but will think himself amply rewarded for his labour if the verdict passed upon his work shall be, that the present edition is, in some respects at least, an improvement upon that by Mr. Laing in 1844.

To Samuel Laing belongs the imperishable fame of having made the first translation of the Heimskringla into the English tongue, and his work has been of great service to many a historian, scholar, and poet. Carlyle is chiefly indebted to him for his "Early Kings of Norway," and Laing's Heimskringla inspired several of Longfellow's best poems. We should not fail to give honour to whom honour is due. We must not forget that Bishop Percy published a translation of Mallet's Northern Antiquities in 1770, that A. T. Cottle attempted a translation of the Elder Edda in 1797, but bear in mind that these men were pioneers, pathfinders, and laboured under immense difficulties. The modern Icelandic scholar has Vigfusson's Icelandic-English Dictionary, and many other excellent aids which were unknown in former generations. All the more credit is therefore due to such pioneers as George Stephens, George Webbe Dasent, Samuel Laing, Benjamin Thorpe, Dr. Carlyle, Sir Edmund Head, Robert Lowe, and William Morris, in England; and to such men as Longfellow, George P. Marsh, and Willard Fiske, in America, for mastering the spirit of Norse history and literature.

From this point of view it seemed to the present publisher and editor peculiarly fitting to make Laing's translation the basis of a new edition of the Heimskringla, to retain his honoured name in connection therewith, and to make at the same time such eliminations, additions and corrections, as the lapse of time and the progress of knowledge have made necessary.

In the performance of this task the reviser has made more or less use of all the later editions and translations of the Heimskringla, and he is under special obligations to Prof. C. R. Unger's edition of the original text, and to P. A. Munch's and H. O. Hildebrand's translations into Norwegian and Swedish. Many of Laing's foot-notes which seemed obsolete or irrelevant have been omitted, and an L. has been added to all those which have been retained. A considerable number of new notes have been substituted and added, and for the greater number of these the editor is indebted to Hildebrand, whose chronology has also invariably been adopted. Anent the fixing of the dates of events in old Scandinavian history, all Norse scholars owe much to "TIMATAL," a treatise on the chronology of the earlier Icelandic history by the late Gudbrand Vigfusson, one of the foremost old Norse scholars of this century. His "Lives of the Early Bishops of Iceland" (1858), his edition of Cleasby's "Icelandic-English Dictionary" (1874), his "Sturlunga Saga with Prolegomena" (1878); and his "Corpus Poeticum Boreale" (1883) are enduring monuments of Vigfusson's great learning, and mark a new epoch in the study of Scandinavian history and literature in the world generally, and in England and America particularly.

The chief features of the revision may be classified and described as follows :—

1. Chapters, paragraphs, sentences, and words have been eliminated here and there, and others have been substituted or added, in order to make the translation as now presented correspond with Prof. C. R. Unger's text-edition published in Christiania in 1868. From Vigfusson a thoroughly revised text-edition was expected, but Unger's is the best hitherto published.

2. Laing's irrelevant notes have been omitted, and many new notes have been added. The new ones have been gleaned from various sources, but chiefly from Hildebrand's Swedish translation of the Heimskringla.

3. The orthography of names of persons and places has been thoroughly revised. With this part of the work the editor is not himself entirely satisfied. He admits that he is guilty of several inconsistencies; but the jewel of consistency in regard to the orthography of old Norse names has not yet been discovered by any of the old Norse scholars. Even Vigfusson abounds in inconsistencies, and he frequently writes the same name in several ways. In this edition a large number of superfluous consonants have been dropped (*e.g.*, Fin for Finn, Hal for Hall; Trygve for Tryggve, &c.); final "i" has been changed to "e;" one "s" has been substituted for "ss" (*e.g.*, Olafson for Olafsson); sometimes the present form

of the name of a place has been preferred to the old Norse (*e.g.*, Throndhjem, Upsala, Jerusalem, Spain, England; but the editor admits that if he should ever have the privilege of revising the Heimskringla again, he would write Trygvason (like Arnason, Skulason, &c.) instead of Trygveson.

4. The dates of events have been inserted in bracket throughout the text, and a new chronological table has been added at the end of Vol. IV.

5. To the liberality of the publisher the reader is indebted for two maps showing approximately how the world, and particularly the North of Europe, looked to Norse eyes in the tenth and eleventh centuries.

6. Two elaborate indexes have been prepared, one of persons, and peoples, and another of geographical names. In both these indexes the old Norse nominative form of the name has been added, so that the reader may in each case see what liberties the editor has taken in regard to the orthography. In the index of places the modern geographical name is also given whenever it is known. Thus the index of places supplies at a glance any geographical information desired by the reader. For this valuable feature the editor is chiefly indebted to Prof. C. R. Unger's text-edition, which is furnished with similar indexes.

7. Samuel Laing's Preface and Preliminary Dissertation have been retained *in extenso*, partly as a memorial of their author's deep interest in his subject, and partly as a record of the high-water mark of Norse scholarship in England and America in the

early part of this century. The critical reader will find fault with many of Mr. Laing's bold statements, but he surely cannot fail to admire the glowing enthusiasm which attended the first introduction of the Heimskringla among the Anglo-American descendants and kinsmen of the gods and heroes, kings, earls, and simple bondes of the grand old North. The conspectus or list of Icelandic literature incorporated in the Preliminary Dissertation had to be thoroughly revised, but aside from that the emendations in the Preface and Preliminary Dissertation are confined chiefly to dates and orthography.

8. The skaldic verses being reproduced by Mr. Laing and his son in rhyme and metre, the reviser could not mend them. They are not translations, but rather original songs or ballads in modern measures. The most that can be said for them as reproductions of the Heimskringla verses, is that they are written on the same themes and celebrate the same events. They do not even paraphrase the thought of the original Icelandic texts. The present editor, ready to confess his own inability to reproduce these old skaldic songs in suitable English translations or paraphrases, or to better Mr. Laing's poetry, was at first inclined to follow the example of P. A. Munch, and omit the most of the quotations from the skalds altogether. The fact is that these verses rarely contain any additional historical matter. They are simply quoted by Snorre in corroboration of what he states in prose, and the reader will lose nothing if he skips them. It is fair to presume that

if printing had been invented in Snorre Sturlason's time, and these poems consequently accessible in other books, he would not have quoted them, but simply informed his readers in a preface or in a foot-note where they might find them. While the Norwegian translator, P. A. Munch, omits the poems, the Danish translator, Bishop Grundtvig, has paraphrased them into modern ballads, and Mr. Laing has attempted to imitate the latter. Inasmuch as Mr. Laing appears to have devoted much time and labour to these songs (see Vol. I., pp. 251-261) and inasmuch as they are not essential to Snorre's prose narrative, and can be skipped by the reader without any interruption in the thread of the story, the editor finally decided to leave them as a monument to Mr. Laing's indefatigable industry. In one or two cases where the skaldic verses seemed to be of considerable importance, Dr. Gudbrand Vigfusson's paraphrases have been added from "Corpus Poeticum Boreale," a magnificent work, in which the poetry of the old Northern tongue from the earliest times to the thirteenth century, is edited, classified, and translated.

In regard to the life of Samuel Laing, I am happy to be able to give the following brief account, based on notes kindly furnished me by his son, Samuel Laing, Esq., and by his daughter, Mrs. Elizabeth Baxter.

The author and traveller Samuel Laing was born in Kirkwall, Orkney, October 4, 1780. He received his early education at the Grammar School there, and entered the college in Edinburgh at the age of seventeen. It appears that he visited Kiel in his youth, but the year of his journey thither is not given. Nor do we know the date of his entering the British army, where he served in what was then called the staff corps. He was stationed at Hythe in Kent, where a canal was being constructed at the foot of the high ground bordering on Romney Marsh, and intended as a defence against a French invasion. This must have been in 1806 or 1807, and there he became engaged to Miss Agnes Kelly, a daughter of Captain Kelly, of an old Devonshire family. Mr. Laing was on the staff of Sir John Moore, and accompanied him on his expedition to Spain, and returned to England after the battle of Corunna, at which he was present. In 1809 he left the army, married Miss Kelly in March, and settled in Edinburgh, where his brother, who had lately succeeded to a large fortune, found him a situation as manager of large mines at Wanlockhead in the south of Scotland. His daughter, Mrs. Baxter, was born in December 1809, and two years later his son Samuel (the lady and gentleman who have kindly supplied me with notes about their father), and in November 1812 his wife died. A few years later we find Mr. Laing in Leith, and in 1818 he was in Orkney, engaged by a firm in London to establish the herring fishing, a matter in which he was wholly successful. At the close of 1818, his brother, W. Malcolm Laing, died and left him the landed property in Orkney on which he lived with his children and sister-in-law until 1834, when, on the marriage of his daughter, he left Orkney and never returned. The next years were spent in Norway, and from his travels there and in Sweden we have his books "A Residence in Norway," "A Tour in Sweden," "Notes of a Traveller," &c. In 1837 his daughter became a widow, and from that time he made his home with her. He continued making excursions abroad, gathering materials for his books, and while at home he wrote constantly. At the age of eighty both his body and mind were still in full vigour, but at eighty-three he broke down both intellectually and physically. He died without any illness, of old age, in his daughter's house in Edinburgh, April 23, 1868, and was buried in the Dean Cemetery near that city. Mr. Laing was of an Orkney family, descended on the male side from a lowland Scotch ancestor, and on the female from a pure Norse stock. His older brother, Malcolm Laing, from whom he inherited the estate, was a distinguished man, author of a "History of Scotland," and a friend and correspondent of Fox, Macintosh, and other leading men of the day. Succeeding as a younger brother to a heavily encumbered estate, he was ruined, like most of the old Orkney proprietors, by the failure of kelp in 1830, and lived for the rest of his life on a small income. But with characteristic energy he took to literature and wrote several works, which were much read and admired. His "Notes of a Traveller" became very popular. His last years were devoted mainly to Norse literature, and in 1844 he published his translation of the Heimskringla, a work which greatly delighted Thomas Carlyle. While residing in Orkney he was the first to introduce the herring fishery and agricultural improvements, which saved the population from destitution on the failure of kelp. He was universally beloved and respected; and for many years he was provost of his native town, Kirkwall. Samuel Laing was a man of singular acuteness and geniality.

In conclusion, I desire to render my personal thanks to Mr. Birket Smith, its chief, and to the other officials at the Copenhagen University Library. I am indebted to them for many services and particularly for placing at my command those works which it was necessary to consult in connection with this revision. I also wish to acknowledge once more my great obligations to the scholarly works of Gudbrand Vigfusson, H. O. Hildebrand, P. A. Munch, Gustaf Storm, and Theodore Möbius. It goes without saying that Möbius' "Catalogus" and "Verzeichniss" have supplied the materials for the Heimskringla bibliography published below. I also avail myself of this opportunity of offering my hearty thanks to the venerable George Stephens, Johannes Steenstrup, to Ernst Sars, Viktor Rydberg, to the Royal Danish Society of Antiquaries, and to all the Scandinavian scholars and authors who have in so many ways assisted and befriended me during my pleasant sojourn in the North.

For the rare opportunity of spending four years in Copenhagen, where these and other volumes have been prepared under far more favourable auspices than would elsewhere have been possible, I am particularly indebted to my friend and neighbour the Honourable William F. Vilas, Secretary of the Interior at Washington, who recommended me to the President of the United States for the diplomatic mission to Denmark, and it affords me the greatest pleasure to be permitted to dedicate these Heimskringla volumes to him as a slight token of my gratitude, and as a *souvenir* of his distinguished and disinterested kindness to an ambitious student of Scandinavian antiquities, history and literature.

RASMUS B. ANDERSON.

COPENHAGEN, DENMARK,
February 11, 1889.

HEIMSKRINGLA BIBLIOGRAPHY.

A.—MANUSCRIPTS.

THE original manuscript is not known. The oldest one extant is—

1. *Kringla*, so called from the first words *Kringla heimsins*, "the world's circle," whence the whole work has received the well-known name Heimskringla. *Kringla* is supposed to be written about the year 1260; in the sixteenth century it was in Norway; in 1633 it came to Copenhagen, was destroyed by the fire of 1728, but had then been copied several times, and an excellent copy, made by Asgeir Jonsson in 1682, is now preserved in the Arne Magnusson collection. The preface is wanting, but in other respects this manuscript is nearly complete.

2. *Jöfraskinna*, written 1260–1270, was brought to Copenhagen

1665, was destroyed by the fire of 1728, but had previously been transcribed by various persons, and also by Asgeir Jonsson. The beginning is wanting.

3. *Gullinskinna*, probably from 1270-1280; was brought to Copenhagen before 1682; was lost in the fire of 1728, but had been copied by Asgeir Jonsson.

4. *Eirspennil*, probably from 1280, is still to be found in the Arne Magnusson collection. It was brought to Copenhagen from Norway, where it had been kept from the fourteenth to the sixteenth century.

5. *Codex Frisianus*, written about the year 1300, came from Norway to Denmark; was owned in the seventeenth century by Otto Friis (hence its name). It is now in the Arne Magnusson collection.

6. Several fragments found in Iceland, and probably dating from the year 1300, are also preserved in the Arne Magnusson collection in Copenhagen.

B.—Text-Editions.

1. J. Peringsköld's, published in Stockholm, with a translation into Swedish, in the year 1697.

2. A folio edition, in three volumes, published in Copenhagen 1777-1783, with Latin and Danish translations by G. Schöning, S., and B. Thorlacius, and E. C. Werlauff. The Danish translation was made by the Icelander Jon Olafsson.

3. C. R. Unger's, published in Christiania in 1868.

4. N. Linder's and K. A. Haggson's, published in three volumes in Upsala, 1870-1872.

C.—Translations.

1. A translation into Norwegian, made 1550 by Laurents Hanssön (never printed).

2. A translation into Danish by Jens Mortensön, published in Copenhagen, 1594.

3. A translation into Norwegian by Peter Claussön, made 1599, published by Ole Worm in Copenhagen, 1633.

4. A translation into Swedish in Peringsköld's edition, published in Stockholm in 1697 (*see above*).

5. A translation into Danish in Schöning's edition, published in Copenhagen in 1777-1783 (*see above*).

6. A translation into Latin in Schöning's edition (*see above*).

7. A translation into Swedish by G. Richert, published in Stockholm, 1816-1829.

8. A translation into Danish by N. F. S. Grundtvig, published in Copenhagen, 1818-1822.

9. A translation into German by F. Wachter, published in Leipzig, 1835-1836.

10. A translation into German by G. Mohnike, published in Stralsund, 1837.

11. A translation into Norwegian by J. Aal, published in Christiania, 1838-1839.

12. A translation into English by Samuel Laing, published in London, 1844.

13. A translation into Norwegian by P. A. Munch, published in Christiania, 1859.

14. A translation into Swedish by H. O. Hildebrand-Hildebrand, published in Orebro, 1869-1871.

D.—Commentaries.

1. Müller, P. E., Undersögelse om Kilderne, til Snorre's Heimskringla, og disses Troværdighed, Copenhagen, 1823.

2. Cronholm, Abr., De Snorronis Sturlonidis Historia, Lundæ, 1841.

3. Maurer, Konrad, Ueber Heimskringla und ihre Entstehung, published in his "Ueber die Ausdrücke: altnordische, altnorwegische, und isländische Sprache," Munich, 1867.

4. Storm, Gust., Snorre Sturlasson's Historieskrivning, en Kritisk Undersögelse, Copenhagen, 1873.

5. Möbius, Th., Über die Heimskringla *in*. Zeitschr f. deutsche Philologie V. 141-146, 1874.

PREFACE TO THE FIRST EDITION.

It is of importance to English history to have, in the English language, the means of judging of the social and intellectual state—of the institutions and literature—of a people who during 300 years bore an important, and for a great portion of that time a predominant part, not merely in the wars, but in the legislation of England; who occupied a very large proportion of the country, and were settled in its best lands in such numbers as to be governed by their own, not by Anglo-Saxon laws; and who undoubtedly must be the forefathers of as large a proportion of the present English nation as the Anglo-Saxons themselves, and of a much larger proportion than the Normans. These Northmen have not merely been the forefathers of the people, but of the institutions and character of the nation, to an extent not sufficiently considered by our historians. Civilised or not in comparison with the Anglo-Saxons, the Northmen must have left the influences of their character, institutions, barbarism or culture, among their own posterity. They occupied one-third of all England for many generations, under their own Danish laws; and for half a century nearly, immediately previous to the Norman conquerors, they held the supreme government of the country. It is doing good service in the fields of literature to place the English reader in a position to judge for himself of the influence which the social arrangements and spirit of these Northmen may have had on the national character, and free institutions which have grown up among us from elements planted by them, or by the Anglo-Saxons. This translation of Snorre Sturlason's Chronicle of the Kings of Norway will place the English reader in this position. He will see what sort of people these Northmen were who conquered and colonised the kingdoms of Northumberland, East Anglia, and other districts, equal to one-third of all England at that time, and who lived under their own laws in that portion of England; and he will see what their institutions and social spirit were at home, whether these bear any analogy to what sprang up in England afterwards, and whether to them or to the Anglo-Saxon race we are most indebted for our national character and free constitution of government. The translator of Snorre Sturlason's Chronicle hopes, too, that his labour will be of good service in the fields of literature, by bringing before the English public a work of great literary merit,—one which the poet, or the reader for amusement, may place in his library, as well as the antiquary and reader of English history.

The translator can lay claim to no considerable knowledge of or great familiarity with the Icelandic. To get at the meaning and spirit of the text in any way was his main object; and where he met difficulties, which generally lay only in his own ignorance, he spared no labour in collating the passages he was in doubt about with the Swedish translation in Peringskiold's edition of the work,—with the Danish translation in the edition begun 1777 by Schöning under the auspices of the Danish Government, and finished in 1826 by Thorlacius and Werlauff, in 6 vols. folio,—and with the excellent translation of it into Norse by M. Jacob Aal, published in quarto, in 1838, at Christiania. His notes and explanations are derived mostly from these sources, and principally from M. Jacob Aal's work: and where from his imperfect acquaintance with the Icelandic he found difficulties in the text, especially in the skaldic poetry, which is often very obscure, he had recourse to M. Jacob Aal's translation as the best guide to the meaning and spirit of the original. That gentleman, as the last effort of a long life spent in commercial and literary pursuits, has translated Snorre Sturlason's Chronicle, and the Sagas of the succeeding times down to the end of Hakon Hakonsson's reign in 1263, for the use of the Norwegian peasantry. He remembered in his youth that these histories, although in the old and almost obsolete language of Peter Clausson's translation of 1590,* were a house-book read at the fireside of almost every peasant in Norway; and at a great

* Peter Clausson's translation remained unprinted until 1633, when Prof. Ole Worm in Copenhagen had it published, together with a review by himself of that period of the history of Norway which lies between King Hakon Hakonsson and Queen Margrete, and a brief description of Norway by Clausson.

expense he has published a new translation of them into Norse, and has placed the book, at a merely nominal price considering its magnificence, again within reach of his countrymen. In the present translation the object has been to make it, like M. Jacob Aal's, not merely a work for the antiquary, but for the ordinary reader of history,—for the common man.

The translator believes, also, that it opens up a new and rich field of character and incident, in which the reader who seeks amusement only will find much to interest him. The adventures, manners, mode of living, characters, and conversations of these sea-kings are highly dramatic, in Snorre's work at least; and are told with a racy simplicity and truthfulness of language which the translator cannot flatter himself with having attained or preserved. All he can say for his work is, that any translation is better than none; and others may be stimulated by it to enter into the same course of study, who may do more justice to a branch of literature scarcely known among us.

EDINBURGH, 1844.

CONTENTS.

Introduction vii
Hemiskringla Biography xviii
Preface xxi

PRELIMINARY DISSERTATION.

CHAPTER I.

PAGE

Of the Literature and Intellectual Condition of the Northmen 1

CHAPTER II.

Of the Religion of the Northmen 5

CHAPTER III.

Of the Social Condition of the Northmen . . . 11

CHAPTER IV.

State of the Useful Arts among the Northmen . .

CHAPTER V.

Of the Discovery of Greenland and America by the Northmen 176
Memoir of Snorre Sturlason 233

THE HEIMSKRINGLA.

	PAGE
PREFACE OF SNORRE STURLASON	262
OF THE PRIEST ARE FRODE	265

I.

YNGLINGA SAGA; OR, THE STORY OF THE YNGLING FAMILY FROM ODIN TO HALFDAN THE BLACK 269

CHAPTER I.
OF THE SITUATION OF COUNTRIES 269

CHAPTER II.
OF THE PEOPLE OF ASIA 270

CHAPTER III.
OF ODIN'S BROTHERS 271

CHAPTER IV.
OF ODIN'S WAR WITH THE VANS 272

CHAPTER V.
ODIN DIVIDES HIS KINGDOM: ALSO CONCERNING GEFION . 273

CHAPTER VI.
OF ODIN'S ACCOMPLISHMENTS 276

CHAPTER VII.
ODIN'S FEATS 277

CHAPTER VIII.
OF ODIN'S LAWGIVING 279

CHAPTER IX.
OF NIORD'S MARRIAGE 280

CHAPTER X.
Of Odin's Death 281

CHAPTER XI.
Of Niord 281

CHAPTER XII.
Frey's Death 282

CHAPTER XIII.
Of Freyja and her Daughters 283

CHAPTER XIV.
Of King Fiolner's Death 284

CHAPTER XV.
Of Svegder 285

CHAPTER XVI.
Of Vanlande, Svegder's Son 286

CHAPTER XVII.
Of Visbur's Death 287

CHAPTER XVIII.
Of Domalde's Death 289

CHAPTER XIX.
Of Domar's Death 289

CHAPTER XX.
Of Dygve's Death 290

CHAPTER XXI.
Of Dag the Wise 291

CHAPTER XXII.
Of Agne 293

CHAPTER XXIII.
Of Alrek and Eirik. 294

CHAPTER XXIV.
Of Yngve and Alf 295

CHAPTER XXV.
Of Hugleik. 297

CHAPTER XXVI.
King Gudlaug's Death 298

CHAPTER XXVII.
Of King Hake 298

CHAPTER XXVIII.
Jorund's Death 299

CHAPTER XXIX.
Of King Ane's Death 300

CHAPTER XXX.
Of Egil and Tunne 302

CHAPTER XXXI.
Of King Ottar 305

CHAPTER XXXII.
Of King Adils' Marriage 307

CHAPTER XXXIII.
Of King Adils' Death 308

CHAPTER XXXIV.
Rolf Krake's Death 309

CONTENTS.

CHAPTER XXXV.
Of Eystein and the Jutland King Solve 310

CHAPTER XXXVI.
Of Yngvar's Fall 311

CHAPTER XXXVII.
Of Onund the Land-clearer 311

CHAPTER XXXVIII.
Of Ingiald Illrade 312

CHAPTER XXXIX.
Of King Onund's Death 314

CHAPTER XL.
The Burning in Upsala 315

CHAPTER XLI.
Of Hiorvard's Marriage 317

CHAPTER XLII.
War between Ingiald and Granmar and Hiorvard . . 318

CHAPTER XLIII.
Death of the Kings Granmar and Hiorvard . . . 320

CHAPTER XLIV.
Of Ingiald Illrade's Death 321

CHAPTER XLV.
Of Ivar Vidfadme 322

CHAPTER XLVI.
Of Olaf the Tree-feller 322

CHAPTER XLVII.
Olaf the Tree-feller Burned 323

CHAPTER XLVIII.
Halfdan Hvitbein made King 324

CHAPTER XLIX.
Of Halfdan Hvitbein 324

CHAPTER L.
Of Ingiald, Brother of Halfdan 325

CHAPTER LI.
Of King Eystein's Death 325

CHAPTER LII.
Of Halfdan the Mild 327

CHAPTER LIII.
Of Gudrod the Hunter 327

CHAPTER LIV.
Of King Olaf's Death 329

CHAPTER LV.
Of Ragnvald the Mountain-high 330

II.
HALFDAN THE BLACK'S SAGA 331

CHAPTER I.
Halfdan Fights with Gandalf and Sigtryg . . . 331

CHAPTER II.
Battle between Halfdan and Eystein 332

CHAPTER III.
Halfdan's Marriage 333

CHAPTER IV.
Halfdan's Strife with Gandalf's Sons 334

CHAPTER V.
King Halfdan's Last Marriage with Sigurd Hiort's Daughter 335

CHAPTER VI.
Of Ragnhild's Dream 337

CHAPTER VII.
Of Halfdan's Dream 338

CHAPTER VIII.
Halfdan's Meat Vanishes at a Feast 339

CHAPTER IX.
Halfdan's Death 341

III.

HARALD HARFAGER'S SAGA 342

CHAPTER I.
Harald's Strife with Hake and his Father Gandalf . 342

CHAPTER II.
King Harald Overcomes Five Kings 344

CHAPTER III.
Of Gyda, Daughter of Eirik 345

CHAPTER IV.
King Harald's Vow 346

CHAPTER V.
The Battle in Orkadal 346

CHAPTER VI.
Of King Harald's Laws for Land Property . . . 347

CHAPTER VII.
Battle in Gaulardal 348

CHAPTER VIII.
King Harald Seizes all Naumudal District . . . 349

CHAPTER IX.
King Harald's Home Affairs 350

CHAPTER X.
Battle of Solskel 351

CHAPTER XI.
Fall of the Kings Arnvid and Audbiorn 353

CHAPTER XII.
King Vemund Burnt to Death 355

CHAPTER XIII.
Death of Earl Hakon and of Earl Atle the Mjove . 356

CHAPTER XIV.
Of King Harald and the Swedish King Eirik . . . 357

CHAPTER XV.
King Harald at a Feast of the Peasant Ake, and the Murder of Ake 358

CHAPTER XVI.
King Harald's Journey to Tunsberg 360

CONTENTS.

CHAPTER XVII.
The Battle in Gautland 361

CHAPTER XVIII.
Hrane the Gautlander's Death 362

CHAPTER XIX.
Battle in Hafersfiord 363

CHAPTER XX.
King Harald the Supreme Sovereign in Norway. Of the Settlement of Distant Lands 365

CHAPTER XXI.
King Harald's Marriage and his Children . . . 366

CHAPTER XXII.
King Harald's Voyage to the West 368

CHAPTER XXIII.
King Harald has his Hair Clipped 370

CHAPTER XXIV.
Rolf Ganger is Driven into Banishment . . . 370

CHAPTER XXV.
Of the Fin Svase and King Harald 372

CHAPTER XXVI.
Of Thiodolf of Hvin, the Skald 374

CHAPTER XXVII.
Of Earl Torf-Einar's obtaining Orkney . . . 375

CONTENTS.

CHAPTER XXVIII.
King Eirik Eymundson's Death 377

CHAPTER XXIX.
Guthorm's Death in Tunsberg 377

CHAPTER XXX.
Earl Ragnvald Burnt in his House 377

CHAPTER XXXI.
Halfdan Haleg's Death 378

CHAPTER XXXII.
King Harald and Earl Einar Reconciled . . . 380

CHAPTER XXXIII.
Death of Guthorm. Death of Halfdan the White . 381

CHAPTER XXXIV.
Marriage of Eirik, the Son of King Harald . . . 381

CHAPTER XXXV.
Harald Divides his Kingdom among his Sons . . . 384

CHAPTER XXXVI.
Death of Ragnvald Rettilbeine 386

CHAPTER XXXVII.
Of Gudrod Liome 387

CHAPTER XXXVIII.
King Biorn the Merchant's Death 387

CONTENTS.

CHAPTER XXXIX.
Of the Reconciliation of the Kings 389

CHAPTER XL.
Birth of Hakon the Good 390

CHAPTER XLI.
King Athelstan's Message 392

CHAPTER XLII.
Hauk's Journey to England 392

CHAPTER XLIII.
Hakon, the Foster-son of Athelstan, is Baptized . 394

CHAPTER XLIV.
Eirik is brought to the Sovereignty 394

CHAPTER XLV.
King Harald's Death 395

CHAPTER XLVI.
The Death of Olaf and of Sigurd 397

THE HEIMSKRINGLA;

OR,

CHRONICLE OF THE KINGS OF NORWAY.

PRELIMINARY DISSERTATION.

CHAPTER I.

OF THE LITERATURE AND INTELLECTUAL CONDITION OF THE NORTHMEN.

SNORRE STURLASON'S Heimskringla is a work known to few English readers. Heimskringla—the world's circle—being the first prominent word of the manuscript that catches the eye, has been quaintly used by the northern antiquaries to designate the work itself. One may well imagine that the librarian, or the scholar, in the midst of the rolls and masses of parchments of the great public and private libraries of Copenhagen and Stockholm, has found his advantage in this simple way of directing an unlettered assistant to the skin he wishes to unfold. It is likely that the illuminated initial letters of ancient manuscripts, and of the early printed books, may have had their origin in a similar use or convenience in the monastic libraries of the Middle Ages. Snorre himself is guiltless of this pedantic conceit; for he calls his work the Saga or Story of the Kings of Norway. It is in reality a chronicle, or rather a connected series of memoirs, of kings and other personages, and of the events in which they have been engaged in. Norway, Denmark, Sweden, England, and other countries, from those early ages in which mythology and history are undistinguishably blended together, down to the period nearly of Snorre Sturlason's own birth, to 1177. Snorre begins with Odin and the half-fabulous tales of the Yngling dynasty, and, showing more judgment than many of the modern Saga scholars and antiquaries, passes rapidly over these as an unavoidable introduction to authentic historical times and narratives. From the middle of the ninth century, from Halfdan the Black, who reigned from about the year 827 to about 860, down to Magnus Erlingson, who reigned from about 1162 to 1184, he gives a continuous narrative of events and incidents in public and private life, very descriptive and characteristic of the men and manners of those times,—of the deeds of bold and bloody sea-kings,—of their cruises, of their forays, of their adventures, battles, conquests in foreign lands, and of their home fireside lives also: and he gives, every now and then, very graphic delineations of the domestic manners, way of thinking, acting, and living in those ages; very striking traits of a semi-barbarous state of mind, in which rapacity, cruelty, and bloody ferocious doings, are not unfrequently lightened up by a ray of high and generous feeling;

and he gives too, every now and then, very natural touches of character, and scenes of human action, and of the working of the human mind, which are, in truth, highly dramatic. In rapid narrative of the stirring events of the wild Viking life,—of its vicissitudes, adventures, and exploits,—in extraordinary yet not improbable incidents and changes in the career of individuals,—in touches true to nature,—and in the admirable management of his story, in which episodes, apparently the most unconnected with his subject, come in by-and-by, at the right moment, as most essential parts of it,— Snorre Sturlason stands as far above Ville Hardouin, Joinville, or Froissart, as they stand above the monkish chroniclers who preceded them. His true seat in the Valhal of European literature is on the same bench—however great the distance between —on the same bench with Shakspeare, Carlyle, and Scott, as a dramatic historian; for his Harald Harfager, his Olaf Trygveson, his Olaf the Saint, are in reality great historical dramas, in which these wild energetic personages, their adherents and their opponents, are presented working, acting, and speaking before you.

This high estimate of the literary merit of Snorre Sturlason's work will scarcely pass unquestioned by English readers,—accustomed indeed to hear of the Anglo-Saxon literature, language, and institutions, as of great importance to the historian and antiquary, and as a study necessary for those who wish to become perfectly acquainted with our own, but who would never discover from the pages of Hume, or of any other of our historical writers, that the northern pagans who, in the ninth and tenth centuries, ravaged the coasts of Europe, sparing neither age, sex, nor condition—respecting neither churches, monasteries, nor their inmates—conquering Normandy, Northumberland (then reckoned, with East-Anglia, equal to one-third of all England)—and, under Svein and Canute the Great, conquering and ruling over the whole of England,—were a people possessing any literature at all, or any laws, institutions, arts, or manners connecting them with civilised life. Our historians have confined themselves for information entirely to the records and chronicles of the Anglo-Saxon monks, who, from their convent walls, saw with horror and dismay the bands of these bloodthirsty pagans roving through the country, ravaging, burning, and murdering; and who naturally represent them as the most ferocious and ignorant of barbarians, and without any tincture of civilisation. Our historians and their readers are apt to forget altogether that, pagan and barbarian as these Danes or Northmen of the ninth and tenth centuries undoubtedly were, they were the same people, only in a different stage of civilisation, as the Anglo-Saxons themselves, and were in the tenth century, in their social state, institutions, laws, religion, and language, what the Anglo-Saxons had been in the fifth century, when they first landed on the Isle of Thanet. They forget, too, that the introduction of Christianity, and with it of the Latin language, and of the learning which had a reference only to the Church, and the introduction of social arrangements, establishments, and ideas of polity and government, cast in one mould for all countries of Christendom by the Romish Church, had during these five centuries altered, exhausted, and rendered almost effete, the original spirit and character of Anglo-Saxon social institutions. They do not sufficiently consider the powerful moral influence of this fresh infusion, in the tenth century, of the same spirit, from the same original source, upon the character, ideas, and even forms of government and social arrangements of the whole English population in the subsequent generations, and through them upon the whole of modern society. They do not sufficiently appreciate the social effects of the settlements of these Northmen in England immediately previous to the Norman conquest, when for four generations of kings, viz., Svein, Canute, Harald, and Hardicanute, they had been sole masters of the country, and had possessed and held under their own Danish laws, for many previous generations, what was reckoned equal to one-third of all England. The renovation of Anglo-Saxon institutions, the revival of principles and social spirit which were exhausted in the old Anglo-Saxon race, may be traced to this fresh infusion from the cognate northern people. This subject is very curious and important.

Two nations only have left permanent impressions of their laws, civil polity, social arrangements, spirit, and character, on the civilised communities of modern times—the Romans, and the handful of northern people from the countries beyond the Elbe which had never submitted to the Roman yoke, who, issuing in small piratical bands from the fifth to the tenth century, under the name of Saxons, Danes, Northmen, plundered, conquered, and settled on every European coast from the White Sea to Sicily. Under whatever name, Goths, Visigoths, Franks, Anglo-Saxons, Danes, or Northmen, these tribes appear to have been all of one original stock,—to have been one people in the spirit of their religion, laws, institutions, manners, and languages, only in different stages of civilisation, and the same people whom Tacitus describes. But in Germany the laws and institutions derived from the Roman power, or formed under it after the Roman empire became Christianised, had buried all the original principles of Teutonic arrangements of society as described by Tacitus; and in France the name was almost all that remained of Frank derivation. All the original and peculiar character, spirit, and social institutions of the first inundation of this Germanic population, had become diluted and merged under the church government of Rome,—when a second wave of populations from the same pagan north inundated again, in the ninth and tenth centuries, the shores of Christendom. Wheresoever this people from beyond the pale and influence of the old Roman empire, and of the later church empire of Rome, either settled, mingled, or marauded, they have left permanent traces in society of their laws, institutions, character, and spirit.

Pagan and barbarian as they were, they seem to have carried with them something more natural, something more suitable to the social wants of man, than the laws and institutions formed under the Roman power. What traces have we in Britain of the Romans? A few military roads, and doubtful sites of camps, posts, and towns,—a few traces of public works, and all indicating a despotic military occupation of the country, and none a civilised condition of the mass of the inhabitants,—alone remain in England to tell the world that here the Roman power flourished during four hundred years.

In every province of the ancient Roman empire, even in Italy itself, the remains of Roman power are of the same character—whether those remains be of material objects, as edifices, public works, roads, temples, statues—or of moral objects, as law, government, religion, and social arrangement; and that character is of a hard iron despotism, in which all human rights, all individual existence in wellbeing, all the objects for which man enters into social union with his fellow-man, are disregarded in favour of ruling classes or establishments in the social body, noble, military, or clerical. The Saxon occupation of England lasted for a similar period to the Roman, for about four hundred years. This first wave of the flood of northern populations has left among us traces of laws and institutions, and of a social character and spirit, in which many outlines of freedom and of just principles of social union are distinguishable; and left the influences on the social body of ideas, manners, language, which still exist. But these traces were nearly obliterated, and it is not to be denied that their influence on society was effete,—that in Anglo-Saxon England, as in the rest of Europe, all social arrangement, character, and spirit were assuming one shape and hue under the pressure of superstition, and of the Roman power, institutions, and ascendency, revived through the influence of the Church of Rome which had been in full operation for four centuries and a half, assimilating everything to one form and principle,—when the second wave of the northern populations, the Danes or Northmen, came, under Svein and Canute the Great, to invigorate and renew the social elements left by the first. The moral power of this people—the Anglo-Saxons and Northmen being essentially the same people—has left deeper impressions on society, and of a nobler character, than the despotic material power of the Romans. It is in activity at the present hour in European society, introducing into every country more just ideas than those which grew up amidst the ruins of the Roman empire, of the social relations of the governing and the governed. The history of modern civilisation resolves itself, in reality, into the history of the moral influences of these two nations. All would have been Roman in Europe at this day in principle and social arrangement,—Europe would have been, like Russia or Turkey, one vast den of slaves, with a few rows in its amphitheatre of kings, nobles, and churchmen, raised above the dark mass of humanity beneath them, if three boats from the north of the Elbe had not landed at Ebbsfleet in the Isle of Thanet fourteen hundred years ago, and been followed by a succession of similar boat expeditions of the same people, marauding, conquering, and settling, during six hundred years, viz., from 449 to 1066. All that men hope for of good government and future improvement in their physical and moral condition—all that civilised men enjoy at this day of civil, religious, and political liberty—the British constitution, representative legislature, the trial by jury, security of property, freedom of mind and person, the influence of public opinion over the conduct of public affairs, the Reformation, the liberty of the press, the spirit of the age—all that is or has been of value to man in modern times as a member of society, either in Europe or in the New World, may be traced to the spark left burning upon our shores by these northern barbarians.

Our English writers and readers direct their attention too exclusively to the Anglo-Saxon branch of this great Teutonic race of people, and scarcely acknowledge the social influence of the admixture of their Danish conquerors,—of that fresh infusion in the tenth century, from the same original stock, of the original spirit, character, and social institutions. The schoolman and the political antiquary find it classical or scholarlike to trace up to obscure intimations, in the treatise of Tacitus on the ancient Germans, the origin of parliaments, trial by jury, and all other free institutions, assuming somewhat gratuitously that the seafaring Saxons, who, four hundred years after the days of Tacitus, crossed the sea from the countries north of the Elbe, and conquered England, were identical in laws and social institutions with the forest Germans on the Rhine whom Tacitus describes; and forgetting that a much nearer and more natural source of all the social elements they are tracing back to the forests of Germany in the time of Agricola, was to be found in full vigour among the people who had conquered and colonised the kingdoms of Northumberland and East Anglia, reckoned equal then to one-third of England, and had held them for several generations, and who conquered and ruled over all England for nearly half a century immediately previous to its final conquest by their own Norman kinsmen. The spirit, character, and national vigour of the old Anglo-Saxon branch of this people, had evidently become extinct under the influence and pressure of the Church of Rome upon the energies of the human mind. This abject state of the mass of the old Christianised Anglo-Saxons is evident from the trifling resistance they made to the small piratical bands of Danes or Northmen who infested and settled on their coasts. It is evident that the people had neither energy to fight, nor property, laws, or institutions to defend, and were merely serfs on the land of nobles, or of the Church, who had nothing to lose by a change of masters. It is to the renewal of the original institutions, social condition, and spirit of Anglo-Saxon society, by the fresh infusion of these Danish conquerors into a very large proportion of the whole population in the eleventh century —and not to the social state of the forest Germans in the first century—that we must look for the actual origin of our national institutions, character, and principles of society, and for that check of the popular opinion and will upon arbitrary rule which grew up by degrees, showing itself even in the first generation after William the Conqueror, and which slowly but necessarily produced the English constitution, laws, institutions, and character. The same seed was no doubt sown by the old Anglo-Saxons, and by the Northmen—for they were originally the same people; but the seed of the former had perished under Romish superstition and Church influence, during five centuries in which the mind and property in every country were subjugated to the priesthood whose home was at Rome; and the seed of the latter flourished, because it was fresh from a land in which all were proprietors with interests at stake, and accustomed, although in a very rude and violent way, to take a part, by Things, or assemblies of the people, in all the acts of their government.

Some German, Anglo-American, and English writers, with a silly vanity, and a kind of party feeling, claim a pre-eminence of the Anglo-Saxon race among the European people of our times, in the social, moral, political, and religious elements of society, and even in physical powers—in intellect and in arms. This is the echo of a bray first heard in the forgotten controversy about the authenticity of Ossian's Poems. Pinkerton contended stoutly for the natural intellectual superiority of the Gothic over the Celtic race, insisting that no intellectual achievement, not even the almost physical achievement of the conquest of a country by force of arms, was ever accomplished by Celts. The black hair, dark eye, and dusky skin of the small-sized Celt, were considered by those philosophers to indicate an habitation for souls less gifted than those which usually dwell under the yellow hair, blue eye, and fair skin of the bulky Goth. This conceit has been revived of late in Germany, and in America; and people talk of the superiority of the Gothic, Germanic, or Anglo-Saxon race, as if no such people had ever existed as the Romans, the Spaniards, the French—no such men as Cæsar, Buonaparte, Cicero, Montesquieu, Cervantes, Ariosto, Raphael, Michael Angelo. If the superiority they claim were true, it would be found not to belong at all to that branch of the one great northern race which is called Teutonic, Gothic, Germanic, or Anglo-Saxon—for that branch in England was, previous to the settlements of the Danes or Northmen in the tenth and eleventh centuries, and is at this day throughout all Germany, morally and socially degenerate, and all distinct and distinguishing spirit or nationality in it dead; but to the small cognate branch of the Northmen or Danes, who, between the ninth and twelfth centuries brought their paganism, energy, and social institutions, to bear against, conquer, mingle with, and invigorate the priest-ridden, inert descendants of the old Anglo-Saxon race. It was not, perhaps, so much an overwhelming number of these Northmen, as the new spirit they brought with them, that mixed with and changed the social elements of the countries they settled in. A spark will set fire to a city, if it find stuff to kindle. This stuff was in human nature; and these Northmen, a handful as they were of mere barbarians, did kindle it with their spark of a free social existence, in which all men had property or interests, and a right to a voice in the affairs of their government and in the enactment of their laws. It must be admitted, whatever we think of the alleged superiority of the Teutonic race over the Celtic or Slavonic, that this Northern branch has been more influential than the older Anglo-Saxon branch of their common race on the state of modern society in Europe. We have only to compare England and the United States of America with Saxony, Prussia, Hanover, or any country calling itself of ancient Germanic or Teutonic descent, to be satisfied that from whatever quarter civil, religious, and political liberty, independence of mind, and freedom in social existence may have come, it was not from the banks of the Rhine, or the forests of Germany.

The social condition, institutions, laws, and literature of this vigorous, influential branch of the race, have been too much overlooked by our historians and political philosophers; and this work of Snorre

Sturlason gives us very different impressions of this branch, in its pagan and barbarous state, from the impressions which the contemporary Anglo-Saxon writers, and all our historians on their authority, afford us. Let us first look at their literature, and compare it with that of the Anglo-Saxon of the same ages.

Our early historians, from the Venerable Bede downwards, however accurate in the events and dates they record, and however valuable for this accuracy, are undeniably the dullest of chroniclers. They were monks, ignorant of the world beyond their convent walls, recording the deaths of their abbots, the legends of their founders, and the miracles of their sainted brethren, as the most important events in history; the facts being stated without exercise of judgment, or inquiry after truth, the fictions with a dull credulity unenlivened by a single gleam of genius. The Historia Ecclesiastica venerabilis Bedæ, and Asser's Life of Alfred, embrace the earlier portion of the same period, viz., the latter half of the eighth century, of which the first Sagas of the Heimskringla of Snorre Sturlason treat. The Saxon Chronicle is a dry record of facts and dates, ending about 1155, or about the same period (within twenty years) at which the Heimskringla ends. Matthew Paris begins his history about 1057, and carries it down to about 1250, which is supposed to be about the period of his own death. He was a contemporary of Snorre, who was born in 1178, and murdered in Iceland in 1241. Matthew Paris was no unlettered, obscure monk. He was expressly selected by the Pope, in 1248, for a mission to Norway to settle some disputes among the monks of the order of Saint Benedict, in the monastery of Nidarholm, or Monkholm, in the diocese of Throndhjem; and after accomplishing the object of his mission he returned to his monastery at St. Albans. It is not to be denied that all this connected series of Anglo-Saxon and Anglo-Norman history, from the dissolution of the Roman empire in Britain in the middle of the fifth century down to the middle of the thirteenth century, although composed by such writers of the Anglo-Saxon population as Bede and Matthew Paris, men the most eminent of their times for learning and literary attainments among the Anglo-Saxons and their descendants, is of the most unmitigated dulness, considered as literary or intellectual production; and that all the historical compositions of the old Anglo-Saxon branch during those eight centuries, either in England or in Germany, are, with few if any exceptions, of the same leaden character. They are also, with the exception of the Saxon Chronicle, and of the translation into Anglo-Saxon of Bede by the great King Alfred, all, or almost all, composed in the Latin tongue, not in the native national tongue of the country in which they were composed and of which they treat;—composed not for the people, and as part of the literature of the country, but for a tribe of cloistered scholars spread over the country, yet cut off by their profession from all community of interests, feelings, or views, with the rest of the nation; a class centralised in Rome, and at home only in her church establishment. It was their literature, not the literature of the nation around them, that these writers composed; and its influence, and even all knowledge of its existence, was confined to their own class. It was not until the thirteenth century that Ville Hardouin composed his Memoirs in the vernacular tongue of his countrymen; and he and Joinville, who wrote about the end of the thirteenth century, are considered the earliest historical writers who emancipated history from the Latinity and dulness of the monkish chroniclers.

When we turn from the heavy Latin records of the Anglo-Saxon monks to the accounts given of themselves in their own language, during the very same ages, by the Northmen, we are startled to find that these wild, bloody sea-kings, worshippers of Thor, Odin, and Frigg, and known to us only from the Anglo-Saxon monks as ferocious pagans, overthrowing kings, destroying churches and monasteries, ravaging countries with fire and sword, and dragging the wretched inhabitants whom they did not murder into slavery, surpassed the cognate Saxon people they were plundering and subduing, in literature as much as in arms—that poetry, history, laws, social institutions and usages, many of the useful arts, and all the elements of civilisation and freedom, were existing among them in those ages in much greater vigour than among the Anglo-Saxons themselves. We cling to the early impression given us by Hume, and all our best historians, upon the authority of our monkish chroniclers, that these Pagan Danes or Northmen were barbarians of an almost brutal ignorance and ferocity, without a spark of civilisation or literature. We see that these Vikings, or marauders from the North, were bloody, daring, capable of incredible enterprises and exertion, and of incredible outrages and cruelty when successful—and that a few hundreds of them landing from row-boats, could daunt and subdue extensive tracts of country and all their inhabitants; yet we do not draw the natural conclusion from these facts, that this terrifying, conquering few must have been superior in mental power, energy, and vigour of action, to the daunted, conquered many. All conquests that history tells of will be found to resolve themselves into the superior mental powers of the conquerors. The Romans conquered nations armed in the same way as themselves by superior tactics, discipline, military arrangement, and perseverance; that is, by superior mental power applied to the same material means. The moderns in America, India, and in Europe, conquer by the superiority of firearms, or of what belongs to the efficiency of firearms, in a campaign. This too is the superiority of mental power in the invention, construction, or application of material means. The Northmen, armed with the same weapons as the inhabitants of England, men of the same physical powers as the Anglo-Saxons, land in small piratical bands, altogether insignificant in numbers, on the coasts of England and France, and terrify, paralyse, and conquer, as the Spaniards with their firearms and horses did in Mexico or Peru. What is this but the superiority of mind, of intellectual power, energy, spirit, over the inert passive Anglo-Saxon inhabitants, tamed down by the Church influence and superstition of five centuries into a state of listless existence, without spirit or feeling as a nation, or confidence and self-dependence as individuals, and looking for aid from saints, prayers, and miracles? It was the human mind in a state of barbarous energy and action, and with the vitality of freedom, conquering the human mind in a state of slavish torpidity and superstitious lethargy. The paucity of numbers of these Danes or Northmen was not compensated by any superiority of the weapons, discipline, or tactics they used; but they were men fighting to acquire property by plunder or conquest, who had laws and institutions which secured to them its enjoyment; and they had as opponents only a population of serfs or labourers, with no property in the soil, no interests to fight for, nothing to lose or to defend but what they could save as well by flying or submitting as by fighting.

It might be surmised by a philosophic reader of the history of those times, that all the vigorous action and energy of mind of these barbarous Danes or Northmen could not be showing itself only in deeds of daring enterprise abroad,—that some of it must be expending itself at home, and in other arts and uses than those of a predatory warfare. It will not, at least, surprise such a reader that some of this mental power was applied at home in attempts, however rude, at history and poetry; but he will be surprised to find that those attempts surpass, both in quality and quantity, all that can be produced of Anglo-Saxon literature during the same ages, either in the Anglo-Saxon language or in the Latin. These literary attempts also, or, to give them their due title, this body of literature, is remarkably distinguished from that of the Anglo-Saxons, or of any other people of the same period, by being composed entirely in the native national tongue, and intended to instruct or amuse an audience of the people; and not in a dead language, and intended merely for the perusal of an educated class in the monasteries. With the exception of Thjodrek the Monk, who wrote in Latin in the time of King Sverre, viz., between 1177 and 1202, a history of the kings of Norway down to the end of the reign of Sigurd the Crusader in 1130, and who appears to have been a foreigner, all the literary attempts among this northern branch of the one great race, during the five centuries in which the other branch, the Teutonic or Anglo-Saxon, was confining all intellectual communication in history or poetry to Latin, and within the walls of the cloisters, were composed in the vernacular tongue of the country, intelligible to, and indeed altogether addressed to, the people of all classes.* This singular instance in Europe

---

* A "Historia Norvagia" also seems to have been written in Latin by a Norwegian in the twelfth century. An attempt to make Latin the literary language was also made in Iceland, when the monks Od

of a national literature diffused among a barbarous and rude people, who had not even received the civilisation which accompanies the Christian religion under every form, before the beginning of the twelfth century, who were Pagans in short fully five hundred years after every other part of Europe was, with the exception of some districts perhaps on the coasts of the Baltic, fully Christianised, has not been sufficiently considered by historians in estimating the influence of literature on national mind, character, and social arrangement. To the influence of this rude national literature we probably owe much of what we now pride ourselves upon as the noblest inheritance from our forefathers,—that national energy, activity, independence of mind, and value for civil and political freedom, which distinguish the population of England from that of all other countries, and have done so ever since the admixture of the Northmen with the Old Anglo-Saxons. It may be said that the influence of sagas or songs, of the literature, such as it may be, upon the spirit and character of a people, is overstated, and that it is but a fond exaggeration, at any rate, to dignify with the title of a national, influential literature, the rude traditionary tales and ballads of a barbarous Pagan population. But a nation's literature is its breath of life, without which a nation has no existence, is but a congregation of

Snorreson (died 1200) and Gunlaug Leifson (died 1218) in the Thingeyra cloister produced works on Olaf Trygveson in the Latin language. The originals are lost, and the works are known only through Icelandic translations and adaptations.

individuals. However low the literature may be in its intellectual merit, it will nationalise the living materials of a population into a mass animated with common feeling. During the five centuries in which the Northmen were riding over the seas, and conquering wheresoever they landed, the literature of the people they overcame was locked up in a dead language, and within the walls of monasteries. But the Northmen had a literature of their own, rude as it was; and the Anglo-Saxon race had none, none at least belonging to the people. The following list will show the reader that in the five centuries between the days of the Venerable Bede and those of Matthew Paris, that is from the ninth to the end of the thirteenth century, the northern branch of the common race was not destitute of intellectuality, notwithstanding all their Paganism and barbarism, and had a literature adapted to their national spirit, and wonderfully extensive. The list is taken from that given by Thormod Torfæus, in his "Series Dynastarum et Regum Daniæ," from that given by Müller in his "Sagabibliothek," and from that of Biorn Haldorson. The notes on the date and contents are extracted chiefly from Müller's work. The words *historical* or *fabulous* indicate only that the work is founded on facts apparently, or is a work of fiction." *

* The list has been thoroughly revised by the present Editor, who has consulted in this connection "Catalogus Librorum," and the "Verzeichniss," by the learned Prof. Theodor Möbius of Kiel, and Gudbrand Vigfusson's edition of Cleasby's Icelandic-English Dictionary and his edition of Sturlunga Saga.

Adonius Saga (of a king and duke in Syria). Fabulous.
Alafleks Saga (of a son of a King Richard of England). Fabulous.
Alexander Mikla Saga (of Alexander the Great, translated by Bishop Brand Jonson, by order of Hakon Hakonson). Historical.
Amicus Saga ok Amilius (of Amicus and Amilius—belongs to the story of the Seven Wise Men). Fabulous.
Amloda Saga (of Hamlet, freely translated from Saxo). Fabulous.
Andra Rimur (Rhymes of or concerning Andreas).
Ans Saga (of An Buesvinger). Mythologico-Historical.
Arna Biskups Saga (of Bishop Arne, flourished 1260). Historical.
Arons Saga Hiorleifssonar (of Aron, son of Hiorleif). Historical.
Asmundar Saga vikings ins Irska.

Bærings Saga fagra (of the beautiful Bæring, a Saxon king). Fabulous.
Bandamanna Saga (of the Confederates—account of an Icelandic law-process in the eleventh century). Local history.
Bardar Saga Snæfelsass (of Bard, son of King Dumo, a giant). Fabulous.
Barlaams Saga ok Josaphats.
Bevus Saga (of Bevis, son of an English Count Ginar). Fabulous.
Biskupa Sögur (Sagas of the Bishops). Of these two large volumes have been published by the Icelandic Literary Society.
Bjarnar Saga Hitdælakappa (of Biorn of Hitdale, a contemporary of King Olaf the Saint). Historical.
Blomstrvalla Saga (a translation from the German by Biorn, in Hakon Hakonson's time). The name Blomstrvalla is from a place near Alexandria, where the scene is laid.
Bodvars Biarka Saga. Historical.
Bose ok Herauts Saga (of Bose and Heraut). Fabulous.
Bragda-Magus Saga. Mythical.
Brandkrossa Thattr (Traits of Helge Asbiornson and Helge Droplaugson). Fabulous.
Breta Sögur (Saga of Wales, called Bretland; the parts of England occupied by the Anglo-Saxons were called Saxland by the Northmen). This is from Geoffrey of Monmouth's work.
Broddhelga Saga (of a chief who died about 974). Historical.
Bua Saga (of Bue Andredson). Fabulous.

Damusta Saga (of a Damusta who killed Ion, king of a country south of France, and became king of Greece). Fabulous.
Dinus Saga Dromblata (of Dionysius the Proud, son of King Ptolemy, in Egypt).
Draplaugarsona Saga (of the sons, Helge and Grim, of Draplaug). History and fable mixed; the period, the tenth century.
Drauma Jons Saga (of John the Dreamer and Earl Henry). Fabulous.

Edda Sæmunds (the elder Edda). Mythological. English translation by Benjamin Thorpe. London, 1866.

Edda Snorres (the younger Edda). Mythological. Translated into English by R. B. Anderson. Chicago, 1880.

Edvardar Saga hins helga (of Saint Edward of England).

Egils Saga Einhenda ok Asmundar (of Egil the One-handed, and Asmund). Fabulous.

Egils Saga Skallagrimssonar (of Egil, son of Skallagrim). Historical; period from the middle of the ninth to the end of the tenth century. Translated into English by Daniel Kilham Dodge, Ph.D.

Eiriks Saga Rauda (of Eirik Red, who discovered Greenland, and Vinland or America). Historical; period from near the end of the ninth to the beginning of the tenth century.

Eiriks Saga Vidforla (of Eirik the Wanderer, who goes in search of the land of immortality). Mythological.

Elis Saga (of Elis or Julius and Rosamund). Translated from the French, 1226, by Monk Robert, by order of Hakon Hakonson.

Eyrbyggia Saga (of Thorgrim, whose forefather, Rolf, came from the Isle of Moster in the west of Norway, and first planted Iceland with people from his island (*eyrbyggia*, isle-settlers) to escape Harald Harfager). Historical; period from the first colonising of Iceland to the middle of the eleventh century.

Færeyinga Saga (of the Fareys). Historical.

Fertrams Saga ok Plato (of Fertram and Plato, sons of King Arthur). Fabulous.

Finnboga Saga hins ramma (of Finboge the Strong). Fable and history, from middle of tenth to eleventh century.

Flateyar-bok (the Flatey Codex, so called from the Isle of Flatey in Breidafiord in Iceland, in which the manuscript was discovered in 1650. The annals end in 1395. It contains many sagas transcribed into it, and is considered a most important historical collection).

Floamanna Saga (of a Thorgil and his ancestors, original settlers in Iceland, and of his adventures in Greenland. Thorgil died 1033). Historical.

Flores Saga ok Blankiflur.

Flovents Saga (of Flovent King of the Franks, invented by Master Simon in Lyons).

Fostbrœdra Saga. Historical.

Fridthiofs Saga (of Fridthiof the Bold). This beautiful story has been the groundwork of several poetic and dramatic imitations, of which Bishop Tegner's, in Swedish, has been translated into English. *

---

* See Anderson's "Viking Tales of the North," which contains Tegner's poem in English, and a translation of the original sagas.

Gautreks Saga. Mythical.

Gibbons Saga (of Gibbon, son of the French King William).

Gisla Saga Surssonar (of Gisle, the son of Sur. Events of the tenth century in Iceland). Historical. Translated into English by G. W. Dasent. Edinburgh, 1866.

Gongu-Hrolfs Saga (of Rolf Ganger, the conqueror of Normandy). Historical.

Grænlendinga Thattr (events in Greenland from 1122, and a list of nine bishops and fifteen churches). Historical.

Gragas (Gray Goose). A collection of the laws of Iceland. Edited and translated into Danish by V. Finsen.

Grettis Saga (of Gretter the Strong). Adventures, fabulous and historical, mixed, of Gretter and his forefathers, in the ninth, tenth, and eleventh centuries. Translated into English by Magnusson and Morris. London, 1869.

Grims Saga lodinkinna (the Saga of Grim Shaggy-chin).

Gudmundar Biskups Saga (of Bishop Gudmund); being part of the third book of the Sturlunga Saga, or account of the Sturlung family, which ends 1264, and of which the first books are supposed to have been written 1201.

Guimars Saga (of Guimar, an English knight).

Gullthoris Saga (of Gold Thorer, or Torskfindinga Saga). Fabulous.

Gunlaugs Saga Ormstungu (of Gunlaug the Serpent-tongued). Historical; the period about 1006. Translated by Eirikr Magnusson and William Morris, in "Three Northern Love Stories." London, 1875.

Gunnars Saga keldugnupsfifls (of Gunnar the Idiot). Fabulous.

Gunnars Saga Thidrandabana (of Gunnar who killed Thidrande). Historical; supposed to be written about the end of the twelfth century.

Hænsa Thoris Saga (of Thorer the hen-merchant). Historical.

Hakonar Konungs Saga Hakonarsonar (of King Hakon Hakonson, who was born 1203, and died 1264). Historical; by Sturle Thordson, a contemporary.

Hakonar Saga Ivarssonar (of Hakon Ivarson). Historical.

Halfdanar Saga Bronufostra (of Halfdan, foster-son of Bran). Fabulous.

Halfdanar Saga Eysteinssonar (of Halfdan, son of Eystein). Fabulous.

Halfs Saga (of Half, who, if not altogether a fabulous personage, lived about the eighth century; or in the sixth, according to others).

Hallfredar Saga Vandrædaskalds (of Halfred "the skald, desperate or difficult to deal with," who lived in King Olaf the Saint's time). Historical.

Haralds Rimur Hringsbana (of Harald, who slew Hring).
Haralds Rimur Kvingiarna (Rhymes of or concerning Harald the Woman-lover.)
Havardar Saga Isfirdings (a tragic tale). Historical.
Hemings Thattr (of Heming, a fabulous personage of Olaf the Saint's time).
Hervarar Saga (of Hervar). Mythological.
Hialmters ok Olvis Saga (of Hialmther and Olver). Fabulous.
Hogne ok Hedins Saga. Mythological.
Holmverja Saga. Mixed fable and historical fact regarding Iceland.
Hrafnkels Saga Freysgoda. Historical; of Harald Harfager's times.
Hrafns Saga Sveinbiarnarsonar (of Hrafn, son of Sveinbiorn).
Heidarviga Saga.
Hrims ok Tryggva Rimur.
Heimskringla (the work by Snorre Sturlason now published). Historical.
Hrolfs Saga Kraka (a collection of Sagas, some historical, some fabulous).
Hrolfs Saga Gautrekssonar (of Hrolf, son of Gautrek). Mythological. The story of the battle on the heath. Historical.
Hrolfs Saga Skuggafifls (of Hrolf, son of Skugge the Idiot).
Hrolfs Saga Kraka ok kappa hans (the Saga of Hrolf Kraka and his heroes).
Hromundar Saga Greipssonar. Fabulous.
Hungrvaka (the Hunger-waking is the name of a saga of the Bishops of Skalholt down to 1178; the author supposing it would raise an appetite for more).

Illuga Saga Gridarfostra (of Illugo, foster-son of Grid). Fabulous.
Isfirdinga Saga (of a division of Iceland called Isfirding). Historical.
Islendingabok Ara Froda (Book of Iceland—concerning the first colonisation of Iceland, the introduction of Christianity, &c., usually called Are Frode Schedæ; written about 1120). Historical.
Ivents Saga Artaskappa. Fabulous; translated from the French by order of Hakon Hakonson.

Jarlmanns Saga ok Hermanns (of Jarlman and Herman). Fabulous.
Jarnsida (the law of Iceland from A.D. 1272-1280).
Jokuls Thattr Buasonar (of Jokul, son of Bue). Fabulous.
Jomsvikinga Saga (of the Vikings of Jomsburg, in the island of Wollin or Jom). Historical.
Jonales Rimur (Rhymes of Jonales).
Jons Biskups Saga (of John the Bishop, viz., Jon Ogmundson, who died 1121, Bishop of Skalholt). Historical.
Jons Saga Leiksveins (of John the Juggler). Fabulous.
Jons Saga Baptista (of John the Baptist).

Jonsbok (the Icelandic code of laws of A.D. 1280, and still in use in Iceland.

Kallinius Rimur (Rhymes of Callinius).
Karlamagnus Saga (of Charlemagne).
Ketils Saga Hængs (of Ketil Hæng). Fable and history.
Kirialax Saga (of the Emperor Alexis, viz., Kurios Alexis; but this is a fabulous emperor).
Klarus Saga Keysarasonar (of Clarus, son of the Emperor). Fabulous.
Knytlinga Saga (of the Danish kings of the Canute dynasty, from Harald Gormson to Canute VII., supposed to be by Olaf Thordson, who died 1259). Historical.
Konrads Saga Keysarasonar (of Konrad, son of the Emperor).
Konungs-skuggsja (the King's Mirror). A didactic scholastic work.
Kormaks Saga (of Kormak the Skald). Fable and history.
Kraks Spa (Prophecy of Krak).
Kristinrettr (Ecclesiastic Laws, of which there are several collections).
Kristni Saga (of the introduction of Christianity into Iceland, from 981 to 1000). Historical.
Kroka Refs Saga (of Ref the Cunning). Fabulous.

Landnamabok (events in Iceland from the original settlement in the ninth to the end of the tenth century; with names of the first settlers, and of their lands, to the number of about 3000 names of persons, and 1400 of places; supposed to have been written in the last half of the thirteenth century). Historical.
Langfedgatal (series of dynasties and kings in the North). Historical.
Laurentius Biskups Saga (of Bishop Laurence, who was born 1267). Historical, by a contemporary.
Laxdæla Saga (of the descendants of Aud, who settled in Laxdale). Historical.
Liosvetninga Saga (Lives of the Descendants of Thorgeir and Gudmund, and their own Lives, between the middle of the tenth and end of the twelfth century). Historical; written about the end of the twelfth century.

Magnus Saga Orkneya Jarls (of Saint Magnus, Earl of Orkney, who was killed 1110). Historical.
Margretar Saga (of Margaret and Sigurd, in Magnus the Good's time).
Mariu Saga (of Mary, viz., the Virgin).
Mirmans Saga (of Mirman, a king in Sicily). Fabulous.
Mottuls Saga (of the magic cloak at the court of King Arthur).

Nikolaus Saga Erkibiskups (of Nicholas, Archbishop of Lucca).
Njals Saga (of Nial). Historical; and supposed to be written by

Sæmund Frode, in the eleventh century. The Saga of Burnt Njal is translated into English by G. W. Dasent. The title is "The Story of Burnt Njal; or, Life in Iceland."

Nornagests Thattr. A mythical story.

Œlkofra-thattr. A comical tale telling how Thorkel, nicknamed Alehood, brewed the beer at the althing. Historical.
Œrverodds Saga (of Od the Archer; literally, Arrow-Od). Fabulous.
Orkneyinga Saga (Saga of the Orkney Isles). Translated into English by Jon Hjaltalin and Gilbert Goudie, and edited, with notes and introduction, by Joseph Anderson. Edinburgh, 1873.

Pals Byskups Saga (of Bishop Paul, the seventh bishop of Skalholt, who died in 1211; probably by a contemporary). Historical.
Parcevals Saga (of Parceval, one of King Arthur's worthies). Fabulous.
Partalopa Saga.
Petrs Saga Postola (of Peter the Apostle).

Ragnars Saga Lodbrokar (of Ragnar Lodbrok). History with fable.
Reinalds Rimur Rhymes of (Reinald and Rosa).
Reykdæla Saga. A story of the feud between the good chief Axel and the evil Vemund Koger. Historical.

Salusar Saga ok Nikanors (of Saul and Nicanor, two foster brothers, one of Galatia, and one of Italy). Fabulous.
Samsons Saga Fagra (of Samson the Fair). Fabulous.
Sigurdar Saga snarfara.
Sigurdr Saga Thegla (of Sigurd the Silent, son of King Lodver in Saxland). Fabulous.
Skaldhelga Rimur (Rhymes of the Skald Helge).
Skida Rimar (Rhyme of Skide).
Stiornu Odda Draumr (Star Odde, viz., the Astrologer Odde's Dream).
Stufs Thattr (Traits of Stuf the Skald, who lived in the time of Harald Sigurdson, about 1050). Historical.
Sturlunga Saga (of the family of Sturla, of which Snorre Sturlason was a descendant, from the beginning of the twelfth century to 1284). Historical. Edited, with Prolegomena, Appendices, Tables, Indices, and Maps, by Dr. G. Vigfusson. Oxford, 1878. A superb edition.
Svarfdæla Saga (of Thorstein, who first settled in Svarfdal in Iceland; and fabulous adventures of his successors). History and fable.
Sveins Thattr ok Finns.
Sverris Saga (of King Sverre, from 1177, when Snorre Sturlason's Heimskringla ends, to King Sverre's death). Historical.
Svinfellinga Saga (the story of the sons of Orm, the noted chief of the Svinfell family). Biographical.

Thidreks Saga (of Dietrik of Bern). The same as the German story.

Thomas Saga Erkibyskups (of Archbishop Thomas of Canterbury). Edited, with English translation, by Eirikr Magnusson. London, 1875.

Thordar Saga Hredu (of Thord the Terrible, who, in 975, left Norway, and settled in Iceland). Historical.

Thorleifs Thattr Jarlaskalds (of Thorleif the Skald of the Earls of Orkney). Historical.

Thormodar Saga Kolbrunarskalds (of Thormod Kolbrunarskald). Historical.

Thorsteins Saga Sidu-Hallssonar (of Thorstein, son of Hal o' Side). Historical.

Thorsteins Saga Vikingssonar (of Thorstein, son of Viking). Fabulous. Translated into English by R. B. Anderson in his "Viking Tales of the North." Chicago, 1877.

Thorvalds Saga Vidforla (tells how Thorvald Kodranson, the far-travelled fellow missionary and companion of the Saxon Bishop Frederick, preached the new faith to the Icelanders for four years, but in vain). Historical.

Ulfhams Rimur (Rhymes of Ulfham).

Valdimars Saga Konungs (of Valdemar, son of King Philip of Saxland).

Vallaliots Saga (of Ljot o' Vall, an Icelander; the story of the twelfth century). Historical.

Valvers Thattr (Traits of the life of Valver).

Vapnfirdinga Saga (tells of the feuds between the men of Hof and the men of Crosswick). Historical.

Vatnsdœla Saga (of Ketil Thrumr, his son Thorstein, Ingemund and Sæmund, his grandsons, who settled in Vatnsdal in Iceland). Historical.

Vigaglums Saga (of Glum, son of Eyjolf, who went to settle in Iceland 922). Historical. Translated, with notes and an introduction, by Edmund Head. London, 1866.

Viktors Saga ok Blaus (of Victor and Blaus). Fabulous.

Vilhialms Sago Siods (of William of the Treasure, a son of King Richard in England). Fabulous.

Vilkina Saga (History of the Vilkins). Mythological, and belonging to the Niblung literature.

Vilmundar Saga (of Vilmund and Hierande, a son of a king in Frankland). Fabulous.

Volsunga Saga. Mythological. Translated into English by E. Magnusson and W. Morris. London, 1870.

The lives of Saints (*Heilagra Manna Sögur*), many of which are mentioned in the above list, constitute two large volumes, published by C. R. Unger in Christiania. The *Postula Sögur*, legendary accounts of the lives of the Apostles, have appeared in a large volume, edited by Prof. C. R. Unger. A large number of the *Riddara Sögur*, that is, Romantic Sagas, have been published by Dr. Engen Kölbing, Strasburg, and by Dr. E. Cederschiold, of Lund, Sweden. The old Icelandic literature also abounds in so-called *Rimur*, or Ballads, founded on written stories. Many of these *rimurs* have not yet been published. The most of the mythical sagas are published, collected in three volumes, by Prof. C. C. Rafn, Copenhagen.

It does not appear that any saga-manuscript * now existing has been written before the fourteenth century, however old the saga itself may be. The Flatey manuscript is of 1395. Those supposed to have been written in the thirteenth century are not ascertained to be so on better data than the appearance and handwriting. It is known that in the twelfth century Are Frode, Sæmund, and others began to take the sagas out of the traditional state, and fix them in writing; but none of the original skins appear to have come down to our times, but only some of the numerous copies of them. Bishop Müller shows good reasons for supposing that before Are Frode's time, and in the eleventh

---

* Fagrskinna, Morkinskinna, Hrokkinskinna—fair skin, dark skin, wrinkled skin—are names applied by Torfæus to manuscripts on parchment, probably to designate, when he resided at Stavanger in Norway, to his friend and correspondent, Arne Magnussen at Copenhagen, the particular skin he wanted to refer to, in a compendious way understood between themselves. Arne Magnussen, whose collection of manuscripts is so often quoted under the name of the Arnœ-Magnœi, was the greatest antiquary who *never wrote*. Although he wrote no books, his judgment and opinions are known from notes, selections, and correspondence, and are of great authority at this day in the Saga literature. Torfæus consulted him in his researches, which gives great weight to the views of Torfæus on many points, as we have in them the combined judgment of two of the greatest northern antiquaries.—L.

century, sagas were committed to writing; but if we consider the scarcity of the material in that age—parchment of the classics, even in Italy, being often deleted, to be used by the monks for their writings—these must have been very few. No well-authenticated saga of ancient date in Runic is extant, if such ever existed; although Runic letters occur in Gothic, and even in Anglo-Saxon manuscripts, mixed with the other characters.

To these Torfæus adds,—Historical Fragments concerning Ivar Vidfadme, Hrærek Slongvanbauge, Helge Hvasse (the Sharp), and the Battle of Bravalla: also the Codex Flateyensis, as above noticed,—a manuscript so called from the island Flatey, on the west side of Iceland, in which it was discovered, containing the genealogies and annals of the Norwegian kings and chiefs: also a manuscript called by him the Fair Skin—Fagrskinna; being a breviary of the history of Norway, or chronological compendium from Halfdan the Black to Sverre's reign; and also several ancient annals, which, being without titles, he cannot cite in his catalogue. Besides these, the following works, no longer extant in any known manuscripts, are referred to in the ancient histories, viz.: The history of Einar the son of Gisle, who killed Giafald, one of the court of King Magnus Barefoot, is cited in the end of the "Life of Saint Jon Bishop of Holar." The history of Sigurd Hjort (Cervus) is cited by Snorre Sturlason in his "Life of Halfdan the Black." The life of Alfgeir is cited in the "Hardar Saga Grimkelssonar."

The history of Grim the son of the widow on the gard Krop, who killed Eid the son of Skegge of Midfiord, is mentioned in the "Life of Grette the Strong." The life of Thorgils the son of Hall, and the history of the people of Niardvik, are cited in the "History of the Laxdale People." The "Landnama" mentions histories of Bodmod, of Gerper, of Grimolf, and the life of Thord Geller. The same work mentions also the history of the Thorskfiord people, and a life of Vebiorn, who was one of the original settlers in Iceland when it was uninhabited. The history of the Sturlung family shows that formerly there were extant a history of the Berserk and Viking Hraungrid, and lives of Olaf king of the Lidmen or army, of Hrok the Black, and of Orm the Poet. Snorre appears to have read a history of the Skioldung family, that is, of the progenitors of the Danish dynasty. The "Life of Hrolf Krake" cites a life of Thorer the Dog-footed, and a life of Agnar son of Hroar king of Denmark. The "Life of Rolf the Walker,* the Conqueror of Normandy," cites a history of the Gautland people. The history of Skiold the son of Dag, and of Herman, is cited in the "Life of Illuge Grid's Foster-son." The "Life of Bose" mentions a life of Sigurd Hring. Mention of the histories of Ulf, son of Sebbe, and of Earl Kvik, is made in the historical relation of

* Hrolf Gangr appears to have been a name in the family; and one of the forefathers of the conqueror of Normandy bore it. The popular tale of his being so stout or corpulent that no horse could carry him, and he was obliged to walk, may therefore be doubted; as such a habit of body would scarcely be consistent with the personal activity of great warriors in those days.—L.

some incidents by the skalds of Harald Harfager. The "History of the Liosvatn People" cites a history of the people of Espihol. The writings of Are, who lived about the year 1117, and first committed to writing the Icelandic compositions, and of Sæmund, who flourished about the year 1083, and had studied at universities in Germany and France, and of Od the Monk, who flourished in the twelfth century, are almost entirely lost. Kolskeg, a contemporary of Are, and, like him, distinguished by the surname of Frode—the wise, or the much knowing—Brand, who lived about the year 1163, Eirik, the son of Od, and his contemporary Karl, abbot of the monastery of Thringey, in the north of Iceland, and several others, appear to have been collectors, transcribers, and partly continuators of preceding chronicles; and all these flourished between the time of Bede in the end of the seventh and beginning of the eighth century, when the devastations of these piratical Vikings were at the worst, and the time of Snorre Sturlason in the middle of the thirteenth century, when the Viking life was given up, invasions of Northmen even under their kings had ceased, and the influence of Christianity and its establishments was diffused.

Now we have here a vast body of literature, chiefly historical, or intended to be so, and all in the vernacular tongue of the Northmen. It is for our Anglo-Saxon scholars and antiquaries to say, whether in the Anglo-Saxon tongue, or in the Anglo-Saxon and the Latin together, such a body of national literature was produced,—whether such intellectual activity existed between the days of the Venerable Bede, our earliest historian, in the beginning of the eighth century, and the days of Matthew Paris, the contemporary of Snorre Sturlason, in the first half of the thirteenth? And these were Pagans, these Northmen, whether in Denmark, Norway, or Iceland, for more than half of these five centuries! This body of literature may surely be called a national literature; for on looking over the subjects it treats of, it will be found to consist almost entirely of historical events, or of the achievements of individuals, which, whether real or fabulous, were calculated to sustain a national spirit among the people for whom they were composed; and scarcely any of it consists of the legends of saints, of homilies, or theological treatises, which constitute the greater proportion of the literature of other countries during the same ages, and which were evidently composed only for the public of the cloisters. It is distinguished also from any contemporary literature, and indeed from any known body of literature, by the peculiar circumstance of its having been for many centuries, and until the beginning of the twelfth century, or within 120 years of Snorre Sturlason's own times, an oral not a written literature, and composed and transmitted from generation to generation by word of mouth, and by memory, not by pen, ink, and parchment. This circumstance may affect the historical value of these documents, if the authenticity of what they relate be not supported by internal or collateral evidence, but does not affect their literary value as the compositions, during five centuries, of the Northmen, and as such to be compared with the compositions, during the same five centuries, of the cognate Anglo-Saxon people. It is of great importance, however, to examine the value, as historical documents, of these compositions.

The early history of every people can only have been preserved by traditionary stories, songs, ballads, until the age when they were fixed by writing. The early history of Rome, for many centuries, has had no other foundation than such a saga-literature as this of the Northmen. Homer, whether the Iliad and Odyssey be the works of one mind or of several, has had traditionary accounts as the historical foundation and authority for the events and personages he celebrates. Snorre Sturlason has done for the history of the Northmen what Livy did for the history of the Romans. The traditionary works of the predecessors of Livy in his historical field, the sagas of the Romans, have unfortunately not reached us. The ancient Roman writers themselves regret that the songs and legends, the sagas from which the historical accounts of their ancestors are derived, and which it appears from two passages in Cicero*

---

* Gravissimus auctor in 'Originibus' dixit Cato, morem apud majores hunc epularum fuisse, ut deinceps, qui accubarent, canerent ad tibiam clarorum virorum laudes atque virtutes.—*Cicero, Tusc. Quæst.* iv. 3.

Utinam exstarent illa carmina quæ multis sæculis ante suam ætatem in epulis esse cantitata a singulis convivis de clarorum virorum laudibus in 'Originibus' scriptum reliquit Cato.—*Cicer. Brutus*, cap. xix.

See on this subject the Preface to "Lays of Ancient Rome." By Thomas Babington Macaulay. London: Longman and Co., 1842.—L.

were extant in the time of the elder Cato, and, like the sagas of the Northmen, were sung or recited at feasts, had fallen into oblivion. Such documents in verse or prose are common to the early history of every people, and on such and on the similar transmission of them by memory, the historical Scriptures of the Old Testament themselves rest. These sagas have been preserved among the Northmen, or at least have not perished so entirely but that the sources from which their historian Snorre drew his information may be examined. They constitute the body of literature of which the list of sagas given above is an imperfect catalogue—imperfect because many sagas, songs, or other compositions referred to in those which are extant no longer exist, and probably never had been taken out of the traditionary state, in which they existed then as matter of memory, and been fixed in writing. If we consider the scarcity of the material—parchment—in the Middle Ages, even in the oldest Christianised countries of Europe, and the still greater scarcity of scribes, and men of learning and leisure, who would bestow their time and material on any subjects but monastic legends in the Latin language, we must wonder that so many of these historical tales had been committed to writing in Iceland; not that so many which once were extant in the traditionary state have not been preserved.

Every intelligent reader of English history who is startled at this view of the comparative literature and intellectual condition of the two branches, the

Pagan and the Christian, of the one great northern race, between the eighth and the thirteenth centuries, will desire information on the following points:—Who were the scribes, collectors, or compilers, who preceded Snorre Sturlason in writing down, gathering, or reducing to history, those traditionary narratives called sagas which had floated down on the memory, in verse or in prose, from generation to generation? Who were the original authors of these compositions; and what was the condition of the class of men, the skalds, who composed them? What were the peculiar circumstances in the social condition of the Northmen in those ages, by which such a class as the skalds was kept in bread, and in constant employment and exertion among them, and even with great social consideration; while among the Anglo-Saxons, a cognate branch of the same people, the equivalent class of the Bards, Troubadours, Minstrels, Minnesingers, was either extinct, or of no more social influence than that of the Court Jesters or the Jougleurs?

Snorre Sturlason tells us, in the preface to his work, that "the priest Are *hinn Frodi* (*hinn Frodi* is applied to several writers, and means the Wise, the Learned; le Prud'homme perhaps of the Norman-French, although antiquaries render it into the more assuming Latin appellative, Polyhistor), was the first man in Iceland who wrote down in the Norse tongue both old and new narratives of events." The Landnama-bok (Liber Originum Islandiæ), which treats of the first occupation of Iceland by the Norwegians, and of their descendants; the Islendinga-bok, or Book of Iceland, usually quoted by the title of the Latin translation, "Schedæ Arii Polyhistoris," which is an account of the introduction of Christianity, and of other affairs in Iceland, and of the judges and other considerable personages; and the Flateyiar-bok, forming part of the important manuscript on parchment quoted so often by northern antiquaries by the name of the Codex Flateyensis,—are works of Are still extant. The Flateyiar-bok appears to have been a chronicle begun by Are, and continued by his successors in his parochial charge. It does not appear that any writing of Are upon parchment is extant, and his labours as a compiler appear to be known from the testimony only of Snorre Sturlason, or from copies such as those in the Codex Flateyensis, made from his writings. Are Frode is reckoned by Torfæus to have been born about the year 1067, and to have written "the old and new narratives of events," which Snorre tells us he did, "two hundred and forty years after the first occupation of Iceland by the Norwegians;" about the year 1117. A manuscript of Biorn of Skardsa, which Torfæus says was once in his possession, speaks of an older compiler than Are. Isleif, the first bishop of Iceland, who was consecrated by Adalbert, archbishop of Bremen, in 1056, and who died in 1080, is stated to have written a life of Harald Harfager and his successors, down to Magnus the Good, who died about 1047, compiled from the current sagas; and his son, Bishop Gissur, is stated to have also collected and written down histories in the common tongue. Are the Learned was brought up as a fosterson in the house of Teit, another son of this Bishop Isleif, and, Torfæus supposes, may have used the materials collected by Isleif; and thus the labours of the two, as compilers or scribes of the ancient sagas, may have been attributed to the one of most celebrity. The celibacy of the clergy appears not to have been regarded in the northern countries in the eleventh or twelfth centuries. We read of the wives and sons of priests down to a late period; and Bishop Isleif was not singular in having sons.

Sæmund, also designated as the Learned (*Frodi*), was a contemporary of Are. He was born in 1056, and after travelling and studying in Germany and France returned to Iceland, and settled as priest of the parish of Odde, in the south of Iceland, and commenced the Annals, which were continued by his successors in the clerical charge of Odde, and are hence called "Annales Oddenses" by the northern antiquaries. The older Edda, of which the Edda of Snorre Sturlason is but an epitome for explaining the mythological language and allusions of the poetical saga, is attributed to him; but unfortunately it is almost entirely lost, so that we know little of the doctrines or establishments of the ancient Odin-worship. Od the Monk, who lived in the following century, refers to an historical work of Sæmund, which is also lost. Sæmund died in 1133. His contemporary Are survived him, and died in 1148.

Kolskeg, also styled Frode, was another contemporary of Are, whose name is known as a compiler, or scribe, but his works are not extant.

Brand, bishop of the diocese of Holar in Iceland, and who died 1264, was also a diligent transcriber of sagas from the memory to parchment. He was a contemporary of Saxo Grammaticus, the Danish historian. Saxo himself, in the preface to his work, gives the strongest testimony to the diligence and importance of the historical researches and traditional records of the Icelanders. "Nor is the industry of the Tylenses (by which name Saxo designates the people of Tyle, Thule, or Iceland) to be passed over in silence, who, from the sterility of their native soil, being deprived of every luxury of food, exercise a perpetual sobriety, and turn every moment of their lives to the cultivation of a knowledge of the affairs of other countries, and, compensating their poverty by their ingenuity, consider it their pleasure to become acquainted with the transactions of other nations, and hold it to be not less honourable to record the virtues of others than to exhibit their own; and whose treasures in the records of historical transactions I have carefully consulted, and have composed no small portion of the present work according to their relations, not despising as authorities those whom I know to be so deeply imbued with a knowledge of antiquity." Saxo appears to have had access to many sagas, either in manuscript, or in *vivâ voce* relation, which are not now extant. Thjodrek the Monk, a contemporary also of Saxo, who flourished about the year 1161, and wrote a history of the kings of Norway in Latin, and almost the only historical work of the Middle Ages composed in that language in Norway, gives a similar testimony to the great amount of historical knowledge among the Icelanders transmitted through their songs and sagas. The causes of this peculiar turn among the Icelanders will be inquired into afterwards.

Erik, the son of Od, wrote a history of King Harald Gille's sons, Sigurd and Inge, who succeeded him, as joint kings of Norway, about 1136, to the death of each of them; and gives also the history of King Magnus the Blind, and of Sigurd Slembedegn. As King Inge fell in battle in the year 1161, the work of Erik is to be placed after that date. Karl Jonsson, abbot of the monastery of Thingeyre in the north of Iceland, who was ordained in 1169, and died in 1212, wrote a life of his contemporary King Sverre, who reigned from 1177 to 1202. His work is highly esteemed.

Od the Monk, also called Frode, was next to, or perhaps contemporary with, these writers, and composed a life of King Olaf Trygveson, containing circumstances not found in other accounts of that reign; from which it is supposed that he had access to sagas not now extant.

These are the principal historical writers who compiled or composed from the ancient unwritten sagas, between the days of Are the Learned in 1117, and the days of Snorre Sturlason in the beginning of the following century. In these hundred or hundred and twenty years between Are and Snorre, the great mass of literature in the vernacular tongue committed to parchment proves a state of great intellectual activity among these Northmen. It is not the literary or historical value, or the true dates or facts of these traditionary pieces called sagas, written down for the first time within those hundred and twenty years, that is the important consideration to the philosophical reader of history; but the extraordinary fact, that before the Norman conquest of England here was a people but just Christianised, whose fathers were Pagans, and who were still called barbarians by the Anglo-Saxons, yet with a literature in their own language diffused through the whole social body, and living in the common tongue and mind of the people. The reader would almost ask if the Anglo-Saxons were not the barbarians of the two,—a people, to judge from their history, without national feeling, interests, or spirit, sunk in abject superstition, and with no literature among them but what belonged to a class of men bred in the cloister, using only the Latin language, and communicating only with each other, or with Rome. In the same period in which the intellectual powers of the Pagan or newly Christianised Northmen were at work in the national tongue upon subjects of popular interest, what was the amount of literary production among the Anglo-Saxons? Gildas, the earliest British writer, was of the ancient British, not of the Anglo-Saxon people, and wrote about the year 560, or a century after the arrival of the Anglo-Saxons in England. Gildas Albanius, or Saint Gildas, preceded him by about a century; and both wrote in Latin, not in the British or the Saxon tongue. The "Historia Ecclesiastica Venerabilis Bedæ" was written in Latin about the year 731; and King Alfred translated this work of the Venerable Bede into Anglo-Saxon about 858, or by other account some time between 872 and 900. Asser wrote "De Vita et Rebus Gestis Alfredi" about the same period, for he died 910. Nennius, and his annotator Samuel, are placed by Pinkerton about the year 858. Florence of Worcester wrote about 1100; Simeon of Durham about 1164; Giraldus Cambrensis in the same century. The "Saxon Chronicle" appears to have been the work of different hands from the eleventh to the twelfth century. Roger of Hovedon wrote about 1210; Matthew Paris, the contemporary of Snorre Sturlason, about 1240. These are the principal writers among the Anglo-Saxons referred to by our historians, down to the age of Snorre Sturlason; and they all wrote in Latin, not in the language of the people—the Anglo-Saxon.

This separation of the mind and language, and of the intellectual influence of the upper educated classes, from the uneducated mass of the Anglo-Saxon people, on the Continent as well as in England, by the barrier of a dead language, forms the great distinctive difference between the Anglo-Saxons and the Northmen; and to it may be traced much of the difference in the social condition, spirit, and character of the two branches of the Teutonic or Saxon race at the present day. It is but about a century ago, about 1740, that this barrier was broken down in Germany, and men of genius or science began to write for the German mind in its own German language. With the exception of Luther's translation of the Bible, little or nothing had been written before the eighteenth century for the German people in the German tongue. That beautiful language itself had become so Latinised by the use and application of Latin in all business and intellectual production—a circumstance which both Goethe and Jean Paul Richter, its greatest masters, deplore—that it was, and to a considerable degree remains in the present times, a different language in writing from the spoken vernacular tongue of the people of Germany. They have to acquire it, as, in some sort, a dead language to them, to understand and enter into the meaning and spirit of their own best writers. Their Plat Deutch, the spoken tongue of the mass of the people, does not merely differ as our Scotch, Yorkshire, or Somersetshire dialects differ from English, only in tone of voice, pronunciation, and in the use of a few obsolete words; but in construction and elements, from the too great admixture of foreign elements from the Latin into the cultivated German. A striking proof of this is, that no sentiment, phrase, popular idea, or expression from the writings of Lessing, Goethe, Schiller, Richter, or any other great German writer, is ever heard among the lower classes in Germany, the peasants, labouring people, and uneducated masses; while, with us, sentiments, expressions, phrases, from Shakspeare, Pope, Burns, Swift, De Foe, Cobbett,—from Cervantes, Le Sage, Moliere,—have crept into common use and application, as proverbial sayings circulating among our totally uneducated classes, who certainly never read those authors, but have caught up from others what is good and natural, because the thought is expressed in language which they are as familiar with as the writer was himself. In our branch of the Saxon race, the intellectuality of the educated class has always worked downwards through a language common to all. The moral influence of this uninterrupted circulation of ideas from the highest to the lowest is very striking in our social condition, and in that of all the people descended from the Northmen, the younger branch of the great Anglo-Saxon race. Under every form of government, whether despotic as in Denmark, aristocratic as in Sweden, democratic as in the United States, or mixed as in England, they are, under all circumstances, distinguished from the other, the old Anglo-Saxon branch, by their strong nationality and distinct national characters. What is this but the diffusion of one mind, one spirit, one mode of thinking and doing, through the whole social body of each of these groups, by a common language and literature, such as it may be, giving one shape and tone to the mind of all? Turn from these groups of the European population, and look at the nationality or national character of the other branch of the race—or rather look for it. Where is it? Have Prussians, Saxons, Hanoverians, Hessians, Baden-Badenians, or whatever their rulers call them, any jot of this national feeling, any national existence at all? Have the Germans as a whole mass, or has any one group of them, any national character at this day, any common feeling among all classes upon any one subject? There is a want of that circulation of the same mind and intelligence through all classes of the social body, differing only in degree, not in kind, in the most educated and the most ignorant, and of that circulation and interchange of impressions through a language and literature common to all, which alone can animate a population into a nation. It would be a curious subject for the political philosopher to examine, what have been the effects of the literature of a people upon their social condition. English literature works much more powerfully upon the great mass of the English people, although uneducated, and unable to receive its influence and impression direct, than German literature, although much more abundant, works upon the people of Germany. The circulation of ideas stops there at a certain class, and the mass remains unmoved by, impenetrable to, and unintelligent of the storms that may be raging on the surface among the upper educated people. The literature of the Northmen in their own tongue undoubtedly kept alive that common feeling and mind—that common sense on matters of common interest, which in England grew up into our national institutions. They had a literature of their own, however rude, a history of their own, however barbarous,—had laws, institutions, and social arrangements of their own; and all these through a common language influencing and forming a common mind in all; and when men, or the children of men whose minds had been so formed, came to inhabit, and not merely to conquer, but to colonise a very large proportion of the surface of England, we may safely assume that what we call the Anglo-Saxon institutions of England, and the spirit and character on which alone free institutions can rest, were the natural productions of this national mind, reared by the Northmen in England, and not by the Anglo-Saxons.

What were the peculiar circumstances, in the social condition of this branch of the Saxon race, which kept alive a national literature, history, spirit, and character, and peculiar laws and institutions, while all that was peculiar to or distinctive of the other branch had long been extinguished in Germany, and in a great measure in England? This question can only be answered by looking at the original position of this northern branch of the same stock, on the European soil.

The race of men who under Odin established themselves in the countries north of the Baltic were undoubtedly of Asiatic origin.* The date of this

---

* On the origin of the Teutonic race, and particularly on the earliest Germanic settlement of Scandinavia, see Viktor Rydberg's remarkable work, recently published in an English translation by R. B. Anderson, and entitled "Researches in Teutonic Mythology." London, Swan, Sonnenschein and Co., 1889.

inundation may have been 400 years before or 400 years after the Christian era (antiquaries have their theories for both periods), or there may have been different Odins, or the name may have been generic and applied to all great conquerors; and the causes, as well as the dates of this vast movement, are lost in the night of antiquity. The fact itself admits of no doubt; for it rests not only on the concurrent traditions and religious belief of the people, but upon customs retained by them to a period far within the pale of written history, and which could only have arisen in the country from which they came, not in that to which they had come. The use, for instance, of horse-flesh could never have been an original indigenous Scandinavian custom, because the horse there is an animal too valuable and scarce ever to have been an article of food, as on the plains of Asia; but down to the end of the eleventh century the eating of horse-flesh at the religious feasts, as commemorative of their original country, prevailed, and was the distinctive token of adhering to the religion of Odin; and those who ate horse-flesh were punished with death by Saint Olaf. A plurality of wives also, in which the most Christian of their kings indulged even so late as the twelfth century, was not a custom which, in a poor country like Scandinavia, was likely to prevail, and appears more probably of Asiatic origin. But what could have induced a migrating population from the Tanais (the Don), on which traditionary history fixes their original seat, after reaching the southern coasts of the Baltic, to have turned to the north and crossed the sea to establish themselves on the bleak inhospitable rocks, and in the severe climate of Scandinavia, instead of overspreading the finer countries on the south side of the Baltic? The political causes from preoccupation, or opposition of tribes as warlike as themselves, cannot now be known from any historical data; but from physical data we may conjecture that such a deviation from what we would consider the more natural run of the tide of a population seeking a living in new homes, may have been preferable to any other course in their social condition. We make a wrong estimate of the comparative facilities of subsisting, in the early ages of mankind, in the northern and southern countries of Europe. If a tribe of red men from the forests of America had been suddenly transported in the days of Tacitus to the forests of Europe beyond the Rhine, where would they, in what is called the hunter state, that is, depending for subsistence on the spontaneous productions of nature, have found in the greatest abundance the means and facilities of subsisting themselves? Unquestionably on the Scandinavian peninsula, intersected by narrow inlets of the sea teeming with fish, by lakes and rivers rich in fish, and in a land covered with forests, in which not only all the wild animals of Europe that are food for man abound, but from the numerous lakes, rivers, ponds, and precipices in this hunting-field, are to be got at and caught with much greater facility than on the boundless plains, on which, from the Rhine to the Elbe, and from the Elbe to the Vistula, or to the steppes of Asia, there is scarcely a natural feature of country to hem in a herd of wild animals in their flight, and turn them into any particular tract or direction to which the hunters could resort with advantage, and at which they could depend on meeting their prey. At this day Norway is the only country in Europe in which men subsist in considerable comfort in what may be called the hunter state,—that is, upon the natural products of the earth and waters, to which man in the rudest state must have equally had access in all ages,—and derive their food, fuel, clothing, and lodging from the forest, the mountains, the fiords, and rivers, without other aid from agriculture, or the arts of civilised life, than is implied in keeping herds of reindeer in a half tame state, or a few cows upon the natural herbage of the mountain glens. We, in our state of society, do not consider that the superior fertility of the warmer climes and better soils of southern countries, adds nothing to the means of subsistence of those who do not live upon those products of the earth which are obtained by cultivation. A hermit at the present day could subsist himself, from the unaided bounty of nature, much better at the side of a fiord in Norway, than on the banks of the Tiber, or of the Tagus, or of the Thames. Iceland, which we naturally think the last abode to which necessity could drive settlers, had in its abundance of fish, wild fowl, and pasturage for sheep and cows, although the country never produced corn, such advantages that it was the earliest of modern colonies, and was a favourite resort of emigrants in the ninth century. The Irish monk Dicuil, who wrote in 825 his work "De Mensura Orbis Terræ," published by C. A. Walckenaer in Paris in 1828, says that for 100 years, that is from 725, the desire for the hermit life had led many Irish clerks to the islands to the north of the British sea, which, with a fair wind, may be reached in two days' sail from the most northerly British isles. These were most likely the Farey Isles, or Westman Isles. "These isles," he says, "from the creation of the world uninhabited, and unnamed, are now, in 825, deserted by the hermits on account of the northern sea robbers. They have innumerable sheep, and many sorts of sea fowl." This would show that even before the settlement of the Northmen in Iceland about 825 (and in one of the sagas it is said the first settlers found in the Westman Isles books and other articles of Irish priests), the facility of subsistence had drawn some individuals to those rocks in the northern ocean, and they were then known lands. Sweden had a still stronger attraction for the warlike tribes from the interior of Asia, who were pressing upon the population of Europe south of the Baltic, and which has been overlooked by the historians who treat of the migrations of mankind from or to the north in the rude ages. Sweden alone had iron and copper for arms and utensils close to the surface of the earth, and, from the richness of the ores, to be obtained by the simplest processes of smelting. This natural advantage must, in those ages, have made Sweden a rallying point for the Asiatic populations coming into Europe from the north of Asia, and from countries destitute of the useful metals in any abundant or easily obtained supply. To them Sweden was a Mexico or Peru, or rather an arsenal from which they must draw their weapons before they could proceed to Germany. This circumstance itself may account for the apparently absurd opinion of the swarms of Goths who invaded Europe having come from Scandinavia; and for the apparently absurd tradition of Odin, or the Asiatics invading and occupying Scandinavia in preference to the more genial countries and climes to the south of the Baltic; and for the historical fact of a considerable trade having existed, from the most remote times, between Novgorod and Sweden, and of which, in the very earliest ages, Wisby, in the Isle of Gotland, was the *entrepôt* or meeting-place for the exchange of products. The great importance of this physical advantage of Scandinavia in the abundance of copper and iron, to an ancient warlike population, will be understood best if we take the trouble to calculate what quantity of iron or copper must have been expended in those days as ammunition, in missile weapons, by an ordinary army in an ordinary battle. We cannot reckon less than one ounce weight of iron, on an average, to each arrowhead, from twenty to twenty-four *drop*, or an ounce and a quarter to an ounce and a half, being considered by modern archers the proper weight of an arrow; and we cannot reckon that bowmen took the field with a smaller provision than four sheaves of arrows, or heads for that number. A sheaf of twenty-four arrows would not keep a bowmen above ten or twelve minutes; and in an ordinary battle of three or four hours, allowing that arrows might be picked up and shot back in great numbers, we cannot suppose a smaller provision belonging to and transported with a body of bowmen than ninety-six rounds each, which, for a body of 4000 men only, would amount to above fourteen tons' weight of iron in arrowheads alone. For casting spears or javelins, of which in ancient armies, as in the Roman, more use was made than of the bow, we cannot reckon less than six ounces of iron to the spear head, or less than two spears to each man; and this gives us nearly two tons' weight more of iron for 4000 men as their provision in this kind of missile. Of hand-weapons, such as swords, battleaxes, halberds, spears, and of defensive armour, such as head-pieces and shields, which every man had, and coats of mail or armour, which some had, it is sufficient to observe that all of it would be lost iron to the troops who were defeated, or driven from the field of battle leaving their killed and wounded behind, and all had to be replaced by a fresh supply of iron. We see in this great amount of iron or bronze arms, to be provided and transported with even a very small body of men in ancient times, why a single battle was almost always decisive, and every thing was staked upon the issue of a single day;

and we see why defeat, as in the case of the battle of Hastings and many others, was almost always irrecoverable with the same troops. They had no ammunition on the losing side after a battle. We may judge from these views how important and valuable it must have been for an invading army of Goths, or whatever name they bore, coming from Asia to Europe, to have got possession of Sweden; so important, indeed, that it is reasonable to believe that if ever an Asiatic people invaded Europe north of the Carpathian mountains, the invaders would first of all proceed north along the Vistula and other rivers falling into the Baltic, and put themselves in communication, by conquest or commerce, with the country which supplied their ammunition, and would then issue armed from the north, and break into the Roman empire, and be considered as a people coming originally from some northern hive. Scandinavia certainly never had food for more human beings than its present inhabitants, and could never have poured out the successive multitudes who, by all accounts, are said to have come in from the north upon the Roman provinces; but in this view it is likely that the flood of people actually did pour in from the north, to which the march must of necessity have been first directed from Asia. It may be objected to these views, that iron or metal was not of such prime necessity as we make it to these barbarians in their warfare; that flint or other stones were much used for arrow-heads, and that we find such commonly in museums.

and even stones that have evidently been intended for javelins or battleaxes. If we look, however, at what exists out of museums, we find that stones which admit of being chiselled, sharpened, or brought to an edge or point that would pierce cloth, leather, or any defensive covering, and inflict a deadly wound, are among the rarest productions. Granite, gneiss, sandstone, limestone, all rise in lumps and cubical masses, scarcely to be reduced by any labour or skill to shapes suitable for a spear or arrowhead. Countries of vast extent are without stone at all near the surface of the earth, and many without such a kind of stone as could be edged or pointed, without such skill and labour as would make stone arrowheads more scarce and valuable than metal ones. Of such stones as might be substituted for metal in missile weapons it happens, singularly enough, that Scandinavia itself is more productive than any part of the north of Europe, if we except perhaps the districts of England abounding in flint. Our ordinary museum arrowheads of stone, or what our country people, when they turn them up by the plough, call elf-bolts, from an obscure impression that they do not belong to the soil, but are, from the regularity of their shape, an artificial production, are in reality the organic fossil called by geologists the Belemnite, which, tapering to a point at both ends from regular equally poised sides, is, in its natural fossil state, an arrowhead. This fossil, and the sharp schists, which could easily be formed into effective points for missile weapons, abound particularly in that great indenture of the Norwegian coast called the Skager Rack, and in the Middle Ages called Viken, or the Wick, or Vik, between the Naze of Norway and the Sound or the coast of Jutland, and from which Pinkerton conjectures the Scottish Picts or Victi, if they were a Gothic tribe, originally proceeded. He founds his conjecture on the similarity of name; and the Vikings or pirates probably derived their name from this district of Viken in which they harboured, and for the obvious reason that here the means of replenishing their ships with the missile arms of the age abounded. Hardsteinagriot, or small hard stones, appear to have been even an article of export at a very early date from Thelemark, and to have been shipped from the coast to which they were transported in quantities of 1500 loads at a time from the interior.* Stones for throwing by hand (the sling, on account of the space required around the slinger, seems never to have been in use) were so important an article in the sea fights of those times, that the ships of war, or long-ships, were always accompanied on the viking cruises by transports or ships of burden, to carry the plunder, clothes, and provisions, the ships of war being loaded with arms and stones. We find two transport vessels to ten ships of war in the Saga of Saint Olaf, as the number with him when he left his ships of war at the mouth of the Humber,

---

* Krafts Beskrivelse, III. 154. Kong Sverre's Saga, by Jacob Aal, note on cap. 91.—L.

after a long viking expedition, and returned to Norway, with 220 men, in his two transport ships. Earl Ragnvald, the son of Kol, invaded Earl Paul in Orkney with six ships of war, five boats of a size to cross the sea from Norway, and three ships of burden;* and in all their expeditions ships of burden were required in some proportion to the ships of war, owing to the great stowage necessary for their weapons. In the Færeyinga Saga, in which the exploits of a viking† called Sigmund Bresteson are related minutely, we read of his walking across a small island on the Swedish coast, and discovering five ships of another viking at anchor on the opposite side, and he returned to his own ships, passed the whole night in landing his goods and plunder, and breaking up stones on shore, and loading his vessel with them, and at daylight he went to attack the other viking, and captured his vessels. In the engagement of Earl Paul in Orkney with the friends of Earl Ragnvald he refused the assistance of men from Erling of Tankerness (Tanskaruness), off which place the battle was fought,

---

* Olaf's Saga, cap. 27. Orkneyinga Saga.—L.

† Viking and sea-king are not synonymous, although, from the common termination in *king*, the words are used, even by our historians, indiscriminately. The sea-king was a man connected with a royal race either of the small kings of the country, or of the Harfager family, and who by right received the title of king as soon as he took the command of men, although only of a single ship's crew, and without having any land or kingdom. The viking is a word not connected with the word kongr or king. Vikings were merely pirates, alternately peasants and pirates, deriving the name of viking from the viks, wicks, or inlets on the coast in which they harboured with their long ships or rowing galleys. Every sea-king was a viking, but every viking was not a sea-king.—L.

because he had as many men as could find room to fight in his vessels, but required his assistance in carrying out stones from the shore to his vessels as long as the enemy would allow it to be done safely. Stones could not be transported or distributed in a conflict on land; and on this account the Northmen appear generally to have kept to their ships in their battles, and, even when marauding on land, to have had their ships far up the rivers to retire upon. This circumstance, namely, the great bulk in stowage, and in transport by land, of the usual arms of the age, arrows, casting spears, and stones, in any considerable quantities for a body of troops, and the difficulty of concentrating stores of them just at the spot where they are needed on land, accounts in a great measure for the success of comparatively small bodies of invaders landing on the coasts of England, or Normandy, in those ages. The invaders had the advantage of a supply of weapons in their vessels to retire upon, or to advance from; while their opponents having once expended what they carried with them, which could scarcely exceed the consumption in one ordinary battle of a few hours' duration, would be totally without missiles.

In the settlement of an Asiatic population in Scandinavia, which, whatever may have been the cause or inducement for preferring that side of the Baltic, undoubtedly did take place at an unascertained date, under a chief called Odin, we find a remarkable difference of social arrangement—and a sufficient cause for it—from that social arrangement which grew up among the people who invaded and seized on the ancient Roman empire. The latter were settling in countries of which the land was already appropriated; and however warlike and numerous we may conceive these invaders to have been, they could be but a handful compared to the numbers of the old indigenous inhabitants. They of necessity, and for security, had to settle as they had conquered, in military array, under local military chiefs whose banners they had followed in war, and were, for safety and mutual protection, obliged to rally around in peace. The people had the same military duties to perform to their chiefs, and their chiefs to the general commander or king, as in the field. They were, in fact, an army in cantonments in an enemy's country; and this, which is the feudal system, is the natural system of social arrangement in every country taken possession of by invaders in spite of the indigenous original inhabitants. It is found in several provinces of India, in several of the South Sea Islands, and wheresoever men have come into a country and seized the land of the first occupants. But where there is none to disturb the invaders—where they are themselves the first occupants, this military arrangement is unnecessary, and therefore unnatural. The first invaders of Scandinavia have entered into an uninhabited or unappropriated country, or if inhabited, it has been by a wandering or very unwarlike population, like the present Laplanders, or the Fenni of Tacitus. We are entitled to draw this conclusion from the circumstance that these invaders did not occupy and sit down in the country feudally. Each man possessed his lot of land without reference to or acknowledgment of any other man,—without any local chief to whom his military service or other quit-rent for his land was due,—without tenure from, or duty or obligation to any superior, real or fictitious, except the general sovereign. The individual settler held his land, as his descendants in Norway still express it, by the same right as the king held his crown—by udal right, or odel,—that is, noble right; subject to no feudal burden, servitude, escheat, or forfeiture to a superior from any feudal casualty. This was the natural arrangement of society, and the natural principle of possession in a country not previously occupied, and in which the settlers had consequently no reason for submitting to feudal obligations and to a military organisation. When the very same people, these unfeudalised Northmen, came to conquer and settle in Normandy, in a country appropriated and peopled, and which they had to defend as well as to invade and occupy, they naturally adopted the feudal social arrangement necessary for their security, and maintained it in all its rigour. In the very same century the kinsman of the same chief, Rolf Ganger, who was conquering and feudally occupying Normandy, came to settle in Iceland, where they had no occasion for the military organisation and principle of the feudal system in the unappropriated, uninhabited island; and they occupied it not feudally, but, as their ancestors had occupied the mother country itself, udally. The udal landowners, although exempt from all feudal services, exactions, or obligations to any other man as their local chief, or, in feudal language, the superior of their lands, were by no means exempt from services or taxes to the king or general chief, who was udal-born to the sovereignty of the whole or of a part of the country, and was acknowledged by the Thing or assembly of the landowners of the district. The kingly power was as great as in any feudally constituted country, either for calling out men and ships for his military expeditions abroad or at home, or for raising taxes. The scat was a fixed land-tax, paid to the king either in money or in kind, that is, in natural products of the land, and was collected by his officers yearly in each district, or even let for a proportion of the amount to his friends or lendermen during life or pleasure. This class of lendermen appears to have been the nearest approach to a feudal class in their social arrangements; but the *len* was a temporary, not an hereditary holding, and was not accompanied by any feudal privileges or baronial powers. The kings also received in their royal progresses through the country free lodging and entertainment for themselves and a certain fixed number of *hirdmen*, that is a court, for a certain fixed number of days in each district. All the most minute particulars of the supplies which each farm or little estate—for each little farm was a distinct udal estate—had to furnish, the turns in which each locality was liable to this entertaining of the king and court, the time and numbers of the court followers to be entertained, were matters of fixed law, and settled by the Things of each district. In these circuits the kings assembled the district Things, and with the assistance of the lawman, who appears to have been a local judge, either hereditary or appointed by the Thing, settled disputes between parties, and fixed the amount of money compensation or fine to be paid to the injured party. All offences and crimes, from the murder of the king himself down to the very slightest injury, or infraction of law, were valued and compensated for in money, and divided in certain portions between the party injured (or his next of kin if he was murdered,) and the king. The offender was an outlaw until he, or his friends for him, had paid the mulct or compensation, and could be slain, without any mulct or fine for his murder. The friends of the injured or murdered party could refuse to accept of any compensation in money, but could lawfully wait an opportunity, and take their revenge in kind. The king could only remit his own share of the mulct, but not that of the friends of the murdered party; and not to revenge an injury received and not compromised by a compensation, appears to have been considered highly dishonourable. The revenues of the kings appear to have been drawn, in some considerable proportion, from this source. When not engaged in warfare they appear to have been subsisted, as their ordinary mode of living, on these royal circuits or progresses through the country. The kings had no fixed residence or palace in Norway; but had estates or royal domains in every district, and houses on them in which they could lodge for a time, and receive what was due for their entertainment in victuals from the neighbourhood; but these houses appear to have been no better mansions than the houses on any other estates, and the kings were usually lodged, with their courts, as well as subsisted, by the land-owners or bondes. This usage of royal progresses for the subsistence of the royal household appears to have been introduced into England at the Norman, or rather at the previous Danish conquest; and the purveyance for it was a royal right, which continued to be exercised down to the end of Queen Elizabeth's reign.

Before the introduction or general diffusion of writing, it is evident that a class of men whose sole occupation was to commit to memory and preserve the laws, usages, precedents, and details of all those civil affairs and rights, and to whose fidelity in relating former transactions implicit confidence could be given, must of necessity have existed in society—must have been in every locality; and from the vast number and variety of details in every district, and the great interests of every community, must have been esteemed and recompensed in proportion to their importance in such a social state. This class were the skalds—the men who were the living books, to be referred to in every case of law or property in which the past had to be applied to the present. Before the introduction of Christianity, and with Christianity the use of written documents, and the diffusion, by the church establishment, of writing in every locality, the skald must have been among the pagan landowners what the parish priest and his written record were in the older Christianised countries of Europe. In these all civil affairs were in written record either of the priest or the lawyer; and the skalds, in these Christianised countries, were merely a class of wandering troubadours, poets, story-tellers, minnesingers, entertained, like the dwarfs, court-jesters, or jugglers, by the great barons at their castles, for the entertainment which their songs, music, stories, or practical jokes might afford. Here, in this pagan country, they were a necessary and most important element in the social structure. They were the registrars of events affecting property, and filled the place and duty of the lawyer and scribe in a society in which law was very complicated; the succession to property, through affinity and family connection, very intricate, from the want of family surnames, and the equal rights of all children; and in which a priesthood like that of the Church of Rome, spread over the country, and acquainted more or less with letters, the art of writing, and law, was totally wanting. The skalds of the north disappeared at once when Christian priests were established through the country. They were superseded in their utility by men of education, who knew the art of writing; and the country had no feudal barons to maintain such a class for amusement only. We hear little of the skalds after the first half of the twelfth century; and they are not quoted at all in the portion of

Magnus Erlingson's reign given by Snorre Sturlason within the twelfth century.

Besides the payment of scat, and the maintenance of the king's household in the royal progresses, the whole body of the landowners were bound to attend the king in arms, and with ships, whenever they were called upon to serve him either at home or abroad. The king appears, in fact, not only not to have wanted any prerogative that feudal sovereigns of the same times possessed, but to have had much more power than the monarchs of other countries. The middle link in the feudal system—a nobility of great crown vassals, with their sub-vassals subservient to them as their immediate superiors, not to the crown —was wanting in the social structure of the Northmen. The kingly power working directly on the people was more efficient; and the kings, and all who had a satisfactory claim to the royal power, had no difficulty in calling out the people for war expeditions. These expeditions, often merely predatory in their object, consisted either of general levies, in which all able-bodied men, and all ships, great and small, had to follow the king; or of certain quota of men, ships, and provisions, furnished by certain districts according to fixed law. All the country along the coasts of Norway, and as far back into the land "as the salmon swims up the rivers," was divided into ship-districts or ship-rathes; and each district had to furnish ships of a certain size, a certain number of men, and a certain equipment, according to its capability; and other inland districts had to furnish cattle and other provision in fixed numbers. This arrangement was made by Harald Harfager's successor, Hakon, who reigned between 935 and 961; and as Hakon was the foster-son of Athelstan of England, and was bred up to manhood in his court, it is not improbable that this arrangement may have been borrowed from the similar arrangement made by King Alfred for the defence of the English coast against the Northmen; unless we take the still more probable conjecture that Alfred borrowed it himself from them, as they were certainly in all naval and military affairs superior to his own people in that age. It is to be observed, that, for the Northmen, these levies for predatory expeditions were by no means unpopular or onerous. "To gather property" by plundering the coasts of cattle, meal, malt, wool, slaves, was a favourite summer occupation. When the crops were in the ground in spring, the whole population, which was seafaring as well as agricultural in its habits, was altogether idle until harvest; and the great success in amassing booty, as vikings, on the coasts, made the Leding, as it was called, a favourite service during many reigns: and it appears that the service might be commuted sometimes into a war tax, when it was inconvenient to go on the levy. Every man, it is to be observed, who went upon these expeditions, was udal born to some portion of land at home; that is, had certain udal rights of succession, or of purchase, or of partition, connected with the little estate of the family of which he was a member.

All these complicated rights and interests connecting people settled in Northumberland, East Anglia, Normandy, or Iceland, with landed property situated in the valleys of Norway, required a body of men, like the skalds, whose sole occupation was to record in their stories trustworthy accounts, not only of the historical events, but of the deaths, intermarriages, pedigrees, and other family circumstances of every person of any note engaged in them. We find, accordingly, that the sagas are, as justly observed by Pinkerton, rather memoirs of individuals than history. They give the most careful heraldic tracing of every man's kin they speak of, because he was kin to landowners at home, or they were kin to him. In such a social state we may believe that the class of skalds were not, as we generally suppose, merely a class of story-tellers, poets, or harpers, going about with gossip, song, and music; but were interwoven with the social institutions of the country, and had a footing in the material interests of the people. To take an interest in the long-past events of history is an acquired intellectual taste, and not at all the natural taste of the unlettered man. When we are told of the Norman baron in his castle-hall, or the Iceland peasant's family around their winter fireside in their turf-built huts, sitting down in the tenth or eleventh century to listen to, get by heart, and transmit to the rising generation, the accounts of historical events of the eighth or ninth century in Norway, England, or Denmark, we feel that, however pleasing this picture may be to the fancy, it is not true to nature, — not consistent with the human mind in a rude illiterate social state. But when we consider the nature of the peculiar udal principle by which land or other property was transmitted through the social body of these Northmen, we see at once a sufficient foundation in the material interests, both of the baron and the peasant, for the support of a class of traditionary relaters of past events. Every person in every expedition was udal born to something at home,—to the kingdom, or to a little farm; and this class were the recorders of the vested rights of individuals, and of family alliances, feuds, or other interests, when written record was not known. For many generations after the first Northmen settled in England or Normandy, it must, from the uncertain issue of their hostilities with the indigenous inhabitants, have been matter of deep interest to every individual to know how it stood with the branch of the family in possession of the piece of udal land in the mother-country to which he also was udal born, that is, had certain eventual rights of succession; and whether to return and claim their share of any succession which may have opened up to them in Norway must have been a question with settlers in Northumberland, Normandy, or Iceland, which could only be solved by the information derived from such a class as the skalds. Before the clergy by their superior learning extinguished the vocation of this class among the Northmen, the skalds appear to have been frequently employed also as confidential messengers or ambassadors; as, for instance, in the proposal of a marriage between Olaf King of Norway and the daughter of King Olaf of Sweden, and of a peace between the two countries to be established by this alliance. The skalds, by their profession, could go from court to court without suspicion, and in comparative safety; because, being generally natives of Iceland, they had no hereditary family feuds with the people of the land, no private vengeance for family injuries to apprehend; and being usually rewarded by gifts of rings, chains, goblets, and such trinkets, they could, without exciting suspicion, carry with them the tokens by which, before the art of writing was common in courts, the messenger who had a private errand to unfold was accredited. When kings or great people met in those ages they exchanged gifts or presents with each other, and do so still in the East; and the original object of this custom was that each should have tokens known to the other, by which any bearer afterwards should be accredited to the original owner of the article sent with him in token, and even the amount of confidence to be reposed in him denoted. We, with writing at command, can scarcely perhaps conceive the shifts people must have been put to, when even the most simple communication or order had to be delivered *vivâ voce* to some agent who was to carry it, and who had to produce some credential or token that he was to be believed. Every act of importance between distant parties had to be transacted by tokens. Our wonder and incredulity cease when we consider that such a class of men as those who composed and transmitted this great mass of saga literature were evidently a necessary element in the social arrangements of the time and people, and, together with their literature or traditional songs and stories, were intimately connected with the material interests of all, and especially of those who had property and power. They were not merely a class of wandering poets, troubadours, or story-tellers, living by the amusement they afforded to a people in a state too rude to support any class for their intellectual amusement only. The skalds, who appear to have been divided into two classes,—poets, who composed or remembered verses in which events were related, or chiefs and their deeds commemorated; and saga-men, who related historical accounts of transactions past or present,—were usually, it may be said exclusively, of Iceland.

It is usually considered a wonderful and unaccountable phenomenon in the history of the Middle Ages, that an island like Iceland, producing neither corn nor wood, situated in the far north, ice-bound in part even in summer, surrounded by a wild ocean, and shaken and laid waste by volcanic fire, should, instead of being an uninhabited land, or inhabited only by rude and ignorant fishermen, have been the centre of intelligence in the north, and of an extensive literature. It is wonderful; but, if we consider the causes, the phenomenon is naturally and soberly accounted for. Iceland was originally colonised by the most cultivated and peaceful of the mother-country; the nobility and people of the highest civilisation then in the north flying, in the ninth century, and especially after the battle of Hafersfiord, from what they considered the tyranny of Harald Harfager, and the oppression of the feudal system which he was attempting to establish in Norway. It was an emigration from principle. The very poor and ignorant, and those who merely sought gain without any higher motive for their emigration, could not go to Iceland; because a suitable vessel, with the necessary outfit and stock, could only be afforded by people of the highest class, and they only had to dread the jealousy and power of Harfager. Their friends, retainers, housemen, and servants attached to their families, went with them; but the *landnammen*, the *origines gentis*, were the sons and brothers of the nobles and kings, as they were called, who from the very same cause, the dread and hatred of Harfager's power, went out to plunder and conquer on the coasts of England and France. At the very same period that Rolf Ganger set out on his expedition, which ended in the conquest of Normandy, one of his brothers sought a peaceful asylum in the uninhabited Iceland; and the more peaceful of the higher class in those days were, we may presume, the most civilised and cultivated of their age. New England, perhaps, and Iceland, are the only modern colonies ever founded on principle, and peopled at first from higher motives than want or gain; and we see at this day a lingering spark in each of a higher mind than in populations which have set out from a lower level. The original settlers in Iceland carried with them whatever there was of civilisation or intelligence in Norway; and for some generations at least were free from the internal feuds, and always were free from the external wars and depredations on their coasts, which kept other countries in a state of barbarism. They enjoyed security of person and property. The means of subsistence in Iceland were not so very different from the means in Norway, nor of so much more difficult attainment, as might on a hasty view be supposed. The south coast of Iceland is not higher north than the country about Throndhjem fiord, and the most northerly part is barely within the Arctic Circle. A large proportion of the population of Norway lived in those ages, and live now, in as high a latitude; and, from not being surrounded by the ocean on all sides, in a severer climate; and under the local disadvantage, from the shape of the country, that the Fielde or mountain ridges in Norway approach much closer to the shore, and leave much less flat level pasture land between them and the sea than the mountains of Iceland. The cultivation of corn is as much out of the question in a great proportion of Norway as in Iceland. The people in the upland districts of every province of Norway, and almost all the population north of the Namsen river, draw the main part of their subsistence at present from the natural products of the land and water,—the pasture for their cattle, and the fishing in the rivers, the lakes, and the sea.

These natural products are as abundant in Iceland as in Norway; and the butter, cheese, wool, dried meat, fish, oil, feathers, skins, the wadmal or coarse woollen cloth, and the coarse linen spun and woven in their households, would be more in demand, more readily exchangeable, and of higher comparative value in former times, than such Icelandic products are now. With the surplus of such articles beyond their own consumpt the Icelanders could supply their own most pressing wants. These were for corn and wood—articles of first necessity, which did not admit of the population sinking into indolence and apathy in providing them. An intercourse and regular trade with England and Denmark for meal and malt, and with Norway for wood, tools of metal, and other necessaries of life, must have existed from the first years of the colonisation of Iceland. The Icelanders had consequently from the first more easy and regular opportunities of visiting foreign countries, and returning again to their own, than the natives of any other country in the north in those ages. They appear also to have traded without molestation, and never to have molested others. No Icelandic viking is mentioned in the sagas, even in the ages when a viking cruise was deemed an honourable occupation. Iceland men are mentioned in the sagas, occasionally, as being in the service of vikings of Norway, as hired men; but no long-ship, or viking belonging to Iceland, is mentioned. The necessity of trading in peace across the sea, and of giving no pretext for capture or

retaliation on Iceland vessels, may have been one
cause for this remarkable abstinence from the
favourite pursuit of the nobles of those ages in other
northern countries. It could not be from the cause
to which it is usually attributed, the want of wood
in the country to build long-ships. The Icelanders
had to buy merchant ships in Norway of a size to
cross the sea, and appear to have had them in abundance; and the same class of people who fitted out
viking expeditions in other countries could have
purchased long-ships as easily as ships of burden.
Their neighbours in the Farey Islands were equally
destitute of wood; yet they had a very celebrated
viking, Sigmund Bresteson. The Orkney Islands
had their Svein, a renowned viking, so late as the
twelfth century. But in none of the sagas in which
the exploits of these vikings are related, is there
any mention of any Iceland viking at any period.
The fair inference is, that the men who emigrated
from Norway to Iceland, and who were of the class
and had the means to fit out long-ships for piracy,
were men more advanced in civilisation and intelligence, and of higher principle, than men of the
same class in that age in the other northern countries. In all the sagas there appears a kind of
reluctance to dwell upon or approve of that part of
the hero's life passed in viking expeditions, or in
"gathering property" by piracy. One imagines, at
least, that in the Saga of Olaf Trygveson, of Olaf
the Saint, and of other great chiefs, the saga-man
shows a disposition to hurry over this part of their

lives, to throw it into the years of extreme youth, and not to approve himself of that part of his tale. The comparatively safe intercourse which the Icelanders undoubtedly had with other countries gave them a higher education, that is, the means of acquiring a greater stock of information on what was doing in other countries, than any other people of those times. When we consider that these Icelandic colonists were connected by the udal law of succession with the principal families and estates in Norway, Denmark, Sweden, England, and France, and were deeply interested in the conquests, revolutions, battles, or changes going on, and in which their friends and relations were the chief actors, we can understand that an historical spirit must have grown up in such a population—a great desire to know, and a great talent to remember and relate. Heritable interests and rights of families in Iceland were involved in what was going on in Normandy and Northumberland as much as in Throndhjem; and the consideration in which the skald or sagaman who could give accounts of such events was held, may not be exaggerated in the sagas. In a community of such colonists, the class of skalds remembering and relating past transactions was an essential element, and must have been held in the highest honour. To return home to Iceland appears, indeed, to have been the end which the most favoured skald at the courts of kings proposed to himself. From their opportunities of visiting various countries, the Icelandic skalds were undoubtedly the educated men of the times when books did not in any way contribute to intelligence, or to forming the mind; but only extensive intercourse with men, and the information gathered from it. Having by the lapse of time no family feuds even with the people of Norway, no injuries, national or private, to avenge or to fear vengeance for from others, the Icelander could travel through other countries on private or public affairs with a degree of personal security which people of the highest rank and power belonging to the country were strangers to in those unhappy ages. This advantage was sufficient of itself to make them a useful class in every court. They were not only neutral men in every strife; but, from their travel and experience, men of intelligence, prudence, and safe counsel, compared to men of no intellectual culture at all, and acquainted only with arms and violence. They had also the advantage of speaking in its greatest purity what was the court language in Norway, Sweden, Denmark, England, and at Rouen.* The moral influence which the skalds enjoyed, as counsellors and personal friends and advisers of many of the kings, may not be exaggerated in the sagas; for it appears to be that which knowledge and education would naturally obtain amidst ignorance and barbarity.

* Normandy, it is to be always remembered in reading the history of those ages, was conquered, but not colonised, as Iceland and Northumberland were colonised, by the introduction of a totally new population, with their own laws, manners, and language. In Normandy, so early as the time of Duke Richard, the second in descent from Rolf Ganger, his son had to be sent to Bayeux to acquire the pure northern language, it having been already corrupted at the court.—L.

The class of skalds and saga-men, supported by intellectual labour in the north of Europe, may not have been very numerous at any one time; but owing to the favourable circumstances peculiar to the Icelanders, the profession centred in Iceland. We hear of no skalds of any other country, not even of Norway. All the intellectual labour of the kind required in the north of Europe was derived from Iceland. We may surely reckon the population in the north of Europe using a common tongue in those times,—of Scandinavia, Denmark, Jutland, and Schlesvig; of the kingdom of Northumberland, East Anglia, and of parts of Mercia; of Normandy, in some proportion of its inhabitants; of the Hebrides and Isle of Man, in some proportion; of the Orkney, Shetland, and Farey Islands, altogether,— to have amounted to two millions of people. Small as the demand might be for intellectual gratification among these two millions, yet they were scattered over countries widely apart, and they used a common tongue, and had a real and effective relationship of families among them all; and the desire for news of what was doing in other lands, and for narratives of events which might be of importance in their family interests, would be sufficient to give an impulse to such a small population as that of Iceland, which never exceeded sixty or sixty-four thousand people, to give employment to all the surplus talent of such a population, and to keep up a literary tone, if it may be so called, of the public mind in such a handful of people. Men of any talent would naturally endeavour to qualify themselves for that profession in which several, and probably a considerable number, attained distinction, wealth, or high consideration. It was better than the chance of advantage from embarking as a private seaman or man-at-arms in viking forays and cruises under a sea-king; better than staying at home tending cattle, cutting peats, making hay, and catching and curing fish. The same motives operate in the same way at this day in the social economy of Iceland. The youth of talents and ambition study, come to the university of Copenhagen, become often men of very great attainments and learning, and with as few chances or examples before them of substantial reward for their labours as the skalds, their predecessors, could have had. The impulse to mind in any community being once given, either by accidental or physical circumstances, the movement in the same direction goes on and seems to be permanent — never to cease. The perpetuity of intellectual movement, of the direction of mind and mental energy in the same way, even when laws, government, and all social arrangements, and even religion itself, are altered, and the old forms not even remembered, is one of the most singular and interesting of the phenomena in the nature of man. It is strikingly illustrated in Iceland. The Icelandic youth prepare themselves now for a learned profession, as the skalds did 800 years ago, exactly from the same intellectual impulse, although in a different field; and the movement of the public mind towards intellectual occupation appears to have remained in this small community unchanged, undiminished, and only less visible because it is not now the only community in the north with the same movement. The continued tendency of mind in Iceland to literary pursuits appears when we compare them, in numbers not exceeding at present 56,000[*] individuals, with any equal number of the British population. The Icelanders had a printing-press among them in the first half of the sixteenth century; and many works in Latin and in Icelandic have been printed at Skalholt, Holar, and other places. The counties of Orkney and Shetland, with an equal population,—of Caithness, Sutherland, Ross, with probably double the population,—have not at this day any such intellectual movement, or any press that could print a book, or any book produced within themselves to print. The whole Celtic population in Scotland, since the beginning of time, never produced in their language a tithe of the literature that has been composed and printed in Icelandic by the Icelanders for their own use within this century. The modern literature of Iceland, or even its saga literature, may not be considered by the critic of a very high class or value, or of merit in itself; but, in judging of the intellectuality of a people, the philosopher will regard its amount and diffusion as of much more importance than its quality. That belongs to the author, and measures merely the genius and talents of the individual: the

[*] The present population of Iceland is about 70,000 individuals.

amount and diffusion measure the intellectual condition of the society. Apply this measure to any town or county in Great Britain of 56,000 inhabitants, and we will find little reason to boast of a more advanced intellectual condition among us than that which the Icelanders appear to retain at this day from bygone times, when an intellectual character was impressed on the public mind in their small community by the skalds; and little reason to believe that the monkish historians of the Anglo-Saxons, in the same ages in which the skalds flourished, have left more deep or influential traces of their literature in the parts of Europe in which they were the only men engaged in those ages in intellectual occupation, than the skalds have done in the narrow circle in which alone they could have influence on posterity.

In these observations on the saga literature, nothing is said of or allowed for the Runic writing inscribed on rocks, monumental stones, wooden staves, drinking-cups of horn or metal, arms, or ornaments, for which at one period a high antiquity was claimed. It seems to be now admitted that a Runic character, apparently borrowed from the Gothic and Roman, and adapted to the material on which it was usually cut, viz., hard stone or wood, by converting all the curves of the letters into straight lines for the facility of cutting, has existed from a very early age among the Northmen. It would, indeed, be absurd to suppose that an intelligent people roaming over the world, who had appeared in the Mediterranean in the days of Charlemagne, and had a regular body of troops, the Varings, in the pay of the Greek Emperors at Constantinople, should not have adopted, or imitated, what would be useful at home, as far as applicable to their means. This appears to have been precisely the extent to which Runic writing was applied. From want of means to write,—that is, the want of the parchment, paper, ink, and writing tools,—the writing in Runic was almost entirely confined to short monumental inscriptions recording the death of an individual, the name of the person who erected the stone to his memory, and also the name of the person who cut the letters—a proof that the use of the Runic characters was rare, and confined to a few. Of these Runic inscriptions, of which a thousand or more have been examined by antiquaries, few can be placed before the introduction of Christianity in the eleventh century. The sign of the cross may, in the dreams of the zealous antiquary, appear the sign of Thor's hammer; but there is no evidence that the pagans used such a symbol, and the obvious interpretation of such a mark upon a tombstone is that it belongs to the age of Christianity. Torfæus, whose antiquarian zeal was tempered with a love of truth, and whose antiquarian knowledge has not been surpassed, says[*] not only that the Runic inscriptions throw no light upon history, but are so intricate and confused, that what you may imagine you catch by the eye you cannot by the understanding; and in proof of his remark he refers

[*] Torfæus, Series Regum Dan., cap. viii.—L.

to conflicting interpretations of the two greatest Runic antiquaries, Worm and Verelius, of the meaning of Runic inscriptions, on which they both agree perfectly as to the strokes or incisions in the stone. Bartholinus* also says, in his Danish Antiquities, that excepting four or five, none of these Runic inscriptions are in any way illustrative of history, and in general are so obscure that the names of the persons for whom the stones are erected can scarcely be extracted, and much is matter of mere conjecture. The opinions of these great antiquaries are singularly confirmed by the recent discovery made by chemical science, that one of the few Runic inscriptions supposed to be illustrative of history,—one upon a rock at Hoby, near Runamo, in the Swedish province of Bleking, which is mentioned by Saxo Grammaticus as being, in his time (namely, about 1160), considered inexplicable, and which modern Runic scholars interpreted a few years ago to relate to the battle of Bravalla fought about the year 680,—is in reality no inscription at all, but a mere *lusus naturæ*; merely veins of one substance interspersed in the body of another substance, and forming marks which resemble Runic letters in the fancy of the antiquary, but which is an appearance in rocks of granitic formation with veins of chlorite interspersed, not unfamiliar to the eye of the mineralogist. Another of the Runic inscriptions, supposed to be illustrative of history, is that on a rock called Korpeklinte, in

* Bartholinus, Antiq. Dan., l. i. cap. 9.—L.

the island of Gotland, which, in Runic characters, told that

"Aar halftridium tusanda utdrog Helge med Gutanum sinum;"*

that is, "in the year half three thousand,—videlicet, two thousand five hundred,—went out Helge with his Goths." This inscription must be, as Worm himself admits, a gross fabrication; for the pagan Northmen did not reckon by years, but by winters, and could have known nothing of the computation of time from the creation of the world, which is derived from the Bible, and was unknown to them in the year of the world 2500. But before the year 1636 somebody had been at the trouble to attempt to impose upon the world by this inscription in Runic letters, although in modern language, and of modern conception. We may believe that inscriptions on stones in memory of the dead,—rude calendars cut in wood,—charms on amulets, rings, shields, or swords,—and tokens of recognition to be sent by messengers to accredit them to friends at a distance, may have existed among the Northmen from their first arrival in Europe; and Odin himself may have invented or used the Runic character in this way: but we have no ground for believing that any distinct use of writing,† *currente*

---

* I have not been able to find this Runic inscription so as to verify it.

† A remarkable proof how little Runic was known, or used, is, that a certain Od Sveinbiornson gave notice to Snorre Sturlason of the conspiracy against his life in September, 1240, in Runic.[1] But neither

[1] In Sturlunga Saga, ch. 154, it is stated that Snorre had a letter, which Od Sveinbjornson at Alptanes had sent him, that it was written in *stafkarla-letr* (a kind of Runic letters), and that it was not read. A *stafkarl* is a poor beggar (a

calamo, applicable to the transmission of historical events, was known before the introduction of Christianity, and of letters with Christianity, in the eleventh century, or was diffused before the diffusion of Church establishments over the north. If the Runic had been a written character among the Northmen of the ninth century, it must have been transported to Iceland, in which the first settlers were not of the rude and ignorant, but of the most cultivated of their age in Norway; but few, if any, Runic inscriptions of a date prior to the introduction of Christianity are found in Iceland. If they had possessed the use of written characters, as they had unquestionably a literature in Iceland, it would be absurd to believe that they had not applied the one to the other; but for two hundred and forty years —that is, until the time of Are—should have committed the sagas to memory, instead of to parchment or paper. Are himself would have used the Runic character, if writing Runic had been diffused among the Northmen; and although no manuscript of the time of Are exists, but only early copies of his writings, yet among the mass of sagas in manuscript some must have been in Runic characters, if Runic writing had been diffused among the Icelanders. No

Snorre (certainly not one of the unlearned of his age in the saga or Icelandic literature), nor any of those with him at the time, could read the Runic characters; and Snorre in consequence fell a victim to the conspiracy, and was murdered in his house on the 22d September, 1240. —*Schoning, Pref. to Heimskringla.*—L.

"staff carle"), and *stafkarla-letr* must be a peculiar kind of secret runes. This is, however, no evidence that Snorre and those with him could not read the ordinary runes.

Runic manuscript, however, on parchment or paper, of unquestionable antiquity and authenticity, has ever been discovered. A fragment, entitled "Historia Hialmari Regis Biornlandiæ atque Thulemarkiæ ex Fragmento Runici MS. literis recentioribus descripta cum genuina versione Johannis Peringskoldi," without date, place of publication, or reference to where the original Runic manuscript on skin or paper is to be found, is evidently a translation of a part of the Saga of Hialmar into Runic letters, for the purpose of imposing on the public, and is to be classed with the Korpeklinte inscription. The controversy concerning the antiquity and historical value of the Runic character and inscriptions ran high in the latter half of the seventeenth century, and unjustifiable means were used to establish opinions as facts. This fragment of ancient Runic writing on parchment was ascribed by Rudbeck to the seventh century, by Stiernman to the tenth, by Biorner to the eleventh or twelfth. It was incorporated into Hicks's Thesaurus as a specimen of written Runic. But Archbishop Benzelius, Celsius the elder and Celsius the younger, Erichson, and Ihre, antiquaries of great note and authority in Sweden, expressed their doubts of the authenticity of this fragment at the time it appeared, —about 1690; and Nardin, in an Academical Dissertation, published at Upsala 1774, proves from the language that this Runic manuscript is an impudent forgery.*

---

* The earliest runes were not writing in proper sense, but fanciful signs, possessing a magical power; such runes have, through vulgar supersti-

tion, been handed down even to the present time. The phrase in the old Danish Ballads, *kaste runer,* "to throw runes," *i.e.*, chips, may be compared to the *Lat. sortes,* Mommsen's History of Rome, vol. i. p. 187, footnote (Eng. Ed.), or the Sibylline leaves in the Æneid. In regard to runes in writing, the word was first applied to the original alphabet, which at an early time was derived from the common Phœnician, probably through Greek and Roman coins, in the first centuries of our era. From these runes were subsequently formed two alphabets, the old Scandinavian (whence again the Anglo-Saxon) as found on the golden horn and the stone in Tune, and the later Scandinavian, in which the inscriptions in the greater number of the Swedish and Danish stone monuments are written, most being of the tenth (ninth?) and following centuries. A curious instance of the employment of runes is their being written on a *kefli* (a round piece of wood) as messages. It is doubtful whether poems were ever written in this way, for almost the only authority for such a statement is Egla, 605, where we read that the Sonatorrek was taken down on a Runic stick, the other instances being mostly from romances or fabulous sagas. This writing on *kefli* is mentioned in the Latin line, "Barbara 'fraxineis' sculpatur runa 'tabellis'" (Capella, fifth century). In later times (from the thirteenth century) Runic writing was practised as a sort of curiosity. Thus calendars used to be written on sticks, of which there is a specimen in the Bodleian Library in Oxford; they were also used for inscriptions on tombstones, spoons, chairs, and the like. There even exists in the Arna-Magn. Library a Runic MS. of an old Danish law, and there is a Runic letter in Sturl. (1241).—G. Vigfusson in Cleasby's Icelandic-English Dictionary, *sub voce.* An abbreviated quotation by A.

Recent authors to be consulted in regard to the runes are:—

Bugge, Soph., Om Runeskriftens Oprindelse. Christiania, 1873.

Wimmer, Ludv., Runeskriftens Oprindelse og Udvikling i Norden. Copenhagen, 1874.

Stephens, Geo., The Old-Northern Runic Monuments of Scandinavia and England, now first collected and deciphered, by Geo. Stephens; with many hundreds of fac-similes and illustrations. 3 vols. London, 1866, and later.

Thorsen, P. G., Om Runernes Brug til Skrift udenfor det monumentale. Copenhagen, 1877.

CHAPTER II.

OF THE RELIGION OF THE NORTHMEN.

It must strike every reader of saga literature how very little we can gather from the sagas of the doctrines and usages of the paganism which existed among the Northmen down to a comparatively late period, and for five hundred years after the cognate Anglo-Saxon branch, both on the Continent and in England, had been entirely Christianised, and had been long under the full influence of the Church and priesthood. The Anglo-Saxons landed in England about the year 450. They appear at that time to have had a religion cognate to, if not identical with, that of the Northmen who landed in England three hundred years afterwards, or about the year 787. Odin, Thor, Frigg, were among their deities; Yule and Easter were religious festivals; and the eating of horse-flesh was prohibited in a council held in Mercia in 785, as "not done by Christians in the East"—which implies that among the Anglo-Saxons also it was a pagan custom, derived from their ancestors. In about a century after the landing of the Saxons, viz., about 550, the Heptarchy was in existence; and in about another century, viz., about 640, Christianity was generally established among them. It was not till a century after their first expeditions, about 787, that the pagan Northmen made a complete and permanent conquest of the kingdom of Northumberland, which they held under independent Danish princes until 953, when independent earls, only nominally subject to the English crown, succeeded; and even at the compilation of Doomsday Book by William the Conqueror, the lands of Northumberland, Westmoreland, Cumberland, and part of Lancashire, are omitted, as not belonging to England. Of these Anglo-Northmen the conversion cannot well be fixed to a date, because they had no scruple apparently of nominally adopting Christianity when it suited their interests; and they appear to have had no desire to convert, or to be converted, in their predatory expeditions. As late, however, as the beginning of the eleventh century, the Northmen and their chiefs were still pagans. Svein, indeed, and his son Canute, who in 1017 became sole monarch of England, were zealous Christians; but they and their contemporaries, Olaf Trygveson and Olaf the Saint, and the small kings in Norway, were born pagans; and their conversion, and the introduction of the Christian religion and religious institutions into Norway and its dependencies, cannot be dated higher than the first half of the eleventh century. It seems surprising that we know so little of a pagan religion existing so near our times,—of this last remnant of paganism among the European people, existing in vigour almost five hundred years after Christianity and the Romish

Church establishment were diffused in every other country! What we know of it is from the Edda compiled by Sæmund* the priest, a contemporary of Are who compiled the historical sagas. Sæmund was born in 1056, and had travelled and studied in Germany and France. He lived consequently in an age when many who had been bred in and understood the religion of their forefathers were still living, and in a country in which, if anywhere, its original doctrines and institutions would be preserved in purity.

If we may take the account of Tacitus as correct, this ancient religion of the Germanic race must have been eminently spiritual, and free from idolatry. He says, in chapter 35, "De Moribus Germanorum," that they held "regnator omnium Deus, cetera subjecta atque parentia;" and, in chapter 9, "ceterum nec cohibere parietibus Deos, neque in ullam humani oris speciem assimilare ex magnitudine cœlestium arbitrantur." The polytheism of the ancient Romans, and the saint-worship of the Christian Church of Rome, and probably the infirmity of the human mind and language requiring, in an uncultivated state, material forms to represent abstract ideas, had in the course of ten centuries, between Tacitus and Sæmund, undoubtedly mingled with and moulded the forms and ideas of the original religion of this race. Idolatry, in every shape and country, is the result of a struggle of the human mind to attain

---

* Sæmund's name became connected with the elder Edda after the revival of Icelandic literature, but this theory is not supported by a scrap of evidence, and is now abandoned by all eminent Old-Norse scholars.

fixed ideas in religion. It is universal at a certain stage of the development of the intellectual powers of man, because that stage is as necessary to be passed through as infancy in the individual, or barbarism in a society of human beings. A love for religious certainty and truth is at the bottom even of the grossest idolatry, if we analyse it rightly. Idolatry is an attempt to individualise the conception of almighty power, under a strong sense of its existence—to make it more possible, or more easy for the mind, in a certain stage of development, to dwell upon and entertain some present conception of that power. Idols should be considered by the Christian philosopher as the imperfect words of a much more pure religious sentiment than our churchmen generally suppose—words different, indeed, from spoken or written words, but intended to convey the same conception, and used with the same sentiment by the ignorant idolater as the most poetic imagery and most eloquent language of our pulpit orator. The most absurd idols of the Hindu or the South Sea islander, sent home to the museum of the Missionary Society as memorials of the spiritual blindness of the heathens at this day—idols with four or five arms and wings, griffin-footed, made up, in short, of emblems of all real or imaginary living beings known to the ignorant idolater,—are in reality words of which these absurdly combined parts are the syllables, or rather are the expression of a sentiment of which these are the words. We smile or shudder in holy horror at these uncouth representations, forgetting, in the pride of our philosophy and theology, that the sentiment is the same—viz., the innate feeling of divine power; and perhaps not less intensely felt in the mind which expresses it in these shapes, than in the mind which expresses it in the shapes of written or spoken language. It is but the way of expressing it, not what is intended to be expressed, that is different. Idol-language and word-language have the same object—namely, to express the impression or sentiment of almighty power and divine existence innate in the human mind; and who shall say that we approach nearer to the understanding and expressing of this almighty power and divine nature, "which passeth understanding," with our alphabet, or written or spoken words, than the ignorant idolater, without words in his language to express his ideas, in his carved and painted idol-language? Each means may be the best adapted for different states and stages of the development of the human mind. It has a stage in its development at which, in a highly civilised country, a little black stroke with a dot over it presents to the philosopher, and as he believes, in the most clear and distinct manner, all that mind knows or can express of self-existence, of individuality—of I. A wooden idol, representing something like to but different from man, presents to the mind of the pagan all that it can conceive, or with its means express, of superhuman divine existence. Is not this a mere defect in the alphabet used? Is not the real inward sentiment in the mind of man, with regard to the

Divinity, the same, precisely the same, whatever be the mode of expressing it? But the pagan, the idolater, the ignorant even of the Catholic church, worship these stocks and stones; and instead of regarding them as signs only shadowing forth what, in its intellectual state, the human mind cannot otherwise express of its religious sentiments, take the signs for the things they represent, and worship them as such. So do we, in all our pride of knowledge and intellectual development. We too worship our signs—our words. Let any man set himself to the task of examining the state of his knowledge on the most important subjects, divine or human, and he will find himself a mere word-worshipper; he will find words without ideas or meaning in his mind venerated, made idols of—idols different from those carved in wood or stone only by being stamped with printers' ink on white paper. This is perhaps the just view which the philosopher and the humble Christian should take, of all the natural forms of religion which have ever existed beyond the pale of the religion revealed to us in the Scriptures.

This necessity of man's expressing in his uncultivated mental condition, by the material visible means of idols, the innate sentiment of religion which he has no other language for, will, if it do not reconcile, render very unimportant the various speculations of antiquaries relative to Odin. Some find that Odin was a real personage, who, on the fall of Mithridates, migrated with his nation from the borders of the Tanais, to escape the Roman yoke, about seventy years before our era. This is the opinion of Snorre Sturlason; and, as far as regards the Asiatic origin of the Northmen, it is confirmed by all the traditions, mythological or historical, of the skalds. Torfæus, reckoning from skaldic genealogies, finds that there must have been an older Odin; and if we are to admit that the god called Odin was a real historical personage, it is impossible to fix, from the traditionary genealogies of those who claimed to be his descendants, at what period he lived. There are no fixed points in the history of the North before the middle of the ninth century, when, about 853 or 854, the birth of Harald Harfager, who lived to 931, is determined from contemporary history. The skaldic genealogies make this king the twenty-eighth in descent from Odin. If we allow eleven years to each reign, which is the average length of the reigns in the Heptarchy, we must place Odin 550 years after the Christian era. If we take Sir Isaac Newton's computation of eighteen years as the average length of reigns, we bring Odin to the year 368 of our era. If we take lives instead of reigns, we must believe twenty-seven successive persons to have lived so long as to average thirty-five years each, in order to place this god called Odin seventy years before our era. When we turn to the Anglo-Saxon genealogies, we find it still more difficult to place Odin. King Alfred was born 849, or about four years before Harald Harfager, and is only twenty-three generations from the Odin or Wodin of Hengist and Horsa. Offa, king of Mercia, who lived about 793, is only fifteen generations; Ida, king of Northumberland, who lived about 547, only nine generations; and Ella, king of Northumberland, who lived about 559, is eleven generations from Odin or Wodin. The reasonable view is that of Pinkerton, which has more recently been developed by Grimm and other German writers on Scandinavian mythology,—that Odin, Wodin, Godin, were names of the Supreme Divine Power among the Germanic race; and that Thor, Frigg, &c., were merely impersonations of divine attributes; that none of these were ever human heroes, deified by their contemporaries or descendants. It may, indeed, be reasonably doubted whether, in any age or country, any such deification of mortals known to be human beings—any such hero-worship as classical schoolmen and antiquaries suppose, ever did take place among any portion of the human race; for it is contrary to the natural tendency and movement of the human mind. It is a trite observation, that no people have ever been discovered by the traveller in so rude and barbarous a state as to be without any sentiment of a Divinity; and that this universal sentiment is more distinctive of the species man even than his reasoning powers,—for in these the elephant, the dog, the beaver, the bee, partake, and almost vie with human beings in the lowest condition of humanity. The writers who make the most of this trite observation in support of natural religion overlook a powerful argument for the truth of revealed religion, in a sentiment equally innate, and as widely diffused among men in a natural state. No people have ever been discovered by the traveller or the antiquary without a strong and distinct impression of the incarnation, past, present, or to come, as well as the existence, of the Divinity. Our divines turn away from this argument in support of revealed religion. They assume that, in the dark ages of every nation, individuals have taken to themselves the attributes and honours of divinity— have imposed themselves upon their contemporaries as gods, or have been taken by their contemporaries or their posterity for gods; and in this schoolboy way great divines, historians, and philosophers think of, tell of, and account for, idolatry and paganism among rude uncivilised nations. But they do not apply to the subject two universally and permanently ruling principles of the human mind in every stage of development: first, that there is no deceiving a man's own consciousness; and, second, that if a man cannot deceive himself, he cannot deceive others. Alexander the Great, or Odin, or the Roman emperors, or the Roman pontiffs, may have placed themselves at the head of the priesthood or church, and may have allowed their flatterers to place their statues among those of the gods, and to append the title of Divus or Saint to their names; but in all this Church trickery these men no more believed themselves gods, than their people believed them to be of divine nature. The human mind, in a state of sanity, never was discovered in so low a condition of the reasoning power as to approach to any such conclusions. As to a rude and ignorant people elevating their deceased leaders, kings, heroes, to a place among their deities, it is the last thing a rude and ignorant people would think of; for in a rude and ignorant state the natural movement of the human mind is to detract from, not to elevate, the merits of others; and the valued endowments of body or mind in such a state—strength, beauty, valour, or even wisdom in the narrow range of their public or private affairs—are more generally diffused than the intellectual attainments valued by men in a civilised state, and are neither so high nor so rare as to be deified, instead of being subjects of envy and detraction, or of emulation. In no state, barbarous or civilised, are men disposed to yield superiority, and allow divine honours and attributes to mortals who, as natural self-love or vanity will whisper to every one, are not much superior,—not divinely superior, to themselves. The natural movement and tendency of the human mind are equally opposed in every stage of development, in every state of society, to any such hero-worship. Divines overlook the weight of this argument for revealed religion drawn from natural religion—that it is not from man upwards, but from the Divinity downwards, that this universal sentiment of an incarnation proceeds; that the existence of a Supreme Being is not a sentiment more innate in the human mind, than that of the incarnation of the Supreme Being at some period, past, present, or future; that it is not Jupiter, Mars, or Odin, who were men or heroes set up by their fellow-men as gods to be worshipped, but that it is the innate and universal sentiment of the human mind which has set them up. What the mind cannot grasp in one conception, or express in one expression, it necessarily divides; and thus groups of divine attributes have been impersonated as distinct deities, individualised, named, incarnated, by an irresistible instinct to find an incarnation of Divine Power; and these distinct individualisations produced in a rude state of the human mind by the poverty of language, have been made historical personages of. But, looking at the natural movement of the human mind, it may be reasonably doubted whether any historical personage, who really lived as a mortal man, was ever made a god of by his fellow-men. The Christian philosopher, who considers the expected Messiah of the Jews, the incarnation of Jupiter, of Odin, and the living incarnation of a Lama among a great proportion of the present population of Asia, will not hesitate to place the accomplished advent of our Saviour upon the same innate sentiment of the human mind as the existence itself of Supreme Divine Power. Both are universal innate sentiments of the mind of man. The divines and Christian philosophers who are not content with resting the existence of a Supreme Divine Power upon the innate sentiment of the human mind, upon the same ground as the proof of our self-existence rests, but who seek to prove the existence of a Supreme Divine Power from the design, contrivance, and wisdom manifested in the material objects around us,—Paley, and the Bridgewater Bequest writers, who undertook, for a prize of two or three hundred pounds given by an English lord, to prove to all and sundry of God's creatures the existence of a God from the mechanism of the hand, the eye, the movements of the planetary bodies, and other natural objects without us, and not from that which is within us,—who seek to prove the spiritual from the material, and not from the spirituality existing and innate in every man's mind,—are not so immeasurably distant from gross paganism as they suppose. They and the pagan,—the Odin-worshipper, or Jupiter-worshipper, or whatever he may be,—proceed upon the very same material grounds, and the pagan appears the closer and stricter reasoner of the two. An ignorant and barbarous people may be wrong in the grounds from which they reason, but are seldom wrong in the reasoning process itself. Their conclusions are usually very correct, only drawn from false premises. They mark the thunderbolt, and conclude there is a Supreme Divine Cause—a thunder-maker. They mark the ocean,—now calm and smiling, now shaking the earth with its fury,—and conclude there must be an ocean god. This is precisely the reasoning of the Paley and of the Bridgewater Bequest philosophy. The manifest design, contrivance, adaptation of means to an end in a watch, prove the existence of a watch-maker—of the hand, of a handmaker—of the eye, of an eye-maker—of the world, of a world-maker. But from these material-world grounds these material philosophers cannot deduce, in strict reasoning, the unity of the Supreme Divine Power; still less the moral perfections of the Supreme Divine Power. The pagan proceeds upon exactly the same grounds in his religious belief; but reasons much more correctly and logically from the same material grounds, when he concludes there is a separate Divine Power for each separate class of material objects—a god of thunder, a Neptune, and so on; and concludes, from the material world grounds, that the superiority of the intelligence that made it, above his which perceives it, is in degree and power, not in kind; and his material-world grounds give him no reason, in his strict reasoning process, to conclude that the divine intelligence, or intelligences, which he deduces from them, are exempt from the passions or frailties of the intelligences he is acquainted with. He attributes, therefore, to his gods, the passions and motives of men; and the Christian philosophers, who reason from the same grounds to prove the existence of the Deity, do not reason half so correctly as the pagan; for these grounds will carry human reason no farther than the pagan goes,—and there Paley, prize-essay divinity, and paganism stick together neck and neck. The material-world grounds prove the existence only of creative power. The goodness, mercy, omniscience, all the attributes of God, are in the innate sentiment of the human mind of an existence of Supreme Divine Power; and upon this innate sentiment or spirituality the material-world argument has to fall back, to rescue itself from mere paganism. The distinctive characteristics of paganism and Christianity are, that the former rests entirely on material-world grounds—and from these grounds reasons strictly and correctly, its conclusions being correctly drawn, but from imperfect premises; the latter rests on the spiritual evidence of the innate sentiment of a Divine Existence, of which every human mind is as conscious as of its own existence—and on Revelation. Paley, the Bridgewater Bequest philosophers, and all that school of Christian reasoners, have, in fact, done infinite mischief to religion, by throwing out of view the innate sense, the spirituality of the human mind, on which pure religion is founded; and by resting its evidences on material external objects, from which the deist, the polytheist, the Odin-worshipper, if such now existed, might draw conclusions in their own favour more strictly logical than this kind of Christian reasoner can; for divine power, and no other of the attributes of the Deity, can be deduced from the material world without a reference to the intellectual, to the human mind, and to the inspired writings. These philosophers have lowered the tone of religion by their evidences drawn from the material world; and their evidences do not, in strict reasoning, prove their conclusions.

Of the doctrine, institutions, and forms of the religion of Odin, we have but few memorials.

There are two Eddas. The older Edda is that which was composed or compiled by Sæmund,* and of it only three fragments are extant.† The one is called the "Voluspa," or the Prophecy of the Vala. In the Scotch words "spæ-wife," and in the English word "spy," we retain words derived from the same root, and with the same meaning, as the word "spa" of the Voluspa. The second fragment is called "Havamal," or the High Discourse; the third is the Magic, or Song of Odin. The Voluspa gives an account by the prophetess of the actions and operations of the gods; a description of chaos; of the formation of the world; of giants, men, dwarfs; of a final conflagration and dissolution of all things; and of the future happiness of the good, and punishment of the wicked. The Havamal is a collection of moral and economical precepts. The Song of Odin is a collection of stanzas in celebration of his magic powers. The younger Edda, composed 120 years after the older, by Snorre Sturlason, is a commentary upon the Voluspa; illustrating it in a dialogue between Gylfe, the supposed contemporary of Odin, under the assumed name of Ganglere, and three divinities,—Har (the High), Jafnhar (equal to the High), and Thride (the Third)—at Asgard (the abode of the gods, or the original Asiatic seat of Odin), to which Gylfe had gone

---

* See note, page 88.
† There are thirty-nine poems in the elder Edda. They are in no special connection with each other, but may be divided into three classes: purely mythological, mythological-didactic, and mythological-historical. At least twenty of them may be classed as mythological. See Anderson's "Norse Mythology," pp. 116-125.

to ascertain the cause of the superiority of the Asiatics. Both the Eddas appear to have been composed as handbooks to assist in understanding the names of the gods, and the allusions to them in the poetry of the skalds; not to illustrate the doctrine of the religion of Odin. The absurd and the rational are consequently mingled. Many sublime conceptions, and many apparently borrowed by Sæmund and Snorre from Christianity,—as, for instance, the Trinity with which Ganglere converses, —are mixed with fictions almost as puerile as those of the classical mythology. The genius of Snorre Sturlason shines even in these fables. In the grave humour with which the most extravagantly gigantic feats of Thor at Utgard are related and explained, Swift himself is not more happy; and one would almost believe that Swift had the adventures of Thor and the giant Utgard-Loke before him when he wrote of Brobdignac. The practical forms or modes of worship in the religion of Odin are not to be discovered from the Eddas, nor from the sagas which the two Eddas were intended to illustrate. It is probable that much has been altered to suit the ideas of the age in which they were committed to writing, and of the scribes who compiled them. Christianity in Scandinavia seems, in the eleventh century, to have consisted merely in the ceremony of baptism, without any instruction in its doctrines. The wholesale conversion of whole districts by Olaf Trygveson, and King Olaf the Saint, was evidently the mere ceremony of baptism. On the eve of the battle of Stiklestad the mere acceptance and performance of that ceremony, without any instruction, was considered by Saint Olaf himself a sufficient Christianising of the pagan robbers, whose assistance he refused unless they would consent to be Christians. From the high importance attached to the mere ceremony without reference to its meaning, it is not improbable that the Christian transcribers, or relaters of the historical sagas, may have thought it decent to make the ancestors of the kings or great personages they are treating of, although they had lived in pagan times, partake of the important Christian ceremony which of itself had, in the rude conception of the early Christian converts, a saving power; for we find on the birth of every child who is to become a king, and leave descendants, that "water was poured over the child, and a name given him;" that he was baptized, in short, although living in the Odin religion. Harald Harfager is stated in his saga " to have had water poured over him, and a name given him," and his son Hakon also, who succeeded him; but we hear nothing of any such baptism of Eirik Blood-axe, or of any other of his sons, nor of any whose descendants did not succeed to power as kings. It may reasonably be doubted if any such ceremony was used in the Odin religion on the birth of a child;* because these pagans certainly exposed their children—a practice not consistent with dedi-

---

* Dr. Konrad Maurer has shown conclusively that the sprinkling of the infant with water was a pre-Christian ceremony among the Norsemen, and also that it prevailed among the pagan Romans and Greeks.

cating them by a ceremony analogous to baptism to the service of their gods in any way; and if it had no meaning in their religion, it could not be practised, unless in imitation of the Christian ceremony. Marriage also appears not to have been celebrated with any religious form. Polygamy was as fully tolerated as in Asia.* Harald Harfager had nine wives, with several concubines. Saint Olaf had concubines besides his wife, and was succeeded by a natural son; for illegitimacy, where it is not founded on any religious element in the marriage tie, is not considered a natural or just disqualification from inheritance. Marriage appears with the Odin-worshippers to have been altogether a civil tie, subject, consequently, to the disruptions which civil circumstances might produce or excuse.

The churches or temples of Odin appear to have had no consecrated order of men like a priesthood set apart for administering in religious rites. In the historical sagas, in the accounts of the direct collisions between Hakon Athelstan's foster-son, or Olaf the Saint, with the worshippers of Odin, in the temples at More, and in Gudbrandsdal, no mention is made of the presence of any priests. Bondes of eminence, great people, and even district kings, are mutilated or put to death—suffer martyrdom in the cause of Odinism; but no word is there to be found of any man in sacerdotal function. Three great religious festivals appear to have been held by the

* A too broad statement. Polygamy was very rare among our heathen ancestors.

Odin worshippers. One, in honour of Thor, was held in midwinter, about the turn of the day; and from coinciding nearly with the Christmas of the Christian church, the name of Yule, derived from Jolner, one of Odin's names, and the festivity and merry-making of the pagan celebration, were amalgamated with the Christian commemoration. The second, in honour of Frigg, was held at the first quarter of the second moon after the beginning of the year;* and the third, in honour of Odin, in the beginning of spring. The convenience of having snow to travel on, and the leisure and facility of travelling while snow covered the ground, have probably been the cause of all these pagan festivals being crowded together in the winter half-year, or between harvest and seed-time. They were not solely nor principally religious festivals, but assemblies of the people at which the regular Things were held, business transacted, and fairs kept for bartering, and buying, and merry-making. An hereditary priesthood descended from the twelve diar, or priests, or godar, who accompanied Odin from Asia, and who originally were judges as well as priests in the Things held at these great religious festivals, existed at the colonisation of Iceland, and down to the time of Snorre Sturlason himself, who was one of those godars; but the sacerdotal function had become merged in the civil function of judge apparently long before the introduction of Chris-

* The Northmen appear to have reckoned by winters, and the beginning of the year or winter from the 16th of October.—L.

tianity. The judicial functions and emoluments of judge descending by hereditary rights in certain families, as appertaining to their hereditary priesthood, could not be a popular institution, especially with no sacerdotal function to perform. True religion, as we see in Scotland and England, can scarcely maintain itself when it mixes up civil power or great wealth with the religious element in its establishment; and much less can a false religion. We may gather from the silence of the sagas on the point, that the godar had no sacerdotal or religious function in society; and did not, even in the earliest historical period, exist as priests, but as hereditary local judges only, each in his own godard or parish. At the Things at which Hakon Athelstan's fosterson, Olaf Trygveson, and Olaf the Saint, come in collision with the religion of Odin, and threaten and even put to death peasants and chiefs who adhere to Odinism, no priests or godar appear, or are spoken of. Their civil power, jurisdiction, and dues or emoluments, however, were derived from their hereditary succession to the priestly office in their respective godards; and it is not unreasonable to suppose that they, their supporters, and the religion on which they founded their rights, were not the popular side at the introduction of Christianity. Some indications may be perceived of its having been a political movement to adopt Christianity. The supporters of the old religion appear to have been the small kings, the rich bondes, and those who may reasonably be supposed to have been themselves godar, or connected with them. The support of Christianity, on the other hand, appears to have come from the people and the kings, and not from the kings alone. In Iceland, where the godar, with their civil powers, were transplanted from Norway by the first aristocratical settlers, and where Christianity had no royal supporters, the Thing of the people declared Christianity the lawful religion of the land. The institutions of Odinism, as well as its doctrines, were evidently become extinct as religion. The incompatible elements of civil power, wealth, and sacerdotal function, had lost all religious influence. The mixture is at this day as ineffective in its power over the human mind as it was in the eleventh century, and in the Christian religion as it was in Odinism.

The only practices connected with religion mentioned in the sagas, at which a priesthood may have officiated, are sacrifices, at the three great festivals, of cattle, which were killed and feasted upon. The door-post, floor, and people are stated to have been sprinkled with the blood of the sacrifice by a brush; but even this may be a fiction of the saga-maker, taken from the similar sprinkling of holy water by a brush in the Romish church. The best established of the religious practices of the Odin worshippers was the partaking of horse-flesh at those festivals, as commemorative of their ancestors. This practice was transplanted even to Iceland by the pagan settlers, and it held its ground there long after Christianity was adopted. As food, the horse never could have been reared in Iceland; and a religious or popular superstition only must have kept up such a custom there. The eating of horse-flesh at those festivals appears to have been held as decisive a test of paganism as baptism of Christianity, and was punished by death in the eleventh century by Saint Olaf. Public business, however, in the Things, and the ordinary business and pleasures of great country fairs, appear to have occupied the people at those festivals much more than any religious observances. Public worship under any form, or private or household devotion in the Odin religion, cannot be distinctly traced in the sagas. It is to be remembered, however, that it might not have been thought right or safe by the saga-relater or saga-scribe to go far into an account of pagan observances, customs, or doctrines; in case of being considered himself as a believer in them. This may have affected considerably the fidelity of delivery of subjects, both religious and political, in the sagas, when they were still in a traditionary, not a written state. To some cause of this kind we must ascribe the trifling amount of information concerning the Odin-worship to be found in the sagas. Religion may have been very little regarded, and a priesthood to support its observances and doctrines may have become a class connected only with civil power and emolument in their godards, and not thought of as belonging to religious service; but still a very strong religious spirit, among some at least of the pagan population, may be inferred from various details in the sagas.

We read of many individuals in the reigns of Hakon Athelstan's foster-son, of Olaf Trygveson, and of Olaf the Saint, suffering the loss of fortune, mutilation, torture, and death, rather than give up their religion and submit to baptism. The religion of Odin had its martyrs in those days, and consequently must have had its doctrines, its devotions, its observances, its application to the mind of man in some way, its something to suffer for; but the sagas leave us in the dark with regard to the doctrines and observances of a religion for which men were willing to suffer. The machinery of the Odin mythology, the fables, allegories, meanings and no-meanings of the Myths, however interesting, give us little or no information on the really important points,—the amount, quality, and social influence of the religion of the pagan Northmen immediately previous to their conversion to Christianity. The many names of places derived from Thor, and other names given to the Supreme Being in their religion, which are still to be recognised, not only in Scandinavia, but in the north of Scotland, the Farey Islands, Iceland, show that the Northmen carried about with them some knowledge of their religion. The many allusions in the poems and songs of the skalds presuppose even a very intimate knowledge, on the part of the hearers, with a very complicated mythological nomenclature and system. Every one, from the lowest to the highest, must have been familiar with the names, functions, attributes, histories ascribed to these gods, or the skald would have been unintelligible. The great development of the intellectual powers among the Northmen, is indeed one of the most curious inferences to be drawn from the sagas. The descriptions of relative situations of countries, as East, West, North, or South, show generalised ideas and habits of thinking among their seafaring men; and the songs of the skalds, as those of the four who accompanied Saint Olaf at the battle of Stiklestad, seem to have been instantly seized and got by heart by the people, — the Biarkemal to have been instantly recognised, and thought applicable to their situation; and all the mythological, and to us obscure allusions, to have been understood generally in the halls in which the skalds recited or sung their compositions. Their religion must have been taught to them, although we find few traces of the religious establishment or social arrangements by which this was done.

The material remains of this religion of Odin are surprisingly few. We find in the North very few remains of temples;—no statues, emblems, images, symbols. Was it actually more spiritual than other systems of paganism, and therefore less material in its outward expression? If we consider the vast mounds raised in memory of the dead, and their high appreciation of their great men of former ages, we can scarcely doubt but that the Northmen had higher notions of a future state than that of drinking ale in Valhal.

The temples of Odin appear to have been but thinly scattered. We hear but of the one at More, and one at Lade, in the Throndhjem district. A mound of earth alone remains at More, which was the principal temple in the north of Norway: houses or halls, constructed of wood, for receiving the people who came together to eat, drink, and transact their business, have probably been all the structures. The temple at Upsala, or Uppsalir (the up-halls or great halls), should have left some traces of former magnificence; for it was the residence of Odin himself,—the headquarters, the Rome of the Odin religion; and in part, at least, was constructed of stone. Adam of Bremen, who lived about the time Christianity was first introduced into Sweden, namely, about 1064, says, "Nobilissimum illa gens templum habet quod Upsala dicitur, non longe positum a Sictona civitate vel Birka. In hoc templo, quod totum ex auro paratum est, statua trium deorum veneratur populus, ita ut potentissimus eorum Thor in medio solum habeat triclinium, hinc et inde locum possident Woden et Fricco." In this passage from a contemporary Christian writer, who, as canon in the cathedral of Bremen, —under the bishop of which all the northern bishops stood at first,—must have had the best opportunity of becoming acquainted with the paganism of the North, Thor is stated to be seated on the throne as the supreme deity, and Odin and Frigg on each side as the minor deities in this pagan trinity; and the temple is stated to have been most noble, and adorned with gold. This temple was converted into a Christian church by Olaf the Swede about 1026; and Severin, an Englishman, was the first bishop. It was plundered of all its wealth, pagan and Christian, by King Stenkil, the son of King Ingve, about the year 1085; and set fire to, and the stone walls only left. King Sverker I. restored it about 1139, and had it consecrated, and dedicated to Saint Laurence. This church appears to be the only building from which the extent at least, if not the magnificence, of the temples of the pagan religion of the North may be guessed at. It stands at Gamle Upsala (Old Upsala), about two miles north of the present town of Upsala, at the end of an extensive plain. Around Gamle Upsala —now consisting of this church, the minister's house, and two or three cottages—there are, according to Professor Verelius, in his notes on the Herverar Saga, tumuli to the number of six hundred and sixty-nine, besides many which have been levelled for cultivation. Reckoning the chain of such hillocks between the town of Upsala and Gamle Upsala, that, or even a greater number of those tumuli, may be conceded to the antiquary. Three of them, close to Gamle Upsala, are called Kongs-högarne (the king's mounds); and one, oblong and flattened at the top, is Tings-högen (the Thing's mound). The circumference of these mounds at the base is about three hundred and fifty paces, and the ascent on any side takes about seventy-five steps; so that the perpendicular height may be about ninety feet. It may also be conceded to the antiquary that these mounds are works of art, in so far that they have been reduced to regular shape by the hands of man, and have been used as places of interment, and still more as places for addressing a multitude from—the steep slopes close to each other admitting of great numbers sitting or standing within sight and hearing of a person addressing them. But whoever looks over this chain of sand-hills at the end of a plain which has been a lake or mire at no distant geological period, and with a mire or morass, now called Myrby Trask, on the other side of it, will doubt whether these mounds be not originally of natural formation. He is struck, at least, with the conviction that not only in other countries, but in Sweden itself in particular, such formations of small ridges, and hillocks of gravel, sand, and rolled stones, upon a tongue of land which has originally divided two lakes, are of most frequent occurrence. Here, about Upsala, man has availed himself of a chain of mounds formed by nature; and, as a natural feature of ground, they account for the selection of Gamle Upsala in the earliest ages for the seat of government. With a lake or mire on each side, a narrow tongue of land dotted with small eminences behind each other gave the defenders a succession of strong posts to retire upon; and when missiles of very short range, and spear, sword, or battle-axe, and fighting hand to hand, were the only weapons and modes of fighting, the advantage of the higher ground was the great object in tactics. Gamle Upsala would be strong when the country was covered with wood, and the flat ground was a flooded morass. Of the old buildings, or town, no vestiges remain. Of the temple some of the walls are supposed to be included in the present church; and the old foundations have been traced by Rudbeck and Peringskiold. Its extreme length has not exceeded one hundred and twenty feet; and the rough unhewn small stones of such walls as may possibly have been parts of the old structure do not tell of much architectural magnificence. The arches, whether of the pagan structure, or of the re-edification in 1139, are the round Saxon arch; and the whole is less than an ordinary parish church in England. An exterior line is said by antiquaries to have surrounded the building, and to have been the golden ring, chain, or serpent surrounding the temple of Odin in skaldic poetry; but this has had no foundation but in their fancy. A wooden palisade may, no doubt, have surrounded the temple, with the tops of it painted or gilded; for the Northmen appear to have been profuse in gilding, from the descriptions of their war vessels with gilded sides and prows; and the skalds, in their symbolical inflated language, may have called this the serpent or dragon of Thor: but wood-work leaves no trace to posterity, and of stone-work no mark remains of any exterior circumvallation; and it must be confessed that no trace remains in this locality of magnificence belonging to the paganism of the North. The gold chains, bracelets, armlets, anklets—too small for men, and of exquisite workmanship—which have been found in the North, and are preserved in the museums of Copenhagen, Christiania, and Stockholm, if they really belonged to northern idols, and were not rather the hoarded plunder of vikings gathered in more civilised or refined lands,* or of Varings returned from Constantinople, give a much higher idea of the splendour of the pagan religion of Odin than any architectural remains in Scandinavia. The sites of Sigtuna and of Birka—now Old Sigtuna, mentioned by Adam of Bremen in the above extract—are at the head of the Malar lake, and would well deserve the careful examination of the antiquarian traveller. Walls are still standing there which have at least the interest of being among the oldest architectural fragments of the North.

The most permanent remains of the Odin religion are to be found in the usages and language of the descendants of the Odin-worshippers. All the descendants of the great Saxon race retain the names of three days of the week—Wednesday, Thursday, and Friday—from the Odin religion. Tuesday, perhaps, or Diss-day, on which the offerings to fate were made, and the courts of justice held, may belong also to the number.† Yule is a pagan festival kept in the pagan way, with merriment and good cheer, all over the Saxon world. Beltan is kept on Midsummer-day, all over the north of Europe, by lighting fires on the hills, and other festivities. It is but within these

* The archæological relics are so numerous in Scandinavia as to wholly exclude the idea of their importation in any manner from foreign lands.
† Tuesday is named after Tir, the one-handed god.

fifty years that trolls or sea-trows, and finmen and dwarfs, disappeared in the northern parts of Scotland. Mara (the nightmare) still rides the modern Saxon in his sleep, and under the same name nearly as she did the Yngling king Vanland; and the evil one in the Odin mythology, Nokken, keeps his ground, in the speech and invocations of our common people, as Old Nick, in spite of the Society for the Diffusion of Christian Knowledge. It is curious to observe how much more enduring ideas are than things—the intellectual than the material objects that mark the existence of the human species. Stone-work, and gold, and statues, and all material remains of this once general religion of the North, have disappeared from the face of the earth; yet words and ideas belonging to it remain.

It is remarkable that in the religion of Odin, as in that of Mahomet, women appear to have had no part in the future life.* We find no allusion to any Valhal for the female virtues. The Paradise of Mahomet and the Valhal of Odin are the same; only the one offers sensual and the other warlike enjoyments to the happy. They both exclude females. This is not the only coincidence. Odin appears to have stood in the same relation to Thor in Odinism that Mahomet stands in to the Supreme Being in Mahometanism. The family of Mahomet, its semi-sacred character, and its rights, as successors to the prophet, to the throne and

* This is a mistake. The mythology teaches that Odin shared the slain equally with Freyja, who rules in *Folkvang*, that is, in the *human* dwellings, where there are seats enough for all.

supremacy of temporal power over Mahometans, and with equal rights of succession in equal degrees of affinity to this sacred source, is in fact the Yngling dynasty of Odinism. If Mahomet had existed 400 years earlier, he would have been in modern history one of the Odins, perhaps *the* Odin, and the person or persons we call Odin would have merged in him. The coincidence between Odinism and Mahometanism in the ideas of a future state, in the exclusion of females from it, in the hereditary succession of a family to sacred and temporal power and function, show a coincidence in the ideas and elements of society among the people among whom the two religions flourished; and this coincidence is perhaps sufficiently strong to prove that the religion of Odin must have sprung up originally in the East among the same ideas and social elements as Mahometanism. The rapid conquest by Christianity over Odinism, about the beginning of the eleventh century, proves that the latter was not indigenous, but imported, and belonged to different physical circumstances and a different social state. The exclusion of females from a future life, and their virtues from reward, was not suited to the physical circumstances under which men live in the North, although among a people living on horseback in the plains of Asia the female may hold no higher social estimation than the horse. Christianity, by including the female sex in its benefits, could not but prevail in the North over Odinism.

The Odin-worship was not the only form of Paganism in the north of Europe. We find, in chapter 143 of the "Saga of Saint Olaf," an account of an expedition of Karl and Gunstein round the North Cape to Biarmeland, or the coast of the White Sea; and after trading for skins at the mouth of the Dwina, where Archangel now stands, of their proceeding, when the fair was over, to plunder the temple and idol of Jomala. They took a cup of silver coins that rested on his knee, a gold ornament that was round his neck, and treasure that was buried with the chiefs interred there, and retired to their ships. If this Jomala had been Thor or Odin, these vikings would not have plundered his temple, especially as one of them, Thorer Hund, was a zealous Odin-worshipper, and a martyr at last to his faith. We find, on the Baltic side of the country, that the Slavonic tribe who inhabit Esthonia had a Jomala, according to Kohl's "Reise in der Deutch-Russisch Ostsee Provinzen," 1842; and the name of Jomsborg given to the fortress of that singular association the Jomsborg Vikings on the island of Wollin, off the coast of Esthonia, seems to have had the same origin. The Joms-Vikings were a military association of pirates inhabiting the castle of Jomsborg, professing celibacy and obedience to their chief, and very similar to the orders of knights—as the Teutonic order, and that of Rhodes and the Templars —which appeared in Europe a century or two later; but these pirate-knights do not appear to have been in any way connected with the religion of Odin, or of Jomala. The Laplanders and Finlanders appear to have worshipped Jomala* also; and he appears to have been altogether a Slavonic, not a Saxon god. From the account of the expedition to Biarmeland, the temple and idol of this worship must have been as rich, and the attendance of guards or priests on the temple much greater than in the Odin-worship in Norway in that age, viz., the beginning of the eleventh century.†

* Jomala is still the name of the Deity—of God—among the Laplanders and Finlanders, according to Geijer.—*Svea Rikes Häfder*, p. 96.—L.

† On the subject of the old Norse religion the reader is referred to Anderson's "Norse Mythology," and to Viktor Rydberg's "Researches in Teutonic Mythology." The latter has been translated by Anderson, and published by Swan, Sonnenschein, & Co., London.

CHAPTER III.

OF THE SOCIAL CONDITION OF THE NORTHMEN.

If the historical sagas tell us little concerning the religion and religious establishments of the pagan Northmen, they give us incidentally a great deal of curious and valuable information about their social condition and institutions; and these are of great interest, because they are the nearest sources to which we can trace almost all that we call Anglo-Saxon in our own social condition, institutions, national character, and spirit. The following observations are picked up from the sagas. The reader of Snorre Sturlason's "Heimskringla" has before him the facts, or narratives, and can see himself whether the following inferences from them are warranted, and the views given of the singular state of society among the Northmen correctly drawn.

The lowest class in the community were the Thræll (Thralls, slaves). They were the prisoners captured by the vikings at sea on piratical cruises, or carried off from the coasts of foreign countries in marauding expeditions. These captives were, if not ransomed by their friends, bought and sold at regular slave markets. The owners could kill them without any fine, mulct, or manbod to the king, as in the case of the murder or manslaughter of a free man. King Olaf Trygveson, in his childhood, his mother Astrid, and his foster-father Thorolf were captured by an Esthonian viking, as they were crossing the sea from Sweden on their way to Novgorod, and were divided among the crew, and sold. An Esthonian man called Klerkon got Olaf and Thorolf as his share of the booty; but Astrid was separated from her son Olaf, then only three years of age. Klerkon thought Thorolf too old for a slave, and that no work would be got out of him to repay his food, and therefore killed him; but sold the boy to a man called Klerk for a goat. A peasant called Reas bought him from Klerk for a good cloak; and he remained in slavery until he was accidentally recognised by his uncle, who was in the service of the Russian king, and was by him taken to the court of Novgorod, where he grew up. His mother Astrid, apparently long afterwards, was recognised by a Norwegian merchant called Lodin at a slave market to which she had been brought for sale. Lodin offered to purchase her, and carry her home to Norway, if she would accept of him in marriage, which she joyfully agreed to; Lodin being a man of good birth, who sometimes went on expeditions as a merchant, and sometimes on viking cruises. On her return to Norway her friends approved of the match as suitable; and when her son, King Olaf Trygveson, came to the throne, Lodin and his sons by Astrid were in high favour. This account of the capturing, selling, and buying slaves, and killing one worn out, is related, as it would be at present in the streets of Washington, as an ordinary matter.* Slavery among the Anglo-Saxons at this period, namely, in the last half of the tenth century, appears to have become rather an *adscriptio glebæ*—the man sold or transferred with the land—than a distinct saleable property in the person of the slave; at least we hear of no slave markets in England at which slaves were bought and sold. In Norway this class appears to have been better treated than on the south side of the Baltic, and to have had some rights. Lodin had to ask his slave Astrid to accept of him in marriage. We find them also in the first half of the eleventh century, at least under some masters, considered capable of acquiring and holding property of their own. When Asbiorn came from Halogaland in the north of Norway to purchase a cargo of meal and malt, of which articles King Olaf the Saint, fearing a scarcity, had prohibited the exportation from the south of Norway, he went to his relation Erling Skialgson, a peasant or *bondi*, who was married to a sister of the late King Olaf Trygveson, and was a man of great power. Erling told Asbiorn that in consequence of the law he could not supply him, but that his thralls or slaves could probably sell him as much as he required for loading his vessel; adding the remarkable observation, that they, the slaves, are not bound by the law and country regulation like other men,—evidently from the notion that they were not parties, like other men, to

---

* This was written before the emancipation proclamation by Abraham Lincoln.

the making the law in the Thing. It is told of this Erling, who was one of the most considerable men in the country and brother-in-law of King Olaf Trygveson, although of the bonde or peasant class, that he had always ninety free-born men in his house, and two hundred or more when Earl Hakon, then regent of the country, came into the neighbourhood; that he had a ship of thirty-two banks of oars; and when he went on a viking cruise, or in a levy with the king, had two hundred men at least with him. He had always on his farm thirty slaves, besides other workpeople; and he gave them a certain task as a day's work to do, and gave them leave to work for themselves in the twilight, or in the night. He also gave them land to sow, and gave them the benefit of their own crops; and he put upon them a certain value, so that they could redeem themselves from slavery, which some could do the first or second year, and "all who had any luck could do it in the third year." With this money Erling bought new slaves, and he settled those who had thus obtained their freedom on his newly cleared land, and found employment for them in useful trades, or in the herring fishery, for which he furnished them with nets and salt. The same course of management is ascribed in the Saga of Saint Olaf to his stepfather, Sigurd Syr, who is celebrated for his prudence, and wisdom, and skill in husbandry; and it has probably been general among the slave-holders. The slaves who had thus obtained their freedom would belong to what appears to have been a distinct class from the peasants or bondes on the one hand, or the slaves on the other—the class of unfree men.

This class—the unfree—appears to have consisted of those who, not being udal-born to any land in the country, so as to be connected with, and have an interest in, the succession to any family estate, were not free of the Things; were not entitled to appear and deliberate in those assemblies; were not Thingsmen. This class of unfree is frequently mentioned in general levies for repelling invasion, when all men, free and unfree, are summoned to appear in arms; and the term unfree evidently refers to men who had personal freedom, and were not thralls, as the latter could only be collected to a levy by their masters. This class would include all the cottars on the land paying a rent in work upon the farm to the peasant, who was udal-born proprietor; and, under the name of housemen, this class of labourers in husbandry still exists on every farm in Norway. It would include also the house-carls, or freeborn indoor men, of whom Erling, we see, always kept ninety about him. They were, in fact, his bodyguard and garrison, the equivalent to the troop maintained by the feudal baron of Germany in his castle; and they followed the *bondi* or peasant in his summer excursions of piracy, or on the levy when called out by the king. They appear to have been free to serve whom they pleased. We find many of the class of bondes who kept a suite of eighty or ninety men; as Erling, Harek of Thiotta, and others. Svein, of the little isle of Gairsay (Gareksey) in Orkney, kept, we are told in the Orkneyinga Saga, eighty men all winter; and as we see the owner of this farm, which could not produce bread for one-fourth of that number, trusting for many years to his success in piracy for subsisting his retainers, we must conclude that they formed a numerous class of the community. This class would also include workpeople, labourers, fishermen, tradesmen, and others about towns and farms, or rural townships, who, although personally free and freeborn, not slaves, were unfree in respect of the rights possessed by the class of bondes, landowners, or peasants, in the Things. They had the protection and civil rights imparted by laws, but not the right to a voice in the enactment of the laws, or regulation of public affairs in the Things of the country. They were, in their rights, in the condition of the German population at the present day.

The class above the unfree in civil rights, the free peasant-proprietors, or bonde class, were the most important and influential in the community. We have no word in English, or in any other modern language, exactly equivalent to the word *bondi*, because the class itself never existed among us. Peasant does not express it; because we associate with the word peasant the idea of inferior social importance to the feudal nobility, gentry, and landed proprietors of a country, and this bonde class was itself the highest class in the country. Yeoman, or, in Cumberland, statesman, expresses their condition only relatively to the portions of land owned by them; not their social position as the highest class of landowners. If the Americans had a word to express the class of small landowners in their old settled states who live on their little properties, have the highest social influence in the country, and are its highest class, and, although without family aggrandisement by primogeniture succession, retain family distinction and descent, and even family pride, but divide their properties on the udal principle among their children, it would express more justly what the bonde class were than the words landholder, yeoman, statesman, peasant-proprietor, or peasant. In the following translation of the Heimskringla, where the word peasant is used for the word *bondi*,* the reader will have to carry in mind that these peasants were, in fact, an hereditary aristocracy, comprehending the great mass of the population, holding their little estates by a far more independent tenure than the feudal nobility of other countries, and having their land strictly entailed on their own families and kin, and with much family pride, and much regard for and record of their family descent and alliances, because each little estate was entailed on each peasant's whole family and kin. Udal right was, and is to this day in Norway, a

* Bóndi (in the plural bœndr) does not suit the English ear, and there is no reasoning with the ear in matters of language. The word itself, bondi or buandi, is present participle of the verb *búa* to live, abide, dwell, turned into a noun. It is of the same root as the Anglo-Saxon *buan* and the German *Bauer*. The word *bu* is still retained in Orkney and Shetland, to express the principal farm and farm-house of a small township or property, the residence of the proprietor; and is used in Denmark and Norway to express stock, or farm stock and substance. The law distinguishes between a grid-man, a labourer, a budsetu-man, a cottager, and a bondi, a man, who has land or stock.

species of entail, in realty, in the family that is udal-born to it. The udal land could not be alienated by sale, gift to the church, escheat to a superior, forfeiture, or by any other casualty, from the kindred who were udal-born to it; and they had, however distantly connected, an eventual right of succession vested in them superior to any right a stranger in blood could acquire. The udal-born to a piece of land could evict any other possessor, and, until a very late period, even without any repayment of what the new possessor having no udal right may have paid for it, or laid out upon it; and at the present day a right of redemption within a certain number of years is competent to those udal-born to an estate which has been sold out of a family. The right to the crown of Norway itself was udal-born right in a certain family or race, traced from Odin down to Harald Harfager through the Yngling dynasty, as a matter of religious faith; but from Harald Harfager as a fixed legal and historical point. All who were of his blood were udal-born to the Norwegian crown, and with equal rights of succession in equal degrees of propinquity. The eldest son had no exclusive right, either by law or in public opinion, to the whole succession, and the kingdom was more than once divided equally among all the sons. This principle of equal succession appears to have been so rooted in the social arrangement and public mind, that notwithstanding all the evils it produced in the succession to the crown by internal warfare between brothers, it seems never to have been shaken as a principle of right; and the kings who had laboured the most to unite the whole country into one sovereignty, as Harald Harfager, were the first to divide it again among their sons. One cause of this may have been the impossibility, among all classes, from the king to the peasant, of providing otherwise for the younger branches of a family than by giving them a portion of the land itself, or of the products of the land paid instead of money taxes to the crown. Legitimacy of birth was held of little account, owing probably to marriage not being among the Odin-worshippers a religious as well as a civil act; for we find all the children, illegitimate as well as legitimate, esteemed equal in udal-born right even to the throne itself; and although high descent on the mother's side also appears to have been esteemed, it was no obstacle even to the succession to the crown that the mother, as in the case of Magnus the Good, had been a slave. This was the consequence of polygamy, in which, as in the East, the kings indulged. Harald Harfager had nine wives at once, and many concubines; and every king, even King Olaf the Saint, had concubines as well as wives; and we find polygamy indulged in down to about 1130, when Sigurd the Crusader's marriage with Cecilia, at the time his queen was alive and not divorced, was opposed by the Bishop of Bergen, who would not celebrate it; but nevertheless the priest of Stavanger performed the ceremony, on the king's duly paying the church for the indulgence. Polygamy appears not to have been confined to kings and great men; for we find in the old Icelandic law book, called the "Gray Goose," that, in determining the mutual rights of succession of persons born in either country, Norway or Iceland, in the other country, it is provided that children born in Norway in bigamy should have equal right as legitimate children,—which also proves that in Iceland civilisation was advanced so much further than in Norway that bigamy was not lawful there, and its offspring not held legitimate. Each little estate was the kingdom in miniature, sometimes divided among children, and again reunited by succession of single successors by udal-born right vesting it in one. These landowners, with their entailed estates, old families, and extensive kin or clanship, might be called the nobility of the country, but that, from their great numbers and small properties, the tendency of the equal succession to land being to prevent the concentration of it into great estates, they were the peasantry. In social influence they had no class, like the aristocracy of feudal countries, above them. All the legislation, and the administration of law also, was in their hands. They alone conferred the crown at their Things. No man, however clear and undisputed his right of succession, ventured to assume the kingly title, dignity, and power but by the vote and concurrence of a Thing. He was proposed by a bonde; his right explained; and he was received by the Thing before he could levy subsistence, or men and aid, or exert any act of kingly power within the jurisdiction of the Thing. After being received and proclaimed at the Ore

Thing held at Throndhjem as the general or sole king of Norway, the upper king,—which that Thing alone had the right to do,—he had still to present himself to each of the other district Things, of which there were four, to entitle him to exercise royal authority, or enjoy the rights of royalty within their districts. The bondes of the district, who had voice and influence in those Things by family connection and personal merit, were the first men in the country. Their social importance is illustrated by the remarkable fact, that established kings—as, for instance, King Olaf Trygveson—married their sisters and daughters to powerful bondes, while others of their sisters and daughters were married to the kings of Sweden and Denmark. Erling the bonde refused the title of Earl when he married Astrid, the king's sister. Lodin married the widow of a king, and the mother of King Olaf Trygveson. There was no idea of disparagement, or inferiority, in such alliances; which shows how important and influential this class was in the community.

It is here, in these assemblies or Things of the Northmen, the immediate predecessors of the Norman conquerors, and their ancestors also—by which, however rudely, legislation and all parliamentary principles were exercised—that we must look for the origin of our parliaments, and the spirit and character of our people; on which, and not on the mere forms, our constitution is founded. The Wittenagemot of the Anglo-Saxon Heptarchic kings were not, like the Things of the Northmen, existing and influential assemblies of the people meeting *suo jure* at stated times, enacting and administering laws, and so interwoven with the whole social and political idiosyncrasy of the people, that the State could have no movement or existence but through such assemblies. The Wittenagemot, as the name implies, appears to have been merely a council of the wise and important men of the country, selected by the king to meet, consult, and advise with him—which is as different from a Thing as a cabinet council from a parliament. The Northmen who invaded and colonised the kingdom of Northumberland, had entirely expelled other occupants in the ninth century. The Anglo-Saxons had fled before the pagan and barbarous invaders who seized and settled on the lands, and, from the proximity to Norway and Denmark, received a rapid accession to their numbers by the influx of new settlers, as well as by their own increase of population. Normandy was only conquered by the Northmen, but Northumberland was colonised. Their religion, language, and laws were established. They had their own, and [not the Anglo-Saxon laws: a proof that they were a population not Anglo-Saxon in their social institutions. This appears from the laws of Edward the Elder, of Alfred himself, and from the treaties of these kings with Guthrun, the leader or chief of the Northmen who then occupied Northumberland. The kingdom of Northumberland, comprehending the present counties of York, Durham, Northumberland, Cumberland, Westmoreland, and parts of Lancashire; East Anglia also, comprehending the Isle of Ely, Cambridgeshire, Norfolk, and Suffolk; and the country of the former East Saxons, comprehending Essex, Middlesex, and part of Hertfordshire, and also parts of the northern and southern extremities of the Anglo-Saxon kingdom of Mercia— were so entirely occupied by Danes, or people of Danish descent, that they were under Danish, not Anglo-Saxon law. From the first invasions of the Danes in 787, or from the end of the eighth century to the time of the Norman conquest in 1066, or nearly 300 years, the laws and usages of the Northmen had prevailed over this large portion of the island. This kingdom of Northumberland would, at the present day, be more populous and wealthy than either of the kingdoms of Sweden, Denmark, Hanover, Holland, Belgium, Saxony, or Würtemberg, and had no doubt a proportional importance in those times. The Northmen, immediately previous to the Norman conquest, had conquered the whole of England, and held it from 1003 to 1041, for four successive reigns; viz., of Svein, Canute the Great, Harald Harefoot, and Hardicanute. In the laws of Edward the Confessor, as given by Lambart in 1568, and republished by Wheloch at the end of his edition of Bede, 1644, it is stated that for sixty-eight years previous to the Norman conquest, these Anglo-Saxon laws, originally framed by Edgar, had been out of use; and when William the Conqueror, in the fourth year of his reign, renewed these laws of Edward the Confessor, he was more inclined to retain the laws of the Northmen then in general use. If we strike off Wales, Cornwall, the western borders towards Scotland, and all comprehended in the kingdom of Northumberland, East Anglia, and other parts peopled by Northmen and their descendants, it is difficult to believe that the old Anglo-Saxon branch could have been predominant in the island, in numbers, power, and social influence; or could have prevailed to such an extent over the character and spirit of the population as to bury all social movement under the apathy and superstition in which they appear to have been sunk. The rebellions against William the Conqueror and his successors appear to have been almost always raised, or mainly supported, in the counties of recent Danish descent, not in those peopled by the old Anglo-Saxon race. The spirit and character of men having rights in society were undoubtedly renewed, and kept alive in England, by this great infusion into the population of people who had these rights, and the spirit and character produced by them, in their native land. A new and more vigorous branch was planted in the country than the old Anglo-Saxon. In historical research it is surely more reasonable to go to the nearest source of the institutions, laws, and spirit of a people—to the recent and great infusion into England from the north, during the ninth, tenth, and eleventh centuries, of men bred up in a rude but vigorous exercise of their rights in legislation, and in all the acts of their government—than to the most remote, and to trace in the obscure hints of Tacitus of popular and free institutions existing a thousand years before in the forests of Germany, the origin of our parliaments, constitution, and national character. The German people, the true unmixed descendants of the old Saxon race whom Tacitus describes, never, from the earliest date in modern history to the present day, had a single hour of religious, civil, and political liberty, as nations, or as individuals—never enjoyed the rights which the American citizen or the British subject, however imperfectly, enjoy in the freedom of person, property, and mind, at the present day, in their social condition. If the great stock itself of the Anglo-Saxon race has not transmitted to its immediate posterity in its own land the institutions of a free people, nor the spirit, character, independence of mind, on which alone they can be founded with stability, it appears absurd to trace to that stock our free institutions, and the principles in our character and spirit by which they are maintained, when we find a source so much nearer from which they would naturally flow. Our civil, religious, and political rights—the principles, spirit, and forms of legislation through which they work in our social union, are the legitimate offspring of the Things of the Northmen, not of the Wittenagemot of the Anglo-Saxons — of the independent Norse viking, not of the abject Saxon monk.

It would be a curious inquiry for the political philosopher to examine the causes which produced, in the tenth century, such a difference in the social condition of the Northmen, and of the cognate Anglo-Saxon branch in England and Germany. Physical causes connected with the nature of the country and climate, as well as the conventional causes of udal right, and the exclusion of inheritance by primogeniture, prevented the accumulation of land into large estates, and the rise of a feudal nobility like that of Germany. The following physical causes appear not only to have operated directly in preventing the growth of the feudal system in the country of the Northmen, but to have produced some of the conventional causes also which concurred to prevent it.

The Scandinavian peninsula consists of a vast table of mountain land, too elevated in general for cultivation, or even for the pasturage of large herds or flocks together in any one locality; and although sloping gently towards the Baltic or the Sound on the Swedish side, and there susceptible of the same inhabitation and husbandry as other countries, in as far as clime and soil will allow, on the other side, —the proper country of the Northmen,—throwing out towards the sea all round huge prongs of rocky and lofty ridges, either totally bare of soil, or covered with pine forests, growing apparently out of the very rock, and with no useful soil beneath them. The valleys and deep glens between these ridges, which shoot up into lofty pinnacles, precipices, and mountains, are filled at the lower end by the ocean, forming fiords, as these inlets of the sea are called, which run far up into the land, in some cases a hundred miles or more; yet so narrow that the stones, it is said, rolling down from the mountain slope on one side of such a fiord, are often projected from the steep overhanging precipice, in which the slope halfway down ends, across to the opposite shore. These fiords in general, however, are fine expanses or inland lakes of the ocean,—calm, deep, pure blue; and shut in on every side by black precipices and green forests, and with fair wooded islets sleeping on the bosom of the water. These fiords are the peculiar and characteristic feature of Norwegian scenery. Rivers of great volume of water, but generally of short and rapid course, pour into the fiords from the Fielde, or high table-land behind, which forms the body or mass of the country. It is on the flat spots of arable land on the borders of these fiords, rivers, and the lakes into which the rivers expand, that the population lives. In some of these river-valleys and sea-valleys, a single farm of a few acres of land is only found here and there in many miles of country, the bare rock dipping at once into the blue deep water, and leaving no margin for cultivation. In others, narrow slips of inhabitable arable land extend some way, but are hemmed in behind, on the land side, by the rocky ridges which form the valley; and they are seldom broad enough to admit of two rows of little farms, or even of two large fields, in the breadth between the hill-foot and the water; and in the length are often interrupted by some bare prong of rock jutting from the side-ridge into the slip of arable level land,

and dividing it from such another slip. All the land capable of cultivation, either with spade or plough, has been cultivated from the most remote times; and there is little room for improvement, because it is the ground rock destitute of soil, not merely trees or loose rocks encumbering the soil, that opposes human industry. The little estates, not averaging perhaps fifty acres each of arable land, are densely inhabited; because the seasons for preparing the ground, sowing, and reaping, are so brief, that all husbandry work must be performed in the shortest possible time, and consequently at the expense of supporting, all the year, a great many hands on the farm to perform it; and the fishing in the fiord, river, or lake, the summer pasturage for cattle in the distant fielde-glens attached to each little estate in the inhabited country, and a little wood-cutting in the forest, afford subsistence to many more people than the little farm itself would require for its cultivation in a better clime, or could support from its own produce. The extent of every little property has been settled for ages, and want of soil and space prevents any alteration in the extent, and keeps it within the unchangeable boundaries of rock and water. It is highly interesting to look at these original little family estates of the men who, in the ninth and tenth centuries, played so important a part in the finest countries of Europe,—who were the origin of the men and events we see at this day, and whose descendants are now seated on the thrones and in the palaces of Europe, and in the

West are making a new world of social arrangements for themselves. The sites, and even the names, of the little estates or gards on which these men were born, remain unchanged, in many instances, to this day; and the posterity of the original proprietors of the ninth century may reasonably be supposed, in a country in which the land is entailed by udal right upon the family, to be at this day the possessors—engaged, however, now in cutting wood for the French or Newcastle market, instead of in conquering Normandy and Northumberland.

Some of our great English nobility and gentry leave their own splendid seats, parks, and estates in England, to enjoy shooting and fishing in Norway for a few weeks. They are little aware that they are perhaps passing by the very estates which their own ancestors once ploughed,—sleeping on the same spot of this earth on which their forefathers, a thousand years ago, slept, and were at home; men, too, as proud then of their high birth, of their descent, through some seven-and-twenty generations, from Odin, or his followers the Godes, as their posterity are now of having "come in with or before the Conqueror." The common traveller visiting this land destitute of architectural remains of former magnificence, without the temples and classical ruins of Italy, or the cathedrals and giant castles of Germany, will yet feel here that the memorials of former generations may be materially insignificant, yet morally grand. These little farms and houses, as they stand at this day, were the homes of men whose rude, but just and firm sense of their civil and political rights in society is, in the present times, radiating from the spark of it they kindled in England, and working out in every country the emancipation of mankind from the thraldom of the institutions which grew up under the Roman empire, and still cover Italy and Germany, along with the decaying ruins of the splendour, taste, magnificence, power, and oppression of their rulers. Europe holds no memorials of ancient historical events which have been attended by such great results in our times, as some rude excavations in the shore-banks of the island of Viger,* in Sondmore,—which are pointed out by the finger of tradition as the dry docks in which the vessels of Rolf Ganger, from whom the fifth in descent was our William the Conqueror, were drawn up in winter, and from whence he launched them, and set out from Norway on the expedition in which he conquered Normandy. The philosopher might seat himself beside the historian amidst the ruins of the Capitol, and with Rome, and all the monuments of Roman power and magnificence under his eye, might venture to ask whether they, magnificent and imposing as they are, suggest ideas of greater social interest,—are connected with grander moral results on the condition, well-being, and civilisation of the human race in every land, than these rude excavations in the isle of Viger, which once held Rolf Ganger's vessels.

* Vigrey, the isle of Viger, is situated in Haram parish, in the bailiwick of Sondmore.—*Strohm's Biskrivelse over Möre*, and *Kraft's Norge*.—L.

It is evident that such a country in such a climate never could have afforded a rent, either in money or in natural products, for the use of the land, to a class of feudal nobility possessing it in great estates, although it may afford a subsistence to a class of small working landowners, like the bondes, giving their own labour to the cultivation, and helping out their agricultural means of living with the earnings of their labour in other occupations—in piracy and pillage on the coasts of other countries in the ninth century, and in the nineteenth with the cod fishery, the herring fishery, the wood trade, and other peaceful occupations of industry. On account of these physical circumstances—of a soil and climate which afford no surplus produce from land, after subsisting the needful labourers, to go as rent to a landlord—no powerful body of feudal nobility could grow up in Norway, as in other countries in the Middle Ages; and from the same causes, now in modern times, during the 400 years previous to 1814 in which Denmark had held Norway, all the encouragement that could be given by the Danish government to raising a class of nobility in Norway was unavailing. Slavery even could not exist in any country in which the labour of the slave would barely produce the subsistence of the slave, and would leave no surplus gain from his labour for a master; still less could a nobility, or body of great landowners drawing rent, subsist where land can barely produce subsistence for the labour which, in consequence of the shortness of the seasons, is required in very large quantity, in proportion to the area, for its cultivation. We find, accordingly, that when the viking trade, the occupation of piracy and pillage, was extinguished by the influence of Christianity, the progress of civilisation, the rise of the Hanseatic League and of its establishments, which in Norway itself both repressed piracy and gave beneficial occupation in the fisheries to the surplus population formerly occupied in piracy and warfare, that class of people fell back upon husbandry and ordinary occupations which had formerly been engaged all summer and autumn in marauding expeditions; and the class of slaves, the thralls, was necessarily superseded in their utility by people living at home all the year. The last piratical expeditions were about the end of the twelfth century, and in the following century thraldom, or slavery, was, it is understood, abolished by law by Magnus the Law Improver. The labour of the slave was no longer needed at home, and would not pay the cost of his subsistence.

Physical circumstances also, and not conventional or accidental circumstances, evidently moulded the other social arrangements of the Northmen into a shape different from the feudal. The Things or assemblies of the people, which kings had to respect and refer to, may be deduced much more reasonably from natural causes similar to those which prevented the rise of a feudal class of nobles in Norway, than from political institutions or principles of social arrangement carried down from the ancient Germans in a natural state of liberty in remote ages. The same causes will produce the same effects in all ages. It is refining too much in political antiquarianism to refer all liberal social arrangements—our English parliaments, our constitutional checks upon the executive power in the State, our popular representation, and the spirit of our laws—to the Wittenagemots of the Saxons, and to trace these again up to principles of freedom in social arrangement derived from the Germanic tribes in the days of Tacitus. But it is not refining too much to conclude that, in every age and country, there are but two ways in which the governing class of a community can issue their laws, commands, or will, to the governed. One is through writing, and by the arts of writing and reading being so generally diffused that in every locality one individual at least, the civil functionary or the parish priest, is able to communicate the law, command, or will of the governing, to that small group of the governed over which he is placed. The other way, and the only way where, from the nature of the soil and climate, the governed are widely scattered, and writing and reading are rarely attained, and such civil or clerical arrangement not efficient, was to convene Things or general assemblies of the people, at which the law, command, or will of the governing could be made known to the governed. There could be no other way, in poor, thinly inhabited countries especially, by which the governing, however despotic, could get their law, command, or will done; for these must be made known to be executed or obeyed, whether they were for a levy of men or of money, for war or for peace, for rewarding and honouring, or for punishing and disgracing—the law, command, or will must be promulgated. Nor is it refining too much to conclude, that wheresoever men are assembled together in numbers for public business, be it merely to hear the law, command, or will of a despotic ruler, the spirit of deliberating upon, considering, and judging of the decree given out, and of the public interests involved in it, is there in the midst of them. The democratic element of society is there,—the spirit of judging in their own affairs is there, and is let loose; for such an assembly is in effect a parliament, in which public opinion will make itself heard; and coming from the only military force of those ages, the mass of the people, and, in the North, of a people without military subordination to a feudal aristocracy in civil affairs, must predominate over the will of the king supported only by his court retinue. The concurrence of a few great nobles could not here give effect to the royal command, law, or will; because the few, the intermediate link of a powerful aristocracy, which to this day chains the Anglo-Saxon race on the Continent, was from physical causes—the poverty of the soil—totally wanting among the Northmen, and the kings had to deal direct with the people in great general assemblies or Things. The necessity of holding such general meetings or Things for announcing to the people the levies of men, ships, and provisions required of them, and for all public business, and the check given by the Things to all measures not approved of by the public judgment, appear in every page of the Heimskringla, and constitute its great value, in fact, to us, as a record of the state of social arrangement among our ancestors. The necessity of assembling the people was so well established, that we find no public act whatsoever undertaken without the deliberation of a Thing; and the principle was so engrafted in the spirit of the people, that even the attack of an enemy, the course to be taken in dangerous circumstances, to retreat or advance, were laid before a Thing of all the people in the fleet or army; and they often referred it to the king's own judgment, that is, the king took authority from the Thing to act in the emergency on his own plan and judgment. A reference to the people in all that concerned them was interwoven with the daily life of the Northmen, in peace and in war. We read of House Things, of Court Things, of District Things for administering law, of Things for consultation of all engaged in an expedition; and in all matters, and on all occasions, in which men were embarked with common interests, a reference to themselves, a universal spirit of self-government in society, was established. King Sverre, who reigned from 1184 to 1202, after the period when Snorre Sturlason's work ends, although taking his own way in his military enterprises, appears in a saga of his reign never to have omitted calling a Thing, and bringing it round by his speeches, which are often very characteristic, to his own opinion and plans.

So essential were Things considered wheresoever men were acting with a common stake and interest, that in war expeditions the call to a Thing on the war-horn or trumpet appears to have been a settled signal-call known to all men,—like the call to arms, or the call to attack; and each kind of Thing, whether it was a general Thing that was summoned, or a House Thing of the king's counsellors, or a Hird Thing of the court, or of the leaders of the troops, appears to have had its distinct peculiar call on the war-horn known to all men. In the ordinary affairs of the country, the Things were assembled in a simple and effective way. A *bod*, called a budstikke in Norway, where it is still used, was a stick of wood like a constable's baton, with a spike at the end of it, which was passed from house to house, as a signal for the people to assemble. In each house it was well known to which neighbouring house it had to be passed, and the penalties for detaining the bod were very heavy. In modern times, the place, house, and occasion of meeting, are stated on a slip of paper enclosed in the bottom of the budstick; but in former times the Thing-place, and the time allowed for repairing there, were known, and whether to go armed or unarmed was the only matter requiring to be indicated. An arrow split into four parts was the known token for appearing in arms. If the people of a house to which the token was carried were from home, and the door locked, the bearer had to stick it on the door by the spike inserted in one end for this purpose; if the door was open, but the people not at home, the bearer had " to stick it in the house-father's great chair at the fireside ; " and this was to be held a legal delivery of the token, exonerating the last bearer from the penalties for detaining it. The peace token, a simple stick with a spike; the war token, an arrow split into quarters, and sent out in different directions; a token in shape of an axe, to denote the presence of the king at the Thing; and one in shape of the cross, to denote that Church matters were to be considered,—are understood to have been used before writing and reading were diffused. On one occasion, we read of Earl Hakon issuing the usual token for the bondes to meet him at a Thing; and it was exchanged, in its course, for the war token, and the bondes appeared in arms, and overpowered the Earl and his attendants.

The Things appear not to have been representative, but primary assemblies, of all the bondes of the district udal-born to land. In Sweden there appears to have been one general Thing held at Upsala, at the time when the festivals or sacrifices to Thor, Odin, and Frigg, were celebrated. From the proceedings of one of the Things held at Upsala, in February or March, 1018, related in the Saga of Saint Olaf, we may have some idea of the power of those assemblies. King Olaf of Sweden, who had a great dislike to Olaf King of Norway, was forced by this Thing to conclude a peace with, and give his daughter in marriage to, King Olaf of Norway, in order to put an end to hostilities between the two countries; and they threatened, by their lagman, to depose him for misgovernment, if he refused the treaty and alliance which King Olaf of Norway proposed by his ambassador Hialte the Skald. The lagman appears to have been the depositary and expounder of the laws passed by the Things, and to have been either appointed by the people as their president at the Things, or to have held his office by hereditary succession from the gode, and to have been priest and judge, exercising both the religious and judicial function. At this general Thing at Upsala the lagman of the district of Upland was entitled to preside; and his influence and power in this national assembly appear to have been much greater than the king's. It is a picturesque circumstance, mentioned in the Saga of Saint Olaf about this Thing at Upsala in 1018, that when Thorgny the lagman rose after the ambassador from Norway had delivered his errand, and the Swedish king had replied to it, all the bondes, who had been sitting on the grass before, rose up, and crowded together to hear what their lagman Thorgny was going to say; and the old lagman, whose white and silky beard is stated to have been so long that it reached his knees when he was seated, allowed the clanking of their arms and the din of their feet to subside before he began his speech. The Things appear to have been always held in the open air, and the people were seated; and the speakers, even the kings, rose up to address them. In the characters of great men given in the sagas we always find eloquence, ready agreeable speaking, a good voice, a quick apprehension, a ready delivery, and winning manners, reckoned the highest qualities of a popular king or eminent chief. His talent as a public speaker is never omitted. In Sweden this one general Thing appears to have been for the whole country; and besides the religious or civil business, a kind of fair for exchanging commodities arose from the concourse of people to it from all parts of the country. In Norway,—owing no doubt to the much greater difference in the means of subsistence in the different quarters of the country, in some of which fishing-grounds out at sea, and even rocks abounding in sea-fowl eggs at the season, were subjects of property; in others pasturages in distant mountain glens, and in others arable lands only, are of importance,—four distinct Things appear in the oldest times to have been necessary for framing laws suitable to the different circumstances of their respective jurisdictions; and, within their jurisdictions, the smaller district Things appear to have determined law cases between parties according to the laws settled at the great Things; and as the mulcts or money penalties paid for all crimes went partly to the king, and were an important branch of the royal revenue, the kings, on their progresses through the land, with the lagman of each district, appear to have held these Things for administering justice and collecting their revenue. The king's bailiff, or the tacksman or donatory of the revenue of the district, appears to have held these Law

Things in the king's absence. The great Things appear to have been legislative, and the small district Things within their circle of jurisdiction administrative. Of the great Things there were in old times four in different quarters of Norway. The Frosta Thing was held in the Throndhjem country, at a place called Frosta, in the present bailiwick of Frosten; Gula Thing, at Eyvindvik, in the shiprath of Gule, on the west coast of Norway; Eidsivia Thing, at Eidsvold, in Upper Raumarike, for the inland or upland districts of Norway; and Borgar Thing, at the old burgh called Sarpsborg, on the river Glommen, near the great waterfall called Sarpsfors. One or two other Law Things appear to have been added in later times: one in Halogaland for the people living far north, and one on the coast between the jurisdiction or circle of the Sarpsborg Thing and that of the Gula Thing. A special Thing, called the Ore Thing, from being held on the Ore, or isthmus * of the river Nid, on which the city of Throndhjem stands, was considered the only Thing which could confer the sovereignty of the whole of Norway, the other Things having no right to powers beyond their own circles. It was only convened for this special purpose of examining and proclaiming the right to the whole kingdom; and it appears to have been only the kingship *de jure* that the Ore Thing considered and confirmed: the king had still to repair to each Law Thing and small

---

* The narrow slip of land between two waters, as at a river-mouth or outlet of a lake, between it and the sea, is still called an Are or Ayre in the north of Scotland, and is the same as the Icelandic *Eyrr*.—L.

Thing, to obtain their acknowledgment of his right, and the power of a sovereign within their jurisdictions. The scat or land-tax,—the right of guest-quarters or subsistence on royal progresses,—the levy of men, ships, provisions, arms, for defence at home, or war expeditions abroad, had to be adjudged to the kings by the Things; and amidst the perpetual contests between udal-born claimants, the principle of referring to the Things for the right and power of a sovereign, and for the title of king, was never set aside. No class but the bondes appeared at Things with any power. The kings themselves appear to have been but Thing-men at a Thing.

Two circumstances, which may be called accidental, concurred with the physical circumstances of the country, soil, and clime, to prevent the rise of a feudal nobility in Norway at the period, the ninth century, when feudality was establishing itself over the rest of Europe. One was the colonisation of Iceland by that class which in other countries became feudal lords; the other was the conquests in England and in France, by leaders who drew off all of the same class of more warlike habits than the settlers in Iceland, and opened a more promising field for their ambition abroad in those expeditions, than in struggling at home against the supremacy of Harald Harfager. In his successful attempt to reduce all the small kings, or district kings, under his authority, he was necessarily thrown upon the people for support, and their influence would be naturally increased by the suppression through their aid of the small independent kings. This struggle was renewed at intervals until the introduction of Christianity by King Olaf the Saint; and the two parties appear to have supported the two different religions: the small kings and their party adhering to the old religion of Odin, under which the small kings, as godes, united the offices of judge and priest, and levied certain dues, and presided at the sacrificial meetings as judges as well as priests; and the other party, which included the mass of the people, supported Christianity, and the supremacy of King Olaf, because it relieved them from the exactions of the local kings, and from internal war and pillage. The influence of the people, and of their Things, gained by the removal to other countries of that class which at home would have grown probably into a feudal aristocracy. In Iceland an aristocratic republic was at first established, and in Normandy and Northumberland all that was aristocratic in Norway found an outlet for its activity.

A physical circumstance also almost peculiar to Norway, and apparently very little connected with the social state of a people, was of great influence, in concurrence with those two accidental circumstances, in preventing the rise of an aristocracy. The stone of the Peninsula in general, and of Norway in particular, is gneiss, or other hard primary rock, which is worked with difficulty, and breaks up in rough shapeless lumps, or in thin schistose plates; and walls cannot be constructed of such building materials without great labour, time, and command of cement. Limestone is not found in abundance in Norway, and is rare in situations in which it can be made and easily transported; and even clay, which is used as a bedding or cement in some countries for rough lumps of stone in thick walls, is scarce in Norway. Wood has of necessity, in all times and with all classes, been the only building material. This circumstance has been of great influence in the Middle Ages on the social condition of the Northmen. Castles of nobles or kings, commanding the country round, and secure from sudden assault by the strength of the building, could not be constructed, and never existed in Norway. The huge fragments and ruins of baronial castles and strongholds, so characteristic of the state of society in the Middle Ages in the feudal countries of Europe, and so ornamental in the landscape now, are wanting in Norway. The noble had nothing to fall back upon but his war-ship, the king nothing but the support of the people. In the reign of our King Stephen, when England was covered with the fortified castles of the nobility, to the number, it is somewhere stated, of 1500, and was laid waste by their exactions and private wars, the sons of Harald Gille—the kings Sigurd, Inge, and Eystein—were referring their claims and disputes to the decision of Things of the people. In Normandy and England the Northmen and their descendants felt the want in their mother-country of secure fortresses for their power; and the first and natural object of the alien landholders was to build castles, and lodge themselves in safety by stone walls against sudden assaults, and above all against the firebrand of the midnight assailant. In the mother-country, to be surprised and burned by night within the wooden structures in which even kings had to reside, was a fate so common, that some of the kings appeared to have lived on board ships principally, or on islands on the coast.

This physical circumstance of wanting the building material of which the feudal castles of other countries were constructed, and by which structures the feudal system itself was mainly supported, had its social as well as political influences on the people. The different classes were not separated from each other, in society, by the important distinction of a difference in the magnitude or splendour of their dwellings. The peasant at the corner of the forest could, with his time, material, and labour of his family at command, lodge himself as magnificently as the king,—and did so. The mansions of kings and great chiefs were no better than the ordinary dwellings of the bondes. Lade, near Throndhjem, —the seat of kings before the city of Throndhjem, or Nidaros, was founded by King Olaf Trygveson, and which was the mansion of Earl Hakon the Great, and of many distinguished men who were earls of Lade,—was, and is, a wooden structure of the ordinary dimensions of the houses of the opulent bondes in the district. Egge—the seat of Kalf Arneson, who led the bonde army against King Olaf which defeated and slew him at the battle of Stiklestad, and who was a man of great note and social importance in his day—is, and always has been, such a farm-house of logs as may be seen on every ordinary farm estate of the same size. The foundation of a few loose stones, on which the lower tier of logs is laid to raise it from the earth, remains always the same, although all the superstructure of wood may have been often renewed; but these show the extent on the ground of the old houses. The equality of all ranks in those circumstances of lodging, food, clothing, fuel, furniture, which form great social distinctions among people of other countries, must have nourished a feeling of independence of external circumstances—a feeling, also, of their own worth, rights, and importance, among the bondes — and must have raised their habits, character, and ideas to a nearer level to those of the highest. The kings, having no royal residences, were lodged, with their court attendants, on the royal progresses, habitually by the bondes, and entertained by them. At the present day there are no royal mansions, or residences of the great, in Norway, different from the ordinary houses of the bondes or peasant-proprietors. His Majesty Carl Johan has to lodge in their houses in travelling through his Norwegian dominions; and no king in Europe could travel through his kingdom, and be lodged so well every night by the same class. In ancient times the kings lived in guest-quarters—that is, by billet upon the peasant-proprietors in different districts in regular turn; and even this kind of intercourse must have kept alive a high feeling of their own importance in the bonde class, in the times when, from the want of the machinery of a lettered functionary class, civil or clerical, all public business had to be transacted directly with them in their Things. The rise and diffusion of letters, learning, and a learned class, in the Middle Ages, retarded perhaps rather than advanced just principles of government and legislation. The people were more enslaved by the power which the learning of the Middle Ages threw into the hands of their rulers, than they were before in the ages of ignorance of letters, when their rude force was in direct contact, face to face, with the rude power of their rulers. This prejudicial effect of the revival of letters on civil, political, and religious liberty, by doing away with all direct *vivâ voce* communication in assemblies of the people between the rulers and the ruled, may be traced even to the present day in Germany and other countries. The people have no influence in their own concerns, because a lettered body of functionaries, spread over the whole social body, and fixed in every locality, receives, and disseminates to the small groups of the population under their jurisdiction, the law, command, or will of the autocratic government, without that reference to the people which could not be avoided when all had to be convened in a Thing or assembly to hear the promulgation. The period in which the influence of the governed should have been made effective slipped by on the Continent, among the Anglo-Saxon race, without being used;

and probably would have slipped by in England also, but for the recent admixture of a wilder, more ignorant, and more free people, in a great proportion of the island, who could not even be oppressed without collecting them into Things, or Folkmots, to make known to them what they had to submit to. The very ignorance of the half pagan people of mixed or pure Danish descent who occupied so large a portion of the island at the Norman conquest, was the providential means of keeping alive that spirit of self-government in public affairs among the people, on which, and not on the mere forms of representative government, our social economy rests. The forms are useless without the life in the spirit of the people to animate them. France, and some countries of Germany, have got the moulds; but the stuff to fill them with is wanting in the people. We inherit this stuff in the national character from the great intermixture of the rude energetic Northmen, bred up in Things and consultations with their leaders, which took place during the Danish conquest immediately previous to the invasion of William the Conqueror; and in the generation immediately after his conquest this stuff began to show itself in fermentation, and worked out our present social institutions, and the spirit of our national character.

The lendermen, or tacksmen of the king's farms and revenues, could scarcely be called a class. They were temporary functionaries, not hereditary nobles; and had no feudal rights or jurisdiction, but had to plead in the Things like other bondes. As individuals they appear to have obtained power and influence, but not as a class; and they never transmitted it to their posterity.

The earls, or jarls, were still less than the lendermen a body of nobility approaching to the feudal barons of other lands. The title appears to have been altogether personal; not connected with property in land, or any feudal rights or jurisdiction. The Earls of Orkney—of the family of Ragnvald Earl of More, the friend of Harald Harfager, and father of Rolf Ganger—appear to have been the only family of hereditary nobles under the Norwegian crown exercising a kind of feudal power. The Earls of More appear to have been only functionaries or lendermen collecting the king's taxes, and managing the royal lands in the district, and retaining a part for their remuneration. The Earls of Orkney, however, of the first line, appear to have grown independent, and to have paid only military service, and a nominal quitrent, and only when forced to do so. This line appears to have been broken in upon in 1129, when Kale, the son of Kol, was made earl, under the name of Earl Ragnvald. His father Kol was married to the sister of Earl Magnus the Saint; but the direct male descendants of the old line, the sons of Earl Magnus's brothers, appear not to have been extinct. In Norway, from the time of Earl Hakon of Lade, who was regent or viceroy for the Danish kings when they expelled the Norwegian descendants of

Harfager, there appears to have been a jealousy of conferring the title of earl, as it probably implied some of Earl Hakon's power in the opinion of the people. Harald Harfager had appointed sixteen earls, one for each district, when he suppressed the small kings; but they appear to have been merely collectors of his rents.

The churchmen were not a numerous or powerful class until after the first half of the twelfth century. They were at first strangers, and many of them English. Nicolas Breakspear, the son, Matthew Paris tells us, of a peasant employed about the Benedictine monastery of Saint Albans in Hertfordshire, and educated by the monks there, was the first priest who obtained any political or social influence in Norway. He was sent there, when cardinal, on a mission to settle the Church; and afterwards, when elected pope, 1154, under the title of Hadrian IV., he was friendly to the Norwegian people. His influence when in Norway was beneficially exerted in preventing the carrying of arms, or engaging in private feuds, during certain periods of truce proclaimed by the Church. The body of priests in the peninsula until the end of the twelfth century being small, and mostly foreigners from England, both in Sweden and in Norway, shows the want of education in Latin and in the use of letters among the pagan Northmen; and shows also the identity or similarity of the language of a great portion at least of England with that of the Scandinavian peninsula.

Several of the smaller institutions in society, which were transplanted into England by the Northmen or their successors, may perhaps be traced to the mode of living which the physical circumstances of the mother-country had produced. The kings having, in fact, no safe resting-place but on board of ship, being in perpetual danger, during their progresses for subsistence on shore, to be surprised and burnt in their quarters by any trifling force, had no reluctance at all to such expeditions against England, the Hebrides, or the Orkney Islands, as they frequently undertook; and when on shore, and from necessity subsisting in guest-quarters in inland districts, we see the first rudiments of the institution of a standing army, or bodyguard, or body of hired men-at-arms. The kings, from the earliest times, appear to have kept a hird, as it was called, or court. The hirdmen were paid men-at-arms; and it appears incidentally from several passages in the sagas that they regularly mounted guard,—posted sentries round the king's quarters,—and had patroles on horseback, night and day, at some distance, to bring notice of any hostile advance. We find that Olaf Kyrre, or the Quiet, kept a body of 120 hirdmen, 60 guests,* and 60 house-carls, for doing such

* The guests were one division of the king's men; they were a kind of policemen, and had not the full privileges of the king's guardmen or hirdmen, although they were in the king's pay; they had their own *seats* in the king's hall, *the guests' bench*, their own *chief*, their own *banner*, their own *meeting*, and they formed a separate body. As the guests were lower in rank than the hirdmen, a recruit had often to serve his apprenticeship among them. See G. Vigfusson's Cleasby's Icelandic-English Dictionary, *sub voce*.

work as might be required. The standing armed force, or bodyguard, appears to have consisted of two classes of people. The hirdmen were apparently of the class udal-born to land, and consequently entitled to sit in Things at home; for they are called Thingmen, which appears to have been a title of distinction. The guest appears to have been a soldier of the unfree class; that is, not of those udal-born to land, and free of, or qualified to sit in, the Things. They appear to have been the common seamen, soldiers, and followers; for we do not find any mention of slaves ever employed under arms in any way, or in any war expeditions. The guests appear to have been inferior to the thingmen or hirdmen, as we find them employed in inferior offices, such as executing criminals or prisoners. The victories of Svein, and Canute the Great, are ascribed to the superiority of the hired bands of thingmen in their pay. The massacre of the Danes in 1002, by Ethelred, appears to have been of the regular bands of thingmen who were quartered in the towns, and who were attacked while unarmed and attending a Church festival. The hirdmen appear not only to have been disciplined and paid troops, but to have been clothed uniformly. Red was always the national colour of the Northmen, and continues still in Denmark and England the distinctive colour of their military dress. It was so of the hirdmen and people of distinction in Norway, as appears from several parts of the sagas, in the eleventh century. Olaf Kyrre, or the Quiet, appears to have introduced, in this century, some court ceremonies or observances not used before. For each guest at the royal table he appointed a torch-bearer, to hold a candle. The butler stood in front of the king's table to fill the cups, which, we are told, before his time were of deers' horn. The court-marshal had a table, opposite to the king's, for entertaining guests of inferior dignity. The drinking was either by measure, or without measure; that is, in each horn or cup there was a perpendicular row of studs at equal distances, and each guest when the cup or horn was passed to him drank down to the stud or mark below. At night, and on particular occasions, the drinking was without measure, each taking what he pleased; and to be drunk at night appears to have been common even for the kings. Such cups with studs are still preserved in museums, and in families of the bondes. The kings appear to have wanted no external ceremonial belonging to their dignity. They were addressed in forms, still preserved in the northern languages, of peculiar respect; their personal attendants were of the highest people, and were considered as holding places of great honour. Earl Magnus the Saint was, in his youth, one of those who carried in the dishes to the royal table; and torch-bearers, hirdmen, and all who belonged to the court, were in great consideration; and it appears to have been held of importance, and of great advantage, to be enrolled among the king's hirdmen.

We may assume from the above observations,

derived from the facts and circumstances stated in various parts of the Heimskringla, that the intellectual and political condition of this branch of the Saxon race, while it was pagan, was not very inferior to, although very different from, that of the Anglo-Saxon branch which had been Christianised five hundred years before, and had among them the learning and organisation of the Church of Rome. They had a literature of their own; a language common to all, and in which that literature was composed; laws, institutions, political arrangements, in which public opinion was powerful; and had the elements of freedom and constitutional government. What may have been the comparative diffusion of the useful arts in the two branches in those ages? The test of the civilisation of a people, next to their intellectual and civil condition, is the state of the useful arts among them.

Note.—For further information in regard to the social and political condition of the old Northmen, see the Story of Burnt Njal, from the Icelandic of Njals Saga, by G. W. Dasent. This gives a vivid description of life in Iceland at the end of the tenth century, and the work contains a scholarly introduction by the learned translator. Another work of great importance in this connection is Paul B. Du Chaillu's "The Viking Age," just issued by John Murray, London, and by the Scribners in New York.

CHAPTER IV.

STATE OF THE USEFUL ARTS AMONG THE NORTHMEN.

THE architectural remains of public buildings in a country—of churches, monasteries, castles,—as they are the most visible and lasting monuments, are often taken as the only measure of the useful arts in former times. Yet a class of builders, or stone-masons, wandering from country to country, like our civil engineers and railroad contractors at the present day, may have constructed these edifices; and a people or a nobility sunk in ignorance, superstition, and sloth, may have paid for the construction, without any diffusion of the useful arts, or of combined industry, in the inert mass of population around. Gothic architecture in both its branches, Saxon and Norman, has evidently sprung from a seafaring people. The nave of the Gothic cathedral, with its round or pointed arches, is the inside of a vessel with its timbers, and merely raised upon posts, and reversed. No working model for a Gothic fabric could be given that would not be a ship turned upside down, and raised on pillars. The name of the main body of the Gothic church—the nave, navis, or *ship* of the building, as it is called in all the northern languages of Gothic root—shows that the wooden structure of the shipbuilder has given the idea and principles to the architect, who has only translated the wood-work into stone, and reversed it, and raised it to be the roof instead of the bottom of a fabric. The Northmen, however, can lay no claim to any attainment in architecture. The material and skill have been equally wanting among them. From the pagan times nothing in stone and lime exists of any importance or merit as a building; and the principal structure of an early age connected with Christianity, the cathedral of Throndhjem, erected in the last half of the twelfth century, cannot certainly be considered equal to the great ecclesiastical structures of Durham, York, or other English cathedrals, scarcely even to that of the same period erected in Orkney—the cathedral of Saint Magnus. We have, however, a less equivocal test of the progress and diffusion of the useful arts among the Northmen than the church-building of their Saxon contemporaries, for which they wanted the material. When we read of bands of ferocious, ignorant, pagan barbarians, landing on the coasts of England or France, let us apply a little consideration to the accounts of them, and endeavour to recollect how many of the useful arts must be in operation, and in a very advanced state too, and very generally diffused in a country, in order to fit out even a single vessel to cross the high seas, much more numerous squadrons filled with bands of fighting men. Legs, arms, and courage, the soldier and his sword, can do nothing here. We can understand multitudes of ignorant, ferocious barbarians, pressing in by land upon the Roman empire, overwhelming countries like a cloud of locusts, subsisting, as they march along, upon the grain and cattle of the inhabitants they exterminate, and settling, with their wives and children, in new homes; but the moment we come to the sea we come to a check. Ferocity, ignorance, and courage will not bring men across the ocean. Food, water, fuel, clothes, arms, as well as men, have to be provided, collected, transported; and be the ships ever so rude, wood-work, iron-work, rope-work, cloth-work, cooper-work, in short almost all the useful arts, must be in full operation among a people, before even a hundred men could be transported, in any way, from the shores of Norway or Denmark to the coasts of England or France. Fixed social arrangements, too, combinations of industry working for a common purpose, laws and security of person and property, military organisation and discipline, must have been established and understood, in a way and to an extent not at all necessary to be presupposed in the case of a tumultuous crowd migrating by land to new settlements. Do the architectural remains, or the history of the Anglo-Saxon people, or of any other, in the eighth or ninth century, and down to the thirteenth, give us any reasonable ground for supposing among them so wide a diffusion of the arts of working in wood and iron, of raising or procuring by commerce flax or hemp, of the arts of making ropes, spinning and weaving sailcloth, preserving provisions, coopering water-casks, and all the other combinations of the primary arts of civilised life, implied in the building and fitting out vessels to carry three or four hundred men across the ocean, and to be their resting-place, refuge, and home for many weeks, months, and on some of their viking cruises even for years? There is more of civilisation, and of a diffusion of the useful arts on which civilisation rests, implied in the social state of a people who could do this, than can be justly inferred from a people quarrying stones, and bringing them to the hands of a master-builder to be put together in the shape of a church or castle. Historians tell us that when Charlemagne, in the ninth century, saw some piratical vessels of the Northmen cruising at a distance in the Mediterranean, to which they had for the first time found their way, he turned away from the window, and burst into tears. Was it the barbarism of these pirates, or their civilisation, their comparative superiority in the art of navigation, and of all belonging to it, that moved him? None of the countries under his sway, none of the Christian populations of Europe in the seventh, eighth, or ninth centuries, had ships and men capable of such a voyage. The comparative state of shipbuilding and navigation, in two countries with sea-coasts, is a better test of their comparative civilisation and advance in all the useful arts than that of their church-building. Compared to Italy, Sicily, or Bavaria, Great Britain or Scandinavia, or the United States of America, would be utterly barbarous and uncivilised, if structures of stone were a measure of the civilisation and general diffusion of the useful arts among a people. It is to be observed, also, that the ships of the Northmen in those ages did not belong to the king, or to the State, but to private adventurers and peasants, and were fitted out by them; and were gathered by a levy or impressment, from all the country, when required for the king's service. The arts connected with the building and fitting out such ships must have been generally diffused. The fleets were not, like those of King Alfred, created by, and belonging to, the king. We need not have any great notion of the kind and size of the vessels gathered by a levy from the peasantry; but the worst and least of them must have been seaworthy, and of a size to navigate along the coast from the most northerly district, such as Halogaland, to the Baltic,—a distance of twelve degrees of latitude; they must have been of a size to carry and shelter men, with their provisions, clothes, and arms; and the arms of those days required great room for stowage. Stones were an ammunition which it was necessary to carry in every ship; because on the rocky steep coasts of Norway, or on the muddy shores of the Baltic, pebbly beaches at which this kind of ammunition could be replaced are not common. Swords, spears, battle-axes, arrowheads, bows, and bow-strings had all to be kept dry, and out of the sea-spray; for rust and damp would make them useless as weapons. These, consequently, had to be stowed under a deck, or in chests. The shields alone could bear exposure to wet, and they appear to have been hung outside from the rails all round the vessel; so that they would occupy the place of quarter-cloths, or wash-boards, above the gunwale of our shipping. The stowage of their plunder also, which consisted of bulky articles, as malt, meal, grain, cattle, wool, clothes, arms taken in the forays on the coast (and they had transport vessels as well as war vessels with them on their marauding expeditions) required vessels of a considerable size. We need not suppose that, of the 1200 vessels which King Olaf in his last levy to oppose Canute the Great had assembled and brought to the Baltic, the greater number were more than large boats, of perhaps thirty feet of keel, with a forecastle deck, a cabin aft, and the centre open, and merely tilted over at night to shelter the crew. Yet to construct many hundreds of such rude craft as this,—and any kind of boat or ship below this, as a class of vessels, could not have withstood sea and weather along the coast of Norway, and across the Skager Rack to the Sound,—implies a general diffusion of the art of working in iron; a trade in the arts of raising and smelting the ore; and a knowledge, in every district of the country, of the smith-work and carpenter-work, and tools and handicrafts necessary for ship-building and fitting out ships for sea. We have some data in the sagas from which we can arrive at the dimensions and appearance of the larger class of vessels used by the Northmen, allowing that the ordinary vessels of the peasants gathered by a levy could be no larger or better than large herring-boats. We have in the Saga of Olaf Trygveson some details of the building of the Long Serpent and the Crane, some time between the years 995 and 1000. The Long Serpent is called the largest vessel that had ever been built in Norway to that time. These were long-ships, which appear to have been a denomination of ships of war, distinguishing them from last-ships, or ships for carrying cargoes. The long-ship was of much smaller breadth in proportion to the length. The long-ships appear to have been divided into two classes: dragon ships, from the figure-head probably of a dragon being used on them, and which appear to have had from twenty to thirty rowers on each side; and snekias,* or cutters, with from ten to twenty rowers on a side. The Crane had thirty banks for rowers; and the forecastle and poop were high, and the vessel very narrow in proportion to her length. The Long Serpent had thirty-four banks for rowers, and the saga gives some interesting details concerning her. The length of her keel, we are told, that rested upon the grass, was seventy-four ells. This ell is stated by Macpherson, in his "Annals of Commerce," on the authority of Thorkelin, a learned antiquary, who was keeper of the Royal Library at Copenhagen, to have been equal to a foot and a half English measure. We have, therefore, 111 feet at the least as the length of keel of this vessel. This would be within about ten feet of the length of keel of one of our frigates of 42 guns, and of 942 tons burden, and

---

* Anglo-Saxon, snace, English, smack; a kind of *swift-sailing ship* belonging to the kind of "long-ship," thus called from its swift snake-like movement in the water. See Vigfusson's Cleasby, *sub voce*.

of a breadth of 38 feet and a depth of 13 feet; or, taking a steam vessel of 111 feet of keel, the extreme breadth would be 22 feet, the depth 13½ feet, the tonnage 296 tons, and the horse-power 120. These are dimensions and proportions given for 111 feet of keel in the able articles on shipbuilding and on steam navigation in the "Encyclopædia Britannica." The Long Serpent, being a rowing as well as sailing vessel, would have as much rake of stem and stern as a steamer; and would be as long on deck. She is described as of good breadth, but the breadth is not stated; well timbered, for which the saga refers to the knees for supporting the beams, which were then to be seen; and with thirty-four benches or banks for rowers, which would be the beams in a modern vessel. One of our long large steam vessels, with high poop deck and forecastle deck, low waist, and small breadth, would probably have very nearly the same appearance in the water as such a vessel as the Long Serpent; only, instead of paddle-boxes and wheels on each side, there would be thirty-four oars out on each side between the forecastle and the poop. The Northmen appear by the saga to have been lavish in gilding and painting their vessels. One of these long low war-ships of the vikings, with a gilded head representing a dragon on the stem, and a gilded representation of its tail at the stern curling over the head of the steersman, with a row of shining red and white shields hung over the rails all round from stem to stern, representing its scaly sides, and thirty oars on each side giving it motion and representing its legs, must have been no inapt representation of the ideal figure of a dragon creeping over the blue calm surface of a narrow gloomy fiord, sunk deep, like some abode for unearthly creatures, between precipices of bare black rock, which shut out the full light of day. Dragon was a name for a class or size of war-ships, but each had its own name. The Crane, the Little Serpent, the Long Serpent, the Bison, and other vessels of about thirty banks for rowers, are mentioned; and vessels of from twenty to twenty-five banks appear to have been common among the considerable bondes, and cutters of ten or fifteen banks to have been the ordinary class of vessels of all who went on sea. A vessel of thirty or thirty-four banks for rowers would have that number of oars out on each side, and not fifteen or seventeen only on each side; because the breadth of such a vessel would be sufficient to give two rowers, sitting midships, a sufficient length of lever between their hands and the fulcrum at the gunwales on either side, to wield and work any length of oar that could be advantageous: but in the smaller class of vessels of ten or fifteen oars it is likely that one oar only was worked on each bank, as in our men-of-war's boats, the whole breadth of the vessel being required for the portion of the lever or oar within the fulcrum or gunwale. Under the feet of the rowers, in the waist of the vessel, the chests of arms, stones for casting, provisions, clothing, and goods, have been stowed, and protected by a deck of movable hatches. Upon this lower deck the crew appear to have slept at night, sheltered from the weather by a tilt or awning, when not landed and under tents on the beach for the night. Ship-tents are mentioned in the outfit of vessels as being of prime necessity, as much as ship-sails. In the voyages in the sagas, we read of fleets collected in the north of Norway, from Throndhjem, and even from Halogaland, sailing south along the coast every summer as far as the Sound, and thence into the Baltic, or along the coast of Jutland and Slesvik, and thence over to Britain, or to the other coasts. The major part of the vessels appear to have taken a harbour every night, or to have been laid, on the coast of Norway, close to the rocks, in some sheltered spot, with cables on the land, or with the fore-foot of the vessel touching the beach; and the people either landed and set up tents on shore, or made a tilt on board by striking the mast, and laying the tilt cloths or sails over it. The large open vessels which at present carry the dried fish from the Lofoden isles to Bergen, although open for the sake of stowage, are of a size to carry masts of forty feet long which are struck by the crew when not under sail, there being no standing rigging, and only one large square-sail. This appears to have been the rig and description of all the ancient vessels, great or small, of the Northmen. They appear to have had a certain show and luxury about their sails; for we read of them having stripes of white, red, and blue cloths; and we read of Sigurd the Crusader waiting for a fortnight at the mouth of the Dardanelles with a fair wind for going up the strait, until he got a wind with which he could sail up with the sails trimmed fore and aft in his ships, that the inhabitants on shore might see the splendour of his sails. These large rowing vessels had one advantage belonging only to steam vessels in our times, that they could back out of seen dangers; and being under command of oars, and with small draught of water, shallows, rocks, and lee shores were not such formidable dangers to them as to our sailing vessels. Many important towns in those times, as, for instance, all our Cinque Ports, appear to have been situated rather with a reference originally to a good convenience for beaching such vessels, than to good sheltered harbours for riding at anchor in. The whole coast of the peninsula, from the North Cape round to Tornea, is protected from the main ocean by an almost continuous belt of islands, islets, rocks, and half-tide reefs, or skerries, within which the navigation is comparatively smooth, although very intricate for vessels with sails only. This inland passage "within the skerries" is used now, even in winter, by small boats going to the cod-fishing in the Lofoden isles from the Bergen district. It is only at particular openings, as at the mouths of the great fiords, that this continuous chain of sheltering isles and rocks is broken, and that the eye of the ocean, as the Norwegian fishermen express it, looks in upon the land. By waiting opportunities to cross these openings vessels of a small class, we may suppose, have accompanied King Olaf in his foray to the Danish islands, in hopes of booty more profitable than fish; and we need not believe his fleet of 1200 vessels raised by his levy to have been all of a large class. When his son Magnus the Good went to Denmark to claim the crown, upon the death of Hardicanute of England, in consequence of an agreement that the survivor of the two should succeed to the heritage of the other, he is stated to have had seventy large vessels with him, by which we may suppose vessels of twenty banks or upwards, such as the considerable bondes possessed, to be meant; and this number probably expresses more correctly the number of large ships then in the country. The size of the war-vessels appears to have been reckoned by the banks, or by the rooms between two banks of oars. Each room or space, we may gather from the sagas, was the berth of eight men, and was divided into half-rooms, starboard and larboard, of four men for working the corresponding oars. When the ships were advancing two men worked the oar, one covered them with his shield from the enemy's missiles, and one shot at the enemy. When the ships got into line, they were bound together by their stems and sterns; and the forecastles and poops, which were decked, and raised high in the construction of their vessels, and sometimes with temporary stages or castles on them, were the posts of the fighting men. The main manœuvre seems to have consisted in laying the high forecastles and poops favourably for striking down with stones, arrows, and casting spears, upon a lower vessel. They used grappling-irons for throwing into the enemy's ship, and dragging her towards them. But these and similar observations will occur to the reader of the many sea-fights recounted in the Heimskringla.

One of the most indispensable articles for a large vessel,—one for which no substitute can be found, and which cannot be produced single-handed, but requires the co-operation of many branches of industry,—is the anchor. Boats may be anchored by a stone, or a hook of strong wood sunk by a heavy stone attached to it; but vessels of from 50 to 111 feet of keel, such as the war-ships and last-ships of the Northmen, must have carried anchors of from ten to fifteen hundredweight at the least, and we read of their riding out heavy gales. To forge, or procure in any way such anchors, betokens a higher state of the useful arts among these pagan Northmen than we usually allow them.* Iron is the mother of all the useful arts; and a people who could smelt iron from the ore, and work it into all that is required for ships of considerable size, from a nail to an anchor, could not have been in a state of such utter barbarism as they are represented to us. We may fairly doubt of their gross ignorance and want of civilisation in their pagan state, when we find they had a literature of their own, and laws, institutions, social arrangements, a spirit and character very analogous to the English, if not the source from which the English flowed; and

* The Museum of Northern Antiquities at Copenhagen contains many articles, both of ornament and use, which display great ingenuity and good workmanship in metal, and betoken a considerable division of trades and of labour in their production, even in the earliest times.—L.

were in advance of all the Christian nations in one branch at least of the useful arts, in which great combinations of them are required—the building, fitting out, and navigating large vessels.*

* Since the above was written two magnificent viking ships have been unearthed in Norway, and can be seen by the traveller in Christiania. An illustration of the Gokstad ship found in 1880 will be found in P. B. Du Chaillu's "Viking Age," in which the archæology of the North, the articles of ornament and use, are discussed and exhibited in illustrations. In "The Viking Age," Du Chaillu has described from the sagas and the finds in the museums the life of the old Norseman from the cradle to the grave, from his birth to his funeral, and his work is a lasting monument to the high civilisation of the ancient Scandinavians.

CHAPTER V.

OF THE DISCOVERY OF GREENLAND AND AMERICA BY THE NORTHMEN.

The discovery of Greenland by the Icelanders about the year 984, and the establishment of considerable colonies on one or on both sides of that vast peninsula which terminates at Cape Farewell,—in which Christianity and Christian establishments, parishes, churches, and even monasteries, were flourishing, or at least existing to such an extent that from 1112 to 1409 there was a regular succession of bishops, of whom seventeen are named, for their superintendence, —are facts which no longer admit of any reasonable doubt. The documentary evidence of the saga,— which gave not merely vague accounts of such a discovery and settlement, but statistical details, with the names and the distances from each other of farms or townships, of which there were, according to accounts of the fourteenth century, ninety in what was called Vestribygd or the western settlement, with three churches, and 190 in the Eystribygd or eastern settlement, with one cathedral, eleven other churches, two towns, and two monasteries,—bears all the internal evidence of truth, in the consistency and simplicity of the statements. The saga accounts also are supported by the incidental notice of Greenland by contemporary writers. Adam of Bremen mentions that the people of Greenland, among other northern people, sent to his diocesan, Adalbert archbishop of Bremen, who died in 1075, for clergymen, who accordingly were sent to them. The first bishop of Greenland mentioned in the Icelandic accounts was Arnold, who was ordained by the archbishop of Lund, in Scania, in 1124. The bishopric of Greenland was afterwards under the archbishop of Throndhjem; and Alf, or Alfus, who is supposed to have died about 1378 in Greenland, is the last who is known to have officiated there. In 1389, Henry, according to Torfæus, was appointed bishop; and in 1406 Askel was appointed to succeed Henry, in case he was dead. But it does not appear, according to Torfæus, that either of them ever reached Greenland; but, since Torfæus's time, a document is said to have been discovered relative to a marriage settlement executed at Gardar, the name of the town or episcopal seat in Greenland, by the last bishop, whose name was Eindride Andreson, not Askel, three years later, viz., in 1409. In 1261, the Greenland settlements appear to have been regularly annexed to the crown of Norway by King Hakon Hakonson, who sent messengers to the people of Greenland; and in the submission which the messengers brought back, it was agreed "that all fines for murders, whether committed by Norwegian or Greenland people, on inhabited or uninhabited land, or even under the pole itself, should be paid to the king."

The payments for murders, or other capital offences compounded for by mulcts to the king and relations, were then a considerable branch of the royal revenues. In 1388-9, Henry the bishop, on setting out for Greenland received instructions to keep the king's revenues safely warehoused in a certain fixed place, those years in which no vessels came to Greenland; which shows that the communications with Iceland were not yearly or regular. A brief of Pope Nicholas V., in 1448, to the bishops of Skalholt and Holar in Iceland, states "that his beloved children dwelling in an island called Greenland, on the utmost verge of the ocean north of Norway, and who are under the archbishop of Throndhjem, have raised his compassion by their complaint that after having been Christians for 600 years, and converted by the holy Saint Olaf, and having erected many sacred buildings and a splendid cathedral on said island, in which divine service was diligently performed, they had thirty years ago been attacked by the heathens of the neighbouring coast, who came with a fleet against them, and killed and dispersed many, and made slaves of those who were able-bodied; but having now gathered together again, they crave the services of priests and a bishop." The pope therefore desires those bishops, as the nearest, to consult with their diocesan, if the distance permit, and to send the Greenland people a suitable man to be their bishop. The sudden extinction of a colony, which must have attained considerable importance and population to have had a regular succession of bishops for 250 years, is much more extraordinary than its establishment. It vanished, as it were, from the face of the earth, about the end of the fourteenth or beginning of the fifteenth century; and even the memory of its former existence passed away. The Christian colony established in the tenth century in Greenland, with its churches, monasteries, bishops, was considered, notwithstanding the internal and the collateral evidence supporting the sagas, to be a pious delusion of the Middle Ages, founded on a mere saga fable. The fable itself is short, and appears to have nothing fabulous in it. In the beginning of the tenth century, an Icelander or a Norwegian, called Gunbiorn, son of Ulf Krage, was driven by a storm to the west of Iceland, and discovered some rocks, which he called Gunbiorn Skerry, and a great country, of which he brought the news to Iceland. Soon after one Eirik Red, or Eirik the Red, was condemned at Thorsnes Thing, in Iceland, to banishment for a murder he had committed. He fitted out a vessel, and told his friends he would go and find the land which Gunbiorn had seen, and come back and let them know what kind of country it was. Eirik sailed west from the Snowfieldsjokul, in Iceland, to the east coast of Greenland, and then followed the coast southwards, looking for a convenient place for dwelling in. He sailed westward round a cape which he called Hvarf,* and passed the first winter on an island,

* Hvarf appears a name given to extreme capes from which the coast turns or bends in a different direction. Cape Wrath, the extreme westerly

which from him was called Eirik's Isle. After passing three years in examining the coast he returned to Iceland, and gave such a fine account of the country that it was called Greenland; and the following year twenty-five vessels with colonists set out with him to settle in it, but only about one half reached their destination, some having turned back, and some being lost in the ice. About fourteen years after Eirik was settled in Greenland, his son Leif, who afterwards discovered Vinland, went over to Norway to King Olaf Trygveson, who had him instructed in Christianity, or baptized, and sent a priest with him to Greenland, who baptized Eirik and all the colonists. Many came over from Iceland from time to time, and the country was settled wherever it was inhabitable. In this account there is nothing incredible or inconsistent. Greenland was to Iceland what Iceland had been to Norway— a place of refuge for the surplus population, for those who had no land or means of living. Iceland was originally an aristocratic republic,—a settlement made by people of family and wealth, who alone could fit out vessels for emigrating to it; and these landnammen took possession of the land. Of the lower class many in course of time must have become retainers, tenants, or workpeople under the higher class, and have been ready to emigrate to a country where they could get land of their own, and at a distance little more than half of that from

point of the coast of Scotland, has originally been called Hvarf, and in time changed to *Wrath*.—L.

Norway to Iceland. The discovery and colonisation of land within a distance so short, compared to the usual voyage from Iceland to Norway, is not incredible, nor wonderful. The means of subsistence in both countries have probably been very much the same. Seals, whales, fish of various kinds, reindeer, hares, wild fowl, would give subsistence; oil, skins, feathers, furs—which in the middle age were in great estimation for dress,—would give surplus products for exchange. Cattle, if we may believe the sagas, were kept in considerable numbers. Corn was not produced in either country. The balance of the natural products which man may subsist on, —such as game, reindeer, seals, fish, and of furs and feathers for barter,—may have been even in favour of Greenland. The extinction of such a colony, after existing for 400 years, is certainly more extraordinary than its establishment, and almost justifies the doubt whether it ever had existed. Several causes are given for this extraordinary circumstance. One is the gradual accumulation of ice on both sides of this vast peninsula, by which not only the pasturages, and temperature in which cattle could subsist, may have been diminished, and with these one main branch of the subsistence of the population; but also the direct communication with small vessels coasting along the shores and through the sounds of Greenland may have been interrupted, and the voyage round Cape Farewell outside of the isles, and ice, and sounds have been too tempestuous for such vessels as they possessed. Another cause was probably the great pestilence called the black death, which appeared in Europe about 1349, and which seems to have been more universal and destructive than the cholera, the plague, or any other visitation known in the history of the human race. It extinguished entirely populations much more numerous, and more wholesomely fed, clothed, and lodged, than we can suppose a colony in Greenland to have been, and it seems to have raged particularly in the north. It is supposed by some that this pestilence either swept off the whole population of the colony, or weakened it so much that the survivors were at last cut off by the Skrælings or Esquimaux, with whom the colonists appear to have been always in hostility. The inequality between the most contemptible race of savages and the most civilised people of Europe would be but small, or the advantage probably on the side of the uncivilised, in all warfare between them before the use of fire-arms; which, next to Christianity, has been the great means of diffusing and securing civilisation among the human race. The pope's brief of 1448, if it be a genuine document,—and it is said to have been found in the archives of the Vatican by a Professor Mallet some years ago, but how he got there is not shown,—would prove the truth of the conjecture which has been made, that the colonists were overpowered by the Skrælings. The existing traditions among the Esquimaux,* of their having come in

* The Esquimaux appear, by the narrative of his discoveries on the north coast of America by the late Mr. Thomas Simpson (1843), to be

their canoes and surprised and killed all the Kabloon or European people in old times, is not worth much, as evidently it is a tradition only of the moment produced by leading questions put to them. No tradition in any country seems to exist but as an impersonation, as an account of an individual person doing a thing; and it is the individual and his personal feats, not the great act itself, that is delivered by tradition as its principal subject matter. Another cause was, that Queen Margaret, on whom the three northern crowns had devolved in 1387, had made the trade to Greenland, Iceland, the Farey Isles, Halogaland, and Finland, a royal monopoly, which could only be carried on in ships belonging to, or licenced by, the sovereign; and certain merchants who had visited Greenland about that time were accused of a treasonable violation of the royal edict, we are told by Torfæus in his "Grænlandia Antiqua," and only escaped punishment by pleading that stress of weather had driven them to those parts. Her successor, Eirik of Pomerania, was too much engaged in Swedish affairs, and his successor, Christopher of Bavaria, in his contests with the Hanseatic League, to think of the colony of Greenland. Under the monopoly of trade the Icelanders could have no vessels, and no object for sailing to Greenland; and the vessels fitted out by government, or its lessees, to trade with them, would only be ready to leave Denmark or Bergen

by no means the poor physically weak people met with at present in Davis's Straits, and described by Captain Parry.—L.

for Iceland, at the season they ought to have been ready to leave Iceland to go to Greenland. The colony gradually fell into oblivion. Its former existence even had become a matter of disputed or neglected tradition. Christian III., who came to the throne 1534, abolished the prohibition of sailing to Greenland; and a few feeble attempts were made at discovery by him and his successors from time to time, and at last even these were given up. It was not until 1721 that a Norwegian minister, Hans Egede,—one of those rare men who go on to their purpose unmoved by any selfish interest, and to whom fame, wealth, honour, comfort, are neither object nor reward,—resigned his living in Norway, and obtained permission, after much difficulty and many petitions to government, to settle himself as a missionary on the coast of Davis's Straits among the Esquimaux. The general opinion was, that the lost colony of Old Greenland was situated on the east coast of the peninsula, and not within Davis's Straits; and it does not appear that Hans Egede himself, at first, had any idea that he was settling upon the ruined seats of Christian predecessors of the same tongue and mother country. It is a curious paragraph in the history of the human race, showing how true it is that in the tide of time man and his affairs return to where they set out,—that Christian churches, bishops, and consequently people in some numbers, and in some state of civilisation, had existed, been extinguished, and forgotten, and again on the same spot, after the lapse of 400 years, men have attempted to live, colonise, and Christianise. The feeble attempt in our times, the struggle to subsist, and the trifling amount of population in the modern colony after a century, are strangely in contrast with the state of the old colony. There are but about 150 Europeans at present in these Danish colonies; and the whole population of the natives, from Cape Farewell as far north as man can live, is reckoned under 6000 people, and about five or six vessels only are employed in trading with them; and this is in a country which formerly subsisted a population of European descent, which had at least sixteen ecclesiastical establishments or parishes, a bishop, monasteries, and consequently a number greatly exceeding 150 souls. The old colonists do not appear to have ever made converts among the natives, and their numbers, which must at one time have been considerable, appear to have found abundant subsistence; for we read in the sagas of vessels with sixty men arriving in autumn, being subsisted all winter, and fitted out in spring, and victualled for voyages of uncertain and long duration: and now if one of the vessels fitted out by charitable contributions by the sect of the Moravians to carry food to their missionaries be delayed for a season, they are in danger of starving. Is it man or nature that has changed? Are men less vigorous, less energetic, less enduring and hardy, than in those old times of the Northmen? or is the land, the sea, the climate less adapted now for the subsistence of the human animal?

The opinion of almost all antiquaries was, that the main settlement of the old colonists,—the Eystribygd, with its 190 gards (farms), its town of Gardar, its cathedral, bishop's seat, and twelve or thirteen churches,—was on the east coast of Greenland, somewhere on the coast north of Cape Farewell, inaccessible now from ice; and the less important Vestribygd to have been west of Cape Farewell, within Davis's Straits. Others supposed that both settlements were on the east coast of Greenland, and that the old colonists did not know that Greenland had a west coast from Cape Farewell. The opinions were founded on certain ancient sailing directions found in the sagas, especially in a saga of King Olaf Trygveson, in which it is mentioned that from Stad, the westermost part of Norway, it is a voyage of seven days' sailing due west to Hornpoint, the eastermost part of Iceland; and that from Snowfieldnes, the point of Iceland nearest to Greenland, it is a voyage of four days' sailing, also due west, to Greenland: and a rock called Gunbiornskerry is stated to be half-way between Iceland and Greenland; but this course, says one of the ancient accounts of unknown date, but certainly of the fourteenth century, "was the old way of sailing; but now the ice from the northern gulf has set down so near to this skerry, that nobody can take this course without danger of life." This rock, skerry, or isle, midway between the coast of Iceland and that of Greenland, is proved by Scoresby and other navigators to have no existence; and the east coast of

Greenland, as far as it has been possible to explore it, is found to be more inclement, icebound, and in every way less adapted naturally to afford subsistence to man, than the west coast within Cape Farewell; although the Eystribygd is represented in all the sagas as the most populous settlement, having 190, and the other only 90 gards. It is now generally admitted that the east coast of Greenland never was inhabited at all by the old colonists; that their east and west settlements had no reference to being east or west of Cape Farewell, but to being easterly or westerly from some place within Davis's Straits, and which formed their division between the two settlements; and in this view the east settlement would be the country nearest to Cape Farewell, and as at present, the best provided with the natural means of subsistence; and the western and poorer settlement would be the country beyond it to the north; and that Gunbiornskerry was not in the midway, or halfway, as it had been interpreted, in the sea between Iceland and Greenland, but some island on the east coast of Greenland, which was half-way, in point of distance and time, between Iceland and the eastern settlement in Greenland. From it they took a new departure, and coasted along, with sails and oars, round Cape Hvarf, or Cape Farewell, and up Davis's Straits to the eastern settlement. Hans Egede, his son Paul, and others, had from time to time examined and sent home accounts of remains of ancient buildings which they had found on their missionary excursions. Arctander, as early as 1777, had made reports of such remains. The Society of Northern Antiquaries at Copenhagen took up the subject with great zeal in this century; and researches have been made of which the result is a kind of synthetic proof, as it may be called, of the veracity of the saga. The remains of former inhabitation of the country, of houses, paths, walls, stepping-stones, churches, foundations of rows of dwellings, show that the saga accounts have not been exaggerated; and it must give every fair unprejudiced reader a confidence which he had not before in the sagas when he finds in this—the most questionable perhaps of all the saga statements—that a considerable Icelandic colony actually had existed in Greenland from the tenth century. The facts they state are fully supported by the discoveries made on the spot within this century. A similar moral confidence in the sagas is given to the few saga readers who happen to be acquainted with the Orkney Islands, from finding, in the Orkneyinga Saga, a minute and accurate knowledge of places, distances, names, and other details of the localities mentioned. In this case of Greenland the remains discovered carry conviction to all. At Karkortok, a branch of a long fiord called Igalikko, in latitude 60° 50' north, and longitude 44° 37', near to the settlement of Julianahope, is a ruin of a building 51 feet in length by 25 feet in breadth, with well-built stone walls, 4 feet thick, standing to the height of 16 and 18 feet; and with two round arched windows, one in each gable, and four other windows not arched, on each side, and with two door-ways,—evidently intended for a church. This appears the most perfect of the ruins yet discovered. Foundations, with walls in some parts 4 feet high, have been found of buildings 120 feet in length by 100 feet in breadth; and from such rows being found in various places, the families may be supposed to have lived in contiguous houses. But single dwellings also have been used, as foundations overgrown with dwarf-willow, and the berry-bearing shrubs, are found in favourable situations on the sides of the fiords. In what appears to have been a church, the foundations being 96 feet long by 48 feet broad, at the extremity of the fiord Igalikko, latitude 60° 55', a stone with a Runic inscription was found in 1830; and to the readers of Runic the inscription offered no difficulty:—" Vigdis, M. D. Hvilir Her. Glæde Gud Sal Hennar;" that is, "Vigdis rests here: God bless her soul." The meaning of the letters M. D. following the name, and which probably refer to the person's family, as Magnus's Daughter, or some similar distinctive use, form the only obscurity. In 1831 the missionary De Fries found near Igigeitum, in latitude 60°, a tombstone used as a door lintel to a Greenland house, with an inscription in Roman characters—" Her Hvilir Hro Kolgrims;" which is, " Here rests Hroar or Hroaldr Kolgrimson." But the most interesting of these inscriptions is one discovered in 1824, in the island Kingiktorsoak in Baffin's Bay, in latitude 72° 55' north, longitude 56° 5' west of Greenwich; as it shows how bold these

Northmen have been in their seamanship, and how far they had penetrated into regions supposed to have been unvisited by man before the voyages of our modern navigators. It now appears that Captain Parry and Captain Lyon had only sailed over seas which had been explored by these Northmen in the twelfth century. The inscription found in this high latitude was sent to three of the greatest antiquaries and Runic scholars in Europe—Finn Magnusen, Professor Rask, and Dr. Bryniulfson in Iceland; and, without communication with each other, they arrived at the same interpretation, viz. "Erling Sighvatson and Biarne Thordarson and Eindrid Oddson, on Saturday before Ascension Week, raised these marks and cleared ground, 1135." The meaning is, that in token of having taken possession of the land, they had raised marks or mounds of which Kragh and Stephenson observed some vestiges on the spot where the inscription was found, and had cleared a space of ground around, being a symbol of appropriation of the land. The interesting part of this inscription has not been sufficiently noticed and examined. In the Romish church the days of the Ascension Week are of peculiar solemnity. The priests, accompanied by the people, walk in long processions with lighted torches around the churches and consecrated ground, chanting, and sprinkling holy water. From the numerous processions going on at this festival, the Ascension Week was called the Gang Dayis, or Ganging Dayis, in old Scotch, —is still called the Gang Week in some parts of

England,—was called Gang Dagas in Anglo-Saxon, —and Ascension Day, Gagn Dagr in the Icelandic; and the going in procession, not the Gagn or Gain of Spiritual Victory, has given the name to the Dies Victoriæ in the northern languages. It appears that there are two festivals which might be called Gagn Dagr in the Romish church, from their being celebrated by processions: one is the Dies Victoriæ Maximus, about the 24th of April; the other procession day is about the 14th of May; and the Laukardakin fyrir Gakndag of the inscription may be the Saturday before either of these procession days. But, to whichsoever it refers, the people who made these marks at that time of the year must have wintered upon the island. By the accounts of all northern voyagers, the sea in Baffin's Bay is not navigable at or near Ascension Week, or any church festival to which Gagn Dagr applies. We must either suppose that these Northmen, without any of our modern outfit of ships for wintering in such high latitudes, did not only winter there, but found the country so endurable as to take possession of it by a formal act indicating an intention to settle in the island; or we must suppose that the cold, within so recent an historical period as 800 years ago, has increased so much in the northern parts of the globe, that countries are now uninhabitable by man which were formerly not so. Both, perhaps, may be taken into account. The capability of enduring cold or heat in extreme degrees may be acquired by individuals or tribes, and the habits and functions of the body become adapted to the temperature. The advance of ice locally in Davis's Straits, and on the east coast of Greenland, seems also ascertained by the yearly increase of the fields of ice in the neighbouring seas within the experience of our whale fishers.

The discovery of America, or Vinland, in the eleventh century, by the same race of enduring enterprising seamen, is not less satisfactorily established by documentary evidence than the discovery and colonisation of Greenland; but it rests entirely upon documentary evidence, which cannot, as in the case of Greenland, be substantiated by anything to be discovered in America. One or two adventurers made voyages, came to new countries to the south and west of Greenland, landed, repeated their visits, and even remained for one or two years trading for skins with the natives, and felling timber to take home in their ships; but they established no colony, left none behind them to multiply, and, as in Greenland, to construct, in stone, memorials of their existence on the coast of America. All that can be proved, or that is required to be proved, for establishing the priority of the discovery of America by the Northmen, is that the saga or traditional account of these voyages in the eleventh century was committed to writing at a known date, viz., between 1387 and 1395, in a manuscript of unquestionable authenticity, of which these particular sagas or accounts relative to Vinland form but a small portion; and that this known date was eighty years before Columbus visited

Iceland to obtain nautical information, viz., in 1477, when he must have heard of this written account of Vinland; and it was not till 1492 that he discovered America. This simple fact, established on documents altogether incontrovertible, is sufficient to prove all that is wanted to be proved, or can be proved, and is much more clearly and ably stated by Thormod Torfæus, the great antiquary of the last century, than it has been since, in his very rare little tract, "Historia Vinlandiæ Antiquæ, 1707." This, however, has not been thought sufficient by modern antiquaries, and great research and talent have been expended in overlaying this simple documentary fact, on which alone the claim of the Icelanders to the priority of discovery rests, with a mass of documents of secondary importance and no validity. These are of secondary importance; because the circumstances which led to or happened upon these voyages, the family descent, or even identity of the adventurers, and the truth or falsehood of the details related, do not either confirm or shake the simple fact on which everything rests,—that a discovery of a new land to the west and south was made and recorded, taken out of the mere traditionary state, and fixed in writing in 1387, or 100 years before Columbus's first voyage. They are of no validity; because, after Columbus's first voyage in 1492, the seafaring people in every country would be talking of and listening to accounts of discoveries, new or old,—imagination would be let loose,—and old sagas would be filled up and new invented; so that no document relative to this question is of real validity which is not proved at setting out to be older than 1492,—that is to say, not merely an older story which may have circulated in the traditionary state from the tenth or eleventh century, but older than 1492, on paper or parchment. Saga antiquaries are sometimes given to confounding together in their speculations these two very distinct ages of their documents. The only document of this kind is the one pointed out by Torfæus in 1707, which is in itself good and sufficient, and beyond all suspicion; and to link it to documents of uncertain or suspicious date, or to details which may or may not be true, and which require the aid of imagination, prejudice, or good will to believe, as well as of sober judgment, is weakening, not strengthening, the argument. Torfæus kindled a light which the moths have gathered about, and almost put out.

In 1697 Peringskiold published the "Heimskringla" of Snorre Sturlason, with a Latin and Swedish translation of the Icelandic. It was discovered and pointed out by Torfæus, that Peringskiold must have had some inferior manuscript of the work before him, because eight chapters of the Saga of King Olaf Trygveson, viz., from chapter 105 to chapter 113, are interpolated, and are not to be found in any genuine manuscript of Snorre's work. These eight chapters contain the accounts of the voyages of Leif and of Thorfin Karlsefne to Vinland. There is internal evidence in Snorre's work itself that these eight chapters are a clumsy interpolation by Peringskiold, or his authority; for they interrupt Snorre's narrative in the most interesting period of King Olaf Trygveson's life, and have no connection with the transactions or personages preceding or following; whereas all Snorre's episodes are, with surprising art and judgment, connected with what goes before or is to follow, and are brought in exactly at the right place. It may be thought, at first sight, that the very circumstance of a man of Snorre's knowledge and judgment in the sagas not knowing, or knowing not adopting, the account of the discovery of Vinland given in these eight chapters subsequently interpolated in his work, is conclusive against their value and authenticity. But it is to be remembered that although he probably knew of them, the subject was altogether foreign to his work. Vinland was an object of no interest in his days, and had not, like Greenland, Iceland, the Farey Isles, or Orkney, been occupied as a colony, or part of the dominions of Norway, and had not employed any of the historical personages of whom he treats; and therefore it would have been inconsistent with his work to introduce the obscure, and in his time unimportant fact, of the discovery of new land, or the adventures of the discoverers. The eight chapters in question, by whomsoever they were interpolated into Snorre Sturlason's work, proved to be taken, with few variations, and none of any importance, from the eighth chapter of the Saga of King Olaf Trygveson in the "Codex Flateyensis." This saga gives more details of the reign of that king than

Snorre Sturlason's saga of it, and is no doubt the source from which he drew his account, using it often verbatim.

The "Flateyar Annall, or Codex Flateyensis," by far the most important of Icelandic manuscripts, takes its name from the island Flatey, in Breidafiord in Iceland, where it had been long preserved, and where Bishop Brynjolf Sveinson of Skalholt purchased it, about 1650, from the owner, Jonas Torfason, for King Frederic III., giving in exchange for it the perpetual exemption from land-tax of a small estate of the owner. The manuscript is in large folio, beautifully written on parchment. On the first page stands—"This book is owned by Ion Hakonson. Here are, first, songs; then how Norway was inhabited or settled; then of Eirik Vidforle (the far-travelled); thereafter Olaf Trygveson, and all his deeds *; then next the Saga of King Olaf the Saint, with all his deeds, and therewith the sagas of the Orkney Earls; then the saga of Sverre, and thereafter the saga of Hakon the Old, with the sagas of King Magnus his son; then are deeds of Einar Sokkeson of Greenland, thereafter of Helge and Ulf the Bad; then begin annals from the time the world was made, showing all to this present time that is come. The priest Ion Thordarson has written from Eirik Vidforle, and the two sagas of the Olafs; and priest Magnus Thorhalson has written

---

* Thattr does not exactly mean deeds, but excerpts or short accounts of deeds. We use in Scotland the expression, "a tait o' woo,"—a little wool pulled out of a fleece; which corresponds to the Icelandic thattr, an excerpt.—L.

from thence, and also what is written before, and has illuminated the whole. God Almighty and the Holy Virgin Mary bless those who wrote, and him who dictated." The writer of this paragraph says, that the annals written out by the priest Magnus Thorhalson from the beginning of the world come down to the present time, and he has consequently been a contemporary of the scribe Magnus Thorhalson. These annals end with the year 1395, and the time at which the writing was concluded is thus distinctly ascertained. The time at which the writing was commenced is also distinctly ascertained; for in the piece on "how Norway was inhabited," in giving the series of kings, it is said, on coming to King Olaf Hakonson, "He was king when this book was writing; and then were elapsed from the birth of our Lord Jesus Christ 1300 and 80 and 7 years." The dates of the beginning and ending of this beautiful piece of penmanship are thus fixed, and the handwriting of each of the scribes perfectly known. The "Codex Flateyensis" is not an original work of one author, but a collection of sagas transcribed from older manuscripts, and arranged in so far chronologically that the accounts are placed under the reign in which the events they tell of happened, although not connected with it or with each other. Under the saga of Olaf Trygveson are comprehended the sagas of the Farey Islands; of the Vikings of Jomsburg; of Eirik Red, and Leif his son, the discoverers of Greenland and Vinland; and the voyages of Thorfin Karlsefne to Vinland, and all the circumstances, true or false, of their adventures. It is evident that the main fact is that of a discovery of a western land being recorded in writing between 1387 and 1395; and whether the minor circumstances, such as the personal adventures of the discoverers, or the exact localities in America which they visited, be or be not known, cannot affect this fact,—nor the very strong side fact, that eighty years after this fact was recorded in writing, in no obscure manuscript, but in one of the most beautiful works of penmanship in Europe, Columbus came to Iceland* from Bristol, in 1477, on purpose to gain nautical information, and must have heard of the written accounts of discoveries recorded in it. It is as great an error to prove too much as to prove too little. Enough is proved for the purpose of establishing the priority of discovery; but when the northern antiquaries proceed to prove the details,— to establish the exact points in the state of Massa-

* The English trade with Iceland appears to have been very considerable. Annals in manuscript of 1411 and 1413, quoted by Finn Magnusen, in his Treatise on the English Trade to Iceland in the "Nordisk Tidsskrift for Oldkyndighed," mention, besides plundering and piracy committed by the English, proclamations of Eirik of Pomerania against trading with them. In 1413 there were thirty English vessels on the Iceland coast. In 1415, in the harbour of Hafnarfiord there were six English vessels at one time. The trade had been made a monopoly, and the English appear to have forcibly broken through its regulations, in spite of the proclamations of their own and the Norwegian sovereign. The Icelandic bishops at that time—viz., Jon Jonson, bishop of Holar, and his successor in 1429, Jon Williamson—were Englishmen; and also the bishop of Skalholt in 1430, John Garrikson, appears to have come from England. Bristol and Hull appear, in 1474, to have had a great share of the trade to Iceland. It appears, from the Memoir of Columbus by his son Fernando, that in February, 1477, his father visited Tyle (Thule) or Friesland, "an island as large as England with which the English, especially those of Bristol, drive a great trade."

chusetts at which Leif put up his wooden booths, and where Thorfin Karlsefne and his wife Gudrid lived, and Freydis committed her wholesale slaughter, and to make imaginary discoveries of Runic inscriptions and buildings erected by Northmen in Rhode Island,—they are poets, not antiquaries. The subject is of so much interest both in Europe and America, and so much has been written in very expensive books to prove what is not susceptible of proof, and of no importance, if proved, that a few pages must be bestowed on it.

From the adventurous spirit of the Northmen in the eleventh century,—from their habits of living on board ship, on their ordinary viking cruises, for many more weeks and months together than are required for a voyage from Iceland to America,—from their being at home on board, and accustomed on their sea expeditions up the Mediterranean, to the White Sea, and to Iceland direct across the ocean,

It is a curious coincidence that he mentions he came to the island without meeting any ice, and the sea was not frozen; and in an authentic document of March in the same year, 1477, it is mentioned as a kind of testimony of the act of which the document is the protocol, that there was no snow whatever upon the ground at the date it was executed,—a rare circumstance, by which it would be held in remembrance. In the year 1477, Magnus Eyjolfson was bishop of Skalholt: he had been abbot of the monastery at Helgafel, where the old accounts concerning Vinland and Greenland were, it is supposed, originally written and preserved, and the discoverers were people originally from that neighbourhood. Columbus came in spring to the south end of Iceland, where Whalefiord was the usual harbour; and it is known that Bishop Magnus, exactly in the spring of that year, was on a visitation in that part of his see, and it is to be presumed Columbus must have met and conversed with him. These are curious coincidences of small circumstances, which have their weight.—See *Captain Zahrtmann on the Voyage of Zeno*, and *F. Magnusen on the English Trade to Iceland*, 2nd vol. of *Nordisk Tidsskrift*, 1833.—L.

to a sea life,—it is not improbable that they should have undertaken a voyage of discovery to the west and south, and have renewed it when they found a land which produced building timber and skins to repay them. It was certainly not seamanship that was wanting among them in those ages, but science only. The class of vessels in which they sailed made them in a great measure independent of the science of navigation; because their vessels were of an easy draught of water, and they had a command with their oars and their numerous crews over their vessels, which made a lee shore, or other unfavourable positions, of no such importance as to modern ships. In size, and as sea-boats, their vessels in general were probably equal or superior to those in which Columbus made his first voyage. One of Columbus's vessels is understood to have been only a half-decked craft. Sebastian Cabot, and some of the earliest explorers of Baffin's Bay, sailed in vessels under thirty tons. The Anna Pink, which accompanied Lord Anson half round the world, was a vessel of eighteen tons. In their shipping, seamanship, and habits of sea life and endurance, there was certainly nothing to make it, *a priori*, improbable that they should undertake a voyage of discovery to the south and west of Greenland. The details of adventures on such a voyage may not be correct, and yet the fact itself true. The following is an abridgment, as short as possible, of the details, and the conclusions drawn from them as to the localities in America which they visited. The eight chapters themselves are annexed to Olaf Trygveson's Saga.

Eirik Red, in spring, 986, emigrated from Iceland to Greenland with Heriulf Bardson. He fixed his abode at Brattahlid, in Eiriksfiord; and Heriulf at Heriulfsnes. Biarne, the son of the latter, was absent in Norway at the time, and finding on his return that his father was gone, resolved to follow him, and put to sea. As winter was approaching, they had bad weather, northerly winds and fogs, and did not know where they were. When it cleared up they saw a land without mountains, but with many small hills, and covered with wood. This not answering the description of Greenland, they turned about and left it on the larboard hand; and sailing two days they came to another land, flat, and covered with wood. Then they stood out to sea with a south-west wind, and saw a third land, high, and the mountains covered with glaciers; and coasting along it they saw it was an island. Biarne did not land, but stood out to sea with the same south-west wind, and sailing with fresh gales reached, in four days more, Heriulfsnes in Greenland, his father's abode.

Some years after this, supposed to be about 994, Biarne was in Norway on a visit to Earl Eirik, and was much blamed, when he told of his discovery, for not having examined the countries more accurately. Leif, a son of Eirik Red, bought his ship, when Biarne returned to Greenland, and with a crew of thirty-five men set out, about the year 1000, to look for these lands. He came first to the land which

Biarne had seen last, landed, found no grass, but vast icy mountains in the interior, and between them and the shore a plain of flat slaty stones (hella), and called the country Helluland. They put to sea, and came to another country, which was level, covered with woods, with many cliffs of white sand, and a low coast, and called the country Markland (outfield or woodland). They again stood out to sea with a north-east wind, and after two days' sailing made land, and came to an island eastward of the mainland, and entered into a channel between the island and a point projecting north-east from the mainland. They sailed eastward, saw much ground laid dry at ebb tide, and at last went on shore at a place where a river which came from a lake fell into the sea. They brought their vessel through the river into the lake, and anchored. Here they put up some log huts; but, after resolving to winter there, they constructed larger booths or houses. After lodging themselves, Leif divided his people into two companies, to be employed by turns in exploring the country and working about the houses. One of the exploring party, a German by birth, called Tyrker, was one day missing. They went out to look for him, and soon met him, talking German, rolling his eyes, and beside himself. He at last told them in Norse, as they did not understand German, that he had been up the country, and had discovered vines and grapes; adding, "that he should know what vines and grapes were, as he was born in a country in which they were in plenty." They now occupied themselves in hewing timber for loading the vessel, and collecting grapes with which they filled the ship's boat. Leif called the country Vinland. They sailed in spring, and returned to Greenland.

Leif's brother, Thorvald, set out, in the year 1002, to Vinland in Leif's vessel, and came to his booths or houses, and wintered there. In spring Thorvald sent a party in the boat to explore the coast to the south. They found the country beautiful, well wooded, with but little space between the woods and the sea, and long stretches of white sand, and also many islands and shoals; and on one island found a corn barn, but no other traces of people. They returned in autumn to Leif's booths. Next summer Thorvald sailed with the large vessel, first eastward, then northward, past a headland opposite to another headland, and forming a bay. They called the first headland Kialarnes (Keel Ness). They then sailed into the nearest fiord, to a headland covered with wood. Thorvald went on shore, and was so pleased that he said "he should like to stay there." On going on board they observed three hillocks on the sandy shore. They went up to them, and found they were three canoes, with three Skrælings under each. They killed eight of them, and one made his escape in his canoe. A great number afterwards came in skin-canoes and attacked them. They were repulsed; but Thorvald was wounded by an arrow and died, and according to his directions was buried at the promontory where he had expressed his wish to stay, or take up his abode, with a cross at the head and one at the foot of his grave; and the place was called Crossness. His companions returned to Leif's booths, wintered there, and in spring sailed to Greenland.

Thorstein, Eirik's third son, set out in the same ship, with his wife Gudrid, and a crew of twenty-five men, to bring home his brother's body; but after driving about all summer they returned, without making the land, to Lysefiord in Greenland, where Thorstein died, and his wife Gudrid returned to Eiriksfiord.

Next summer, viz., 1006, two ships from Iceland came to Greenland. One was commanded by Thorfin, called Karlsefne (of manly endowment); the other by Biarne Grimolfson. A third ship was commanded by Thorvard. Thorfin Karlsefne had married in the course of the winter Gudrid, the widow of Thorvald, and by her advice resolved on going to Vinland in spring. Thorvard had married Freydis, a natural daughter of Eirik; and the three ships set out with 160 men, and all kinds of live stock, to establish a colony in Vinland. They sailed first to the Vestribygd (within Davis's Straits), and to Biarney (Disco Isle). From thence they sailed in a southerly direction to Helluland, where they found many foxes. From thence, sailing two days to the south, they came to Markland, a wooded country stocked with animals. Then they sailed south-west for a long time, having the land to starboard, until they came to Kialarnes, where there were great deserts, and long beaches and sands. When they had passed these, the land was indented with inlets. They had two Scots with them, Hake and Hekia, whom Leif had formerly received from King Olaf Trygveson, and who were very swift of foot. They were put on shore to explore the country to the south-west, and in three days they returned with some grapes, and some ears of wheat, which grew wild in that country. They continued their course until they came to a fiord which penetrated far into the land. Off the mouth of it was an island with strong currents round it, and also up in the fiord. They found vast numbers of eider ducks on the island, so that they could scarcely walk without treading on their eggs. They called the island Straumey (Stream Isle), and the fiord Straumfiord. A party of eight men, commanded by Thorhal, left them here, and went north to seek for Vinland. Thorfin Karlsefne proceeded with Snorre, Biarne, and the rest, in all 151 men, southwards. Those who went northwards passed Kialarnes; but were driven by westerly gales off the land, and to the coast of Ireland, where, it was afterwards reported they were made slaves. Thorfin Karlsefne and his men arrived at the place where a river issuing from a lake falls into the sea. Opposite to the mouth of the river were large islands. They steered into the lake, and called the place Hop (the Hope). On the low grounds they found fields of wheat growing wild, and on the rising grounds vines. One morning a number of skin-canoes came to them. The people were sallow-coloured, ill-looking, with ugly heads of hair, large eyes, and broad cheeks; and after looking at the strangers they retired round the cape to the south-west. Thorfin Karlsefne put up dwelling-houses a little above the bay, and they wintered there: no snow fell, and their cattle lived in the open field. On the shortest day the sun was above the horizon in the watch before and after midday watch. A number of canoes came again from the south-west, holding up a white shield as a signal of peace, and bartered gray furs for bits of cloth, and for milk soup. The bull belonging to the party happened to bellow, and the Skrælings were terrified, and fled in their canoes. Gudrid, Thorfin Karlsefne's wife, lay in here of a son, who was called Snorre. In the beginning of the following winter, the Skrælings attacked them. They were defeated by the courage of Gudrid (who appears to have been far advanced in pregnancy at the time of this attack); but lost a man, and were so dispirited by the prospect of constant hostilities with the natives, that they resolved to return. They sailed east, and came to Straumfiord. Thorfin Karlsefne then took one of the ships to look for Thorhal, while the rest remained behind. They proceeded northwards round Kialarnes and afterwards to the north-west, the land being to larboard of them, and covered with thick forests. They considered the hills they saw at Hope, and these, as one continuous range. They spent the third winter at Straumfiord. Thorfin Karlsefne's son was now three years old. When they sailed from Vinland they had southerly winds, and came to Markland, where they met five Skrælings, and took two boys whom they taught Norse, and who told them their people had no houses, but lived in holes and caves: that they had kings; one called Avaldamon, and the other Valdidida. Biarne Grimolfson was driven into the Irish Ocean, and came into waters so infested with worms that their ship was in a sinking state. Some of the crew were saved in the boat, which had been smeared over with seal-oil, which is a preventive against worms in wood. Thorfin Karlsefne continued his voyage to Greenland, and arrived at Eiriksfiord.

During the same summer, 1011, a ship from Norway came to Greenland. The vessel belonged to two brothers, Helge and Finboge, who wintered in Greenland. Freydis (the natural daughter of Eirik Red, who had married Thorvard) proposed to them to join in an expedition to Vinland, each party to have thirty men, and to divide the gain equally. They agreed, and set out, and reached Leif's booths, where they spent the winter; but Freydis, who had taken five men more with her than the agreement allowed, quarrelled with the brothers, and murdered them and the whole of their people, and returned in spring (1013) to Greenland.

Thorfin Karlsefne went to Norway with his Vinland cargo next summer, and it was considered very valuable. He sold even a piece of wood used for a door-bar, or a broomstick, to a Bremen merchant for half a mark of gold; for it was of mosur-wood * of Vinland. He returned, and purchased land in Iceland; and many people of distinction are descended

* Supposed to have been bird's-eye maple. Early Engl. *maser*.

from him and his son Snorre, who was born in Vinland. After his death his widow, Gudrid, went to Rome, and on her return lived in religious seclusion in Iceland.

The above is an abridgment of the eight chapters on which the whole accounts of Vinland rest, and which are given at length in the Appendix; and so much fanciful speculation has been reared upon this foundation, that it deserves examination. The main facts—the discovery of various lands to the south and west of Greenland, the repeated voyages to them, and the reasonable motives of such voyages—bear all the internal evidences of simple truth. We may generally believe in the truth of the accounts of men's actions, when we see reasonable and sufficient motives for them so to act. Iceland, although it had wood in those days, and has some still, produced only a scrubby small brushwood of birch or hazel, not fit for ship-building, nor for the large halls which it was the fashion of the age for great people to have for entertaining and lodging their followers in; and the state of society made it necessary for safety to keep large bodies of retainers always at hand, and about them. It is told as a remarkable thing in the Landnama Book, or History of the Original Settlers in Iceland (page 29), that Avang found such large wood where he settled, that he built a long-ship; and in the Kristni Saga it is mentioned that Hialte Skeggjeson built a ship at home, so large that he sailed in it to Norway. In general, however, they had to buy their sea-going vessels in Norway. The drift-wood found about the shores of Iceland in great abundance to a late period, and perhaps even now, would be too much shaken and worm-eaten to be fit for ship-building, even if it were of a sufficient size. To go in quest of the wooded countries to the southwest, from whence drift-wood came to their shores, was a reasonable, intelligible motive for making a voyage in search of the lands from whence it came, and where this valuable material could be got for nothing. So far we see reasonable motives followed by reasonable and perfectly credible acts and results. In the account, however, of the details upon which so much has been built up by modern antiquaries, we find no such consistency, credibility, or internal evidence of truthfulness. Leif and his successors, Thorfin Karlsefne and others, arrive in Vinland in spring—say in May, June, or July. In what climate, or part of the world, are grapes to be found in those months? They can hardly tread on Straum Island —settled by our modern antiquaries to be Egg Island, at the mouth of Plymouth Sound in Massachusetts— for the eggs of eider ducks. It was consequently early in spring, before birds were hatched, and before grapes have the shape of fruit in any climate, that they found ripe grapes and ears of wheat! Do vines, or wheat, or corn of any kind, grow spontaneously in those countries? This is a question by no means satisfactorily ascertained. Tyrker the German, who knew so well grapes and vines, "because he was born in a country in which these are not scarce," comes back to his party after a short absence, rolling his eyes, making faces, talking German, and half drunk. All the grapes in Germany, and Vinland to boot, would not make a man drunk, without their juice undergoing the vinous fermentation. This is clearly the fiction of some saga-maker, who knew no more of wine than that it was the juice of the grape; and all the geographical speculations upon the sites and localities of the Vinland of the Northmen, built upon the natural products of the land, fall to the ground. The eider duck, on our side of the world, is very rarely seen in lower latitudes than 60°. It may be different on the American coast; but the Skrælings, the sallow-complexioned people with skin-canoes with whom they bartered cloth and milk for sable and squirrel or grey skins, are, together with their articles of traffic, of northern origin. The red race of Indians could never have been called Skrælings, and described as such,—viz., with broad cheeks, and sallow complexions,—by Northmen who knew the Skrælings, or Esquimaux race, in Greenland. But we are told the Esquimaux race extended once much farther south, beyond Newfoundland and the Gulf of St. Lawrence, and as far south as we please to have them. It is as easy to tell us that once the juice of the grape would intoxicate without the vinous fermentation,—that wheat would grow without being sown,—and that a barn, or more properly a kiln-barn, might be found in a land without dwelling-houses. All the geographical knowledge that can be drawn from the accounts of the natural products of Vinland in these eight chapters, points clearly to the Labrador coast, or Newfoundland, or some places north of the Gulf of St. Lawrence. The terror of the Skrælings at the bellowing of Thorfin Karlsefne's bull points rather to an island people, as the natives of Prince Edward's Island, or of Newfoundland; for a continental people in that part of America could not be strangers to the much more formidable bison, or musk ox, or buffalo. The piece of mosur-wood from Vinland, which Thorfin Karlsefne sold to a Bremen merchant for half a mark of gold, must have derived its value either from its intrinsic worth or beauty as wood or dye-wood, or as a stick coming from a distant unknown land. In the latter case the kind or quality of the wood, and whether it grew south or north, were circumstances of no consequence to the buyer: it was a curiosity from an unknown land. In the former case, Thorfin Karlsefne must be supposed to have gone to Honduras to cut his broomstick. The maple, or whatever wood for furniture grows more to the north in America, is not more beautiful than birch wood or other European wood. If it had been logwood, fustic, mahogany, that was meant by mosur-wood, it would be a proof that the saga-writer was drawing upon his own imagination in the details of his account of Thorfin Karlsefne; for vines and wheat growing spontaneously, mahogany trees or dye-woods, and Esquimaux in skin-canoes trading with sable skins and grey skins, and furs described to be white or all grey,— "grávara ok safali ok alskonar skinnavara," and "algra skinn"—never met in one locality: for the former are products of a very southern latitude, and the people and animals described belong to a northern climate. The account of the time from land to land in the voyages of Biarne, Leif, or Thorfin Karlsefne, leads to no satisfactory result as to the land they came to; because we neither know their rate of sailing in a day, nor whether by a day's sailing they meant sailing day and night, or that they took down and stowed their great square-sail at night, and lay-to with a little try-sail aft till daylight, as similarly rigged vessels on the fishing banks do at the present day. The lying-to all night, as they were in an unknown sea, was the better seamanship, and we may suppose it was their way of sailing. In their ordinary voyages they appear always to have put up their tent-cloths at night, brought their vessel to the land or to an anchor, and to have gone to rest, leaving only a watch on deck. It is usually mentioned in the saga when they sail night and day, as a special circumstance. It does not appear probable they would run with all sail in the night through an unknown sea; and if they took down sail at night, and lay-to in the Gulf stream, all conjecture founded on a number of days' sailing from Helluland to Markland, or from Markland to Vinland is quite arbitrary, and without guide. The description of the land is equally unsatisfactory as a means of discovering the localities in Vinland they visited, without more precise data. A country of stony soil, with little vegetation among the slaty fragments that cover it, applies to all the country from

Hudson's Bay to Newfoundland; and Helluland, so called from this circumstance, is a name that would suit any part of Labrador as well as Newfoundland. Markland, so called because low or level, and covered with thick forests, as a description may be applied to any part of America as well as to Nova Scotia. An island with a sound between it and the main, or a low shore with remarkably white sand cliffs and shallow water, a fiord or inlet of the sea, a river running out of a lake, a bay between two headlands, one of them of a conspicuous figure, are good landmarks for identifying a country of which the position is known, but are good for nothing as data for fixing that position itself; because these are features common to all sea-coasts, and, on a small or great scale, to be found within every hundred miles of a run along the sea-board of a country. It is evident from the personal adventures ascribed by the saga-maker to the personages, that the details are imaginary, and only the general outline true. The revival of Thorstein Eirikson's body, and its prophesying what was to befall Gudrid in her lifetime, are within the ordinary belief of those times, and therefore do not lessen the confidence in other circumstances related; nor the appearance to her alone of another Gudrid who spoke Norse to her in Vinland, and whom nobody else saw. But the adventures of Freydis, her murder of the two brothers, thirty men, and the women, is an improbable, not to say an impossible circumstance; as her thirty-five men had no motive for such a butchery of their comrades, in a country in which they needed all their strength for their safety, and for the objects of their voyage. All the details seem merely the filling up of imagination, to make a story of a main fact, the discovery of Vinland by certain personages, whose names, and the fact of their discovering unknown lands south-west of Greenland, are alone to be depended upon.

But two facts are stated by our modern antiquaries, which are held to be quite conclusive as to the locality in America discovered by the Icelanders. One is, that in the details of Leif's voyage and residence in Vinland, it is stated that on the shortest day the sun was above the horizon from half-past seven o'clock in the morning to half-past four o'clock in the afternoon, or nine hours, which gives the latitude of the place 41° 24′ 10″, and which brings it to between Seaconnet Point in 41° 26′, and Judith Point in 41° 23′, and which two points form the entrance into Mount Hope Bay; which corresponds, even to the name Hop or Hope, with the description of a river, now called the Taunton, running from a lake into the sea, and with all the other landmarks or accounts of the appearance of the coast given in the saga. The other fact, not less striking, is, that in this very neighbourhood,—viz., at Assonet Point, on the shore of the river Taunton, in latitude 41° 44′, near the town of Berkley in the district of Massachusetts,— a stone covered with Runic inscriptions is still to be seen, and is known by the name of the Dighton Writing (written) Rock, and was an object of curiosity to the early English settlers as far back as 1680. These two happy coincidences are so happy—so like finding a box, and 800 years afterwards finding the key that of all the keys in the world can alone open it—that people almost doubt, at the first hearing of it, whether the news be not too good to be true. The first question that arises to the doubting reader is, how, in Leif Eirikson's time,—that is, about the year 1000, when Christianity was scarcely introduced, and Church festivals, Church time, and the knowledge and prayers of churchmen unknown,—did the Icelanders divide time? The whole circle of the horizon appears to have been divided by them into four quarters, each subdivided into two, making eight divisions, or *áttir* (from which our old word airths applied to the winds, seems derived); and these eight watches, each of three of our hours, made up the day, which we divide into 24 parts. It was not until 120 years after Leif's voyage, viz., in 1123, that Bishop Thorlak established in Iceland a code of Church regulations or laws, by which time was more minutely ascertained for Church prayers and observances. For all secular business, among a seafaring and labouring population, the division of time into eight watches was sufficiently minute for all their practical purposes. Now the saga says, "Sol hafdi thar Eyktarstad ok Dagmalastad um skammdegi;" which clearly means that, on the shortest day, they had the sun in the watches called the Dagmalastad and the Eyktarstad; that the sun rose in the former, and set in the latter, and not as in Iceland, where the rising and setting were, on the shortest day, included in one watch.

The Dagmalastad was the watch immediately before the mid-day watch (Middegi), and the Eyktarstad that immediately after. Now if we reckon from noon, the middle of the mid-day watch, it would begin at half-past ten o'clock of our time, and end at half-past one o'clock; Dagmalastad would begin at half-past seven, and end at half-past ten; and Eyktarstad begin at half-past one, and end at half-past four in the afternoon. Now if the sun rose any time within the Dagmalastad, and set any time within the Eyktarstad watch,—that is to say, any time between half-past seven and half-past ten for its rising, and any time between half-past one and half-past four of our time for its setting, — it would answer all the conditions of the text of the saga, which merely says they had the sun in these watches, not during the whole of these watches; and the precision of ideas and expression which characterises the Icelanders would undoubtedly have expressed, if that had been the meaning, that the sun rose at the beginning of Dagmalastad, and set at the end of Eyktarstad. Torfæus, certainly not inferior in judgment and knowledge to any antiquary of our times, and who, as a contemporary and friend, had on every doubtful point the opinion of Arne Magnæus, the first Icelandic antiquary who has ever appeared, makes out, from the same text, that the sun may be considered to have been above the horizon from the middle of Dagmalastad to the middle of Eyktarstad,—that is, for about six hours, —which would correspond to a latitude of 49° instead of 41°; and he, and Arne Magnæus we may presume with him, bring Vinland to some place in Newfoundland, or in the St. Lawrence, which certainly would agree better with the description of the people and products, excepting the ready-made wine, the spontaneous wheat, and the fine wood, than Taunton river in Massachusetts. With regard to the Dighton Writing Rock, upon which so much has been built in vast and expensive publications, such as the "Antiquitates Americanæ" (Hafniæ, 1837), and other works, the following observations may lead to a true estimate of its historical value. The rock or stone is a boulder or transported mass, not a stone belonging to the ground-rock of the country. It is about $11\frac{1}{2}$ feet long by $5\frac{1}{2}$ feet high, running up to an edge, and the surface, or side on which the Runic inscription is found, sloping at an angle of 60° from its base. It is one of that class of detached masses of primary rock scattered over the whole northern hemisphere of our globe—the evidences of some vast convulsion beyond human knowledge or conjecture. Whoever has examined this class of stones must have observed that it is almost a characteristic, distinguishing them from fixed ground rock of similar formation, that they are more interspersed with black or greenish veins or marks of a different substance from the component parts of the rock, and in short, with lines which often assume the appearance of sea-weed or other fossil plant, enclosed in the crystallised matrix of the stone, but which are in reality small veins, or rather lines of chlorite. The Runic inscription at Runamo, in Bleking in Sweden, which from the days of Saxo Grammaticus to the present times, was considered to be an inscription of real but unintelligible letters on the ground-rock, and which antiquaries but a few years ago supposed they had deciphered, and actually published their explanation of it, is now discovered, and admitted to be nothing but veins of one substance interspersed in another. Chemistry settled the historical value of this Runic inscription. The Dighton Writing Rock would perhaps be the better of a certificate from the mineralogist, as well as the antiquary. Supposing it beyond all doubt a stone with artificial characters, letters or

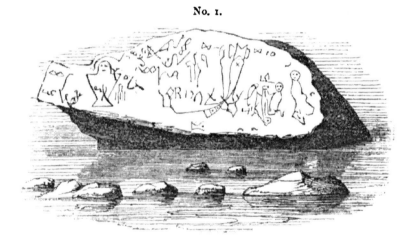

No. 1.

figures inscribed upon it, the first question that occurs to every inquirer must be, what is there to prove that these marks are the work of the Northmen, and not of the natives, or of the first European settlers about the year 1620? The stone, of which No. 1 is a delineation (No. 2 is a copy of the marks or inscription in 1790, and No. 3 in 1830), bears nothing to show by

KINGS OF NORWAY.

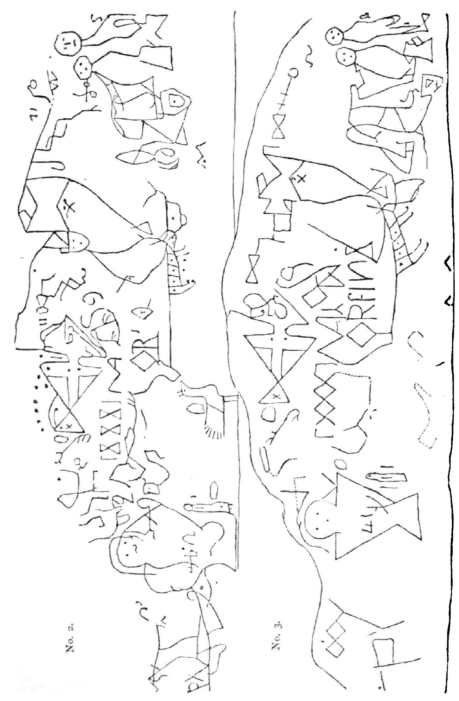

No. 2.—A copy of the inscription on the Dighton Stone, as given by Baylie and Goodwin in 1790.
No. 3.—A copy of the same, as given by the American antiquaries in 1830.

whom, or when, the marks in question were scratched upon it. The native tribes of America, the Hottentots, even the natives of Australia according to Captain Gray's narrative of his travels, have a propensity to delineate rude figures and marks upon the sides of caves and remarkable rocks, to indicate that they have been there, and even to show their tribe, numbers, and the direction they have taken. This stone is, by the description, quite tempting to indulge the propensity common to all men, savage or civilised, to leave some mark after them of their having existed; for it is said to be conspicuous from its position, flat surface, and different texture from the common rock of the country around.* It is evident, on referring to No. 2 and No. 3, that there is no sequence of letters, either Runic or Roman, upon the Dighton Writing Rock, but only detached unconnected marks, belong-

* "The Dighton stone is a fine-grained greywakke, and the rock of the neighbourhood a large conglomerate. It is situated 6½ miles south of Taunton, on the Taunton river, a few feet from the shore, and is covered with water at flood tide, on the west side of Assonet Point, in the town of Berkley, county of Bristol, and state of Massachusetts, and in a parish or district called Dighton. The marks are described as 'showing no method in the arrangement of them.' The lines are from half an inch to an inch in width, and in depth sometimes one-third of an inch, though in general very superficial. They were, inferring from the rounded elevations and intervening depressions, pecked in upon the rock and not chiselled or smoothly cut out."—(*Communication of the Rhode Island Historical Society to the Society of Northern Antiquaries at Copenhagen*, 1830.) Other rocks, similarly marked with rude hieroglyphics, or figures of animals, are found in various parts of the interior of America, far from the coast,—as on the Alleghany river, the Connecticut river, about Lake Erie, on Cumberland river, about Rockcastle Creek,—and similar, as sculptured work, to the Dighton stone. Are these too memorials of Thorfin Karlsefne, left in Vinland by his party of woodcutters? or are they the rude memorials of the wandering Indians, left, if they have a meaning, to show those of their own tribe who may follow that they have been on the spot some time before?—L.

ing to any people or period one may please to fancy. What is there to prove that these are not the scratches of some idle sailor boy, or of some master Dighton of the first settlers in 1620? Every Runic inscription given by Olaus Wormius, in his "Literatura Runica," is in regular columns of letters from right to left, or from top to bottom, or going round the stone; but still in regular rows, letter after letter. Here all the scratches are detached marks, such as a child would make on the smooth side of a stone, without meaning. The only semblance to letters is in the middle of the stone, in which antiquaries discover the name of Thorfin,—viz., Thorfin Karlsefne, the leader of the expedition. In the older copy (No. 2 of the inscription) we see a lozenge-shaped mark, a Roman letter R, a stroke, and a triangular mark. In the later copy of 1830 (No. 3), the lozenge has got a tail to it, and the Roman letters R F I N are distinct. The first copy was taken in 1790, by Dr. Baylie and Mr. Goodwin; the latter in 1830, by the Rhode Island Historical Society. Both copies coincide; except that the figure of a cock, and of some animal apparently, and some unintelligible marks delineated in the older, have in the course of forty years become obliterated, and are not given in the later copy. But by some strange process, although it is one not at all uncommon in stones that have attracted the antiquary's notice, the thing sought for —the letters of the word Thorfin—has in the course of the forty years gained wonderfully in distinctness, instead of becoming obliterated or less legible. Let any one look at the upper copy (No. 2), and make out, if he can, any thing approaching to the word Thorfin, except a lozenge and R, such as one may see on a box or package in a ship's cargo; but let him look at No. 3—the copy taken since the Icelandic origin of the inscription was broached,—and there to be sure he will see without spectacles a lozenge with a tail, and the Roman letters F I N Z, making Thorfinz. In the tables of the various forms of Runic letters given by Wormius, in his "Literatura Runica," there is no such lozenge-shaped letter to express Th or Tho; but as in many districts Runic letters appear to have had different shapes from those used in other parts, this circumstance is of little importance. The letter R may have been common to both alphabets, the Roman and Runic: the letters F I N are decidedly Roman; so that in this Runic inscription there is but one letter that may possibly be Runic, if it be a letter at all, and the rest are all Roman characters. In both copies, just over the lozenge letter, is a mark, also in Roman characters, which may be N A, or M A; the letter A being formed by the last branch of the M. Either will do; because, if it be N A, it may be part of the word Landnam; and if it be M A, it will surely be part of the word Madr: and Landnammadr signifies the first settler of a country,—the *origines gentis*,—and is so used to denote the original settlers in Iceland, of whom the Landnama Saga treats. Close to this N A, in both copies, are marks of three tens and a one, in Roman numerals, viz., XXXI.; and before the first is something like a Greek gamma, but which may possibly be intended for a Roman C. Now if this Roman C be intended for a hundred, it would not be for a Roman hundred or centum of five score, but a long hundred of six score, by which the Icelanders always counted; and CXXXI. would in reality mean 151, not 131. Now, Thorfin Karlsefne had lost nine of his original party, who had gone northwards under Thorhal; and this number 9 added to the 151 so clearly and satisfactorily made out on the stone, just makes up the 160 men, the original number of Landnammen of Vinland who embarked with Thorfin Karlsefne. It would be puerile to dwell on such puerilities. To believe that Thorfin Karlsefne, or any of his party, was acquainted not only with the Runic and Roman letters, but with the Roman numerals, yet without knowing the use of those numerals, and the number of units they express; and should leave a Runic inscription, as it is called, without a Runic letter in it, and so rude as to show—if the marks are letters at all, and not merely scratches, marks, or initials, made at various times by various hands — a complete ignorance of the collocation of letters in a row so as to form words, and a complete ignorance of the value of the Roman numerals he was using,—would require the antiquarian credulity of a Jonathan Oldbuck.

The northern antiquaries are misled in their speculations about Vinland by the singular case of the ancient Greenland colony. By the rarest coincidences of new and old colonisation, a kind of double evidence has come out to prove the veracity of the saga accounts of that old Icelandic colony. First is the documentary evidence of the saga, bearing no inconsistency or internal evidence of deviation from truth, and supported by collateral documentary evidence, from Adam of Bremen and other writers of the eleventh and twelfth centuries incidentally mentioning Greenland and its bishops, and which is evidence precisely similar in kind to the documentary evidence relative to Vinland. But a second mass and kind of evidence substantiating the first has come out in our times, by the discovery in Greenland of remains of buildings, churches, and of inscriptions and other material proofs, corroborating the documentary proofs of the existence and state of this ancient colony in Greenland given in the sagas. Our modern antiquaries want to substantiate the documentary evidence of the saga relative to Vinland by a similar kind of material evidence to be discovered in America, without considering that the cases are totally distinct and different. Greenland was a colony with communications, trade, civil and ecclesiastical establishments, and a considerable population, for 300 years at least before it was lost sight of. Vinland was only visited by flying parties of wood-cutters, remaining at the utmost two or or three winters, but never settling there permanently as colonists, nor as far as can be seen from the sagas, with any intention of settling. No division and occupation of the land, no agricultural preparations are mentioned.

Cattle they would have taken for milk, or food probably, at any rate, as salt to preserve meat must have been scarce in Greenland, where it could only be obtained by evaporating sea-water. Cattle taken with them, if the circumstance be true, are the only indication of any intention to settle; and a settlement or colony was not established. Three winters are the longest period any of these wood-cutting parties stayed in Vinland. To expect here, as in Greenland, material proofs to corroborate the documentary proofs, is weakening the latter by linking them to a sort of evidence which, from the very nature of the case,—the temporary visits of a ship's crew,—cannot exist in Vinland, and, as in the case of Greenland, come in to support them. It would be quite as judicious and consistent with sound principle of investigation to go to New Zealand, or the Sandwich Isles, to search for material proofs (old shoes, cocked hats, or pen-knives) of Captain Cook's having visited those places, and to link the documentary proofs of his discoveries to the authenticity of the material proofs — of the old shoes, cocked hats, and pen-knives—left by him on those shores. This is precisely the kind of investigation and reasoning, with regard to the discovery of Vinland by the Northmen, which antiquaries are pursuing; and to be sure it does lead them into laughable discoveries—quite as ridiculous as that of the Runic inscription on the Dighton Writing Stone, or as Oldbuck's Roman Prætorium on the Kaim of Kinprunes. Here is another specimen of the development of the imaginative faculty among antiquaries.

In the town of Newport, near to the south end of Rhode Island, stands the circular stone-work of an old windmill, of about 18 feet in diameter within walls, and raised upon eight pillars of about 7 feet high and 5 to 6 feet apart, arched over so as to admit carts to come under the floor of the mill, and the corn-sacks to be hoisted up or lowered down through a hatch in the wooden floor above. This is the ordinary plan in large well-arranged windmills, as it takes the horses and carts out of the way of the wings of the mill, and of the lever on the ground by which the moveable wooden superstructure or head of the mill was formerly turned to the wind. The pillars supported the beams of the floor; and windows and a fireplace, corresponding to the floor or platform of the mill, are in the wall, which is about 24 feet high, built of rough stone very substantially, and with lime-mortar, and has been harled or roughcast with lime. The situation is at the summit, or nearly so, of the principal eminence in the neighbourhood, open to the sea breezes, and with no out-walls or anything near it to intercept the wind. It is universally called by the inhabitants of the neighbourhood "the old stone mill." These are pretty good proofs that the building has been a mill; but there is also documentary proof of it. Rhode Island was first settled by the English in 1636, and two years afterwards (1638) Newport and the south end were occupied. In 1678, that is forty years afterwards, Benedict Arnold, who appears to have been governor of the settlement at one time, in his last will and testament calls this very building his stone mill. This is not all. One of the first settlers, a Mr. Peter Easton, had the laudable custom of marking in his pocket-book whatever notable event occurred in his township; and under the year 1663 he makes the memorandum, "This year we built the first windmill." Now we have here, first, the documentary evidence of Governor Arnold's will, calling it, in 1678, his stone mill, and bequeathing it as such; and of Mr. Peter Easton's pocket-book, giving posterity the information that "the first windmill in the township was built in 1663;" and as they could scarcely have required two mills at once if they had none before, we may fairly presume that the mill built in 1663 was that bequeathed in 1678. And, secondly, we have the material proof of the building, with its modern walls, built with mortar of lime and sand, and harled, and with a chimney-place, windows, and beam supports for a mill platform or floor, being altogether fit for and on the plan of a mill, and of nothing else, and its situation also being adapted for that purpose; and the only name given to it by the neighbouring inhabitants being "the old stone mill." Don Quixote himself could not have resisted such evidence of this having been a windmill. But those sly rogues of Americans dearly love a quiet hoax. With all gravity they address a solemn communication to the

Royal Society of Northern Antiquaries at Copenhagen, respecting those interesting remains of "a structure bearing an antique appearance"—"a building possibly of the ante-Columbian times"—"a relic, it may be, of the Northmen, the first discoverers of Vinland!" After describing the situation of the mill, they go on to say that this "dilapidated structure" has long attracted the attention of the numerous strangers who come in the fine season, from all parts of the Union, to enjoy the sea-bathing and pure air of Newport, and they often question the inhabitants concerning its origin; but the only answer they receive is, that it has always been known by the name of "the old stone mill." It has excited the most lively interest among the learned in those parts, and many conjectures have been hazarded about its origin and object; but these, say the wags, with great solemnity of phrase, "are shrouded with mystery;" and all that can be learned from the inhabitants is, that as long as people can remember it has been called "the old stone mill." But whether this structure could have been built for a mill, although no doubt it is so well adapted for a mill that it may have been used for such purpose at some period, is matter of grave doubt to many; because no similar building, of old or new date, for any purpose, exists in the neighbourhood, or in all the country. They send, along with their communication concerning this interesting structure of the original Scandinavian discoverers of Vinland in the eleventh century, drawings of the exterior and interior, a ground plan and an elevation of the old stone mill; all which they submit to the consideration of the Royal Society of Northern Antiquaries at Copenhagen. It must be allowed that these Rhode Island wags have played off their joke with admirable dexterity. They conceal nothing that fixes the building to have been beyond all doubt a mill; neither the name it has always gone by,—nor its windmill plan and site,—nor its modern walls built with lime and sand, and roughcast,—nor General Arnold's will calling it his stone mill,—nor Mr. Peter Easton's memorandum of the year in which it was built; but they cunningly keep all these circumstances in the background, and bring to the front " the dilapidated structure," — " the wonder of strangers from all parts of the United States,"—" the structure bearing an antique appearance,"—its origin and use " shrouded with mystery," — " but possibly ante-Columbian," — " a remain, possibly, of the Scandinavian discoverers of Vinland in the eleventh century." The bait took; and no doubt these comical fellows at Newport are chuckling in their club-room at seeing their "old stone mill" figuring in the Annals of the Northern Antiquarian Society, with arches and pillars like a Grecian temple. It is only when one comes, compass in hand, to a scale of feet and inches, that one finds this magnificent structure, with pillars and arches, and of which an exterior and interior view is given in the Annals of the Antiquarian Society of Copenhagen, is in reality the bottom of a mill of the very ordinary size of eighteen feet within walls, standing on pillars six or seven feet high and five or six feet apart, and arched over,—like to and on the scale of the pillars and arches of a cart-shed, or a horse-course of a thrashing-mill, instead of a structure, as the plates, of which no less than three are given, would lead you to believe, on the plan of the Coliseum, and of the size of the Temple of Vesta. This is very amusing; but it is not quite so amusing to have to pay heavy prices for magnificent books, got up in two or three languages, superb in size, paper, and type, decorated with fac-simile specimens of the writing of illuminated beautifully executed saga manuscripts, illustrated with splendid copper-plates, and published in the name and under the auspices of a great and learned antiquarian society; and to find you have been paying gold for such old-wifery as this Dighton writing rock and Newport old stone mill. It would, in fact, be difficult to point out any fact or observation of value relative to the discovery of Vinland which has not been brought to light and weighed by Torfæus, in 1707, in his little tract on Vinland. Torfæus was an antiquary of great judgment. He came first into the field, and seized upon the only fact with respect to the discovery of Vinland that there was to seize upon—the documentary proof, from a manuscript of fixed date, of a discovery of Vinland, recorded in writing a hundred years before Columbus's first voyage; and that record known to Columbus, or Columbus in a situation to know of it, a few years before he undertook that voyage. Torfæus left nothing behind to glean with respect to Vinland of any value in the question of the discovery of America by the Northmen.

The legend of Gudleif Gudlaugson, like this saga of Thorfin Karlsefne, gives a discovery not unlikely to have taken place, and much more to the south; but with adventures which border on the incredible. It is contained in the Eyrbyggia Saga. Towards the end of King Olaf the Saint's time, about 1030, Gudleif, on his voyage westward from Iceland to Ireland, was overtaken by a heavy storm from east and north-east, which drove him far out to sea to the south-west, so that none of them knew where they were. After driving about the greater part of the summer they came to land; but were seized by the natives, who came in crowds to the vessel, and spoke a language they did not understand, but it appeared to them like Irish. They observed that the natives were disputing whether to make slaves of them or put them to death. In the meantime an old grand-looking man, with white hair, came riding along, and all the natives received him with the greatest respect. He accosted the Icelanders in the Norse tongue, and asked them if Snorre the Gode (one of the most important personages in Iceland) was alive still, and his sister Thurid. He would not tell his name, and forbade his countrymen to come there again, as the people were fierce, and attacked strangers, and the country had no good harbours. He gave them a gold ring to deliver to Thurid, whom, he said, he liked better than her brother Snorre, and a sword for Thurid's son. Gudleif brought these things home, and people concluded that the man must have been Biorn Breidvikingakappe, a skald, who was much respected, and who had fallen in love with Thurid, on which account her husband and her brother had persecuted him, and he had left Iceland in a vessel about the year 998, and had never afterwards been heard of: "and this is the only truth known concerning Biorn." This saga is supposed to have been written or composed in the beginning of the thirteenth century; as it mentions one Gudny telling him, the saga-writer, of taking up and interring in a church the bones of some of Snorre the Gode's predecessors, and this Gudny is known to have died about 1220. The legend has a value independent of the truth or falsity of the details. These are at least improbable. The man could have no object in concealing his name, which the tokens he was sending to Iceland would at once reveal, and no intelligible motive for not returning with his countrymen. But it is valuable, because, whatever may be the truth of the filling up, or even of the main event of a vessel being driven to an unknown land, it shows an existing rumour or idea among seafaring men, long before Columbus's discoveries, that a north-east wind would bring a vessel sailing from Iceland to Ireland to a new land on the south-west, if she ran before it; and not into an uninhabitable region of fire, as the Romans appear to have conceived of the world. Some obscure knowledge of a western land must have been circulating as a foundation for this legend. The White Man's Land, the Great Ireland, a country in the west peopled by Christians originally of Ireland, has the same kind of value of showing that men, either from the reason in the supposition, which is the most likely, or from some actual chance discovery, had come to the conclusion that there was land in the west opposite the shores of Europe; and it also has the same kind of worthlessness as the other two legends—that the details are evidently fictitious and improbable.*

MEMOIR OF SNORRE STURLASON.

SNORRE STURLASON was born in the year 1178, at Hvam, in the present bailiwick of Dale, in the western province of Iceland. His father, Sturla Thordson, was a man of consequence, descended from the royal stock of Odin to which the Northern kings, and all the great families among the Northmen, traced their lineage; and he held by hereditary right the dignity of a Gode, which in the times of the Odin worship was hereditary in certain families descended from the twelve Diar, Drottar, or Godar, who accompanied Odin from Asgard. The office of Gode appears to have combined the functions of priest and judge

---

* A full account of the Norse voyages to Vinland, with a discussion of the relation of Columbus to them, will be found in my little work "America not discovered by Columbus," third edition, Chicago, 1883.

originally; and long after the sacerdotal function had ceased the judicial remained, and was exercised as an hereditary jurisdiction over the locality or godord, even long after the establishment of Christianity. Snorre was sent in his infancy to Jon Loptson, of Odde, to be fostered. It was the custom of the age for people of consequence to send their children to be fostered by others, sometimes of higher and sometimes of lower station; but always of a station, connection, or influence that would be of use afterwards to the foster-child. This fostering was not merely nursing the child until he was weaned, but implied bringing him up to the age of manhood; and the ties of foster-father, foster-son, and foster-brother, appear to have been as strong and influential as the natural ties of blood relationship. The custom has arisen in turbulent times, from the policy of not giving an opportunity to hereditary enemies to cut off an entire family at one swoop, leaving no heir and avenger, and of strengthening the family by collateral alliances through the new ties. We read of many instances of the kings sending their infants to influential bondes to be fostered; by which, no doubt, a great local interest and connection was secured to the foster-child. In the social state of those ages each family was a distinct dynasty, beholden for its security to its own strength in friends and followers, and its own power to avenge its wrongs, rather than to the guardianship and force of law. The system of fosterage was a consequence of this social state; and the custom lingered in

England for a long time in the form of sending children to be brought up as pages in the families of distinguished personages. Jon Loptson appears to have been a person of more distinction than Snorre's own father. His grandfather was Sæmund Frode, the contemporary of Are who first committed the historical sagas to writing; Jon Loptson's mother, Thora, was an illegitimate daughter of King Magnus Barefoot. In such a family, we may presume the literature of the country would be cultivated, and the sagas of the historical events in Norway, and of the transactions of her race of kings, would be studied with great interest.

One would like to know how people of distinction in that age lived and were lodged in Iceland? What kind of house and housekeeping the daughter of a king would have there? We have no positive data to judge from; but we may infer from various circumstances that this class would be at least as well off as in Norway; that comparatively the comforts, luxuries, and splendour of life in the poor countries, would not be so much inferior to those of the rich countries as in our own days. Sugar, coffee, tea, silks, cotton, and all foreign articles, were almost equally out of reach and enjoyment in all the countries of the North. From the natural products, or crops of the land, all that was enjoyed had to be obtained. Iceland enjoyed the advantage of more security of property and person; and the natural products of Iceland,—fish, oil, skins, butter, wool, and before the introduction of cotton as a clothing material, the wadmal, or coarse woollen cloth manufactured in Iceland, in which rent and taxes were paid, and which circulated as money through all the North, and in which even other goods were valued as a medium of exchange,—would all be of much higher comparative value than in after ages, when commerce and manufactures gave people a greater supply of better and cheaper articles for the same uses. The market for wood of Norway being confined to such islands as produced none for building purposes, the houses would probably be much the same in size and conveniences as those common among all classes in Norway, and little more expensive. The trade of bartering their products for those of other countries would probably be much more extensive than now, because their kind of products were much more generally used in other countries. In Throndhjem, Bergen, and Tunsberg, several merchant vessels at the same time are often spoken of in the sagas; and Torfæus, in his "Vinlandia," page 69, mentions a Hrafnus Limiricepeta, so called from his frequent voyages to Limeric in Ireland—a Limeric trader—who had related to Thorfin, Earl of Orkney, some accounts of a Great Ireland in the Western Ocean. In the Færeyinga Saga, we read of merchants frequenting the Farey Isles to purchase the products of the country, and of the people sending off cargoes of their wool to Norway. The commercial intercourse of those times has probably been greater than we suppose, although dealings were only in the rude products of one land bartered against those of another. Matthew Paris tells us of his being at Bergen in the year 1248, and of there being more than 200 vessels in that port at the same time. The poorer lands and countries of Europe, and the employment of their inhabitants, have in fact undergone a great depreciation in value, and which is still going on, by the introduction and general diffusion of better articles for food, clothing, and enjoyment, from better climes, and by the diffusion of more refined tastes and habits than the products of their soil and industry can gratify. When wadmal, or coarse woollen cloth, was the ordinary wear; stockfish, or salt fish, in great use even in royal households; fish oil the only means in the North for lighting rooms,—the poorest countries, such as Iceland, Greenland, or the north of Norway, which produced these, must have been much more on a par with better countries, such as Denmark or England, which did not produce them, and must have been comparatively much better to live in, and the inhabitants nearer to the general condition of the people of other countries, than they are now. The daughter of King Magnus Barefoot would probably be as well lodged, fed, clothed, and attended, as she would have been in Scotland in that age.

Jon Loptson died when Snorre was nineteen years of age. Snorre continued to live with his foster-brothers, his own father being dead, and his patrimony inconsiderable and much wasted by his mother. At twenty-one years of age he married Herdis, the daughter of a wealthy priest called Berse, who lived at Berg, in the bailiwick of Myre, where he also took up his abode. He got a considerable fortune with his wife,* by whom he had several children, but only two who grew up; a son called Jon Murt, and a daughter called Halbera. He had also several illegitimate children; a son called Urokia, and a daughter called Ingibjorg. After being twenty-five years married to Herdis, he married, some time about 1224, she being still alive, another wife, Halveig, a rich widow, with whom he got also a large fortune. He quarrelled with the children of his first wife about their fortunes to which they were entitled when he parted from their mother. He was in enmity also with the husbands of both his daughters, each of whom had been divorced, or had had two husbands; and these sons-in-law, and his own brother Sighvat, were the parties who finally murdered him in their family feud. What is known of Snorre Sturlason is derived from an account of the Sturla family, called the "Sturlunga Saga," composed evidently by one of the descendants of the kinsmen with whom he had been in enmity. His bad actions are probably exaggerated, and his good concealed. With every allowance, however, for the false colouring which hatred and envy may have given to the picture, Snorre appears from it to have been a man of violent disposition,—greedy, selfish, ambitious, and under no restraint of principle in gratifying his avarice and evil passions. He is accused of amassing great wealth by unjust litiga-

* Four thousand dollars it is reckoned to have been by antiquaries—a large fortune before silver became plentiful by the discovery of America.—L.

tion with his nearest kindred, and by retaining unjustly the property which of right belonged to them on his parting with his first wife; and of appearing at the Things with an armed body of 600 or 800 men, and obtaining by force the legal decisions he desired. He is accused also of having, on his visits to Norway, betrayed the independence of his country, and contributed to reduce Iceland to the state of a province of Norway. It is probable that much of the vices of the age, and of the inevitable events in history prepared by causes of remote origin, is heaped up by the saga-writer on Snorre's head. He was clearly guilty of the two greatest charges which, in a poor country and ignorant age, can be brought against a man—he was comparatively rich and comparatively learned. Of his wealth we are told that he possessed six considerable farms, on which his stock of cattle was so great that in one year, in which fodder was scarce, he lost 120 head of oxen without being seriously affected by it in his circumstances. He employed much of his wealth in improving and fortifying his main residence at Reykholt, to which he had removed from Berg. At Reykholt, he constructed a bathing room of cut freestone, into which the water from a warm spring in the neighbourhood was conducted by a covered drain or pipe. Stone buildings in the North being rare, this structure was considered magnificent, and is spoken of as a proof at once of Snorre's wealth and extravagance. In this age it will rather be considered a proof that Snorre was a man of habits far more refined than those of the people around him; that, trifling as the structure may have been, it shows a mind of great energy and activity to have executed it, and of some refinement and improved habits to have felt the want of accommodations for personal cleanliness in his house. Snorre's first journey to Norway appears to have been about the year 1221, when he was forty-three years of age, and was still married to his first wife Herdis. He appears to have come to Norway on a visit to Earl Hakon Galen, who was married to Lady Christina, the daughter of King Sigurd the Crusader. We are told in the Sturlunga Saga, that Snorre had composed a poem in honour of the earl, who in return had sent him a sword and a suit of armour. On his arrival he found that the earl was dead, and his widow was married again to Askel, the Lagman of Gautland. He remained the first winter at the court of King Hakon and Earl Skule, who then ruled over Norway, and proceeded in summer to visit Lady Christina, by whom he was well received; and it may be supposed that on this journey he collected the information relative to former transactions in Sweden and Denmark, as well as in Norway, that he gives in his Chronicle. The Lady Christina was a daughter of King Sigurd the Crusader by Malmfrid, a daughter of King Harald of Novgorod, whose mother was Gyda, a daughter of the English King Harald, the son of Earl Godwin, who fell at the battle of Hastings. This Lady Christina appears to have been married first to Erling Skakke, by whom she had a son who was King of Norway, Magnus Erlingson, in the middle of whose reign Snorre's Chronicle ends. She was then married to Earl Hakon Galen, after whom she married the Lagman Askel. On his return from this visit Snorre remained two years with Earl Skule in Norway. It is evident that, as a chronicler, Snorre Sturlason had thus enjoyed opportunities of collecting or correcting the accounts of transactions of former times, which few contemporary writers possessed. He was made a cup-bearer, or dish-bearer, equivalent to the modern dignity of chamberlain, by King Hakon; and is accused by his enemies of having entered into a private agreement with the King and Earl Skule that he should use his influence to subvert the independence which Iceland had hitherto enjoyed, and to persuade the Thing to submit to the Government of the King of Norway; and that he should be made the King's lenderman, or even earl over the country, in reward of this service. Whatever may have been Snorre Sturlason's ambition or want of principle, no grounds for this charge appear in his life. The subjection of Iceland to the crown of Norway was, on the contrary, carried into effect two years after his murder by his personal enemies; and the event may rather be considered the inevitable result of the changes which had taken place in the social condition, military spirit, and arrangements and relative importance of different countries, about the middle of the thirteenth century, than the consequence of any conspiracy or treachery. Snorre returned to Reykholt, and, divorcing his first wife, married his second, for the sake, it is alleged, of her large fortune, and became the richest, and probably the most unhappy, man of his day in Iceland. He was involved in disputes and lawsuits with his sons and his wife's family, who appear to have had just and legal claims to their shares of the properties which he continued to keep in his own possession. He appears to have visited Norway once, if not twice again, before or about the year 1237, and to have attached himself to the party of Duke Skule, who had claims on the succession to the crown of Norway. In 1237 Snorre returned to Iceland, and Duke Skule assumed the title of king at Throndhjem, in opposition to his son-in-law, King Hakon Hakonson; but in the following year he and his son were slain. Snorre Sturlason, as a friend or adherent of Duke Skule, was declared a traitor by King Hakon. As the king's chamberlain, he might in that age, although not a Norwegian subject, be considered a traitor. Letters from the king were issued to his enemies to bring him prisoner to Norway, or to put him to death; and on this authority his relations, with whom he was in enmity in a family feud,—his three sons-in-law, Gissur, Kolbein, and Arne,—came by night, in September 1241, to his residence at Reykholt, and murdered him in the 63rd year of his age. The same party, two years afterwards, brought Iceland under subjection to the crown of Norway. It seems unjust to throw upon the memory of Snorre Sturlason, as far as the circumstances can be made out, the imputation of having sought to betray the independence of his country, when no overt act of his appears to have tended to that result, and when his enemies, who assassinated him, and from whom alone any account of his life proceeds, were avowedly the parties who brought it about. But it cannot be denied that their accounts, and even their enmity, prove that Snorre has been a man unjust to and hated by his family,—selfish, rapacious, and without restraint from principle or natural affection.

The judgment for posterity to come to probably is, that Snorre Sturlason, and even his relations who murdered him, were rather a type of the age in which they lived than individuals particularly prominent for wickedness in that age. The moral influences of Christianity had not yet taken root among the Northmen, while the rude virtues of their barbarous pagan forefathers were extinct. The island of Iceland had never contained above sixty-three or sixty-four thousand inhabitants—the population of an ordinary town. The providing of food, fuel, and of winter provender for their cattle, and such employments, have necessarily at all times occupied a much greater proportion of the population than in more favoured climes. The enterprising, energetic, and restless spirits found occupation abroad in the roving viking expeditions of the Norwegians, for the Icelanders themselves fitted out no viking expeditions; while the equally ambitious, but more peaceful and cultivated, appear to have acquired property and honour, as skalds, in no inconsiderable number. But the rise of the Hanseatic League, and the advance of the south and west of

Europe in civilisation, trade, and naval power, had extinguished the vikings on the sea. They were no longer, in public estimation, exercising an allowable or honourable profession; but were treated as common robbers, and punished. The diffusion of Christianity, and of a lettered clergy over the Scandinavian peninsula, had in the same age superseded the skalds, even as recorders of law or history. The skald, with his saga and his traditional verses, gave way at once before the clerk, with his paper, pen, and ink. Both occupations—that of the viking and of the skald—fell as it were at once, and in one generation—in the end of the twelfth and beginning of the thirteenth century; and the wild, unquiet, ambitious spirits, in the small Icelandic population, which were formerly absorbed by them, were thrown back into their native island, and there, like tigers shut up together in a den, they preyed on or worried each other. In Scandinavia itself the same causes produced in that age the same effects. The Birkibeins and the Baglers, who, from the middle of Magnus Erlingson's reign, raised their leaders alternately to the Norwegian crown, were in reality the vikings, driven from the seas to the forests,—were the daring, the idle, the active of society, who could find no living or employment in the ordinary occupations of husbandry, which were preoccupied by the ordinary agricultural population, nor in the few branches of manufacture or commerce then exercised as means of subsistence; and whose former occupations of piracy at sea, or marauding expeditions on land under foreign vikings, was cut off by the progress of Christian influences on conduct,—of the power of law, and of the naval, military, and commercial arrangements in all other countries. The employments and means of living peaceably were not increased so rapidly as the employment given by private warfare on sea and land had been put down; and in all Europe there was an overpopulation, in proportion to the means of earning a peaceful livelihood, which produced the most dreadful disorders in society. This was probably the main cause of the unquiet, unsettled state of every country, from the eleventh century to the fifteenth. The Crusades even appear to have been fed not more by fanaticism than by this want of employment at home in every country. Law and social order were beginning to prevail, and to put down private wars, and the claim of every petty baron to garrison his robber nest and pillage the weak; but this growing security had not advanced so far that trade and manufacture could absorb, and give a living to, the men not wanted in agricultural and thrown out of military employment. It takes a long time, apparently, before those tastes and habits of a nation on which manufactures and commerce are founded, can be raised. Society was in a transition state. The countries which took but little part in the Crusades,—such as Scotland, Norway, Sweden, Denmark, and this little population of Iceland,—and which had no outlet for the unquiet spirits reared in private wars or piracy, present a deplorable state of society for many generations. A bad, unquiet, cut-throat spirit, was transmitted to succeeding generations, and kept those countries in a half-barbarous state to a much later period than the other countries which had got rid of a prior turbulent generation in the Crusades. The Sturlunga Saga, or account of Snorre and his family, contains little else but a recital of private feuds in Iceland,— of murders, burning of houses, treachery, and a social disorganisation among this handful of people, which might well excuse Snorre Sturlason if he had wished and attempted to obtain the common benefit of all social union—the security of life and property—by the surrender of a nominal independence, but a real anarchy, into the hands of a government strong enough to make laws respected.

Snorre Sturlason must be measured, not by our scale of moral and social worth, but by the scale of his own times. Measured by that scale, he will be judged to have been a man of great but rough energy of mind,—of strong selfishness, rapacity, and passions unrestrained by any moral, religious, or social consideration,—a bold, bad, unprincipled man, of intellectual powers and cultivation far above any of his contemporaries whose literary productions have reached us,—a specimen of the best and worst in the characters of men in that transition-age from barbarism to civilisation,—a type of the times, a man rough, wild, vigorous in thought and deed, like the men he describes in his Chronicle.

How, it may fairly be asked, could a work of such literary merit as the translator claims for

Snorre Sturlason's Chronicle, have lain hid so long from English readers, and have been valued, even on the Continent, only by a few antiquaries in search of small facts connected with Danish history? The Heimskringla has been hardly used by the learned men of the period in which it was first published. It appeared first in the literary world in 1697, frozen into the Latin of the Swedish antiquary Peringskiold. A Swedish translation, indeed, as well as a Latin, accompanied the Icelandic text; but the Swedish language was then, and is now, scarcely more known than the Icelandic in the fields of European literature. Modern Latin, or Latin applied to subjects beyond its own classical range, is a very imperfect medium for conveying realities to the mind, and, like algebra, presents only equivalents for things or words, — not the living words and impressions themselves. It may be an advantage in science, law, metaphysics, to work with the dead terms of a dead fixed language; but in all that addresses itself to the fancy, taste, or sympathy of men, the dead languages are dead indeed, and do not convey ideas vividly to the mind like the words of a living tongue belonging to existing realities. Conceive Shakespeare translated into Latin, or Schiller, or Sir Walter Scott! Would the scholar the most versed in that language have the slightest idea of those authors, or of their merits? About the time also when Peringskiold published the Heimskringla, antiquarian research was, and still continues to be, the principal literary occupation of the educated classes in Sweden and Denmark, and that which led, more than any other branch of literature, to distinction and substantial reward from government. Peringskiold, Torfæus, Arne Magnusen, Schöning, and many other antiquaries of great learning, research, and talent in their own antiquarian pursuits, dug for celebrity in this mine of the Heimskringla, and generally threw away the sterling ore to bring home the worthless pebble. Dates were determined, localities ascertained, royal genealogies put to rights,—the ancestor of the Danish dynasty proved, to the satisfaction of all men, to have been a descendant of Odin called Skiold, and not Dan,—and a great deal of such learned dust was raised, swept into a heap, and valued as dust of gold; but the historical interest, the social condition, the political institutions of the Northmen, as delineated in the Heimskringla, were not laid before the public by those great antiquaries: and possibly these were subjects of which they could not safely treat. These profound scholars, so laboriously and successfully occupied, appear to have forgotten altogether, in their zeal to do each other justice, and amidst the compliments they were interchanging on their own merits, that there was a Snorre Sturlason entitled to his share in the honours. His work was treated as some of the classics have been by their learned commentators—the text overwhelmed, buried, and forgotten, under annotations and unimportant explanations of it. It is pleasing to observe how the natural taste of a people selects what is good in their literature, what is adapted to the mind of all, with more just tact than even the educated classes among them. While the merit of Snorre was hid from the educated under a mass of learned rubbish, the people both in Norway and Denmark had a true feeling for it; and in 1594 a translation into Danish of parts of the Heimskringla* was published in Denmark by Mortensen.

In 1590 a priest, Peter Claussen,—himself as wild a manslaying priest as the priest Thangbrand, or any other of the rough energetic personages in the work of Snorre,—translated the Heimskringla for the benefit of his countrymen in Norway, the language of Snorre having become obsolete, or at least obscure, even to the Norwegian peasantry. His translation was published in 1633 by Ole Worm, and it became a house-book among the Norwegian bondes.

At the present day, in the dwellings of the remote valleys, especially of the Throndhjem district,—such

---

* The copy of the Heimskringla made in 1230 by Snorre's nephew, Sturla, is considered the ground text from which all the other manuscripts have been made; and copies in writing of his work have been made as late as 1567. The exact date of any of the manuscripts used by Mortensen in 1594, or by Claussen in 1590, printed by Wormius in 1663, or by Peringskiold in 1697, is not ascertained. They appear to have all had different manuscripts before them; some better, apparently, in some parts, and in others not so perfect. The Heimskringla of Schöning, in folio,—the first volume published in 1777, the last in 1826, in Icelandic, Latin, and Danish at Copenhagen,—is the best.—L.

Since the above was written a text edition has been published in Christiania by C. R. Unger, 1868, and another in Upsala by N. Linder and K. A. Hagson, 1870. Several translations have appeared in various languages. See editor's preface.

as Stordal, Veerdal, Indal,—a well-used copy of some saga, generally that of King Olaf the Saint, reprinted from Peter Claussen's work, will be found along with the Bible, Prayer Book, Christian the Fourth's Law Book, and the Storthing's Transactions, to be the housefather's library. During a winter passed in one of those valleys, the translator, in the course of acquiring the language of the country, borrowed one of those books from his neighbour Arne of Ostgrunden, a bonde or peasant-proprietor of a farm so called. It was the saga of King Olaf the Saint. Reading it in the midst of the historical localities, and of the very houses and descendants of the very men presented to you in the stirring scenes of this saga at the battle of Stiklestad, he may very probably have imbibed an interest which he cannot impart to readers unacquainted with the country, the people, and their social state. He read with delight the account of old manners and ways of living given in the saga,—old, yet not without much resemblance to what still exists in ordinary family life among the bondes. He found, from knowing the localities, the charm of truth from internal evidence in the narratives of that saga. It is not unlikely that these favourable circumstances may have given the translator a higher impression of the literary merit of the Heimskringla than others may receive from it. He was not aware at the time that the volume which delighted him was but a translation of a single saga from Snorre Sturlason's work into a Norse which itself was becoming obsolete, and like the Scotch of Lindsay of Pitscottie's Chronicle, was in some degree a forgotten language even among the peasantry. It has since been the occasional and agreeable occupation of his leisure hours to study the work of Snorre in the original. To much knowledge of, or familiarity with the Icelandic he cannot lay any claim. To get at the meaning and spirit of the text, helping himself over the difficulties, which generally only lay in his own ignorance of the language, by collating every passage he was in doubt about with the meaning given to it in the translations of Peringskiold, Schöning, and Aal, and to give a plain faithful translation into English of the Heimskringla, unencumbered with antiquarian research, and suited to the plain English reader, has been his object.

The short pieces of skaldic poetry which Snorre intermixes with his narrative, and quotes as his authorities for the facts he is telling, are very difficult to deal with in a translation. They are not without a rude grandeur of imagery, and a truthfulness in description of battles and sea-fights; and they have a simplicity which, although often flat, is often natural and impressive. They have probably been originally delivered *vivâ voce* in recitative, so that the voice, adroitly managed, would form a measure. Icelandic poetry does not, like the Greek or Latin, differ from prose by certain measures or feet in a verse, but has a formation peculiar to itself. All Icelandic poems, or almost all, are divided into strophes consisting of eight lines. The strophe is further subdivided into two half-strophes, and each of these again into two parts. Each part is a fourth of the whole strophe, and contains two verses or lines. The first of these lines is called the fore line and the second the back line; and the two are connected together, as verses, by rhyme-letters, or rhyme-syllables. This rhyme-letter, or alliteration, consists in having two words in the fore line beginning with the same initial letter; and a third word, that which is the most important in the meaning, in the second or back line, and beginning with the same letter. For example:—

"Farvel fagnadar
Fold og heilla."

"Farewell, favoured
Fold (land) and holy."

The letter F in the word Fold is the head letter of the alliteration, and the same letter in Farvel and fagnadar are the two subsidiary alliterative sounds in the first line. In the use of this alliteration there are several subdivisions, from exceptions or limitations to the general principle. Besides this alliteration or letter-rhyme, there are syllable-rhymes, in which the first syllables of words, instead of the first letters only, form, by their collocation in the fore and back lines, the versification; and if the first syllables rhyme together, the last may be different sounds. Thus, merki and sterka, or gumar and sumir, are perfect syllable-rhymes in a line. End-rhymes, as in the other Gothic languages, are also used in Icelandic versification, connected also always with the alliterative and syllabic rhymes. Thus :—

"Nu er hersis hefnd,
Vid hilmi efnd.
Gengr ulfr ok ærn
Af ynglings bærn."

The two lines only rhyme together, in Icelandic versification, which are connected by the rules of alliterative verse,—viz., the first and second, and the third and fourth; but the first and third, or second and fourth, are never made to rhyme together. Longer verse-lines than of eight syllables are not used, and lines of three or six appear more common. A short measure, admitting of no pause or cæsura in the middle of the line, appears to have been most agreeable to the Icelandic ear, or mode of recitative in which the skalds have chanted their verses. These observations are taken from Rask's "Veiledning til det Islandske Sprog, 1811;" in which there is a valuable dissertation on the Icelandic versification, with examples of the different kinds of verses. Some later Icelandic scholars are of opinion that what Rask has treated as two lines, on the supposition that the Icelandic versification had no cæsura, had in reality been one line, with the cæsura marked by a rhyme corresponding to the end-rhyme of the line, which middle rhyme is of common occurrence in old English verses.

For example, in the following old English verses on the bee, the line is not concluded at the rhyme in the middle, which marks a strong cæsura or pause, not a total want of it:—

> "In winter daies, when Phœbus' raies
>   Are hid with misty cloud,
> And stormy showers assault her bowers,
>   And cause her for to crowd."
>
> *Baret's Alvearia*, 1580.

And also in the Latin rhymes of the monks in the Middle Ages, as for instance in these—

> "Omnia terrena per vices sunt aliena.
>   Nunc mei nunc hujus, post mortem nescio cujus,"

the rhyme by no means concludes the line. The mode of writing on parchment or paper has, for economy of the material, been in continuous lines, like prose, without any division, in the manuscripts of old date; so that nothing can be concluded from the writing concerning the length or forms of the verses. Whether the skalds adapted their verses to music, or tunes, seems not well ascertained. Little mention, if any, is made in any of the sagas of tunes, or musical instruments; yet they have had songs. All their pieces are called songs, and are said to be sung, and many of them evidently were intended to be sung. We find mention also of old songs; for instance, the "Biarkamal," was instantly recognised by the whole army at Stiklestad. They must have had tunes for these songs. We find also a refrain, or chorus to songs, mentioned. All, perhaps, that can be safely said of Icelandic versification is, that the system has been very artificial, and full of technical difficulties in the construction; and, independently of the beauties of poetic spirit and ideas, may have had the merit of technical difficulties in the verse adroitly overcome by the skald,—a merit which it would be going too far to contemn, because we, with minds and ears not trained in the same way, cannot feel it. How much of our own most esteemed poetry gives us pleasure from similar conventional sources distinct altogether from poetical imagery, or ideas which all men of all countries and ages would relish and feel pleasure from? There may also have been a harmony and measured cadence given by the voice in reciting or chanting such verses—and they were composed to be recited, not silently read—which are lost to us. All we can judge of them is, that if such verses could be constructed in the English language, they would be without harmony or other essential property of verse to us; as our minds, ears, and the genius of our language are formed in a different mould. Besides, the peculiarities of the construction of the verse, the poetical language, and the allusions to the Odin mythology, are so obscure, involved, and far-fetched, that volumes of explanation would be necessary almost for every line in any verbatim translation. Torfæus, who was himself an Icelander, and was unquestionably the first of northern antiquaries, declares that much of the skaldic poetry is so obscure, that no meaning at all can be twisted out of it by the most intense study. The older and younger Eddas were in fact handbooks composed expressly for explaining the mythological allusions and metaphors occurring in the poetry of the skalds; so that this obscurity and difficulty appear to have been felt even before the Odin worship was totally extinct, and its mythology forgotten. Examples will best illustrate the obscurity of allusion. In the verses composed by Berse, quoted in the forty-eighth chapter of Olaf the Saint's Saga, in the sixth line, the literal translation of the text would be, "Giver of the fire of the ship's out-field." The "out-field" of the ship is the ocean which surrounds a ship, as the out-field surrounds a farm. The fire of the ocean is gold; because Ægir, when he received the gods into his hall in the depths of the ocean, lighted it with gold hung round instead of the sun's rays; and hence the ocean's rays is a common poetical term for gold in the skaldic poetry. Now "the giver of the fire of the ship's out-field" means the giver of gold, the generous king. Another example of the obscurity of allusion is in the first line of the verses quoted in the twenty-first chapter of Olaf Trygveson's Saga. In the original the expression is literally, "Hater of the bow-seat's fire." Now the bow-seat is the hand which carries the bow; its fire is the gold which adorns the hand in rings or bracelets; and the hater of this fire is the man who hates to keep it, who gives it away,—the generous man. Every piece almost of the skaldic poetry quoted in the Heimskringla has allusions of this obscure kind, which would be unintelligible without voluminous explanation; and yet the character of these short poetical pieces does not consist in these, which seem to be but expletives for filling out their artificial structure of verses, but in their rude simplicity and wild grandeur. The translator intended at first to have left out these pieces of skaldic poetry altogether. They are not essential to Snorre's prose narrative of the events to which they refer. They are not even authorities for the facts he details, although he quotes them in that view; for they only give the summary or heads of events of which he gives the particular minute accounts. They appear to be catch-words, or preliminary verses, for aiding the memory in recurring to some long account or saga in prose of which they are the compendium or text. The oldest translator of Snorre's work, Peter Claussen, who is supposed to have had, in 1590, a manuscript to translate from which is now lost, omits altogether the verses. The translator consulted a literary friend,—his son, Mr. S. Laing, late Fellow of St. John's College, Cambridge, now of the Railroad Department of the Board of Trade,—and went over with him the translation of the prose narrative of Snorre, and translations into prose of the poetical pieces connected with it. They came to the conclusion that although these pieces of skaldic poetry are not essential to Snorre's prose narrative of the historical events to which they refer, they are essential to the spirit and character of Snorre's work. However obscure, unpoetical, monotonous in the ideas, or uninteresting and flat they may be, they show the mind, spirit, and intellectual state of the age and people,—show what it was they considered poetry; and the poorest of these compositions have, in this view, great historical interest. Many of them are, especially in the descriptions and imagery connected with the warfare of those times, highly poetical; and, under any forms of verse and language, the "Hakonarmal," chapter 33 of Hakon the Good's Saga, the "Biarkamal," chapter 220 of Saint Olaf's Saga, and many of the pieces of Sigvat the Skald and others, would be acknowledged as genuine poetry. On examining more closely these pieces of skaldic poetry it appears, in general, that the second half of the strophe of eight lines, which their rules of versification required as the length of their poetical pieces, is but a repetition of the idea of the first half, and the second two lines but an echo of the first two. The whole meaning—all that the skald has to say in the strophe, is very often comprehended within the first two lines, the fore line and back line, which are connected together by the alliterative letters or syllables; and the one idea is expanded, only in other words, over the whole surface of the rest of the strophe of eight lines. The extraordinary metaphors and mythological allusions, the epithets so long-winded and obscure, the never-ending imagery of wolves glutted and ravens feasted by the deeds of the warriors, arise evidently from the necessity imposed on the skald of finding alliteratives, and conforming to the other strict rules of their versification. The beauty of this artificial construction is lost even upon the best Icelandic scholars of our times; and it appears to have been the only beauty many of these pieces of poetry ever pretended to, for the ideas so expressed are often not in any way poetical. Grundtvig, in his translation into Danish of the Heimskringla, and some German translators of skaldic poems, have cut the loop of this difficulty. They have taken only the most poetical of the pieces of the skalds, and have freely translated, or freely paraphrased them into modern ballads, or songs, in modern measures. Grundtvig has done so with great poetic genius and spirit; and his translations have justly placed him in the first rank of Danish poets. Many of his translations might be placed by the side of the best pieces of Burger or of Scott in the ballad style; but then they are Grundtvig's, not the skalds'. They are no more a translation of the verses of the skalds quoted by Snorre Sturlason, than Shakespeare's Hamlet is a translation of the story of Hamlet in Saxo Grammaticus.

The translator and Mr. S. Laing have rendered into English verse these skaldic pieces of poetry, from prose translations of them laboriously made out. The ideas in each strophe, the allusions, and imagery, were first ascertained by collating the Norse translation of them in M. Jacob Aal's excellent translation of the Heimskringla published in 1838, and those in the folio edition of 1777, and the Latin prose translation of them by Thorlacius and Werlauff, in the sixth volume of that edition, published in 1826, with the Icelandic text.* The ideas, allu-

---

* The versions of Jacob Aal and Thorlacius and Werlauff into the

sions, and imagery are, much oftener than could be expected, obtained, and rendered line for line; and the meaning of each half strophe is always, it is believed, given in the corresponding four English verses. The English reader, it is hoped, will thus be better able to form an idea of the poetry of the skalds, than if the translators had been more ambitious, and had given a looser paraphrase of these pieces according to their own taste or fancy. Some of these pieces of skaldic poetry, it will be seen even by this dim reflection, have very considerable poetical merit; many, again, are extremely flat and prosaic, and are merely prose ideas cut into the shape of verse by the skald. These, it must be recollected, may have had their beauty and merit in the technical construction of the verse, and may have been very pleasing and harmonious, although such merit is lost upon us in a different language. The ideas are all we can get at; not the forms and technical beauties of the expression of those ideas. It will not escape the observation of the English reader that in the ideas there is a very tedious monotony, in the descriptions of battles and bloodshed, in the imagery of war, in the epithets applied to the warriors and kings; and in general there is a total want of sentiment or feeling. The spirit is altogether material. The skalds deal only in description of material objects, and mainly of those connected with warfare by sea

cognate Northern tongue are much more graphic than the Latin, and more true to the spirit of the Icelandic. These versions have been referred to for the meaning of the skald in all cases in which the Icelandic was obscure.—L.

or land. But this, no doubt, belongs to the spirit of the state of society and times; and it will be considered of some importance to know what the ideas were which were then considered poetical, and which pleased the cultivated classes for whom the skalds composed. The English public will be able, in some degree, from these translations, to judge what the poetry of the skalds was,—what may have been its real poetic merit: of the labour and difficulty of presenting these pieces to the public, even in this imperfect way, none can judge but those who will try the same task.

THE HEIMSKRINGLA;

OR,

CHRONICLE OF THE KINGS OF NORWAY.

PREFACE OF SNORRE STURLASON.*

In this book I have had old stories written down, as I have heard them told by intelligent people, concerning chiefs who have held dominion in the northern countries,† and who spoke the Danish

---

* Gudbrand Vigfusson has shown, in his *Prolegomena* to his edition of Sturlunga Saga (Oxford, 1878), that Are Frode and not Snorre is the author of this Preface. That part only which gives the life of Are is written by Snorre. Indeed, Vigfusson takes the Ynglinga Saga to be "the very work of Are, abridged here and there, but still preserving in many chapters (especially those which depict the life and rites of the heathen days) his characteristic words." He believes the Ynglinga Saga to be a reproduction of Are's Konunga-bok (Book of Kings).

"It (the Book of Kings) has perished," says Vigfusson, "except so far as it is embodied in Snorre's work, in which we can detect some fragments of it apparently verbally cited—*e.g.*, the preface, 'á bók þessi' . . . which certainly cannot be ascribed to Snorre, as Gisli Brynjulfsson long ago maintained. The writer repeatedly speaks of *viva voce* sources, never of books: 'As I have heard wise men say,' 'As I have been told,' 'old traditions' (fornar frásagnir), 'poems' (kvæði), 'epic lays' (söguljoð) used for entertainment—these are his sources. He also speaks of *Langfeðgatal*, by which we take him to mean genealogical lays, which indeed were especially styled *tal* (*Ynglingatal, Haleygjatal*). All this is in good keeping with Are and his age; when Snorre wrote a century later, a whole cycle of written sagas had sprung up, whilst tradition had at the same rate died away, or was becoming extinct."

† That is to say, in all the three northern countries, Denmark, Sweden, and Norway.

tongue; and also concerning some of their family branches, according to what has been told me. Some of this is found in ancient family registers, in which the pedigrees of kings and other personages of high birth are reckoned up, and part is written down after old songs and ballads which our forefathers had for their amusement. Now, although we cannot just say what truth there may be in these, yet we have the certainty that old and wise men held them to be true.

Thiodolf * of Hvin † was the skald of Harald Harfager, and he composed a poem for King Ragnvald the Mountain-high, which is called "Ynglingatal." This Ragnvald was a son of Olaf Geirstada-Alf, the brother of King Halfdan the Black. In this poem thirty of his forefathers are reckoned up, and the death and burial-place of each are given. He begins with Fiolner, a son of Ingvefrey, whom the Swedes, long after his time, worshipped and sacrificed to, and from whom the race or family of the Ynglings take their name.

* Family surnames were not in use, and scarcely are so now, among the Northmen. Olaf the son of Harald was called Olaf Haraldson; Olaf's son Magnus, Magnus Olafson; and his son Hakon, Hakon Magnuson: thus dropping altogether any common name with the family predecessors. This custom necessarily made the tracing of family connection difficult, and dependent upon the memory of skalds or others. The appellations Fair-haired, Black, &c., have been given to help in distinguishing individuals of the same name from each other. Hinn Frodi the Wise, the Much-knowing,—the Polyhistor, as it is translated into Latin by the antiquarians,—is applied to many persons; and is possibly connected with the old Norman French appellative Prud-Prud'homme.—L.

† Hvin, now Kvinesdal, in Norway. Thjodolf composed, besides Ynglingatal, also the poem Haustlong. He is mentioned in Harald Harfager's Saga.

Eyvind Skaldaspiller * also reckoned up the ancestors of Earl Hakon the Great in a poem called "Haleygiatal," composed about Hakon; and therein he mentions Saming, a son of Ingvefrey,† and he likewise tells of the death and funeral rites of each. The lives and times of the Yngling race were written from Thiodolf's relation enlarged afterwards by the accounts of intelligent people.

As to funeral rites, the earliest age is called the Age of Burning; because all the dead were consumed by fire, and over their ashes were raised standing stones.‡ But after Frey was buried under a cairn at Upsala,§ many chiefs raised cairns, as commonly as stones, to the memory of their relatives.

The Age of Cairns began properly in Denmark after Dan Mikillate ‖ had raised for himself a burial-cairn, and ordered that he should be buried in it on his death, with his royal ornaments and armour, his horse and saddle-furniture, and other valuable goods; and many of his descendants followed his example. But the burning of the dead continued, long after that time, to be the custom of the Swedes and

---

* Skaldaspiller = skald-spoiler, a poetaster or plagiarist. Vigfusson, *s.v.*, believes this nickname was given to Eyvind because two of his chief poems were modelled after contemporary poets, the Haleygiatal after the Ynglingatal, and the Hakonarmal after the Eiriksmal.

† Doubtless an error in copying, as chapter 9 of the Ynglinga Saga names Njord as Saming's father.

‡ Bauta-Steinar are in Scotland called standing stones by the common people, and we have no other word in our language for those monuments.—L.

§ Uppsalir, the High Halls, was not the present city of Upsala; but Gamle Upsala, two miles north of the present Upsala.—L.

‖ Mikil-lati—the Magnificent (*superbus*).—L.

Northmen.* Iceland was occupied in the time that Harald Harfager was the King of Norway.† There were skalds in Harald's court whose poems the people know by heart even at the present day, together with all the songs about the kings who have ruled in Norway since his time; and we rest the foundations of our story principally upon the songs which were sung in the presence of the chiefs themselves or of their sons, and take all to be true that is found in such poems about their feats and battles: for although it be the fashion with skalds to praise most those in whose presence they are standing, yet no one would dare to relate to a chief what he, and all those who heard it, knew to be a false and imaginary, not a true account of his deeds; because that would be mockery, not praise.

OF THE PRIEST ARE FRODE.

The priest Are Frode ‡ (the learned), a son of Thorgils the son of Geller, was the first man in this country § who wrote down in the Norse language narratives of events both old and new. In the beginning of his book he wrote principally about the first settlements in Iceland, the laws and government, and next of the lagmen,‖ and how long each

---

* As Vigfusson has pointed out, this statement is incorrect, and is refuted by Havamal and by the monuments. The great bulk of the bauta stones belong to the eleventh and even the twelfth century.

† The occupation of Iceland is usually given as taking place in the year 874.

‡ Are Frode was born in Iceland 1068, and died Nov. 9, 1148.—L.

§ That is, in Iceland.

‖ Lagmen were district judges appointed by the Things to administer the law.—L.

had administered the law; and he reckoned the years at first, until the time when Christianity was introduced into Iceland, and afterwards reckoned from that to his own times. To this he added many other subjects, such as the lives and times of kings of Norway* and Denmark, and also of England; besides accounts of great events which have taken place in this country itself. His narratives are considered by many men of knowledge to be the most remarkable of all; because he was a man of good understanding, and so old that his birth was as far back as the year after Harald Sigurdson's fall.† He wrote, as he himself says, the lives and times of the kings of Norway from the report of Od Kolson, a grandson of Hal of Sida. Od again took his information from Thorgeir Afradskol, who was an intelligent man, and so old that when Earl Hakon the Great was killed he was dwelling at Nidarnes—the same place at which King Olaf Trygveson afterwards laid the foundation of the merchant town of Nidaros (*i.e.* Throndhjem) which is now there. The priest Are came, when seven years old, to Haukadal ‡ to Hal Thorarinson, and was there fourteen years. Hal was a man of great knowledge and of excellent memory; and he could even remember being baptized, when he was three years old, by the priest Thangbrand, the year before Christianity was established by law in Iceland. Are was twelve years

---

* The ancient name of Norway was *Norvegr* or *Noregr*, probably originally Nordvegr or Nordrvegr, the North way.

† Consequently in the year 1067.

‡ In the south of Iceland.

of age when Bishop Isleif* died, and at his death eighty years had elapsed since the fall of Olaf Trygveson. Hal died nine years later than Bishop Isleif,† and had attained nearly the age of ninety-four years. Hal had traded between the two countries, and had enjoyed intercourse with King Olaf the Saint, by which he had gained greatly in reputation, and he had become well acquainted with the kingdom of Norway. He had fixed his residence in Haukadal when he was thirty years of age, and he had dwelt there sixty-four years, as Are tells us. Teit, a son of Bishop Isleif, was fostered in the house of Hal at Haukadal, and afterwards dwelt there himself. He taught Are the priest, and gave him information about many circumstances which Are afterwards wrote down. Are also got many a piece of information from Thurid, a daughter of the gode ‡ Snorre.§ She was wise and intelligent, and remembered her father Snorre, who was nearly thirty-five years of age when Christianity was introduced into Iceland, and died a year after King Olaf the Saint's fall.‖ So it is not wonderful that Are the priest had good information about ancient events both here in Iceland, and abroad, being a man

* Isleif was the first bishop of Iceland, and had studied at Erfurth in Germany, and died 1080.—L.

† Consequently in the year 1089.

‡ Godes were priests and judges, and an hereditary class, apparently, in Iceland in the heathen time. But we hear little or nothing of such a priesthood in Norway; nor is it clear what their civil jurisdiction may have been in Iceland compared to that of the lagmen, or whether the godes, originally the priests by hereditary right, as descendants of Odin's twelve diar, were not ex officio the lagmen or judges also.—L.

§ Snorre Gode was born 964 and died 1031.

‖ This happened 1030.—L.

anxious for information, intelligent and of excellent memory, and having besides learned much from old intelligent persons. But the songs seem to me most reliable if they are sung correctly, and judiciously interpreted.

I.

YNGLINGA SAGA;

OR, THE STORY OF THE YNGLING FAMILY FROM ODIN TO HALFDAN THE BLACK.

CHAPTER I.—*Of the Situation of Countries.*

IT is said that the earth's circle which the human race inhabits is torn across into many bights, so that great seas run into the land from the out-ocean. Thus it is known that a great sea goes in at Niorvasund,[*] and up to the land of Jerusalem. From the same sea a long sea-bight stretches towards the north-east, and is called the Black Sea, and divides the three parts of the earth; of which the eastern part is called Asia, and the western is called by some Europa, by some Enea. Northward of the Black Sea lies Svithiod the great,[†] or the Cold. The Great Svithiod is reckoned by some not less than the great Serkland;[‡] others compare it to the Great Blueland.[§] The northern part of Svithiod lies

---

[*] Niorvasund, the Straits of Gibraltar; Niorvasund was for the first time passed by a Norseman (Skopte) A.D. 1099.

[†] Svithiod the Great, or the Cold, is the ancient Scythia; and is also called Godheim in the mythological sagas, or the home of Odin and the other gods. Svithiod the Less is Sweden proper, and is called Manheim, or the home of the kings the descendants of these gods.—L.

[‡] Serkland means North Africa and Spain, and the countries of the Saracens in Asia.—L.

[§] Blaland, the country of the blacks in Africa.—L.

uninhabited on account of frost and cold, as likewise the southern parts of Blueland are waste from the burning of the sun. In Svithiod are many great domains, and many wonderful races of men, and many kinds of languages. There are giants, and there are dwarfs, and there are also blue men. There are wild beasts, and dreadfully large dragons. On the north side of the mountains which lie outside of all inhabited lands runs a river through Svithiod, which is properly called by the name of Tanais,* but was formerly called Tanaquisl,† or Vanaquisl, and which falls into the ocean at the Black Sea. The country of the people on the Vanaquisl was called Vanaland, or Vanaheim;‡ and the river separates the three parts of the world, of which the eastermost part is called Asia, and the westermost Europe.

CHAPTER II.—*Of the People of Asia.*

The country east of the Tanaquisl in Asia was called Asaland, or Asaheim, and the chief city in that land was called Asgard.§ In that city was a chief called Odin, and it was a great place for sacrifice. It was the custom there that twelve temple godes ‖

---

* Tanais is the river Don.

† Quisl, means a branch of a river at its mouth.

‡ Vanaheim belongs exclusively to the domain of mythology. See Anderson's "Norse Mythology."

§ Asgard is supposed by those who look for historical fact in mythological tales to be the present Assor; others that it is Chasgar in the Caucasian ridge, called by Strabo Aspurgum—the Asburg or castle of Aas; which word Aas still remains in the northern languages, signifying a ridge of high land. The word belongs exclusively to mythology—L.

‖ Hof godes, whose office of priests and judges continued hereditary in Scandinavia.—L.

should both direct the sacrifices, and also judge the people. They were called Diar, or Drotnar, and all the people served and obeyed them. Odin was a great and very far-travelled warrior, who conquered many kingdoms, and so successful was he that in every battle the victory was on his side. It was the belief of his people that victory belonged to him in every battle. It was his custom when he sent his men into battle, or on any expedition, that he first laid his hand upon their heads, and called down a blessing* upon them; and then they believed their undertaking would be successful. His people also were accustomed, whenever they fell into danger by land or sea, to call upon his name; and they thought that always they got comfort and aid by it, for where he was they thought help was near. Often he went away so long that he passed many seasons on his journeys.

CHAPTER III.—*Of Odin's Brothers.*

Odin had two brothers, the one called Ve, the other Vile, and they governed the kingdom when he was absent. It happened once when Odin had gone to a great distance, and had been so long away that the Asas doubted if he would ever return home, that his two brothers took it upon themselves to

---

* The word in the original is *bianak*, a word foreign to the Scandinavian tongues, and supposed to mean *blessing*; compare the Latin *benedictio*; the Scot. *bannock*, from Gael. *banagh*, an oat-cake.

divide his estate; but both of them took his wife Frigg to themselves. Odin soon after returned home, and took his wife back.*

CHAPTER IV.—*Of Odin's War with the Vans.*

Odin went out with an army against the Vans; but they were well prepared, and defended their land, so that victory was changeable, and they ravaged the lands of each other, and did damage. They tired of this at last, and on both sides appointed a meeting for establishing peace, made a truce, and exchanged hostages. The Vans sent their best men, Niord the Rich, and his son Frey. The Asas sent a man called Hœner, whom they thought well suited to be a chief,† as he was a stout and very handsome man, and with him they sent a man of great understanding called Mimer; and on the other side the Vans sent the wisest man in their community, who was called Kvaser. Now, when Hœner came to Vanaheim he was immediately made a chief, and Mimer came to him with good counsel on all occasions. But when Hœner stood in the Things or other meetings, if

---

* Much of the Ynglinga Saga will be found to belong to the domain of mythology. Ve and Vile (sanctity and will), Odin, the Asas, &c., are simply divinities made human. The editor refers the reader, in regard to all these mythological names, to his "Norse Mythology" and to his translation of Snorre's Younger Edda. The discrepancy between the mythological statements of the Ynglinga Saga and those of the Younger Edda, is properly noted by Vigfusson as a confirmation of his view that the Ynglinga is mainly the work of Are Frode.

† These exchanges appear not to have been of hostages, but of chiefs to be incorporated with the people to whom they were sent, and thus to preserve peace.—L.

Mimer was not near him, and any difficult matter was laid before him, he always answered in one way, —"Now let others give their advice;" so that the Vans got a suspicion that the Asas had deceived them in the exchange of men. They took Mimer, therefore, and beheaded him, and sent his head to the Asas. Odin took the head, smeared it with herbs so that it should not rot, and sang incantations over it. Thereby he gave it the power that it spoke to him, and discovered to him many secrets. Odin placed Niord and Frey as priests of the sacrifices, and they became deities of the Asas. Niord's daughter Freyja was priestess of the sacrifices, and first taught the Asas the magic art, as it was in use and fashion among the Vans. While Niord was with the Vans he had taken his own sister in marriage, for that he was allowed by their law; and their children were Frey and Freyja. But among the Asas it was forbidden to come together in so near relationship.

CHAPTER V.—*Odin Divides his Kingdom: Also concerning Gefion.*

There goes a great mountain barrier from north-east to south-west, which divides the Greater Svithiod from other kingdoms. South of this mountain ridge it is not far to Tyrkland,* where

---

* Tyrkland was the country of which the chief city was Troy. Tyrk may be a corruption of Teukrer. The tradition anent a descent from Troy was widely diffused in the Teutonic world, but had no foundation in fact.

Odin had great possessions. In those times the Roman chiefs went wide around in the world, subduing to themselves all people; and on this account many chiefs fled from their domains. But Odin having foreknowledge, and magic-sight, knew that his posterity would come to settle and dwell in the northern half of the world. Odin then set his brothers Ve and Vile over Asgard; and he himself, with all the gods and a great many other people, wandered out, first westward to Gardarike,* and then south to Saxland.† He had many sons; and after having subdued an extensive kingdom in Saxland, he set his sons to defend the country. He himself went northwards to the sea, and took up his abode in an island which is called Odinse, in Fyen. Then he sent Gefion across the sound to the north, to discover new countries; and she came to King Gylfe, who gave her a ploughgate of land. Then she went to Jotunheim, and bore four sons to a giant, and transformed them into a yoke of oxen, and yoked them to a plough, and broke out the land into the ocean right opposite to Odinse, which land was called Seeland, where she afterwards settled and dwelt. Skiold, a son of Odin, married her, and they dwelt at Leire.‡ Where the ploughed land was is a lake or sea called Lag.§ In the Swedish

---

* Gardarike is Russia.   † Saxland is Germany.

‡ Leidre, or Hleidre, or Leire, at the end of Isefiord, in the county of Ledreborg, is considered the oldest royal seat in Denmark. It is situated about one and a half Danish miles south west of Roeskilde.—L.

§ Lag (Icelandic, lögr) means simply water or lake, but is also used as a proper noun to designate Lake Mœlar in Sweden.

land the fiords of Lag correspond to the nesses in Seeland. Brage the Old * sings thus of it : †—

> "Gefion from Gylfe drove away,
> To add new land to Denmark's sway,—
> Blythe Gefion ploughing in the smoke
> That steamed up from her oxen-yoke:
> Four heads, eight forehead stars had they,
> Bright gleaming, as she ploughed away;
> Dragging new lands from the deep main
> To join them to the sweet isle's plain."

Now when Odin heard that things were in a prosperous condition in Gylfe's land to the east, he went thither, and Gylfe made a peace with him, for Gylfe thought he had no strength to oppose the Asas. Odin and Gylfe had many tricks and enchantments against each other; but the Asas had always the superiority. Odin took up his residence at the lake Lag, at the place now called Old Sigtuna.‡ There he erected a large temple, where there were sacrifices according to the customs of the Asas. He appropriated to himself the whole of that district of country, and called it Sigtuna.§ To the temple gods he gave also domains. Niord dwelt in Noatun, Frey in Upsala, Heimdal in Himinbjorg, Thor in Thrudvang, Balder in Breidablik; to all of them he gave good domains.‖

* Brage the Old is supposed to have died about the year 800. He was skald at the court of King Bjorn at Hauge in Sweden.

† This fable is possibly the echo of some tradition of a convulsion in which the ocean broke into the Baltic through the Sound and Belts, or in which the island of Seeland was raised from the deep.—L.

‡ Situated near the present Sigtuna, across the lake, close by Signildsberg.

§ Sigtuna = Odin's town, Sige being one of Odin's names.

‖ *All these names except Upsala are found in Grimnismal in the Elder Edda.

CHAPTER VI.—*Of Odin's Accomplishments.*

When Asa-Odin came to the north, and the gods with him, it is truthfully stated that he began to exercise and teach others the arts which the people long afterwards have practised. Odin was the cleverest of all, and from him all the others learned their magic arts; and he knew them first, and knew many more than other people. But now, to tell why he is held in such high respect, we must mention various causes that contributed to it. When sitting among his friends his countenance was so beautiful and dignified, that the spirits of all were exhilarated by it; but when he was in war he appeared fierce and dreadful. This arose from his being able to change his colour and form in any way he liked. Another cause was, that he conversed so cleverly and smoothly, that all who heard were persuaded. He spoke everything in rhyme, such as now composed, and which we call skald-craft. He and his temple gods were called song-smiths, for from them came that art of song into the northern countries. Odin could make his enemies in battle blind, or deaf, or terror-struck, and their weapons so blunt that they could no more cut than a willow twig; on the other hand, his men rushed forwards without armour, were as mad as dogs or wolves, bit their shields, and were strong as bears or wild bulls, and killed people at a blow, and neither fire nor iron told upon them. This was called Berserk-gang.*

---

* *Ber-serkr.* We will here condense G. Vigfusson's explanation as

CHAPTER VII.—*Of Odin's Feats.*

Odin could transform his shape: his body would lie as if dead, or asleep; but then he would be in shape of a fish, or worm, or bird, or beast, and be off in a twinkling to distant lands upon his own or other people's business. With words alone he could quench fire, still the ocean in tempest, and turn the wind to any quarter he pleased. Odin had a ship which was called Skidbladner,* in which he sailed over wide seas, and which he could roll up like

---

given in Cleasby's Icelandic-English Dictionary. "The etymology of the word has been much contested. Some, upon the authority of Snorre, 'his men rushed forward without armour,' derive it from '*berr*' (bare) and '*serkr*' (shirt), but this etymology is inadmissible, because '*serkr*' is a noun not an adjective. Others derive it from '*berr*' (bear=*ursus*), which is greatly to be preferred, for in olden ages athletes and champions used to wear hides of bears, wolves, and reindeer. The old poets so understood the word. In battle the berserks were subject to fits of frenzy called *berseks-gangr* (*furor bersercicus*), when they howled like wild beasts, foamed at the mouth, and gnawed the iron rim of their shields. During these fits they were, according to popular belief, proof against steel and fire, and made great havoc in the ranks of the enemy; but when the fever abated they were weak and tame. A somewhat different sort of berserk is also recorded in Norway, as existing in gangs of professional bullies, roaming about from house to house, challenging husbandmen to *holmgang* (duel), extorting ransom, and in case of victory carrying off wives, sisters, or daughters; but in most cases the damsel is happily rescued by some travelling Icelander, who fights and kills the berserk. No berserk is described as a native of Iceland. The historians are anxious to state that those who appeared in Iceland were born Norse (or Swedes); and they were looked upon with fear and execration. That men of the heathen age were taken with fits of the '*furor athleticus*' is recorded in the case of Kveldulf in the Egla, and proved by the fact that the law set a penalty upon it. The author of Ynglinga Saga attributes the berserk-gang to Odin and his followers; but this is a sheer misinterpretation, or perhaps the whole passage is a rude paraphrase of Havamal, 149 *seq.* With the introduction of Christianity this championship disappeared altogether."

* According to the Younger Edda, this ship belonged not to Odin but to Frey.

a cloth.* Odin carried with him Mimer's head, which told him all the news of other countries. Sometimes even he called the dead out of the earth, or set himself under the gallows; whence he was called the ghost-sovereign, and lord of the gallows. He had two ravens,† to whom he had taught the speech of man; and they flew far and wide through the land, and brought him the news. From all such things he became pre-eminently wise. He taught all these arts in Runes, and songs which are called incantations, and therefore the Asas are called incantation-smiths. Odin understood also the art in which the greatest power is lodged, and which he himself practised; namely, what is called magic. By means of this he could know beforehand the pre-destined fate ‡ of men, or their not yet completed lot; and also bring on the death, ill luck, or bad health of people, and take the strength or wit from one person and give it to another. But after such witchcraft followed such weakness, that it was not thought respectable for men to practise it; and therefore the priestesses were brought up in this art. Odin also knew where all missing things were concealed under the earth, and understood the songs by which the earth, the hills, the stones, and mounds were opened to him; and he bound those who dwell in them by the power of his word, and went in and

---

* This possibly refers to boats covered with skin or leather—the coracle of the Welsh and Irish.—L.

† Hugin and Munin.

‡ Örlög—the original law, the primæval law fixed from the beginning. It is curious that this idea of a predestination existed in the religion of Odin.—L.

took what he pleased. From these arts he became very celebrated. His enemies dreaded him; his friends put their trust in him, and relied on his power and on himself. He taught the most of his arts to his priests of the sacrifices, and they came nearest to himself in all wisdom and witch-knowledge. Many others, however, learned much thereof, and from that time witchcraft spread far and wide, and continued long. People sacrificed to Odin, and the twelve chiefs from Asaland,—called them their gods, and believed in them long after. From Odin's name came the name Audun, which people gave to his sons; and from Thor's name comes Thorer, also Thorarin; and also it is sometimes augmented by other additions, as Steinthor, or Hafthor, and many kinds of alterations.

Chapter VIII.—*Of Odin's Lawgiving.*

Odin established the same law in his land that had been in force among the Asas. Thus he established by law that all dead men should be burned, and their property laid with them upon the pile, and the ashes be cast into the sea or buried in the earth. Thus, said he, every one would come to Valhal with the riches he had with him upon the pile; and he would also enjoy whatever he himself had buried in the earth. For men of consequence a mound should be raised to their memory, and for all other warriors who had been distinguished for manhood a standing stone; which custom remained long after Odin's time. Towards winter there should be blood-sacrifice for a good year, and in the middle of winter for a good crop; and the third sacrifice should be in summer, for victory in battle.* Over all Svithiod † the people paid Odin a tax—so much on each head; but he had to defend the country from enemy or disturbance, and pay the expense of the sacrifice feasts for a good year.

CHAPTER IX.—*Of Niord's Marriage.*

Niord took a wife called Skade; ‡ but she would not live with him, but married afterwards Odin, and had many sons by him, of whom one was called Saming; and of this Eyvind Skaldaspiller sings thus:—

> "To Asa's son Queen Skade bore
> Saming, who dyed his shield in gore,—
> The giant-queen of rock and snow,
> Who loves to dwell on earth below,
> The iron pine-tree's daughter, she
> Sprung from the rocks that rib the sea,
> To Odin bore full many a son,
> Heroes of many a battle won."

To Saming Earl Hakon the Great reckoned up his pedigree. This Svithiod they called Manheim, but the Great Svithiod they called Godheim; and of Godheim great wonders and novelties were related.

---

* Towards winter (October 14) for the coming year, in the middle of winter (January 12), for the growing crops, and in summer (April 14), for the warlike expeditions to be undertaken. The old Norsemen knew only two seasons: winter (October 14, April 14), and summer.

† Svithiod is the present Sweden or Svealand.

‡ Skade was the daughter of the giant Thjasse. See Younger Edda.

CHAPTER X.—*Of Odin's Death.*

Odin died in his bed in Sweden; and when he was near his death he made himself be marked with the point of a spear,* and said he was going to Godheim, and would give a welcome there to all his friends, and all brave warriors should be dedicated to him; and the Swedes believed that he was gone to the ancient Asgard, and would live there eternally. Then began anew the belief in Odin, and the calling upon him. The Swedes believed that he often showed himself to them before any great battle.† To some he gave victory; others he invited to himself; and they reckoned both of these to be well off in their fate. Odin was burnt, and at his pile there was great splendour. It was their faith, that the higher the smoke arose in the air, the higher he would be raised whose pile it was; and the richer he would be, the more property that was consumed with him.

CHAPTER XI.—*Of Niord.*

Niord of Noatun was then the sole sovereign of the Swedes; and he continued the sacrifices, and was called the drot or sovereign by the Swedes, and

---

* The meaning seems to be, that he was marked with the sign of the head of a spear; that is, with the sign of the cross. The sign of Thor's hammer, viz., the head of a battle-axe or halberd, was said to be used as the sign of the cross was after Christianity was introduced: it was a kind of consecration by a holy sign. But this is probably a pious interpolation.—L.

† Thus Odin appeared before the battle near Lena, in 1208, see Fornmanna Sögur ix. 175.

he received scat and gifts from them. In his days were peace and plenty, and such good years, in all respects, that the Swedes believed Niord ruled over the growth of seasons and the prosperity of the people. In his time all the diar or gods died, and blood-sacrifices were made for them. Niord died on the bed of sickness, and before he died made himself be marked for Odin with the spear-point. The Swedes burned him, and all wept much over his grave-mound.

Chapter XII.—*Frey's Death.*

Frey took the kingdom after Niord, and was called drot by the Swedes, and they paid taxes to him. He was, like his father, fortunate in friends and in good seasons. Frey built a great temple at Upsala,* made it his chief seat, and gave it all his taxes, his land, and goods. Then began the Upsala domains,† which have remained ever since. Then began, in his days, the Frode-peace; and then there were good seasons in all the land, which the Swedes ascribed to Frey; so that he was more worshipped than the other gods, as the people became much richer in his days by reason of the peace and good seasons. His wife was called Gerd, daughter of Gymer, and their son was called Fiolner. Frey was called

---

* Upsala (properly Uppsala) is in fact genitive plural of Uppsalir, but the old preposition *till* governed genitive, and in course of time the place got the nominative form Upsala, as it is now called.

† The Upsala domains were certain estates for the support of the sovereign, and of the temple and rites of worship; which after the introduction of Christianity remained with the crown, and constituted a large portion of the crown property in Sweden.—L.

by another name, Yngve; and this name Yngve was considered long after in his race as a name of honour, so that his descendants have since been called Ynglings. Frey fell into a sickness; and as his illness took the upper hand, his men took the plan of letting few approach him. In the meantime they raised a great mound, in which they placed a door with three holes in it. Now when Frey died they bore him secretly into the mound, but told the Swedes he was alive; and they kept watch over him for three years. They brought all the taxes into the mound, and through the one hole they put in the gold, through the other the silver, and through the third the copper money that was paid. Peace and good seasons continued.

CHAPTER XIII.—*Of Freyja and her Daughters.*

Freyja alone remained of the gods, and she became on this account so celebrated that all women of distinction were called by her name, whence they now have the title Frue; so that every woman is called frue, or mistress over her property, and the wife is called the house-frue. Freyja continued the blood-sacrifices. Freyja was rather fickle-minded. Her husband was called Od, and her daughters Hnos and Gerseme. They were so very beautiful, that afterwards the most precious jewels were called by their names.

When it became known to the Swedes that Frey was dead, and yet peace and good seasons continued, they believed that it must be so as long as Frey remained in Sweden; and therefore they would not burn his remains, but called him the god of this world, and afterwards offered continually blood-sacrifices to him, principally for peace and good seasons.

CHAPTER XIV.—*Of King Fiolner's Death.*

Fiolner, Yngve Frey's son, ruled thereafter over the Swedes and the Upsala domains. He was powerful, and lucky in seasons and in holding the peace. Fredfrode ruled then in Leire, and between them there was great friendship and visiting. Once when Fiolner went to Frode in Seeland, a great feast* was prepared for him, and invitations to it were sent all over the country. Frode had a large house, in which there was a great vessel many ells high, and put together of great pieces of timber; and this vessel stood in a lower room. Above it was a loft, in the floor of which was an opening through which liquor was poured into this vessel. The vessel was full of mead, which was excessively strong. In the evening Fiolner, with his attendants, was taken into the adjoining loft to sleep. In the night he went out on the balcony outside to seek the privy of the house, and he was very sleepy, and

---

* The old Norse word is *veizla*, a grant, gift, or allowance. It may mean simply a feast or banquet, but in Snorre's Sagas of the Kings it usually refers to the receptions or entertainments to be given, according to law, to the Norse kings or to the king's landed men or his stewards; for, as will appear in the Heimskringla, the king in olden times used to go on regular circuits through his kingdom, taking each county in turn; his retinue, the places of entertainment, and the time of his staying at each place, being regulated by law. See Vigfusson, Dict., *s.v.*

exceedingly drunk. As he came back to his room he went along the balcony to the door of another loft, went into it, and his foot slipping he fell into the vessel of mead, and was drowned. So says Thiodolf of Hvin :—

> " In Frode's hall the fearful word,
> The death-foreboding sound was heard :
> The cry of fey * denouncing doom,
> Was heard at night in Frode's home.
> And when brave Frode came, he found
> Svithiod's dark chief, Fiolner, drowned.
> In Frode's mansion drowned was he,
> Drowned in a waveless, windless sea."

CHAPTER XV.—*Of Svegder.*

Svegder took the kingdom after his father, and he made a solemn vow to seek Godheim and Odin the old. He went with twelve men through the world, and came to Tyrkland, and the Great Svithiod, where he found many of his connections. He was five years on this journey; and when he returned home to Svithiod he remained there for some time. He had got a wife in Vanheim, who was called Vana, and their son was Vanland. Svedger went out afterwards to seek again for Godheim, and came to a mansion on the east side of Svithiod † called

---

* Fey, feig, (Icelandic, *feigr*, doomed to die), is used in the same sense in the northern languages as in Scotland, denoting the acts or words or sounds preceding, and supposed to be portending, a sudden death. "The gauger is fey," in Sir Walter Scott's novel "Guy Mannering," is an expression seized by that great painter of Scottish life from the common people, and applied in its true meaning.—L.

† It is not clear whether the Greater or Minor Svithiod is here meant. In Upland in Sweden are found gards by this name, but there is also a place called Stein, in Estland.

Stein, where there was a stone as big as a large house. In the evening after sunset, as Svegder was going from the drinking-table to his sleeping-room, he cast his eye upon the stone, and saw that a dwarf was sitting under it. Svegder and his man were very drunk, and they ran towards the stone. The dwarf stood in the door, and called to Svegder, and told him to come in, and he should see Odin. Svedger ran into the stone, which instantly closed behind him, and Svedger never came back. Thiodolf of Hvin tells of this:—

> " By Durner's* elfin race,
> Who haunt the cliffs and shun day's face,
> The valiant Svegder was deceived,
> The elf's false words the king believed.
> The dauntless hero rushing on,'
> Passed through the yawning mouth of stone:
> It yawned—it shut—the hero fell,
> In Sokmimer's † hall, where giants dwell."

CHAPTER XVI.—*Of Vanlande, Svegder's Son.*

Vanlande, Svegder's son, succeeded his father, and ruled over the Upsala domain. He was a great warrior, and went far around in different lands. Once he took up his winter abode in Finland with Snow the Old, and got his daughter Drifa in marriage; but in spring he set out leaving Drifa behind, and although he had promised to return within three years he did not come back for ten. Then Drifa

---

* Durnr, the second chief of the dwarfs or elves, in the Scandinavian mythology.—L.

† Sokmimer—the giant of the deep, the destructive maelstrom of the ocean.—L.

sent a message to the witch Huld; and sent Visbur, her son by Vanlande, to Sweden. Drifa bribed the witch-wife Huld, either that she should bewitch Vanlande to return to Finland, or kill him. When this witch-work was going on Vanlande was at Upsala, and a great desire came over him to go to Finland; but his friends and counsellors advised him against it, and said the witchcraft of the Fin people showed itself in this desire of his to go there. He then became very drowsy, and laid himself down to sleep; but when he had slept but a little while, he cried out, saying, "Mara* was treading upon him." His men hastened to him to help him; but when they took hold of his head she trod on his legs, so that they nearly broke, and when they laid hold of his legs she pressed upon his head; and it was his death. The Swedes took his body and burnt it at a river called Skut, where a standing stone was raised over him. Thus says Thiodolf:—

> "And Vanlande, in a fatal hour,
> Was dragg'd by Grimhild's daughter's power,
> The witch-wife's, to the dwelling-place
> Where men meet Odin face to face.
> Trampled to death, to old Skut's shore
> The corpse his faithful followers bore;
> And there they burnt, with heavy hearts,
> The good chief killed by witchcraft's arts."

CHAPTER XVII.—*Of Visbur's Death.*

Visbur inherited after his father Vanlande. He married the daughter of Aude the Rich, and gave

* Mara, the nightmare. We retain the name, and the notion that it is a demon riding or treading on the sleeper.—L.

her as her bride-gift three large farms, and a gold ornament. They had two sons, Gisl and Ondur; but Visbur left her and took another wife, whereupon she went home to her father with her two sons. Visbur had a son who was called Domalde, and his stepmother used witchcraft to give him ill-luck. Now, when Visbur's sons were, the one twelve, the other thirteen years of age, they went to their father's place, and desired to have their mother's dower; but he would not deliver it to them. Then they said that the gold ornament should be the death of the best man in all his race; and they returned home. Then they began again with enchantments and witchcraft, to try if they could destroy their father. The volva* Huld said that by witchcraft she would bring about not only this, but also that a murderer of his own kin should never be wanting in the Yngling race; and they agreed to have it so. Thereafter they collected men, came unexpectedly in the night on Visbur, and burned him in his house. So sang Thiodolf:—

> "Have the fire-dogs' fierce tongues yelling
> Lapt Visbur's blood on his own hearth?
> Have the flames consumed the dwelling
> Of the hero's soul on earth?
> Madly ye acted, who set free
> The forest foe, red fire, night thief,
> Fell brother of the raging sea,†
> Against your father and your chief."

---

* The volva, sometimes erroneously written vola, is a word of uncertain etymology meaning a prophetess, sibyl, wise woman. Volva, seid-kona, and spakona are synonymous. They are all skilled in sorcery.

† Forniot was father of Loge, Hler, and Kare; or Fire, the Sea, and the Wind; and hence fire is called by the skalds the brother of the sea.

CHAPTER XVIII.—*Of Domalde's Death.*

Domalde took the heritage after his father Visbur, and ruled over the land. As in his time there was great famine and distress in Svithiod, the Swedes made great offerings of sacrifice at Upsala. The first autumn they sacrificed oxen, but the succeeding season was not improved by it. The following autumn they sacrificed men, but the succeeding year was rather worse. The third autumn, when the offer of sacrifices should begin, a great multitude of Swedes came to Upsala; and now the chiefs held consultations with each other, and all agreed that the times of scarcity were on account of their king Domalde, and they resolved to offer him for good seasons, and to assault and kill him, and sprinkle the altar of the gods with his blood. And they did so. Thiodolf tells of this:—

> "It has happened oft ere now,
> That foeman's weapon has laid low
> The crowned head, where battle plain
> Was miry red with the blood-rain.
> But Domalde dies by bloody arms,
> Raised not by foes in war's alarms,—
> Raised by his Swedish liegeman's hand,
> To bring good seasons to the land."

CHAPTER XIX.—*Of Domar's Death.*

Domalde's son, called Domar, next ruled over the land. He reigned long, and in his days were good

---

Loge is a word still retained n the northern parts of Scotland to signify fire. The *lowe*, for the blaze or flame of fire, is indeed in general use in Scotland.—L.

seasons and peace. Nothing is told of him but that he died in his bed in Upsala, and was transported to the Fyrisvols,* where his body was burned on the river-bank, and where his standing stone still remains. So says Thiodolf:—

> "I have asked wise men to tell
> Where Domar rests, and they knew well.
> Domar, on Fyri's wide-spread ground,
> Was burned, and laid on Yngve's mound."

CHAPTER XX.—*Of Dygve's Death.*

Dygve was the name of his son, who succeeded him in ruling the land; and about him nothing is said but that he died in his bed. Thiodolf tells of it thus:—

> "Dygve the Brave, the mighty king,
> It is no hidden secret thing,
> Has gone to meet a royal mate,
> Riding upon the horse of Fate.
> For Loke's daughter † in her house
> Of Yngve's race would have a spouse;
> Therefore the fell-one snatched away
> Brave Dygve from the light of day."

Dygve's mother was Drot, a daughter of King Danp, the son of Rig, who was first called king in the Danish tongue.‡ His descendants always afterwards considered the title of king the title of highest dignity. Dygve was the first of his family called

---

* The plains near Upsala, now called by the Swedes Fyrisvall.

† Loke's (the evil principle) daughter was Hel, who received in the under world those who, not having fallen in battle, were not received by Odin in Valhal. Our word "hell" is connected with the name of this goddess apparently.—L.

‡ That is, throughout the North.

king, for his predecessors had been called *Drottnar*, and their wives *Dróttningar*, and their court *Drott*.* Each of their race was called Yngve, or Yngune, and the whole race together Ynglings.† The queen Drot was a sister of King Dan Mikillate, from whom Denmark took its name.

Chapter XXI.—*Of Dag the Wise.*

King Dygve's son, called Dag, succeeded to him, and was so wise a man that he understood the language of birds. He had a sparrow which told him much news, and flew to different countries. Once the sparrow flew to Reidgotaland,‡ to a farm called Vorve, where he flew into the peasant's corn-field and took his grain. The peasant came up, took a stone, and killed the sparrow. King Dag was ill pleased that the sparrow did not come home; and

---

* See further on this point in Rigsmal in the Elder Edda.

† Is it possible that the Ingævones of Tacitus can have any relation to this of tribe Yngune or Yngve? The passage, cap. 2, "De Moribus Germaniæ," has a remarkable coincidence with the saga story of these Northmen. "Celebrant carminibus antiquis (quod unum apud illos memoriæ et annalium genus est) Tuisconem Deum, terra editum, et filium Mannum, originem gentis conditoresque. Manno tres filios e quorum nominibus proximi oceano Ingævones." Here is a tribe of Ingæve deriving their origin from the gods, like the Yngve or Yngune of the saga.—L.

‡ Reidgotaland is understood to mean Jutland,[1] and Eygotaland the islands inhabited by the same people. It is by no means clear that the so called Gotlanders on the Baltic coast have any connection with the great population called the Goths, unless a fortuitous similarity of name and a common origin. That the vast hordes called Goths who overwhelmed Italy came from these Gotlands, it is inconsistent with common sense to suppose. The whole coasts of the Baltic could furnish no such masses of armed men now even, when they furnish more subsistence for man.—L.

---

[1] Reidgotaland is supposed by some authorities to mean the island Gotland. The *reid* means waggon or chariot.

as he, in a sacrifice of expiation, inquired after the sparrow, he got the answer that it was killed at Vorve. Thereupon he ordered a great army, and went to Gotland; and when he came to Vorve he landed with his men and plundered, and the people fled away before him. King Dag returned in the evening to his ships, after having killed many people and taken many prisoners. As they were going across a river at a place called Skiotan's Ford, or Weapon Ford, a labouring thrall came running to the river-side, and threw a hay-fork into their troop. It struck the king on the head, so that he fell instantly from his horse and died, and his men went back to Svithiod. In those times the chief who ravaged a country was called Gram,* and the men-at-arms under him Grams. Thiodolf sings of it thus:—

> "What news is this that the king's men,
> Flying eastward through the glen,
> Report? That Dag the Brave, whose name
> Is sounded far and wide by Fame,—
> That Dag, who knew so well to wield
> The battle-axe in bloody field,
> Where brave men meet, no more will head
> The brave—that mighty Dag is dead!
>
> "Vorve was wasted with the sword,
> And vengeance taken for the bird,—
> The little bird that used to bring
> News to the ear of the great king.
> Vorve was ravaged, and the strife
> Was ended when the monarch's life
> Was ended too—the great Dag fell
> By the hay-fork of a base thrall!"

---

* Gram is equivalent to grim, fierce.—L.

CHAPTER XXII.—*Of Agne.*

Agne was the name of Dag's son, who was king after him,—a powerful and celebrated man, expert, and exercised in all feats. It happened one summer that King Agne went with his army to Finland, and landed and marauded. The Fins gathered a large army, and proceeded to the strife under a chief called Froste. There was a great battle, in which King Agne gained the victory, and Froste fell there with a great many of his people. King Agne proceeded with armed hand through Finland, subdued it, and made enormous booty. He took Froste's daughter Skialf, and her brother Loge, and carried them along with him. When he sailed from the east he came to land at Stoksund,[*] and put up his tent to the south on the ness, where then there was a wood. King Agne had at the time the gold ornament which had belonged to Visbur. He now married Skialf, and she begged him to make a burial feast in honour of her father. He invited a great many distinguished guests, and made a great feast. He had become very celebrated by his expedition, and there was a great drinking match. Now when King Agne had got drunk, Skialf bade him take care of his gold ornament which he had about his neck; therefore he took hold of the ornament, and bound it fast about his neck before he went to sleep. The land-tent stood at the wood side, and a high

---

[*] Stoksund is the sound or stream at Stockholm, between the Mælar lake and the sea.

tree over the tent protected it against the heat of the sun. Now when King Agne was asleep, Skialf took a noose, and fastened it under the ornament. Thereupon her men threw down the tent-poles, cast the loop of the noose up in the branches of the tree, and hauled upon it, so that the king was hanged close under the branches and died; and Skialf with her men ran down to their ships, and rowed away. King Agne was cremated upon the spot, which was afterwards called Agnafit;* and it lies on the east side of the Taurrin, and west of Stoksund. Thiodolf speaks of it thus:—

> "How do ye like the high-souled maid,
> Who, with the grim Fate-goddess' aid,
> Avenged her sire?—made Svithiod's king
> Through air in golden halter swing?
> How do ye like her, Agne's men?
> Think ye that any chief again
> Will court the fate your chief befell,
> To ride on wooden horse to hell?"

CHAPTER XXIII.—*Of Alrek and Eirik.*

The sons of Agne were called Alrek and Eirik, and were kings together after him. They were powerful men, great warriors, and expert at all feats of arms. It was their custom to ride and break in horses both to walk and to gallop, which nobody understood so well as they; and they vied with each other who could ride best, and keep the best horses. It happened one day that both the brothers rode out to-

---

* Icelandic, *fit*, genitive *fitjar*, means literally the webbed foot of water-birds, but metaphorically meadow-land on the banks of a firth, lake, or river, and is here used in the latter sense.

gether alone, and at a distance from their followers, with their best horses, and rode on to a field; but never came back. The people at last came out to look after them, and they were both found dead with their heads crushed. As they had no weapons, except it might be their horses' bridles, people believed that they had killed each other with them. So says Thiodolf:—

> "Alrek fell, by Eirik slain,
> Eirik's life-blood dyed the plain.
> Brother fell by brother's hand;
> And they tell it in the land,
> That they worked the wicked deed
> With the sharp bits that guide the steed.
> Shall it be said of Frey's brave sons,
> The kingly race, the noble ones,
> That they have fought in deadly battle
> With the head-gear of their cattle?"

CHAPTER XXIV.—*Of Yngve and Alf.*

Alrek's sons, Yngve and Alf, then succeeded to the kingly power in Svithiod. Yngve was a great warrior, always victorious; handsome, expert in all exercises, strong and very sharp in battle, generous and full of mirth; so that he was both renowned and beloved. Alf his brother was a silent, harsh, unfriendly man, and sat at home in the land, and never went out on war expeditions. They called him Elfse. His mother was called Dageid, a daughter of King Dag the Great, from whom the family of Daglings are descended. King Alf had a wife named Bera,* who was the most agreeable of women, very

* A female bear.

brisk and gay. One autumn Yngve, Alrek's son, had arrived at Upsala from a viking cruise by which he was become very celebrated. He often sat long in the evening at the drinking table; but Alf went willingly to bed very early. Queen Bera sat often till late in the evening, and she and Yngve conversed together for their amusement; but Alf often told her that she should not sit up so late in the evening, but should go first to bed, so as not to waken him. She replied, that happy would be the woman who had Yngve instead of Alf for her husband; and as she often repeated the same, he became very angry. One evening Alf went into the hall, where Yngve and Bera sat on the high-seat speaking to each other. Yngve had a short sword upon his knees, and the guests were so drunk that they did not observe the king's coming in. King Alf went straight to the high-seat, drew a sword from under his cloak, and pierced his brother Yngve through and through. Yngve leaped up, drew his short sword, and gave Alf his death-wound; so that both fell dead on the floor. Alf and Yngve were buried under mounds in Fyrisvols. Thus tells Thiodolf of it:—

> "I tell you of a horrid thing,
> A deed of dreadful note I sing,—
> How by false Bera, wicked queen,
> The murderous brother-hands were seen
> Each raised against a brother's life;
> How wretched Alf with bloody knife
> Gored Yngve's heart, and Yngve's blade
> Alf on the bloody threshold laid.
> Can men resist Fate's iron laws?
> They slew each other without cause.'

CHAPTER XXV.—*Of Hugleik.*

Hugleik was the name of King Alf's son, who succeeded the two brothers in the kingdom of the Swedes, the sons of Yngve being still children. King Hugleik was no warrior, but sat quietly at home in his country. He was very rich, but had still more the reputation of being very greedy. He had at his court all sorts of players, who played on harps, fiddles, and viols; and had with him magicians, and all sorts of witches. Hake and Hagbard were two brothers, very celebrated as sea-kings, who had a great force of men-at-arms. Sometimes they cruised in company, sometimes each for himself, and many warriors followed them both. King Hake came with his troops to Svithiod against King Hugleik, who, on his side, collected a great army to oppose him. Two brothers came to his assistance, Svipdag and Geigad, both very celebrated men, and powerful combatants. King Hake had about him twelve champions, and among them Starkad the Old;* and King Hake himself was a murderous combatant. They met on Fyrisvols, and there was a great battle, in which King Hugleik's army was soon defeated. Then the combatants, Svipdag and Geigad, pressed forward manfully; but Hake's champions went six against one, and they were both taken prisoners. Then King Hake penetrated within the shield-circle †

---

* Full accounts of Starkad and of Hagbard can be found in Saxo Grammaticus.

† A bulwark or covering of shields—the testudo of the Romans—seems always to have been formed round the king's person in battle.—L.

around King Hugleik, and killed him and two of his sons within it. After this the Swedes fled; and King Hake subdued the country, and became king of Svithiod. He then sat quietly at home for three years; but during that time his combatants went abroad on viking expeditions, and gathered property for themselves.

CHAPTER XXVI.—*King Gudlaug's Death.*

Jorund and Eirik, the sons of Yngve Alrekson, lay all this time in their war-ships, and were great warriors. One summer they marauded in Denmark, where they met a King Gudlaug, king of the Haleygians, and had a battle with him, which ended in their clearing Gudlaug's ship and taking him prisoner. They carried him to the land at Straumeynes, and hanged him there, and afterwards his men raised a mound over him. So says Eyvind Skaldaspiller:—

> "By the fierce East-kings'* cruel pride,
> Gudlaug must on the wild horse ride—
> The wildest horse you e'er did see:
> 'Tis Sigar's steed—the gallows tree.
> At Straumeynes the tree did grow,
> Where Gudlaug's corpse waves on the bough.
> A high stone stands on Straumey's heath,
> To tell the gallant hero's death."

CHAPTER XXVII.—*Of King Hake.*

The brothers Eirik and Jorund became much celebrated by this deed, and appeared to be much

---

* The Swedish kings Jorund and Eirik, of Yngve's race, are said to be of the East—as relative to Norway, from which Gudlaug came.—L.

greater men than before. When they heard that King Hake in Svithiod had sent from him his champions, they steered towards Svithiod, and gathered together a strong force. As soon as the Swedes heard that the Ynglings were come to them, they flocked to the brothers in multitudes. The brothers proceeded up the Lag (Mælar) lake, and advanced towards Upsala against King Hake, who came out against them on the Fyrisvols with far fewer people. There was a great battle, in which King Hake went forward so bravely that he killed all who were nearest to him, and at last killed King Eirik, and cut down the banner of the two brothers. King Jorund with all his men fled to their ships. King Hake had been so grievously wounded that he saw his days could not be long; so he ordered a war-ship which he had to be loaded with his dead men and their weapons, and to be taken out to the sea; the tiller to be shipped, and the sails hoisted. Then he set fire to some tar-wood, and ordered a pile to be made over it in the ship. Hake was almost if not quite dead, when he was laid upon this pile of his. The wind was blowing off the land,—the ship flew, burning in clear flame, out between the islets, into the ocean. Great was the fame of this deed in after times.

CHAPTER XXVIII.—*Jorund's Death.*

Jorund, King Yngve's son, remained king at Upsala. He ruled the country; but was often, in summer, out on war expeditions. One summer he went with his forces to Denmark; and having plundered all around in Jutland, he went into Limfiord in autumn, and marauded there also. While he was thus lying in Oddasund with his people, King Gylaug of Halogaland, a son of King Gudlaug, of whom mention is made before, came up with a great force, and gave battle to Jorund. When the country people saw this they swarmed from all parts towards the battle, in great ships and small; and Jorund was overpowered by the multitude, and his ships cleared of their men. He sprang overboard, but was made prisoner and carried to the land. Gylaug ordered a gallows to be erected, led Jorund to it, and had him hanged there. So ended his life. Thiodolf talks of this event thus:—

> "Jorund has travelled far and wide,
> But the same horse he must bestride
> On which he made brave Gudlaug ride.
> He too must for a necklace wear
> Hagbard's * fell noose in middle air.
> The army leader thus must ride
> On Odin's † horse, at Limfiord's side."

CHAPTER XXIX.—*Of King Ane's Death.*

Aun or Ane was the name of Jorund's son, who became king of the Swedes after his father. He was a wise man, who made great sacrifices to the gods; but, being no warrior, he lived quietly at

---

* Hagbard's noose—the gallows rope by which Hagbard was hanged.—L.
† Odin was the god of the hanged; and Odin's horse was a name for the gallows. The gallows are (as on p. 298) called the horse of Sigar, from the love tale of the hero by that name.

home. In the time when the kings we have been speaking of were in Upsala, Denmark had been ruled over by Dan Mikellate, who lived to a very great age; then by his son, Frode Mikellate, or the Peace-loving, who was succeeded by his sons Halfdan and Fridleif, who were great warriors. Halfdan was older than his brother, and above him in all things. He went with his army against King Aun to Svithiod, and was always victorious. At last King Aun fled to West Gautland when he had been king in Upsala about twenty-five years, and was in Gautland twenty-five years, while Halfdan remained king in Upsala. King Halfdan died in his bed, and was buried there in a mound; and King Aun returned to Upsala when he was sixty years of age. He made a great sacrifice, and in it offered up his son to Odin. Aun got an answer from Odin, that he should live sixty years longer; and he was afterwards king in Upsala for twenty-five years. Now came Ale the Bold, a son of King Fridleif, with his army to Svithiod against King Aun, and they had several battles with each other; but Ale was always the victor. Then Aun fled a second time to West Gautland; and for twenty-five years Ale reigned in Upsala, until he was killed by Starkad the Old. After Ale's fall, Aun returned to Upsala and ruled the kingdom for twenty-five years. Then he made a great sacrifice again for long life, in which he sacrificed his second son, and received the answer from Odin, that he should live as long as he gave him one of his sons every tenth year, and also that he should name one of the districts of his country after the number of sons he should offer to Odin. When he had sacrificed the seventh of his sons he continued to live for ten years; but so that he could not walk, but was carried on a chair. Then he sacrificed his eighth son, and lived thereafter ten years, lying in his bed. Now he sacrificed his ninth son, and lived ten years more; but so that he drank out of a horn like an infant. He had now only one son remaining, whom he also wanted to sacrifice, and to give Odin Upsala and the domains thereunto belonging, under the name of the Tenth Land, but the Swedes would not allow it; so there was no sacrifice, and King Aun died, and was buried in a mound at Upsala. Since that time it is called Aun's sickness when a man dies, without pain, of extreme old age. Thiodolf tells of this :—

> "In Upsala town the cruel king
> Slaughtered his sons at Odin's shrine—
> Slaughtered his sons with cruel knife,
> To get from Odin length of life.
> He lived until he had to turn
> His toothless mouth to the deer's horn;
> And he who shed his children's blood
> Sucked through the ox's horn his food.
> At length fell Death has tracked him down,
> Slowly, but sure, in Upsala town."

CHAPTER XXX.—*Of Egil and Tunne.*

Egil was the name of Aun the Old's son, who succeeded as king in Svithiod after his father's death. He was no warrior, but sat quietly at home. Tunne was the name of a slave who had been the counsellor and treasurer of Aun the Old; and when Aun died Tunne took much treasure and buried it in the earth. Now when Egil became king he put Tunne among the other slaves, which he took very ill, and ran away with others of the slaves. They dug up the treasures which Tunne had concealed, and he gave them to his men, and was made their chief. Afterwards many malefactors flocked to him; and they lay out in the woods, but sometimes fell upon the domains, pillaging and killing the people. When King Egil heard this, he went out with his forces to pursue them; but one night when he had taken up his night quarters, Tunne came there with his men, fell on the king's men unexpectedly, and killed many of them. As soon as King Egil perceived the tumult, he prepared for defence, and set up his banner; but many people deserted him, because Tunne and his men attacked them so boldly, and King Egil saw that nothing was left but to fly. Tunne pursued the fugitives into the forest, and then returned to the inhabited land, ravaging and plundering without resistance. All the goods that fell into Tunne's hands he gave to his people, and thus became popular and strong in men. King Egil assembled an army again, and hastened to give battle to Tunne. But Tunne was again victorious, and King Egil fled with the loss of many people. Egil and Tunne had eight battles with each other, and Tunne always gained the victory. Then King Egil fled out of the country, and went to Seeland in Denmark, to Frode the Bold, and promised him a scat from the Swedes to obtain help. Frode gave him an army, and also his champions, with which force King Egil repaired to Svithiod. When Tunne heard this he came out to meet him; and there was a great battle, in which Tunne fell, and King Egil recovered his kingdom, and the Danes returned home. King Egil sent King Frode great and good presents every year, but he paid no scat to the Danes; but notwithstanding the friendship between Egil and Frode continued without interruption. After Tunne's fall, Egil ruled the kingdom for three years. It happened in Svithiod that an old bull, which was destined for sacrifice, was fed so high that he became dangerous to people; and when they were going to lay hold of him he escaped into the woods, became furious, and was long in the forest committing great damage to the people. King Egil was a great hunter, and often rode into the forest to chase wild animals. Once he rode out with his men to hunt in the forest. The king had traced an animal a long while, and followed it in the forest, separated from all his men. He observed at last that it was the bull, and rode up to it to kill it. The bull turned round suddenly, and the king struck him with his spear; but it tore itself out of the wound. The bull now struck his horn in the side of the horse, so that he instantly fell flat on the earth with the king. The king sprang up, and was drawing his sword, when the bull struck his horns right into the king's breast. The king's men then came up and killed the bull. The king lived but a short time, and was buried in a mound at Upsala. Thiodolf sings of it thus:—

> "The fair-haired son of Odin's race,
> Who fled before fierce Tunne's face,
> Has perished by the demon-beast
> Who roams the forest of the East.
> The hero's breast met the full brunt
> Of the wild bull's shaggy front;
> The hero's heart's asunder torn
> By the fell Jotun's spear-like horn."

Chapter XXXI.—*Of King Ottar.*

Ottar* was the name of King Egil's son who succeeded to the domains and kingdom after him. He did not continue friendly with King Frode, and therefore King Frode sent messengers to King Ottar to demand the scat which Egil had promised him. Ottar replied, that the Swedes had never paid scat to the Danes, neither would he; and the messengers had to depart with this answer. Frode was a great warrior; and he came one summer with his army to Svithiod, and landed and ravaged the country. He killed many people, took some prisoners, burned all around in the inhabited parts, made a great booty, and made great devastation. The next summer King Frode made an expedition to the eastward;† and when King Ottar heard that Frode was

---

* The Beowulf poem names among the Shilfings the old Ongenþew and his sons Onela and Ohthere (Ottar). Ohthere's sons, one of whom was named Eadgils, and the other Eánmund, came into conflict with the ruling king, a son of Ongenþeow (Angantyr?) and had to take flight from him to the Gauts.

† Icelandic, *i Austrveg;* that is, to the east side of the Baltic.

not at home in his own country, he went on board his own ships, sailed over to Denmark, and ravaged there without opposition. As he heard that a great many people were collected at Seeland, he proceeds westward to the Sound, and sails north about to Jutland; lands at Limfiord; plunders the Vendel district;* burns, and lays waste, and makes desolate the country he goes over with his army. Vot and Faste were the names of the earls whom Frode had appointed to defend the country in Denmark while he was abroad. When the earls heard that the Swedish king was laying waste Denmark, they collected an army, hastened on board their ships, and sailed by the south side to Limfiord. They came unexpectedly upon Ottar, and the battle began immediately. The Swedes gave them a good reception, and many people fell on both sides; but as soon as men fell in the Danish army other men hastened from the country to fill their places, and also all the vessels in the neighbourhood joined them. The battle ended with the fall of Ottar and the greater part of his people. The Danes took his body, carried it to the land, laid it upon a mound of earth, and let the wild beasts and ravens tear it into pieces. Thereafter they made a figure of a crow out of wood, sent it to Sweden, and sent word with it that their king, Ottar, was no better than it; and from this he was called Ottar Vendelcrow. Thiodolf tells so of it:—

"By Danish arms the hero bold,
Ottar the Brave, lies stiff and cold.

---

* Vendel, the part of Jutland north of Limfiord, now called Vendsyssel.

> To Vendel's plain the corpse was borne;
> By eagles' claws the corpse is torn,
> Spattered by ravens' bloody feet,
> The wild bird's prey, the wild wolf's meat.
> The Swedes have vowed revenge to take
> On Frode's earls, for Ottar's sake;
> Like dogs to kill them in their land,
> In their own homes, by Swedish hand."

CHAPTER XXXII.—*Of King Adils' Marriage.*

Adils was the name of King Ottar's son and successor. He was a long time king, became very rich, and went also for several summers on viking expeditions. On one of these he came to Saxonland with his troops. There a king was reigning called Geirthiof, and his wife was called Alof the Great; but nothing is told of their children. The king was not at home, and Adils and his men ran up to the king's house and plundered it, while others drove a herd of cattle down to the strand.* The herd was attended by slave-people, carls, and girls, and they took all of them together. Among them was a remarkably beautiful girl called Yrsa. Adils returned home with this plunder. Yrsa was not one of the slave girls, and it was soon observed that she was intelligent, spoke well, and in all respects was well

---

* The ordinary way, with the vikings, of victualling their ships, was driving cattle down to the strand and killing them, without regard to the property of friends or enemies; and this was so established a practice, that it was expressed in a single word, "strandhug." King Harald Harfager had prohibited the strandhug being committed in his own dominions by his own subjects on their viking cruises; and Rolf Ganger, the son of the Earl of More, having, notwithstanding, landed and made a strandhug in the South of Norway where the king happened to be, was outlawed; and he in consequence set out on an expedition, in which he conquered and settled in Normandy.—L.

behaved. All people thought well of her, and particularly the king; and at last it came to so far that the king celebrated his wedding with her, and Yrsa became queen of Svithiod, and was considered an excellent woman.

CHAPTER XXXIII.—*Of King Adils' Death.*

King Halfdan's son Helge ruled at that time over Leire. He came to Svithiod with so great an army, that King Adils saw no other way than to fly at once. King Helge landed with his army, plundered, and made a great booty. He took Queen Yrsa prisoner, carried her with him to Leire, took her to wife, and had a son by her called Rolf Krake. When Rolf was three years old, Queen Alof came to Denmark, and told Queen Yrsa that her husband, King Helge, was her own father, and that she, Alof, was her mother. Thereupon Yrsa went back to Svithiod to King Adils, and was queen there as long as she lived. King Helge fell in a war expedition; and Rolf Krake, who was then eight years old, was taken to be king in Leire. King Adils had many disputes with a king called Ale of the Uplands; he was from Norway; and these kings had a battle on the ice of the Vener lake, in which King Ale fell, and King Adils won the battle. There is a long account of this battle in the Skioldunga Saga,* and also about Rolf Krake's coming to Adils, and sowing gold upon the Fyrisvols. King Adils was a great lover of

* The Skjoldunga Saga is lost, but a reference to it is also found in Skaldskaparmal of the Younger Edda.

good horses, and had the best horses of these times. One of his horses was called Slongver, and another Hrafn. This horse he had taken from Ale on his death, and bred from him a horse, also called Hrafn, which the king sent as a present to King Godgest in Halogaland. When Godgest mounted the horse he was not able to manage him, and fell off, and was killed. This accident happened at Omd in Halogaland.* King Adils was at a Disa † sacrifice; and as he rode around the Disa hall his horse Hrafn stumbled and fell, and the king was thrown forward upon his head, and his skull was split, and his brains dashed out against a stone. Adils died at Upsala, and was buried there in a mound. The Swedes called him a great king. Thiodolf speaks thus of him :—

> " Witch-demons, I have heard men say,
> Have taken Adils' life away.
> The son of kings of Frey's great race,
> First in the fray, the fight, the chase,
> Fell from his steed—his clotted brains
> Lie mixed with mire on Upsala's plains.
> Such death (grim Fate has willed it so)
> Has struck down Ale's deadly foe."

CHAPTER XXXIV.—*Rolf Krake's Death.*

Eystein, King Adils' son, ruled next over Svithiod, and in his lifetime Rolf Krake of Leire fell. In those days many kings, both Danes and Northmen,

---

* Halogaland is the province of Norway now called Nordland, extending from the Namsen river north to Westfiord, where it joins the province of Finmark.—L.

† The dises (Icel. *dis*, pl. *disir*) of Teutonic mythology were goddesses or female guardian angels, who followed every man from his birth, and only left him in the hour of his death.

ravished the Swedish dominions; for there were many sea-kings who ruled over many people, but had no lands, and he might well be called a sea-king who never slept beneath sooty roof-timbers, and never drank near the hearthstone.

CHAPTER XXXV.—*Of Eystein and the Jutland King Solve.*

There was a sea-king called Solve,* a son of Hogne of Niardey,† who at that time plundered in the Baltic, but had his dominion in Jutland. He came with his forces to Svithiod, just as King Eystein was at a feast in a district called Lofund.‡ Solve came unexpectedly in the night on Eystein, surrounded the house in which the king was, and burned him and all his court. Then Solve went to Sigtuna and desired that the Swedes should receive him, and give him the title of king; but they collected an army, and tried to defend the country against him, on which there was a great battle, that lasted, according to report, eleven days. There King Solve was victorious, and was afterwards king of the Swedish dominions for a long time, until at last the Swedes betrayed him, and he was killed. Thiodolf tells of it thus:—

> "For a long time none could tell
> How Eystein died—but now I know
> That at Lofund the hero fell;
> The branch of Odin was laid low,

---

* Solve is in all later documents written *Salve*.
† Niardey, an island in North Throndhjem district.—L.
‡ Lofund, an isle in the Mælar lake, on which the palace of Drottningholm now stands.—L.

> Was burnt by Solve's Jutland men.
> The raging tree-devourer fire
> Rushed on the monarch in its ire;
> First fell the castle timbers, then
> The roof-beams—Eystein's funeral pyre."

CHAPTER XXXVI.—*Of Yngvar's Fall.*

Yngvar, who was King Eystein's son, then became king of Sweden. He was a great warrior, and often lay out with his war-ships; for the Swedish dominions were much ransacked then by Danes and East-country men. King Yngvar made a peace with the Danes; but betook himself to ravaging the East country in return. One summer he went with his forces to Esthonia, and plundered at a place called Stein. The Esthonians came down from the interior with a great army, and there was a battle; but the army of the country was so brave that the Swedes could not withstand them, and King Yngvar fell, and his people fled. He was buried close to the sea-shore under a mound, in Adalsysla*, and after this defeat the Swedes returned home. Thiodolf sings of it thus:—

> "Certain it is the Estland foe
> The fair-haired Swedish king laid low.
> On Estland's strand, o'er Swedish graves,
> The East Sea sings her song of waves;
> King Yngvar's dirge is ocean's roar
> Resounding on the rock-ribbed shore."

CHAPTER XXXVII.—*Of Onund the Land-clearer.*

Onund was the name of Yngvar's son who succeeded him. In his days there was peace in Svithiod,

* In Esthonia or Estland, written by Snorre *Eistland*.

and he became rich in valuable goods. King Onund went with his army to Esthonia to avenge his father, and landed and ravaged the country round far and wide, and returned with a great booty in autumn to Svithiod. In his time there were fruitful seasons in Svithiod, so that he was one of the most popular of kings. Svithiod is a great forest land, and there are such great uninhabited forests in it that it is a journey of many days to cross them. Onund bestowed great diligence and expense on clearing the woods and cultivating the cleared land. He made roads through the desert forests; and thus cleared land is found all through the forest country, and great districts are settled. In this way extensive tracts of land were brought into cultivation, for there were country people enough to occupy the land. Onund had roads made through all Svithiod, both through forests and morasses, and also over mountains; and he was therefore called Onund Roadmaker. He had a house built for himself in every district of Svithiod, and went over the whole country in guest-quarters.*

CHAPTER XXXVIII.—*Of Ingiald Illrade.*

Onund had a son called Ingiald, and at that time Yngvar was king of the district of Fiadrundaland.

---

* This continued to be the ordinary way of subsisting the kings and court in Norway for many generations. In Sweden the kings appear to have had a fixed residence at Upsala, and in Denmark at Leire and Odinse; while in Norway they appear to have lived always in royal progresses through the districts in turns, without any palace, castle, or fixed abode.—L.

It is somewhat strange to find Onund building roads in every district of Sweden, inasmuch as he ruled over only a part of it. See chapter 43.

Yngvar had two sons by his wife,—the one called Alf, the other Agnar,—who were about the same age as Ingiald. Onund's district-kings were at that time spread widely over Svithiod, and Svipdag the Blind ruled over Tiundaland, in which Upsala is situated, and where all the Swedish Things are held. There also were held the mid-winter sacrifices, at which many kings attended. One year at mid-winter there was a great assembly of people at Upsala, and King Yngvar had also come there with his sons. Alf, King Yngvar's son, and Ingiald, King Onund's son, were there,—both about six years old. They amused themselves with child's play, in which each should be leading on his army. In their play Ingiald found himself not so strong as Alf, and was so vexed that he cried bitterly. His foster-brother Gautved came up, led him to his foster-father Svipdag the Blind, and told him how ill it appeared that he was weaker and less manly than Alf, King Yngvar's son. Svipdag replied that it was a great shame. The day after Svipdag took the heart of a wolf, roasted it on the tongs, and gave it to the king's son Ingiald to eat, and from that time he became a most ferocious person, and of the worst disposition. When Ingiald was grown up, Onund applied for him to King Algaute for his daughter Gauthild. Algaute was a son of Gautrek the Mild, and grandson of Gaut; and from them Gautland *

---

* This derivation of the name Gautland, given to the small kingdoms in Sweden called East and West Gautland, from the name of a chief, does away with a great deal of absurd speculation that these small districts were the original seats of the mighty people called Goths

took its name.* King Algaute thought his daughter would be well married if she got King Onund's son, and if he had his father's disposition; so the girl was sent to Svithiod, and King Ingiald celebrated his wedding with her in due time.

CHAPTER XXXIX.—*Of King Onund's Death.*

King Onund one autumn, travelling between his mansion-houses, came over a road called Himinheath, where there are some narrow mountain valleys, with high mountains on both sides. There was heavy rain at the time, and before there had been snow on the mountains. A landslip of clay and stones came down upon King Onund and his people, and there he met his death, and many with him.† So says Thiodolf; namely,—

> "We all have heard how Jonaker's † sons,
> Whom weapons could not touch, with stones
> Were stoned to death—in open day,
> King Onund died in the same way.
> Or else perhaps the wood-grown land,
> Which long had felt his conquering hand,
> Uprose at length in deadly strife,
> And pressed out Onund's hated life."

---

who overwhelmed the Roman empire. The name is written by Snorre *Gautland.*—L.

* Gautland (Swedish, Götaland) is the land of the Gauts, and is to be distinguished from the island Gotland in the Baltic Sea. Gaut was one of the many names of Odin.

† Jonaker was a king in the Edda whose sons were stoned to death, because steel weapons could not wound them. The meaning is, that Onund was killed in the same way by stones—which the earth may have showered down upon him for his cutting down wood and improving land.—L.

CHAPTER XL.—*The Burning in Upsala.*

Then Ingiald, King Onund's son, came to the kingdom. The Upsala kings were the highest in Svithiod among the many district-kings who had been since the time that Odin was chief. The kings who resided at Upsala had been the supreme chiefs over the whole Swedish dominions until the death of Agne, when, as before related, the kingdom came to be divided between brothers. After that time the dominions and kingly powers were spread among the branches of the family as these increased; but some kings cleared great tracks of forest-land, and settled them, and thereby increased their domains. Now when Ingiald took the dominions and the kingdom of his father, there were, as before said, many district-kings. King Ingiald ordered a great feast to be prepared in Upsala, and intended to enter at it on his heritage after King Onund his father. He had a large hall made ready for the occasion,—one not less, nor less sumptuous, than that of Upsala; and this hall was called the Seven Kings' Hall, and in it were seven high-seats, for kings. Then King Ingiald sent men all through Svithiod, and invited to his feast kings, earls, and other men of consequence. To this heir-feast came King Algaute, his father-in-law; Yngvar king of Fiadrundaland, with his two sons, Alf and Agnar; King Sporsnial of Nerike; King Sigvat of Attundaland: but Granmar king of Sudermannaland did not come.

Six kings were placed in the seats in the new hall; but one of the high-seats which Ingiald had prepared was empty. All the persons who had come got places in the new hall; but to his own court, and the rest of his people, he had appointed places at Upsala. It was the custom at that time that he who gave an heirship-feast after kings or earls, and entered upon the heritage, should sit upon the footstool in front of the high-seat, until the full bowl, which was called the Brage-bowl, was brought in.* Then he should stand up, take the Brage-bowl, make solemn vows to be afterwards fulfilled, and thereupon empty the bowl. Then he should ascend the high-seat which his father had occupied; and thus he came to the full heritage after his father. Now it was done so on this occasion. When the full Brage-bowl came in, King Ingiald stood up, grasped a large bull's horn, and made a solemn vow to enlarge his dominions by one half, towards all the four corners of the world, or die; and thereupon pointed with the horn to the four quarters. Now when the guests had become drunk towards evening King Ingiald told Svipdag's sons, Folkvid and Hulvid, to arm themselves and their men, as had before been settled; and accordingly they went out, and came up to the new hall, and set fire to it. The hall was soon in a blaze, and the six kings, with all their people, were burned in it. Those who tried to come

---

* The bowl drunk in honour of Brage, the god of poetry. At banquets the gods were first toasted. After the introduction of Christianity, the first toasts were consecrated to Christ and to the saints. On the vows taken see Olaf Trygveson's Saga, chapter 39.

out were killed. Then King Ingiald laid all the dominions these kings had possessed under himself, and took scat from them.

CHAPTER XLI.—*Of Hiorvard's Marriage.*

When King Granmar heard the news of this treachery, he thought the same lot awaited him if he did not take care. The same summer King Hiorvard, who was surnamed Ylfing,* came with his fleet to Svithiod, and went into a fiord called Myrkva-fiord.† When King Granmar heard this he sent a messenger to him to invite him and all his men to a feast. He accepted it willingly; for he had never committed waste in King Granmar's dominions. When he came to the feast he was gladly welcomed. In the evening, when the full bowls went round, as was the custom of kings when they were at home, or in the feasts they ordered to be made, they drank together, the men and women with each other in pairs, and the rest of the company drank all round in one set. But it was the law among the vikings that all who were at the entertainment should drink together in one company all round. King Hiorvard's high-seat was placed right opposite to King Granmar's high-seat, and on the same bench sat all his men. King Granmar told his daughter Hildigun, who was a remarkably beautiful girl, to make ready to carry ale

* The Ylfings were an ancient royal family, mentioned both in the Beowulf poem and in the Elder Edda. They appear to have resided in southern Sweden. An Ylfing is a descendant of Ulf.
† Now Morköfiord, in Sodermanland province.—L.

to the vikings. Thereupon she took a silver goblet, filled it, bowed before King Hiorvard, and said, "Success to all Ylfings: this cup to the memory of Rolf Krake,"—drank out the half, and handed the cup to King Hiorvard. He took the cup, and took her hand, and said she must sit beside him. She says, that is not viking fashion to drink two and two with women. Hiorvard replies, that it were better for him to make a change and leave the viking law and drink in company with her. Then Hildigun sat down beside him, and both drank together, and spoke a great deal with each other during the evening. The next day, when King Granmar and Hiorvard met, Hiorvard spoke of his courtship, and asked to have Hildigun in marriage. King Granmar laid his proposal before his wife Hild, and before people of consequence, saying they would have great help and trust in Hiorvard; and all approved of it highly, and thought it very advisable. And the end was, that Hildigun was promised to Hiorvard, and the wedding followed soon after; and King Hiorvard stayed with King Granmar, who had no sons, to help him to defend his dominions.

CHAPTER XLII.—*War between Ingiald and Granmar and Hiorvard.*

The same autumn King Ingiald collected a war-force with which he intended to fall upon these two relations; he had warriors from all the realms which he had conquered. But when they heard it they also collected a force, and Hogne, who ruled over East Gautland, together with his son Hilder, came to their assistance. Hogne was father of Hilder, who was married to King Granmar. King Ingiald landed with his army, which was by far the most numerous. A battle began, which was very sharp; but after it had lasted a short time, the chiefs who ruled over Fiadrundaland, West Gautland, Nerike, and Attundaland, took to flight with all the men from those countries, and hastened to their ships. This placed King Ingiald in great danger, and he received many wounds, but escaped by flight to his ships. Svipdag the Blind, Ingiald's foster-father, together with his sons, Gautvid and Hulved, fell. Ingiald returned to Upsala very ill satisfied with his expedition; and he thought the army levied from those countries he had acquired by conquest had been unfaithful to him. There was great hostility afterwards between King Ingiald and King Granmar, and his son-in-law King Hiorvard; and after this had continued a long time the friends of both parties brought about a reconciliation. The king appointed a meeting, and concluded a peace. This peace was to endure as long as the three kings lived, and this was confirmed by oath and promises of fidelity. The spring after King Granmar went to Upsala to make offering, as usual, for a steady peace. Then the foreboding* turned out for him so that it did not promise him long life, and he returned to his dominions.

* That is, resulting from the offering.

CHAPTER XLIII.—*Death of the Kings Granmar and Hiorvard.*

The autumn after, King Granmar and his son-in-law Hiorvard went to a feast at one of their farms in the island Sile.* When they were at the entertainment, King Ingiald came there in the night with his troops, surrounded the house, and burnt them in it, with all their men. Then he took to himself all the country these kings had possessed, and placed chiefs over it. King Hogne and his son Hilder often made inroads on horseback into the Swedish dominions, and killed King Ingiald's men, whom he had placed over the kingdom which had belonged to their relation Granmar. This strife between King Ingiald and King Hogne continued for a long time; but King Hogne defended his kingdom against King Ingiald to his dying day. King Ingiald had two children by his wife;—the eldest called Asa, the other Olaf Wood-carver. Gauthild, the wife of Ingiald, sent the boy to his foster-father Bove, in West Gautland, where he was brought up along with Saxe, Bove's son, who had the surname of Fletter.† It was a common saying that King Ingiald had killed twelve kings, and deceived them all under pretence of peace; therefore he was called Ingiald the Evil-adviser (Illrade). He was king over the greater part of Sweden. He married his daughter Asa to Gudrod king of Scania; and she was like her father in disposition. Asa

---
* Now Sela isle, in the Mælar lake.—L.
† One who braids his hair.

brought it about that Gudrod killed his brother Halfdan, father of Ivar Vidfadme; and also she brought about the death of her husband Gudrod, and then fled to her father; and she thus got the name also of Asa the Evil-adviser.

CHAPTER XLIV.—*Of Ingiald Illrade's Death.*

Ivar Vidfadme came to Scania after the fall of his uncle Gudrod, and collected an army in all haste, and moved with it into Svithiod. Asa had gone to her father before. King Ingiald was at a feast in Raning,* when he heard that King Ivar's army was in the neighbourhood. Ingiald thought he had not strength to go into battle against Ivar, and he saw well that if he betook himself to flight his enemies would swarm around him from all corners. He and Asa took a resolution which has become celebrated. They drank until all their people were dead drunk, and then put fire to the hall; and it was consumed, with all who were in it, including themselves, King Ingiald, and Asa. Thus says Thiodolf:—

> "With fiery feet devouring flame
> Has hunted down a royal game
> At Raning, where King Ingiald gave
> To all his men one glowing grave.
> On his own hearth the fire he raised,
> A deed his foeman even praised;
> By his own hand he perished so,
> And life for freedom did forego."

---

* Ranninge, a village in Fogd isle, in the Mælar lake, is supposed to have been the Raning of the saga. A large circle of stones, or a wall, remains, still called Ranningeborg, on a heath.—L.

CHAPTER XLV.—*Of Ivar Vidfadme.*

Ivar Vidfadme subdued the whole of Sweden. He brought in subjection to himself all the Danish dominions, a great deal of Saxland, all the East country,* and a fifth part of England. From his race the kings of Sweden and Denmark who have had the supreme authority in those countries, are descended. After Ingiald the Evil-adviser the Upsala dominion fell from the Yngve race, notwithstanding the length of time they could reckon up the series of their forefathers.

CHAPTER XLVI.—*Of Olaf the Tree-feller.*

When Olaf, King Ingiald's son, heard of his father's end, he went, with the men who chose to follow him, to Nerike; for all the Swedish community rose with one accord to drive out Ingiald's family and all its friends. Now, when the Swedes got intelligence of him he could not remain there, but went on westwards, through the forest, to a river which comes from the north and falls into the Vener lake, and is called the Elf.† There they sat themselves down, turned to, and cleared the woods, burnt, and then settled there. Soon there were great districts, which altogether were called Vermaland; and a good living was to be made there. Now when it

---

* By East country is meant the eastern shores of the Baltic, *i.e.*, Esthland and Kurland.

† Elf means river, and is the old name of the Klara river, flowing through Vermaland, and called at its source in Norway Trysil.

was told of Olaf, in Sweden, that he was clearing the forests, they laughed at his proceedings, and called him the Tree-feller. Olaf got a wife called Solva, or Solveig, a daughter of Halfdan Gulltan, westward in Soley Islands. Halfdan was a son of Solve Solveson, who was a son of Solve the Old, who first settled on these islands. Olaf Tree-feller's mother was called Gauthild, and her mother was Alof, daughter of Olaf Skygne, king in Nerike. Olaf and Solveig had two sons, Ingiald and Halfdan. Halfdan was brought up in Soley Isles, in the house of his mother's brother Solve, and was called Halfdan Hvitbein.

CHAPTER XLVII.—*Olaf the Tree-feller Burned.*

There were a great many people who fled the country from Svithiod, on account of King Ivar; and when they heard that King Olaf had got good lands in Vermaland, so great a number came there to him that the land could not support them. Then there came dear times and famine, which they ascribed to their king; as the Swedes used always to reckon good or bad crops for or against their kings. The Swedes took it amiss that Olaf was sparing in his sacrifices, and believed the dear times must proceed from this cause. The Swedes therefore gathered together troops, made an expedition against King Olaf, surrounded his house, and burnt him in it, giving him to Odin as a sacrifice for good crops. This happened at the Vener lake. Thus tells Thiodolf of it:—

> "The temple wolf,* by the lake shores,
> The corpse of Olaf now devours.
> The clearer of the forests † died
> At Odin's shrine by the lake side.
> The glowing flames stripped to the skin
> The royal robes from the Swedes' king.
> Thus Olaf, famed in days of yore,
> Vanished from earth at Vener's shore."

Chapter XLVIII.—*Halfdan Hvitbein made King.*

Those of the Swedes who had more understanding found that the dear times proceeded from there being a greater number of people on the land than it could support, and that the king could not be blamed for this. They took the resolution, therefore, to cross the Eid forest‡ with all their men, and came quite unexpectedly into Soleys, where they put to death King Solve, and took Halfdan Hvitbein prisoner, and made him their chief, and gave him the title of king. Thereupon he subdued Soleys, and proceeding with his army into Raumarike, plundered there, and laid that district also in subjection by force of arms.

Chapter XLIX.—*Of Halfdan Hvitbein.*

Halfdan Hvitbein became a great king. He was married to Asa, a daughter of Eystein the Severe, who was king of the Upland people, and ruled over Hedemark. Halfdan and Asa had two sons, Eystein and Gudrod. Halfdan subdued a great part of Hede-

---

* The temple wolf—the fire which devoured the body of Olaf.—L.

† Olaf was called the Tree-feller.—L.

‡ Eydaskogr, a great uninhabited forest, which then, and to a late period, covered the frontier of Norway towards Sweden on the south.—L.

mark, Thoten, Hadeland, and much of Vestfold.* He lived to be an old man, and died in his bed at Thoten, from whence his body was transported to Vestfold, and was buried under a mound at a place called Skareid, at Skiringsal.† So says Thiodolf :—

> " Halfdan, esteemed by friends and foes,
> Receives at last life's deep repose :
> The aged man at last, though late,
> Yielded in Thoten to stern fate.
> At Skiringsal hangs o'er his grave
> A rock, that seems to mourn the brave.
> Halfdan, to chiefs and people dear,
> Received from all a silent tear."

Chapter L.—*Of Ingiald, Brother of Halfdan.*

Ingiald, Halfdan's brother, was king of Vermaland; but after his death King Halfdan took possession of Vermaland, raised scat from it, and placed earls over it as long as he lived.

Chapter LI.—*Of King Eystein's Death.*

Eystein, Halfdan Hvitbein's son, became king after in Raumarike and Vestfold. He was married to

---

* Hedemark, Thoten, Hadeland, Vestfold, and the Uplands or Highlands, are all districts in Norway, and in the south of Norway; except the Uplands, which apparently included the upper parts of the valleys of which the waters flow northwards from the dividing ridge, the Dovrefield.—L.

† Skiringssalr is rather a place of note. It is called "Sciringeshael" in the Voyage of Ottar of Halogaland, written by our King Alfred in the end of the ninth century, and the most learned antiquaries have been puzzled where to look for it. Scania, the neighbourhood of Stockholm, and even Prussia, have been considered the true locality of this ancient seat of trade. The Norwegian antiquary Jacob Aal, in his translation of Snorre, places Skiringsal in Vestfold, in Tiolling parish, in the bailiwick of Laurvig; and the situation, access, ancient names, and remains of tumuli around, make this the probable site of the merchant town of Sciringshael.—L.

Hild, a daughter of Eirik Agnarson, who was king in Vestfold. Agnar, Eirik's father, was a son of Sigtryg, king in the Vendel district. King Eirik had no son, and died while King Halfdan Hvitbein was still in life. The father and son, Halfdan and Eystein, then took possession of the whole of Vestfold, which Eystein ruled over as long as he lived. At that time there lived at Varna a king called Skiold, who was a great warlock. King Eystein went with some ships of war to Varna, plundered there, and carried away all he could find of clothes or other valuables, and of peasants' stock, and killed cattle on the strand for provision, and then went off. King Skiold came to the strand with his army, just as Eystein was at such a distance over the fiord that King Skiold could only see his sails. Then he took his cloak, waved it, and blew into it. King Eystein was sitting at the helm as they sailed within the Earl Isle, and another ship was sailing at the side of his, when there came a stroke of a wave, by which the boom of the other ship struck the king and threw him overboard, which proved his death. His men fished up his body, and it was carried into Borro, where a mound was thrown up over it, upon a cleared field out towards the sea at Vadla.* So says Thiodolf:—

"King Eystein sat upon the poop
Of his good ship: with sudden swoop

---

* Now the farm Vold, on which the mounds of Eystein and his son Halfdan and others still remain. It adjoins Borre, about six miles from Tunsberg.—L.

> The swinging boom dashed him to hell,
> And fathoms deep the hero fell
> Beneath the brine. The fury whirl
> Of Loke,* Tempest's brother's girl,
> Grim Hel, clutched his soul away;
> And now where Vadla's ocean bay
> Receives the ice-cold stream, the grave
> Of Eystein stands,—the good, the brave!"

CHAPTER LII.—*Of Halfdan the Mild.*

Halfdan was the name of King Eystein's son who succeeded him. He was called Halfdan the Mild, but the Bad Entertainer; that is to say, he was reported to be generous, and to give his men as much gold as other kings gave of silver, but he starved them in their diet. He was a great warrior, who had been long in viking cruises, and had collected great property. He was married to Hlif, a daughter of King Dag of Vestmarar. Holtar, in Vestfold, was his chief house; and he died there on the bed of sickness, and was buried at Borro under a mound. So says Thiodolf:—

> "By Hel's summons, a great king
> Was called away to Odin's Thing:
> King Halfdan, he who dwelt of late
> At Holtar, must obey grim Fate.
> At Borro, in the royal mound,
> They laid the hero in the ground."

CHAPTER LIII.—*Of Gudrod the Hunter.*

Gudrod, Halfdan's son, succeeded. He was called Gudrod the Magnificent, and also Gudrod the Hunter.

---

* Loke (the evil principle) was brother of Byleist, the god of tempests; and Loke's daughter was Hel—from which probably our word Hell, the abode of evil spirits, is derived.—L.

He was married to Alfhild, a daughter of King Alfarin of Alfheim, and got with her half the district of Vingulmark. Their son Olaf was afterwards called Geirstada-Alf. Alfheim, at that time, was the name of the land between the Raum and Gaut rivers. Now when Alfhild died, King Gudrod sent his men west to Agder to the king who ruled there, and who was called Harald Redbeard. They were to make proposals to his daughter Asa upon the king's account; but Harald declined the match, and the ambassadors returned to the king, and told him the result of their errand. Soon after King Gudrod hove down his ships into the water, and proceeded with a great force in them to Agder. He immediately landed, and came altogether unexpectedly, at night, to King Harald's house. When Harald was aware that an army was at hand, he went out with the men he had about him, and there was a great battle, although he wanted men so much. King Harald and his son Gyrd fell, and King Gudrod took a great booty. He carried away with him Asa, King Harald's daughter, and had a wedding with her. They had a son by their marriage called Halfdan; and the autumn that Halfdan was a year old Gudrod went upon a round of feasts. He lay with his ship in Stiflusund, where they had been drinking hard, so that the king was very tipsy. In the evening, about dark, the king left the ship; and when he had got to the end of the gangway from the ship to the shore,*

---

* The ships appear generally to have been laid all night close to or at the shore, with a gangway to land by; and the crew appear to have had tents on shore to pass the night in.—L.

a man ran against him, thrust a spear through him, and killed him. The man was instantly put to death, and in the morning when it was light the man was discovered to be Asa's footboy: nor did she conceal that it was done by her orders. Thus tells Thiodolf of it:—

> "Gudrod is gone to his long rest,
> Despite of all his haughty pride,—
> A traitor's spear has pierced his side:
> For Asa cherished in her breast
> Revenge; and as, by wine opprest,
> The hero staggered from his ship,
> The cruel queen her thrall let slip
> To do the deed of which I sing:
> And now the far-descended king,
> At Stiflusund, in the old bed
> Of the old Gudrod race, lies dead."

CHAPTER LIV.—*Of King Olaf's Death.*

Olaf came to the kingdom after his father. He was a great warrior, and an able man; and was besides remarkably handsome, very strong, and large of growth. He had Vestfold; for King Alfgeir took all Vingulmark to himself, and placed his son Gandalf over it. Both father and son made war on Raumarike, and subdued the greater part of that land and district. Hogne was the name of a son of the Upland king, Eystein the Great, who subdued for himself the whole of Hedemark, Thoten, and Hadeland. Then Vermaland fell off from Gudrod's sons, and turned itself, with its payment of scat, to the Swedish king. Olaf was about twenty years old when Gudrod died; and as his brother Halfdan now had the kingdom with him, they divided it between them; so that Olaf got the eastern, and Halfdan the southern part. King Olaf had his main residence at Geirstad.* There he died of a disease in his foot, and was laid under a mound at Geirstad. So sings Thiodolf:—

> "Long while this branch of Odin's stem
> Was the stout prop of Norway's realm;
> Long while King Olaf with just pride
> Ruled over Vestfold far and wide.
> At length by cruel gout oppressed,
> The good King Olaf sank to rest:
> His body now lies underground,
> Buried at Geirstad, in the mound."

CHAPTER LV.—*Of Ragnvald the Mountain-high.*

Ragnvald was the name of Olaf's son who was king of Vestfold after his father. He was called "Mountain-high," and Thiodolf of Hvin composed for him the "Ynglinga-tal;"† in which he says:—

> "Under the heaven's blue dome, a name
> I never knew more true to fame
> Than Ragnvald bore; whose skilful hand
> Could tame the scorners of the land,—
> Ragnvald, who knew so well to guide
> The wild sea-horses‡ through the tide:
> The 'Mountain-high' was the proud name
> By which the king was known to fame."

---

* Geirstadir. This ancient seat of royalty in small is now supposed to have been a farm called Gierrestad, in the same parish, Tiolling, in which Skiringsal was situated.—L.

† Ynglinga-tal—the succession of the Yngling race. Our word tale applied to numbers, as things told over one by one, appears connected with this word.—L.

‡ The wild sea-horses—ships, which are generally called the horses of the ocean in skaldic poetry.—L.

II.

HALFDAN THE BLACK'S SAGA.*

CHAPTER I.—*Halfdan Fights with Gandalf and Sigtryg.*

HALFDAN was a year old when his father † was killed, and his mother Asa set off immediately with him westwards to Agder, and set herself there in the kingdom which her father Harald had possessed. Halfdan grew up there, and soon became stout and strong; and, by reason of his black hair, was called Halfdan the Black. When he was eighteen years old he took his kingdom in Agder, and went immediately to Vestfold, where he divided that kingdom, as before related, with his brother Olaf. The same autumn he went with an army to Vingulmark against King Gandalf. They had many battles, and sometimes one, sometimes the other gained the victory; but at last they agreed that Halfdan should have half of Vingulmark, as his father Gudrod had had it before.

---

* Halfdan the Black reigned from about the year 821 to about 860. In the preceding Saga of the Yngling race, there are but few points to be fixed down as historical by dates and coincidences with other history; and the earlier part of it belongs to mythology, not to history. Facts there are—we hold them to be facts only because they are not extravagant enough to be fables—intermingled with the mythological accounts of Odin and his times; but Snorre with great judgment goes over this period rapidly, and comes as quickly as possible to the period when authentic history begins to dawn,—to the reigns of Halfdan and Harald Harfager. Their royal derivation from the Yngve race (the Ynglings) could not be omitted; but Snorre hastens over it, as only a necessary preface to his more authentic narratives.—L.

† King Gudrod, son of Halfdan. See Ynglinga Saga, chapter 53.

Then King Halfdan proceeded to Raumarike, and subdued it. King Sigtryg, son of King Eystein, who then had his residence in Hedemark, and who had subdued Raumarike before, having heard of this, came out with his army against King Halfdan, and there was a great battle, in which King Halfdan was victorious; and just as King Sigtryg and his troops were turning about to fly, an arrow struck him under the left arm, and he fell dead. Halfdan then laid the whole of Raumarike under his power. King Eystein's second son, King Sigtryg's brother, was also called Eystein, and was then king in Hedemark. As soon as Halfdan had returned to Vestfold, King Eystein went out with his army to Raumarike, and laid the whole country in subjection to him.

CHAPTER II.—*Battle between Halfdan and Eystein.*

When King Halfdan heard of these disturbances in Raumarike, he again gathered his army together; and went out against King Eystein. A battle took place between them, and Halfdan gained the victory, and Eystein fled up to Hedemark, pursued by Halfdan. Another battle took place, in which Halfdan was again victorious; and Eystein fled northwards, up into the Dales* to the herse † Gud-

---

* The Dales (*Dalir*) included what is now called Gudbrandsdalen and Osterdalen.

† The herse (Icelandic, *hersir*) was the political name of the Norse chiefs or lords of the oldest age, especially before Harald Harfager and the settlement of Iceland. The records respecting the office are scanty, as they chiefly belonged to pre-historical times. They were probably not liegemen, but resembled the godes of the old Icelandic common-

brand. There he was strengthened with new people, and in winter he went towards Hedemark, and met Halfdan the Black upon a large island which lies in the Mjosen lake. There a great battle was fought, and many people on both sides were slain, but Halfdan won the victory. There fell Guthorm, the son of the herse Gudbrand, who was one of the finest men in the Uplands. Then Eystein fled north up the valley, and sent his relation Halvard Skalk to King Halfdan to beg for peace. On consideration of their relationship, King Halfdan gave King Eystein half of Hedemark, which he and his relations had held before; but kept to himself Thoten, and the land so called. He likewise appropriated to himself Hadeland; and he plundered far and wide around, and was become a mighty king.

CHAPTER III.—*Halfdan's Marriage.*

Halfdan the Black got a wife called Ragnhild, a daughter of Harald Goldbeard, who was a king in Sogn. They had a son, to whom Harald gave his own name; and the boy was brought up in Sogn, by his mother's father, King Harald. Now when this Harald had lived out his days nearly, and was become weak, having no son, he gave his dominions to his daughter's son Harald, and gave him his title of king; and he died soon after. The same winter his daughter Ragnhild died; and the following spring

wealth, being a kind of patriarchal and hereditary chiefs. The old Norse herses were no doubt the prototypes of the barons of Normandy and Norman England. See Vigfusson, *s.v.*

the young Harald fell sick, and died at ten years of age. As soon as Halfdan the Black heard of his son's death, he took the road northwards to Sogn with a great force, and was well received. He claimed the heritage and dominion after his son; and no opposition being made, he took the whole kingdom. Earl Atle Mjove (the Slender), who was a friend of King Halfdan, came to him from Gaular; and the king set him over the Sogn district, to judge in the country according to the country's laws, and collect scat upon the king's account. Thereafter King Halfdan proceeded to his kingdom in the Uplands.

CHAPTER IV.—*Halfdan's Strife with Gandalf's Sons.*

In autumn, King Halfdan proceeded to Vingulmark. One night when he was there in guest quarters, it happened that about midnight a man came to him who had been on the watch on horseback, and told him a war force was come near to the house. The king instantly got up, ordered his men to arm themselves, and went out of the house and drew them up in battle order. At the same moment, Gandalf's sons, Hysing and Helsing, made their appearance with a large army. There was a great battle; but Halfdan being overpowered by the numbers of people, fled to the forest, leaving many of his men on this spot. His foster-father, Olver the Wise, fell here. The people now came in swarms to King Halfdan, and he advanced to seek Gandalf's sons. They met at Eid, near Eyna,* and fought there. Hysing and Helsing fell, and their brother Hake saved himself by flight. King Halfdan then took possession of the whole of Vingulmark, and Hake fled to Alfheim.

CHAPTER V.—*King Halfdan's last Marriage with Sigurd Hiort's Daughter.*

Sigurd Hiort was the name of a king in Ringerike, who was stouter and stronger than any other man, and his equal could not be seen for a handsome appearance. His father was Helge the Sharp; and his mother was Aslaug, a daughter of Sigurd the Worm-eyed, who again was a son of Ragnar Lodbrok. It is told of Sigurd, that when he was only twelve years old he killed in single combat the berserk Hildebrand, and eleven others of his comrades; and many are the deeds of manhood told of him in a long saga about his feats.† Sigurd had two children, one of whom was a daughter, called Ragnhild, then twenty years of age, and an excellent brisk girl. Her brother Guthorm was a youth. It is related that Sigurd had a custom of riding out quite alone in the uninhabited forest to hunt the wild beasts that are hurtful to man, and he was always very eager at this sport. One day he rode out into the forest as usual, and when he had ridden a long way he came out at a piece of cleared land near to

* The old name for the present Lake Oyeren, in the south-eastern part of Norway, written also Oieren.

† This saga is lost, but it is quoted in a fragment concerning Ragnar's sons. (*Script. rerum danicarum*, ii. 284.)

Hadeland. There the berserk Hake came against him with thirty men, and they fought. Sigurd Hiort fell there, after killing twelve of Hake's men; and Hake himself lost one hand, and had three other wounds. Then Hake and his men rode to Sigurd's house, where they took his daughter Ragnhild and her brother Guthorm, and carried them, with much property and valuable articles, home to Hadeland, where Hake had many great farms. He ordered a feast to be prepared, intending to hold his wedding with Ragnhild; but the time passed on account of his wounds, which healed slowly; and the berserk Hake of Hadeland had to keep his bed, on account of his wounds, all the autumn and beginning of winter. Now King Halfdan was in Hedemark at the Yule entertainments when he heard this news; and one morning early, when the king was dressed, he called to him Harek Gand, and told him to go over to Hadeland, and bring him Ragnhild, Sigurd Hiort's daughter. Harek got ready with a hundred men, and made his journey so that they came over the lake to Hake's house in the grey of the morning, and beset all the doors and stairs of the places where the house-servants slept. Then they broke into the sleeping-room where Hake slept, took Ragnhild, with her brother Guthorm, and all the goods that were there, and set fire to the house-servants' place, and burnt all the people in it. Then they covered over a magnificent waggon, placed Ragnhild and Guthorm in it, and drove down upon the ice. Hake got up and went after them a while; but when he came to the ice on the lake, he turned his sword-hilt to the ground and let himself fall upon the point, so that the sword went through him. He was buried under a mound on the banks of the lake. When King Halfdan, who was very quick of sight, saw the party returning over the frozen lake, and with a covered waggon, he knew that their errand was accomplished according to his desire. Thereupon he ordered the tables to be set out, and sent people all round in the neighbourhood to invite plenty of guests; and the same day there was a good feast which was also Halfdan's marriage-feast with Ragnhild, who became a great queen. Ragnhild's mother was Thorny, a daughter of Harald Klak, king in Jutland, and a sister of Thyre Dannebod, who was married to the Danish king, Gorm the Old, who then ruled over the Danish dominions.

CHAPTER VI.—*Of Ragnhild's Dream.*

Ragnhild, who was wise and intelligent, dreamt great dreams. She dreamt, for one, that she was standing out in her herb-garden, and she took a thorn out of her shift; but while she was holding the thorn in her hand it grew so that it became a great tree, one end of which struck itself down into the earth, and it became firmly rooted; and the other end of the tree raised itself so high in the air that she could scarcely see over it, and it became also wonderfully thick. The under part of the tree was red with blood, but the stem upwards was beautifully green, and the branches white as snow. There were many and great limbs to the tree, some high up, others low down; and so vast were the tree's branches that they seemed to her to cover all Norway, and even much more.*

Chapter VII.—*Of Halfdan's Dream.*

King Halfdan never had dreams, which appeared to him an extraordinary circumstance; and he told it to a man called Thorleif the Wise, and asked him what his advice was about it. Thorleif said that what he himself did, when he wanted to have any revelation by dream, was to take his sleep in a swinesty, and then it never failed that he had dreams. The king did so, and the following dream was revealed to him. He thought he had the most beautiful hair, which was all in ringlets; some so long as to fall upon the ground, some reaching to the middle of his legs, some to his knees, some to his loins or the middle of his sides, some to his neck, and some were only as knots springing from his head. These ringlets were of various colours; but one ringlet surpassed all the others in beauty, lustre, and size. This dream he told to Thorleif, who interpreted it thus:—There should be a great posterity from him, and his descendants should rule over countries with great, but not all with equally great honour; but one of his race should be more celebrated than

---

* See Chapter XLV. of Harald Harfager's Saga, where this dream is interpreted as applying to Harald Fairhair.

all the others. It was the opinion of people that this ringlet betokened King Olaf the Saint.

King Halfdan was a wise man, a man of truth and uprightness,—who made laws,* observed them himself, and obliged others to observe them. And that violence should not come in place of the laws, he himself fixed the number of criminal acts in law, and the compensations, mulcts, or penalties, for each case, according to every one's birth and dignity.†

Queen Ragnhild gave birth to a son, and water was poured over him, and the name of Harald given him, and he soon grew stout and remarkably handsome. As he grew up he became very expert at all feats, and showed also a good understanding. He was much beloved by his mother, but less so by his father.

CHAPTER VIII.—*Halfdan's Meat vanishes at a Feast.*

King Halfdan was at a Yule-feast in Hadeland, where a wonderful thing happened one Yule ‡ even-

---

* He was the author of the so-called Eidsiva-law. See Preliminary Dissertation.

† The penalty, compensation, or manbod for every injury, due to the party injured, or to his family and next of kin if the injury was the death or premeditated murder of the party, appears to have been fixed for every rank and condition, from the murder of the king down to the maiming or beating a man's cattle or his slave. A man for whom no compensation was due was a dishonoured person, or an outlaw. It appears to have been optional with the injured party, or his kin if he had been killed, to take the mulct or compensation, or to refuse it, and wait an opportunity of taking vengeance for the injury on the party who inflicted it, or on his kin. A part of each mulct or compensation was due to the king; and these fines or penalties appear to have constituted a great proportion of the king's revenues, and to have been settled in the Things held in every district for administering the law with the lagman.—L.

‡ The feast of Jolner, one of the names of Odin, was celebrated

ing. When the great number of guests assembled were going to sit down to table, all the meat and all the ale* disappeared from the table. The king sat alone very confused in mind; all the others set off, each to his home, in consternation. That the king might come to some certainty about what had occasioned this event, he ordered a Fin to be seized who was particularly knowing, and tried to force him to disclose the truth; but however much he tortured the man, he got nothing out of him. The Fin sought help particularly from Harald, the king's son; and Harald begged for mercy for him, but in vain. Then Harald let him escape against the king's will, and accompanied the man himself. On their journey they came to a place where the man's chief had a great feast, and it appears they were well received there. When they had been there until spring, the chief said, "Thy father took it much amiss that in winter I took some provisions from him,—now I will repay it to thee by a joyful piece of news: thy father is dead; and now thou shalt return home, and take possession of the whole kingdom which he had, and with it thou shalt lay the whole kingdom of Norway under thee."

by the pagan Northmen in mid-winter; and the name of Yule and the festivity were made to coincide with the Christmas of the Church of Rome, which is called Yule all over the North, from Jolner. In Scotland, as well as in Scandinavia, Yule is the name given to the Christmas holidays.—L.

* The Icelandic word is *mungat*, which means a finer sort of ale. There are many accounts in Scandinavia of the disappearance of the food and drink through the agency of spirits.

CHAPTER IX.—*Halfdan's Death*.

Halfdan the Black was driving from a feast* in Hadeland, and it so happened that his road lay over the lake called Rand.† It was in spring, and there was a great thaw. They drove across the bight called Rykinsvik, where in winter there had been a pond broken in the ice for cattle to drink at, and where the dung had fallen upon the ice the thaw had eaten it into holes. Now as the king drove over it the ice broke, and King Halfdan and many with him perished. He was then forty years old. He had been one of the most fortunate kings in respect of good seasons. The people thought so much of him, that when his death was known, and his body was floated to Ringerike to bury it there, the people of most consequence from Raumarike, Vestfold, and Hedemark, came to meet it. All desired to take the body with them to bury it in their own district, and they thought that those who got it would have good crops to expect. At last it was agreed to divide the body into four parts. The head was laid in a mound at Stein in Ringerike, and each of the others took his part home and laid it in a mound; and these have since been called Halfdan's Mounds.‡

* According to *Fagrskinna* the feast was given at a gard called Brandabo.

† The lake now called Rands-fiord; and the bight called Rykinsvik is at a farm called Röken.—L.

‡ The Flatey-bok gives Vingulmark instead of Hedemark. One MS. of *Fagrskinna* states that the entrails were buried in Tingelstad, the body at Stein, and the head at Skiringsal in Vestfold.

III.

HARALD HARFAGER'S SAGA.*

CHAPTER I.—*Harald's Strife with Hake and his Father Gandalf.*

HARALD † was but ten years old when he succeeded his father (Halfdan the Black). He became a stout, strong, and comely man, and withal prudent and

---

* The first twenty chapters of this saga refer to Harald's youth and his conquest of Norway. This portion of the saga is of great importance to the Icelanders, as the settlement of their isle was a result of Harald's wars. The second part of the saga (chaps. 21-46) treats of the disputes between Harald's sons, of the jarls of Orkney, and of the jarls of More. With this saga we enter the domain of history.

The following Icelandic sagas treat more or less of Harald Harfager's time:—

1. Egil Skallagrimsons Saga.
2. Vatnsdæla Saga (chaps. 1-16).
3. Floamanna Saga (chaps. 1-7).
4. Eyrbyggia Saga (chaps. 1-9).
5. Laxdæla Saga (chaps. 1-6).
6. Gullthoris Saga.
7. Kormaks Saga.
8. Gisla Saga Surssonar.
9. Svarfdæla Saga.
10. Grettis Saga.
11. Landnama-bok.
12. Islendinga-bok.
13. Sturlunga Saga. First part.
14. Fagrskinna (chaps. 5-24).

The skalds in Harald's time were Odin Ondskald, Thorbjorn Hornklofe, Alver Nufva, Thjodolf of Hvin, Ulf Sebbason, Guthorm Sindre, Torf-Einar, and Jorun.

Harald Harfager reigned from about the year 860 to about the year 930. Pinkerton thinks Torfæus dates his reign thirty years too far back, and that Harald Harfager's reign began in 900 or 910. As he agrees, however, in placing his death in 931 or 936, the only difference between the two antiquaries is, that Torfæus begins to reckon Harald's reign from his father's death, and Pinkerton from the subjugation of the small kings, by which he became sole king of Norway.—L.

manly. His mother's brother, Guthorm, was leader of the hird,* at the head of the government, and commander (*hertogi*)† of the army. After Halfdan the Black's death, many chiefs coveted the dominions he had left. Among these King Gandalf was the first; then Hogne and Frode, sons of Eystein, king of Hedemark; and also Hogne Karuson came from Ringerike. Hake, the son of Gandalf, began with an expedition of 300 men against Vestfold, marched by the main road through some valleys, and expected to come suddenly upon King Harald; while his father Gandalf sat at home with his army, and prepared to cross over the fiord into Vestfold. When Guthorm heard of this he gathered an army, and marched up the country with King Harald against Hake. They met in a valley, in which they fought a great battle, and King Harald was victorious; and there fell King Hake and most of his people. The place has since been called Hakadale. Then King Harald and Guthorm turned back, but they found King Gandalf had come to Vestfold. The two armies marched against each other, and met, and had a great battle; and it ended in King Gandalf flying, after leaving most of his men

---

* A court or *hird* about the king's person were men-at-arms of the court or hird, kept in pay, and holding guard by night, even on horseback (see chapter 4 of the preceding saga); and appear to have been an establishment coeval with the kingly power itself. This kind of paid standing army must have existed from the earliest period, where no feudal rights over vassals or retainers could give the king or his nobles a constant command of armed followers.—L.

† Hertogi (Anglo-Saxon, *heretoga*; German, *herzog*), originally a leader or commander. As a title, *duke*, it was first borne by Skale, created duke in the year 1237. In Sweden the title was introduced by the Folkung dynasty.

dead on the spot, and in that state he came back to his kingdom. Now when the sons of King Eystein in Hedemark heard the news, they expected the war would come upon them, and they sent a message to Hogne Karuson and to Herse Gudbrand, and appointed a meeting with them at Ringsaker in Hedemark.

CHAPTER II.—*King Harald overcomes Five Kings.*

After the battle King Harald and Guthorm turned back, and went with all the men they could gather through the forests towards the Uplands. They found out where the Upland kings had appointed their meeting-place, and came there about the time of midnight, without the watchmen observing them until their army was before the door of the house in which Hogne Karuson was, as well as that in which Gudbrand slept. They set fire to both houses; but King Eystein's two sons slipped out with their men, and fought for a while, until both Hogne and Frode fell. After the fall of these four chiefs, King Harald, by his relation Guthorm's success and power, subdued Hedemark, Ringerike, Gudbrandsdal, Hadeland, Thoten, Raumarike, and the whole northern part of Vingulmark. King Harald and Guthorm had thereafter war with King Gandalf, and fought several battles with him; and in the last of them King Gandalf was slain, and King Harald took the whole of his kingdom as far south as the river Raum.*

* The present name of the river Raum is Glommen.

CHAPTER III.—*Of Gyda, Daughter of Eirik.*

King Harald sent his men to a girl called Gyda, a daughter of King Eirik of Hordaland, who was brought up as foster-child in the house of a great bonde in Valders. The king wanted her for his concubine; for she was a remarkably handsome girl, but of high spirit withal. Now when the messengers came there, and delivered their errand to the girl, she answered, that she would not throw herself away even to take a king for her husband, who had no greater kingdom to rule over than a few districts.* "And methinks," said she, "it is wonderful that no king here in Norway will make the whole country subject to him, in the same way as Gorm the Old did in Denmark, or Eirik at Upsala." The messengers thought her answer was dreadfully haughty, and asked what she thought would come of such an answer; for Harald was so mighty a man, that his invitation was good enough for her. But although she had replied to their errand differently from what they wished, they saw no chance, on this occasion, of taking her with them against her will; so they prepared to return. When they were ready, and the people followed them out, Gyda said to the messengers, "Now tell to King Harald these my words. I will only agree to be his lawful wife upon the condition that he shall first, for my sake, subject to himself the whole of Norway, so that he may rule over that kingdom as freely and

* The word for districts in the original is *fylki*. Norway was divided into *fylkis*, and each *fylki* was governed by a fylkir, that is, a kinglet or king. Before the time of Harald Fairhair there were thirty-one *fylkis* in Norway.

fully as King Eirik over the Swedish dominions, or King Gorm over Denmark; for only then, methinks, can he be called the king of a people."

Chapter IV.—*King Harald's Vow.*

Now came the messengers back to King Harald, bringing him the words of the girl, and saying she was so bold and foolish that she well deserved that the king should send a greater troop of people for her, and inflict on her some disgrace. Then answered the king, "This girl has not spoken or done so much amiss that she should be punished, but rather she should be thanked for her words. She has reminded me," said he, "of something which it appears to me wonderful I did not think of before. And now," added he, "I make the solemn vow, and take God to witness, who made me and rules over all things, that never shall I clip or comb my hair until I have subdued the whole of Norway, with scat,* and duties, and domains; or if not, have died in the attempt." Guthorm thanked the king warmly for his vow; adding, that it was royal work to fulfil royal words.

Chapter V.—*The Battle in Orkadal.*

After this the two relations gather together a great force, and prepare for an expedition to the Uplands,

---

* Scat was a land-tax, paid to the king in money, malt, meal, or flesh-meat, from all lands; and was adjudged by the Thing to each king upon his accession, and being proposed and accepted as king.

In Orkney, where the land in general has been feudalised since the annexation in 1463 of the islands to the Scotch crown, the old udal tax of scat remains as an item in the feu-duties payable to the crown.—L.

and northwards up the valley (Gudbrandsdal), and north over Dovrefield; and when the king came down to the inhabited land he ordered all the men to be killed, and everything wide around to be delivered to the flames. And when the people came to know this, they fled every one where he could; some down the country to Orkadal, some to Gaulardal, some to the forests. But some begged for peace, and obtained it, on condition of joining the king and becoming his men. He met no opposition until he came to Orkadal. There a crowd of people had assembled, and he had his first battle with a king called Grÿting. Harald won the victory, and King Gryting was made prisoner, and most of his people killed. He took service himself under the king, and swore fidelity to him. Thereafter all the people in Orkadal district went under King Harald, and became his men.

CHAPTER VI.—*Of King Harald's Laws for Land Property.*

King Harald made this law over all the lands he conquered, that all the udal property should belong to him; and that the bondes, both great and small, should pay him land dues for their possessions.*
Over every district he set an earl to judge according to the law of the land and to justice, and also to collect the land dues and the fines; and for this each earl received a third part of the dues, and services, and fines, for the support of his table and other

---

* This appears to have been an attempt to introduce the feudal system.—L.

expenses.* Each earl had under him four or more herses, each of whom had an estate of twenty marks yearly income bestowed on him and was bound to support twenty men-at-arms, and the earl sixty men, at their own expenses. The king had increased the land dues and burdens so much, that each of his earls had greater power and income than the kings had before; and when that became known at Throndhjem, many great men joined the king, and took his service.†

CHAPTER VII.—*Battle in Gaulardal.*

It is told that Earl Hakon Griotgardson came to King Harald from Yrjar,‡ and brought a great crowd of men to his service. Then King Harald went into Gaulardal, and had a great battle, in which he slew two kings, and conquered their dominions; and these were Gaulardal district and Strind district. He gave Earl Hakon Strind district to rule over as earl. King Harald then proceeded to Stjoradal, and had a third battle, in which he gained the victory, and took that district also. Thereupon the Throndhjem people assembled, and four kings met together with their troops. The one ruled over Veradal,§ the second over

---

* This system of compensating officials prevailed in the North throughout the Middle Ages.

† Hakon the Good restored the udal right to the bondes. In regard to the details of King Harald's conduct in this matter, the reader is referred to Egils Saga, chap. 4.

‡ Yrjar is the present Orland, in Fossum, north of Throndhjem.

§ Veradal (Værdal), Skaun (Skogn), the Sparbyggja district (Sparbu), and Indriey (Inderöen), are small districts or parishes on the side of the Throndhjem fiord.

Skaun, third over the Sparbyggja district, and the fourth over Indriey; and this latter had also Eyna district. These four kings marched with their men against King Harald, but he won the battle; and some of these kings fell, and some fled. In all, King Harald fought at the least eight battles, and slew eight kings, in the land of Throndhjem, and laid the whole of it under him.

CHAPTER VIII.—*King Harald seizes all Naumudal District.*

North in Naumudal were two brothers, kings,— Herlaug and Hrollaug; and they had been for three summers raising a mound or tomb of stone and lime and of wood. Just as the work was finished, the brothers got the news that King Harald was coming upon them with his army. Then King Herlaug had a great quantity of meat and drink brought into the mound, and went into it himself, with eleven companions, and ordered the mound to be covered up.* King Hrollaug, on the contrary, went upon the summit of the mound, on which the kings were wont to sit, and made a throne to be erected, upon which he seated himself. Then he ordered feather-beds to be laid upon the bench below, on which the earls were wont to be seated, and threw himself down from his

---

* On the gard Skei, in Leko parish, north of Throndhjem, a mound answering this description was opened. There was an inner wall made of stone and wood, dividing it into two chambers. In one chamber were found bones of cattle, and in the other two human skeletons, one of which seemed to be in a sitting posture. This has been believed to be Herlaug's cairn. Snorre says the mound was raised of stone and *lime* and wood. Lime was not known in the heathen time.

high seat or throne into the earls' seat, giving himself the title of earl. Now Hrollaug went to meet King Harald, gave up to him his whole kingdom, offered to enter into his service, and told him his whole proceeding. Then took King Harald a sword, fastened it to Hrollaug's belt, bound a shield to his neck, and made him thereupon an earl, and led him to his earl's seat; and therewith gave him the district of Naumudal, and set him as earl over it.*

CHAPTER IX.—*King Harald's Home Affairs.*

King Harald then returned to Throndhjem, where he dwelt during the winter, and always afterwards called it his home. He fixed here his head residence, which is called Lade. This winter he took to wife Asa, a daughter of Earl Hakon Griotgardson, who then stood in great favour and honour with the king. In spring the king fitted out his ships. In winter he had caused a great frigate (a dragon) to be built, and had it fitted out in the most splendid way,

---

* Before writing was in general use, this symbolical way of performing all important legal acts appears to have entered into the jurisprudence of all savage nations; and according to Gibbon, chap. 44, "the jurisprudence of the first Romans exhibited the scenes of a pantomime: the words were adapted to the gestures, and the slightest error or neglect in the *forms* of proceeding was sufficient to annul the *substance* of the fairest claims." This ceremony of demission from the seat of a king, and assumption of the rank and seat of an earl, and the subsequent investiture of Hrollaug by the ceremony of binding a sword and shield on him, and leading him to the earl's seat, have probably been ceremonies adopted from the feudal countries. Harald Harfager's object appears to have been to feudalise the dominions he conquered from the small kings; but the subsequent partition of the country among his descendants, and their feuds with each other, prevented the permanency of feudal tenures under the crown; and the holdings being only personal not hereditary, were of less value than the udal rights to land.—L.

and brought his house-troops and his berserks on board. The forecastle men were picked men, for they had the king's banner. From the stem to the mid-hold was called rausn,* or the fore-defence; and there were the berserks.† Such men only were received into King Harald's house-troop as were remarkable for strength, courage, and all kinds of dexterity; and they alone got place in his ship, for he had a good choice of house-troops from the best men of every district. King Harald had a great army, many large ships, and many men of might followed him. Hornklofe, in his poem called "Glymdrapa," tells of this; and also that King Harald had a battle with the people of Orkadal, at Opdal forest, before he went upon this expedition.

> "O'er the broad heath the bowstrings twang,
> While high in air the arrows sang;
> The iron shower drives to flight
> The foeman from the bloody fight.
> The warder of great Odin's shrine,
> The fair-haired son of Odin's line,
> Raises the voice which gives the cheer,
> First in the track of wolf or bear.
> His master voice drives them along
> To Hel—a destined, trembling throng;
> And Nokve's ship, with glancing sides,
> Must fly to the wild ocean's tides,—
> Must fly before the king who leads
> Norse axe-men on their ocean steeds."

CHAPTER X.—*Battle at Solskel.*

King Harald moved out with his army from Thrond-

---

* Rausn is explained by Schöning to have been that part of the vessel where the rise begins to form the bow—the forecastle-deck.—L.

† Berserk. See note, chapter 6, "Ynglinga Saga."

hjem, and went southwards to More.* Hunthiof was the name of the king who ruled over the district of More. Solve Klofe was the name of his son, and both were great warriors. King Nokve, who ruled over Raumsdal,† was the brother of Solve's mother. Those chiefs gathered a great force when they heard of King Harald, and came against him. They met at Solskel,‡ and there was a great battle, which was gained by King Harald [A.D. 867]. Hornklofe tells of this battle:—

> "Thus did the hero known to fame,
> The leader of the shields, whose name
> Strikes every heart with dire dismay,
> Launch forth his war-ships to the fray.
> Two kings he fought; but little strife
> Was needed to cut short their life.
> A clang of arms by the sea-shore,—
> And the shields' sound was heard no more."

The two kings were slain, but Solve escaped by flight; and King Harald laid both districts under his power. He stayed here long in summer to establish law and order for the country people, and set men to rule them, and keep them faithful to him; and in autumn he prepared to return northwards to Thrond-hjem. Ragnvald Earl of More, a son of Eystein Glumra, had the summer before become one of Harald's men; and the king set him as chief over these two districts, North More and Raumsdal;

---

* Mæri appears derived from the old northern word mar, the sea; the same as the Latin mare, and retained by us in moor or morass. It is applied to a flat bordering on the sea; and possibly our Murrayshire may have a common root with the two districts of Norway called South and North More.—L.

† Raumsdal is the present Romsdal.—L.

‡ Solskel is an island in the parish of Ædo, in North More.—L.

strengthened him both with men of might and bondes, and gave him the help of ships to defend the coast against enemies. He was called Ragnvald the Mighty, or the Wise; and people say both names suited well. King Harald came back to Throndhjem about winter.

CHAPTER XI.—*Fall of the Kings Arnvid and Audbiorn.*

The following spring [A.D. 868], King Harald raised a great force in Throndhjem, and gave out that he would proceed to South More. Solve Klofe had passed the winter in his ships of war, plundering in North More, and had killed many of King Harald's men; pillaging some places, burning others, and making great ravage: but sometimes he had been, during the winter, with his friend King Arnvid in South More. Now when he heard that King Harald was come with ships and a great army, he gathered people, and was strong in men-at-arms; for many thought they had to take vengeance of King Harald. Solve Klofe went southwards to the Fiord, which King Audbiorn ruled over, to ask him to help, and join his force to King Arnvid's and his own. "For," said he, "it is now clear that we all have but one course to take; and that is to rise, all as one man, against King Harald, for we have strength enough, and fate must decide the victory: for as to the other condition of becoming his servants, that is no condition for us, who are not less noble than Harald. My father thought it better to fall in battle for his kingdom, than to go willingly into King Harald's service, or not to abide the chance of weapons like the Naumudal kings." King Solve's speech was such that King Audbiorn promised his help, and gathered a great force together, and went with it to King Arnvid, and they had a great army. Now, they got news that King Harald was come from the north, and they met within Solskel. And it was the custom to lash the ships together, stem to stem; so it was done now. King Harald laid his ship against King Arnvid's, and there was the sharpest fight, and many men fell on both sides. At last King Harald was raging with anger, and went forward to the fore-deck, and slew so dreadfully that all the forecastle men of Arnvid's ship were driven aft of the mast, and some fell. Thereupon Harald boarded the ship, and King Arnvid's men tried to save themselves by flight, and he himself was slain in his ship. King Audbiorn also fell; but Solve fled. So says Hornklofe:—

> "Against the hero's shield in vain
> The arrow-storm fierce pours its rain.
> The king stands on the blood-stained deck,
> Trampling on many a stout foe's neck;
> And high above the dinning stound
> Of helm and axe, and ringing sound
> Of blade and shield, and raven's cry,
> Is heard his shout of 'Victory!'"

Of King Harald's men, fell his earls Asgaut and Asbiorn, together with his brothers-in-law Griotgard and Herlaug, the sons of Earl Hakon of Lade. Solve became afterwards a great sea-king, and often did great damage in King Harald's dominions.

CHAPTER XII.—*King Vemund Burnt to Death.*

After this battle [A.D. 868] King Harald subdued South More; but Vemund, King Audbiorn's brother, still had the Firda-district. It was now late in harvest, and King Harald's men gave him the counsel not to proceed southwards round Stad.* Then King Harald set Earl Ragnvald over South and North More and also Raumsdal, and he had many people about him. King Harald returned to Throndhjem [A.D. 869]. The same winter Ragnvald went over Eid, and southwards to the Firda district. There he heard news of King Vemund, and came by night to a place called Naustdal, where King Vemund was living in guest-quarters. Earl Ragnvald surrounded the house in which they were quartered, and burnt the king in it, together with ninety men. Then came Berdlukare to Earl Ragnvald with a completely armed long-ship, and they both returned to More. The earl took all the ships Vemund had, and all the goods he could get hold of. Berdlukare † proceeded north to Throndhjem to King Harald, and became his man; and a dreadful berserk he was.

---

* Stad is often mentioned in the sagas, being the most westerly part of the mainland of Norway; and vessels coasting along from the north or south had to steer a new course along the coast after passing Stad. It is now called Statland.—L.

† Kare of Berdla, called Berdlukare, lived in the Firda district. His sons Eyvind and the skald Alver were in the king's hird and enjoyed his favour. Berdlakare was related to Kveldulf, who, with his sons Thorulf and Grim, is mentioned in the *Egla*.

CHAPTER XIII.—*Death of Earl Hakon and of Earl Atle the Mjove.**

The following spring [869] King Harald went southwards with his fleet along the coast, and subdued the Firda-district. Then he sailed eastward along the land until he came to Viken;† but he left Earl Hakon Griotgardson behind, and set him over the Firda-district. Earl Hakon sent word to Earl Atle the Mjove that he should leave Sogn district, and be earl over Gaular district, as he had been before, alleging that King Harald had given Sogn district to him. Earl Atle sent word that he would keep both Sogn district and Gaular district, until he met King Harald. The two earls quarrelled about this so long, that both gathered troops. They met at Fialar, in Stavanger fiord, and had a great battle, in which Earl Hakon fell, and Earl Atle got a mortal wound, and his men carried him to the island of Atley,‡ where he died. So says Eyvind Skaldaspiller:—

> "He who stood a rooted oak,
> Unshaken by the swordsman's stroke,
> Amidst the whiz of arrows slain,
> Has fallen upon Fialar's plain.

---

* *Mjove* means the Slender.
† The statements in this chapter that Harald sailed eastward, and that he had given Hakon the Sogne district, are in conflict with the *Egla* and the *Flateyar-bok*, where we read that Harald went to Throndhjem this year, and that Earl Hroald received Firda-fylke after it had been conquered. Snorre is probably in error.
‡ Atle isle in Fialar, now included in Söndfiord, has probably got its name from Atle. Three standing stones at Velnes church, supposed to have been erected to his memory, still remain.—L.

> There, by the ocean's rocky shore,
> The waves are stained with the red gore
> Of stout Earl Hakon Griotgard's son,
> And of brave warriors many a one."

CHAPTER XIV.—*Of King Harald and the Swedish King Eirik.*

King Harald came with his fleet eastward to Viken, and landed at Tunsberg, which was then a trading town. He had then been four years in Throndhjem, and in all that time had not been in Viken.* Here he heard the news that Eirik Eymundson, king of Sweden, had laid under him Vermaland, and was taking scat or land-tax from all the forest settlers; and also that he called the whole country north to Svinasund, and west along the sea, West Gautland; and which altogether he reckoned to his kingdom, and took land-tax from it. Over this country he had set an earl, by name Hrane Gauzke, who had the earldom between Svinasund and the Gaut river, and was a mighty earl. And it was told to King Harald that the Swedish king said he would not rest until he had as great a kingdom in Viken as Sigurd Ring, or his son Ragnar Lodbrok, had possessed; and that was Raumarike and Vestfold, all the way to the isle Grenmar, and also Vingulmark, and all that lay south of it. In these districts many chiefs, and many other people, had given obedience to the Swedish king. King Harald was very angry at this, and summoned the bondes to a

* In reference to the chronology, the reader is referred to the note in chapter 13, and to the *Egla*, chapters 3-19.

Thing at Folden, where he laid an accusation *
against them for treason towards him. Some bondes
defended themselves from the accusation, some paid
fines, some were punished. He went thus through
the whole district during the summer, and in harvest
he did the same in Raumarike, and laid the two districts under his power. Towards winter he heard
that Eirik king of Sweden was, with his court,
going about in Vermaland in guest-quarters.

CHAPTER XV.—*King Harald at a Feast of the Peasant
Ake, and the Murder of Ake.*

King Harald takes his way across the Eid forest
eastward, and comes out in Vermaland, where he also
orders feasts to be prepared for himself. There was
a man, by name Ake, who was the greatest of the
bondes of Vermaland, very rich, and at that time
very aged. He sent men to King Harald, and invited
him to a feast, and the king promised to come on the
day appointed. Ake invited also King Eirik to a
feast, and appointed the same day. Ake had a great
feasting hall, but it was old; and he made a new
hall, not less than the old one, and had it ornamented
in the most splendid way. The new hall he had
hung with new hangings, but the old had only its old
ornaments. Now when the kings came to the feast,
King Eirik with his court was taken into the old hall;
but Harald with his followers into the new. The

---

* A reference to a Thing, and an accusation before it, appears to
have been a necessary mode of proceeding, even to authorise the king to
punish for treason the udal landholders.—L.

same difference was in all the table furniture, and King Eirik and his men had the old-fashioned vessels and horns, but all gilded and splendid; while King Harald and his men had entirely new vessels and horns adorned with gold, all with carved figures, and shining like glass: and both companies had the best of liquor. Ake the bonde had formerly been King Halfdan the Black's man. Now when daylight came, and the feast was quite ended, and the kings made themselves ready for their journey, and the horses were saddled, came Ake before King Harald, leading in his hand his son Ubbe, a boy of twelve years of age, and said, "If the goodwill I have shown to thee, sire, in my feast, be worth thy friendship, show it hereafter to my son. I give him to thee now for thy service." The king thanked him with many agreeable words for his friendly entertainment, and promised him his full friendship in return. Then Ake brought out great presents, which he gave to the king, and they gave each other thereafter the parting kiss. Ake went next to the Swedish king, who was dressed and ready for the road, but not in the best humour. Ake gave to him also good and valuable gifts; but the king answered only with few words, and mounted his horse. Ake followed the king on the road, and talked with him. The road led through a wood which was near to the house; and when Ake came to the wood, the king said to him, "How was it that thou madest such a difference between me and King Harald as to give him the best of everything, although thou knowest thou art my man?" "I think," answered

Ake, "that there failed in it nothing, king, either to you or to your attendants, in friendly entertainment at this feast. But that all the utensils for your drinking were old, was because you are now old; but King Harald is in the bloom of youth, and therefore I gave him the new things. And as to my being thy man, thou art just as much my man." On this the king out with his sword, and gave Ake his death-wound. King Harald was ready now also to mount his horse, and desired that Ake should be called. The people went to seek him; and some ran up the road that King Eirik had taken, and found Ake there dead. They came back, and told the news to King Harald, and he bids his men to be up, and avenge Ake the bonde. And away rode he and his men the way King Eirik had taken, until they came in sight of each other. Each for himself rode as hard as he could, until Eirik came into the wood which divides Gautland and Vermaland. There King Harald wheels about, and returns to Vermaland, and lays the country under him, and kills King Eirik's men wheresoever he can find them. In winter King Harald returned to Raumarike, and dwelt there a while.

CHAPTER XVI.—*King Harald's Journey to Tunsberg.*

King Harald went out in winter to his ships at Tunsberg, rigged them, and sailed away eastward over the fiord, and subjected all Vingulmark to his dominion. All winter he was out with his ships, and marauded in Ranrike;* so says Thorbiorn Hornklofe :—

> "The Norseman's king is on the sea,
> Tho' bitter wintry cold it be,—
> On the wild waves his Yule keeps he.
> When our brisk king can get his way,
> He'll no more by the fireside stay
> Than the young sun : he makes us play
> The game of the bright sun-god † Frey.
> But the soft Swede loves well the fire,
> The well-stuffed couch, the downy glove,
> And from the hearth-seat will not move."

The Gautlanders gathered people together all over the country.

CHAPTER XVII.—*The Battle in Gautland.*

In spring, when the ice was breaking up, they drove stakes into the Gaut river to hinder King Harald with his ships from coming to the land. But King Harald laid his ships alongside the stakes, and plundered the country, and burnt all around; so says Hornklofe :—

> "The king, who finds a dainty feast
> For battle-bird and prowling beast,
> Has won in war the southern land
> That lies along the ocean's strand.
> The leader of the helmets, he
> Who leads his ships o'er the dark sea,
> Harald, whose high-rigged masts appear
> Like antlered fronts of the wild deer,
> Has laid his ships close alongside
> Of the foe's piles with daring pride."

---

* Ranrike was the present Bahuus province, between the Gota and Glommen river-mouths.—L.

† In northern mythology Frey, the god of the sun, is supposed to have been born at the winter solstice; and the return of the lengthening day was celebrated by a feast called Yule, which coinciding with Christmas, was transferred to the Christian festival.—L.

Afterwards the Gautlanders came down to the strand with a great army, and gave battle to King Harald, and great was the fall of men. But it was King Harald who gained the day. Thus says Hornklofe:—

> "Whistles the battle-axe in its swing,
> O'er head the whizzing javelins sing,
> Helmet and shield and hauberk ring ;
> The air-song of the lance is loud,
> The arrows pipe in darkening cloud ;
> Through helm and mail the foemen feel
> The blue edge of our king's good steel.
> Who can withstand our gallant king ?
> The Gautland men their flight must wing."

CHAPTER XVIII.—*Hrane the Gautlander's Death.*

King Harald went far and wide through Gautland, and many were the battles he fought there on both sides of the river, and in general he was victorious. In one of these battles fell Hrane Gauzke; and then the king took his whole land north of the river and west of the Vener, and also Vermaland. And after he turned back therefrom, he set Guthorm as chief to defend the country, and left a great force with him. King Harald himself went first to the Uplands, where he remained a while, and then proceeded northwards over the Dovrefield to Throndhjem, where he dwelt for a long time. Harald began to have children. By Asa he had four sons. The eldest was Guthorm.* Halfdan the Black and Halfdan the White were twins. Sigfrod was the fourth. They were all brought up in Throndhjem with all honour.

* According to the *Flatey-bok*, Guthorm was the son of Gyda.

CHAPTER XIX.—*Battle in Hafersfiord.*

News came in from the south land that the people of Hordaland and Rogaland, Agder and Thelemark, were gathering, and bringing together ships and weapons, and a great body of men. The leaders of this were Eirik king of Hordaland; Sulke king of Rogaland, and his brother Earl Sote; Kiotve the Rich, king of Agder, and his son Thor Haklang; and from Thelemark two brothers, Hroald Hryg and Had the Hard. Now when Harald got certain news of this, he assembled his forces, set his ships on the water, made himself ready with his men, and set out southwards along the coast, gathering many people from every district. King Eirik heard of this when he came south of Stad; and having assembled all the men he could expect, he proceeded southwards to meet the force which he knew was coming to his help from the east. The whole met together north of Jadar,* and went into Hafersfiord, where King Harald was waiting with his forces. A great battle began, which was both hard and long; but at last King Harald gained the day. There King Eirik fell, and King Sulke, with his brother Earl Sote. Thor Haklang, who was a great berserk, had laid his ship against King Harald's, and there was above all measure a desperate attack, until Thor Haklang fell, and his whole ship was cleared of men. Then King Kiotve fled to a little isle outside, on which there was a good place of strength. Thereafter all his

---
* The present Jaderen, near Stavanger.

men fled, some to their ships, some up to the land; and the latter ran southwards over the country of Jadar. So says Hornklofe, viz. :—

> "Has the news reached you?—have you heard
> Of the great fight at Hafersfiord,*
> Between our noble king brave Harald
> And King Kiotve rich in gold?
> The foemen came from out the East,
> Keen for the fray as for a feast.
> A gallant sight it was to see
> Their fleet sweep o'er the dark-blue sea;
> Each war-ship, with its threatening throat
> Of dragon fierce or ravenous brute †
> Grim gaping from the prow; its wales
> Glittering with burnished shields,‡ like scales;
> Its crew of udal men of war,
> Whose snow-white targets shone from far;
> And many a mailed spearman stout
> From the West countries round about,
> English and Scotch, a foreign host,
> And swordsmen from the far French coast.§
> And as the foemen's ships drew near,
> The dreadful din you well might hear;
> Savage berserks roaring mad,
> And champions fierce in wolf-skins clad,‖
> Howling like wolves; and clanking jar
> Of many a mail-clad man of war.
> Thus the foe came; but our brave king
> Taught them to fly as fast again.
> For when he saw their force come o'er,
> He launched his war-ships from the shore;
> On the deep sea he launched his fleet,
> And boldly rowed the foe to meet.

---

* Hafrafiördr, now Hafsfiord, north of Jaderen district, near Stavanger.—L.

† The war-ships were called dragons, from being decorated with the head of a dragon, serpent, or other wild animal; and the word "draco" was adopted in the Latin of the Middle Ages to denote a ship of war of the larger class. The snekke was the cutter or smaller war-ship.—L.

‡ The shields were hung over the side-rails of the ships.—L.

§ It is curious to find that English, Scotch, and French men-at-arms, from the West countries, were in Kiotve's army.—L.

‖ The wolf-skin pelts were nearly as good as armour against the sword.—L.

> Fierce was the shock, and loud the clang
> Of shields, until the fierce Haklang,
> The foeman's famous berserk, fell.
> Then from our men burst forth the yell
> Of victory; and the King of Gold
> Could not withstand our Harald bold,
> But fled before his flaky locks
> For shelter to the island rocks.
> All in the bottom of the ships
> The wounded lay, in ghastly heaps;
> Backs up and faces down they lay,
> Under the row-seats stowed away;
> And many a warrior's shield, I ween,
> Might on the warrior's back be seen,
> To shield him as he fled amain
> From the fierce stone-storm's pelting rain.
> The mountain-folk, as I've heard say,
> Ne'er stopped as they ran from the fray,
> Till they had crossed the Jadar sea,
> And reached their homes—so keen each soul
> To drown his fright in the mead bowl."

CHAPTER XX.—*King Harald the Supreme Sovereign in Norway. Of the Settlement of Distant Lands.*

After this battle King Harald met no opposition in Norway, for all his opponents and greatest enemies were cut off. But some, and they were a great multitude, fled out of the country, and thereby great districts were peopled. Jemtaland and Helsingjaland were peopled then, although some Norwegians had already set up their habitation there. In the discontent that King Harald seized on the lands of Norway,[*] the out-countries of Iceland and the Farey[†] Isles were discovered and peopled. The Northmen had also a

---

[*] This taking the land appears to have been an attempt to introduce the feudal tenures and services.—L.

[†] Icelandic, *Færeyar;* literally sheep-isles.

great resort to Shetland,* and many men left Norway, flying the country on account of King Harald, and went on viking cruises into the West sea. In winter they were in the Orkney Islands and Hebrides;† but marauded in summer in Norway, and did great damage. Many, however, were the mighty men who took service under King Harald, and became his men, and dwelt in the land with him.

CHAPTER XXI.—*King Harald's Marriage and his Children.*

When King Harald had now become sole king over all Norway, he remembered what that proud girl had said to him; so he sent men to her, and had her brought to him, and took her to his bed. And these were their children: Alof—she was the eldest; then was their son Hrorek; then Sigtryg, Frode, and Thorgils. King Harald had many wives ‡ and many children. Among them he had one wife, who was called Ragnhild the Mighty, a daughter of King Eirik, from Jutland; and by her he had a son, Eirik Blood-axe. He was also married to Svanhild, a daughter of Earl Eystein; and their sons were Olaf

---

* Called Hjaltland in the Icelandic.
† Called in the Icelandic Sudreyar.
‡ Polygamy—possibly brought with them from their original seats in Asia—appears to have been a privilege of the royal race, among the Northmen, down to the thirteenth century. The kings had concubines as well as a plurality of wives; and the children appear to have been equally udal-born to the kingdom, whether born in marriage or not. It does not appear from the sagas what forms or ceremonies constituted a marriage before the introduction of Christianity. A marriage feast or wedding is mentioned, and one of the wives appears to have been the drottning or queen; but we are not told of any religious ceremony besides the feast.—L.

Geirstade-Alf, Biorn and Ragnar Rykkil. Lastly, King Harald married Ashild, a daughter of Hring Dagson, up in Ringerike; and their children were, Dag, Hring, Gudrod Skiria, and Ingigerd. It is told that King Harald put away nine wives when he married Ragnhild the Mighty. So says Hornklofe:—

> "Harald, of noblest race the head,
> A Danish wife took to his bed;
> And out of doors nine wives he thrust,—
> The mothers of the princes first,
> Who in Holmryger hold command,
> And those who rule in Hordaland.
> And then he packed from out the place
> The children born of Holge's race."

King Harald's children were all fostered and brought up by their relations on the mother's side. Guthorm the Duke had poured water over King Harald's eldest son,* and had given him his own name. He set the child upon his knee,† and was

---

* See note, page 102. According to ancient Scandinavian laws, the right of inheritance on the part of the children depended on their having received baptism. Jacob Grimm has shown in his "Deutsche Rechts-Alterthümer" (page 457, edition of 1828) that in the old Norse heathendom the father was permitted to expose his own child only before it had been sprinkled with water. A few years ago the distinguished scholar, Dr. Konrad Maurer, of Munich, took pains to collect in a pamphlet all the references to infant baptism in heathen times. He there shows how extensively baptism was practised among the heathen Teutons, and what its real significance was. He also points out its connection with similar ancient rites among the Greeks and Romans. In his scholarly work, Dr. Maurer has sifted every passage bearing on this subject not only in the old Icelandic and Norwegian literary monuments, but also in the historical documents of the Swedes, Danes, South Germans, and Anglo-Saxons. Dr. Maurer shows that this heathen baptism was a naming ceremony, that a peculiar bond was established between the person baptizing and the person baptized, and also between the child and the witnesses present, a relation corresponding to that of god-father and god-mother among Christians, and that it made the child an heir.

† This appears to have been a generally used symbol of adoption of a child.—L.

his foster-father, and took him with himself eastward to Viken, and there he was brought up in the house of Guthorm. Guthorm ruled the whole land in Viken, and the Uplands, when King Harald was absent.

CHAPTER XXII.—*King Harald's Voyage to the West.*

King Harald heard that the vikings, who were in the West sea in winter, plundered far and wide in the middle part of Norway; and therefore every summer he made an expedition to search the isles and out-skerries * on the coast. Wheresoever the vikings heard of him they all took to flight, and most of them out into the open ocean. At last the king grew weary of this work, and therefore one summer he sailed with his fleet right out into the West sea. First he came to Hjaltland (Shetland), and he slew all the vikings who could not save themselves by flight. Then King Harald sailed southwards, to the Orkney Islands, and cleared them all of vikings. Thereafter he proceeded to the Sudreys (Hebrides), plundered there, and slew many vikings who formerly had had men-at-arms under them. Many a battle was fought, and King Harald was always victorious. He then plundered far and wide in Scotland itself, and had a battle there. When he was come westward as far as the Isle of Man, the report of his exploits on the land had gone before him; for all the inhabitants had fled over to Scotland, and the island was left entirely bare both

* Skerries are the uninhabited dry or half-tide rocks of a coast.—L.

of people and goods, so that King Harald and his men made no booty when they landed. So says Hornklofe :—

> "The wise, the noble king, great Harald,
> Whose hand so freely scatters gold,
> Led many a northern shield to war
> Against the town upon the shore.
> The wolves soon gathered on the sand
> Of that sea-shore; for Harald's hand
> The Scottish army drove away,
> And on the coast left wolves a prey."

In this war fell Ivar, a son of Ragnvald, Earl of More; and King Harald gave Ragnvald, as a compensation for the loss, the Orkney and Shetland isles, when he sailed from the West; but Ragnvald immediately gave both these countries to his brother Sigurd, who remained behind them; and King Harald, before sailing eastward, gave Sigurd the earldom of them. Thorstein the Red, a son of Olaf the White and of Aud the Wealthy, entered into partnership with him; and after plundering in Scotland, they subdued Caithness and Sutherland, as far as Ekkjalsbakke.* Earl Sigurd killed Melbridge Tooth, a Scotch earl, and hung his head to his stirrup-leather; but the calf of his leg was scratched by the teeth, which were sticking out from the head, and the wound caused inflammation in his leg, of which the earl died, and he was laid in a mound at Ekkjalsbakke. His son Guthorm ruled over these countries

---

* Ekkjalsbakke, the Ekkial, is now the Oickel, a river falling into the Frith of Dornoch; and the banks or braes on its borders are the Ekkjalsbakke of the saga—not the Ochil hills, as some have imagined; and the burial mound may be still remaining possibly.—L.

for about a year thereafter, and died without children. Many vikings, both Danes and Northmen, set themselves down then in those countries.

CHAPTER XXIII.—*King Harald has his Hair Clipped.*

After King Harald had subdued the whole land, he was one day at a feast in More, given by Earl Ragnvald. Then King Harald went into a bath, and had his hair dressed. Earl Ragnvald now cut his hair, which had been uncut and uncombed for ten years; and therefore the king had been called Lufa (*i.e.*, with rough matted hair). But then Earl Ragnvald gave him the distinguishing name—Harald Harfager (*i.e.*, fair hair); and all who saw him agreed that there was the greatest truth in that surname, for he had the most beautiful and abundant head of hair.

CHAPTER XXIV.—*Rolf Ganger is driven into Banishment.*

Earl Ragnvald was King Harald's dearest friend, and the king had the greatest regard for him. He was married to Hild, a daughter of Rolf Nefia, and their sons were Rolf and Thorer. Earl Ragnvald had also three sons by concubines,—the one called Hallad, the second Einar, the third Hrollaug; and all three were grown men when their brothers born in marriage were still children. Rolf became a great viking, and was of so stout a growth that no horse could carry him, and wheresoever he went he must go on foot;

and therefore he was called Gange-Rolf.* He plundered much in the East sea.† One summer, as he was coming from the eastward on a viking's expedition to the coast of Viken, he landed there and made a cattle foray.‡ As King Harald happened, just at that time, to be in Viken, he heard of it, and was in a great rage; for he had forbid, by the greatest punishment, the plundering within the bounds of the country. The king assembled a Thing, and had Rolf declared an outlaw over all Norway. When Rolf's mother, Hild, heard of it she hastened to the king, and entreated peace for Rolf; but the king was so enraged that her entreaty was of no avail. Then Hild spake these lines:—

> "Think'st thou, King Harald, in thy anger,
> To drive away my brave Rolf Ganger,
> Like a mad wolf, from out the land?
> Why, Harald, raise thy mighty hand?
> Why banish Nefia's gallant name-son,
> The brother of brave udal-men?
> Why is thy cruelty so fell?
> Bethink thee, monarch, it is ill
> With such a wolf at wolf to play,
> Who, driven to the wild woods away,
> May make the king's best deer his prey."

Gange-Rolf went afterwards over sea to the West to the Hebrides, or Sudreys;§ and at last farther

---

* Gange-Rolf, Rolf Ganger, Rolf the Walker, was the conqueror of Normandy. He appears to have had among his ancestors a Rolf Ganger; so that the popular story of his great obesity, which seems scarcely consistent with his great military activity, may not be literally true.—L.

† Austrvegr, the lands on the south side of the Baltic.—L.

‡ A strandhögg, or foray for cattle to be slaughtered on the strand for his ships.—L.

§ Sudreyar,—of which we still retain the name Sodor, applied to the bishopric of Sodor and man,—was the southern division of the Hebrides, or Hebudes.—L.

west to Valland,* where he plundered and subdued for himself a great earldom, which he peopled with Northmen, from which that land is called Normandy. Gange-Rolf's son was William, father to Richard, and grandfather to another Richard, who was the father of Robert Longspear, and grandfather of William the Bastard, from whom all the following English kings are descended. From Gange-Rolf also are descended the earls in Normandy. Queen Ragnhild the Mighty lived three years after she came to Norway; and, after her death, her son and King Harald's was taken to Thorer Hroaldson, and Eirik was fostered by him.†

CHAPTER XXV.—*Of the Fin Svase and King Harald.*

King Harald, one winter, went about in guest-quarters in the Uplands, and had ordered a Christmas feast to be prepared for him at the farm Thoptar.‡ On Christmas eve came Svase to the door, just as the king went to table, and sent a message to the king to ask if he would go out with him. The king was angry at such a message, and the man who had brought it in took out with him a reply of the king's displeasure. But Svase, notwithstanding, desired that his message should be delivered a second time;

---

* Valland was the name applied to all the west coast of France, but more particularly to Bretagne, as being inhabited by the Valer or inhabitants of Wales and Cornwales (Cornwall), expelled by the Saxons from Great Britain in the last half of the fifth century. The adjective Valskr (Welsh) was used to denote what belonged to this Valland.—L.

† Gange-Rolf was baptized in 912, and died in 931. He probably left Norway about the year 890.

‡ Now Tofte, near the head of Gudbrandsdal.—L.

adding to it, that he was the Fin whose hut the king had promised to visit, and which stood on the other side of the ridge. Now the king went out, and promised to follow him, and went over the ridge to his hut, although some of his men dissuaded him. There stood Snowfrid, the daughter of Svase, a most beautiful girl; and she filled a cup of mead for the king. But he took hold both of the cup and of her hand. Immediately it was as if a hot fire went through his body; and he wanted that very night to take her to his bed. But Svase said that should not be unless by main force, if he did not first make her his lawful wife. Now King Harald made Snowfrid his lawful wife, and loved her so passionately that he forgot his kingdom, and all that belonged to his high dignity. They had four sons: the one was Sigurd Hrise; the others Halfdan Haleg, Gudrod Liome, and Rognvald Rettilbeine. Thereafter Snowfrid died; but her corpse never changed, but was as fresh and red as when she lived. The king sat always beside her, and thought she would come to life again. And so it went on for three years that he was sorrowing over her death, and the people over his delusion.* At last Thorleif the Wise succeeded, by his prudence, in curing him of his delusion by accosting him thus:—"It is nowise wonderful, king, that thou grievest over so beautiful and noble a wife, and bestowest costly coverlets and beds of down on her corpse, as she desired; but these honours fall short of what is due, as she

* There is a similar story about Charlemagne.

still lies in the same clothes. It would be more suitable to raise her, and change her dress." As soon as the body was raised in the bed all sorts of corruption and foul smells came from it, and it was necessary in all haste to gather a pile of wood and burn it; but before this could be done the body turned blue, and worms, toads, newts, paddocks, and all sorts of ugly reptiles came out of it, and it sank into ashes. Now the king came to his understanding again, threw the madness out of his mind, and after that day ruled his kingdom as before. He was strengthened and made joyful by his subjects, and his subjects by him, and the country by both.

CHAPTER XXVI.—*Of Thiodolf of Hvin, the Skald.*

After King Harald had experienced the cunning of the Fin woman, he was so angry that he drove from him the sons he had with her, and would not suffer them before his eyes. But one of them, Gudrod Liome, went to his foster-father Thiodolf of Hvin, and asked him to go to the king, who was then in the Uplands; for Thiodolf was a great friend of the king. And so they went, and came to the king's house late in the evening, and sat down together unnoticed near the door. The king walked up and down the floor casting his eye along the benches; for he had a feast in the house, and the mead was just mixed. The king then murmured out these lines:—

"Tell me, ye aged grey-haired heroes,
Who have come here to seek repose,

> Wherefore must I so many keep
> Of such a set, who, one and all,
> Right dearly love their souls to steep,
> From morn till night, in the mead-bowl?"

Then Thiodolf replies:—

> "A certain wealthy chief, I think,
> Would gladly have had more to drink
> With him, upon one bloody day,
> When crowns were cracked in our sword-play."

Thiodolf then took off his hat, and the king recognised him, and gave him a friendly reception. Thiodolf then begged the king not to cast off his sons; "for they would with great pleasure have taken a better family descent upon the mother's side, if the king had given it to them." The king assented, and told him to take Gudrod with him as formerly; and he sent Halfdan and Sigurd to Ringerike, and Ragnvald to Hadeland, and all was done as the king ordered. They grew up to be very clever men, very expert in all exercises. In these times King Harald sat in peace in the land, and the land enjoyed quietness and good crops.

CHAPTER XXVII.—*Of Earl Torf-Einar's obtaining Orkney.*

When Earl Ragnvald in More heard of the death of his brother Earl Sigurd, and that the vikings were in possession of the country, he sent his son Hallad westward, who took the title of earl to begin with, and had many men-at-arms with him. When he arrived at the Orkney Islands, he established himself in the country; but both in harvest, winter, and spring, the vikings cruised about the isles, plundering the headlands, and committing depredations on the coast. Then Earl Hallad grew tired of the business, resigned his earldom, took up again his rights as an allodial owner,* and afterwards returned eastward into Norway. When Earl Ragnvald heard of this he was ill pleased with Hallad, and said his sons were very unlike their ancestors. Then said Einar, "I have enjoyed but little honour among you, and have little affection here to lose: now if you will give me force enough, I will go west to the islands, and promise you what at any rate will please you—that you shall never see me again." Earl Ragnvald replied, that he would be glad if he never came back; "For there is little hope," said he, "that thou will ever be an honour to thy friends, as all thy kin on thy mother's side are born slaves." Earl Ragnvald gave Einar a vessel completely equipped, and he sailed with it into the West sea in harvest. When he came to the Orkney Isles, two vikings, Thorer Treskeg and Kalf Skurfa, were in his way with two vessels. He attacked them instantly, gained the battle, and slew the two vikings. Then this was sung :—

> "Then gave he Treskeg to the trolls,
> Torf-Einar slew Skurfa."

He was called Torf-Einar, because he cut peat for fuel, there being no firewood, as in Orkney there are

---

* The Icelandic word in Snorre is *hauldsrett*, that is, the rights of a *hauldr*. This word is not derived from *halda* (to hold), but is identical with Anglo-Saxon *hæle*, German, *held* (a hero). As a law term it means the owner of allodial land, a kind of higher yeomen, like the Westmoreland statesman. It is identical in meaning with the modern Norwegian *Odelsbonde* (udal farmer). See Vigfusson, *s.v.*

no woods. He afterwards was earl over the islands, and was a mighty man. He was ugly, and blind of an eye, yet very sharp-sighted withal.

CHAPTER XXVIII.—*King Eirik Eymundson's Death.*

Duke Guthorm * dwelt principally at Tunsberg, and governed the whole of Viken when the king was not there. He defended the land, which, at that time, was much plundered by the vikings. There were disturbances also up in Gautland as long as King Eirik Eymundson lived; but he died when King Harald Fairhair had been ten years king of all Norway.

CHAPTER XXIX.—*Guthorm's Death in Tunsberg.*

After Eirik, his son Biorn was king of Svithiod for fifty years. He was father of Eirik the Victorious, and of Olaf the father of Styrbiorn. Guthorm died on a bed of sickness at Tunsberg, and King Harald gave his son Guthorm the government of that part of his dominions, and made him chief of it.

CHAPTER XXX.—*Earl Ragnvald Burnt in his House.*

When King Harald was forty years of age many of his sons were well advanced, and indeed they all came early to strength and manhood. And now they began to take it ill that the king would not give

* Duke Guthorm, Harald Harfager's uncle.—L.

them any part of the kingdom, but put earls into every district; for they thought earls were of inferior birth to them. Then Halfdan Haleg and Gudrod Liome set off one spring with a great force, and came suddenly upon Earl Ragnvald, earl of More, and surrounded the house in which he was, and burnt him and sixty men in it. Thereafter Halfdan took three long-ships, and fitted them out, and sailed into the West sea; but Gudrod set himself down in the land which Ragnvald formerly had. Now when King Harald heard this he set out with a great force against Gudrod, who had no other way left but to surrender, and he was sent to Agder. King Harald then set Earl Ragnvald's son Thorer over More, and gave him his daughter Alof, called Arbot, in marriage. Earl Thorer, called the Silent, got the same territory his father Earl Ragnvald had possessed.

Chapter XXXI.—*Halfdan Haleg's Death.*

Halfdan Haleg came very unexpectedly to Orkney, and Earl Einar immediately fled; but came back soon after, about harvest time, unnoticed by Halfdan. They met, and after a short battle Halfdan fled the same night. Einar and his men lay all night without tents, and when it was light in the morning they searched the whole island, and killed every man they could lay hold of. Then Einar said "What is that I see upon the isle of Rinansey?* Is it a man or a bird? Sometimes it raises itself up, and sometimes

---

* North Ronaldsay in the Orkneys.

lies down again." They went to it, and found it was Halfdan Haleg, and took him prisoner.

Earl Einar sang the following song the evening before he went into this battle:—

> "Where is the spear of Hrollaug?* where
> Is stout Rolf Ganger's bloody spear!
> I see them not; yet never fear,
> For Einar will not vengeance spare
> Against his father's murderers, though
> Hrollaug and Rolf are somewhat slow,
> And silent Thorer sits and dreams
> At home, beside the mead-bowl's streams."

Thereafter Earl Einar went up to Halfdan, and cut a spread eagle upon his back,† by striking his sword through his back into his belly, dividing his ribs from the back-bone down to his loins, and tearing out his lungs; and so Halfdan was killed. Einar then sang:—

> "For Ragnvald's death my sword is red:
> Of vengeance it cannot be said
> That Einar's share is left unsped.
> So now, brave boys, let's raise a mound,—
> Heap stones and gravel on the ground
> O'er Halfdan's corpse: this is the way
> We Norsemen our scat duties pay."

Then Earl Einar took possession of the Orkney Isles as before. Now when these tidings came to Norway, Halfdan's brothers took it much to heart, and thought that his death demanded vengeance;

---

* Hrollaug, Rolf Ganger, Thorer the Silent, and Einar were all sons of that Earl Ragnvald whom Harald Harfager's sons, and among them Halfdan, had surprised and burnt in his house. They ought, according to the opinion of the times, to have taken vengeance as well as Einar on the murderers.—L.

† This kind of punishment was called *rista örn*—to cut an eagle.—L.

and many were of the same opinion. When Einar heard this, he sang :—

> " Many a stout udal-man, I know,
> Has cause to wish my head laid low ;
> And many an angry udal knife
> Would gladly drink of Einar's life.
> But ere they lay Earl Einar low,—
> Ere this stout heart betrays its cause,
> Full many a heart will writhe, we know,
> In the wolf's fangs, or eagle's claws."

Chapter XXXII.—*King Harald and Earl Einar Reconciled.*

King Harald now ordered a levy, and gathered a great force, with which he proceeded westward to Orkney; and when Earl Einar heard that King Harald was come, he fled over to Caithness. He made the following verses on this occasion :—

> " Many a bearded man must roam,
> An exile from his house and home,
> For cow or horse ; but Halfdan's gore
> Is red on Rinansey's wild shore.
> A nobler deed—on Harald's shield
> The arm of one who ne'er will yield
> Has left a scar. Let peasants dread
> The vengeance of the Norsemen's head ;
> I reck not of his wrath, but sing,
> ' Do thy worst !—I defy thee, king !' "

Men and messages, however, passed between the king and the earl, and at last it came to a conference; and when they met the earl submitted the case altogether to the king's decision, and the king condemned the earl and the Orkney people to pay a fine of sixty marks of gold. As the bondes thought this was too heavy for them to pay, the earl offered to pay the whole if they would surrender their udal lands to him.* This they all agreed to do: the poor because they had but little pieces of land; the rich because they could redeem their udal rights again when they liked. Thus the earl paid the whole fine to the king, who returned in harvest to Norway. The earls for a long time afterwards possessed all the udal lands in Orkney, until Sigurd Hlodvison † gave back the udal rights.‡

CHAPTER XXXIII.—*Death of Guthorm. Death of Halfdan the White.*

While King Harald's son Guthorm had the defence of Viken, he sailed outside of the islands on the coast, and came in by one of the mouths of the Gaut river. When he lay there Solve Klofe came upon him, and immediately gave him battle, and Guthorm fell. Halfdan the White and Halfdan the Black went out on an expedition, and plundered in the East sea, and had a battle in Eistland,§ where Halfdan the White fell.

CHAPTER XXXIV.—*Marriage of Eirik, the Son of King Harald.*

Eirik, Harald's son, was fostered in the house of the herse Thorer, son of Hroald, in the Fiord district.

---

* See the Saga of Olaf the Saint, chapter 99, where the story is told somewhat differently.

† He fell in the Brian battle, 1014, after a reign of about 30 years. He was one of the greatest of the Orkney jarls.

‡ There are still a few udal properties in Orkney, and many which are described in the feudal charters as having been udal lands of old.—L.

§ Eistland is Esthonia.—L.

He was the most beloved and honoured by King Harald of all his sons. When Eirik was twelve* years old, King Harald gave him five long-ships, with which he went on an expedition,—first in the Baltic; then southwards to Denmark, Friesland,† and Saxonland; on which expedition he passed four years. He then sailed out into the West sea, and plundered in Scotland, Bretland,‡ Ireland, and Valland,§ and passed four years more in this way. Then he sailed north to Finmark,‖ and all the way to Biarmaland,¶ where he had many a battle, and won many a victory. When he came back to Finmark, his men found a girl in a Lapland hut, whose equal for beauty they never had seen. She said her name was Gunhild, and that her father dwelt in Halogaland, and was called Ozur Tote. "I am here," she said, "to learn Lapland-art, from two of the most knowing Laplanders in all Finmark, who are now out hunting. They both want me in marriage. They are so skilful that they can hunt out traces either upon the frozen or the thawed earth, like dogs; and

---

* In the saga-age, young men became of age at twelve, and could then enter public life.

† Friesland appears to have been the name given to the whole coast from the Eyder in Slesvik to North Holland, and to have been called Saxonland (Saxland) or Friesland.—L.

‡ Bretland (Britton land) was that part of Britain inhabited by the ancient inhabitants. The sagas give the name of England only to the parts inhabited by the Anglo-Saxons. Wales, Cornwall, and the west coast of the island, are always called Bretland.—L.

§ Valland, the west coast of France, in which the inhabitants of Wales and Cornwall settled in the fifth century.—L.

‖ Finmark is the country we call Lapland in the north of Norway and Sweden.—L.

¶ Biarmaland was the coast of the White Sea about the mouth of the Dwina, and now the Russian province of Archangel.—L.

they can run so swiftly on skees,* that neither man nor beast can come near them in speed. They hit whatever they take aim at, and thus kill every man who comes near them. When they are angry the very earth turns away in terror, and whatever living thing they look upon then falls dead. Now ye must not come in their way; but I will hide you here in the hut, and ye must try to get them killed." They agreed to it, and she hid them, and then took a leather bag, in which they thought there were ashes which she took in her hand, and strewed both outside and inside of the hut. Shortly after the Laplanders came home, and asked who had been there; and she answered, "Nobody has been here." "That is wonderful," said they; "we followed the traces close to the hut, and can find none after that." Then they kindled a fire, and made ready their meat, and Gunhild prepared her bed. It had so happened that Gunhild had slept the three nights before, but the Laplanders had watched the one upon the other, being jealous of each other. "Now," she said to the Laplanders, "come here, and lie down one on each side of me." On which they were very glad to do so. She laid an arm round the neck of each, and they went to sleep directly. She roused them up; but they fell to sleep again instantly, and so soundly that she scarcely could waken them. She even raised them up in the bed, and still they slept. Thereupon she took two great seal-skin bags, and put their heads in them, and tied them fast under their arms;

* A kind of long snow-shoes or snow-skates.

and then she gave a wink to the king's men. They run forth with their weapons, kill the two Laplanders, and drag them out of the hut. That same night came such a dreadful thunder-storm that they could not stir. Next morning they came to the ship, taking Gunhild with them, and presented her to Eirik. Eirik and his followers then sailed southwards to Halogaland; and he sent word to Ozur Tote, the girl's father, to meet him. Eirik said he would take his daughter in marriage, to which Ozur Tote consented, and Erik took Gunhild, and went southwards with her [A.D. 922].

CHAPTER XXXV.—*Harald divides his Kingdom among his Sons.*

When King Harald was fifty years of age many of his sons were grown up, and some were dead. Many of them committed acts of great violence in the country, and were in discord among themselves. They drove some of the king's earls out of their properties, and even killed some of them. Then the king called together a numerous Thing* in the south part of the country, and summoned to it all the people of the Uplands. At this Thing he gave to all his sons the title of king, and made a law that his descendants in the male line should each succeed to the kingly title and dignity; but his descendants by the female side only to that of earl. And he divided the country among them thus:—Vingulmark,

* This must have been the Eidsiva-thing, at the present Eidsvold, where the present constitution of Norway was adopted, May 17, 1814.

Raumarike, Vestfold, and Thelemark, he bestowed on Olaf, Biorn, Sigtryg, Frode, and Thorgils. Hedemark and Gudbrandsdal he gave to Dag, Hring, and Ragnar. To Snowfrid's sons he gave Ringerike, Hadeland, Thoten, and the lands thereto belonging. His son Guthorm, as before mentioned, he had set over the country from Svinesund to the Glommen, and to defend the country eastwards. King Harald himself generally dwelt in the middle of the country, and Hrorek and Gudrod were generally with his court, and had great estates in Hordaland and in Sogn. King Eirik was also with his father King Harald; and the king loved and regarded him the most of all his sons, and gave him Halogaland and North More, and Raumsdal. North in Throndhjem he gave Halfdan the Black, Halfdan the White, and Sigrod land to rule over. In each of these districts he gave his sons the one half of his revenues, together with the right to sit on a high-seat,—a step higher than earls, but a step lower than his own high-seat. His king's seat each of his sons wanted for himself after his death, but he himself destined it for Eirik. The Throndhjem people wanted Halfdan the Black to succeed to it. The people of Viken, and the Uplands, wanted those under whom they lived. And thereupon new quarrels arose among the brothers; and because they thought their dominions too little, they drove about in piratical expeditions. In this way, as before related, Guthorm fell at the mouth of the Gaut river, slain by Solve Klofe; upon which Olaf took the kingdom he had possessed. Halfdan the White fell in Eistland, Halfdan Haleg in Orkney. King Harald gave ships of war to Thorgils and Frode, with which they went westward on a viking cruise, and plundered in Scotland, Ireland, and Bretland. They were the first of the Northmen who took Dublin.* It is said that Frode got poisoned drink there; but Thorgils was a long time king over Dublin, until he fell into a snare of the Irish, and was killed.

CHAPTER XXXVI.—*Death of Ragnvald Rettilbeine.*

Eirik Blood-axe expected to be head king over all his brothers and King Harald intended he should be so; and the father and son lived long together. Ragnvald Rettilbeine governed Hadeland, and allowed himself to be instructed in the arts of witchcraft, and became a great warlock. Now King Harald was a hater of all witchcraft. There was a warlock in Hordaland called Vitgeir; and when the king sent a message to him that he should give up his art of witchcraft, he replied in this verse:—

> "The danger surely is not great
> From wizards born of mean estate,
> When Harald's son in Hadeland,
> King Ragnvald, to the art lays hand."

But when King Harald heard this, King Eirik Blood-axe went by his orders to the Uplands, and

* Thorgils and Frode founded a Norman kingdom in Ireland about the year 840, or even earlier, according to the Irish annals, consequently they could not be Harald's sons. Olaf the White came there in 852, and restored the Norse kingdom.—Vigfusson. Snorre calls Dublin *Dyflin.*

came to Hadeland, and burned his brother Ragnvald in a house, along with eighty other warlocks; which work was much praised.

CHAPTER XXXVII.—*Of Gudrod Liome.*

Gudrod Liome was in winter on a friendly visit to his foster-father Thiodolf in Hvin, and had a well-manned ship, with which he wanted to go north to Rogaland. It was blowing a heavy storm at the time; but Gudrod was bent on sailing, and would not consent to wait. Thiodolf sang thus:—

> "Wait, Gudrod, till the storm is past,—
> Loose not thy long-ship while the blast
> Howls over-head so furiously,—
> Trust not thy long-ship to the sea,—
> Loose not thy long-ship from the shore:
> Hark to the ocean's angry roar!
> See how the very stones are tost,
> By raging waves high on the coast!
> Stay, Gudrod, till the tempest's o'er—
> Deep runs the sea off Jadar's shore."

Gudrod set off in spite of what Thiodolf could say; and when they came off the Jadar the vessel sunk with them, and all on board were lost.

CHAPTER XXXVIII.—*King Biorn the Merchant's Death.*

King Harald's son, Biorn, ruled over Vestfold at that time, and generally lived at Tunsberg, and went but little on war expeditions. Tunsberg at that time was much frequented by merchant vessels, both from Viken and the north country, and also from the south, from Denmark, and Saxonland. King Biorn had also merchant ships on voyages to other lands, by which he procured for himself costly articles, and such things as he thought needful; and therefore his brothers called him the Seaman, and the Chapman (merchant). Biorn was a man of sense and understanding, and promised to become a good ruler. He made a good and suitable marriage, and had a son by his wife, who was named Gudrod. Eirik Blood-axe came from his Baltic cruise with ships of war, and a great force, and required his brother Biorn to deliver to him King Harald's share of the scat and incomes of Vestfold. But it had always been the custom before, that Biorn himself either delivered the money into the king's hands, or sent men of his own with it; and therefore he would continue with the old custom, and would not deliver the money. Eirik again wanted provisions, tents, and liquor. The brothers quarrelled about this; but Eirik got nothing, and left the town. Biorn went also out of the town towards evening up to Seaheim. In the night Eirik came back after Biorn, aud came to Seaheim just as Biorn and his men were seated at table drinking. Eirik surrounded the house in which they were; but Biorn with his men went out and fought. Biorn, and many men with him, fell. Eirik, on the other hand, got a great booty, and proceeded northwards. But this work was taken very ill by the people of Viken, and Eirik was much disliked for it; and the report went that King Olaf would avenge his brother

Biorn, whenever opportunity offered. King Biorn lies in the Seaman's mound at Seaheim.*

CHAPTER XXXIX.—*Of the Reconciliation of the Kings.*

King Eirik went in winter northwards to More, and was at a feast in Solve, within the point Agdanes;† and when Halfdan heard of it he set out with his men, and surrounded the house in which they were. Eirik slept in a room which stood detached by itself, and he escaped into the forest with five others; but Halfdan and his men burnt the main house, with all the people who were in it. With this news Eirik came to King Harald, who was very wroth at it, and assembled a great force against the Throndhjem people. When Halfdan the Black heard this he levied ships and men, so that he had a great force, and proceeded with it to Stad, within Thorsberg. King Harald lay with his men at Reinplain. Now people went between them, and among others a clever man called Guthorm Sindre, who was then in Halfdan's army, but had been formerly in the service of King Harald, and was a great friend of both. Guthorm was a great skald, and had once composed a song both about the father and the son, for which they had offered him a reward. But he would take nothing; but only asked that, some day or other, they should grant him any request

---

* Seaheim, called afterwards Seim, is a farm in the present Sems parish in Jarlsberg, about two miles from the town of Tunsberg. The Seaman's mound is still to be seen.—L.

† Agdanes is the south point of land at the entrance of the Throndhjem fiord.—L.

he should make, which they promised to do. Now he presented himself to King Harald, — brought words of peace between them, and made the request to them both that they should be reconciled. So highly did the king esteem him, that in consequence of his request they were reconciled. Many other able men promoted this business as well as he; and it was so settled that Halfdan should retain the whole of his kingdom as he had it before, and should let his brother Eirik sit in peace. After this event Jorund, the skald-maid, composed some verses in *Sendibit* (The Biting Message):—

> "I know that Harald Fairhair
> Knew the dark deed of Halfdan.
> To Harald Halfdan seemed
> Angry and cruel." *

Chapter XL.—*Birth of Hakon the Good.*

Earl Hakon Griotgardson of Lade had the whole rule over Throndhjem when King Harald was anywhere away in the country; and Hakon stood higher with the king than any in the country of Throndhjem. After Hakon's death his son Sigurd succeeded to his power in Throndhjem, and was the earl, and had his mansion at Lade. King Harald's sons, Halfdan the Black and Sigrod, who had been before in the house of his father Earl Hakon, continued to be brought up in his house. The sons of Harald and Sigurd were about the same age. Earl Sigurd was one of the wisest men of his time, and

---

* Nothing is known of the poetess Jorund.

married Bergliot, a daughter of Earl Thorer the Silent; and her mother was Alof Arbot, a daughter of Harald Harfager. When King Harald began to grow old he generally dwelt on some of his great farms in Hordaland; namely, Alrekstad, or Seaheim, Fitjar, Utstein, or Augvaldsnes in the island Karmt. When Harald was seventy years of age he begat a son with a girl called Thora Mosterstang, because her family came from Moster. She was descended from good people, being connected with Horda-Kare; and was moreover a very stout and remarkably handsome girl. She was called the king's servant-girl; for at that time many were subject to service to the king who were of good birth, both men and women. Then it was the custom, with people of consideration, to choose with great care the man who should pour water over their children, and give them a name. Now when the time came that Thora, who was then at Moster, expected her confinement, she would go to King Harald, who was then living at Seaheim; and she went northwards in a ship belonging to Earl Sigurd. They lay at night close to the land; and there Thora brought forth a child upon the land, up among the rocks,* close to the ship's gangway, and it was a man child. Earl Sigurd poured water over him, and called him Hakon, after his own father, Hakon earl of Lade. The boy soon grew handsome, large in size, and very like his father King Harald. King Harald let him follow his mother, and they were both in the king's house as long as he was an infant.

* In this place Hakon the Good also died. See his saga, chapter 32.

CHAPTER XLI.—*King Athelstan's Message.*

At this time a king called Athelstan [*] had taken the kingdom of England. He was called victorious and faithful. He sent men to Norway to King Harald, with the errand that the messengers should present him with a sword, with the hilt and handle gilt, and also the whole sheath adorned with gold and silver, and set with precious jewels. The ambassador presented the sword-hilt to the king, saying, "Here is a sword which King Athelstan sends thee, with the request that thou wilt accept it." The king took the sword by the handle; whereupon the ambassador said, "Now thou hast taken the sword according to our king's desire, and therefore art thou his subject, as thou hast taken his sword." King Harald saw now that this was a jest, for he would be subject to no man. But he remembered it was his rule, whenever anything raised his anger, to collect himself, and let his passion run off, and then take the matter into consideration coolly. Now he did so, and consulted his friends, who all gave him the advice to let the ambassadors, in the first place, go home in safety.

CHAPTER XLII.—*Hauk's Journey to England.*

The following summer King Harald sent a ship westward to England, and gave the command of it

---

[*] King of the Anglo-Saxons, 925-941, grandson of Alfred the Great. No English chronicle tells of Hakon's fostering in England.

to Hauk Habrok.* He was a great warrior, and very dear to the king. Into his hands he gave his son Hakon. Hauk proceeded westward to England, and found King Athelstan in London,† where there was just at the time a great feast and entertainment. When they came to the hall, Hauk told his men how they should conduct themselves; namely, that he who went first in should go last out, and all should stand in a row at the table, at equal distance from each other; and each should have his sword at his left side, but should fasten his cloak so that his sword should not be seen. Then they went into the hall, thirty in number. Hauk went up to the king and saluted him, and the king bade him welcome. Then Hauk took the child Hakon, and set it on the king's knee. The king looks at the boy, and asks Hauk what the meaning of this is. Hauk replies, "Harald the king bids thee foster his servant-girl's child." The king was in great anger, and seized a sword which lay beside him, and drew it, as if he was going to kill the child. Hauk says, "Thou hast borne him on thy knee, and thou canst murder him if thou wilt; but thou wilt not make an end of all King Harald's sons by so doing." On that Hauk went out with all his men, and took the way direct to his ship, and put to sea,—for they were ready,—and came back to King Harald. The king was highly pleased with this; for it is the common observation of all people, that the man who fosters another's

---

* Habrok = High Breeches.
† Called in the sagas *Lundún* and *Lundúnaborg*.

children is of less consideration than the other. From these transactions between the two kings, it appears that each wanted to be held greater than the other; but in truth there was no injury to the dignity of either, for each was the upper king in his own kingdom till his dying day.

CHAPTER XLIII.—*Hakon, the Foster-son of Athelstan, is Baptized.*

King Athelstan had Hakon baptized, and brought up in the right faith, and in good habits, and all sorts of exercises, and he loved Hakon above all his relations; and Hakon was beloved by all men. Athelstan was a man of understanding and eloquence, and also a good Christian. King Athelstan gave Hakon a sword, of which the hilt and handle were gold, and the blade still better; for with it Hakon cut down a mill-stone to the centre eye, and the sword thereafter was called the Kvernbite.* Better sword never came into Norway, and Hakon carried it to his dying day.

CHAPTER XLIV.—*Eirik is brought to the Sovereignty.*

When King Harald was eighty years of age he became very heavy, and unable to travel through the country, or do the business of a king. Then he

---

* Quern is still the name of the small hand mill-stones still found in use among the cottars in Orkney, Shetland, and the Hebrides.—L.

This sword is mentioned in the Younger Edda. There were many excellent swords in the olden time, and many of them had proper names.

brought his son Eirik to his high-seat, and gave him the power and command over the whole land. Now when King Harald's other sons heard this, King Halfdan the Black also took a king's high-seat, and took all Throndhjem land, with the consent of all the people, under his rule as upper king. After the death of Biorn the Chapman, his brother Olaf took the command over Vestfold, and took Biorn's son, Gudrod, as his foster-child. Olaf's son was called Trygve; and the two foster-brothers were about the same age, and were hopeful and clever. Trygve, especially, was remarkable as a stout and strong man. Now when the people of Viken heard that those of Hordaland had taken Eirik as upper king, they did the same, and made Olaf the upper king in Viken, which kingdom he retained. Eirik did not like this at all. Two years after this, Halfdan the Black died suddenly at a feast in Throndhjem, and the general report was that Gunhild had bribed a witch to give him a death-drink. Thereafter the Throndhjem people took Sigrod to be their king.

CHAPTER XLV.—*King Harald's Death.*

King Harald lived three years after he gave Eirik the supreme authority over his kingdom, and lived mostly on his great farms which he possessed, some in Rogaland, and some in Hordaland. Eirik and Gunhild had a son, on whom King Harald poured water, and gave him his own name, and the promise that he should be king after his father Eirik.

King Harald married most of his daughters within the country to his earls, and from them many great families are descended. King Harald died on a bed of sickness in Rogaland, and was buried under a mound at Hauge in Karmtsund. In Haugasund is a church, now standing; and not far from the churchyard, at the north-west side, is King Harald Harfager's mound; but his grave-stone stands west of the church, and is thirteen feet and a half high, and two ells broad. One stone was set at the head and one at the feet; on the top lay the slab, and below on both sides were laid small stones. The grave,* mound, and stone, are there to the present day.† Harald Harfager was, according to the report of men of knowledge, of remarkably handsome appearance, great and strong, and very generous and affable to his men. He was a great warrior in his youth; and people think that this was foretold by his mother's dream before his birth, as the lowest part of the tree she dreamt of was red as blood. The stem again was green and beautiful, which betokened his flourishing kingdom; and that the tree was white at the top showed that he should reach a grey-haired old age. The branches and twigs showed forth his posterity, spread over the whole land: for of his race, ever since, Norway has always had kings.‡

* A monument was built there in 1872, 1000 years after the battle of Hafersfjord.
† The stone and some remains of the mound are still to be seen at Gar, the principal farm-house in the parish of Karmtsund.—L.
‡ The last male descendant of Harald Fairhair that ruled in Norway, Hakon Magnuson V., died in the year 1319. There are peasants in Gudbrandsdal who claim to be descended from Harald.

CHAPTER XLVI.—*The Death of Olaf and of Sigrod.*

King Eirik took all the revenues which the king had in the middle of the country, the next winter after King Harald's decease. But Olaf took all the revenues eastward in Viken, and their brother Sigrod all that of the Throndhjem country. Eirik was very ill pleased with this; and the report went that he would attempt with force to get the sole sovereignty over the country, in the same way as his father had given it to him. Now when Olaf and Sigrod heard this, messengers passed between them; and after appointing a meeting place, Sigrod went eastward in spring to Viken, and he and his brother Olaf met at Tunsberg, and remained there a while. The same spring King Eirik levied a great force, and ships, and steered towards Viken. He got such a strong steady gale that he sailed night and day, and came faster than the news of him. When he came to Tunsberg, Olaf and Sigrod, with their forces, went out of the town a little eastward to a ridge, where they drew up their men in battle order; but as Eirik had many more men he won the battle. Both brothers, Olaf and Sigrod, fell there; and both their grave-mounds are upon the ridge where they fell. Then King Eirik went through Viken, and subdued it, and remained far into summer. Gudrod and Trygve fled to the Uplands. Eirik was a stout handsome man, strong, and very manly,—a great and fortunate man of war; but bad-minded, gruff, unfriendly, and silent. Gunhild, his wife, was the most beautiful of women,—clever, with much knowledge, and lively; but a very false person, and very cruel in disposition. The children of King Eirik and Gunhild were, Gamle, the oldest; then Guthorm, Harald, Ragnfrod, Ragnhild, Erling, Gudrod, and Sigurd Sleva. All were handsome, and of manly appearance.*

* Of Eirik, his wife, and children, see the following sagas.

END OF VOL. I.